RABIES: SCIENTIFIC BASIS OF THE DISEASE AND ITS MANAGEMENT

THIRD EDITION

RABIES: SCIENTIFIC BASIS OF THE DISEASE AND ITS MANAGEMENT

THIRD EDITION

Edited by

ALAN C. JACKSON
University of Manitoba, Winnipeg, MB, Canada

AMSTERDAM • BOSTON • HEIDELBERG • LONDON • NEW YORK
OXFORD • PARIS • SAN DIEGO • SAN FRANCISCO
SINGAPORE • SYDNEY • TOKYO

Academic Press is an imprint of Elsevier

Academic Press is an imprint of Elsevier
The Boulevard, Langford Lane, Kidlington, Oxford OX5 1GB, UK
525 B Street, Suite 1800, San Diego, CA 92101-4495, USA

First edition 2002
Second edition 2007
Third edition 2013

Notice
No responsibility is assumed by the publisher for any injury and/or damage to persons
or property as a matter of products liability, negligence or otherwise, or from any use or
operation of any methods, products, instructions or ideas contained in the material herein.
Because of rapid advances in the medical sciences, in particular, independent verification of
diagnoses and drug dosages should be made.

British Library Cataloguing-in-Publication Data
A catalogue record for this book is available from the British Library

Library of Congress Cataloging-in-Publication Data
A catalog record for this book is available from the Library of Congress

ISBN: 978-0-12-396547-9

For information on all Academic Press publications
visit our website at elsevierdirect.com

Typeset by MPS Ltd, Chennai, India
www.adi-mps.com

Printed and bound in the United States

13 14 15 16   10 9 8 7 6 5 4 3

Contents

4 Molecular Epidemiology   123

SUSAN A. NADIN-DAVIS

5 Rabies in Terrestrial Animals   179

CATHLEEN A. HANLON

6 Bat Rabies   215

ASHLEY C. BANYARD, DAVID T. S. HAYMAN, CONRAD M. FREULING, THOMAS MÜLLER, ANTHONY R. FOOKS, AND NICHOLAS JOHNSON

7 Human Disease   269

ALAN C. JACKSON

8 Pathogenesis   299

ALAN C. JACKSON AND ZHEN F. FU

9 Pathology   351

JOHN P. ROSSITER AND ALAN C. JACKSON

TIZIANA LEMBO

ALAN C. JACKSON

Preface

Although rabies is an ancient disease, today it remains a very important, neglected disease that continues to cause considerable human suffering and deaths. Progress has been slow in reducing the rabies global death rate, although the methods to do so are now already well established. Momentum is clearly developing for commitments from countries for global elimination of canine rabies by 2025 or earlier, so the next decade or two could be a very exciting time in the rabies field. This volume describes developments and progress in many areas of the rabies field since the second edition was published in 2007, including in the areas of epidemiology, molecular epidemiology, prevention and therapy of human rabies, and control of wildlife and dog rabies. This volume is truly a multidisciplinary effort by authors in many different fields. I am sure that this edition will give a diverse audience—including physicians, veterinarians, virologists, immunologists, wildlife biologists, and students in different fields—a much better understanding of the disease.

I would like to thank Elsevier for allowing me to serve as the editor of a third edition. I thank all of the authors for sharing their expertise in their insightful and up-to-date chapters. All of the contributors have done a superb job, and I very much appreciate their efforts. Finally, special thanks to Dr. William Wunner, who co-edited the first two editions of this book with me, and to Dr. Frederick Murphy for writing the foreword in all three editions and for providing advice to me over the years.

Alan C. Jackson
Winnipeg, Manitoba, Canada
August 2012

Foreword

Rabies research continues to be a lively, important part of world of infectious disease sciences, made so by the continuing importance of rabies as a human disease, the financial and resource burden of the disease to public health agencies worldwide, and the ever-intriguing nature of the viruses (the lyssaviruses), their molecular biology and genetics, their pathogenesis and pathophysiology, and their ecology and epidemiology. Since the publication of the first edition of this book in 2002 and the second edition in 2007, research has continued at a laudable pace, across broad areas of basic and public health sciences, again clearly warranting the decision of the editor, Alan Jackson, to carry out all the work necessary to construct and edit this third edition.

In some settings, among competing human and zoonotic viral diseases, rabies is marginalized; it is said, after all, that there are only a few rabies-related human deaths each year in developed countries. However, a global perspective quickly changes this viewpoint, and when the shortcomings in the reporting of human cases in so many parts of the less developed world are considered, the importance of the disease becomes clear, indeed. It is from this global perspective that this volume is derived, and Jackson has captured this perspective in his superb choice of authors for the various chapters.

Rabies, almost uniquely—having separate disease control establishments in animal disease, wildlife disease, and human disease prevention/control agencies—has often been said to fall between the cracks, not getting the attention and resources warranted from any public agency. In some countries, it is difficult to determine just who is in charge of rabies control. Until recently, this situation has defied the notion of significant disease control/prevention action, much less national/regional disease elimination. However, new ideas and new funding streams, as well as new research such as that reported here, make me think that we may be entering a new optimistic era where much more is possible. At the same time, there are regions of the world where prevention/control strategies are embryonic and where zoonotic reservoir host niches pose problems that will continue to defy efforts to interrupt transmission far into the future. So, it seems clear that the integration of recent breakthrough research into the continuum of research progress made over the past century is most important. That integration is accomplished in this volume, but it also indicates that there will be a continuing need for further editions in the future. Alan Jackson and the authors of the chapters in this volume will have their work cut out for them as new editions are called for.

As he retires from his duties as editor of the first two editions of this book, congratulations are due to William H. Wunner. Bill Wunner has been a pillar of rabies research for many years, coming on the scene at the time of the rise of molecular virology and continuing as a leader in the incredible advance in our understanding of the nature of the viruses themselves. Bill has always had great perspective using the advances in basic rabies virology to drive advances in disease prevention and control. His role in the initial concept of using recombinant vaccines for the control of wildlife rabies is exemplary.

As I did in writing the forewards for the previous two editions, I have reviewed this edition with much interest—I must admit that I am interested in every facet of rabies science, from the viruses themselves, to the pathology of disease they cause, to the natural history and ecology of the viruses in their reservoir hosts, to all aspects of epidemiology and public health disease prevention and control. It is all here: updated with wonderful new research bases for application in the field. There does not seem to be a simple way to organize any overarching thoughts (this is not a book review, it is a foreward), but here are a few categories that always lead to discussion among the members of the world's "rabies club":

- *History* — I still never get tired of Louis Pasteur's charge: *"Dans les champs de l'observation le hasard ne favorise que les esprits préparés."* ("In matters of observation chance favors only the prepared mind.") We inherit not only a grand beginning from Pasteur and his colleagues, but a memorable continuum that always surprises the scientist–historian who not only sees the insights in early experiments (in the face of so much false dogma) but also sees long periods of stasis and entrenchment. For example, at the time of the development of the first rabies vaccine by Pasteur, Emile Roux, Charles Chamberland, and Louis Thuillier, as well as the immediate development of many vaccination centers at the branches of the Institut Pasteur around the world, there was a surge of enthusiasm for research, but seemingly through the first half of the twentieth century there was little further innovation to make vaccines ever better. Alan Jackson's chapter serves as a reminder that progress is not automatic, that it must be pushed by innovative scientists who understand the barriers—technical and psychological—and are anxious to move past them.
- *The viruses* — Now that it is understood that each lyssavirus has evolved to fit a particular econiche and that continuing evolutionary progression yields new variants and new threats to human and animal health, the need to unravel the finest details of the structure/function of the viruses becomes increasingly important. Progress here has

been spectacular, but so much new detail has been added since the last edition of this volume that it seems clear that there is still more to come. Descriptive virology gives way to functional understanding of the parts of the virus that represent the viral armamentarium in the complex battle between virus and host or host population. The perpetuation of the viruses proves their worthiness as adversaries (the universal drift of the pathogen toward commensalism does not seem to pertain in rabies unless some unrealistic timeline is in play). As they illustrate in their chapter, William Wunner and Karl-Klaus Conzelmann show that many of the chinks in the armor of the viruses seem to lie in the details of their molecular structure/function.

- *The disease, its pathology and pathogenesis* — As Alan Jackson, Zhen Fu, and John Rossiter show in their chapters, most of what we know about the events that take place during rabies infection has been learned from experimental animal models, where we are fortunate to have seemingly good concordance between the models and hosts of medical and veterinary importance. Still, we have a long way to go to understand the neuroanatomical, neuropathological, and neurophysiological bases for the behavioral changes seen in clinical rabies. One basis for optimism here, in this incredibly complex field, is the introduction of remarkable new research technologies— molecular probes, labels and tracers, imaging systems, behavioral correlative systems, and so forth. I never tire of Richard Johnson's 1971 observation: *"It is difficult to harmonize the dramatic clinical signs and lethal outcome of street rabies virus infection with the paucity of pathological changes in neurons. The greater localization to the limbic system with relative sparing of the neocortex provides a fascinating clinicopathologic correlate with the alertness, loss of natural timidity, aberrant sexual behavior and aggressiveness that may occur in clinical rabies. No other virus is so diabolically adapted to selective neuronal populations that it can drive the host in fury to transmit the virus to another host animal."* This statement contains the driver for so much future research.

- *The host response to infection* — Monique Lafon, in her chapter, deals with the seeming enigma of a virus composed of proteins adequate to evoke effective innate and adaptive immune responses, which in such a slowly progressing infection should outrace the virus, versus the exceptionally inadequate functionality of the response that underpins the fatal outcome of infection. The key, of course, is the early entry of the neurotropic virus into the unique, immunologically sequestered environment of the nervous system, together with the intrinsic capacity of rabies virus to specifically inhibit the innate immune response and also to inhibit the migration into the nervous system of protective T cells (normally part of a fundamental system to protect the nervous system, here turned to the advantage of the virus).

In addition, the central control of immunological homeostasis operated by the nervous system results in an inappropriate downregulation of immune, inflammatory, and interferon responsiveness in the periphery. Again, this seems to represent a diabolical survival mechanism that the virus has evolved. Much of our understanding of these complex immunological events is rather new—but, with an appreciation that in yet another way rabies is unique, we can clearly see that unraveling these events holds the promise of new targets for post-exposure intervention.

- *The natural history, ecology and epidemiology of the viruses in nature* — the chapters by Cathleen Hanlon, James Childs, and Susan Nadin-Davis remind us that molecular ecology/epidemiology has become the central theme in the study of virus–host relationships and the co-adaptation of particular hosts and particular virus genotypes. Phylogenic analysis continues to provide powerful insights into the natural history of each virus genotype, especially now as genotypic analyses have been extended to most reservoir host econiches, even those in remote parts of the world. Recent discoveries of increasingly important "spillover events" are now seen as a source of new, emerging rabies threats, especially when a new reservoir host niche impacts human and livestock habitats. In this work, which relies on field and laboratory research, there is contained an opportunity to better focus intervention actions, such as bait-delivered wildlife vaccines.

- *The special issue of bat rabies* — In recent years, about 70% of human rabies cases in the United States have been caused by bat-associated virus genotypes, making bat rabies and its control a high priority for research. Yet, few ideas have come forward to minimize human exposure other than educating the public (and in some instances vaccinating high-risk individuals). Clearly much more research on the natural history of rabies in bats has been called for, and in their chapter, Ashley Banyard, David Hayman, Conrad Freuling, Thomas Műller, Anthony Fooks, and Nicholas Johnson have shown that much has been learned in the past few years. It has not hurt that so many other emerging viral zoonoses (caused by coronaviruses, filoviruses, and henipaviruses) have derived from bats serving as reservoir hosts, but it is the long-running research on rabies and other lyssavirus infections of bats that has become the foundation for dealing with these other zoonoses. Recent research emphasis seems to have centered on the bats themselves, the complexity of their natural history and population biology, their diversity (there are over 1,100 recognized species) and their many lifestyles (e.g., solitary and colonial lifestyles, diverse migration patterns, etc.). Although great progress has been made in linking the phylogeny of the various bat species to the phylogeny of associated lyssavirus isolates/ genotypes (the result of long-standing evolutionary progression in each econiche), it is still the case that too little is known to answer key questions about variations in the disease in various species, variations in disease pathogenesis and pathophysiology, and variations in host response and immunity. Since bat conservation is so broadly supported, worldwide, it is hard to envision novel intervention strategies that will minimize human risk of infection—but it is extremely important to try, in the name of human health and bat conservation.

- *Diagnostics* — In their chapters, Cathleen Hanlon, Susan Nadin-Davis, Susan Moore, and Chandra Gordon present a review of all aspects of rabies diagnostic medicine, as well as a detailed manual of the methods currently employed in diagnostic laboratories. They conclude that the direct immunofluorescence technique, which has served as the cornerstone of rabies diagnosis for the past half century, will be with us for some time. They point out clear shortcomings in adapting methods such as PCR, which have slower turn-around times, higher costs, increased need for staff expertise, added risks of cross-contamination, and the confounding effects of often poorly preserved specimens. The notion of a "dip-stick" rabies test that might be used anywhere seems to have been debunked, even though some non-PCR technologies look promising at research/development stages of testing (e.g., direct rapid immunohistochemical tests using light microscopy, several enzyme immunoassays, including dot-blot enzyme immunoassays and several lateral-flow immunodiagnostic assays). Since the direct immunofluorescence technique is very sensitive and specific, it seems that the only ongoing worry about its continued use concerns the infrastructure and staffing of the public health laboratories that do the testing, which in the United States involves ~100 laboratories testing ~100,000 specimens per year. The fiscal crisis facing public health laboratories makes this concern all too real. Perhaps the most worrisome prospect here is that in some localities in the United States potentially exposed persons or animal owners have to pay for testing (stated to cost from US $65.00 to $145.00 per standard direct fluorescent antibody test). Even so, as the authors point out, it is not uncommon for there to be only a single individual trained to do the testing, and/or for testing to be available only on certain days of the week. Here is a subject where basic or field research is not the solution—but where the solution lies in fixing failing support levels for public disease control and prevention systems.
- *Post-exposure prophylaxis of humans* — Deborah Briggs, Thirumeni Nagarajan, Charles Rupprecht, Zhi Quan Xiang, Hildegund Ertl,

Louise Taylor, and Peter Costa, in their chapters, cover the large interconnected subjects grounded in the disparate reality that in developed countries the goal of post-exposure vaccination is perfect efficacy and safety, cost notwithstanding, whereas in poor developing countries (where most human rabies cases occur), the requirement of post-exposure vaccination is low cost and practicality, such that more and more poor people may be treated. As stated by the authors: "In regions where canine rabies is enzootic and there are no control programs in place, human rabies vaccines are the only defense against contracting rabies." Of course, the role of potent anti-rabies immunoglobulin is recognized in both settings, in the latter still awaiting some kind of inexpensive substitute for human (or humanized) immunoglobulin, some kind of inexpensive antiviral drug with immediate efficacy to fill the window before the vaccine-evoked immune response takes over. The authors review all the present choices in human vaccine production, from the seed virus, the cell substrate, the production technology, the adjuvanting, and the potency and quality assurance/control testing systems. Alas, it is possible to see some of the kinds of streamlining that might be used to lower overall costs, and such has been done in vaccine production facilities in some Asian and South American countries, but it must be possible to go further without compromising safety. After all, foot-and-mouth disease vaccines, produced in the most modern facilities in South America, are safe, efficacious and extremely inexpensive—why cannot there be more manufacturing innovation?

- *Vaccination of domestic animals and wildlife* — Although only a few rabies vaccines have been used in the very effective bait-delivered wildlife vaccination programs in place in North America and Europe, many vaccine production innovations have prospered in the domestic animal vaccine industry. The less restrictive regulatory standards for animal vaccines are clear enough, but the downstream production technologies used in their production do seem to offer ideas for adaptation to the need for new vaccines for extending wildlife vaccination programs to species that have been difficult to reach and/ or difficult to vaccinate with current vaccines. Again, the role of more applied research seems clear and the authors, Zhi Quan Xiang and Hildegund Ertl, offer several tempting ideas.

- *Rabies control: public health organization and action, locally, nationally, and globally* — Darryn L. Knobel, Tiziana Lembo, Michelle Morters, Sunny Townsend, Sarah Cleaveland, Katie Hampson, Richard Rosatte and Tiziana Lembo, in their chapters, cover the large subject of organized prevention actions. Under the heading of doing first things first, within the realities of the places in the world where rabies is a substantial problem, focus is rightly centered on dog rabies: on post-exposure treatment of humans and mass vaccination of dogs. There are several elements in successful programs dealing with these two elements, all of which have been understood for many years, but that need to be organized holistically. The World Health Organization (WHO), a few non-governmental organizations (NGOs), and now the Global Alliance for Rabies Control have been active in more and more countries. The latter has organized its activities under the banner, "Blueprint for Rabies Prevention and Control," which extends its lessons effectively in online materials to help governments plan, implement, evaluate, and sustain canine rabies elimination programs. Of particular note in this volume is the attention drawn to the notorious lack of surveillance and disease-burden data for some rabies-endemic countries. Certainly, sustaining control/elimination programs will require that this failing be corrected, so that the economic impact of endemic dog rabies can better drive long-term government support. From an international perspective, recent studies that human rabies incidence in parts of Africa and Asia is far higher than suggested from WHO data should be getting more coverage in mainstream science/medicine media. In this regard, the annual celebration of World Rabies Day, the work of the Global Alliance for Rabies Control, is a wonderful start.

This third edition of *Rabies: Scientific Basis of the Disease and Its Management* not only brings to the reader a wealth of new information representing the major advances in research on rabies virus and the other lyssaviruses and on the disease of rabies itself, it also provides an updated scientific base for guiding rabies prevention and control programs. The practical application of research breakthroughs, such as improved vaccines and diagnostics, is one key to dealing with rabies as a global health problem. It is gratifying to see the updating and application of the other elements of animal rabies control in so many of the advanced-developing countries of the world: no matter the vernacular, we still must emphasize vaccination and control of the movement of pet dogs and cats, stray dog removal, laboratory diagnosis, surveillance, and public education. More and more countries will also have to employ vaccination of wildlife, but this seems far down the road in the poorest countries where wildlife rabies reservoirs present the risk of introducing virus into domestic animal populations. There is still much to be done, and Alan Jackson and his colleagues in "the rabies club" are the right people to carry on Pasteur's legacy.

Frederick A. Murphy
Galveston, Texas
August 2012

List of Contributors

Ashley C. Banyard Wildlife Zoonoses and Vector Borne Diseases Research Group, Department of Virology, Animal Health and Veterinary Laboratories Agency, Weybridge, New Haw, Addlestone, Surrey, KT15 3NB, UK.

Deborah J. Briggs Global Alliance for Rabies Control, 529 Humboldt St., Suite One, College of Veterinary Medicine, Kansas State University, Manhattan, KS 66502, USA.

James E. Childs Department of Epidemiology and Public Health Yale University School of Medicine, 60 College Street, P.O. Box 208034, New Haven, CT 06520, USA.

Sarah Cleaveland Boyd Orr Centre for Population and Ecosystem Health, Institute of Biodiversity, Animal Health and Comparative Medicine, College of Medical, Veterinary and Life Sciences, University of Glasgow, Glasgow G12 8QQ UK.

Karl-Klaus Conzelmann Max von Pettenkofer-Institute and Gene Center, Ludwig-Maximilians-University, 81377 Munich, Germany.

Peter Costa Global Alliance for Rabies Control, 529 Humboldt St., Suite One, Manhattan, KS 66502, USA.

Hildegund C.J. Ertl The Wistar Institute, 3601 Spruce St, Philadelphia, PA 19104, USA.

Anthony R. Fooks Wildlife Zoonoses and Vector Borne Diseases Research Group, Department of Virology, Animal Health and Veterinary Laboratories Agency, Weybridge, New Haw, Addlestone, Surrey, KT15 3NB, UK; National Consortium for Zoonosis Research, University of Liverpool, Leahurst, Chester High Road, Neston, Wirral, CH64 7TE, UK.

Conrad M. Freuling Institute of Molecular Biology, Friedrich-Loeffler-Institut, Federal Research Institute for Animal Health, D-17493 Greifswald - Insel Riems, Germany.

Zhen F. Fu Department of Pathology, University of Georgia, 501 D. W. Brooks Dr., Athens, GA 30602, USA; State-Key Laboratory of Agricultural Microbiology, Huazhong Agricultural University, 1 Shizishan Road, Wuhan, Hubei 430070, China.

Chandra R. Gordon Kansas State University Rabies Laboratory, 2005 Research Park Circle, Manhattan, KS 66502, USA.

Katie Hampson Boyd Orr Centre for Population and Ecosystem Health, Institute of Biodiversity, Animal Health and Comparative Medicine, College of

Medical, Veterinary and Life Sciences, University of Glasgow, Glasgow G12 8QQ UK.

Cathleen A. Hanlon Kansas State University Rabies Laboratory, 2005 Research Park Circle, Manhattan, KS 66502, USA.

David T.S. Hayman Wildlife Zoonoses and Vector Borne Diseases Research Group, Department of Virology, Animal Health and Veterinary Laboratories Agency, Weybridge, New Haw, Addlestone, Surrey, KT15 3NB, UK; Department of Biology, Colorado State University, Fort Collins, CO 80523, USA.

Alan C. Jackson Departments of Internal Medicine (Neurology) and Medical Microbiology, University of Manitoba, Health Sciences Centre, GF-543, 820 Sherbrook Street, Winnipeg, MB R3A 1R9, Canada.

Nicholas Johnson Wildlife Zoonoses and Vector Borne Diseases Research Group, Department of Virology, Animal Health and Veterinary Laboratories Agency, Weybridge, New Haw, Addlestone, Surrey, KT15 3NB, UK.

Darryn L. Knobel Department of Veterinary Tropical Diseases, Faculty of Veterinary Science, University of Pretoria, Onderstepoort, 0110, South Africa.

Monique Lafon Viral Neuroimmunology, Department of Virology, Institut Pasteur, 25 rue du Dr Roux 75724 Paris cedex 15, France.

Tiziana Lembo Boyd Orr Centre for Population and Ecosystem Health, Institute of Biodiversity, Animal Health and Comparative Medicine, College of Medical, Veterinary and Life Sciences, University of Glasgow, Glasgow G12 8QQ, UK; Partners for Rabies Prevention, Global Alliance for Rabies Control, 529 Humboldt St., Suite One, Manhattan, KS 66502, USA.

Susan M. Moore Kansas State University Rabies Laboratory, 2005 Research Park Circle, Manhattan, KS 66502, USA.

Michelle Morters Cambridge Infectious Diseases Consortium, Department of Veterinary Medicine, Cambridge University, Madingley Road, Cambridge, CB3 0ES, UK.

Thomas Müller Institute of Molecular Biology, Friedrich-Loeffler-Institut, Federal Research Institute for Animal Health, D-17493 Greifswald - Insel Riems, Germany.

Susan A. Nadin-Davis Animal Health Microbiology Research, Ottawa Laboratory (Fallowfield), Canadian Food Inspection Agency, Ottawa, ON K2J 4S1, Canada.

Thirumeni Nagarajan Research and Development Department, Biological E. Limited, Plot No. 1, Phase II, SP Biotech Park, Genome Valley, Kolthur Village, Shameerpet, Hyderabad, India.

Richard C. Rosatte Ontario Ministry of Natural Resources, Wildlife Research and Development Section, Trent University, DNA Building, Peterborough, ON K9J 7B8, Canada.

John P. Rossiter Department of Pathology and Molecular Medicine, Queen's University and Kingston General Hospital, Kingston, Ontario K7L 3N6, Canada.

Charles E. Rupprecht Global Alliance for Rabies Control, 529 Humboldt St., Suite One, Manhattan, KS 66502, USA.

Louise H. Taylor Global Alliance for Rabies Control, 529 Humboldt St., Suite One, Manhattan, Kansas, 66502, USA.

Sunny E. Townsend Boyd Orr Centre for Population and Ecosystem Health, Institute of Biodiversity, Animal Health and Comparative Medicine, College of Medical, Veterinary and Life Sciences, University of Glasgow, Glasgow G12 8QQ, UK.

William H. Wunner The Wistar Institute, 3601 Spruce St., Philadelphia, PA 19104-4268, USA.

Zhi Quan Xiang The Wistar Institute, 3601 Spruce St, Philadelphia, PA 19104, USA.

1

History of Rabies Research

Alan C. Jackson

Departments of Internal Medicine (Neurology) and Medical Microbiology
University of Manitoba, Winnipeg, Manitoba R3A 1R9, Canada

1 ANCIENT GREEK AND ROMAN TIMES

The ancient Greeks coined the word *lyssa* for rabies in dogs, apparently from the root *lud*, meaning "violent." The writings of Democritus, Aristotle, Hippocrates, and Celsus described the clinical features of rabies (Fleming, 1872). For example, Aristotle wrote "the disease is fatal to the dog itself and to any animal that it may bite, man excepted," but it is unclear whether or not he thought humans were susceptible to rabies (Wilkinson, 1977). Fracastoro (1930) has suggested that Aristotle was merely emphasizing that humans may not develop disease after a bite. Hippocrates probably refers to rabies when he wrote that persons in a frenzy drink very little, are disturbed and frightened, tremble at the least noise, or are seized with convulsions (Fleming, 1872). In 100 AD, the physician Celsus described human rabies and used the term hydrophobia (*hudrophobian*), which is derived from Greek words meaning "fear of water." Celsus recognized that the saliva of the biting animal contained the poisonous agent: "every bite has mostly some venom" (*aiitem omnis morsus hahetfere quoddam virus*), and he recommended the practice of using caustics, burning, cupping, and also sucking the wounds due to rabid dog bites (Fleming, 1872). Hence, rabies was prevalent and reasonably well understood in these ancient times.

Thomsen & Blaisdell (1994) have postulated that these early ideas about rabies may have come from writings about Cerberus, the multi-headed dog of Hades. A variety of literary sources in the period 800–700 BC, including Homer's *Iliad*, contained passages describing this creature, whose task it was to guard the underworld. Homer is also thought to refer to rabies when he refers to the "dog-star, or Orion's dog, as exerting a malignant influence upon the health of mankind" (Fleming, 1872). The myth likely originated in oral folklore. Cerberus manifested mad behavior and emitted poisonous substances from his jaws. In this myth, Cerberus has features of rabies, and the myth may have influenced classical beliefs about rabies (Thomsen & Blaisdell, 1994).

2 GIROLAMO FRACASTORO

Girolamo Fracastoro (1478–1553) was an Italian physician who proposed a scientific germ theory of disease (theory of contagion) more than 300 years before the experimental studies of Robert Koch and Louis Pasteur laid the foundations of modern microbiology. In 1546, Fracastoro wrote: "Since, then, this contagion is not communicated by fomes, and is not produced in the skin by simple contact, but requires laceration of the skin, we must suppose that its germs are not very viscous, and that they are perhaps too thick to be able to establish themselves in pores" (Fracastoro, 1930). Although the nature of viruses was not understood until centuries later, Francastoro had great insight into the nature of the viral agent causing rabies. Francastoro also described clinical rabies in humans.

3 JOHN MORGAGNI

In 1769, the pathologist John Morgagni (1735–1789), who has been described as the father of pathological anatomy, speculated that rabies "virus does not seem to be carried through the veins, but by the nerves, up to their origins" (Morgagni, 1960). He apparently concluded this because he noted that the early symptom of paresthesias in rabies occurred in the area of the original wound (Johnson, 1965)

4 EARLIEST PATHOGENESIS STUDIES

In 1804, Georg Gottfried Zinke (1771–1813) published a volume that was designed to prove that the infective agent of rabies was transmitted in infected saliva (Zinke, 1804). This work contained the first planned experimental studies on transmission of a viral disease. He took saliva from a rabid dog and painted it with a small brush into incisions he had made in healthy animals, including other dogs, cats, and rabbits, and the animals subsequently developed rabies (Wilkinson, 1977). These experiments proved that rabies was an infectious disease. A few years earlier, in 1793, John Hunter (1754–1809) suggested evaluating the transmissibility of rabies between species and indicated that transmission by incision and transfer of infected saliva on the point of a lancet should be feasible (Hunter, 1793). However, it is unclear if Zinke had read this article and whether or not it provided the inspiration for his experiments (Wilkinson, 1977). In 1821, the French neurophysiologist François Magendie (1783–1855) reported the transmission of rabies to a dog by inoculation of saliva from a human case of rabies (Magendie, 1821), but there was no indication that he was familiar with the previous writings of either John Hunter or Georg Zinke (Wilkinson, 1977).

5 LOUIS PASTEUR AND RABIES VACCINATION

In 1879, Pierre-Victor Galtier (1846–1908), who was a professor at a veterinary school in Lyon, France, used rabbits in his rabies experiments, which was technically much less difficult and less dangerous than experiments using dogs and cats, and he noted the paralytic and convulsive features of the disease in this species (Galtier, 1879; Wilkinson, 1977). Experimental transmission from dogs to rabbits was also associated with a marked reduction in the incubation period of rabies to an average of 18 days in rabbits (versus about a month in dogs), in addition to the advantages of being relatively cheap, easy to handle, and easy to keep (Geison, 1995). Shortly afterwards, the experimental model in rabbits was taken up by Pasteur.

Louis Pasteur (1822–1895) experimentally transmitted rabies virus by inoculating central nervous system (CNS) material of rabid animals into the brains of other animals (Figure 1.1). In 1881, Pasteur published his first paper on rabies (Pasteur, Chamberland, Roux, & Thuillier, 1881). A technique was developed to transmit the disease from one animal to another. Brain tissue was extracted from a rabid dog under sterile conditions and then inoculated subdurally onto the surface of the brain of a healthy dog via trephination, which reduced the incubation period from a month in dogs with transmission from bites to less than three weeks. Subsequently, Pasteur noted progressive shortening of the incubation period to a limit of 6 to 7 days, in which the virus and the incubation period became "fixed" (Pasteur, 1885), and the term "fixed" has persisted and has become standard terminology for laboratory strains that have been generated in this manner. The fixed virus was found to be more neurovirulent than the street (wild-type) virus, and it had a shorter and more reproducible incubation period. Pasteur also observed that sequential passage in the CNS led to attenuation after peripheral inoculation (Pasteur, Chamberland, & Roux, 1884). In 1885, Pasteur successfully immunized the first patient, Joseph Meister, with his rabies vaccine.

Gerald Geison published information acquired from Pasteur's private laboratory notebooks, which did not become available until after the death of Pasteur's grandson in 1971. There was no printed catalogue of

FIGURE 1.1 **Portrait of Louis Pasteur working in his laboratory.** *From Images from the History of Medicine (National Library of Medicine) (http://ihm.nlm.nih.gov/images/B30055).*

the collection until 1985 (Geison, 1995). Pasteur had actually treated two patients with suspected rabies with his rabies vaccine prior to Meister (Figure 1.2), which never appeared in any of his publications. The first case was Girard, a 61-year-old Parisian man who had been bitten by a dog in March and was admitted to the Necker Hospital with a suspicion of rabies (Geison, 1995). On May 2, 1885, Girard was injected with a dose of Pasteur's vaccine prepared from desiccated rabbit spinal cords, but no further doses were given because of concerns by hospital authorities and the absence of Girard's attending physician. Girard deteriorated through May 6th, then improved and was discharged as cured on May 25th without further follow-up. In retrospect, it seems clear that Girard did not actually have rabies. The second case was Julie-Antoinette Poughon, an 11-year-old girl who had been bitten on the lip by her puppy in May and had clinical features of rabies and was admitted to the Hospital of St. Denis (Geison, 1995). She was given two doses of Pasteur's vaccine beginning on June 22, 1885, and she died on June 23rd. Interestingly, Pasteur had not done any experimental studies using the rabies vaccine on animals that had already exhibited clinical signs of rabies, with the exception of an unsuccessful treatment of a rabbit with rabies a few days after Girard's treatment had begun (Geison, 1995).

On July 6, 1885 Pasteur initiated treatment of a 9-year-old boy from a village in Alsace, Joseph Meister, who had been severely bitten two days earlier by a rabid dog with a dozen bites or more (some deep) involving his finger, thighs, and calves. He was treated with 13 inoculations of infected rabbit spinal cord material over 11 days, and physician

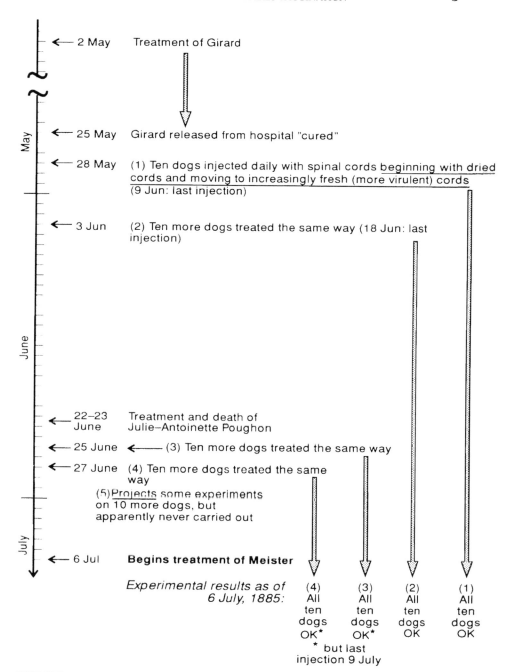

FIGURE 1.2   **Timeline of therapy of Girard, Julie-Antoinette Poughon, and Joseph Meister and of experimental studies on dogs with rabies vaccine by Louis Pasteur.** *(Reproduced with permission from GL Geison in The Private Science of Louis Pasteur, 1995, Princeton University Press, Princeton, NJ, pp 234–256; Copyright Princeton University Press.)*

Dr. Jacques-Joseph Grancher actually administered the vaccine. Pasteur used spinal cord tissue because this tissue was associated with a higher viral titer than brain tissue. The spinal cord tissue contained previously passaged virus that had been partially inactivated with progressively shorter periods of desiccation, which ranged from 15 days (first dose) to 1 day (last dose) (Pasteur, 1885). Joseph Meister never developed rabies. Émile Roux was Pasteur's leading collaborator on rabies studies, and, unlike Pasteur, he was qualified to practice medicine. He later became the third director of Institut Pasteur (1904–1933). Roux had an intimate knowledge of the experimental evidence with respect to the safety and efficacy of the vaccine. Roux and Pasteur had conflicts over rabies research, and it is likely that Roux refused to participate in Meister's treatment because he considered Pasteur's treatment a form of unjustified human experimentation because the available evidence at the time did not justify the treatment (Geison, 1990).

Next, Pasteur successfully treated Jean-Baptiste Jupille, who was a 15-year-old shepherd boy been bitten by a rabid dog from Villers-Farlay, which was near Pasteur's home town (Figure 1.3), beginning on October 20, 1885 (6 days after the bites). A year later, in 1886, Pasteur reported the results of treatment of 350 cases, and only one person developed rabies (Steele, 1975). This individual was a child bitten on October 3rd, but who was not brought in for treatment until November 9th. Over the next few decades hundreds of thousands of people with potential rabies exposures were immunized with nervous system vaccinations at the Institut Pasteur in Paris, which was founded in November, 1888, and in other

FIGURE 1.3 Statue of 15-year-old shepherd boy Jean-Baptiste Jupille and a rabid dog on the grounds of Institut Pasteur in Paris, France.

locations throughout the world. Later in life Meister became a caretaker of the Institut Pasteur. Meister committed suicide in 1940, apparently because he preferred to die rather than open the tomb of Pasteur to the invading Nazi forces (Haas, 1998).

Neuroparalytic complications of early rabies vaccines became recognized. Improved methods of preparing vaccine were developed with inactivation using carbolic acid. In 1911 Sir David Semple reported on a method of preparing an inactivated carbolized vaccine that was derived from nerve tissues of infected animals and became known as the Semple vaccine, and this vaccine was widely used in many countries for several decades (Semple, 1911).

6 EARLIEST PATHOGENESIS STUDIES FOCUSED ON PATHWAYS OF NEURAL SPREAD TO CNS

In 1889, DiVestea & Zagari (1889) showed that inoculation of rabies virus into the sciatic nerve of a rabbit and a dog caused rabies, and that death could be prevented by sectioning and cauterizing the nerve after injection. They also noted that the clinical signs depended on the location of the inoculated nerve and the site that it entered the CNS.

Goodpasture (1925) inoculated street rabies virus (using dog brain homogenates) into the masseter muscle of rabbits in order to investigate whether rabies virus spreads through "axis-cylinders" (within axons) or through perineural lymph spaces. He observed lesions involving the sensory division of the fifth cranial nerve and necrosis of neurons in the corresponding Gasserian ganglia and also in the pons, which were also associated with changes involving neurofibrillar material. He concluded that rabies virus passes in axons and also produces degenerative changes within these structures.

7 ADELCHI NEGRI AND NEGRI BODIES

Adelchi Negri (1875–1912) (Figure 1.4) attended medical school at the University of Pavia and worked with Camillo Golgi. After graduation in 1900, he became Golgi's assistant and began performing independent pathological studies on rabies. He used the Mann's staining method, which contained eosin, and observed eosinophilic cytoplasmic inclusion bodies in neurons from Ammon's horn in 50 of 75 brains from rabid dogs. Negri (1903) described these pathognomonic inclusion bodies detected with the newly developed histological techniques (Figure 1.5). He thought that the inclusions represented protozoan organisms. The putative parasitic organism was named *Neurocytes hydrophobiae*,

FIGURE 1.4 **Portrait of Adelchi Negri.** *From Images from the History of Medicine (National Library of Medicine) (http://ihm.nlm.nih.gov/images/ B019731).*

probably as a result of the similarity with Golgi's studies on malaria with the demonstration of *Plasmodium* in red blood cells (Kristensson, Dastur, Manghani, Tsiang, & Bentivoglio, 1996; Negri, 1909). Negri proposed that the detection of the bodies could be used for the diagnosis of rabies. Similar findings were subsequently demonstrated in different species.

8 PAUL REMLINGER

Paul Remlinger (1871–1964) worked in the Rabies Vaccine Laboratory in Constantinople, Turkey. In 1903, he was able to demonstrate the filterability of rabies virus. He and his co-workers mixed a fixed rabies virus brain homogenate with a virulent culture of the fowl cholera agent (*Pasteurella multocida*), and put the mixture through a Berkefeld V ultrafilter and inoculated the filtrate into rabbits; the absence of Pasteurella in the inoculated rabbits confirmed the success of the ultrafiltration, whereas their later death from rabies made clear that the etiologic agent of rabies was ultrafilterable (Murphy, 2012; Remlinger, 1903).

9 EARLY STUDIES USING ELECTRON MICROSCOPY

Electron microscopic studies of Negri bodies in the hippocampus of street rabies virus—infected mice were reported in 1951 (Hottle, Morgan,

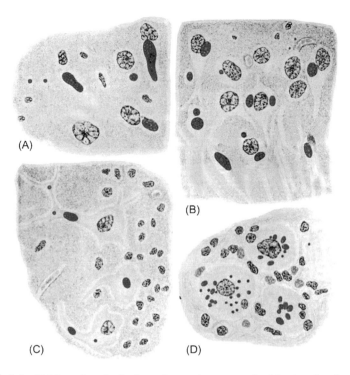

FIGURE 1.5   **Rabies virus inclusions shown in camera lucida drawing by Adelchi Negri from his original paper (reproduced from Negri A. Contributo allow studio dell'eziologia della rabia. Bollettino della Societa Medico-Chirurgica Pavia 2:88–115, 1903).** (A) Cerebral cortex of rabbit infected with street rabies virus with death occurring after 19 days. (B) Ammon's horn. (C) Cerebellum of a dog 19 days after subdural inoculation with street rabies virus. (D) Spinal ganglion of a rabbit infected in the eye with street rabies virus with death occurring after 17 days. Mann's staining.

Peers, & Wyckoff, 1951). They observed that Negri bodies contain small, dense granules and multiple oval vacuoles filled with less dense material. They clarified that there was no evidence that nucleolar constituents contribute directly to the development of Negri bodies. They did not observe any viral particles. In 1962, Seiichi Matsumoto (1918–1983) reported electron microscopic findings in neurons from mouse brain infected with street rabies virus (Matsumoto, 1962) (Figure 1.6). Dense homogeneous masses composed of filamentous aggregates (matrix) were observed in the cytoplasm in association with elongated (rod-like) particles with a width of 120 μm, which they thought were rabies virus virions. Three years later, Miyamoto & Matsumoto (1965) studied the nature of the Negri body by comparing Epon- and paraffin-embedded brain tissues from street virus-infected mice. They found that Negri bodies corresponded to cytoplasmic matrix, providing evidence that they are related

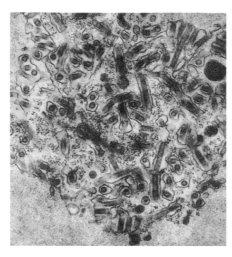

FIGURE 1.6   **Thin section electron microscopy of a rabies virus-infected mouse neuron showing an inner body composed of a large number of virus particles and some cytoplasmic organelles.** Adapted with permission. ©1967, Rockefeller University Press. From Miyamoto, K, Matsumoto, S. Comparative studies between the pathogenesis of street and fixed rabies virus infection. Originally published in The Journal of Experimental Medicine. 125:447–456. doi: 0.1084/jem.125.3.447.

to viral replication, rather than being degenerative structures. They also studied the ultrastructural pathology of street versus fixed virus strain infections and showed that the main difference was in the cellular damage that occurs in association with fixed viruses, rather than anything fundamentally different in viral morphogenesis (Murphy, 2012).

10 FLUORESCENT ANTIBODY STAINING BY GOLDWASSER AND KISSLING

In 1958, Goldwasser & Kissling (1958), who were working in the virology section of the Communicable Disease Center (later became the Centers for Disease Control) in Montgomery, Alabama, used indirect fluorescent antibody staining with human rabies virus antiserum as primary antibody and rabbit-antihuman globulins conjugated with fluorescein in order to demonstrate rabies virus antigens in impression smears of brain tissues infected with both street and fixed rabies viruses. They noted that the number and distribution of the larger fluorescent bodies correlated well with Negri bodies in the street rabies virus infection. This technique subsequently become important for rabies diagnosis and was also used as a powerful tool to follow the spread of the virus in experimental animal hosts at serial time intervals in pioneering rabies pathogenesis studies by Richard Johnson and Frederick Murphy and their colleagues (see below).

11 PATHOGENESIS STUDIES IN RODENTS BY RICHARD T. JOHNSON AND FREDERICK A. MURPHY AND CO-WORKERS

The fluorescent antibody technique of immunostaining played an important role in the first relatively modern experimental pathogenesis studies of rabies in mice by Richard Johnson. Johnson & Mercer (1964) first reported, in studies performed in Australia, fluorescent antibody staining of cerebellar Purkinje cells after intracerebral inoculation of mice with the CVS strain of fixed rabies virus. Purkinje cells were the only neural cell type noted to be infected in the cerebellum, and antigen was observed in the perikaryon and dendrites with an increase in the amount of staining from 2 to 6 days after post-inoculation. A year later, Johnson (1965) published the first detailed pathogenesis studies in experimental rabies in which the sequential infection of cells was studied using fluorescent antibody staining. This technique was very powerful in demonstrating the sequential spread of infections in hosts over time and provided important insights into our understanding of the pathogenesis of many viral infections. Brain and spinal cord tissues from adult or suckling mice were studied at serial time points after both hind limb foot pad and intracerebral inoculation of virus. These studies demonstrated that rabies virus was highly neuronotropic and sequential spread of the virus infection was demonstrated in the nervous system after the two different routes of inoculation. In this report, Johnson also studied the infection after hind limb foot pad inoculation. He was not able to detect rabies virus in the blood or viscera, and he was able to prevent the development of the disease with amputation of the feet and by ligation or sectioning of the sciatic nerve. This work showed that rabies virus spreads in the peripheral and central nervous systems by a neuronal route.

Later, Frederick Murphy and co-workers at the Centers for Disease Control in Atlanta performed further classical pathogenesis studies in suckling hamsters, which was a highly susceptible host, using street rabies virus and rabies-related virus (e.g., Mokola virus) strains (Murphy, Bauer, Harrison, & Winn, 1973; Murphy, Harrison, Winn, & Bauer, 1973). They found early infection in striated muscle cells near the site of inoculation (Figure 1.7A), and then infection of nearby neuromuscular and neurotendinal spindles; later there was involvement of peripheral nerves (Murphy, Bauer, et al., 1973). They found viral antigen in dorsal root ganglion neurons (Figure 1.7B) and detected virus and viral nucleocapsid inclusions in axons by electron microscopy, and they concluded that rabies virus spreads to the CNS in the axoplasm of peripheral nerves. After CNS infection was established, they found evidence of spread of the infection to the eye and multiple organs, including the intestines, adrenal medulla, pancreas, brown adipose tissue, myocardium, taste

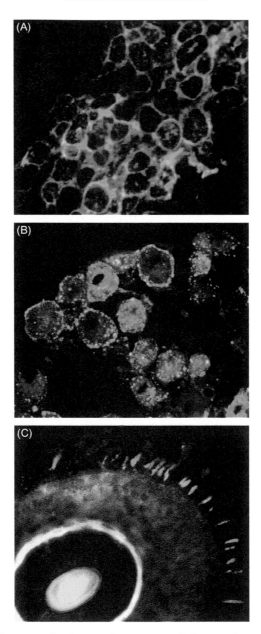

FIGURE 1.7   **Frozen section immunofluorescence of a hamster inoculated in a hindlimb muscle with Mokola virus.** Striated muscle of the inoculated limb showing rabies virus antigen primarily peripherally in myocytes in a transverse section (A). Lumbar dorsal root ganglion showing abundant rabies virus antigen in numerous neuronal cell bodies at 6 days post-infection (B). Rabies virus antigen in nerve endings at the root of a tactile hair (whisker) at 5 days post-infection (C), resulting from centrifugal spread of rabies virus. *Courtesy of Dr. Frederick Murphy, University of Texas Medical Branch, Galveston, Texas.*

buds, olfactory epithelium, salivary glands, and nerve endings associated with tactile hairs (Figure 1.7C), indicating centrifugal peripheral
neural spread of the viral infection (Murphy, Harrison, et al., 1973).

12 FIRST CLONE OF A RABIES VIRUS GENE

In 1983, Yelverton and co-workers cloned the mature full length rabies
virus glycoprotein gene of the CVS strain into plasmids for its direct
expression in *Escherichia coli* (Yelverton, Norton, Obijeski, & Goeddel,
1983). This allowed the amino acid sequence of the viral glycoprotein
to be deduced. This was one of the important early steps in developing
methods for a variety of molecular diagnostic studies. It also lead the way
to further developments, including the generation of recombinant viruses
for use as oral rabies vaccines for wildlife (e.g., vaccinia recombinant virus
containing the rabies virus glycoprotein) (Rupprecht et al., 1986; Wiktor et
al., 1984) and for the development of reverse genetics technology by Karl-
Klaus Conzelmann and colleagues in Munich (Conzelmann, 1998), which
have proved to be very powerful tools for rabies pathogenesis studies and
for the development of novel rabies vaccines (Schnell, Tan, & Dietzschold,
2005).

References

Conzelmann, K. K. (1998). Nonsegmented negative-strand RNA viruses: genetics and manipulation of viral genomes. *Annual Review of Genetics, 32*, 123–162.

DiVestea, A., & Zagari, G. (1889). Sur la transmission de la rage par voie nerveuse. *Annales
de L'Institut Pasteur (Paris), 3*, 237–248.

Fleming, G. (1872). *Rabies and hydrophobia: Their history, nature, causes, symptoms, and prevention*. London: Chapman and Hall.

Fracastoro, G. (1930). Rabies: *Hieronymi Fracastorii, De contagione et conlugiosis mor-
bis et eorum curatione, libri III [English translation by W. C. Wright]*. New York: Putnam.
pp. 124–133.

Galtier, V. (1879). Étude sur la rage. *Comptes Rendus Hebdomadaires des Seances de l Academie
des Sciences, 89*, 444–446.

Geison, G. L. (1990). Pasteur, Roux, and rabies: scientific versus clinical mentalities. *Journal
of the History of Medicine and Allied Sciences, 45*, 341–365.

Geison, G. L. (1995). *The private science of Louis Pasteur*. Princeton, N.J.: Princeton University
Press.

Goldwasser, R. A., & Kissling, R. E. (1958). Fluorescent antibody staining of street and fixed
rabies virus antigens. *Proceedings of the Society for Experimental Biology and Medicine, 98*,
219–223.

Goodpasture, E. W. (1925). A study of rabies, with reference to a neural transmission of the
virus in rabbits, and the structure and significance of Negri bodies. *American Journal of
Pathology, 1*, 547–584.

Haas, L. F. (1998). Louis Pasteur (1822–95). *Journal of Neurology, Neurosurgery, and Psychiatry,
64*(3), 330.

Hottle, G. A., Morgan, G., Peers, J. H., & Wyckoff, R. W. G. (1951). The electron microscopy of rabies inclusion (Negri) bodies. *Proceedings of the Society for Experimental Biology and Medicine, 77*, 721–723.

Hunter, J. (1793). Observations and heads of inquiry on canine madness. *Transactions of a Society for the Improvement of Medical and Chirurgical Knowledge, 1*, 294–329.

Johnson, R. T. (1965). Experimental rabies: studies of cellular vulnerability and pathogenesis using fluorescent antibody staining. *Journal of Neuropathology and Experimental Neurology, 24*, 662–674.

Johnson, R. T., & Mercer, E. H. (1964). The development of fixed rabies virus in mouse brain. *Australian Journal of Experimental Biology and Medical Science, 42*, 449–456.

Kristensson, K., Dastur, D. K., Manghani, D. K., Tsiang, H., & Bentivoglio, M. (1996). Rabies: interactions between neurons and viruses: a review of the history of Negri inclusion bodies. *Neuropathology and Applied Neurobiology, 22*, 179–187.

Magendie, F. (1821). Expérience sur la rage. *Journal de Physiologie Expérimentale, 1*, 40–46.

Matsumoto, S. (1962). Electron microscopy of nerve cells infected with street rabies virus. *Virology, 17*, 198–202.

Miyamoto, K., & Matsumoto, S. (1965). The nature of the Negri body. *Journal of Cell Biology, 27*, 677–682.

Morgagni, J. B. (1960). Wherein madness, melancholy, and hydrophobia, are treated of [English translation by B. Alexander] *The seats and causes of diseases investigated by anatomy* (vol. 1). New York: Hafner Publishing Co.. pp. 144–187.

Murphy, F. A. (2012). *The foundations of virology*. West Conshohocken, PA: Infinity Publishing.

Murphy, F. A., Bauer, S. P., Harrison, A. K., & Winn, W. C. (1973). Comparative pathogenesis of rabies and rabies-like viruses: viral infection and transit from inoculation site to the central nervous system. *Laboratory Investigation, 28*, 361–376.

Murphy, F. A., Harrison, A. K., Winn, W. C., & Bauer, S. P. (1973). Comparative pathogenesis of rabies and rabies-like viruses: infection of the central nervous system and centrifugal spread of virus to peripheral tissues. *Laboratory Investigation, 29*, 1–16.

Negri, A. (1903). Beitrag zum studium der aetiologie der tollwuth. *Zeitschrift fur Hygiene und Infektionskrankheiten, 43*, 507–528.

Negri, A. (1909). Uber die morphologie und der entwicklungszyklus des parasiten der tollwut (Neurocytes hydrophobiae Calkins). *Zeitschrift fur Hygiene und Infektionskrankheiten, 63*, 421–440.

Pasteur, L. (1885). Méthode pour prévenir la rage après morsure. *Comptes Rendus de l'Académie des Sciences, 101*, 765–774.

Pasteur, L., Chamberland, C. E., Roux, E., & Thuillier, L. (1881). Sur la rage. *Comptes Rendus de l'Académie des Sciences, 92*, 1259–1260.

Pasteur, L., Chamberland, C. E., & Roux, E. (1884). Nouvelle communication sur la rage. *Comptes Rendus de l'Académie des Sciences, 98*, 457–463.

Remlinger, P. (1903). Le passage du virus rabique à travers les filtres. *Annales de L'Institut Pasteur (Paris), 101*, 765–774.

Rupprecht, C. E., Wiktor, T. J., Johnston, D. H., Hamir, A. N., Dietzschold, B., Wunner, W. H., et al. (1986). Oral immunization and protection of raccoons (*Procyon lotor*) with a vaccinia-rabies glycoprotein recombinant virus vaccine. *Proceedings of the National Academy of Sciences of the United States of America, 83*, 7947–7950.

Schnell, M. J., Tan, G. S., & Dietzschold, B. (2005). The application of reverse genetics technology in the study of rabies virus (RV) pathogenesis and for the development of novel RV vaccines. *Journal of Neurovirology, 11*(1), 76–81.

Semple, D. (1911). *The preparation of a safe and efficient antirabic vaccine*. Calcutta: Superintendent Government Printing, India. (Scientific memoirs by officers of the medical and sanitary departments of the government of India. New series, no. 44).

Steele, J. H. (1975). History of rabies. In G. M. Baer (Ed.), *The natural history of rabies* (vol. 1, pp. 1–29). New York: Academic Press.

Thomsen, A. G., & Blaisdell, J. (1994). From the fangs of Cerberus: the possible origin of classical beliefs about rabies. *Argos, 11,* 5–8.

Wiktor, T. J., Macfarlan, R. I., Reagan, K. J., Dietzschold, B., Curtis, P. J., Wunner, W. H., et al. (1984). Protection from rabies by a vaccinia virus recombinant containing the rabies virus glycoprotein gene. *Proceedings of the National Academy of Sciences of the United States of America, 81,* 7194–7198.

Wilkinson, L. (1977). The development of the virus concept as reflected in corpora of studies on individual pathogens: rabies—two millennia of ideas and conjecture on the aetiology of a virus disease. *Medical History, 21,* 15–31.

Yelverton, E., Norton, S., Obijeski, J. F., & Goeddel, D. V. (1983). Rabies virus glycoprotein analogs: biosynthesis in *Escherichia coli. Science, 219,* 614–620.

Zinke, G. G. (1804). *Neue Ansichten der Hundswuth; ihrer Ursachen und Folgen, nebst einer sichern Behandlungsart der von tollen Thieren gebissenen Menschen.* Jena, Germany: Gabler.

2

Rabies Virus

William H. Wunner[1] and Karl-Klaus Conzelmann[2]

[1]The Wistar Institute, Philadelphia, Pennsylvania, 19104-4268, USA
[2]Max von Pettenkofer-Institute and Gene Center, Ludwig-Maximilians-University, 81377 Munich, Germany

1 INTRODUCTION

Rabies virus (RABV) is the prototype virus of the genus *Lyssavirus* (from the Greek *lyssa,* meaning "rage") in the family *Rhabdoviridae* (from the Greek *rhabdos,* meaning "rod") of the order *Mononegavirales* (MNV). RABV, the causative agent of classic rabies in animals and humans, is a highly neurotropic virus in the mammalian host invariably causing a fatal encephalomyelitis once the infection is established and has reached the brain. RABV is distributed worldwide among specific mammalian reservoir hosts comprising various carnivore and bat species. Fourteen lyssaviruses, which share certain morphological and structural characteristics with RABV and with the exception of Lagos bat virus (LBV) have caused a rabies-like encephalomyelitis in humans are currently recognized within the *Lyssavirus* genus. Twelve of the fourteen lyssaviruses, which differentiate genetically, segregate into two phylogroups based on phylogenetic analyses. Phylogroup I includes the classic (prototype) RABV, Duvenhage virus (DUVV), European bat lyssavirus, type 1 (EBLV-1), and type 2 (EBLV-2), and Australian bat lyssavirus (ABLV), the putative species Aravan virus (ARAV), Khujand virus (KHUV), Irkut virus (IRKV), and Bokeloh bat lyssavirus (BBLV), the newest species (Freuling et al., 2011) to be characterized and considered an independent species within phylogroup I (Hanlon et al., 2005; Kuzmin, Niezgoda, et al., 2008). Phylogroup II includes LBV, Mokola virus (MOKV), and Shimoni bat virus (SHIBV), the newest species to be characterized and considered an independent species within phylogroup II (Kuzmin et al., 2010). West Caucasian bat virus (WCBV) and Ikoma Lyssavirus (IKOV), isolated from an African civet (Marston et al., 2012), which do not cross-react serologically with any members of the two phylogroups, could tentatively belong to a third phylogroup; see Chapters 3 and 4. Phylogenetic analyses suggest that all lyssaviruses have originated from a precursor bat virus. Lyssavirus species share many of the biologic and physico-chemical features that are associated with other viruses of the *Rhabdoviridae* family. These include the bullet-shaped virus morphology, helical nucleocapsid (NC) or ribonucleoprotein (RNP) core, and general organization of the viral RNA (vRNA) genome and structural proteins. In contrast to all other rhabdoviruses, however, lyssaviruses are not transmitted by insect vectors and have adapted to direct transmission. The five structural proteins of the lyssavirus particle (virion) include a nucleocapsid protein (N), phosphoprotein (P), matrix protein (M), glyco-protein (G) and RNA-dependent RNA polymerase or large protein (L). These lyssavirus proteins generally share many of the biologic functions that the same viral proteins have in other rhabdoviruses. Some of the structural proteins of lyssaviruses, on the other hand, can differ dramati-cally in their antigenic properties and in their posttranslational modifi-cations to convey different, often specific properties that distinguish lyssaviruses from other rhabdoviruses. Lyssaviruses, like other rhabdo-viruses, also use similar mechanisms to enter susceptible cells (although they may use different receptors), express and replicate their genome RNA, and assemble and release mature progeny virions from the plasma membrane or internal membranes of infected cells.

2 RABIES VIRUS ARCHITECTURE

2.1 Virus Structure and Composition

Lyssaviruses, like other rhabdoviruses, consist mainly of RNA (2–3%), protein (67–74%), lipid (20–26%), and carbohydrate (3%) as integral com-ponents (percent of total mass) of their structure. The vRNA genome forms the backbone of the tightly coiled helical RNP (RNA plus protein) core, which extends along the longitudinal axis of the bullet-shaped virus particle. Included in the RNP core are the protein components, N, P, and L, which are surrounded by the viral membrane proteins, M and G, and a mixture of lipoprotein components derived from the cell mem-brane that form the outer envelope or "membrane matrix" of the virion (Figure 2.1). The RNA genome is single-stranded and non-segmented, and has a negative-sense (minus-strand) polarity. This implies that the genome RNA is not infectious.

The five viral genes, which encode the structural proteins of the virus (Figure 2.2), are arranged in a strictly conserved order (3′-N-P-M-G-L-5′) and are flanked by short terminal regulatory sequences (see Section 3.1).

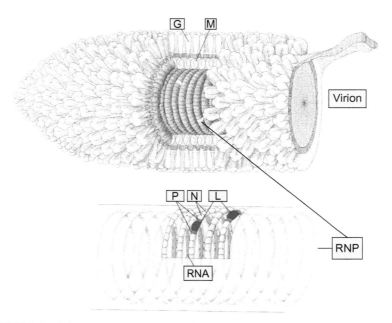

FIGURE 2.1   **Schematic representation of the rabies virion.**
The drawing shows the internal ribonucleoprotein (RNP) core consisting of the single-strand, negative-sense genome RNA encapsidated with nucleocapsid protein (N), the virion-associated RNA polymerase (L) and polymerase cofactor phosphoprotein (P). The RNP core in association with the matrix protein (M) is condensed into the typical bullet-shape particle that is characteristic of rhabdoviruses. A lipid bilayer envelope (or membrane) in which the surface trimeric glycoprotein (G) spikes are anchored surrounds the RNP-M structure. The membrane 'tail' depicted in the drawing represents the trailing piece of envelope that is frequently observed in the electron microscope attached to the virus as it buds from the plasma membrane of the infected cell. [*Reproduced from Wunner, W. H., Larson, J. K., Dietzschold, B., and Smith, C. L., Review of Infectious Diseases* **10**, *Supplement 4, S771–S784, 1988, with permission.*]

Rabies Virus Genome Structure

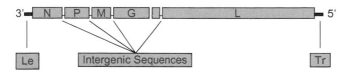

FIGURE 2.2   **Organization of the rabies virus genome.**
The nucleoprotein (N), phosphoprotein (P), matrix protein (M), glycoprotein (G), and large RNA-polymerase protein (L) genes are separated by intergenic di- (at N/P gene border), penta- (at P/M and M/G gene borders), and 19 to 28-nucleotide (at G/L border) sequences (white bars) and are flanked by the leader (Le) and trailer (Tr) sequences (blue bars) at the 3′ and 5′ ends, respectively.

2.2 Morphology and Core Structure of Standard and Defective Virus Particles

RABV particles are best described as bacilliform, rod or bullet-shaped, with one end rounded (hemispherical) and the other flattened (planar), the morphological hallmark of most rhabdoviruses (Figure 2.1). The average length of standard-size, infectious rabies virions measures 180 nm (130–250 nm), and the average diameter is 75 nm (60–110 nm) (Davies, Englert, Sharpless, & Cabasso, 1963; Hummeler, Koprowski, & Wiktor, 1967). Cryo-electron microscopy (cryo-EM) more recently revealed that the outermost lipid bilayer leaflet of the bullet-shaped rhabdovirus, vesicular stomatitis virus (VSV), measures 700 Å (Ge et al., 2010). The exact length of the virion varies: 1960 ± 80 Å, from the conical end, which comprises ~25% of the total length, through the cylindrical (helical) trunk, comprising ~75%. The conical end contains approximately seven turns of a spiral before reaching the cylindrical (helical) trunk. The trunk of a typical virion contains approximately 29 spiral turns (Ge et al., 2010). Docking of the "ring" structure of N and RNA (for details, see Section 3.6. below) into the cryo-EM structure of the RNP establishes the directionality of vRNA in the virion. The docked ring structure shows that the 5' end is at the conical tip of the bullet and the 3' end is at the base of the trunk. Therefore, it follows that the bullet tip defines the origin and that assembly starts at the apex of the virion tip, with the helical trunk forming "downward" from the origin. Cryo-EM tomography of RABV confirms the canonical bullet-shaped morphology of the lyssavirus with a diameter of 81 nm and a length of 188 nm but showed this for only 33% of the total particle number (Guichard et al., 2011).

Beside standard-size bullet-shaped rabies virions, other shorter (truncated), often cone-shaped 'defective' virions are sometimes co-produced, particularly in cell culture. They contain RNA genomes, which are typically shorter than full-size genomes as a result of genome truncations or internal sequence deletions from the genome (Conzelmann, Cox, & Thiel, 1991; Lazzarini, Keene, & Schubert, 1981; Marriott & Dimmock, 2010).

Some defective rabies virions, like defective virions of other rhabdoviruses and other RNA viruses, replicate at the expense of standard 'helper' virus. Defective virions can grow to become the dominant particle type(s) in infected cells and do so by interfering with production of standard infectious virus. These are therefore known as defective-interfering (DI) virions (Clark, Parks, & Wunner, 1981; Wiktor, Dietzschold, Leamnson, & Koprowski, 1977). DI virions of RABV are readily generated in standard cell cultures infected with laboratory-adapted (fixed) strains of RABV (Clark et al., 1981; Wunner & Clark, 1980). They have not been described in rabies virus infections *in vivo*, although their role in controlling production of infectious virions *in vivo* has been suggested (Marriott & Dimmock, 2010).

At the core of all standard and defective virions is a ribbon of tightly coiled yet flexible, right-handed helical RNP that has a periodicity of approximately 7.5 nm per turn. The tightly coiled RNP core in the standard-size infectious virion measures approximately 165 nm in length and 50 nm in diameter. During virus assembly, the RNP core is associated with M to form the 'skeleton' structure of the virus (Mebatsion, Weiland, & Conzelmann, 1999). As virus particles mature and bud through the cellular membrane, the skeleton structure acquires a lipid bilayer envelope (7.5 to 10 nm thick) that surrounds the mature virion. Located on the external surface of the viral envelope are the surface projections that measure 8.3 to 10 nm in length. Each projection or spike contains three molecules (a trimer) of the viral G (Gaudin, Ruigrok, Tuffereau, Knossow, & Flamand, 1992). These have been described when viewed in the EM as the short spikes extending outward with the appearance of hollow knobs at their distal ends. It is estimated that the height of the hollow knobs or 'heads' of the spike is about 4.8 nm; the rest of the spike is made up of the thin ~3.5 nm-'stalk' on which the head rests (Gaudin et al., 1992).

2.3 Viral Proteins

The three viral proteins located in the viral RNP core (Figure 2.1) are the N, the noncatalytic polymerase-associated P, and the catalytic L component of the virion-associated RNA polymerase. All three proteins are involved in the RNA polymerase activity of the virion. Both the N and P are phosphorylated in RABV, unlike in other rhabdoviruses, including VSV, in which only the P is phosphorylated (Gupta, Blondel, Choudhary, & Banerjee, 2000; Sokol et al., 1974). The most abundant protein in the RNP core is N (1325 or 1800 copies) followed by P (691 or 950 copies) and L (25 or 72 copies) (Flamand, Raux, Gaudin, & Ruigrok, 1993; Madore & England, 1977). The stoichiometric relationship that emerges from these independent estimates, however, indicates that the N:P ratio in the RNP complex is 2:1 per virion, which is supported by recent structural analyses. The proteins of the RNP core and the association of N with P and L will be discussed in more detail later (see Section 3.2).

The remaining two structural proteins of the RABV, G and M, are associated with the lipid-bilayer envelope that surrounds the RNP core. The M is a small-size protein that lines the viral envelope. A part of the M is attached at one end to the viral envelope, and the rest sticks into the interior of the virion, where it interacts with N to form an inner leaflet between the envelope and RNP core (Guichard et al., 2011). Interestingly, the multifunctional M associates with both the RNP and the viral G, collaborating with G in infected cells, to produce progeny virions in the budding process at the cell membrane (Mebatsion et al., 1999; Nakahara et al., 1999).

The G is the only glycosylated protein. It produces the trimeric spike-like projections on the surface of the viral envelope (Gaudin et al., 1992). The viral G molecule is glycosylated with branched-chain oligosaccharides, which account for 10–12% of the total mass of the protein. The number of G and M molecules per virion has been estimated to be 1,205 and 1,148 (Flamand et al., 1993), respectively, and 1,800 and 1,547 (Madore & England, 1977), respectively. This calculates to approximately 450 trimeric G spikes distributed on the outer surface of each virion. The rabies virion envelope contains other host-derived minor protein components such as actin and heat shock proteins of the Hsp70 type, CD44 and CD99-related glycoprotein (VAP21) similar to other negative-strand RNA viruses (Lahaye, Vidy, Fouquet, & Blondel, 2012; Naito & Matsumoto, 1978; Sagara et al., 1998). It is possible that the molecular chaperones, such as the heat shock protein calnexin, that associate with the viral proteins during synthesis are incorporated into virions after binding to and assisting in G folding (Gaudin, 1997). In a similar manner, cytoskeleton proteins normally expressed underneath the host cell surface may be incorporated into virions as a consequence of their proximal location and function in virus budding (Sagara et al., 1998). Cellular kinases that activate the transcriptional function of P in RABV may also be packaged into rabies virions (Gupta et al., 2000).

2.3.1 Nucleoprotein (N)

The RABV N (Pasteur virus (PV) strain) contains 450 amino acids, one of which, the serine residue at position 389 (S389), is phosphorylated, and has a molecular weight of ~57 kDa. The N in RABV appears to be phosphorylated by a cellular casein kinase II (Gupta et al., 2000; Wu, Lei, & Fu, 2003). The amino acid sequence of N is the most conserved of the viral proteins among the lyssaviruses (Marston et al., 2007). Despite the conserved nature of the N, there is a relatively high degree of genetic diversity within short segments of the N gene between the genotypes (Bourhy, Kissi, & Tordo, 1993; Bourhy et al., 1999; Kissi, Tordo, & Bourhy, 1995). An important reason for the high level of amino acid sequence conservation within specific regions in N is that it must retain certain key functions that are dependent on specific protein-RNA genome interactions (see Section 3.2). On the other hand, the noted amino acid differences provide unique, genotype-specific epitopes on the N that define antigenic relationships between virus strains within and between genotypes on the basis of their reactivity patterns (antigenicity) with a panel of anti-N monoclonal antibodies (MAbs) (Dietzschold, Lafon, et al., 1987; Flamand, Wiktor, & Koprowski, 1980a, 1980b; Smith, 1989). The exploitation of the qualitative diversity in the N gene at the nucleotide level has also lead to an extensive analysis of phylogenetic relationships of lyssaviruses and suggested quantitative criteria for lyssavirus genotype definition using the polymerase chain reaction (PCR) and nucleotide sequencing technologies; see Chapters 3 and 10.

The binding site on N for its interaction with vRNA was initially localized, using peptide fragments of N that bound leader RNA probe, to a region of N between amino acid residues 298 and 352 (Kouznetzoff, Buckle, & Tordo, 1998). After binding to vRNA, the N undergoes sufficient conformational change to acquire a number of conformation-dependent epitopes, one of which enables S389 of N to be phosphorylated (Dietzschold, Lafon, et al., 1987; Kawai et al., 1999; Toriumi & Kawai, 2004). Phosphorylation of N Ser389 stabilizes the interaction between N and P in the rabies viral RNP complex (Toriumi & Kawai, 2004). It has also been suggested that the phosphorylation of N during vRNA or complementary viral RNA (cRNA) encapsidation is important to regulate vRNA transcription and replication (Liu, Yang, Wu, & Fu, 2004; Wu, Gong, Foley, Schnell, & Fu, 2002; Yang, Koprowski, Dietzschold, & Fu, 1999).

The N is the second most extensively analyzed of the RABV proteins (after the G) with respect to its antigenic and immunogenic structure and function. The immunological interest in N stems from the observation that the RNP of RABV induces protective immunity against a peripheral challenge of lethal rabies virus in animals (Dietzschold, Wang, et al., 1987). Three linear epitopes (antibody binding sites) on N were mapped to amino acids 358 to 367 (antigenic site I), and three linear epitopes (antigenic site IV) were mapped to two independent regions, amino acids 359 to 366 and 375 to 383 (Goto et al., 2000). Although the stretch of amino acids between residues 359 and 366 is shared by the two independent antigenic sites I and IV, the MAbs that recognize epitopes within these sites do not compete with each other for binding to the N antigen. Thus, it would appear that the respective epitopes are detected on different forms of N, one that represents N that is diffusely distributed in the cytoplasm, and the other that is associated with viral RNPs (Goto et al., 2000; Jiang et al., 2010). The fact that the N associated with cytoplasmic inclusion bodies (IBs) or Negri bodies (NBs) (Lahaye et al., 2009) may represent the N in its mature form is suggested by a MAb specific for antigenic site II, which only recognizes a conformation-specific epitope on the IB-associated N antigen. These and the conformation-dependent epitopes present in antigenic sites II and III and also at the phosphorylation site (serine 389) of N-RNA, and others yet to be mapped for which MAbs are available, provide valuable diagnostic tools (Kawai et al., 1999; Lahaye et al., 2009).

The N is also a major target for T helper (Th) cells that cross-react among rabies and rabies-related viruses (Celis, Karr, Dietzschold, Wunner, & Koprowski, 1988; Ertl et al., 1989). Several Th cell epitopes in the RABV N were identified and mapped using a series of overlapping synthetic peptides corresponding to N sequences of approximately 15 amino acids in length (Ertl et al., 1989).

Finally, the RABV N functions as an exogenous superantigen (Lafon et al., 1992). It is perhaps the only viral superantigen that has been identified in humans. Some of the properties and responses found not only in humans but also in mice that are attributable to the RABV N in the role of superantigen include (1) its potent activation of peripheral blood lymphocytes in human vaccines, (2) its ability to produce a more rapid and heightened virus-neutralizing antibody (VNA) response upon injection of inactivated rabies vaccines, (3) its induction of early T cell activation steps and expansion and mobilization of CD4$^+$ Vb8 T cells to trigger and support production of VNA, and (4) its ability to bind to HLA class II antigens expressed on the surface of cells; (see Lafon et al., 1992).

2.3.2 Phosphoprotein (P)

The RABV P (PV strain) contains 297 amino acids (38 kDa) and is the least conserved of the five RABV proteins. The diversity of P sequences is also greatest (43–97%) among certain pairings of lyssaviruses with two regions of greatest variability between residues 52–78 and 155–178 (Marston et al., 2007). The P is a multifunctional and multifaceted protein. It interacts with the N to form N-P complexes and acts as a chaperone for newly synthesized N, preventing its polymerization (self-assembly) and nonspecific binding to cellular RNA (Mavrakis et al., 2003) and specifically directs N encapsidation of the vRNA (Chenik, Chebli, Gaudin, & Blondel, 1994; Fu, Zheng, Wunner, Koprowski, & Dietzschold, 1994; Gigant, Iseni, Gaudin, Knossow, & Blondel, 2000). As a subunit of the RNA polymerase (P-L) complex, the P plays a pivotal role as a noncatalytic cofactor in transcription and replication of the viral genome. The P stabilizes the L and places the P-L complex on the RNA template, which the L alone is unable to do (Chenik et al., 1994; Chenik, Schnell, Conzelmann, & Blondel, 1998; Fu et al., 1994).

The P of lyssaviruses, like the P in other negative-strand viruses, exists in a variety of phosphorylated forms. Two prominent forms of P are present in both rabies virions and in virus-infected cells. One is a major hypophosphorylated 37-kDa form, and the other is a minor hyperphosphorylated 40-kDa form (Toriumi & Kawai, 2004). RABV P is phosphorylated in the N-terminal portion by two distinct types of protein kinases, one of which is a unique heparin-sensitive protein kinase (Gupta et al., 2000; Takamatsu et al., 1998). This unique 71-kDa kinase, designated RABV protein kinase (RABV-PK), phosphorylates recombinant P (36 kDa, expressed in *E. coli*) at S63 and S64 (sequence of the challenge virus standard (CVS) strain) and nascent P (37 kDa) expressed in baby hamster kidney (BHK)-21 cells infected with the high egg passage (HEP)-Flury strain of RABV. In both cases, hyperphosphorylation alters the mobility of P

(36 kDa and 37 kDa) in sodium dodecyl sulfate-polyacrylamide gel electrophoresis (SDS-PAGE), causing it to migrate more slowly, as a protein of 40 kDa (Toriumi & Kawai, 2004). The other phosphorylating enzyme is protein kinase C, which has several isomers (PKCα, β, γ, and ζ). In contrast to the RABV-PK, phosphorylation of P by the PKC isoforms, dominated by PKCγ activity, did not alter the migration of P in SDS-PAGE (Gupta et al., 2000). Upon analyzing the PK activity in rabies virions for the presence of these two types of enzymes, it was concluded the RABV-PK is selectively packaged in mature rabies virions along with a smaller amount of the predominant PKCγ isoform as the rabies virion-associated PKs.

RABV P interacts with the nascent soluble N (N°) and L via domains of P that are specific for each of these two proteins (Castel et al., 2009; Chenik et al., 1998). At least two independent N-binding sites have been found on P. One binds nascent RABV P to nascent N° to maintain N° in a competent form for RNA encapsidation, and the other binds to the C-terminal part of N in the RNA-N complex (Chenik et al., 1994; Fu et al., 1994; Schoehn, Iseni, Mavrakis, Blondel, & Ruigrok, 2001). One site is located within the C-terminal 30 amino acids (between amino acids 267 and 297) of P, and the other is located in the N-terminal portion of the protein between amino acids 69 and 177 (Chenik et al., 1994). Both sites interact with N in a manner that is mutually independent (Fu et al., 1994). The interaction involving the N-terminal binding site (first 40 amino acids) of P requires that the P interact with N° soon after the two proteins are synthesized *in vivo* (Castel et al., 2009; Mavrakis et al., 2006), but that it may compete with endogenous RNA (Albertini, Ruigrok, & Blondel, 2011). The P is able to form elongated dimers, (Gerard et al., 2007; Mavrakis et al., 2003).

In the formation of progeny RNP, the P is required to bind L, the large catalytic subunit of the viral RNA polymerase, to produce a virus-encoded RNA polymerase complex that is fully active. The P subunit has a major binding site for L within the first 19 amino acids of P (Chenik et al., 1998), which interacts through the N-terminal unfolded region of P (Castel et al., 2009; Gerard et al., 2007; Gigant et al., 2000; Spadafora, Canter, Jackson, & Perrault, 1996). Oligomerization of RABV P does not require phosphorylation, nor is the N-terminal domain (first 52 amino acids) necessary for oligomerization or binding to the N-RNA (cRNA and vRNA) template (Gigant et al., 2000; Mavrakis et al., 2003). This is in contrast to the P of VSV, which requires phosphorylation for oligomer formation, to be fully active and available for binding both to L and to the RNA template (Albertini et al., 2011; Ding, Green, Lu, & Luo, 2006; Gao et al., 1996; Gerard et al., 2007).

Other protein–protein interactions of RABV P involve cellular proteins. The 10-kDa cytoplasmic dynein light chain (LC8), which is involved in the intracellular transport of organelles was found to interact strongly with the P of RABV and MOKV (Jacob, Badrane, Ceccaldi, & Tordo, 2000; Raux, Flamand, & Blondel, 2000). The P domain that interacts with dynein LC8 was mapped to the N-terminal half of the P, between amino acids 138 and 172 in the P. Speculation initially that binding of P to LC8 could facilitate axonal transport of the RNP complex via a microtubule-mediated process appears not to be the model for axonal transport of RABV, since deletion of the dynein LC8 binding domain within the P did not change RABV entry into the central nervous system (CNS) (Rasalingam, Rossiter, Mebatsion, & Jackson, 2005). Deletion of the dynein LC8 binding domain however significantly suppressed viral transcription and replication in the CNS resulting in loss of viral infectivity in the CNS (Tan, Preuss, Williams, & Schnell, 2007).

RABV P is also responsible for inhibiting the activation of latent interferon regulatory factor (IRF-3), the key factor for initiating an interferon (IFN) response (Honda & Taniguchi, 2006). It prevents the phosphorylation of serine 386 at the C terminus of the IRF-3 protein, which allows dimers of IRF-3 to form and be recruited to the IFN-β enhancer as part of a larger protein complex that includes transcription factors and co-activators (Brzozka, Finke, & Conzelmann, 2005). Thus, RABV P is necessary and sufficient to prevent a critical IFN response in virus-infected cells.

2.3.3 Virion-Associated RNA Polymerase or Large Protein (L)

The RABV L (PV strain) contains 2,142 amino acids (2,127 amino acids in SAD-B19 strain) (244 kDa) and is encoded in the fifth gene, which comprises more than half (54%) of the coding potential of the RABV genome. The L is the catalytic component of the polymerase complex, which, along with the noncatalytic cofactor P, is responsible for the majority of enzymatic activities involved in vRNA transcription and replication. Many of the activities of this multifunctional enzyme have been identified in genetic and biochemical studies using VSV, the prototype virus and model for studying the virion-associated RNA polymerase of negative-strand RNA viruses (Banerjee & Chattopadhyay, 1990; Poch, Blumberg, Bougueleret, & Tordo, 1990). As in all negative-strand RNA viruses, the RABV virion-associated viral RNA polymerase plays a unique role at the start of infection by initiating the primary transcription of the genome RNA after the RNP core is released into the cytoplasm of the infected cell. The enzymatic steps of transcription include initiation and elongation of the leader RNA and mRNA transcripts, as well as the cotranscriptional modifications of the mRNAs that include 5′-capping, methylation, and 3′-polyadenylation. Comparisons of L sequences from different negative-strand RNA viruses have helped to map the functionally homologous and unique sequences in attempts to locate the ascribed enzyme activities (Barik, Rud, Luk, Banerjee, & Kang, 1990; Poch et al., 1990; Tordo, Poch, Ermine, Keith, & Rougeon, 1988).

One of the main features of L that comes out of the sequence comparison is the number of clusters of conserved amino acid residues that appear to be purposefully aligned in blocks (or boxes), I-VI, along the protein (Lij, Rahmeh, Morelli, & Whelan, 2008; Poch et al., 1990). Within these boxes, some residues form strongly conserved domains, with a high proportion of amino acids either strictly or conservatively maintained in identical positions, while other domains are more variable, consistent with the multifunctional nature of L (Banerjee & Chattopadhyay, 1990; Poch et al., 1990; Tordo et al., 1988). One of the blocks (box III), the catalytic domain, in the central part of the RABV L, between residues 530 and 1177, contains four motifs, A, B, C and D, that represent regions of highest similarity (Poch, Sauvaget, Delarue, & Tordo, 1989; Tordo et al., 1988). These motifs, which are thought to constitute the 'polymerase module' of L, maintain the same linear arrangement and location in all viral RNA-dependent RNA and DNA polymerases (Barik et al., 1990; Delarue, Poch, Tordo, Moras, & Argos, 1990; Poch et al., 1990). Among the conserved sequences in these four motifs is the tri-amino acid catalytic center sequence GDN (glycine, aspartic acid, and asparagine) in motif C, which is extensively conserved in all non-segmented negative-strand RNA viruses, (Poch et al., 1989). At least two other sequences between amino acid residues 754 to 778 and 1332 to 1351 in the VSV L have been identified as consensus sites for binding and utilization of ATP, similar to those found in cellular kinases (Barik et al., 1990; Canter, Jackson, & Perrault, 1993). Three essential activities encoded by L are involved in the binding and utilization of ATP. These are (1) the transcriptional activity that requires binding to substrate ribonucleoside triphosphates (rNTPs), (2) polyadenylation, and (3) protein kinase activity for specific phosphorylation of the P in transcriptional activation (Banerjee & Chattopadhyay, 1990; Sanchez, De, & Banerjee, 1985). Many of the putative functions of this multifunctional protein, including mRNA capping, methylation, and polyadenylation, remain to be delineated and mapped within the RABV L. The process of mapping active sites in the RABV L using the mutational and deletion approach will be helped considerably now that it is feasible to apply the powerful technique of reverse genetics (Schnell & Conzelmann, 1995). Using reverse genetics it is possible to express the RABV L with selected amino acid deletions and point mutations in the cDNA of the L gene to map the locations in L that carry out the various functional activities.

2.3.4 Matrix Protein (M)

The RABV M (PV strain) is the smallest of the virion proteins, with 202 amino acids (25 kDa) (Tordo, Poch, Ermine, & Keith, 1986), and exists in two isoforms. The major 24-kDa form, Mα, accounts for 70–75% of total M in rabies virions, and the minor 23-kDa form, Mβ, accounts for

25–30% of total M in the virion and only 10–15% of total M in the cell (Ameyama et al., 2003). The Mβ isoform, in particular, is found in the Golgi apparatus of the cell, where it co-localizes with viral G. In addition, the M forms dimers (54 kD, 10–20% of total M in the cell) that form a strong association with G (Nakahara et al., 2003). M binds to and condenses the nascent NC core into a tightly coiled, helical RNP-M protein complex, forming a sheath around the NC core and producing the bullet-shaped 'skeleton' structure of the virion. Approximately 1,200 to 1,500 copies of M molecules bind to RABV RNP core. At the same time M mediates binding of the viral core structure to the host membrane, where it initiates virus budding (Mebatsion et al., 1999). A proline-rich motif (PPEY) in the M located at residues 35–38 within the highly conserved 14-amino acid sequence near the N-terminus of the RABV M was the first motif to be associated with virus budding from the cell membrane surface (Harty et al., 2001; Harty, Paragas, Sudol, & Palese, 1999). Other proline-rich (PPxY, PT/SAP, YxxL, and FPIV) motifs and related core sequences are found in the M of VSV and in other viruses; (see Irie, Licata, McGettigan, Schnell, & Harty, 2004). The PPEY motif (but not the overlapping YxxL motif (residues 38–41) in RABV M), more recently referred to as a late-budding domain (L-domain) and the L-domains identified in other viral proteins are all involved in a late step of the virus budding process and conceivably might provide a potential target for VNA and/or antiviral drug interactions (references in Irie et al., 2004). Although the exocytotic release of progeny virions requires the RABV M in the RNP-M 'skeleton' complex, the efficiency of virus budding is greatly enhanced by the interaction of the RNP-M skeleton complex with the envelope G (Mebatsion, Konig, & Conzelmann, 1996; Mebatsion et al., 1999). Increased virion production as a result of direct interaction of the cytoplasmic domain of the transmembrane spike G, and the viral RNP-M core in RABV assembly suggests that a concerted action of both core and spike proteins is necessary for efficient recovery of virions (Mebatsion et al., 1999). However, the interaction of M with the cytoplasmic domain of G does not need to be optimal, that is, the interaction is sufficient if the G of related viruses is substituted for the homologous G in budding virions (Mebatsion, Schnell, & Conzelmann, 1995).

2.3.5 Glycoprotein (G)

The RABV G (all genotype 1 strains) is translated from a G-mRNA transcript that encodes 524 amino acids, which includes an N-terminal 19-amino acid signal peptide (SP) domain. The SP provides a membrane insertion signal, which transports the nascent protein into the rough Endoplasmic Reticulum (ER)-Golgi-plasma membrane pathway before it is cleaved from the N-terminus of the G molecule in the Golgi apparatus. (Conzelmann et al., 1990; Tordo et al., 1988). The mature G (minus the SP) of all RABV strains for which the amino acid sequence as been determined has 505 amino acids (~65 kDa), and that of MOKV has 503 amino acids (Bourhy et al., 1993). The G is a type I membrane glycoprotein with an N-terminal ectodomain (ED), which extends outward on the plasma membrane and surface of mature virus particles, a 22-amino acid transmembrane (TM) (anchoring) domain, and a 44-amino acid long C-terminal domain (CD) locating in the cytoplasm. The resulting transmembrane G is organized into trimers (three monomers of 65 kDa each) in the Golgi apparatus, which later form the G (trimeric) spikes embedded in the plasma membrane and on the virion surface (Gaudin et al., 1992; Whitt, Buonocore, Prehaud, & Rose, 1991). The G spikes in the viral envelope extend 8.3 nm from the virus surface and represent the major surface protein of the virion. The CD (last 44 amino acids) of the G extends inward from the plasma membrane into the cytoplasm of the infected cell, where it interacts with M of the skeleton particle to complete the virion assembly. The ED of G (residues 1 to 439 of the mature RABV G) in each spike is the business end of the molecule for a variety of functional virus interactions. It is responsible for interaction with cellular binding sites (receptors), and therefore, is important in viral pathogenesis by targeting the appropriate cells for infection (Sissoeff, Mousli, England, & Tuffereau, 2005). It is also responsible for low pH-induced fusion of the viral envelope with plasma and endosomal membranes in the cell in the early phase of the RABV life cycle (Albertini, Baquero, Ferlin, & Gaudin, 2012). And it is critical for the induction of a host humoral immune response to RABV infection and as the target of VNAs, as well as for virus specific helper and cytotoxic T cells (Celis, Karr, et al., 1988; Macfarlan, Dietzschold, & Koprowski, 1986).

Appropriate glycosylation of RABV G is important for its proper expression and function. The oligosaccharides that are associated with the G are linked to asparagine (N, in single letter code) residues in the tripeptide sequence (sequon) asparagine-X-serine (NXS) or asparagine-X-threonine (NXT), where "X" is any amino acid other than proline (Shakin-Eshleman, Remaley, Eshleman, Wunner, & Spitalnik, 1992). RABV G molecules can have one, two or three (and sometimes four) sequons or N-linked carbohydrate (N-glycan) sites per molecule depending on the virus strain. Even if an N-glycan site is present, it may be inefficiently glycosylated or not occupied at all depending on the amino acid composition of the sequon (Dietzschold, Wiktor, Wunner, & Varrichio, 1983; Kasturi, Eshleman, Wunner, & Shakin-Eshleman, 1995; Shakin-Eshleman, Wunner, & Spitalnik, 1993; Wunner, Dietzschold, Smith, Lafon, & Golub, 1985). A virus can be heterogeneous with regard to the number of sites that are occupied with N-glycans (macroheterogeneity) and with regard to the types of glycan structures (microheterogeneity) present at each site (Wojczyk et al., 2005). Clearly, various factors can affect the glycosylation of individual sequons in Gs, some influence efficiency of N-glycosylation and others influence processing of N-glycans into a variety of oligosaccharide structures. The sugar residues of the N-glycan are presynthesized in the cell and transferred from a lipid precursor by the enzyme oligosaccharyl transferase as a precursor core-oligosaccharide unit (Glc3Man9GlcNAc2) to the specific sequons of the nascent G molecule. Typically, the transfer occurs as the protein, which is synthesized on the membrane-bound ribosomes (rough ER), begins to fold cotranslationally, and during translocation to the cytoplasmic ER membrane. After the precursor core-oligosaccharide is transferred, the high-mannose triglucoslylated oligosaccharide is trimmed and processed in the lumen of the rough ER and Golgi stacks to form the 'complex type' N-linked monoglucosylated oligosaccharide of the mature G molecule (see references in Gaudin, 1997). The molecular chaperone calnexin recognizes the partially trimmed monoglucosylated glycan and binds to it, assisting the G to fold correctly and completely in order to achieve its biological activity, stability, and antigenicity (Gaudin, 1997; Okazaki, Ohno, Takase, Ochiai, & Saito, 2000; Shakin-Eshleman et al., 1992). This dependence of the RABV G on the molecular interaction with calnexin explains why it is critical that at least one of the asparagine residues, that is, N319 in the RABV G, is conserved in all virus genotypes. N319-glycosylation is essential for correct and complete folding of the nascent RABV G and required for subsequent transport to the cell surface (Wojczyk et al., 2005).

The RABV G is also modified by the addition of palmitic acid (referred to as fatty acid acylation or palmitoylation) at cysteine 461, located in the intra-CD on the C-terminal side of its TM region (Gaudin, Tuffereau, Benmansour, & Flamand, 1991). Although the functional significance of palmitoylation is not entirely clear, it is presumed to have a stabilizing effect on the trimeric G spike anchored in the membrane. Palmitoylation may also play a role in the virus budding process by facilitating the interaction between the CD 'tail' of the G and the M in the RNP-M complex at the cell membrane.

Another important function of the RABV G that is critical in establishing the phenotype of the virus is its role in defining the pathogenicity and determining the neuroinvasive pathway of the virus (Ito, Takayama, Yamada, Sugiyama, & Minamoto, 2001; Kucera, Dolivo, Coulon, & Flamand, 1985; Yan, Mohankumar, Dietzschold, Schnell, & Fu, 2002). While the neurotropism of a particular RABV strain is primarily a function and major defining characteristic of its G, it is relevant to note that other viral components and attributes are also important in altering the viral pathogenesis of rabies (Faber et al., 2004; Morimoto, Foley, McGettigan, Schnell, & Dietzschold, 2000b). Whether or not the virus will cause a lethal infection (follow a specific pathway to induce a fatal disease) is determined first by the interaction of the G spike with a specific receptor on neuronal cells *in vivo* (Lafon, 2005; Sissoeff et al., 2005). An attempt to map the p75NTR receptor binding site on RABV G has suggested that the receptor binds to a region of the G (within residues 318–352) on both sides of antigenic site III and site 'a' that is not neutralized by anti-G antibody (Langevin & Tuffereau, 2002). Lentiviral vectors pseudotyped with RABV G have demonstrated that the G not only allows entry into the nervous system but also, upon infection of neurons at distal connected sites within the nervous system, it facilitates and enhances the retrograde axonal transport of the virus to the CNS (Mazarakis et al., 2001). Second, the pathogenic or virulence phenotype of the virus correlates with a single amino acid at a specific position in the RABV G. For example, arginine (R) 333 or lysine (K) 333 of wildtype (virulent) G determines the virulence phenotype or neuroinvasive pattern of RABV in the CNS. Virus variants that substitute glutamine (Q), isoleucine (I), glycine (G), methionine (M) or serine (S) for R333 in the rabies virus G express a phenotype that is either less pathogenic or totally avirulent compared to the parental wildtype virus in adult immunocompetent mice (Dietzschold, Wunner, et al., 1983; Seif, Coulon, Rollin, & Flamand, 1985; Tao et al., 2010; Tuffereau et al., 1989). While these particular mutations in the G correlate with reduced or abolished neuroinvasiveness, they do not impair the ability of the virus to multiply in cell culture. The same substitutions (e.g., Q333 for R333 or R333Q) in the G can, however, affect the rate of virus spread from cell to cell (Dietzschold et al., 1985; Faber et al., 2005) and the ability to infect motor neurons *in vivo* and *in vitro* (Coulon, Ternaux, Flamand, & Tuffereau, 1998). They can modify the host range spectrum of the virus and determine the choice of neuronal pathways the virus uses to reach the CNS (Etessami et al., 2000; Kucera et al., 1985) as well as the distribution of virus to different areas of the brain (Yan et al., 2002). The R333Q substitution is also associated with a greater ability of the virus to induce apoptosis in neuronal cells (Tao et al., 2010). Still, other amino acid substitutions at other positions on the RABV G appear to confer or influence viral pathogenicity. For example, amino acid substitutions located between positions 34 and 42 and at positions 198 and 200 (related to epitopes specific to antigenic site II) of the CVS strain G reduced the pathogenicity of the CVS strain when inoculated intramuscularly in adult mice (Prehaud, Coulon, Lafay, Thiers, & Flamand, 1988). One or more amino acids between residues 164 and 303 in the G of the pathogenic parental Nishigahara Ni-CE strain compared with the avirulent Ni-CE variant, RC-HL, also appear to define the lethality that is characteristic of this virus because R333 is present in the G of both strains (Takayama-Ito, Ito, Yamada, Minamoto, & Sugiyama, 2004).

Finally, the RABV G is a potent immunogen with major importance immunologically for the induction of the host immune response against virus infection. Because it is such an important antigen for the immune system to mount a response against, it is probably the most extensively studied RABV protein in terms of structure in relation to its immunogenicity and antigenicity for VNAs, that is, as a target for VNAs. The G induces VNAs that recognize both conformational and linear epitope (antibody binding) sites and stimulates helper, as well as cytotoxic T-cell, activity.

2.4 Viral Lipids and Carbohydrate

A lipoprotein bilayer forms the viral envelope (or membrane matrix) surrounding the helical RNP core. The lipids, which constitute 20–26% of the viral lipoprotein envelope, are derived entirely from the host cell and, depending on where the virus buds through the cellular membrane, concentrations of certain lipids may be higher in the viral envelope than are represented in the rest of the plasma membrane. In general, the RABV membrane contains a mixture of lipids including phospholipids (mainly sphingomyelin, phosphatidylethanolamine, and phosphatidylcholine), neutral lipids (mainly triglycerides and cholesterol), and glycolipids.

3 GENOME AND RNP STRUCTURES

One of the characteristic features of MNV like RABV is the stable helical (supercoiled) RNP complex in which the genome RNA is tightly encapsidated by N (N-vRNA) and associated with P and L. RNP structure and function is not only important for progeny virus morphogenesis but also is directly interrelated with the typical mode of gene expression. This involves sequential transcription of non-encapsidated subgenomic mRNAs from genome RNPs, as well as replication of full-length antigenome and progeny genome RNPs. Release of the RNP from infecting virions into the cytoplasm of cells and dissociation of M surrounding the RNP results in transition of the RNP from the supercoiled state to a relaxed structure, in which the N-vRNA complex can serve as a template for RNA synthesis (Iseni, Barge, Baudin, Blondel, & Ruigrok, 1998). In fact, exclusively N-vRNA (and N-cRNA), rather than naked vRNA, is a suitable template for the viral polymerase, both for synthesis of subgenomic non-encapsidated mRNAs (transcription) and of full-length progeny as well as novel DI RNPs (replication) (Arnheiter, Davis, Wertz, Schubert, & Lazzarini, 1985).

Details of the molecular structure of the helical N-RNA complexes of RABV and VSV were revealed by X-ray diffraction of crystals obtained from ring structures of N-vRNA complexes containing 11 or 10N molecules (Albertini et al., 2006; Green, Zhang, Wertz, & Luo, 2006). The RNA in these structures is tightly sequestered in a cavity at the interface between the N- and C-terminal lobes of the N, which appear to clamp down onto the bound RNA. The characteristics of RNP folding, RNA binding, and assembly is highly conserved in VSV and RABV, despite their lack of significant homology in amino acid sequences of N (Luo, Green, Zhang, Tsao, & Qiu, 2007). The complete occlusion of the vRNA by N explains the excellent protection of the vRNA against high salt, attack by RNases, and silencing by siRNAs (Albertini et al., 2006; Albertini, Schoehn, Weissenhorn, & Ruigrok, 2008). The tight N-vRNA clamping appears to open exclusively and transiently during genome transcription and replication to specifically give access for the polymerase complex (L/P or L-P/N) to the vRNA template. This requires conformational changes in N caused by the polymerase and/or P alone (Albertini et al., 2008). Easiest access to the N-vRNA would be at the 3′-end of the genome and antigenome. Indeed, exclusive entry at or close to the 3′-terminus of the genome and (obligatory) sequential transcription of the downstream succession of genes is the characteristic feature of all MNV viruses.

3.1 Genome Primary Structure

The single-strand negative-sense RNA genomes (vRNAs) of lyssaviruses represent the prototypical MNV genomes. They are about 12 kilobases in length and have unmodified 3′-hydroxyl (3′-OH) and 5′-triphosphate (5′-PPP) ends. They consist of a succession of 5 genes in the strictly conserved order 3′-N-P-M-G-L-5′, flanked by short 3′- and 5′-terminal regulatory extragenic regions known as leader (Le) and trailer (Tr) regions, respectively (Figure 2.3).

The 3′-terminal Le region can be regarded as the genomic promoter (GP) that directs the transcription of monocistronic mRNAs, as well as the synthesis of full-length antigenome RNA (cRNA). The 3′ end of the cRNA (i.e., the complement of the genomic Tr region) functions as the antigenome promoter (AGP), but unlike the GP, only directs synthesis of full-length N-encapsidated cRNAs, and not of unencapsidated RNAs. GP and AGP share a similar composition with a high A and U content, and in RABV the 11 3′-terminal nucleotides are identical (3′-UGCGAAUUGUU…5′) (Tordo, Poch, Ermine, & Keith, 1986; Tordo, Poch, Ermine, Keith, & Rougeon, 1986; Tordo et al., 1988). Such terminal complementarity is a common feature of MNV genomes and provides strong evidence that these terminal sequences are important for promoter function (Tordo et al., 1988; Whelan, Barr, & Wertz, 2004).

3.2 Lyssavirus Genes

Most of the genome (>99%) is comprised of the protein-encoding genes. They are transcribed by the viral polymerase to yield mRNAs that can be efficiently translated by the host cell machinery. The mRNAs have a typical 5′-terminal 7-methylguanylate cap structure (m^7G) facilitating

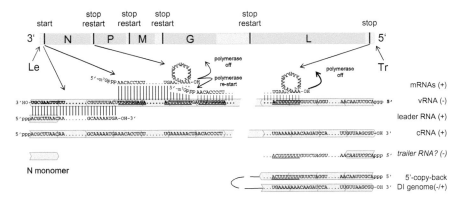

FIGURE 2.3   **RABV genome transcription and replication products.**
On top, the 12-kb genome (vRNA) of RABV is schematically shown. The localization of open reading frames (gray) and transcription signals (black bars) are indicated. In the lower part, details of the vRNA sequence are shown, including terminal sequences and transcription signals (underlined) of the N and L genes. Transcription of 5'-capped mRNAs is obligatory sequential and starts at the 3' proximal N start signal. At the stop/polyadenylation signal the polymerase may resume transcription of the downstream gene at the nearby start signal or dissociate from the template, which leads to the transcription gradient. The intergenic regions between the stop/polyadenylation and restart signals are probably not transcribed. Replication involves concurrent encapsidation of RNA by N. Incomplete encapsidation of the growing leader RNA might lead to abortive replication at the leader/N gene border or a switch to transcription. Whether a corresponding regulatory product of the cRNA (trailer RNA) exists in RABV is not clear. Defective interfering RNAs of RABV comprise the cRNA 5'-end and an exact complementary 3'-end, which is generated by back-copying of the nascent positive strand by the polymerase.

translation initiation, which in all lyssaviruses invariantly starts with the sequence 5'-m^{7}G-AACAYYNCU, where Y = C or U, N = any nucleotide (nt). The mRNA sequences also have very short 5'-noncoding regions in the range of only 20–30 nts, and longer 3'-noncoding regions in the range of 100–500 nts. A particularly long 3'-noncoding region is observed in the G mRNA of RABV, corresponding to the region in the genome that was initially described as a pseudogene (ψ).

Differences in the 3'-noncoding region of genes mostly account for the observed differences in length of lyssavirus genomes. Among the lyssaviruses, RABVs seem to have the shortest genomes, which are in the range of 11,923–11,932 nts. The genome lengths of other lyssavirus species seem to increase somewhat, though the overall variation remains low. The longest lyssavirus genome reported is that of WCBV and still it comprises only 12,178 nts (Kuzmin, Wu, Tordo, & Rupprecht, 2008). For more details on the genetic variability of lyssavirus genomes and proteins, refer to (Banyard, Hayman, Johnson, McElhinney, & Fooks, 2011; Nadin-Davis & Real, 2011); see also Chapter 6.

3.3 Transcription of Genes

According to the widely accepted stop/start model of transcription of MNV viruses, RABV transcription initiates exclusively at the 3′ terminal GP (Figure 2.3), and the genes are transcribed in an obligatory sequential manner (Flamand & Delagneau, 1978; Holloway & Obijeski, 1980). The observed synthesis of an abundant 58nt-long 5′-PPP- and non-polyadenylated leader RNA from the 3′-terminal Le region of rhabdoviruses (Colonno & Banerjee, 1978; Leppert, Rittenhouse, Perrault, Summers, & Kolakofsky, 1979) originally argued in favor of a model in which leader RNA synthesis is absolutely required for transcription of the downstream protein-encoding genes. However, recent data from several virus systems argue as well in favor of transcription initiation mechanisms, which are independent of leader RNA synthesis (Banerjee, 2008; Curran & Kolakofsky, 2008; Whelan, 2008).

The protein-encoding genes downstream of the Le region are separated by conserved gene border signals, which direct the activity of the polymerase during vRNA transcription, but which must be ignored during vRNA replication. They specify the end of the upstream mRNA by a stop/polyadenylation (poly(A)) signal and the beginning of the downstream mRNA (start signal). Typically, stop and start signals are separated by short, most likely non-transcribed intergenic sequences (IGSs). While in VSV these IGSs invariably comprise two nucleotides, lyssaviruses mostly have IGSs of 2 (at the N/P gene border), 5 (at P/M and M/G borders), and 19–28nts (at G/L borders), though some variation (16 nts) is observed at the M/G border in some lyssaviruses.

The stop/poly(A) signals are characterized by a run of mostly 7 U residues (in RABV 3′-..ACUUUUUUU...-5′), on which the polymerase repeatedly stutters back and forth to synthesize a 50–150nt-long poly(A) tail (Barr, Whelan, & Wertz, 1997). Without leaving the N-vRNA template, the polymerase is thought to release the newly generated polyadenylated mRNA and to initiate transcription of the downstream gene at the conserved restart signal, (3′...UUGURRNGA...-5′, where R=G or A, N=any nt) leaving the border signals between sequential genes untranscribed. As was shown for VSV, restart involves the addition of a typical 5′-m^{7}G cap by the L protein through an unconventional capping mechanism (Lij et al., 2008; Ogino & Banerjee, 2007). The stop/poly(A) signal in the L gene is not followed by a restart signal, which implies that the downstream Tr sequence, comprising about 70nts of the lyssavirus genome RNA, is not transcribed.

As a consequence of the exclusive entry of the polymerase at the 3′-end of the genome and eventual dissociation of the polymerase at every gene junction, upstream (promoter proximal) mRNAs are produced more abundantly than downstream mRNAs. This type of transcriptional attenuation is a very simple and efficient method to achieve differential expression levels of gene products. The conserved gene order of MNV genomes therefore seems to directly reflect the relative protein amounts needed for completion of the viral life cycle.

The steepness of the mRNA transcript gradient depends on the rate of dissociation of the polymerase from the template at the gene borders. For RABV, there is evidence that re-initiation is influenced by the length of the intergenic (IG) region, that is, the distance between stop/poly(A) and restart signal. A conserved feature of lyssaviruses is the elongated IGS of the G/L gene border, which was interpreted to reflect a need for downregulation of L expression. Indeed, the short 2 or 5nt-long IGSs of the N/P-, P/M-, and M/G-gene borders directed high level expression of reporter genes from artificial bicistronic RABV mini-genomes, whereas the 24-nt IGS from the G/L border was poorly active in mediating expression of the downstream reporter gene (Finke, Mueller-Waldeck, & Conzelmann, 2003). Moreover, replacement of the long G/L IGS with the 2nts of the N/P IG region in recombinant RABV resulted in more abundant transcription of the L mRNA. This was correlated with higher accumulation of L and an overall increase in vRNA synthesis (Finke, Cox, & Conzelmann, 2000). The resulting viruses showed pronounced cytopathic effects *in vitro*, a phenotype predicted to be detrimental for viruses like RABV, which rely on a stealth strategy for pathogenesis. Thus, attenuating L expression through transcriptional attenuation by a long IG region between the G and L genes seems to be of advantage for RABV and other lyssaviruses.

Another observation strongly supporting the need for specifically limiting the expression of the RABV polymerase is the presence of extra stop signal or stop signal-like sequences in the noncoding region downstream of the G ORF in several virus strains, including Evelyn-Rokitniki-Abelseth (ERA) and PV strains. These signals comprise 5 or 6 U residues and can lead to transcription stop and polyadenylation in a certain percentage of G mRNAs, such that, in addition to the standard long G mRNAs, shorter G mRNAs are generated (Conzelmann et al., 1990; Tordo, Poch, Ermine, Keith, et al., 1986). It is assumed that a polymerase complex that terminates transcription at the upstream signal is not able to reinitiate at the start signal of the L gene, because it cannot cope with an "intergenic region" of several hundred nucleotides. Therefore, only those polymerase complexes terminating at the downstream G/L border with the 24nt-long IGS are supposed to contribute to L mRNA transcription. Rather than representing the remnants of a former gene, as proposed initially (Tordo, Poch, Ermine, Keith, et al., 1986), the extra transcription stop signals between G and L genes may serve the goal of limiting L mRNA transcription. The noncoding region of the G gene is not critical for virus growth and has been used preferably for substitution with single or multiple extra genes (Finke & Conzelmann, 2005; Gomme, Wanjalla, Wirblich, & Schnell, 2011; Schnell, Mebatsion, & Conzelmann, 1994).

The simple and modular organization of rhabdovirus genomes, where (1) cistrons are defined by 'short' gene border signals and (2) gene expression primarily depends on the position of the gene, has been exploited in recent years by numerous reverse genetics approaches. These involved addition of extra single or multiple genes and rearrangement of gene order (Conzelmann, 2004). Indeed, the artificial shifting of viral genes to other positions in the genome has become a versatile tool to study the impact of gene dosage on the rhabdovirus life cycle and pathogenesis (Ball, Pringle, Flanagan, Perepelitsa, & Wertz, 1999; Brzozka, Finke, & Conzelmann, 2005). For RABV, for example, moving the P gene from the second to the most downstream fifth position has revealed a requirement for high protein levels in order to successfully counteract the cellular antiviral interferon response (Brzozka et al., 2005). Moving the RABV M gene to the most promoter-distal position has revealed its contribution to the regulation of viral transcription (Finke et al., 2003), and insertion of multiple G genes has been used to enhance antigen expression for vaccine purposes (Faber et al., 2009).

3.4 Replication of RABV N-vRNA Genomes

Replication of genome N-vRNA complexes involves the synthesis of complementary full-length antigenome N-cRNA. Consequently, the viral polymerase here acts as a RNP-dependent RNP polymerase (a replicase), rather than a RNP-dependent RNA polymerase (a transcriptase). Like N-vRNA, the nascent N-cRNA is encapsidated concurrently with elongation. The coupling of cRNA elongation and N-encapsidation also dictates that replication of full-length N-cRNA is assured only upon prior accumulation of large quantities of N. In addition, transcriptional signal sequences must be ignored by the replicase. In contrast to the transcriptase, which is composed of L-P heteromers, the replicase is considered to be composed of L associated with P-N heteromers in which the P chaperones N for specific encapsidation of nascent cRNA (Chenik et al., 1994; Fu et al., 1994; Schoehn et al., 2001).

N-cRNA synthesis starts at the 3'-end of the N-vRNA genome by primer-independent synthesis to produce the 5'-end of the antigenome (i.e., producing a sequence corresponding to that of the short leader RNA) (Figure 2.3). The finding of N-encapsidated leader RNA in rhabdovirus-infected cells (Blumberg, Giorgi, & Kolakofsky, 1983) indicated that encapsidation of nascent antigenome sequences is initiated by a replicase, or that the Le RNA sequence comprises a specific N-packaging signal that further directs the polymerase to the replicase mode. As revealed by X-ray analyses, every RABV N protomer covers 9–10 nts of RNA, and, as suggested by studies revealing the importance of flush-end N phasing in other MNV viruses, should be formed in a way that the first 9–10 nts are covered by the first N protomer. In the presence of high levels of N-P (or L-P-N) heteromers, synthesis of encapsidated cRNA (or vRNA) is occurring in a processive mode until the 5′-end of the template is reached. The resulting antigenome N-cRNA has a 3′-end identical in the first 11 nts to that of the genome vRNA and can efficiently direct the synthesis of abundant amounts of genome N-vRNAs, provided that N is available (Figure 2.3).

Notably, the AGP of rhabdoviruses is much more active in directing replication than the GP, such that the overwhelming amount of full-length N-RNAs generated in infected cells comprise negative-strand vRNAs, which are useful for further transcription, replication, and assembly of virus progeny. Specifically, vRNAs make up 98% and 90% of full-length RNAs in RABV- and VSV- infected cells, respectively (Finke & Conzelmann, 1999; Whelan et al., 2004). In contrast, recombinant RABV, in which both genome and antigenome RNAs have identical promoters derived from the AGP, produce equal amounts of genome and antigenome N-RNAs. The viral AGP therefore seems to effectively out-compete the GP for replicase activity. This is most impressively illustrated by DI particles (Clark et al., 1981; Wiktor et al., 1977; Wunner & Clark, 1980). The genomes of VSV DI particles, which most efficiently interfere with the replication and propagation of VSV, belong to the so-called 5′ copy-back type (Marriott & Dimmock, 2010); see Figure 2.3. These are generated during RNA synthesis initiating from the viral AGP by dissociation of the polymerase from the template N-cRNA and back-copying of the nascent N-cRNA, and therefore positive and negative strand RNAs have an identical 3′ terminal AGP promoter. It appears that copy-back DI genomes of RABV are less readily generated compared to VSV. Internal deletion-type DI genomes, whose ends are derived from the 3′ and 5′ termini of the genomes of parental viruses have been identified for both VSV and RABV. Their ability to interfere with replication of the parental viruses stems from mutations in the promoters (Conzelmann et al., 1991).

3.5 Regulation of Transcription and Replication

The mechanisms and regulation of genome transcription and replication involved in the postulated switch between the two modes of RNA synthesis are insufficiently known. Transcription is obviously required throughout infection, initially for providing the structural proteins for replication like N and P, and later for providing the proteins, M and G, for assembly of progeny virions. In fact, transcription remains the major mode of RNA synthesis throughout infection, even in the presence of high amounts of N and P.

The term "primary transcription" has been used to describe transcription from the incoming genome RNPs, when the amount of N produced is insufficient to support replication. According to the conventional view, transcription involves the synthesis and release of a positive strand 5'-PPP leader RNA and re-initiation of transcription at the N gene start signal. A switch from the transcription mode to the replication mode is thought without direct evidence to be caused by the presence of sufficient N-P- or L-P-N complexes. For leader-dependent transcription an intriguing model was proposed (Vidal & Kolakofsky, 1989), in which complete encapsidation of the nascent Le RNA prevents the polymerase from recognizing the transcription stop/start signal at the Le region-N gene junction. In this case, the polymerase would not release the leader RNA from the 'leader RNA/N-mRNA' transcript and continue RNA encapsidation in the replication mode. In case of insufficient N-P availability, naked or incompletely encapsidated leader RNA would not prevent the polymerase from switching to the transcription mode. Once in the transcription mode and downstream of the Le RNA/N gene junction, the polymerase is thought not to be able to convert back to the replicase mode. In this model, the synthesis of leader RNA might be regarded as abortive replication.

The alternative models of leader-independent transcription suggest the initiation of transcription by a transcriptase form of the polymerase. They suggest that the transcriptase either enters at the 3'-end of the genome and scans the GP without synthesizing a leader RNA until it encounters the start signal for N mRNA synthesis or that the transcriptase (in contrast to the replicase) can enter directly at the N start signal (Banerjee, 2008; Curran & Kolakofsky, 2008; Whelan, 2008). Any of these models would yield unencapsidated mRNAs and is compatible with the stop/restart model of MNV transcription. Whether the mechanisms are mutually exclusive or may coexist remains to be determined.

Interestingly, rather than the concentration of N alone, it is the level of M that seems to influence the mode of RNA synthesis. Early *in vitro* work suggested that purified M inhibits transcription of RABV (Ito, Nishizono, Mannen, Hiramatsu, & Mifune, 1996). Reverse genetics experiments involving M gene-shifting in recombinant RABV or complementation of mini-genomes with M confirmed an inhibitory effect on transcription, while replication remained unaffected (Finke et al., 2003; Ito, Nishizono, Mannen, Hiramatsu, & Mifune, 1996). Moreover, specific mutations introduced in the M resulted in viruses with a "high transcription" phenotype, which produced much more mRNAs relative to genome templates than the parental viruses (Finke & Conzelmann, 2003). The molecular mechanisms behind such models of transcription regulation are not yet resolved but may involve association of M with the template N-vRNA or different types of the polymerase complex.

3.6 Structural Aspects of RABV RNA Synthesis

Obviously, any RNA synthesis of RABV requires elaborate interplay of the proteins involved, namely N, P, and L and their interaction with the vRNA genome. A better understanding of RABV RNA synthesis requires insight into the structure and dynamics of these protein interactions. Great progress has been achieved regarding RABV RNA synthesis by resolving the atomic structure of the RABV N-RNA complex, as a template for RNA synthesis, but atomic details of how this template is used for RNA synthesis is limited so far. The N-vRNA structure was determined from short N-RNA rings that formed after expression of recombinant N^o and encapsidation of cellular RNA in cells expressing N^o in the absence of other RABV proteins; reviewed in (Albertini et al., 2011). The N^o forms two domains with a positively charged cleft, in which 9 nts of the RNA are buried. Terminal extensions of the protein extend to neighboring protomers, thereby stabilizing the N-RNA complex. Binding of P to N^o (see Section 2.3.2) probably shields the N-RNA cleft and the terminal N extensions to prevent unspecific RNA binding and aggregation of N^o. Simultaneous binding of P to L via its N-terminal domain and to N-RNA via the C-terminal domain is probably involved in bringing the L in contact with the N-RNA template. The contact to N-RNA is probably mediated via the C-terminal lobe of N^o (Schoehn et al., 2001), in a way that P covers two N^o-protomers, resulting in a 1:2 ratio of P:N in the artificial rings (Ribeiro et al., 2008; Ribeiro et al., 2009). In the infected cell, binding of P to N-vRNA is enhanced by phosphorylation of N at Ser389 (Toriumi & Kawai, 2004). It is assumed that several N protomers might dissociate locally to provide sufficient space for the large RNA polymerase (L) molecule to bind.

4 LIFE CYCLE OF RABIES VIRUS INFECTION

The sequence of events in the RABV life cycle, that is, replication *in vitro* and *in vivo* (in cell culture or animal) can be divided into three phases. The first or early phase includes virus attachment to receptors on susceptible host cells, entry via direct virus fusion externally with the plasma membrane and internally with endosomal membranes of the cell, and uncoating of virus particles and liberation of the helical RNP in the cytoplasm. The second or middle phase includes transcription and replication of the viral genome and viral protein synthesis, and the third or late phase includes virus assembly and egress from the infected cell. The early phase of the RABV life cycle, often regarded as the most difficult of the events in RABV infection to fully understand, has been studied in many different cell culture systems. These include neuronal and non-neuronal cell lines and primary, dissociated cell cultures derived from dissected pieces of nervous tissue. One caveat that overshadows the use of experimental cell culture systems is that the cells appear to behave differently *ex vivo* in their susceptibility for RABV infection compared with their susceptibility to infection *in vivo*. That is, once cells are removed from their *in vivo* environment, particularly neuronal cells, they lose their natural control over susceptibility (or resistance) to RABV infection. Nevertheless, many studies using *in vitro* cell culture systems describe how virus enters the host cell by direct fusion with the plasma and endosomal membranes (Iwasaki, Wiktor, & Koprowsk, 1973; Le Blanc et al., 2005) or by receptor-mediated endocytosis (Hummeler et al., 1967; Iwasaki et al., 1973; Lyckc & Tsiang, 1987; Superti, Derer, & Tsiang, 1984; Tsiang, Delaporte, Ambroise, Derer, & Koenig, 1986). No *in vitro* system has yet provided a detailed explanation of how RABV enters muscle cells *in vivo* to support the experimental infections in hamster (Murphy & Bauer, 1974; Murphy, Bauer, Harrison, & Winn, 1973) and skunk (Charlton & Casey, 1979) that show virus replication in striated muscle cells near the site of inoculation.

4.1 Early Phase Events: Role of the RABV Receptor, Endocytosis, and G-mediated Fusion

RABV infection starts with virus attachment to the surface of a target cell and penetration into the cell through an endosomal transport pathway (Le Blanc et al., 2005). Most likely the virus attaches itself to a 'receptor' molecule or cellular receptor unit (CRU) on the cell surface that leads to or permits direct virus entry into susceptible cells in culture (*in vitro*) or specific target cells at the site of inoculation (*in vivo*) (Figure 2.4). Studies using various cell culture systems have implicated various lipids, gangliosides, carbohydrate, and protein of the plasma membrane in RABV binding to cells in culture (Broughan & Wunner, 1995; Conti, Superti, & Tsiang, 1986; Conti et al., 1988; Superti, Seganti, Tsiang, & Orsi, 1984; Superti et al., 1986; Wunner, Reagan, & Koprowski, 1984), but none have proven to be 'specific' receptors. Others have focused on specific cellular receptor molecules or CRUs *in vivo* that appear to correlate with the defined neurotropism of the virus (Lafon, 2005); see also Chapter 8. One of these CRUs is the nicotinic acetylcholine receptor (nAChR) found at neuromuscular junctions, where RABV can also be found co-localized *in situ* (Lewis, Fu, & Lentz, 2000). Not all cell lines infected with RABV *in vitro*, however, express the

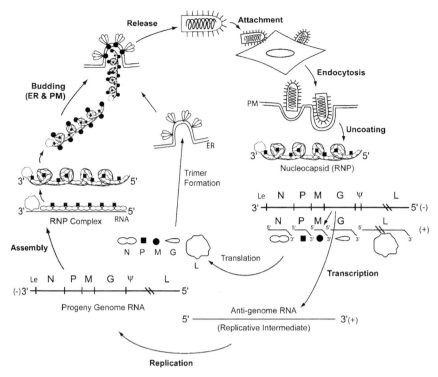

FIGURE 2.4   Rabies virus life cycle in the cell.
Virus enters the cell following attachment through coated pits (viropexis) or via cell surface receptors, mediated by the viral glycoprotein (G) fusing with the cellular membrane (endocytosis). After internalization, the viral G mediates low pH-dependent fusion with the endosomal membrane and the virus is uncoated, releasing the helical nucleocapsid (NC) of the ribonucleoprotein (RNP) core. The five structural genes (N, P, M, G, and L) of the genome RNA (vRNA) in the NC are transcribed into five positive (+) strand monocistronic messenger (m)RNAs and a full-length + strand (anti-genome) replicative intermediate RNA (cRNA), which serves as the template for replication of progeny genome (– strand) vRNA. The proteins (N, P, M, and L) are synthesized from their respective mRNAs on membrane-free ribosomes in the cytoplasm and the G is synthesized from the G-mRNA on membrane-bound ribosomes (rough endoplasmic reticulum). Some of the N-P molecular complexes produce cytoplasmic inclusion bodies (Negri bodies) *in vivo* and some N-P complexes encapsidate the + strand and – strand viral RNAs. After progeny genome vRNA is encapsidated by N-P protein complex, and L is incorporated to form progeny RNP (both full-length standard and shorter defective) structures, the M binds to the RNP and condenses the RNP into the 'skeleton' structures. The skeleton structures interact with the trimeric G structures anchored in the plasma membrane and assemble into virus particles (virions) that bud from the plasma membrane of the infected cell into adjacent extracellular or interstitial space.

nAChR (Reagan & Wunner, 1985; Tsiang, 1993; Tsiang et al., 1986), and some neuronal cells infected with rabies virus *in vivo* may not express the nAChR (Kucera et al., 1985; Lafay et al., 1991; Tsiang et al., 1986). Two other possibilities are the neural cell adhesion molecule (NCAM) CD56 on the cell surface of RABV-susceptible cell lines (Thoulouze et al., 1998) and the low-affinity neurotrophin receptor (p75NTR), a nerve growth factor (Langevin & Tuffereau, 2002; Tuffereau, Benejean, Blondel, Kieffer, & Flamand, 1998). Some doubt has been cast as well on the importance of the p75NTR as an obligatory RABV receptor because the RABV Challenge Virus Standard (CVS) strain infects p75NTR-deficient mice (Jackson & Park, 1999). Whatever receptor is present on the cell surface, it seems the virus is not limited to choosing only one type of receptor in order to initiate the virus life cycle in the infected animal.

After targeted binding of virus to its receptor(s) on host cells, virus is internalized by endocytosis. RABV, like VSV, may also enter the cell through coated pits and uncoated vesicles (viropexis or pinocytosis), which often incorporate several (two to five) virions per vesicle (Tsiang, Derer, & Taxi, 1983). As part of the internalization process, whether by receptor-mediated endocytosis (via the endocytic pathway) or through coated pits, fusion between the viral and endosomal membranes is activated in the acidic environment (pH 6.3–6.5) within the endosomal compartment. At the threshold pH for fusion activation, a series of specific and discrete conformational changes in RABV G takes place whereby it assumes at least three structurally distinct 'conformational' states; for review see (Albertini et al., 2012).

Prior to binding to the cellular receptor, the G on the virion surface is in its 'native' state. After the virus attaches to the receptor and is internalized, the G is 'activated' to a hydrophobic state, which enables it to interact with the hydrophobic endosomal membrane. Upon entering the endosomal compartment and low pH environment of the cellular compartment, the fusion capacity of the G is activated via a major structural change in the G that exposes the fusion domain, which interacts with the target cell membrane. By further rearrangements in the G, the viral and cell membrane are brought into close vicinity such that (via hemifusion) a fusion pore may develop (Albertini et al., 2012).

The low pH-induced exposition of the fusion domain, which is thought to lie between amino acids 102 and 179 (Durrer, Gaudin, Ruigrok, Graf, & Brunner, 1995), is not to be confused with the proposed fusogenic domain (amino acids 360–386) on the RABV G (Morimoto, Ni, & Kawai, 1992). The functional state of the RABV G in which it acquires fusogenic activity is correlated with at least one specific conformational epitope. This epitope, which appears to be formed by combining two separate regions, the neurotoxin-like region (residues 189–214) of RABV G and the conformational antigenic site III (residues 330–340), is abrogated when the G is exposed to acidic conditions (Kankanamge, Irie, Mannen, Tochikura, & Kawai, 2003; Sakai et al., 2004). The transition from virus interaction to generating the fusion pore is a high-cost energy step and depends on the integrity and correct folding of the G trimers directly involved in the fusion process. It has been shown that more than one trimer of G is required to build a competent fusion site.

After low-pH fusion, the G assumes a reversible 'fusion-inactive' conformation, which makes the G monomer appear longer than the 'native' conformation and assume selective antigenic distinctions (Kankanamge et al., 2003). The fusion-inactivated G, which is no longer relevant to the fusion process is highly sensitive to cellular proteases and appears to be in a dynamic equilibrium with the 'native' G that is regulated by lowering and raising the pH (Gaudin, Tuffereau, Segretain, Knossow, & Flamand, 1991; Gaudin, Raux, Flamand, & Ruigrok, 1996). Interestingly, the fusion-inactive conformation serves the G in another capacity. During nascent viral protein synthesis, the G assumes an 'inactive state-like' conformation, protecting the G posttranslationally from fusing with the acid nature of Golgi vesicles while it is transported through the Golgi stacks to the cell surface. At the cell surface, the G acquires its 'native' conformation and structure (Gaudin et al., 1999). The MAbs that recognize specific low pH-sensitive conformational epitopes of the G can identify certain acid-induced conformational changes, as well as detect the various stages of nascent G monomer folding and its association with molecular chaperones like BiP and calnexin (Gaudin, 1997; Kankanamge et al., 2003; Maillard & Gaudin, 2002).

4.2 Middle Phase Events: Transcription, Replication and Nascent Protein Synthesis

4.2.1 *Viral RNP Release, Initiation of Genome RNA Transcription, Replication and Protein Synthesis*

In the second phase of the RABV life cycle, vRNA genome transcription is initiated in the cytoplasm of the infected cell after the tightly coiled transcriptionally 'frozen' RNP core is released from endosomal vesicles (Figure 2.4). With the dissociation of M from the RNP during the 'uncoating' process, the tightly coiled RNP structure relaxes to form a loosely coiled helix (Iseni et al., 1998), conceivably to facilitate the ensuing vRNA transcription and replication events in the cell as described in detail in Section 3.3. The incoming RNP serves as a template for RNA synthesis by approximately 50 RNP-associated RNA polymerase (L-P) complexes. The location of these complexes on the RNP is not known. It is assumed that the polymerase either initiates transcription at the 3'-end of the genome RNA, or it resumes transcription at the next downstream internal mRNA start site on the viral genome, close to where the polymerase complex was 'frozen' in place during progeny virus assembly in a previously infected cell. This part of the replication process has been termed 'primary transcription' (see Section 3.5). According to the stop/restart model of transcription (see Section 3.3), 5'-capped and poly-adenylated monocistronic mRNA transcripts are sequentially produced from the genes and eventually are translated into one of the viral proteins (Flamand & Delagneau, 1978; Holloway & Obijeski, 1980). Re-entry of polymerase released after transcription of the most downstream L gene, or of newly produced polymerase, leads to the typical transcript gradient in which the mRNAs of 3' proximal genes are transcribed more abundantly than those of distal genes.

The proteins of the virus are synthesized from the viral mRNAs using the protein synthesis machinery of the host cell. The G-mRNA is translated on membrane-bound polyribosomes (polysomes) and inserted cotranslationally into the lumen of the ER where disulfide bond formation occurs and the molecular chaperones are available to assist in the folding of G monomers before the molecule is transported out of the ER (Gaudin, 1997). While in the lumen of the ER, the G monomers undergo modification at specific asparagine (N) residues by core glycosylation and N-glycan processing (see Section 2.3.5) and form homotrimers (Gaudin et al., 1992; Whitt et al., 1991). The final processing of the N-linked carbohydrate side chains takes place in the Golgi apparatus of the intracellular membrane network. The other four viral mRNAs (N-, P-, M-, and L-mRNA) are translated on 'free' polysomes in the cytoplasm. As the M accumulates in the cytoplasm of the infected cell it is likely to specifically interact with the eukaryotic translation initiation factor, eIF3h, and inhibit translation of cellular (host) mRNAs that have a Kozak-like 5'-UTR and seize control of the host translational machinery in infected cells for the translation of viral mRNAs (Komarova et al., 2007). Accumulation of other viral proteins, specifically N and P, allows replication to initiate, which involves N-encapsidation-dependent synthesis of full-length antigenome N-cRNAs. These serve as templates for amplification of genome N-vRNAs, which can serve in secondary transcription and protein expression.

Accumulation of proteins leads to formation of cytoplasmic IBs, which increase in size over time (Lahaye et al., 2009; Menager et al., 2009). These IBs, long known as NBs in neurons (Kristensson, Dastur, Manghani, Tsiang, & Bentivoglio, 1996), which are readily detectable by staining, have been used as diagnostic markers for RABV infection. The IBs consist predominantly of viral N and P, but they also contain L (Finke, Brzozka, & Conzelmann, 2004), as well as newly synthesized viral RNAs, including vRNA, cRNA and mRNAs (Lahaye et al., 2009). The latter finding suggested that IBs represent active viral factories, rather than just a compartment for storage or degradation of excess viral proteins. In addition, cellular proteins accumulate inside IBs, such as the toll-like receptor 3 (TLR3), which was found to be required for regular IB formation (Menager et al., 2009) and the 70-kDa Hsp70, which interacts with the viral N accumulated in IBs as a protein-folding chaperone and is found in rabies virions in addition to being involved in other cellular processes (Mayer, 2005). Artificial downregulation of Hsp70 results in decrease viral mRNAs, proteins, and virus particles, suggesting that Hsp70 has a proviral function during RABV infection (Lahaye et al., 2012). RABV IBs are further often associated with cellular membranes, which would provide the possibility to interact with M and G and associated lipids involved in virus envelopment.

Viral mRNAs are always much more abundant than genome and antigenome RNPs, suggesting that transcription of mRNAs remains the major mode of RNA synthesis throughout infection. While initially the proteins produced are required for vRNA replication, at later stages they are needed in abundant amounts for assembly of progeny virus. Genome N-vRNA templates, which must serve both processes, are produced at higher levels than antigenome cRNAs. Quantification with strand-specific probes revealed a 50-fold excess of genomic vRNAs over anti-genome cRNAs. This biased replication could be attributed to a stronger activity of the GP, which successfully competes with the AGP for polymerase (Finke & Conzelmann, 1997).

4.3 Late Phase: Assembly and Budding of Progeny Virus

The process of virus assembly (RABV morphogenesis) actually begins in the middle phase of the life cycle with encapsidation of vRNA and formation of vRNA-N-P, that is, RNP, which requires accumulation of sufficient pools of viral N, P, and L to support RNP formation (Iseni et al., 1998; Liu et al., 2004; Mavrakis et al., 2003). RABV morphogenesis seems to be associated with the formation of the IBs and NBs as sites of virus assembly commonly found particularly in neurons in brain tissue, as well as in tissue culture (Hummeler et al., 1967; Hummeler, Tomassini, Sokol, Kuwert, & Koprowski, 1968; Matsumoto, 1962); see Chapter 9. Assembly of progeny RNP and virions continues into the late phase of the life cycle as long as the cells remain metabolically active.

The M is the next viral protein to associate with RNP complexes. From the time M enters the virus assembly pathway as a soluble protein in the cytoplasm it is involved in all steps that lead to virus budding. M is a multifunctional protein, and it plays several key roles in the dynamics of the formation of progeny virus. First, the association of M with newly formed transcriptionally active RNP changes the balance between viral RNA transcription and replication by inhibiting transcription and stimulating replication (Finke et al., 2003; Ito et al., 1996). M imparts this differential effect in RNA synthesis on encapsidated vRNA, either by binding to the vRNA + N template or the L + P polymerase in the complex (Finke et al., 2003). It is conceivable that the different regulatory functions of M in RNA synthesis may also be provided by the different conformational forms of the monomer M, Mα and Mβ (see Section 2.3.4). The M will then localize the RNP coil at the cellular membrane where the glycosylated trimeric G is concentrated and where the M is able to interact with G (Mebatsion et al., 1999). Finally, the M alters the structure of the RNP by condensing it into a tightly coiled 'skeleton-like' form in which the polymerase activity is 'frozen'. In this function of M, which may involve M as a dimer, M initiates virus budding at cellular membranes where the M dimers become enveloped by the cellular membrane and enter the virus budding process (Mebatsion et al., 1996; Mebatsion et al., 1999; Nakahara et al., 2003). In the mature rabies virion that buds from the cell membrane, the M lies between the lipid bilayer envelope (formed by interaction with the host cell membrane) and the helical RNP that it covers (Mebatsion et al., 1999). Thus, the M covering and condensing the helical RNP is thought to play an important role in virion morphogenesis, that is, giving the particle its bullet-shape morphology. If the M is missing from RABV particles that contain G spikes on the surface, a morphological variation in budded particles is observed suggesting that the particles contain uncondensed RNP. M-deficient RABV also causes increased cell-cell fusion and enhanced cell death, in contrast to the relatively benign cytopathic effect that is observed with wildtype virus. While assembly and budding of bullet-shaped particles can occur in the absence of the transmembrane G spikes that are normally associated with infectious particles (Mebatsion et al., 1996, 1999; Robison & Whitt, 2000), more significantly, virus budding is much less efficient in the absence of M.

In the final stages of RABV assembly, the mature virions acquire their lipid bilayer envelope as the assembled skeleton (RNP + M) structure buds through the host cell plasma membrane. Mature virions that bud through the plasma membrane into the extracellular space are frequently observed in extraneural tissue cells *in vivo* and in a variety of *in vitro* tissue culture systems (Davies et al., 1963; Hummeler et al., 1967; Iwasaki et al., 1973; Matsumoto & Kawai, 1969; Tsiang et al., 1983). Virions can also mature intracellularly by budding through the cytoplasmic ER or Golgi apparatus (Finke, Granzow, Hurst, Pollin, & Mettenleiter, 2010). This is often observed in infected neuronal cells of brain (Gosztonyi, 1994; Murphy, Harrison, Winn, & Bauer, 1973).

If virion budding occurs at a site in the cell membrane (or cytoplasmic ER) where the glycosylated trimeric RABV transmembrane G is also targeted, then infectious virions will be produced bearing the G molecules arranged as trimeric spike-like structures tightly packed and anchored in the viral envelope (Whitt et al., 1991). The C-terminal tail of the G molecule is then free to interact with the M of the RNP + M skeleton structure. Interaction of G with M is essential for stabilization of the G in trimers on the virion surface and for efficient budding of RABV. This does not exclude the possibility that skeleton structures may bud from membrane regions where no G exists (Mebatsion et al., 1996; Mebatsion et al., 1999; Robison & Whitt, 2000). In this case, budding would be inefficient, producing low levels, and the bullet-shaped particles produced would be spikeless (free of G) and non-infectious (Mebatsion et al., 1996). If skeletons bud through ER or Golgi membranes, they bud into the lumen of vesicles produced from these membranes and may be secreted from the cell through the normal secretory pathway.

4.4 Cell-to-Cell Spread (transport) of Progeny Virus

Progeny rabies virions that bud from infected cells are able to spread from cell-to-cell in cell culture (*in vitro*), presumably as they do in the animal (*in vivo*). They have the option of spreading to contiguous cells (direct cell-to-cell spread) or to non-contiguous cells, which are surrounded by interstitial space, although transmission *in vivo*, at least in the initial phase of infection seems to be exclusively via synapses (Wickersham et al., 2007). Successful spread *in vitro* and *in vivo* requires the G (Etessami et al., 2000), and the G determines the phenotypic differences of the viruses with respect to neuroinvasiveness and virulence as discussed further in Section 2.3.5.

In the case of direct cell-to-cell spread, RABV spreads despite a continuous presence of serum VNA (Dietzschold et al., 1985). Alternatively, virus that buds from an infected cell into the surrounding interstitial space must find another cell to infect. In this case, the spread of virus is limited by the presence, *in vitro* and *in vivo*, of VNAs that block virus attachment to cellular receptors, fusion and subsequent virus entry into a susceptible cell (Dietzschold et al., 1985; Flamand et al., 1993). *In vivo*, RABV can travel long distances within a cell, particularly within cells of peripheral nerves and neuronal cells of the CNS, within intra-axonal transport vesicles in a microtubule network-dependent process. Virus that moves intra-axonally can cover great distances, particularly in bipolar neurons, before membrane fusion and RNP release occurs (Ceccaldi, Gillet, & Tsiang, 1989; Coulon et al., 1989; Gillet, Derer, & Tsiang, 1986; Klingen, Conzelmann, & Finke, 2008; Kucera et al., 1985; Lafay et al., 1991); see Chapter 8. Evidence for axonal transport of naked viral NCs (Gosztonyi, 1994) is lacking so far. This hypothesis of transsynaptic transfer of naked NCs, however, was weakened considerably, if not negated, by recent studies that demonstrate an absolute requirement for the virion G in transsynaptic spread of RABV, both *in vivo* and *in vitro* (Etessami et al., 2000).

By spreading directly cell-to-cell, RABV is able to survive at least the humoral arm of the immune protection mechanisms during infection (Dietzschold et al., 1985). When cultures of mouse neuroblastoma (NA) and baby hamster kidney-21 (BHK-21) BHK cells were infected with the virulent phenotype of RABV and maintained in the presence of anti-rabies VNA, the virus was able to spread throughout the NA and the BHK cell cultures. Under similar conditions, the avirulent RABV failed to spread from cell-to-cell in the NA cell culture as if there was some impairment in the G-mediated cell fusion process in the NA cells. The avirulent virus spread cell-to-cell in the BHK cell culture equally well in comparison to the virulent virus (Dietzschold et al., 1985). In many ways, the pathogenic and avirulent viruses behaved *in vitro* in a manner that reflects their ability to spread *in vivo*, after direct inoculation into the brain of the mouse. The pathogenic virus spread *in vivo* more rapidly in the CNS and infected more neurons than the avirulent virus. Could it be that the G was not 'activated' to a hydrophobic state due to the amino acid substitution at position 333, enabling it to interact with the hydrophobic endosomal membrane? Others using NA cells and BHK cells, which constitutively express the RABV G in the absence of other viral proteins, have shown that only G with R333 as the determinant for its pathogenic phenotype demonstrated an ability to induce syncytium formation (cell-cell fusion) at neutral pH (pH-independent fusion) in the NA cell culture (Morimoto et al., 1992). Thus, the RABV G with R333, but not with Q333, has the ability to mediate virus spread among neuronal cells in culture. Since some Q333 variants can kill adult immunocompetent mice when infected by stereotaxic inoculation (Yang & Jackson, 1992), it appears that R333 in the G is required for the neuropathogenicity of the virus from a peripheral site of inoculation and for its transsynaptic and axonal spread *in vivo*. It is also apparent that the pH-independent cell fusion induced by the RABV G may involve the interaction of one or more neuronal cell-specific host cell factors, which are expressed in the NA cells but not in BHK-21 cells.

References

Albertini, A. A., Baquero, E., Ferlin, A., & Gaudin, Y. (2012). Molecular and cellular aspects of rhabdovirus entry. *Viruses*, 4(1), 117–139. doi:10.3390/v4010117viruses-04-00117. [pii].

Albertini, A. A., Wernimont, A. K., Muziol, T., Ravelli, R. B., Clapier, C. R., Schochn, G., et al. (2006). Crystal structure of the rabies virus nucleoprotein-RNA complex. *Science*, 313(5785), 360–363. doi: 1125280 [pii] 10.1126/science.1125280.

Albertini, A. A. V., Ruigrok, R. W. H., & Blondel, D. (2011). Rabies virus transcription and replication. *Advances in Virus Research: Research Advances in Rabies*, 79, 1–22. doi:10.1016/B978-0-12-387040-7.00001-9.

Albertini, A. A. V., Schoehn, G., Weissenhorn, W., & Ruigrok, R. W. H. (2008). Structural aspects of rabies virus replication. *Cellular and Molecular Life Sciences*, 65(2), 282–294. doi:10.1007/S00018-007-7298-1.

Ameyama, S., Toriumi, H., Takahashi, T., Shimura, Y., Nakahara, T., Honda, Y., et al. (2003). Monoclonal antibody #3-9-16 recognizes one of the two isoforms of rabies virus matrix protein that exposes its N-terminus on the virion surface. *Microbiology Immunology, 47*(9), 639–651.

Arnheiter, H., Davis, N. L., Wertz, G., Schubert, M., & Lazzarini, R. A. (1985). Role of the nucleocapsid protein in regulating vesicular stomatitis-virus RNA-synthesis. *Cell, 41*(1), 259–267.

Ball, L. A., Pringle, C. R., Flanagan, B., Perepelitsa, V. P., & Wertz, G. W. (1999). Phenotypic consequences of rearranging the P, M, and G genes of vesicular stomatitis virus. *Journal of Virology, 73*(6), 4705–4712.

Banerjee, A. K. (2008). Response to "Non-segmented negative-strand RNA virus RNA synthesis in vivo". *Virology, 371*(2), 231–233. doi: S0042-6822(07)00796-9 [pii] 10.1016/j. virol.2007.11.026.

Banerjee, A. K., & Chattopadhyay, D. (1990). Structure and function of the RNA polymerase of vesicular stomatitis virus. *Advances in Virus Research, 38*, 99–124.

Banyard, A. C., Hayman, D., Johnson, N., McElhinney, L., & Fooks, A. R. (2011). Bats and lyssaviruses. *Advances in Virus Research: Research Advances in Rabies, 79*(79), 239–289. doi:10.1016/B978-0-12-387040-7.00012-3.

Barik, S., Rud, E. W., Luk, D., Banerjee, A. K., & Kang, C. Y. (1990). Nucleotide sequence analysis of the L gene of vesicular stomatitis virus (New Jersey serotype): Identification of conserved domains in L proteins of nonsegmented negative-strand RNA viruses. *Virology, 175*(1), 332–337.

Barr, J. N., Whelan, S. P. J., & Wertz, G. W. (1997). Cis-acting signals involved in termination of vesicular stomatitis virus mRNA synthesis include the conserved AUAC and the U7 signal for polyadenylation. *Journal of Virology, 71*(11), 8718–8725.

Blumberg, B. M., Giorgi, C., & Kolakofsky, D. (1983). N-Protein of vesicular stomatitis-virus selectively encapsidates leader Rna invitro. *Cell, 32*(2), 559–567.

Bourhy, H., Kissi, B., Audry, L., Smreczak, M., Sadkowska-Todys, M., Kulonen, K., et al. (1999). Ecology and evolution of rabies virus in Europe. *The Journal of General Virology, 80*(Pt 10), 2545–2557.

Bourhy, H., Kissi, B., & Tordo, N. (1993). Molecular diversity of the Lyssavirus genus. *Virology, 194*(1), 70–81. doi: S0042-6822(83)71236-5 [pii] 10.1006/viro. 1993, 1236.

Broughan, J. H., & Wunner, W. H. (1995). Characterization of protein involvement in rabies virus binding to bhk-21-cells. *Archives of Virology, 140*(1), 75–93.

Brzozka, K., Finke, S., & Conzelmann, K. K. (2005). Identification of the rabies virus alpha/beta interferon antagonist: Phosphoprotein P interferes with phosphorylation of interferon regulatory factor 3. *Journal of Virology, 79*(12), 7673–7681. doi:10.1128/Jvi.79.12.7673-7681.2005.

Canter, D. M., Jackson, R. L., & Perrault, J. (1993). Faithful and efficient in vitro reconstitution of vesicular stomatitis virus transcription using plasmid-encoded L and P proteins. *Virology, 194*(2), 518–529. doi: S0042-6822(83)71290-0 [pii] 10.1006/viro.1993.1290.

Castel, G., Chteoui, M., Caignard, G., Prehaud, C., Mehouas, S., Real, E., et al. (2009). Peptides that mimic the amino-terminal end of the rabies virus phosphoprotein have antiviral activity. *Journal of Virology, 83*(20), 10808–10820. doi:10.1128/Jvi.00977-09.

Ceccaldi, P. E., Gillet, J. P., & Tsiang, H. (1989). Inhibition of the transport of rabies virus in the central nervous system. *Journal of Neuropathology and Experimental Neurology, 48*(6), 620–630.

Celis, E., Karr, R. W., Dietzschold, B., Wunner, W. H., & Koprowski, H. (1988). Genetic restriction and fine specificity of human T cell clones reactive with rabies virus. *Journal of Immunology, 141*(8), 2721–2728.

Charlton, K. M., & Casey, G. A. (1979). Experimental rabies in skunks: Immunofluorescence light and electron microscopic studies. *Laboratory Investigation, 41*(1), 36–44.

Chenik, M., Chebli, K., Gaudin, Y., & Blondel, D. (1994). In vivo interaction of rabies virus Phosphoprotein (P) and Nucleoprotein (N) - Existence of 2 N-Binding Sites on P-Protein. *Journal of General Virology, 75*, 2889–2896.

Chenik, M., Schnell, M., Conzelmann, K. K., & Blondel, D. (1998). Mapping the interacting domains between the rabies virus polymerase and phosphoprotein. *Journal of Virology, 72*(3), 1925–1930.

Clark, H. F., Parks, N. F., & Wunner, W. H. (1981). Defective interfering particles of fixed rabies viruses: Lack of correlation with attenuation or auto-interference in mice. *The Journal of General Virology, 52*(Pt 2), 245–258.

Colonno, R. J., & Banerjee, A. K. (1978). Complete nucleotide-sequence of leader Rna synthesized invitro by vesicular stomatitis-virus. *Cell, 15*(1), 93–101.

Conti, C., Hauttecoeur, B., Morelec, M. J., Bizzini, B., Orsi, N., & Tsiang, H. (1988). Inhibition of rabies virus-infection by a soluble membrane-fraction from the rat central nervous-system. *Archives of Virology, 98*(1-2), 73–86.

Conti, C., Superti, F., & Tsiang, H. (1986). Membrane carbohydrate requirement for rabies virus binding to chicken-embryo related cells. *Intervirology, 26*(3), 164–168.

Conzelmann, K. K. (2004). Reverse genetics of mononegavirales. *Biology of Negative Strand Rna Viruses: The Power of Reverse Genetics, 283*, 1–41.

Conzelmann, K. K., Cox, J. H., Schneider, L. G., & Thiel, H. J. (1990). Molecular cloning and complete nucleotide sequence of the attenuated rabies virus SAD B19. *Virology, 175*(2), 485–499.

Conzelmann, K. K., Cox, J. H., & Thiel, H. J. (1991). An L (Polymerase)-deficient rabies virus defective interfering particle RNA is replicated and transcribed by heterologous helper Virus L-Proteins. *Virology, 184*(2), 655–663.

Coulon, P., Derbin, C., Kuccra, P., Lafay, F., Prehaud, C., & Flamand, A. (1989). Invasion of the peripheral nervous systems of adult mice by the CVS strain of rabies virus and its avirulent derivative Avo1. *Journal of Virology, 63*(8), 3550–3554.

Coulon, P., Ternaux, J. P., Flamand, A., & Tuffereau, C. (1998). An avirulent mutant of rabies virus is unable to infect motoneurons in vivo and in vitro. *Journal of Virology, 72*(1), 273–278.

Curran, J., & Kolakofsky, D. (2008). Nonsegmented negative-strand RNA virus RNA synthesis in vivo. *Virology, 371*(2), 227–230. doi:10.1016/J.Virol.2007.11.022.

Davies, M. C., Englert, M. E., Sharpless, G. R., & Cabasso, V. J. (1963). The electron microscopy of rabies virus in cultures of chicken embryo tissues. *Virology, 21*, 642–651.

Delarue, M., Poch, O., Tordo, N., Moras, D., & Argos, P. (1990). An attempt to unify the structure of polymerases. *Protein Engineering, 3*(6), 461–467.

Dietzschold, B., Lafon, M., Wang, H., Otvos, L., Jr., et al., Celis, E., Wunner, W. H., et al. (1987). Localization and immunological characterization of antigenic domains of the rabies virus internal N and NS proteins. *Virus Research, 8*(2), 103–125. doi: 0168-1702(87)90023-2 [pii].

Dietzschold, B., Wang, H. H., Rupprecht, C. E., Celis, E., Tollis, M., Ertl, H., et al. (1987). Induction of protective immunity against rabies by immunization with rabies virus ribonucleoprotein. *Proceedings of the National Academy of Sciences of the United States of America, 84*(24), 9165–9169.

Dietzschold, B., Wiktor, T. J., Trojanowski, J. Q., Macfarlan, R. I., Wunner, W. H., Torresanjel, M. J., et al. (1985). Differences in cell-to-cell spread of pathogenic and apathogenic rabies virus invivo and invitro. *Journal of Virology, 56*(1), 12–18.

Dietzschold, B., Wiktor, T. J., Wunner, W. H., & Varrichio, A. (1983). Chemical and immunological analysis of the rabies soluble glycoprotein. *Virology, 124*(2), 330–337.

Dietzschold, B., Wunner, W. H., Wiktor, T. J., Lopes, A. D., Lafon, M., Smith, C. L., et al. (1983). Characterization of an antigenic determinant of the glycoprotein that correlates with pathogenicity of rabies virus. *Proceedings of the National Academy of Sciences of the United States of America, 80*(1), 70–74.

Ding, H. T., Green, T. J., Lu, S. Y., & Luo, M. (2006). Crystal structure of the oligomerization domain of the phosphoprotein of vesicular stomatitis virus. *Journal of Virology, 80*(6), 2808–2814. doi:10.1128/Jvi.80.6.2808-2814.2006.

Durrer, P., Gaudin, Y., Ruigrok, R. W. H., Graf, R., & Brunner, J. (1995). Photolabeling identifies a putative fusion domain in the envelope glycoprotein of rabies and vesicular stomatitis viruses. *Journal of Biological Chemistry, 270*(29), 17575–17581.

Ertl, H. C., Dietzschold, B., Gore, M., Otvos, L., Jr., et al., Larson, J. K., Wunner, W. H., et al. (1989). Induction of rabies virus-specific T-helper cells by synthetic peptides that carry dominant T-helper cell epitopes of the viral ribonucleoprotein. *Journal of Virological*, *63*(7), 2885–2892.

Etessami, R., Conzelmann, K. K., Fadai-Ghotbi, B., Natelson, B., Tsiang, H., & Ceccaldi, P. E. (2000). Spread and pathogenic characteristics of a G-deficient rabies virus recombinant: An in vitro and in vivo study. *Journal of General Virology*, *81*, 2147–2153.

Faber, M., Faber, M. L., Papaneri, A., Bette, M., Weihe, E., Dietzschold, B., et al. (2005). A single amino acid change in rabies virus glycoprotein increases virus spread and enhances virus pathogenicity. *Journal of Virological*, *79*(22), 14141–14148. doi: 79/22/14141 [pii] 10.1128/JVI.79.22.14141-14148.2005.

Faber, M., Li, J., Kean, R. B., Hooper, D. C., Alugupalli, K. R., & Dietzschold, B. (2009). Effective preexposure and postexposure prophylaxis of rabies with a highly attenuated recombinant rabies virus. *Proceedings of the National Academy of Sciences of the United States of America*, *106*(27), 11300–11305. doi: 0905640106 [pii] 10.1073/pnas.0905640106.

Faber, M., Pulmanausahakul, R., Nagao, K., Prosniak, M., Rice, A. B., Koprowski, H., et al. (2004). Identification of viral genomic elements responsible for rabies virus neuroinvasiveness. *Proceedings of the National Academy of Sciences of the United States of America*, *101*(46)), 16328–16332. doi: 0407289101 [pii] 10.1073/pnas.0407289101.

Finke, S., Brzozka, K., & Conzelmann, K. K. (2004). Tracking fluorescence-labeled rabies virus: Enhanced green fluorescent protein-tagged phosphoprotein P supports virus gene expression and formation of infectious particles. *Journal of Virology*, *78*(22), 12333–12343. doi:10.1128/Jvi.78.22.12333-12343.2004.

Finke, S., & Conzelmann, K. K. (1997). Ambisense gene expression from recombinant rabies virus: Random packaging of positive- and negative-strand ribonucleoprotein complexes into rabies virions. *Journal of Virology*, *71*(10), 7281–7288.

Finke, S., & Conzelmann, K. K. (1999). Virus promoters determine interference by defective RNAs: Selective amplification of mini-RNA vectors and rescue from cDNA by a 3′ copyback ambisense rabies virus. *Journal of Virology*, *73*(5), 3818–3825.

Finke, S., & Conzelmann, K. K. (2005). Recombinant rhabdoviruses: Vectors for vaccine development and gene therapy. *World of Rhabdoviruses*, *292*, 165–200.

Finke, S., & Conzelmann, K. M. (2003). Dissociation of rabies virus matrix protein functions in regulation of viral RNA synthesis and virus assembly. *Journal of Virology*, *77*(22), 12074–12082. doi:10.1128/Jvi.77.22.12074-12082.2003.

Finke, S., Cox, J. H., & Conzelmann, K. K. (2000). Differential transcription attenuation of rabies virus genes by intergenic regions: Generation of recombinant viruses overexpressing the polymerase gene. *Journal of Virology*, *74*(16), 7261–7269.

Finke, S., Granzow, H., Hurst, J., Pollin, R., & Mettenleiter, T. C. (2010). Intergenotypic replacement of lyssavirus matrix proteins demonstrates the role of lyssavirus M proteins in intracellular virus accumulation. *Journal of Virology*, *84*(4), 1816–1827. doi: JVI.01665-09 [pii] 10.1128/JVI.01665-09.

Finke, S., Mueller-Waldeck, R., & Conzelmann, K. K. (2003). Rabies virus matrix protein regulates the balance of virus transcription and replication. *Journal of General Virology*, *84*, 1613–1621. doi:10.1099/Vir.0.19128-0.

Flamand, A., & Delagneau, J. F. (1978). Transcriptional mapping of rabies virus in vivo. *Journal of Virology*, *28*(2), 518–523.

Flamand, A., Raux, H., Gaudin, Y., & Ruigrok, R. W. (1993). Mechanisms of rabies virus neutralization. *Virology*, *194*(1), 302–313. doi: S0042-6822(83)71261-4 [pii] 10.1006/viro.1993.1261.

Flamand, A., Wiktor, T. J., & Koprowski, H. (1980a). Use of hybridoma monoclonal-antibodies in the detection of antigenic differences between rabies and rabies-related virus proteins. 1. The Nucleocapsid Protein. *Journal of General Virology*, *48*(May), 97–104.

Flamand, A., Wiktor, T. J., & Koprowski, H. (1980b). Use of hybridoma monoclonal-antibodies in the detection of antigenic differences between rabies and rabies-related virus Proteins. 2. The Glycoprotein. *Journal of General Virology, 48*(May), 105–109.

Freuling, C. M., Beer, M., Conraths, F. J., Finke, S., Hoffmann, B., Keller, B., & Muller, T. (2011). Novel Lyssavirus in Natterer's bat, Germany. *Emerging Infectious Diseases, 17*(8), 1519–1522. doi:10.3201/Eid1708.110201.

Fu, Z. F., Zheng, Y. M., Wunner, W. H., Koprowski, H., & Dietzschold, B. (1994). Both the N-Terminal and the C-Terminal domains of the nominal phosphoprotein of rabies virus are involved in binding to the nucleoprotein. *Virology, 200*(2), 590–597.

Gao, Y., Greenfield, N. J., Cleverley, D. Z., & Lenard, J. (1996). The transcriptional form of the phosphoprotein of vesicular stomatitis virus is a trimer: Structure and stability. *Biochemistry, 35*(46), 14569–14573.

Gaudin, Y. (1997). Folding of rabies virus glycoprotein: Epitope acquisition and interaction with endoplasmic reticulum chaperones. *Journal of Virology, 71*(5), 3742–3750.

Gaudin, Y., Moreira, S., Benejean, J., Blondel, D., Flamand, A., & Tuffereau, C. (1999). Soluble ectodomain of rabies virus glycoprotein expressed in eukaryotic cells folds in a monomeric conformation that is antigenically distinct from the native state of the complete, membrane-anchored glycoprotein. *Journal of General Virology, 80*, 1647–1656.

Gaudin, Y., Raux, H., Flamand, A., & Ruigrok, R. W. H. (1996). Identification of amino acids controlling the low-pH-induced conformational change of rabies virus glycoprotein. *Journal of Virology, 70*(11), 7371–7378.

Gaudin, Y., Ruigrok, R. W. H., Tuffereau, C., Knossow, M., & Flamand, A. (1992). Rabies virus glycoprotein is a trimer. *Virology, 187*(2), 627–632.

Gaudin, Y., Tuffereau, C., Benmansour, A., & Flamand, A. (1991). Fatty acylation of rabies virus proteins. *Virology, 184*(1), 441–444.

Gaudin, Y., Tuffereau, C., Segretain, D., Knossow, M., & Flamand, A. (1991). Reversible conformational-changes and fusion activity of rabies virus glycoprotein. *Journal of Virology, 65*(9), 4853–4859.

Ge, P., Tsao, J., Schein, S., Green, T. J., Luo, M., & Zhou, Z. H. (2010). Cryo-EM model of the bullet-shaped vesicular stomatitis virus. *Science, 327*(5966), 689–693. doi:10.1126/Science.1181766.

Gerard, F. C. A., Ribeiro, E. D., Albertini, A. A. V., Gutsche, I., Zaccai, G., Ruigrok, R. W. H., & Jamin, M. (2007). Unphosphorylated Rhabdoviridae phosphoproteins form elongated dimers in solution. *Biochemistry, 46*(36), 10328–10338. doi:10.1021/Bi7007799.

Gigant, B., Iseni, F., Gaudin, Y., Knossow, M., & Blondel, D. (2000). Neither phosphorylation nor the amino-terminal part of rabies virus phosphoprotein is required for its oligomerization. *Journal of General Virology, 81*, 1757–1761.

Gillet, J. P., Derer, P., & Tsiang, H. (1986). Axonal-transport of rabies virus in the central nervous system of the rat. *Journal of Neuropathology and Experimental Neurology, 45*(6), 619–634.

Gomme, E. A., Wanjalla, C. N., Wirblich, C., & Schnell, M. J. (2011). Rabies virus as a research tool and viral vaccine vector. *Advances in Virus Research: Research Advances in Rabies, 79*(79), 139–164. doi:10.1016/B978-0-12-387040-7.00009-3.

Gosztonyi, G. (1994). Reproduction of lyssaviruses: Ultrastructural composition of lyssavirus and functional aspects of pathogenesis. *Current topics in Microbiology and Immunology, 187*, 43–68.

Goto, H., Minamoto, N., Ito, H., Ito, N., Sugiyama, M., Kinjo, T., & Kawai, A. (2000). Mapping of epitopes and structural analysis of antigenic sites in the nucleoprotein of rabies virus. *Journal of General Virology, 81*(Pt 1), 119–127.

Green, T. J., Zhang, X., Wertz, G. W., & Luo, M. (2006). Structure of the vesicular stomatitis virus nucleoprotein-RNA complex. *Science, 313*(5785), 357–360. doi:10.1126/Science.1126953.

Guichard, P., Krell, T., Chevalier, M., Vaysse, C., Adam, O., Ronzon, F., & Marco, S. (2011). Three dimensional morphology of rabies virus studied by cryo-electron tomography. *Journal of Structural Biology, 176*(1), 32–40. doi:10.1016/J.Jsb.2011.07.003.

Gupta, A. K., Blondel, D., Choudhary, S., & Banerjee, A. K. (2000). The phosphoprotein of rabies virus is phosphorylated by a unique cellular protein kinase and specific isomers of protein kinase C. *Journal of Virology, 74*(1), 91–98.

Hanlon, C. A., Kuzmin, I. V., Blanton, J. D., Weldon, W. C., Manangan, J. S., & Rupprecht, C. E. (2005). Efficacy of rabies biologics against new lyssaviruses from Eurasia. *Virus Res, 111*(1), 44–54. doi: S0168-1702(05)00073-0 [pii] 10.1016/j.virusres.2005.03.009.

Harty, R. N., Brown, M. E., McGettigan, J. P., Wang, G., Jayakar, H. R., Huibregtse, J. M., & Schnell, M. J. (2001). Rhabdoviruses and the cellular ubiquitin-proteasome system: A budding interaction. *Journal of Virological, 75*(22), 10623–10629. doi:10.1128/JVI.75.22.10623-10629.2001.

Harty, R. N., Paragas, J., Sudol, M., & Palese, P. (1999). A proline-rich motif within the matrix protein of vesicular stomatitis virus and rabies virus interacts with WW domains of cellular proteins: Implications for viral budding. *Journal of Virological, 73*(4), 2921–2929.

Holloway, B. P., & Obijeski, J. F. (1980). Rabies virus-induced RNA-synthesis in bhk-21-Cells. *Journal of General Virology, 49*(Jul), 181–195.

Honda, K., & Taniguchi, T. (2006). IRFs: Master regulators of signalling by Toll-like receptors and cytosolic pattern-recognition receptors. *Nature Reviews Immunology, 6*(9), 644–658. doi:10.1038/Nri1900.

Hummeler, K., Koprowski, H., & Wiktor, T. J. (1967). Structure and development of rabies virus in tissue culture. *Journal of Virological, 1*(1), 152–170.

Hummeler, K., Tomassini, N., Sokol, F., Kuwert, E., & Koprowski, H. (1968). Morphology of the nucleoprotein component of rabies virus. *Journal of Virological, 2*(10), 1191–1199.

Irie, T., Licata, J. M., McGettigan, J. P., Schnell, M. J., & Harty, R. N. (2004). Budding of PPxY-containing rhabdoviruses is not dependent on host proteins TGS101 and VPS4A. *Journal of Virological, 78*(6), 2657–2665.

Iseni, F., Barge, A., Baudin, F., Blondel, D., & Ruigrok, R. W. (1998). Characterization of rabies virus nucleocapsids and recombinant nucleocapsid-like structures. *The Journal of General Virology, 79*, 2909–2919. Pt 12.

Ito, N., Takayama, M., Yamada, K., Sugiyama, M., & Minamoto, N. (2001). Rescue of rabies virus from cloned cDNA and identification of the pathogenicity-related gene: Glycoprotein gene is associated with virulence for adult mice. *Journal of Virology, 75*(19), 9121–9128.

Ito, Y., Nishizono, A., Mannen, K., Hiramatsu, K., & Mifune, K. (1996). Rabies virus M protein expressed in Escherichia coli and its regulatory role in virion-associated transcriptase activity. *Archives of Virology, 141*(3-4), 671–683.

Iwasaki, Y., Wiktor, T. J., & Koprowsk, H. (1973). Early events of rabies virus replication in tissue-cultures - electron-microscopic study. *Laboratory Investigation, 28*(2), 142–148.

Jackson, A. C., & Park, H. (1999). Experimental rabies virus infection of p75 neurotrophin receptor-deficient mice. *Acta Neuropathologica, 98*(6), 641–644.

Jacob, Y., Badrane, H., Ceccaldi, P. E., & Tordo, N. (2000). Cytoplasmic dynein LC8 interacts with lyssavirus phosphoprotein. *Journal of Virology, 74*(21), 10217–10222.

Jiang, Y., Luo, Y. H., Michel, F., Hogan, R. J., He, Y., & Fu, Z. F. (2010). Characterization of conformation-specific monoclonal antibodies against rabies virus nucleoprotein. *Archives of Virology, 155*(8), 1187–1192. doi:10.1007/S00705-010-0709-X.

Kankanamge, P. J., Irie, T., Mannen, K., Tochikura, T. S., & Kawai, A. (2003). Mapping of the low pH-sensitive conformational epitope of rabies virus glycoprotein recognized by a monoclonal antibody #1-30-44. *Microbiology and Immunology, 47*(7), 507–519.

Kasturi, L., Eshleman, J. R., Wunner, W. H., & Shakin-Eshleman, S. H. (1995). The Hydroxy Amino-Acid in an Asn-X-Ser/Thr sequon can influence N-Linked core glycosylation efficiency and the level of expression of a cell-surface glycoprotein. *Journal of Biological Chemistry, 270*(24), 14756–14761.

Kawai, A., Toriumi, H., Tochikura, T. S., Takahashi, T., Honda, Y., & Morimoto, K. (1999). Nucleocapsid formation and/or subsequent conformational change of rabies virus nucleoprotein (N) is a prerequisite step for acquiring the phosphatase-sensitive epitope of monoclonal antibody 5-2-26. *Virology, 263*(2), 395–407. doi: 10.1006/viro.1999.99 62S0042-6822(99)99962-2 [pii].

Kissi, B., Tordo, N., & Bourhy, H. (1995). Genetic polymorphism in the rabies virus nucleoprotein gene. *Virology, 209*(2), 526–537. doi: S0042-6822(85)71285-8 [pii] 10.1006/viro.1995.1285.

Klingen, Y., Conzelmann, K. K., & Finke, S. (2008). Double-labeled rabies virus: Live tracking of enveloped virus transport. *Journal of Virological, 82*(1), 237–245. doi: JVI.01342-07 [pii] 10.1128/JVI.01342-07.

Komarova, A. V., Real, E., Borman, A. M., Brocard, M., England, P., Tordo, N., & Jacob, Y. (2007). Rabies virus matrix protein interplay with eIF3, new insights into rabies virus pathogenesis. *Nucleic Acids Research, 35*(5), 1522–1532. doi: gkl1127 [pii] 10.1093/nar/gkl1127.

Kouznetzoff, A., Buckle, M., & Tordo, N. (1998). Identification of a region of the rabies virus N protein involved in direct binding to the viral RNA. *The Journal of General Virology, 79*(Pt 5), 1005–1013.

Kristensson, K., Dastur, D. K., Manghani, D. K., Tsiang, H., & Bentivoglio, M. (1996). Rabies: Interactions between neurons and viruses. A review of the history of Negri inclusion bodies. *Neuropathology and Applied Neurobiology, 22*(3), 179–187.

Kucera, P., Dolivo, M., Coulon, P., & Flamand, A. (1985). Pathways of the early propagation of virulent and avirulent rabies strains from the eye to the brain. *Journal of Virology, 55*(1), 158–162. doi:10.1016/j.virusres.2005.03.008.

Kuzmin, I. V., Mayer, A. E., Niezgoda, M., Markotter, W., Agwanda, B., Breiman, R. F., & Rupprecht, C. E. (2010). Shimoni bat virus, a new representative of the Lyssavirus genus. *Virus Research, 149*(2), 197–210. doi: S0168-1702(10)00044-4 [pii] 10.1016/j.virusres.2010.01.018.

Kuzmin, I. V., Niezgoda, M., Franka, R., Agwanda, B., Markotter, W., Beagley, J. C., & Rupprecht, C. E. (2008). Possible emergence of West Caucasian bat virus in Africa. *Emerging Infectious Diseases, 14*(12), 1887–1889.

Kuzmin, I. V., Wu, X., Tordo, N., & Rupprecht, C. E. (2008). Complete genomes of Aravan, Khujand, Irkut and West Caucasian bat viruses, with special attention to the polymerase gene and non-coding regions. *Virus Research, 136*(1-2), 81–90. doi: S0168-1702(08)00170-6 [pii] 10.1016/j.virusres.2008.04.021.

Lafay, F., Coulon, P., Astic, L., Saucier, D., Riche, D., Holley, A., & Flamand, A. (1991). Spread of the CVS strain of rabies virus and of the Avirulent mutant AvO1 along the Olfactory pathways of the mouse after intranasal inoculation. *Virology, 183*(1), 320–330.

Lafon, M. (2005). Rabies virus receptors. *Journal of Neurovirology, 11*(1), 82–87. doi:10.1080/13550280590900427.

Lafon, M., Lafage, M., Martinez-Arends, A., Ramirez, R., Vuillier, F., Charron, D., & Scott-Algara, D. (1992). Evidence for a viral superantigen in humans. *Nature, 358*(6386), 507–510. doi:10.1038/358507a0.

Lahaye, X., Vidy, A., Fouquet, B., & Blondel, D. (2012). Hsp70 protein positively regulates Rabies Virus infection. *Journal of Virology* doi: JVI.06501-11 [pii] 10.1128/JVI.06501-11.

Lahaye, X., Vidy, A., Pomier, C., Obiang, L., Harper, F., Gaudin, Y., & Blondel, D. (2009). Functional characterization of Negri bodies (NBs) in rabies virus-infected cells: Evidence that NBs are sites of viral transcription and replication. *Journal of Virology, 83*(16), 7948–7958. doi:10.1128/Jvi.00554-09.

Langevin, C., & Tuffereau, C. (2002). Mutations conferring resistance to neutralization by a soluble form of the neurotrophin receptor (p75NTR) map outside of the known antigenic sites of the rabies virus glycoprotein. *Journal of Virology, 76*(21), 10756–10765. doi:10.1128/Jvi.76.21.10756-10765.2002.

Lazzarini, R. A., Keene, J. D., & Schubert, M. (1981). The origins of defective interfering particles of the negative-strand RNA viruses. *Cell, 26*(2), 145–154.

Le Blanc, I., Luyet, P. P., Pons, V., Ferguson, C., Emans, N., Petiot, A., & Gruenberg, J. (2005). Endosome-to-cytosol transport of viral nucleocapsids. *Nature Cell Biology, 7*(7), 653–664. doi:10.1038/Ncb1269.

Leppert, M., Rittenhouse, L., Perrault, J., Summers, D. F., & Kolakofsky, D. (1979). Plus and minus strand leader rnas in negative strand virus-infected cells. *Cell, 18*(3), 735–747.

Lewis, P., Fu, Y. G., & Lentz, T. L. (2000). Rabies virus entry at the neuromuscular junction in nerve-muscle cocultures. *Muscle and Nerve, 23*(5), 720–730.

Lij, J., Rahmeh, A., Morelli, M., & Whelan, S. P. J. (2008). A conserved motif in region V of the large polymerase proteins of nonsegmented negative-sense RNA viruses that is essential for mRNA capping. *Journal of Virology, 82*(2), 775–784. doi:10.1128/Jvi.02107-07.

Liu, P., Yang, J., Wu, X., & Fu, Z. F. (2004). Interactions amongst rabies virus nucleoprotein, phosphoprotein and genomic RNA in virus-infected and transfected cells. *The Journal of General Virology, 85*(Pt 12), 3725–3734. doi: 85/12/3725 [pii] 10.1099/vir.0.80325-0.

Luo, M., Green, T. J., Zhang, X., Tsao, J., & Qiu, S. H. (2007). Conserved characteristics of the rhabdovirus nucleoprotein. *Virus Research, 129*(2), 246–251. doi:10.1016/J. Virusres.2007.07.011.

Lycke, E., & Tsiang, H. (1987). Rabies virus-infection of cultured rat sensory neurons. *Journal of Virology, 61*(9), 2733–2741.

Macfarlan, R. I., Dietzschold, B., & Koprowski, H. (1986). Stimulation of cytotoxic T-Lymphocyte responses by rabies virus glycoprotein and identification of an immunodominant domain. *Molecular Immunology, 23*(7), 733–741.

Madore, H. P., & England, J. M. (1977). Rabies virus protein synthesis in infected BHK-21 cells. *Journal of Virological, 22*(1), 102–112.

Maillard, A. P., & Gaudin, Y. (2002). Rabies virus glycoprotein can fold in two alternative, antigenically distinct conformations depending on membrane-anchor type. *Journal of General Virology, 83*, 1465–1476.

Marriott, A. C., & Dimmock, N. J. (2010). Defective interfering viruses and their potential as antiviral agents. *Reviews in Medical Virology, 20*(1), 51–62. doi:10.1002/Rmv.641.

Marston, D. A., Horton, D. L., Ngeleja, C., Hampson, K., McElhinney, L. M., Banyard, A. C., & Lembo, T. (2012). Ikoma lyssavirus, highly divergent novel lyssavirus in an African civet. *Emerging Infectious Diseases, 18*(4), 664–667. doi:10.3201/eid1804.111553.

Marston, D. A., McElhinney, L. M., Johnson, N., Muller, T., Conzelmann, K. K., Tordo, N., & Fooks, A. R. (2007). Comparative analysis of the full genome sequence of European bat lyssavirus type 1 and type 2 with other lyssaviruses and evidence for a conserved transcription termination and polyadenylation motif in the G-L 3′ non-translated region. *The Journal of General Virology, 88*(Pt 4), 1302–1314. doi: 88/4/1302 [pii] 10.1099/vir.0.82692-0.

Matsumoto, S., & Kawai, A. (1969). Comparative studies on development of rabies virus in different host cells. *Virology, 39*(3), 449–459.

Matsumoto, S. (1962). Electron microscopy of nerve cells infected with street rabies virus. *Virology, 17*, 198–202.

Mavrakis, M., Iseni, F., Mazza, C., Schoehn, G., Ebel, C., Gentzel, M., & Ruigrok, R. W. H. (2003). Isolation and characterisation of the rabies virus N degrees-P complex produced in insect cells. *Virology, 305*(2), 406–414. doi:10.1006/Viro.2002.1748.

Mavrakis, M., Mehouas, S., Real, E., Iseni, F., Blondel, D., Tordo, N., & Ruigrok, R. W. H. (2006). Rabies virus chaperone: Identification of the phosphoprotein peptide that keeps nucleoprotein soluble and free from non-specific RNA. *Virology, 349*(2), 422–429. doi:10.1016/J.Virol.2006.01.030.

Mayer, M. P. (2005). Recruitment of Hsp70 chaperones: A crucial part of viral survival strategies. *Reviews of Physiology, Biochemistry and Pharmacology, 153*, 1–46. doi:10.1007/s10254-004-0025-5.

Mazarakis, N. D., Azzouz, M., Rohll, J. B., Ellard, F. M., Wilkes, F. J., Olsen, A. L., & Mitrophanous, K. A. (2001). Rabies virus glycoprotein pseudotyping of lentiviral vectors enables retrograde axonal transport and access to the nervous system after peripheral delivery. *Human Molecular Genetics, 10*(19), 2109–2121.

Mebatsion, T., Konig, M., & Conzelmann, K. K. (1996). Budding of rabies virus particles in the absence of the spike glycoprotein. *Cell, 84*(6), 941–951. doi: S0092-8674(00)81072-7 [pii].

Mebatsion, T., Schnell, M. J., & Conzelmann, K. K. (1995). Mokola virus glycoprotein and chimeric proteins can replace rabies virus glycoprotein in the rescue of infectious defective rabies virus particles. *Journal of Virological, 69*(3), 1444–1451.

Mebatsion, T., Weiland, F., & Conzelmann, K. K. (1999). Matrix protein of rabies virus is responsible for the assembly and budding of bullet-shaped particles and interacts with the transmembrane spike glycoprotein G. *Journal of Virological, 73*(1), 242–250.

Menager, P., Roux, P., Megret, F., Bourgeois, J. P., Le Sourd, A. M., Danckaert, A., & Lafon, M. (2009). Toll-like receptor 3 (TLR3) plays a major role in the formation of rabies virus negri bodies. *Plos Pathogens, 5*(2) doi: Artn E1000315Doi 10.1371/Journal.Ppat.1000315.

Morimoto, K., Foley, H. D., McGettigan, J. P., Schnell, M. J., & Dietzschold, B. (2000b). Reinvestigation of the role of the rabies virus glycoprotein in viral pathogenesis using a reverse genetics approach. *Journal of Neurovirology, 6*(5), 373–381.

Morimoto, K., Ni, Y. J., & Kawai, A. (1992). Syncytium formation is induced in the murine neuroblastoma cell-cultures which produce pathogenic Type-G proteins of the rabies virus. *Virology, 189*(1), 203–216.

Murphy, F. A., & Bauer, S. P. (1974). Early street rabies virus-infection in striated-muscle and later progression to central nervous-system. *Intervirology, 3*(4), 256–268.

Murphy, F. A., Bauer, S. P., Harrison, A. K., & Winn, W. C. (1973). Comparative pathogenesis of rabies and rabies-like viruses - viral-infection and transit from inoculation site to central nervous-system. *Laboratory Investigation, 28*(3), 361–376.

Murphy, F. A., Harrison, A. K., Winn, W. C., & Bauer, S. P. (1973). Comparative pathogenesis of rabies and rabies-like viruses: Infection of the central nervous system and centrifugal spread of virus to peripheral tissues. *Laboratory Investigation, 29*(1), 1–16.

Nadin-Davis, S. A., & Real, L. A. (2011). Molecular phylogenetics of the lyssaviruses-insights from a coalescent approach. *Advances in Virus Research: Research Advances in Rabies, 79*(79), 203–238. doi:10.1016/B978-0-12-387040-7.00011-1.

Naito, S., & Matsumoto, S. (1978). Identification of cellular actin within the rabies virus. *Virology, 91*(1), 151–163.

Nakahara, K., Ohnuma, H., Sugita, S., Yasuoka, K., Nakahara, T., Tochikura, T. S., et al. (1999). Intracellular behaviour of rabies virus matrix protein (M) is determined by the viral glycoprotein (G). *Microbiology and Immunology, 43*, 259–270.

Nakahara, T., Toriumi, H., Irie, T., Takahashi, T., Ameyama, S., Mizukoshi, M., & Kawai, A. (2003). Characterization of a slow-migrating component of the rabies virus matrix protein strongly associated with the viral glycoprotein. *Microbiology Immunology, 47*(12), 977–988.

Ogino, T., & Banerjee, A. K. (2007). Unconventional mechanism of mRNA capping by the RNA-dependent RNA polymerase of vesicular stomatitis virus. *Molecular Cell, 25*(1), 85–97. doi:10.1016/J.Molcel.2006.11.013.

Okazaki, Y., Ohno, H., Takase, K., Ochiai, T., & Saito, T. (2000). Cell surface expression of calnexin, a molecular chaperone in the endoplasmic reticulum. *The Journal of Biological Chemistry, 275*(46), 35751–35758. doi: 10.1074/jbc.M007476200M007476200 [pii].

Poch, O., Blumberg, B. M., Bougueleret, L., & Tordo, N. (1990). Sequence comparison of five polymerases (L proteins) of unsegmented negative-strand RNA viruses: Theoretical assignment of functional domains. *The Journal of General Virology, 71*(Pt 5), 1153–1162.

Poch, O., Sauvaget, I., Delarue, M., & Tordo, N. (1989). Identification of four conserved motifs among the RNA-dependent polymerase encoding elements. *EMBO Journal, 8*(12), 3867–3874.

Prehaud, C., Coulon, P., Lafay, F., Thiers, C., & Flamand, A. (1988). Antigenic Site-Ii of the rabies virus glycoprotein – structure and role in viral virulence. *Journal of Virology, 62*(1), 1–7.

Rasalingam, P., Rossiter, J. P., Mebatsion, T., & Jackson, A. C. (2005). Comparative pathogenesis of the SAD-L16 strain of rabies virus and a mutant modifying the dynein light chain binding site of the rabies virus phosphoprotein in young mice. *Virus Research, 111*(1), 55–60. doi: S0168-1702(05)00074-2 [pii] 10.1016/j.virusres.2005.03.010.

Raux, H., Flamand, A., & Blondel, D. (2000). Interaction of the rabies virus P protein with the LC8 dynein light chain. *Journal of Virological, 74*(21), 10212–10216.

Reagan, K. J., & Wunner, W. H. (1985). Rabies virus interaction with various cell-lines is independent of the acetylcholine-receptor – brief report. *Archives of Virology, 84*(3-4), 277–282.

Ribeiro, E. A., Favier, A., Gerard, F. C. A., Leyrat, C., Brutscher, B., Blondel, D., & Jamin, M. (2008). Solution structure of the C-terminal nucleoprotein-RNA binding domain of the vesicular stomatitis virus phosphoprotein. *Journal of Molecular Biology, 382*(2), 525–538. doi:10.1016/J.Jmb.2008.07.028.

Ribeiro, E. D., Leyrat, C., Gerard, F. C. A., Albertini, A. A. V., Falk, C., Ruigrok, R. W. H., & Jamin, M. (2009). Binding of rabies virus polymerase cofactor to recombinant circular nucleoprotein-RNA complexes. *Journal of Molecular Biology, 394*(3), 558–575. doi:10.1016/J.Jmb.2009.09.042.

Robison, C. S., & Whitt, M. A. (2000). The membrane-proximal stem region of vesicular stomatitis virus G protein confers efficient virus assembly. *Journal of Virology, 74*(5), 2239–2246.

Sagara, J., Tochikura, T. S., Tanaka, H., Baba, Y., Tsukita, S., & Kawai, A. (1998). The 21-kDa polypeptide (VAP21) in the rabies virion is a CD99-related host cell protein. *Microbiology Immunology, 42*(4), 289–297.

Sakai, M., Kankanamge, P. J., Shoji, J., Kawata, S., Tochikura, T. S., & Kawai, A. (2004). Studies on the conditions required for structural and functional maturation of rabies virus glycoprotein (G) in G cDNA-transfected cells. *Microbiology and Immunology, 48*(11), 853–864.

Sanchez, A., De, B. P., & Banerjee, A. K. (1985). In vitro phosphorylation of NS protein by the L protein of vesicular stomatitis virus. *The Journal of General Virology, 66*(Pt 5), 1025–1036.

Schnell, M. J., & Conzelmann, K. K. (1995). Polymerase activity of in vitro mutated rabies virus L protein. *Virology, 214*(2), 522–530. doi: S0042-6822(85)70063-3 [pii] 10.1006/viro.1995.0063.

Schnell, M. J., Mebatsion, T., & Conzelmann, K. K. (1994). Infectious rabies viruses from cloned cDNA. *EMBO Journal, 13*(18), 4195–4203.

Schoehn, G., Iseni, F., Mavrakis, M., Blondel, D., & Ruigrok, R. W. (2001). Structure of recombinant rabies virus nucleoprotein-RNA complex and identification of the phosphoprotein binding site. *Journal of Virological, 75*(1), 490–498. doi:10.1128/JVI.75.1.490-498.2001.

Seif, I., Coulon, P., Rollin, P. E., & Flamand, A. (1985). Rabies virulence - effect on pathogenicity and sequence characterization of rabies virus mutations affecting antigenic Site-Iii of the glycoprotein. *Journal of Virology, 53*(3), 926–934.

Shakin-Eshleman, S. H., Remaley, A. T., Eshleman, J. R., Wunner, W. H., & Spitalnik, S. L. (1992). N-Linked glycosylation of rabies virus glycoprotein - individual sequons differ in their glycosylation efficiencies and influence on cell-surface expression. *Journal of Biological Chemistry, 267*(15), 10690–10698.

Shakin-Eshleman, S. H., Wunner, W. H., & Spitalnik, S. L. (1993). Efficiency of N-Linked core glycosylation at asparagine-319 of rabies virus glycoprotein is altered by deletions C-Terminal to the glycosylation sequon. *Biochemistry, 32*(36), 9465–9472.

Sissoeff, L., Mousli, M., England, P., & Tuffereau, C. (2005). Stable trimerization of recombinant rabies virus glycoprotein ectodomain is required for interaction with the p75(NTR) receptor. *Journal of General Virology, 86*, 2543–2552. doi:10.1099/Vir.0.81063-0.

Smith, J. S. (1989). Rabies virus epitopic variation: Use in ecologic studies. *Advances in Virus Research, 36*, 215–253.

Sokol, F., Clark, H. F., Wiktor, T. J., McFalls, M. L., Bishop, D. H., & Obijeski, J. F. (1974). Structural phosphoproteins associated with ten rhabdoviruses. *The Journal of General Virology, 24*(3), 433–445.

Spadafora, D., Canter, D. M., Jackson, R. L., & Perrault, J. (1996). Constitutive phosphorylation of the vesicular stomatitis virus P protein modulates polymerase complex formation but is not essential for transcription or replication. *Journal of Virology, 70*(7), 4538–4548.

Superti, F., Derer, M., & Tsiang, H. (1984). Mechanism of rabies virus entry into cer cells. *Journal of General Virology, 65*(Apr), 781–789.

Superti, F., Hauttecoeur, B., Morelec, M. J., Goldoni, P., Bizzini, B., & Tsiang, H. (1986). Involvement of gangliosides in rabies virus-infection. *Journal of General Virology, 67*, 47–56.

Superti, F., Seganti, L., Tsiang, H., & Orsi, N. (1984). Role of phospholipids in rhabdovirus attachment to CER cells. Brief report. *Archives of Virology, 81*(3-4), 321–328.

Takamatsu, F., Asakawa, N., Morimoto, K., Takeuchi, K., Eriguchi, Y., Toriumi, H., & Kawai, A. (1998). Studies on the rabies virus RNA polymerase: 2. Possible relationships between the two forms of the non-catalytic subunit (P protein). *Microbiology and Immunology, 42*(11), 761–771.

Takayama-Ito, M., Ito, N., Yamada, K., Minamoto, N., & Sugiyama, M. (2004). Region at amino acids 164 to 303 of the rabies virus glycoprotein plays an important role in pathogenicity for adult mice. *Journal of Neurovirology, 10*(2), 131–135. doi:10.1080/13550280490279799.

Tan, G. S., Preuss, M. A., Williams, J. C., & Schnell, M. J. (2007). The dynein light chain 8 binding motif of rabies virus phosphoprotein promotes efficient viral transcription. *Proceedings of the National Academy of Sciences of the United States of America, 104*(17), 7229–7234. doi: 0701397104 [pii] 10.1073/pnas.0701397104.

Tao, L., Ge, J., Wang, X., Zhai, H., Hua, T., Zhao, B., & Bu, Z. (2010). Molecular basis of neurovirulence of flury rabies virus vaccine strains: Importance of the polymerase and the glycoprotein R333Q mutation. *J Virol, 84*(17), 8926–8936. doi: JVI.00787-10 [pii] 10.1128/JVI.00787-10.

Thoulouze, M. I., Lafage, M., Schachner, M., Hartmann, U., Cremer, H., & Lafon, M. (1998). The neural cell adhesion molecule is a receptor for rabies virus. *Journal of Virology, 72*(9), 7181–7190.

Tordo, N., Poch, O., Ermine, A., & Keith, G. (1986). Primary structure of leader RNA and nucleoprotein genes of the rabies genome: Segmented homology with VSV. *Nucleic Acids Research, 14*(6), 2671–2683.

Tordo, N., Poch, O., Ermine, A., Keith, G., & Rougeon, F. (1986). Walking along the rabies genome: Is the large G-L intergenic region a remnant gene? *Proceedings of the National Academy of Sciences of the United States of America, 83*(11), 3914–3918.

Tordo, N., Poch, O., Ermine, A., Keith, G., & Rougeon, F. (1988). Completion of the rabies virus genome sequence determination: Highly conserved domains among the L (polymerase) proteins of unsegmented negative-strand RNA viruses. *Virology, 165*(2), 565–576.

Toriumi, H., & Kawai, A. (2004). Association of rabies virus nominal phosphoprotein (P) with viral nucleocapsid (NC) is enhanced by phosphorylation of the viral nucleoprotein (N). *Microbiol Immunol, 48*(5), 399–409.

Tsiang, H. (1993). Pathophysiology of rabies virus-infection of the nervous-system. *Advances in Virus Research, 42*(42), 375–412.

Tsiang, H., Delaporte, S., Ambroise, D. J., Derer, M., & Koenig, J. (1986). Infection of cultured rat myotubes and neurons from the spinal-cord by rabies virus. *Journal of Neuropathology and Experimental Neurology, 45*(1), 28–42.

Tsiang, H., Derer, M., & Taxi, J. (1983). An in vivo and in vitro study of rabies virus infection of the rat superior cervical ganglia. *Archives of Virology, 76*(3), 231–243.

Tuffereau, C., Benejean, J., Blondel, D., Kieffer, B., & Flamand, A. (1998). Low-affinity nerve-growth factor receptor (P75NTR) can serve as a receptor for rabies virus. *EMBO Journal*, *17*(24), 7250–7259. doi:10.1093/emboj/17.24.7250.

Tuffereau, C., Leblois, H., Benejean, J., Coulon, P., Lafay, F., & Flamand, A. (1989). Arginine or lysine in position-333 of Era and Cvs glycoprotein is necessary for rabies virulence in adult mice. *Virology*, *172*(1), 206–212.

Vidal, S., & Kolakofsky, D. (1989). Modified-model for the switch from sendai virus transcription to replication. *Journal of Virology*, *63*(5), 1951–1958.

Whelan, S. P. (2008). Response to "Non-segmented negative-strand RNA virus RNA synthesis in vivo". *Virology*, *371*(2), 234–237. doi: S0042-6822(07)00797-0 [pii] 10.1016/j.virol.2007.11.027.

Whelan, S. P. J., Barr, J. N., & Wertz, G. W. (2004). Transcription and replication of nonsegmented negative-strand RNA viruses. *Biology of Negative Strand Rna Viruses: The Power of Reverse Genetics*, *283*, 61–119.

Whitt, M. A., Buonocore, L., Prehaud, C., & Rose, J. K. (1991). Membrane-Fusion activity, oligomerization, and assembly of the rabies virus glycoprotein. *Virology*, *185*(2), 681–688.

Wickersham, I. R., Lyon, D. C., Barnard, R. J., Mori, T., Finke, S., Conzelmann, K. K., & Callaway, E. M. (2007). Monosynaptic restriction of transsynaptic tracing from single, genetically targeted neurons. *Neuron*, *53*(5), 639–647. doi: S0896-6273(07)00078-5 [pii] 10.1016/j.neuron.2007.01.033.

Wiktor, T. J., Dietzschold, B., Leamnson, R. N., & Koprowski, H. (1977). Induction and biological properties of defective interfering particles of rabies virus. *Journal of Virological*, *21*(2), 626–635.

Wojczyk, B. S., Takahashi, N., Levy, M. T., Andrews, D. W., Abrams, W. R., Wunner, W. H., & Spitalnik, S. L. (2005). N-glycosylation at one rabies virus glycoprotein sequon influences N-glycan processing at a distant sequon on the same molecule. *Glycobiology*, *15*(6), 655–666. doi:10.1093/Glycob/Cwi046.

Wu, X., Lei, X., & Fu, Z. F. (2003). Rabies virus nucleoprotein is phosphorylated by cellular casein kinase II. *Biochemical and Biophysical Research Communications*, *304*(2), 333–338. doi: S0006291X03005941 [pii].

Wu, X. F., Gong, X. M., Foley, H. D., Schnell, M. J., & Fu, Z. F. (2002). Both viral transcription and replication are reduced when the rabies virus nucleoprotein is not phosphorylated. *Journal of Virology*, *76*(9), 4153–4161. doi:10.1128/Jvi.76.9.4153-4161.2002.

Wunner, W. H., & Clark, H. F. (1980). Regeneration of DI particles of virulent and attenuated rabies virus: Genome characterization and lack of correlation with virulence phenotype. *The Journal of General Virology*, *51*(Pt 1), 69–81.

Wunner, W. H., Dietzschold, B., Smith, C. L., Lafon, M., & Golub, E. (1985). Antigenic variants of CVS rabies virus with altered glycosylation sites. *Virology*, *140*(1), 1–12.

Wunner, W. H., Reagan, K. J., & Koprowski, H. (1984). Characterization of saturable binding-sites for rabies virus. *Journal of Virology*, *50*(3), 691–697.

Yan, X. Z., Mohankumar, P. S., Dietzschold, B., Schnell, M. J., & Fu, Z. F. (2002). The rabies virus glycoprotein determines the distribution of different rabies virus strains in the brain. *Journal of Neurovirology*, *8*(4), 345–352. doi:10.1080/13550280290100707.

Yang, C. X., & Jackson, A. C. (1992). Basis of neurovirulence of avirulent rabies virus variant Av01 with stereotaxic brain inoculation in mice. *Journal of General Virology*, *73*, 895–900.

Yang, J., Koprowski, H., Dietzschold, B., & Fu, Z. F. (1999). Phosphorylation of rabies virus nucleoprotein regulates viral RNA transcription and replication by modulating leader RNA encapsidation. *Journal of Virological*, *73*(2), 1661–1664.

Epidemiology

Cathleen A. Hanlon[1] and James E. Childs[2]
[1]Kansas State University Rabies Laboratory, 2005 Research Park Circle, Manhattan, KS 66506, USA, [2]Department of Epidemiology and Public Health Yale University School of Medicine, 60 College Street, P.O. Box 208034, New Haven, CT 06520, USA

1 INTRODUCTION

The word *rabies* may invoke profound indifference, keen attention, or powerful fear among the public, physicians and related human health care providers, public health professionals, veterinarians, pet owners, and caregivers of domestic and wild animals. Looking at the last 50 years of rabies transmission patterns, challenges are clear. In North America, an extensive but economically pressured public health infrastructure continues to provide protection from an overwhelmingly and universally lethal disease for which there is limited and, with recent economic developments, diminishing support for diagnosis, surveillance, and public and professional education essential for disease prevention.

Rabies has the highest mortality of any known infectious agent with only 3 survivors (one was pre-exposure vaccinated) among the 100 human rabies cases in the USA from 1960–2007. In comparison, the second most lethal viral disease, Ebola hemorrhagic fever, resulted in the death of 18 of 26 cases during a 2000 outbreak in Masindi, Uganda (Borchert et al., 2011). The global burden of rabies is still from sustained transmission of canine-adapted rabies virus variants among domestic dog populations, mainly in Asia and Africa (Lembo et al., 2011). The large-scale threat of rabies exposure to humans and domestic animals in Europe and North America was significantly reduced a number of decades ago through coordinated, methodical, population-based vaccination of dogs and control of stray dog populations. Similar to trends in the

Rabies: Scientific Basis of the Disease and its Management
DOI: http://dx.doi.org/10.1016/B978-0-12-396547-9.00003-1

USA and Europe during the 1950s and 1960s, recent substantial improvements in canine rabies control have occurred in Latin America, reducing the incidence of human disease (Schneider et al., 2011; Ruiz & Chavez, 2010; Belotto, 2004). In practicality, these techniques could be applied globally to push remaining canine rabies virus variants to extinction, if sufficient effort and resources were applied in a cooperative and coordinated manner towards this goal.

Due to elimination of dog-to-dog transmission, trends in animal and human rabies in many areas of the world changed substantially during the last half of the 20th century. In developed countries, human rabies was decreased due to fewer exposures from canine rabies virus variants and the availability of better prophylactic biologics but, in many areas, an increase in rabies transmission among wildlife was recognized. In the USA, the increase in reported wildlife rabies between 1970 and 1990 was due to intensive and geographically extensive epizootics among skunk and raccoon populations. However, enhanced surveillance among other wildlife reservoir species—bats, for example—also influenced the number of reports submitted from states for annual summary.

Overall animal rabies trends as determined by legally mandated, but passive, surveillance efforts in North America indicate a decline from over 20,000 cases per year in the late 1980s to the current level of approximately 6,000–8,000 per year, with the vast majority of cases occurring in the USA (Figure 3.1). These trends reflect the recent success in canine

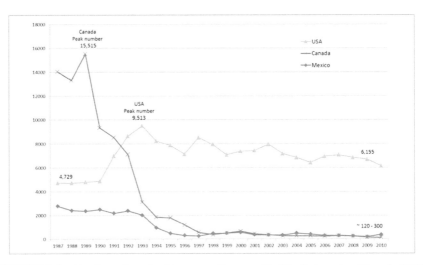

FIGURE 3.1   **Total number of rabies cases diagnosed in North America (Canada, USA, Mexico), 1987–2010.**[1]

[1]Note that the 2009 case total from Mexico is provisional, and the 2010 value is an estimate based on 4 times the first quarter number (n = 92 to date).

rabies control in Mexico, as well as a decline in red fox rabies in Canada due to oral vaccination. The continuing enzootic of raccoon rabies in the USA and its incursions into Canada still pose significant challenges to human and veterinary health (Boyer, Canac-Marquis, Guerin, Mainguy, & Pelletier, 2011; Fehlner-Gardiner et al., 2012; Rosatte et al., 2009).

Despite the historic extinction of canine rabies virus variants in most of North America, numerous rabies virus variants are found among various terrestrial wildlife species (Figure 3.2), and, when considered with the growing number of variants identified among a variety of insectivorous bat species across the Americas and hematophagous bats in Latin America, these constitute an ever-present risk of potential exposure. Sometimes animal owners and the public in the USA are surprised to discover that between 6,000 and 8,000 animal rabies cases are diagnosed every year (Figure 3.3). These cases represent a subset of all naturally occurring cases

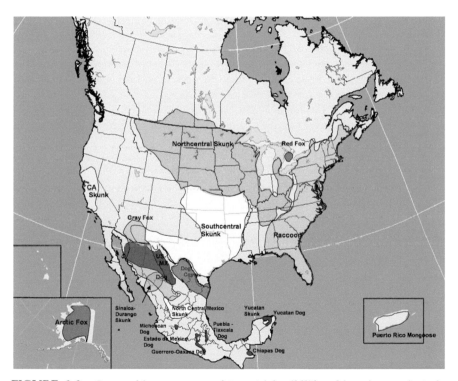

FIGURE 3.2 **Geographic occurrence of terrestrial wildlife rabies virus variants in North America.** The distribution of terrestrial rabies virus variants in North America is delineated by the primary host species or the common name of the group of terrestrial carnivore species serving as reservoir hosts. Numerous rabies virus variants in a number of insectivorous bat species occur throughout the main part of the continent from Alaska and south throughout Mexico and South America. *Data adapted from Velasco-Villa (2006) and USA annual rabies surveillance reports.*

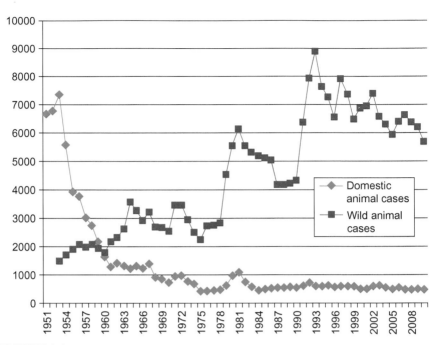

FIGURE 3.3   Cases of rabies in domestic and wild animals in the USA, 1951–2010.

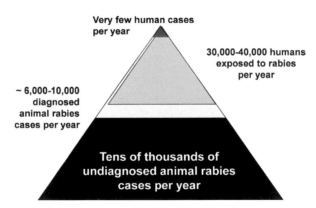

FIGURE 3.4   Measurable impacts of enzootic and epizootic rabies and the "tip-of-the-iceberg" concept.

(Figure 3.4). Moreover, humans travel, domestic, and in some instances wild animals, are moved, and bats, implicated as reservoirs for a variety of rabies virus variants, fly. The risk of exposure to rabies is global. However, understanding global and local transmission patterns and the pathogenesis of the disease will facilitate responsible management of potentially exposed persons and animals towards disease prevention.

2 GLOBAL EPIDEMIOLOGY

The disease rabies is defined as an acute fatal encephalomyelitis in mammals, resulting from infection by any of the viruses in the genus Lyssavirus. As of 2012, the number of recognized species within the genus *Lyssavirus* has grown from seven in 2002 (Childs, Krebs, & Smith, 2002) to 12 (Table 3.1). Since 2003, five novel lyssaviruses were accepted as new species by the International Committee on Taxonomy of Viruses (ICTV) including four unique isolates originating from Eurasia, each from different species of insectivorous bats (Hanlon et al., 2005) and another unique isolate, Shimoni bat virus (SHIBV), from Africa (Kuzmin et al., 2010).

The type species, rabies virus (RABV), is distributed worldwide among mammalian reservoirs consisting of carnivores and bats. RABV, often referred to as genotype 1 (GT 1) and historically as serotype 1 (ST 1), is comprised of a number of related viruses capable of initiating acute fatal neurological disease among mammals. This group is responsible for the majority of human and animal deaths both over the centuries and at present. The term 'genotype' has been applied to lyssaviruses since molecular techniques replaced serotyping as a method of classification (Bourhy, Kissi, & Tordo, 1993). However, the ICTV recognizes viral species rather than genotypes.

Lagos bat virus (LBV), referred to as genotype 2 (GT 2) is found in pteropid bats in sub-Saharan Africa with infrequent spillovers into other species (Markotter et al., 2008). The reservoir host for Mokola virus (MOKV)(genotype 3 (GT 3)) remains elusive but isolates exist from shrews, cats, dogs, a rodent, and two humans in sub-Saharan Africa (Sabeta et al., 2007; Nel et al., 2000). Duvenhage virus (DUVV)(genotype 4 (GT 4)) has been isolated from insectivorous bats and humans who died after bat bites, again, in sub-Saharan Africa (Markotter et al., 2008; van Thiel et al., 2009). The primary host species of European bat lyssavirus, type 1 (EBLV1)(genotype 5 (GT 5)) is the Serotine bat (*Eptesicus serotinus*) and occurs in insectivorous bats across Europe, with occasional infection of humans (Fooks et al., 2003; Kuzmin, Hughes, & Rupprecht, 2006). European bat lyssaviruses, type 2 (EBLV2)(genotype 6 (GT 6)) are isolated primarily from insectivorous bats in the *Myotis* genus and from infected humans following a bite from bats in northwestern Europe (Fooks et al., 2003). The death of a Scottish bat conservationist due to EBL2 infection was the first indigenously acquired case of rabies in the UK in 100 years (Nathwani et al., 2003).

Australian bat lyssavirus (ABLV)(genotype 7 (GT 7)) was first described in 1996 (Speare et al., 1997), is diagnosed with standard rabies diagnostic reagents, prevented with traditional rabies vaccines, and results in fatal human disease that is indistinguishable from classic rabies

TABLE 3.1   Recognized or Proposed Members of the Genus *Lyssavirus*, Family *Rhabdoviridae*

Virus Species Name ICTV Abbreviation[a]	Species Implicated in Maintenance	Distribution	Annual Human Deaths	Reference
Rabies (ST 1/GT 1) RABV	Dogs, wild carnivores, bats	Worldwide (with exception of Australia, Antarctica, and designated rabies-free countries)	>55,000	(Shope, 1982)
Lagos bat (ST 2/GT 2) LBV	Bats-Megachiroptera; *Eidolon helvum*, *Micropterus pusillus*, *Epomophourus wahlbergi*	Africa: Central African Republic, Ethiopia, Nigeria, Senegal, South Africa	not reported	(Bougler and Porterfield, 1958)
Mokola (ST 3/GT 3) MOKV	Uncertain. Shrew-Insectivora; *Crocidura* spp.; Rodentia; *Lopyhromys sikapusi*	Africa: Cameroon, Central African Republic, Ethiopia, Nigeria, South Africa, Zimbabwe	occasional	(Shope et al., 1970)
Duvenhage (ST 4/G 4) DUVV	Bats-Microchiroptera; *Miniopterus schreibersii*, *Nycteris gambiensis*, *N. thebaica*	Africa: South Africa, Guinea, Zimbabwe	occasional	(Meredith et al., 1971)
European bat Lyssavirus 1 (GT 5) EBLV-1	Bats-Microchiroptera; *Myotis dasycneme*, *M. daubentonii*	Europe	occasional	(Bourhy et al., 1993)
European bat Lyssavirus 2 (GT 6) EBLV-2	Bats-Microchiroptera; *Eptesicus serotinus*	Europe	occasional	(Bourhy et al., 1993)

Virus	Host	Location, year	Occurrence in humans	Reference
Australian bat Lyssavirus (GT 7) ABLV	Bats-Megachiroptera; *Pteropus alecto, P. scapulatus*	Australia, 1996; possibly SE Asia mainland	occasional	(Speare et al., 1997)
Aravan virus ARAV	Bat (single isolate) Microchiroptera; *Myotis blythi*	Kyrgyzstan, 1991	not reported	(Arai et al., 2003)
Khujand virus KHUV	Bat (single isolate) Microchiroptera; *Myotis mystacinus*	Tajikistan, 2001	not reported	(Kuzmin et al., 2003)
Irkut virus IRKV	Bat (single isolate) Microchiroptera; *Murina leucogaster*	Eastern Siberia, 2002	not reported	(Kuzmin et al., 2003)
West Caucasian bat virus WCBV	Bat (single isolate) Microchiroptera; *Miniopterus schreibersi*	Caucasus Mountains, 2003	not reported	(Botvinkin, 2003)
Shimoni bat virus SHIV	Bat (single isolate), Leaf-nosed (*Hipposideros commersoni*)	Kenya	not reported	(Kuzmin et al., 2010)
Bokeloh virus[b] BOKV	Bat (single isolate) Natterer's bat (*Myotis nattererii*)	Germany	not reported	(Freuling et al., 2011)
Ikoma virus[b] IKOV	African civet (single isolate) (*Civettictis civetta*)	Tanzania	not reported	Marsten et al., 2012)

[a]ICTV = International Committee on Taxonomy of Viruses, 2011 updates.
[b]As yet unclassified new lyssavirus

(Hanna et al., 2000; Samaratunga, Searle, & Hudson, 1998; Warrilow, 2005; Warrilow, Smith, Harrower, & Smith, 2002). In regard to international travel of domestic animals, Australia is considered "rabies-free" by the World Health Organization, despite these rabies diagnostic and fatal clinical observations and prevention practices with rabies biologics. The Australian bat lyssaviruses circulate among insectivorous and pteropid bats. As previously mentioned, five new species were accepted as new genus lyssavirus members in the 8th report update of the ICTV including Aravan virus (ARAV) and Khujand virus (KHUV) which were each isolated from an insectivorous bat of the *Myotis* genus in Central Asia (Kuzmin et al., 2003), Irkut virus (IRKV) which was isolated from an insectivorous bat, *Murina leucogaster* in eastern Siberia (Kuzmin, Hughes, Botvinkin, Orciari, & Rupprecht, 2005), West Caucasian bat virus (WCBV) which was isolated from an insectivorous bat, *Miniopterus schreibersi* in southeastern Europe (Botvinkin et al., 2003) and Shimoni bat virus identified from an adult female Commerson's leaf-nosed bat (*Hipposideros commersoni*). The latter virus was identified from a dead bat during active disease surveillance in a cave near the southern coast of Kenya (Kuzmin et al., 2010). New lyssaviruses continue to be discovered and present new challenges but the breadth of their occurrence, and the impact of these rabies-related viruses in regard to human and animal mortality appear to be substantially less than those from the type species, RABV.

More recently, Bokeloh bat lyssavirus (BBLV) isolated from a Natterer's bat (*Myotis nattererii*) in Germany (Freuling et al., 2011) and Ikoma virus (IKOV) ioslated from an African civet (*Civettictis civetta*) in Tanzania (Marston et al., 2012) have been proposed as novel members within the genus, although they are not yet formally recognized.

Rabies virus variants can be differentiated by their antigenic characteristics, as measured by differential binding patterns with monoclonal antibodies (Rupprecht, Glickman, Spencer, & Wiktor, 1987; Smith, 1988), and by variations in nucleotide substitutions in their single stranded negative-sense RNA genome (Nadin-Davis, Casey, & Wandeler, 1994; Sacramento, Bourhy, & Tordo, 1991; Smith, Orciari, Yager, Seidel, & Warner, 1992). Based on these tools, it is evident that rabies virus is maintained in diverse mammalian species within the Orders Carnivora and Chiroptera. Humans do not contribute to the maintenance of rabies virus or other lyssaviruses and are considered "dead-end" hosts. Scant evidence exists for natural human-to-human transmission (Fekadu et al., 1996).

Rabies disproportionately affects populations of the species serving as the reservoir host for a specific variant. Transmission of rabies virus is primarily among conspecifics, with cross-species transmission to other mammals usually depending upon direct interactions with infected individuals of the reservoir host species (Blanton, Palmer, Dyer, & Rupprecht, 2011; Blanton, Palmer, & Rupprecht, 2010; McQuiston, Yager, Smith, & Rupprecht, 2001).

Conceptually, rabies virus resembles a metapopulation in which genotypically and phenotypically distinguishable rabies virus variants are adapted to, and maintained by, a single or a few mammalian species, which, *in toto*, constitute the GT 1/ST 1 of genus *Lyssavirus*. The classic definition of a metapopulation is a population of subpopulations that are linked by occasional movements (Barbour & Pugliese, 2004). In the case of rabies virus, movements are represented by the occasional cross-species transmission (i.e., spillover) of rabies virus variants from a reservoir host to a secondary species in which the original variant subsequently becomes adapted to sustained transmission within the secondary species, coincident with the genetic and phenotypic changes indicative of its emergence as a novel and unique sub-variant of rabies.

The epidemiology of human and animal rabies almost invariably reflects the regional terrestrial virus variants maintained within their specific animal reservoirs, and the opportunity for other animal species and human-animal interactions (Anderson, Nicholson, Tauxe, & Winkler, 1984; Feder, Jr., Petersen, Robertson, & Rupprecht, 2012; McQuiston et al., 2001; Noah et al., 1998). Exceptions to this are the occasional spillover of bat-associated rabies virus variants to domestic animals and humans (McQuiston et al., 2001; Messenger, Smith, & Rupprecht, 2002), which are more difficult to spatially map with accuracy due to the ecology of the bat species involved. In addition, "imported" cases of rabies occur following travel or translocation of a human or animal that succumbs to rabies in the new location due to a virus variant from a country of origin.

Animals are also subject to intentional or accidental long distance movement by humans. Translocation of raccoons was responsible for one of the most intense wildlife rabies epizootics on record, the mid-atlantic raccoon rabies epizootic (Nettles, Shaddock, Sikes, & Reyes, 1979). Another very dangerous translocation was responsible for cases of coyote/dog variant rabies among coyotes in Florida, where the virus was introduced with translocated coyotes from Texas (Anonymous, 1995).

Events of viral evolution and adaptation are generally inferred by molecular epidemiologic reconstruction (Hayman et al., 2011; Hughes, Orciari, & Rupprecht, 2005; Kuzmin et al., 2012; Nel et al., 2005; Smith, Lucey, Waller, Childs, & Real, 2002; Talbi et al., 2010). By molecular phylogenetic analyses, a bat ancestry is hypothesized as the origin of rabies virus variants affecting terrestrial carnivores (Badrane & Tordo, 2001). In North America, a molecular clock model suggests the date of divergence of extant bat-associated rabies from the most recent common ancestor occurred about 1651–1660 (Hughes et al., 2005). The bat rabies virus variants found in Latin America among the common vampire bat (*Desmodus rotundus*) and species of the genus *Tadarida*, family *Mollosidae* or free-tailed bats, are closest to the earliest common ancestor. Adaptation of rabies virus variants to species of bats living in medium sized colonies, genera *Eptesicus* and *Myotis*, occurred more rapidly and earlier in time than did adaptation to the more solitary genera, *Lasionycteris*, *Pipistrellus*, and *Lasiuris* (Hughes et al., 2005). In certain instances, as with red foxes on Prince Edward Island (Daoust, Wandeler, & Casey, 1996) and striped skunks in Arizona (Leslie et al., 2006), sustained transmission of bat variants of rabies virus within terrestrial carnivore populations has occurred. Significantly, the identification of specific virus metapopulations within readily identifiable mammalian host species has served as a basis for primary prevention through education of the public and pet owners in regard to risk and prevention, appropriate management of domestic animals at-risk for exposure, and assessment and management of exposed persons. Moreover, this understanding is crucial for experimental approaches towards species-specific design of control measures, such as delivering rabies vaccines by parenteral or oral routes for immunizing sylvatic reservoir hosts and domestic dogs (Charlton et al., 1992; Cliquet et al., 2012; Frontini et al., 1992; Gruzdev, 2008; Hanlon et al., 1989; MacInnes, Tinline, Voigt, Broekhoven, & Rosatte, 1988; Potzsch, Kliemt, Kloss, Schroder, & Muller, 2006; Rupprecht, Hanlon, & Slate, 2004; Wandeler, Capt, Kappeler, & Hauser, 1988) (also see Chapter 13).

3 ROUTES OF RABIES VIRUS TRANSMISSION

3.1 Natural Routes of Transmission

The most common and natural route of rabies virus exposure and transmission is through an animal bite (World Health Organization, 2005; Manning et al., 2008). Contamination of fresh, open bleeding wounds with infectious material is another means of exposure to rabies virus, particularly, for example, if the wounds have been inflicted by the rabid animal through saliva-contaminated claws. Contamination of intact skin with saliva is not considered an exposure to rabies (Manning et al., 2008). Contamination of mucous membranes is a much less effective route of potential exposure to rabies than through bites or open wounds. When exposure occurs by the oral route, rabies occurs infrequently and in some cases, vaccination occurs (Baer, Abelseth, & Debbie, 1971; Baer, Broderson, & Yager, 1975). Certainly ocular exposure through transplantation of corneas from humans dying of rabies generally results in rabies in the recipient (Houff et al., 1979; Javadi, Fayaz, Mirdehghan, & Ainollahi, 1996; Maier et al., 2010; Vetter et al., 2011). However, the risk of transmission through direct contamination of an intact corneal surface is unknown, and no naturally occurring cases have implicated this route. Rabies virus infection possibly acquired by droplets or aerosolized virus has been described in two persons visiting Frio Cave in Texas (Irons, Eads, Grimes, & Conklin, 1957). Millions of Mexican free-tailed bats (*Tadarida brasiliensis*) congregate in this cave, and rabies virus is endemic within the bat population (Humphrey, Kemp, & Wood, 1960). Experimental studies with animals and with an electrostatic precipitation device suggest that air-borne transmission of rabies virus can occur under these exceptional circumstances (Constantine, 1962; Winkler, 1968). It is clear that intranasal exposure to droplets or spray is a potentially dangerous route of exposure because of direct olfactory pathway spread to the brain (Winkler, Fashinell, Leffingwell, Howard, & Conomy, 1973; Anonymous, 1977; Johnson, Phillpotts, & Fooks, 2006), but rabies virus rarely forms a natural aerosol, relegating the inhalation or intranasal route to one of minor importance in the natural history of transmission and disease.

The most important route of rabies virus transmission to humans is animal bite (Manning et al., 2008). Rare reports exist of human rabies following transmission by licks to mucous membranes (Leach & Johnson, 1940), transdermal scratches contaminated with infectious material, and even improperly inactivated rabies vaccines (Para, 1965). Two cases of possible human-to-human transmission of rabies have been described from Ethiopia, although it is not clear if other potential sources of exposure were unlikely (Fekadu et al., 1996). Although transplacental transmission of rabies virus has been reported in a single human case (Sipahioglu & Alpaut, 1985), infants have survived delivery from mothers infected with rabies, when the child received post-exposure prophylaxis (PEP) (Lumbiganon & Wasi, 1990).

There is little documentation for natural rabies transmission by simple contact with virus-infected tissue, although isolated reports suggest infection following butchering of infected carcasses, (Anonymous, 2004; Tariq, Shafi, Jamal, & Ahmad, 1991; Noah et al., 1998) and recently an unvaccinated veterinarian using no personal protective equipment (i.e., gloves, eye protection, etc.) prepared a domestic herbivore for submission for rabies testing and later succumbed to rabies (Brito et al., 2011). In the United States, ingestion of unpasteurized milk from rabid cows has been considered a possible exposure to virus (Centers for Disease Control and Prevention, 1999). Scratches received from a rabid animal could potentially be contaminated with saliva containing rabies virus, and in the United States PEP is considered for persons in these situations (Manning et al., 2008).

Many of the recent human rabies cases in the United States have no documented history of animal bite reported by the patient, relatives, or close companions (Feder, Jr. et al., 2012; Noah et al., 1998). The most likely reason is ignorance regarding the potential rabies exposure risk from a bat encounter and the very minor wound from a bite from a small

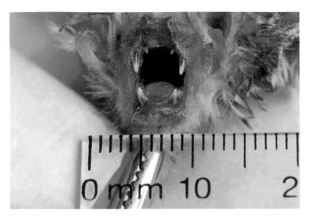

FIGURE 3.5   **Rabies-positive red bat (*Lasuirus borealis*).**

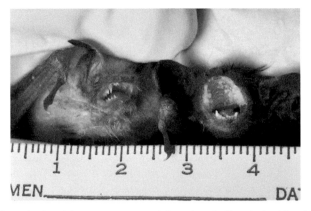

FIGURE 3.6   **Tricolored bat (*Perimyotis subflavus*) and silver-haired bat (*Lasionycterus novtivagans*).**

bat which, unlike bites from carnivores, would not prompt a medical consult solely on the basis of trauma (Figures 3.5 and 3.6).

3.2 Iatrogenic Routes of Transmission

Although rare, iatrogenic human-to-human transmission of rabies has been well documented for recipients of transplanted human tissues. For example, rabies was not recognized in one donor and resulted in deaths among four recipients of kidneys, a liver, and an arterial segment from the index patient in 2004 in the United States (Anonymous, 2004; Noah et al., 1998; Srinivasan et al., 2005). Subsequently, in 2005, three rabies deaths occurred among recipients of organ transplants in Germany (Noah et al., 1998; Maier et al., 2010).

4 RISK AND PREVENTION OF RABIES FOLLOWING AN EXPOSURE

The risk of developing rabies depends on the anatomical site and severity of the bite, the species inflicting the wound, and presumably the rabies virus variant. Published data indicate the risk of developing clinical rabies in unvaccinated persons was 50–80% following multiple, severe head bites; 15–40% following multiple, severe finger, hand, or arm bites; and 3–10% following multiple, severe leg bites inflicted by large terrestrial carnivores, such as wolves or bears (Hattwick et al., 1974). In general, exposure to most body fluids or blood from a rabid animal or a rabid human, with the notable exception of saliva and tears (Anderson et al., 1984; Helmick, Tauxe, & Vernon, 1987), is not regarded as an exposure to rabies (Manning et al., 2008). However, in certain circumstances, laboratory technicians reporting a definite and significant exposure (e.g., a technician cut by a broken specimen container from a rabid patient) to CSF or urine from a human rabies case have been given PEP (Anderson, Williams, Jr., Layde, Dixon, & Winkler, 1984).

In general, modern cell culture derived vaccines, when properly administered with anti-rabies immunoglobulin (RIG), are virtually 100% effective in preventing rabies after an exposure has occurred (Manning et al., 2008). The need for administration of both vaccine and RIG for the prevention of rabies following exposure has been appreciated for more than a half-century (Baltazard & Bahmanyar, 1955; Hemachudha et al., 1999). Vaccine failures have been reported when RIG was not infiltrated around the bite site (Anonymous, 1987; Wilde et al., 1996) or when RIG was omitted in treatment (Gacouin et al., 1999). Potential vaccine failures and failure to adequately seroconvert to nerve tissue origin and cell culture derived vaccines among persons concurrently taking chloriquine for malaria have been documented (Pappaioanou et al., 1986). There are no contraindications for rabies PEP; if one is exposed to rabies, PEP is warranted irrespective of pregnancy, infancy, old-age, or concurrent infections (Manning et al., 2008).

4.1 WHO and the Advisory Committee on Immunization Practices (ACIP) Recommendations for Pre- and Post-exposure Prophylaxis (PEP)

In the United States, all potential rabies virus exposures are treated with HRIG infiltration around the site of exposure and intramuscular vaccination with either the full five doses (Manning et al., 2008) or just the first four (Rupprecht et al., 2009). Two vaccines, Imovax and RabAvert, are licensed in the United States for PEP or pre-exposure

immunization (Manning et al., 2008). The WHO recommends PEP prophylaxis according to categorical grades of potential rabies virus contact
and exposure and lists a number of suitable vaccines for human and animal use (World Health Organization, 2005). Grade II exposures (potential contact with rabies virus through nibbling of uncovered skin or
minor scratches or abrasions with bleeding) require vaccination, unless
or until the animal is determined to be negative for rabies virus infection.
Grade III exposures are potential contact with rabies virus by single or
multiple transdermal bites or scratches, licks, or broken skin, or contamination of mucous membrane with saliva (i.e., licks), or exposure to bats
and requires vaccination and RIG (World Health Organization, 2005). An
additional difference from ACIP recommendations in the United States
is that WHO recommends the use of intradermal vaccination for PEP
and pre-exposure vaccination with two commercially available vaccines,
which permits significant cost savings (World Health Organization, 2005).

As many as 40,000 persons annually may receive PEP in the United
States (Krebs, Long-Marin, & Childs, 1998), although frequently prophylaxis may be administered in situations where it is not indicated (Noah
et al., 1996). Increasingly, mass human exposures to rabid animals have
depleted the local availability of rabies biologicals in the United States
and resulted in expensive episodes (Robbins, Eidson, Keegan, Sackett, &
Laniewicz, 2005; Rotz, Hensley, Rupprecht, & Childs, 1998). Somewhat
to the surprise of public health officials, it appears that when recommendations for PEP are not adhered to in an emergency room setting
in the United States, more often PEP is withheld when it should be
recommended (Moran et al., 2000). Although closer adherence to recommended policies would be ideal, it is unclear whether this would
increase or decrease rabies biological use in the United States.

4.2 PEP for Non-rabies Lyssaviruses

Rabies vaccines and RIG have also been used for post-exposure prophylaxis for other lyssaviruses (Nel, 2005), notably ABLV (Fielding &
Nayda, 2005). In cross-neutralization studies using human sera from
persons vaccinated with human diploid-cell vaccine (HDCV), human
antibodies prevented cell infection with ABLV and ELBV types 1 and 2
in a modified fluorescent antibody virus neutralization assay at titers
$\geq 0.5\,IU/ml$ (Fielding & Nayda, 2005). In experimental trials using PEP
to treat Syrian hamsters and domestic ferrets, standard rabies biologicals
demonstrated reduced, but offered some protection against heterologous
challenge with the four newly described Eurasian bat lyssaviruses (Table I;
ARAV, KHUV, IRKV, and WCBV) (Hanlon et al., 2005). The effectiveness of
vaccine and RIG in animal models was "…inversely related to the genetic
distance between the new isolates and traditional rabies virus," providing

no significant protection against WCBV, the most divergent of the four new bat lyssaviruses from rabies virus (Hanlon et al., 2005).

4.3 Rabies Epizootics and Human Exposure

The tracking and analysis of PEP delivered during successive epizootic and enzootic phases of wildlife rabies are required for understanding the risk of human and animal exposure to rabies from these reservoir hosts (Gordon, Krebs, Rupprecht, Real, & Childs, 2005). The number of human PEPs can remain at elevated levels in regions of the United States where counts of rabid raccoons have greatly diminished in post-epizootic temporal stages (Chang et al., 2002). It is unknown if this is a phenomenon of passive surveillance where the cost and effort to submit a sample for testing is perceived as high such that every sick reservoir species individual is presumed rabid and simply disposed of rather than tested. As data from New York State demonstrate (Table 3.2), the number of animals submitted for rabies testing from a county unit in the years following the first few years of infection declines substantially. The average number of reservoir host species tested per county per year does not appear to constitute an adequate surveillance strategy for measuring rabies intensity among the wildlife species population in a geographic area. In some cases, no reservoir species animals were tested from a county unit for several years. This can be deceptive to the public because when they see zero cases, they interpret the finding as evidence that the disease is no longer present. What may be readily overlooked is that if no animals are tested, then there can be no cases, even though enzootic transmission is indeed highly likely to be occurring in an immediate locality if susceptible populations are present.

5 EPIDEMIOLOGY OF HUMAN RABIES IN NORTH AMERICA AND EUROPE

Human rabies in North America and Europe is now a very rare disease. The history and molecular typing of rabies viruses from cases occurring from 1990–2010 in the United States, Europe, and Japan show that many individuals were exposed in another country (Malerczyk, Detora, & Gniel, 2011; Velasco-Villa et al., 2008). Analyses of human brain material from many imported rabies cases have implicated canine-associated rabies virus variants from countries in which the dog bite was received; imported human rabies in the United States and Europe will continue to occur until Asia and Africa achieve canine rabies control.

Wherever there is or was dog-to-dog transmission of canine rabies virus variants, human rabies cases are or were largely attributable to

TABLE 3.2 Public Health Surveillance Testing of Raccoons, the Reservoir Species, in New York State Over Time, According to Year of Invading Epizootic[a]

Yr of Infection (data from N = ___ # of counties)	Average Positive Cases	# Pos. Range (SD)	Ave. # Tested	# Tested Range (SD)	Overall % Positive	Range of % Pos.[b]	Median Number	
							Positives	Tested
1st (53 counties)	22	2–42 (20)	63	22–104 (41)	35%	7–75 %	17	53
2nd (53 counties)	72	0–177 (105)	112	0–259 (146)	64%	0–91 %	54	79
3rd (50 counties)	23	0–57 (34)	37	0–86 (49)	61%	15–78 %	12	15
4th (44 counties)	8	0–20 (12)	18	0–46 (28)	43%	17–80 %	4	8
5th (32 counties)	7	0–21 (14)	14	0–28 (14)	52%	18–92 %	5	10
6th (18 counties)	7	0–21 (14)	14	0–28 (14)	50%	17–83 %	6	10
7th (6 counties)	6	4–8 (2)	8	5–11 (3)	71%	50–89%	6	9

[a]Data are from 1990–1996, as the raccoon rabies epizootic invaded northward throughout the state.
[b]The range of percent positive among samples tested is based only on data from counties from which >5 raccoons were tested throughout the year.

dog-bite exposure. From 1946, the year the Communicable Disease Center established its national rabies control program, to 1965, 236 cases of human rabies were reported from the United States, of which 70% of the cases were male, 51.3% were ≤15 years of age, and approximately 82% were attributed to dog exposures (Held, Tierkel, & Steele, 1967). Although the risk of rabies has changed substantially due to post-exposure prophylaxis, the age distribution of persons bitten by dogs in the United States is indistinguishable from that occurring in Asia and Africa (Wunner & Briggs, 2010).

Table 3.3 summarizes 111 human rabies cases occuring in the USA from 1960–2011. Over these 52 years, a strong seasonal trend is observed in that the highest number of cases occurs in July through September (Figure 3.7). As previously observed, males are over-represented, with 78 cases among 111 total (70%). The median age of cases was 29 with a range of 2 to 82 years of age. There were 27 cases due to canine rabies virus variants from other countries, including 10 from Mexico and 4 from the Philippines (Figure 3.8). Most of the naturally occurring human rabies cases (40 of 100) in North America from 1960–2011 have been due to variants of rabies virus maintained by insectivorous bats (Table 3.4)(Anderson et al., 1984; Feder, Jr. et al., 2012; Noah et al., 1998).

There has been an inability to elicit a history of animal bite or direct contact with a bat from the patient or family and friends of some human rabies cases due to insectivorous bats (Table 3.5). For example, bites were reported in 12 cases, direct contact with a bat was reported in 11, and a bat in the home or immediate vicinity was reported in 7; for the remaining 8 of 40 cases, no history of a bat or other animal contact could be elicited. The most likely explanation for the lack of history of a bat bite are that the individuals failed to report a bat encounter, perhaps due to perception that it was insignificant (Gibbons, Holman, Mosberg, & Rupprecht, 2002) or because the bat bite went unnoticed (Messenger et al., 2002) (Figures 3.5 and 3.6).

The genotype of rabies virus variants associated with the bat species *Perimyotis subflavus* (formerly *Pipistrellus subflavus*) (4–10 gm) and *Lasionycteris noctivagans* (P.s./L.n. variant) (8–10 gm), is the most frequent associated with human rabies due to bat variants in North America (Messenger, Smith, Orciari, Yager, & Rupprecht, 2003). Experimental evidence suggests that these viruses may have characteristics for enhanced transmission via superficial wounds at peripheral body sites (Dietzschold, Schnell, & Koprowski, 2005; Messenger et al., 2003; Morimoto et al., 1996). The observations that an unnoticed or trivial contact with bats may result in rabies transmission to humans has led to a recommendation that when a bat is physically present, a bite to an individual cannot be ruled out, and rabies cannot be ruled out through testing of the bat, PEP should be considered (Advisory Committee on Immunization Practices, 1999).

TABLE 3.3 Human Rabies Cases in the USA, 1960–2011

	Year	Month of Illness	Outcome	Age	Gender	State	RT-PCR Positive	Diagnostic Method(s)	Location	Source	Exposure Route	Variant
1	1960	5	died	9	M	GA	no sample	>Two	USA	Dog	Bite	Unknown
2	1960	8	died	19	F	OH	no sample	>Two	Guatemala	Cat	Unknown	Unknown
3	1961	1	died	53	F	KY	no sample	>Two	USA	Fox	Bite	Unknown
4	1961	1	died	76	M	CA	no sample	>Two	USA	Dog	Bite	Unknown
5	1961	6	died	74	M	KY	no sample	>Two	USA	Fox	Bite	Unknown
6	1962	7	died	3	M	TX	no sample	>Two	USA	Dog	Direct contact	Unknown
7	1962	10	died	11	M	ID	no sample	>Two	USA	Bat	Bite	Unknown
8	1963	8	died	52	F	AL	no sample	>Two	USA	Dog	Direct contact	Unknown
9	1964	8	died	10	M	MN	no sample	>Two	USA	Skunk	Bite	Unknown
10	1965	5	died	60	M	WV	no sample	>Two	USA	Dog	Bite	Unknown
11	1966	8	died	10	M	SD	no sample	>Two	USA	Skunk	Bite	Unknown
12	1967	7	died	58	F	NY	no sample	>Two	Guinea	Dog	Unknown	Unknown
13	1967	7	died	9	M	OR	yes	>Two	Egypt	Dog	Unknown	Dog, Old World
14	1968	8	died	14	M	KS	no sample	>Two	USA	Dog	Bite	Unknown
15	1969	4	died	2	M	CA	no sample	>Two	USA	Bobcat	Bite	Unknown
16	1970	7	died	11	M	AZ	no sample	>Two	USA	Skunk	Bite	Unknown
17	1970	7	died	4	M	SD	no sample	>Two	USA	Skunk	Bite	Unknown

18	1970	10	**Survived**	6	M	OH	no sample	>Two	USA	Bat	Bite	Unknown
19	1971	3	died	6	M	CA	no sample	>Two	USA	Unknown	Unknown	Unknown
20	1971	11	died	64	M	NJ	no sample	>Two	USA	Bat	Bite	Unknown
21	1972	3	died	70	M	CA	yes	>Two	Philippines	Dog	Unknown	Dog, Phillipines
22	1972	3	died	56	M	TX	no sample	>Two	USA	Aerosol	Laboratory accident	Vaccine strain
23	1973	9	died	26	M	KY	no sample	>Two	USA Bat	Bite	Bat, Perimyotis subflavus	
24	1975	1	died	60	N	MN	no sample	>Two	USA	Cat	Bite	Skunk, North-Central United States
25	1975	7	died	51	M	PR	no sample	>Two	USA, Puerto	Dog	Unknown	Unknown
26	1975	8	died	16	F	CA	yes	>Two	Mexico	Dog	Unknown	Dog, West Mexico/United States
27	1976	6	died	55	F	MD	yes	>Two	USA	Bat	Bite	Bat, Perimyotis subflavus (Ps)
28	1976	8	died	17	M	TX	yes	>Two	Mexico	Dog	Unknown	Dog, Northeast Mexico/United States
29	1977	4	**Survived**	32	M	NY	no sample	>Two	USA	Aerosol	Laboratory accident	Vaccine strain (ERA)

(Continued)

TABLE 3.3 (Continued)

	Year	Month of Illness	Outcome	Age	Gender	State	RT-PCR Positive	Diagnostic Method(s)	Exposure Location	Exposure Source	Exposure Route	Variant
32	1978	6	died	25	M	TX	yes	>Two	Mexico	Dog	Unknown	Dog, Northeast Mexico/United States
30	1978	7	died	39	M	OR	no sample	>Two	USA	Unknown	Unknown	Unknown
31	1978	9	died	37	F	ID	yes	>Two	USA	Corneal transplant	Transplant	Bat, Lasionycteris noctivagans (Ln)
33	1978	12	died	50	M	WV	no sample	>Two	USA	Unknown	Unknown	Unknown
34	1979	5	died	8	M	TX	yes	>Two	Mexico	Dog	Unknown	Dog, Northeast Mexico/United States
35	1979	6	died	7	F	TX	no sample	>Two	USA	Dog	Bite	Unknown
36	1979	7	died	37	M	CA	yes	>Two	Mexico	Dog	Unknown	Dog, West Mexico/United States
37	1979	9	died	24	M	OK	yes	>Two	USA	Unknown	Unknown	Bat, Perimyotis subflavus
38	1979	11	died	45	M	KY	yes	>Two	USA	Ground hog	Direct contact	Bat, Perimyotis subflavus
39	1981	7	died	27	M	OK	yes	>Two	USA	Unknown	Unknown	Skunk, South-Central United States

40	1981	9	died	40	M	AZ	yes	>Two	Mexico	Dog, Mexico	Unknown	Dog, West Mexico/United States
41	1983	1	died	30	M	MA	yes	>Two	Nigeria	Dog, Nigeria	Unknown	Dog, Nigeria
42	1983	3	died	5	F	MI	yes	>Two	USA	Bat	Bite	Bat, Lasionycterus noctivagans
43	1984	8	died	12	F	TX	yes	>Two	Laos	Dog	Unknown	Dog, Southeast Asia (Laos)
44	1984	9	died	12	M	PA	yes	>Two	USA	Unknown	Unknown	Bat, Myotis californicus
45	1984	10	died	72	F	CA	yes	>Two	Guatemala	Dog	Unknown	Dog, Mexico/Guatemala
46	1985	5	died	19	M	TX	yes	>Two	Mexico	Dog	Unknown	Dog, Mexico City
47	1987	12	died	13	M	CA	yes	>Two	Philippines	Unknown	Unknown	Dog, Phillipines
48	1989	2	died	18	M	OR	yes	>Two	Mexico	Unknown	Unknown	Dog, Mexico
49	1990	6	died	22	M	TX	yes	>Two	USA	Bat	Bite	Bat, Tadarida brasiliensis
50	1991	8	died	55	F	TX	yes	>Two	USA	Unknown	Unknown	Dog/coyote, USA
51	1991	8	died	29	M	AR	yes	>Two	USA	Bat	Bite	Bat, L. nocyivagans/P. subflavus

(Continued)

TABLE 3.3 (Continued)

	Year	Month of Illness	Outcome	Age	Gender	State	RT-PCR Positive	Diagnostic Method(s)	Location	Exposure		Variant
										Source	Route	
52	1991	10	died	27	F	GA	yes	>Two	USA	Unknown	Unknown	Bat, L. nocyivagans/ P. subflavus
53	1992	5	died	11	M	CA	yes	>Two	India	Dog	Bite	Dog, India
54	1993	7	died	11	F	NY	yes	>Two	USA	Unknown	Unknown	Bat, L. nocyivagans/ P. subflavus
55	1993	11	died	82	M	TX	yes	>Two	USA	Cow	Direct contact	Bat, L. nocyivagans/ P. subflavus
56	1993	11	died	69	M	CA	yes	>Two	Mexico	Dog	Bite	Dog, Mexico
57	1994	1	died	44	M	CA	yes	>Two	USA	Cat	Direct contact	Bat, L. nocyivagans/ P. subflavus
58	1994	6	died	40	M	FL	yes	>Two	Haiti	Dog	Unknown	Dog, Haiti
59	1994	10	died	24	F	AL	yes	>Two	USA	Unknown	Bat in vicinity	Bat, Tadarida brasiliensis
60	1994	10	died	41	M	WV	yes	>Two	USA	Bat	Bite	Bat, L. nocyivagans/ P. subflavus
61	1994	11	died	42	F	TN	yes	>Two	USA	Unknown	Unknown	Bat, L. nocyivagans/ P. subflavus

62	1994	11	died	14	M	TX	yes	>Two	USA	Dog	Direct contact	Dog/coyote, USA
65	1995	3	died	4	F	WA	yes	>Two	USA	Bat (likely)	Bat in home	Bat, Myotis sp.
63	1995	9	died	27	M	CA	yes	>Two	USA	Bat	Direct contact	Bat, Tadarida brasiliensis
66	1995	10	died	13	F	CT	yes	>Two	USA	Unknown	Bat in home	Bat, L. nocyivagans/P. subflavus
64	1995	11	died	74	M	CA	yes	>Two	USA	Bat	Direct contact	Bat, L. nocyivagans/P. subflavus
67	1996	2	died	26	M	FL	yes	>Two	Mexico	Dog	Bite	Dog, Mexico
68	1996	8	died	32	F	NH	yes	>Two	Nepal	Dog	Bite	Dog, Southeast Asia (Nepal)
69	1996	10	died	42	F	KY	yes	>Two	USA	Unknown	Unknown	Bat, L. nocyivagans/P. subflavus
70	1996	12	died	49	M	MT	yes	>Two	USA	Bat (likely)	Bat in vicinity	Bat, L. nocyivagans/P. subflavus
71	1997	1	died	65	M	MT	yes	>Two	USA	Bat (likely)	Bat in vicinity	Bat, L. nocyivagans/P. subflavus
72	1997	1	died	71	M	TX	yes	>Two	USA	Bat	Direct contact	Bat, L. nocyivagans/P. subflavus

(*Continued*)

TABLE 3.3 (Continued)

	Year	Month of Illness	Outcome	Age	Gender	State	RT-PCR Positive	Diagnostic Method(s)	Location	Exposure		
										Source	Route	Variant
73	1997	10	died	32	M	NJ	yes	>Two	USA	Bat	Direct contact	Bat, L. nocyivagans/P. subflavus
74	1997	10	died	64	M	WA	yes	>Two	USA	Unknown	Bat in vicinity	Bat, Eptesicus fuscus
75	1998	12	died	29	M	VA	yes	>Two	USA	Unknown	Bat in vicinity	Bat, L. nocyivagans/P. subflavus
76	2000	9	died	47	M	MN	yes	>Two	USA	Bat	Bite	Bat, L. nocyivagans/P. subflavus
77	2000	9	died	54	M	NY	yes	>Two	Ghana	Dog	Bite	Dog, A frica (Ghana)
78	2000	10	died	49	M	CA	yes	>Two	USA	Bat	Direct contact	Bat, Tadarida brasiliensis
79	2000	10	died	26	M	GA	yes	>Two	USA	Bat	Direct contact	Bat, Tadarida brasiliensis
80	2000	10	died	69	M	WI	yes	>Two	USA	Bat	Direct contact	Bat, L. nocyivagans/P. subflavus
81	2001	2	died	72	M	CA	yes	>Two	Philippines	Unknown	Unknown	Dog, Phillipines

82	2002	3	died	28	M	CA	yes	>Two	USA	Bat	Direct contact	Bat, L. nocyivagans/P. subflavus
83	2002	8	died	13	M	TN	yes	>Two	USA	Bat	Direct contact	Bat, Tadarida brasiliensis
84	2002	9	died	20	M	IA	yes	>Two	USA	Unknown	Unknown	Bat, L. nocyivagans/P. subflavus
85	2003	2	died	25	M	VA	yes	>Two	USA	Unknown	Unknown	Raccoon, eastern US
86	2003	5	died	64	M	PR	yes	>Two	USA, Puerto	Dog	Bite	Dog/mongoose, Puerto Rico
87	2003	8	died	66	M	CA	yes	>Two	USA	Bat	Bite	Bat, L. nocyivagans/P. subflavus
88	2004	2	died	41	M	FL	yes	>Two	Haiti	Dog	Bite	Dog, Haiti
89	2004	4	died	20	M	AR	no sample	>Two	USA	Bat	Bite	Bat, Tadarida brasiliensis
90	2004	5	died	53	M	OK	yes	>Two	USA	Organ transplant	Transplant	Bat, Tadarida brasiliensis
91	2004	5	died	50	F	TX	yes	>Two	USA	Organ transplant	Transplant	Bat, Tadarida brasiliensis
93	2004	5	died	18	M	TX	yes	>Two	USA	Organ transplant	Transplant	Bat, Tadarida brasiliensis

(Continued)

TABLE 3.3 (Continued)

	Year	Month of Illness	Outcome	Age	Gender	State	RT-PCR Positive	Diagnostic Method(s)	Location	Exposure		Variant
										Source	Route	
92	2004	6	died	55	F	TX	yes	>Two	USA	Artery transplant	Transplant	Bat, Tadarida brasiliensis
94	2004	10	Survived	15	F	WI	neg	>Two	USA	Bat	Bite	no isolate
95	2004	10	died	22	M	CA	yes	>Two	El Salvador	Unknown	Unknown	Dog, El Salvador
96	2005	9	died	10	M	MS	no sample	>Two	USA	Bat	Direct contact	no isolate
97	2006	5	died	16	M	TX	yes	>Two	USA	Bat	Direct contact	Bat, Tadarida brasiliensis
98	2006	9	died	10	F	IN	yes	>Two	USA	Bat	Bite	Bat, Lasionycterus noctivagans
99	2006	11	died	11	M	CA	yes	>Two	Philippines	Dog	Bite	Dog, Phillipines
100	2007	9	died	46	M	MN	yes	>Two	USA	Bat	Bite	Bat, L. nocyivagans / P. subflavus
101	2008	3	died	16	M	CA	yes	>Two	Mexico	Fox	Bite	Bat, Mexico, new Tb
102	2008	11	died	55	M	MO	yes	>Two	USA	Bat	Bite	Bat, Lasionycterus noctivagans
103	2009	2	Survived	17	F	TX	neg	IFA only	USA	Unknown	Unknown	no isolate

104	2009	10	died	43	M	IN	yes	>Two	USA	Unknown	Bat in vicinity	Bat, Perimyotis subflavus
105	2009	10	died	42	M	VA	yes	>Two	India	Dog	Direct contact	Dog, India
106	2009	10	died	55	M	MI	yes	>Two	USA	Bat	Direct contact	Bat, Lasionycterus noctivagans
107	2010	8	died	19	M	LA	yes	>Two	Mexico	Bat	Bite	Bat, Mexico, Desmodus rotundus
108	2010	12	died	70	M	WI	yes	>Two	USA	Unknown	Unknown	Bat, Perimyotis subflavus
109	2011	4	**Survived**	8	F	CA	**neg**	IFA only	USA	Unknown	Unknown	no isolate
110	2011	7	died	73	F	NJ	yes	>Two	Haiti	Dog	Bite	Dog, Haiti
111	2011	8	died	24	M	NY	yes	>Two	Afghanistan	Dog	Bite	Dog, Afghanistan

Source: Anderson, Nicholson, Tauxe, & Winkler, 1984; Feder, Jr., Petersen, Robertson, & Rupprecht, 2012; Noah et al., 1998 and numerous Morbidity and Mortality Weekly Reports published by the Centers for Disease Control and Prevention. Diagnostic methods: antemortem and postmortem diagnostic methods varied throughout the years but include antibody detection in serum, cerebrospinal fluid and sometimes other body fluids by mouse neutralization test (MNT), and then rapid fluorescent focus inhibition test (RFFIT) and more recently augmented with an indirect fluorescent antibody (IFA) test, antigen detection in various samples (i.e., skin, brain, etc.) with direct fluorescent antibody reagents and monoclonal antibodies, virus isolation in suckling mice and cell culture, and genome detection by reverse transcription polymerase chain reaction assay. Rabies virus variant details: Dog/mongoose, Puerto Rico—mongoose-transmitted variant with canine rabies virus variant characteristics and transmissibility Dog/Coyote—variant transmitted among coyotes a unique amino acid change but still with canine rabies virus variant characteristics and transmissibility; Bat Mexico, new Tb—see Velasco-Villa et al., 2008, single human isolate appears to be related to those associated with insectivorous bat species. Highlighted cells in the RT-PCR column emphasize the three cases for which contemporary RT-PCR methods and patient samples were available but tested negative; those in the Exposure column emphasize rabies cases acquired in the laboratory or through iatrogenic surgical interventions; those in the Variant column emphasize cases for which contemporary RT-PCR methods were available but samples were negative or were not available.

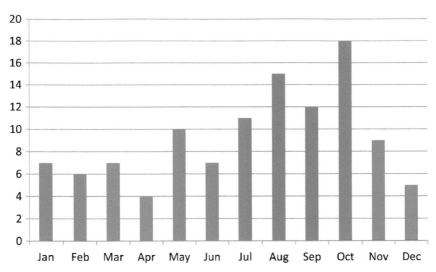

FIGURE 3.7   Monthly occurrence of human rabies cases in the USA, 1960–2011.

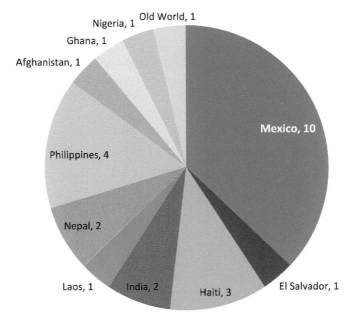

FIGURE 3.8   Human rabies cases in the USA due to canine rabies virus variants from other countries, 1960–2011.

Due to the relative rarity of human rabies cases in some countries, clinical suspicion of rabies can be low and only recognized postmortem when infected material from the index patient is transpanted into recipient patients (Centers for Disease Control and Prevention, 2004; Maier

TABLE 3.4   Variant and RT-PCR Test Status of 111 Human Rabies Cases in the USA, 1960–2011

Rabies Virus Variant	Number of Cases
Lasionycteris noctivagans/Perimyotis subflavus	32
Tadarida brasiliensis	13
Eptescus fuscus	1
Myotis californicus	1
Desmodus rotundus	1
Myotis sp.	1
Total bat cases	49
Dog, imported cases	27
Dog/coyote, USA	2
Skunk	2
Dog/mongoose, Puerto Rico	1
Raccoon	1
Laboratory strain (accidental infection)	2
Total other cases	35
RT-PCR (or variant determination) test status	
No sample for PCR testing	29
RT-PCR positive	79
RT-PCR negative	3
Total	111

et al., 2010). Numerous cases of human rabies have been detected only through postmortem examination of formalin-fixed brain tissue (Noah et al., 1998) or, less frequently, analysis of serum and cerebrospinal fluid for rising antibody titers to rabies virus that neutralize virus as measured by the rapid fluorescent focus inhibition test (RFFIT)(Lewis, 2007; Feder, Jr. et al., 2012). A 2005 case of rabies in a Mississippi resident (Figure 3.9; Table 3.3) was diagnosed retrospectively on the basis of detection of rabies virus-specific antibodies by both RFFIT and the indirect fluorescent antibody (IFA) test (Centers for Disease Control and Prevention, 2006; Lewis, 2007). The diagnosis of rabies in humans, particularly antemortem, and also postmortem when samples from the patient may be limited, can be challenging. A battery of tests is necessary. Ante- and postmortem diagnostic methods have varied through the years, but current approaches consist of detection of neutralizing antibody in serum, cerebrospinal fluid (CSF),

TABLE 3.5   Exposure History of Naturally Occurring Human Rabies Cases in the USA due to Bat-Associated Rabies Virus Variants, 1960–2011

Species of Bat with which the Rabies Virus Variants in Human Cases are Associated[a]	Bite	Direct Contact	Bat in Home/ Vicinity	Unknown	Number of cases
Lasionycteris noctivagans/ Perimyotis subflavus	10	6	5	7	28
Tadarida brasiliensis	2	5	1	0	8
Eptescus fuscus	0	0	1	0	1
Myotis sp.	0	0	1	1	2
Desmodus rotundus (Mexico)	1	0	0	0	1
Total	13	11	8	8	40

[a]Excludes a total of nine bat-associated rabies virus variant cases which occurred in the USA; four cases of rabies due to Lasionycteris noctivagans/Perimyotis subflavus virus variant with one occurring after a corneal transplant and one each following direct contact with a cow, cat, and groundhog; one case of a Tadarada brasiliensis-like variant following a fox bite in Mexico; and four cases of rabies due to the Tadarida brasiliensis rabies virus variant following organ or vascular graft transplantation.

FIGURE 3.9   Human rabies cases in the United States, 1960–2011, by geographic occurrence.

and sometimes other body fluids by the rapid fluorescent focus inhibition test (RFFIT), detection of binding antibody by an indirect fluorescent antibody (IFA) test, antigen detection in various samples (i.e., skin, brain, etc.) with direct fluorescent antibody reagents and monoclonal antibodies, and genome detection by reverse transcription polymerase chain reaction (RT-PCR) assay most often on saliva or brain material. Often, sequential sampling from suspect patients is necessary (Feder, Jr. et al., 2012).

The RT-PCR diagnostic method is exquisitely sensitive. Of the 111 cases reviewed in Table 3.3, 79 had positive findings, although no samples were available for testing from 29 cases. Only 3 cases had negative RT-PCR findings. The first of these cases was the Wisconsin teenager (Figure 3.9; Table 3.3) with virus neutralizing antibody in CSF on the first day of hospitalization. Subsequent sampling revealed increasing neutralizing titers in CSF and serum, but IFA findings were only positive in serum (Feder, Jr. et al., 2012). These findings, along with a high index of suspicion of rabies upon admission to hospital and novel, proactive supportive treatment appears to have contributed to the survival and positive outcome for this patient (Willoughby, Jr. et al., 2005). In contrast, the remaining two cases with RT-PCR negative findings were diagnosed solely on the IFA diagnostic method (Centers for Disease Control and Prevention, 2010; Centers for Disease Control and Prevention, 2012). No neutralizing antibody activity was detected in either patient's serum or CSF, and no rabies virus antigen was detected in skin biopsies. The 2009 Texas case (Figure 3.9; Table 3.3) never required intensive care. Considering the rapid recovery from clinical illness and no neurologic sequelae in these two cases of "survival from rabies," it would seem that further investigation of the source and interpretation of the finding of binding antibodies by IFA would be warranted. Moreover, these methods for diagnosis of rabies in humans are not under the jurisdiction of CLIA but rather are more appropriately considered experimental. Each method as it is applied in unique laboratories requires validation and periodic scrutiny to optimize test performance characteristics.

Although the raccoon rabies epizootic has been one of the most intensive and extensive wildlife epizootics ever recorded, involving 100,000s of animals affected by spillover of this variant (Childs et al., 2000; Gordon et al., 2005), only a single human rabies death has been attributed to this particular variant (Centers for Disease Control and Prevention, 2003). From 1960–2011, two cases of human rabies due to skunk rabies virus variants have been diagnosed in the United States (Tables 3.3 and 3.4). The last human case attributable to exposure to a rabid fox was in Kentucky in 1961 (Anderson et al., 1984).

6 EPIDEMIOLOGY OF HUMAN RABIES IN AFRICA AND ASIA

The vast majority of all human rabies deaths occur in the developing countries where canine rabies virus variants continue to circulate extensively (World Health Organization, 2005). The geographic area of the tropics has accounted for more than 99% of human deaths and ~90% of PEP for rabies (Acha & Arambulo, 1985). Although most rabies deaths are reported from urbanized areas where dog and human populations reach their highest population densities (Beran, 1982), the rate of bites and risk of infection is greater in rural locations. Overall estimates of the annual incidence of bites received from suspect rabid dogs are 100 per 100,000 persons in urban and rural settings in Africa and 100 and 120, respectively, for persons in rural and urban settings in Asia (Cleaveland, Kaare, Tiringa, Mlengeya, & Barrat, 2003; Knobel et al., 2005; Meslin, 2005).

Rabies in Africa and Asia disproportionately afflicts residents in areas of lower socio-economic status in rural and urban settings (Fagbami, Anosa, & Ezebuiro, 1981; Knobel et al., 2005). The estimated direct and indirect costs of rabies PEP is US $39.57 in Africa and US $49.41 in Asia, which represents approximately 5.8% and 3.4% of the annual income for the average person in Africa and Asia, respectively (Knobel et al., 2005). Rabies epizootics among dogs in cities may continue for longer periods in areas of lower socio-economic status (Eng et al., 1993), in part due to the large populations of free-roaming dogs, lack of adequate vaccination coverage of owned or neighborhood dogs, and the lack of resources available for timely control of epizootics (Wilde et al., 2005). Even within the United States, more severe problems associated with higher densities of free-ranging dogs occur in economically challenged urban neighborhoods (Beck, 1975).

In most areas of the world, accurate estimates of human rabies deaths are impossible to obtain as surveillance systems and regional laboratories are inadequate or nonexistent for the systematic detection and laboratory confirmation of human or animal rabies cases. The proportion of rabies cases detected and reported to the WHO in 1999 was estimated to represent only 3% of the total global rabies mortality (Knobel et al., 2005). Novel methods of estimating human rabies deaths have focused on extrapolations from studies estimating domestic dog population densities in different regions in Africa and Asia (Childs et al., 1998), or directly from the incidence of dog-bite in African countries, such as Uganda and Tanzania (Cleaveland, Fevre, Kaare, & Coleman, 2002; Fevre et al., 2005). Risk models based on the likelihood of clinical rabies developing after being bitten by a rabid dog suggest that some 55,000 (90% confidence interval = 24,000–93,000) human rabies deaths occur annually in Africa and Asia (Knobel et al., 2005).

Dog-associated rabies has increased throughout most of sub-Saharan Africa during the last 70 years, although it remains difficult to assess the magnitude of rabies deaths in countries without adequate surveillance or laboratory facilities in the region. In 1997, laboratory confirmation of rabies was available for less than 0.5% of the estimated human rabies cases (World Health Organization, 2005). Sensitive, specific, and, perhaps most important, widely available laboratory testing is an essential element for surveillance for rabies—as is the case with any disease.

7 THE BURDEN AND COST OF RABIES IN AFRICA AND ASIA

An additional health burden to that directly resulting from rabies mortality is the high percentage of adverse reactions to PEP involving nerve tissue origin vaccines. An estimated 200,000 persons in Africa and 7,500,000 persons in Asia (India, China, and other southeastern Asian nations) annually receive PEP for rabies exposures (Knobel et al., 2005). Of those individuals receiving Semple type vaccines, derived from phenol-treated sheep- or goat-brain tissue (Swaddiwuthipong, Weniger, Wattanasri, & Warrell, 1988), or those receiving vaccines derived from suckling-mouse brain tissue (Held & Adaros, 1972), an estimated 360 (CI = 142–586) and 44,525 (CI = 17,585–72,575) disability-adjusted life years (DALY) are lost annually due to adverse reactions among individuals treated in Africa and Asia, respectively (Knobel et al., 2005). The proportion of neuroparalytic complications among persons receiving brain tissue vaccines have been estimated at between 0.3 to 0.8 adverse reactions per 1,000 vaccinees (World Health Organization, 2005).

The global public health cost of rabies is far in excess of metrics limited to the loss of human life. Estimates of the annual burden of canine rabies in Africa and Asia, based on direct medical expenses and costs incurred by patients seeking treatment, amount to US $20.5 million (CI = 19.3–21.8) and US $563 million (CI = 520–605.8), respectively (Knobel et al., 2005). In local studies, such as in Thailand, an estimated >200,000 persons received PEP with cell-culture-derived vaccines in 1997 at a cost of approximately US $10 million (Knobel et al., 2005).

8 EPIZOOTIOLOGY OF RABIES IN LATIN AMERICA

Reports of human rabies deaths in Latin America are substantially lower than rates in Asia and Africa. In large part, the declining rates in Latin America reflect a highly effective regional program of dog rabies control (Organización Panamericana de la Salud (OPS), 1983). Nonetheless, dog-associated rabies remains the principal source of human rabies throughout Latin America. During the interval between 1993 and 2002, 65.2% of the 1,147 human deaths recorded were associated with dogs (Belotto, Leanes, Schneider, Tamayo, & Correa, 2005). In

Mexico, where vampire bat and dog rabies variants co-occur, dog exposures account for ~81% of human rabies deaths (mainly urban) and vampire bats account for ~11% of cases (mostly rural)(de Mattos et al., 1999).

As in Mexico, rabies occurring among domestic animals and humans in Brazil is either of vampire bat origin or associated with dogs (Ito et al., 2003), irrespective of the presence of other variants circulating among other species of bats (Kobayashi et al., 2005).

On islands in the Caribbean, dog-variants of rabies viruses have become established within introduced populations of mongooses and mongooses are responsible for sporadic cases of human rabies such as reported from Puerto Rico (Krebs, Mandel, Swerdlow, & Rupprecht, 2005).

Detailed information on rabies virus variants circulating among sylvatic animal reservoirs is becoming available from Mexico (de Mattos et al., 1999; Velasco-Villa et al., 2002; Velasco-Villa et al., 2005; Velasco-Villa et al., 2008) and several South American countries, such as Bolivia (Favi, Nina, Yung, & Fernandez, 2003), Brazil (Bordignon et al., 2005; Schaefer, Batista, Franco, Rijsewijk, & Roehe, 2005; Shoji et al., 2006), Chile (de Mattos, Favi, Yung, Pavletic, & de Mattos, 2000; Favi, Bassaletti, Lopez, Rodriguez, & Yung, 2011; Favi et al., 2002; Yung, Favi, & Fernandez, 2012), and Colombia (Paez, Nunez, Garcia, & Boshell, 2003; Paez, Saad, Nunez, & Boshell, 2005). In Brazil, molecular sequence data based on the rabies virus N gene have identified four rabies virus variants clustering with four bat genera; one lineage is associated with the common vampire bat, *Desmodus rotundus*, the other three segregate with three families/genera of insectivorous bats, *Eptesicus*, *Molossus*, and *Nyctinomops* (Kobayashi et al., 2005, 2007). A novel variant of rabies isolated from marmosets (*Callithix jacchus*) in Brazil was linked to a human rabies case (Favoretto, de Mattos, Morais, Alves Araujo, & de Mattos, 2001). Rare human rabies deaths in Chile, Colombia, and Brazil have been attributed to rabies virus variants circulating in insectivorous bats (Favi et al., 2002; Paez et al., 2003). In Colombia, variants of rabies virus detected from a case of human rabies and three cases of dog rabies indicated infection by bat-associated variants of rabies virus circulating in two species *Eptesicus brasiliensis* and *Molossus molossus* (Paez et al., 2003). The domestic dog variant of rabies virus appears to have successfully established enzootic maintenance among gray foxes (*Urocyon cineroargenteus*) in Colombia (Paez et al., 2005).

In Mexico, where human rabies is associated with vampire bat and canine rabies virus variants, a newly identified rabies virus variant present among skunks (species not identified) and distinguishable from those rabies virus variants circulating among skunks in North America has been identified (de Mattos et al., 1999).

8.1 Vampire Bat Rabies and Epizootic Cycles of Bovine Rabies

The association of vampire bats and rabies epizootics among cattle in Latin America was first noted in 1910 in Brazil (Carini, 1911). The first human deaths attributed to bites received from vampire bats were documented on the Island of Trinidad (Carini, 1911; Hurst & Pawan, 1959), where vampire bat transmitted rabies continues to be a sporadic disease of cattle (Carini, 1911; Wright, Rampersad, Ryan, & Ammons, 2002). Outbreaks of human rabies due to vampire bats continue to be reported from many South and Central American countries and rabies virus variants originating from vampire bats continue to be isolated from cattle and other species in Argentina (Carini, 1911; Cisterna et al., 2005; Wright et al., 2002), Bolivia (Carini, 1911; Favi et al., 2003; Wright et al., 2002), Brazil (Batista-da-Costa, Bonito, & Nishioka, 1993; Carini, 1911; Ito et al., 2001; Kobayashi et al., 2008; Wright et al., 2002), Peru (Lopez, Miranda, Tejada, & Fishbein, 1992), Venezuela (Caraballo, 1996), Costa Rica (Badilla et al., 2003), Colombia (Badillo, Mantilla, & Pradilla, 2009), and Mexico (de Mattos et al., 1999; Martinez-Burnes et al., 1997).

Vampire bats feed preferentially on livestock. In addition to compromising animal productivity, vampire bats present a significant economic burden through losses due to rabies (Baer, 1991; Delpietro & Russo, 1996; Martinez-Burnes et al., 1997). The use of anticoagulants, applied by topical treatment to the backs of captured bats, subsequently released to return to roosting sites (Linhart, Flores, & Mitchell, 1972), or through systemic treatment of cattle, have been used to achieve reductions in vampire bat biting rates on cattle of 85 to 96% (Flores, Said, De Anda, Ibarra, & Anaya, 1979). Unfortunately, vampire bat control also leads to the death of many non-target species of bats (Mayen, 2003).

The migratory wave of vampire bat rabies has been estimated to travel at 40 to 50 km per year (Brass, 1994), remarkably similar to the rates (30–60 km per year) established for the spread of red fox and raccoon rabies epizootics (Wandeler, 2004; Lucey et al., 2002). Cyclic changes in vampire bat populations could drive cyclic and periodic epizootics of cattle rabies caused by vampire bat transmitted rabies. In regions of Central and South America, areas affected by vampire rabies experience outbreaks every 2 to 3 years (Ruiz & Chavez, 2010).

9 CASE NUMBER DYNAMICS AND MODELING

Rabies epizootics among wildlife reservoir host populations within defined regions frequently follow a distinct course. Intervals of increased disease activity (epizootics) are separated by intervals (inter-epizootics) in which rabies may seem to disappear or reach undetectable levels within a local mammalian community (Anderson, Jackson, May, & Smith, 1981; Childs et al., 2000; Steck & Wandeler, 1980; Wandeler et al., 1974). Following an initial epizootic of rabies, which is typically the largest of a possible series of epizootics that may emerge over time as wildlife rabies enters into a new region to infect previously naive populations, a series of successively smaller epizootics may occur at increasing frequency over time (Anderson et al., 1981; Coyne, Smith, & McAllister, 1989; Smith et al., 2002). The periodic epizootic structure of rabies epizootics may become indistinguishable against a background level of disease. Primary data from New York State during the initial invasion of the raccoon rabies epizootic, reveal the weaknesses of wildlife rabies surveillance based on a passive public health system (Table 3.2). With an estimated average raccoon population in the affected counties, testing fewer than one suspect raccoon per month per county is a poor indicator of the intensity of rabies among the susceptible population and most likely underestimates the risk to domestic animals and humans.

Despite the limitations of a passive surveillance system intended solely for public health purposes, rabies virus maintenance in animal reservoirs and related data have served as a model system for illustrating many important concepts in infectious disease epidemiology and the theoretical modeling of the population biology of a virus and its host. Predictions from model outcomes have been used to identify temporal linkages in the risk of rabies virus spillover from sylvatic reservoirs to domestic species and other wildlife (Gordon et al., 2004; Guerra et al., 2003), the design of rational intervention schemes based on geographic simulators projecting rabies spread following breaches in oral wildlife vaccination zones (Russell, Smith, Childs, & Real, 2005), and the development of analytic methods to inform economic models with finer scale resolution of cost structures associated with different temporal stages of rabies epizootics, providing improved estimates of the savings potentially accrued through active interventions (Gordon et al., 2005).

9.1 Red Fox Rabies

Beginning in the 1940s, an epizootic of red fox rabies began spreading from Russia and Poland towards Western Europe, eventually affecting much of the continent (Steck & Wandeler, 1980; Wandeler, 2008). The epizootic front of red fox rabies in Europe advanced in an irregular wavelike fashion at an estimated 25–60 km per year (Steck & Wandeler, 1980; Wandeler, 2004). From 1978 to 1999, over 151 million vaccine-laden baits were distributed in 18 European countries, which has greatly reduced and even eliminated fox rabies from many previously affected regions (Wandeler, 2008).

Areas of high quality habitat supporting high population densities of red foxes suffer the greatest population depression due to rabies and the highest incidence of disease during the initial epizootic, as based on hunter index estimates of red fox population size (Steck & Wandeler, 1980). Based on population estimates from hunter indeces, rabies reduced populations of red foxes in numbers to 50–60% below pre-epizootic levels (Bogel, Arata, Moegle, & Knorpp, 1974). Estimates of the time in years to recovery of fox populations to pre-rabies densities illustrate how high reproductive rates among these carnivores can rapidly increase densities above the minimum values required to sustain periodic reemergence of disease outbreaks (Bogel et al., 1974; Macdonald & Bacon, 1982).

Prevalence is a difficult attribute to estimate for wildlife disease such as rabies, as the required denominator (the population at a specific time or average population size during an interval of time) is almost never known. However, theoretical estimates and reported estimates from hunter indices of the equilibrium prevalence of rabies suggest the value remains fairly constant at 3–7% during outbreaks (Anderson et al., 1981; Bogel et al., 1974).

9.2 Fox Rabies

Arctic fox rabies virus variants have a near circumpolar distribution, and these fox virus variants remain an occasional source of human rabies in Asia (Kuzmin, 1999; Kuzmin, Hughes, Botvinkin, Gribencha, & Rupprecht, 2008; Shao et al., 2011). These variants were the source for independent cycling of rabies virus among raccoon dogs in Europe and Asia (Bourhy, Dacheux, Strady, & Mailles, 2005; Bourhy et al., 1999; Potzsch et al., 2006).

In North America, a major epizootic of the Arctic fox variant of rabies virus, involving red and arctic foxes (*Alopex lagopus*), began in northern Canada in the 1940s (Tabel, Corner, Webster, & Casey, 1974). In the early 1960s, the epizootic of red fox rabies expanded from Ontario into the northeastern states of New York, New Hampshire, Vermont, and Maine. Red fox rabies and spillover to domestic and wild animals in the United States occurred until the mid-1990s (Gordon et al., 2004), when effective control efforts initiated in Canada brought the fox rabies under control in neighboring Ontario (MacInnes et al., 2001), and coincidently resulted in the disappearance of red fox-associated rabies in the northeastern United States.

The first recorded case of rabies in foxes in the United States occurred in a gray fox (*U. cinereoargenteus*) in Georgia in 1940 (MacInnes et al., 2001). Within years, rabies was endemic among gray foxes in Alabama, Florida, and Tennessee, and from 1940 to 1960 gray and red foxes were the wild carnivore most commonly reported rabid in the United States. Since the 1940s, the endemic area affected by the gray fox-associated variant of rabies virus has diminished in size such that endemic gray fox rabies in the United States is presently known only from relatively small areas in Texas and Arizona (Figure 3.2) (Rohde, Neill, Clark, & Smith, 1997). However, a phylogenetically related virus has been recovered from bobcats in Mexico (de Mattos et al., 1999), suggesting that this rabies virus variant may have a more extensive range than previously appreciated.

10 SKUNK RABIES VIRUS VARIANTS

Rabies among skunks has a long history in North America. The earliest reports of rabies from California in 1826 incriminate spotted skunks (genus *Spilogale*) as the source of human disease (Parker, 1975). Three genera of skunks occur in North American; these are *Mephites*, the striped skunk, and *Conepatus*, the hog-nosed skunks, and *Spilogale*. The striped skunk is by far the most common (Parker, 1975). Currently, two rabies virus biotypes, the South-Cental and North-Central biotypes, circulate among skunks in the central United States (Figure 3.2), with the North-Central Skunk biotype extending into central and western Canada; a third biotype of rabies virus circulates among skunks in California (Crawford-Miksza, Wadford, & Schnurr, 1999). Beginning in the 1960s and continuing until 1990 (Figure 3.10), skunks were the group of terrestrial mammals most frequently reported rabid in the United States.

In the United States and Canada, skunk rabies is common in prairie habitats (Greenwood, Newton, Pearson, & Schamber, 1997; Pool & Hacker, 1982), where periodically epizootic disease has been postulated to be a major factor in driving the cyclic variations in skunk population numbers (Pybus, 1988). The potential impact of rabies on skunk populations was amply demonstrated by Greenwood et al., (1997) who followed a population of radio-collared animals during an epizootic in South Dakota. Estimated rates of skunk density fell from 0.85 skunks per km^2 during April to June, 1991, to 0.17 skunks per km^2 in April to July, 1992, during the rabies epizootic.

11 RACCOON RABIES IN NORTH AMERICA

The epizootic associated with raccoons in the eastern United States is believed to have been initiated in the mid-Atlantic region by the interstate translocation of raccoons incubating rabies from an established focus of raccoon rabies in the southeastern USA for the purpose of restocking dwindling local populations (Nettles, Shaddock et al., 1979; Smith, Sumner et al., 1984). Since the mid-1970s, this raccoon-adapted variant of rabies virus has spread north to Maine and Ontario, Canada,

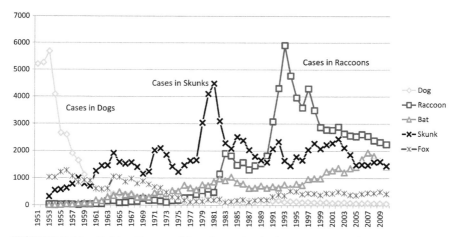

FIGURE 3.10   **Rabies diagnosed in select animals that are reservoir hosts for rabies in the USA, 1951–2010.**

and west to Ohio, causing one of the most intensive outbreaks of animal rabies ever recorded (Childs et al., 2000). Raccoons have been the wild animal species most frequently diagnosed as rabid since the early 1990s (Figure 3.10).

The magnitude of this epizootic was enhanced by the spread of virus through naive raccoon populations of very high density, often in states that had not experienced terrestrial rabies for decades (Hanlon et al., 1998). The interval from the start date of the initial epizootic as raccoon rabies enters a previously unaffected area, and the start date of the second epizootic, as defined by a statistical algorithm (Childs et al., 2000) obtained from time series data from the eastern United States, was 48 months.

There are no published estimates for the critical threshold value of the raccoon population density required to support rabies transmission in areas of the United States. However, direct and indirect estimates of raccoon population size or density indicate significant declines in population size following epizootics of raccoon rabies (Anthony et al., 1990). The magnitude of recorded epizootics, as well as the number of animals tested for possible rabies infection, is correlated with the human population size or density at the level of township and county (Gordon et al., 2005).

The local rate of disease propagation is significantly affected by local environmental heterogeneities, such as those posed by major rivers running orthogonal to the major direction of raccoon rabies spread. In Connecticut, models using a stochastic simulator determined the reduction in local transmission of raccoon rabies to be sevenfold for townships separated by a major river compared to townships without such a physical barrier (Smith et al., 2002). Habitat variation affecting the equilibrium raccoon population density appears to have an influence on the rate of spread of raccoon rabies, as estimated by empirical data on the date of first appearance of rabies in townships of New York State and simulations of the expected rate of rabies spread (Russell, Smith, Waller, Childs, & Real, 2004). Coincident with the rabies wavefront reaching the Adirondack Mountains, advancement of the raccoon epizootic slowed approximately 4 years after entering New York State from the south in 1991. The population density of raccoons in this region of coniferous forests is extremely low (Godin, 2012). The Adirondack region continues to be a formidable barrier to raccoon rabies through 2010 (Blanton et al., 2011), and raccoon rabies has, for the most part, gone around rather than through this region into Canada.

Raccoon rabies was first detected in Ontario, Canada, in 1999, from across the St. Lawrence River border with the United States (Wandeler & Salsberg, 1999). Molecular epidemiologic analyses of raccoon variants from Canada indicate that three independent incursions of raccoon rabies have occurred—two in 1999 in Ontario, and one in 2000 in New Brunwick (Nadin-Davis, Muldoon, & Wandeler, 2006). More recently, confirmed positive raccoon-variant cases in southern Quebec have led to extensive efforts including local depopulation, trapping, and parenteral vaccination and vaccines offered in baits to try to control the incursion (Sterner, Meltzer, Shwiff, & Slate, 2009).

11.1 Rabies Virus Spillover to Domestic Dogs and Cats

With the effective implementation of dog vaccination, spillover of wildlife variants of rabies virus to dogs has declined in North America and Europe to levels below that reported for cats. Thus, in the 1990s, cats have replaced dogs as the companion animal species most commonly reported rabid. In the United States, the annual number of dog rabies cases, primarily due to spillover from rabies virus variants circulating among terrestrial wildlife, has fallen below 100, whereas the average number of cat cases is close to 300 per year (Figure 3.11). In southern Texas, a number of domestic dogs were infected with a dog/coyote variant of rabies virus in the 1990s. However, an aggressive vaccination campaign, both parenteral for dogs and targeted wildlife vaccination, appears to have successfully controlled this variant (Sidwa et al., 2005).

The majority of rabid cats are reported from the eastern United States, where the raccoon-adapted variant of rabies virus is endemic. The large and disproportionate number of rabid cats being identified in the United States presumably reflects poorer vaccine coverage in this animal than is achieved for dogs (Nelson, Mshar, Cartter, Adams, & Hadler, 1998). Required vaccination for cats is still not legally mandated in some states or counties. In 1996, a survey of the 50 states, the District of Columbia, and 3 of 5 territories revealed that 74% of these political units required

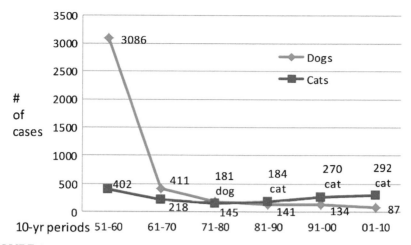

FIGURE 3.11   Ten-year average number of cases of rabies diagnosed in dogs and cats in the USA, 1951–2010.

dog vaccination compared with 52% requiring cat vaccination (Johnston & Walden, 1996). The large number of stray and unvaccinated cats has contributed greatly to the increase in rabies in this species. Cats have been the cause of several large-scale exposures of humans to potentially rabid animals, including one situation involving over 600 PEPs (Noah et al., 1996; Rotz et al., 1998).

In a study assessing the risk of cat cases associated with different temporal stages of the raccoon rabies epizootics, there was a significant urban to rural trend in the increased risk of a cat testing positive for rabies. The risk of a cat testing positive for rabies [Odds Ratio (OR)] in rural counties in the lowest quartile for human population density (<61.6 inhabitants per mi^2) was 2.7-fold above the referent value (odds ratio = 1) for counties in the highest quartile of human population density (>420.2 inhabitants per mi^2); counties with intermediate human population densities, in the second and third quartiles, also showed increasing risk for cat rabies above counties with the highest human density, with odds ratios of 1.7 and 2.0 above the referent value, respectively (Gordon et al., 2004; Noah et al., 1996).

In Western Europe, where red fox rabies is the dominant form of endemic rabies and domestic dog rabies has long been controlled, rabies cases in cats remain a public health problem. Feral cat colonies are recognized as a potential hotspot for spillover of wildlife variants to cats; 14 rabid cats and no rabid dogs occurred between 1979 and 2000 (Mutinelli, 2010; Mutinelli, Stankov, Hristovski, Theoharakou, & Vodopija, 2004; Noah et al., 1996). Between 1979 and 2000, France reported rabies among 1,256 cats and 694 dogs, and between 1966 and 2000, Belgium reported 295 rabid cats and 64 rabid dogs (Aubert et al., 2004).

11.2 Rabies Virus Spillover to Livestock

Although not reservoirs for rabies virus, livestock species are susceptible to infection by variants maintained in other species. These species frequently are not vaccinated against rabies. At least one human death has been documented as a result of contact with a rabid domestic herbivore in Brazil (Brito et al., 2011). Mass human exposures have occurred where animals in public settings developed rabies. For example, in 2004, exposure to a suspect rabid sheep at a Texas wildlife center with an animal petting area resulted in more than 650 people receiving PEP (Star Telegram, 2004; Blanton, Krebs, Hanlon, & Rupprecht, 2006). Further examples include a rabid goat at a New York county fair in 1996, a rabid Wyoming rodeo pony in 1995, and two rabid Massachusetts dairy cows in 1998 and 1996, which resulted in prophylaxis for 465, 12, and 89 people, respectively (Centers for Disease Control and Prevention, 1999; Chang et al., 2002; Compendium of animal rabies prevention and control, 2004). Despite considerations that the oral route of exposure is realtively inefficient for transmission of rabies and a dilution factor of milk from a rabid cow in a bulk tank pool from a herd, 80 of the 89 people exposed to these rabid dairy cows received postexposure prophylaxis due to the ingestion of raw milk. Pasteurization temperatures inactivate rabies virus, therefore drinking pasteurized milk from a rabid animal is not a rabies exposure. Because animal-to-animal transmission is uncommon in livestock, quarantine of the exposed animal may not be warranted (Compendium of Animal Rabies Prevention and Control, 2004).

In the United States, cattle rabies has declined in a similar manner to that of dog rabies (Figure 3.12); from 2001–2010, an average of 85 cattle were reported rabid. Spillover of red fox rabies to cattle in Western Europe has exceeded the levels of dog and cat rabies over the past several decades. Between 1977 and 2000, 6,047 cases of cattle rabies were reported from Germany, 681 cases were reported from Austria (Muller, Cox, & Muller, 2004), and 5 cases were reported from Italy (Mutinelli et al., 2004). Between 1979 and 2000, 2,153 cases of cattle rabies were reported from France, and between 1966 and 2000, 1,629 cases of cattle rabies were reported from Belgium (Aubert et al., 2004).

12 EPIDEMIOLOGY OF MONGOOSE-ASSOCIATED RABIES

Dog-associated variants of rabies virus have been implicated in novel maintenance cycles in wildlife. For example, rabies in the Caribbean in the Asian yellow mongoose (*Herpestes javanicus*; formerly designated

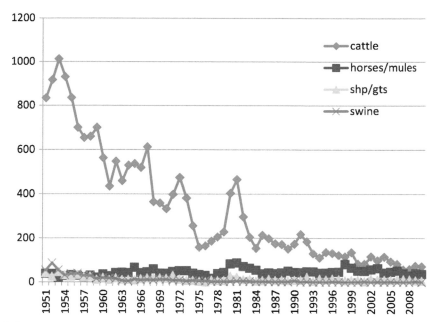

FIGURE 3.12 **Rabies diagnosed in selected species of livestock in the USA, 1951–2010, demonstrating the decline in rabid cattle as canine rabies was eliminated.**

as *Cynicus penicillata*; family *Herpestidae*) has been historically linked to when this species was introduced from Asia. Inter-island variation among the dog-associated rabies virus variants found in the mongooses is substantial and suggests multiple cross-species introductions of virus from dogs (Smith et al., 1992).

All mongooses present today in the Caribbean are descendants of animals brought from India to Jamaica in the 1870s for rodent control on sugar cane plantations (Everard & Everard, 1992). Although it is possible that rabies virus was introduced with these animals, official reports of rabies in mongooses in the Caribbean were not made until 1950 in Puerto Rico (Tierkel, Arbona, Rivera, & De Juan, 1952). Mongooses continue to be a source of human rabies in the Caribbean, where the virus is referred to as the dog/mongoose variant of rabies virus (Messenger et al., 2002).

In sub-Saharan Africa, rabies in the yellow mongoose (*C. penicillata*) appears to involve rabies virus variants distinct from those associated with domestic dogs. Data generated from recent molecular epidemiologic studies indicate an extended history of evolutionary adaptation of rabies virus within yellow mongoose and slender mongoose (*Galerella sanguinea*) populations in South Africa and Zimbabwe, respectively, and suggest that the enzootic area affected by these variants has a larger geographic range than previously suspected (Messenger et al., 2002; Nel et al., 2005).

13 CANINE RABIES VIRUS VARIANTS

Rabies virus variants associated with dog-to-dog transmission throughout the world demonstrate limited antigenic and genetic diversity (Nadin-Davis & Bingham, 2004; Smith, 1989). From these observations, it appears that dog rabies was probably introduced into South America, Africa, and parts of Asia due to translocation of dogs by early European colonialists. The similarity of canine rabies isolates from Latin America, Africa, Asia, and Eastern Europe reflect a global reservoir of rabies in dogs that arose from a common source. The isolate has been termed "the cosmopolitan strain" (Nel & Rupprecht, 2007). The cosmopolitan strain of rabies is believed to have its origins in the Paleartic region that includes Europe, the Middle East, and northern Africa (Nadin-Davis & Bingham, 2004; Badrane & Tordo, 2001). Nonetheless, there remains sufficient sequence variation among variants of rabies solely associated with dogs to detect differences within canine populations. This diversity permits epidemiological tracking of the dispersal of dogs, along with their human hosts, and directly influences patterns of rabies in countries such as Thailand (Denduangboripant et al., 2005).

Although significant numbers of feral, stray, or neighborhood dogs may contribute to free-ranging populations, human behaviors with regard to recruitment of companion animals, in addition to their interactions with stray or neighborhood dogs (Beck, 2000), can rapidly alter the demographic features of free-ranging dog populations.

Quantitative estimates of human mortality due to rabies have been frequently based on dog-bite injuries, estimates of the densities of dog populations, human population density and human-to-dog ratios, and knowledge of the incidence of bite injuries and risk of canine rabies within different populations of dogs (Brooks, 1990; Cleaveland et al., 2002; Cleaveland et al., 2003; Fevre et al., 2005; Knobel et al., 2005; Perry, 1993; Robinson, Miranda, Miranda, & Childs, 1996). The close relationship between reports of rabid dogs and human rabies deaths within countries where the cosmopolitan variant of rabies virus circulates within canines has been demonstrated in Africa, Asia, and South America and was widely appreciated within developed countries prior to canine rabies control (Belotto et al., 2005; Bingham, 2005; Denduangboripant et al., 2005; Tierkel, Graves, Tuggle, & Wadley, 1950).

Although variants of rabies virus introduced by domestic dogs have been implicated as the recent source of rabies virus independently circulating among several populations of wild carnivores, dogs remain the primary rabies threat for humans and animals in Africa (Talbi et al., 2009; Talbi et al., 2010). Similar to the situation in Zimbabwe, rabies virus variants believed to be circulating independently among a gray fox

(*U. cinereoargenteus*) reservoir in Colombia are most closely related to, and group phylogenetically with, domestic dog variants (Paez et al., 2005).

13.1 Management of Canine Rabies

The canine rabies virus variants responsible of dog-to-dog transmission of rabies can be pushed to extinction locally and regionally through comprehensive programs of parenteral vaccination and animal management. Canine rabies control through vaccination, reproduction control, movement restrictions, and habitat modification not only reduces the incidence of human rabies but also provides a cost-effective intervention for reducing the need of human PEP (Bogel & Meslin, 1990). Removal or culling of free-ranging dogs is not recommended as a primary means of dog population reduction or even as a supplementary measure to mass vaccination of dog populations (World Health Organization, 2005). The World Health Organization endorses national legislation in countries where such a program is affordable and enforceable (World Health Organization, 2005)

Mass vaccination campaigns in Africa indicate that a sufficiently high percentage of vaccine coverage can be achieved by the parental route in some urban settings (Kayali et al., 2003; Kayali, Mindekem, Hutton, Ndoutamia, & Zinsstag, 2006). Vaccination rates of 70% among free ranging dog (stray, feral, and neighborhood dogs) populations has been a traditional target promoted as sufficient to establish herd immunity (World Health Organization, 2005). However, theoretical estimates suggest herd immunity levels between 39 and 57%, with an upper 95% confidence interval of 55 and 71%, respectively, may prove sufficient to end epizootic transmission (Coleman & Dye, 1996; Kayali et al., 2006; Kayali et al., 2003).

13.2 The Risk of Pet Travel and Threat to Canine Rabies Prevention

The regulation and control of dog movements within and between countries remains an important strategy for preventing rabies. Rabies-free locations within a country, such as Hawaii in the United States (Fishbein, Corboy, & Sasaki, 1990) and rabies-free nations, such as the UK (rabies-free since 1922) and Japan, had traditionally enforced strict laws requiring 6 months of quarantine (Fishbein et al., 1990). The recent translocation of dogs with canine rabies virus variants into Europe and mainland USA demonstrates the very real risk for reintroduction of canine rabies virus variants into areas where they no longer exist (Castrodale, Westcott, Dobson, & Rupprecht, 2008; Johnson, Freuling, Horton, Muller, & Fooks, 2011).

The United Kingdom (UK) had initiated a Pet Travel Scheme (PETS) requiring imported animals to have an implanted microchip identification tag and documented proof of current vaccination, as evidenced not only by certificate documentation but also by the presence of antibody, within a six-month period prior to presenting proof of procedure at an official UK port (Jones, Kelly, Fooks, & Wooldridge, 2005). Several risk assessments models have been formulated in an attempt to estimate the probability through travel of pet animals of rabies introduction into the UK, which is currently free of terrestrial animal rabies. Having required either a 6-month quarantine for pet animals or evidence of seroconversion following rabies vaccination, the European Food Safety Authority (EFSA) in 2002 and the UK Veterinary Laboratories Agency (VLA) in 2011 specifically attempted to assess the consequence of abandoning the serological test to measure response to rabies vaccination in pet animals (DEFRA, 2011; Goddard et al., 2010; Scientific Committee on Animal Health and Animal Welfare, 2006). The EFSA concluded that serological testing is only beneficial when the waiting time after rabies vaccination exceeds 100 days. In other words, the value of serology would be in detecting low or non-responding animals whose titers may have initially been adequate but which subsequently declined in the interim or "waiting" period. There is also an implicit assumption that, while the animal intended for travel is in the "waiting" period, it should be held in a manner to preclude an exposure to rabies at its home location; for this recommendation to minimize risk, the confinement or quarantine conditions should be clearly stated. In contrast, the VLA concluded that the annual risk of rabies entry into the UK would increase tenfold if the serological test requirement was removed entirely, irrespective of any waiting period after vaccination. It is interesting to note that the model prediction of the number of years before rabies entry is quite long. Specifically, with current practices it is estimated as 13,272 years but declined to 1,152 years without serologic testing. The model identified that removing the serological test and reducing the waiting period to 30 days after vaccination would increase the annual risk of rabies entry such that the estimated number of years between entries was 322 years. Modeling is a powerful tool, but it is inherently limited by the data and design upon which the models are constructed. The power of modeling often comes from iterative refinement. Overall, the risk of rabies entry into the UK is significantly influenced by the inclusion of serologic testing. Moreover, what these models do not consider is that presenting a certificate of vaccination with an appropriately identified animal assumes the authenticity of the document with very little ability to verify this assumption through contact with the responsible veterinary practitioner.

To what level of scrutiny is the information on a rabies certificate form (http://www.nasphv.org/Documents/RabiesVacCert.pdf) held when an animal is presented for travel, and how informative is this information? A measure of rabies immunity through a serum sample tested at a laboratory that is held to quality control practices and performance standard provides a somewhat higher degree of certainty about an animal's response to vaccination. Of course, deception could occur through this system by substitution of a blood sample from a different animal, but this would require collusion between an owner and veterinary practitioner. However, problems occur even with requirements for an official result from a qualified laboratory. Over the past three years, the Kansas State University Rabies Laboratory has documented approximately 75 instances of falsified rabies serology result documents that have been made to appear as if issued by our laboratory, but there is no record of such testing, during which time we tested over 100,000 samples by the fluorescent antibody virus neutralization method for pet travel (Hanlon, unpublished data). At present, regulatory authorities with jurisdiction over animal travel have Internet access to results from our laboratory so that they may verify whether a particular sample was tested at our laboratory. A critical component of prevention of disease translocation is education of owners and the public as to the importance of these requirements for disease prevention.

14 CONCLUSIONS

It is of paramount importance to keep in mind that the first essentials in reducing the risk and consequential damage of any type of disease introduction or novel emergence are to maintain diagnostic capabilities at specialty labs and to assess threats posed by outbreaks in other countries. Potential emergent "hotspots" (Keesing, Belden et al., 2010) that may be identified based on land-use change and biodiversity patterns should be targeted for surveillance of endemic rabies virus variants, and other viruses, that may have the potential to jump host species. Lastly, wherever possible, adequate disease surveillance measures aimed at intensively farmed livestock in newly developed clearings juxtaposed to intact natural habitats and in locations, such as wetmarkets, where wild caught animals are held in close-proximity to domesticated species, could provide early warnings of outbreaks such as Nipah virus and SARS coronavirus (Chua, Goh et al., 1999; Drosten, Gunther et al., 2003). However, managing potential emergence hotspots by attempting to alter the ecology of high-risk areas is likely to be detrimental because the species most resilient to habitat destruction and degradation may be those that amplify pathogen transmission.

Like many zoonoses and other emerging infections, rabies prevention requires the cooperation of animal control, law enforcement, natural

resource personnel, veterinarians, diagnosticians, public health professionals, physicians, and others. The risk of disease translocation can be
mitigated through carefully crafted requirements for animal identification, vaccination, serological monitoring, and advance planning
essentially equivalent to a quarantine period. A critical component is
education of owners and the public as to the importance of requirements.
Historically, rabies diagnosis and prevention has been a core part of public health practice at local and state health agencies. With declining case
numbers and substantial pressure from economic constraints, a number
of public health laboratories are moving away from rabies diagnosis. The
Kansas State University Rabies Laboratory provides rabies diagnostic
testing and virus characterization on positive samples for a number of
states and localities because it is within the public health mission of this
educational, nonprofit, state university laboratory.

Much of the public remains insulated and largely unaware of rabies
transmission patterns either locally or globally. The unpredictability of
human behavior and lack of awareness of the hazards of not complying
with reasonable regulations present the greatest threat of disease translocation and associated exposures and potential mortality. Although rabies
excites the imagination, current vulnerabilities include the potential for
reintroduction of dog-to-dog transmitted rabies, a decline in diagnostic
expertise and capacity, and a lack of basic research, especially to understand recent advances, or lack thereof, towards treatment of clinical
rabies. As we increasingly approach the reality of global community with
rapid and high volume exchange of animate beings, diligent attention and
dedicated effort will be required to maintain and indeed, even advance
emerging and zoonotic disease control, with rabies as a tangible "best-
practices" template, beyond the major advances made in the last 50 years.

References

Anonymous, (1977). Rabies in a laboratory worker – New York. *Morbidity Mortality Weekly Report, Surveillance Summary, 26*, 183–184.

Anonymous, (1987). Human rabies despite treatment with rabies immune globulin and human diploid cell rabies vaccine—Thailand. *Morbidity Mortality Weekly Report, 36*, 759–760, 765.

Anonymous, (1995). Translocation of coyote rabies—Florida, 1994. *Morbidity Mortality Weekly Report, 44*, 580–581, 587.

Anonymous, (2004). Investigation of rabies infections in organ donor and transplant recipients--Alabama, Arkansas, Oklahoma, and Texas, 2004. *Morbidity Mortality Weekly Report, 53*, 586–589.

Acha, P. N., & Arambulo, P. V., III (1985). Rabies in the Tropics-history and current status. In E. Kuwert, C. Merieux, H. Koprowski, & K. Bogel (Eds.), *Rabies in the tropics (pp. 343–359)*. Berlin: Springer-Verlag.

Advisory Committee on Immunization Practices, (1999). Human rabies prevention–United States, 1999. Recommendations of the Advisory Committee on Immunization Practices (ACIP). *Morbidity Mortality Weekly Report Recommendations Report, 48*, 1–21.

Anderson, L. J., Nicholson, K. G., Tauxe, R. V., & Winkler, W. G. (1984). Human rabies in the United States, 1960 to 1979: Epidemiology, diagnosis, and prevention. *Annals of Internal Medicine, 100,* 728–735.

Anderson, L. J., Williams, L. P., Jr., Layde, J. B., Dixon, F. R., & Winkler, W. G. (1984). Nosocomial rabies: Investigation of contacts of human rabies cases associated with a corneal transplant. *American Journal of Public Health, 74,* 370–372.

Anderson, R. M., Jackson, H. C., May, R. M., & Smith, A. M. (1981). Population dynamics of fox rabies in Europe. *Nature, 289,* 765–771.

Anthony, J. A., Childs, J. E., Glass, G. E., Korch, G. W., Ross, L., & Grigor, J. K. (1990). Land use associations and changes in population indices of urban raccoons during a rabies epizootic. *Journal of Wildlife Diseases, 26,* 170–179.

Aubert, M. F., Cliquet, F., Smak, J. A., Brochier, B., Schon, J., & Kappeler, A. (2004). Rabies in France, The Netherlands, Belgium, Luxembourg and Switzerland. In A. A. King, A. R. Fooks, M. Aubert, & A. Wandeler (Eds.), *Historical persective of rabies in europe and the mediteranean basin (pp. 129–145).* Paris: OIE.

Badilla, X., Perez-Herra, V., Quiros, L., Morice, A., Jimenez, E., Saenz, E., et al. (2003). Human rabies: A reemerging disease in Costa Rica? *Emerging Infectious Diseases, 9,* 721–723.

Badillo, R., Mantilla, J. C., & Pradilla, G. (2009). Human rabies encephalitis by a vampire bat bite in an urban area of Colombia. *Biomedica, 29,* 191–203.

Badrane, H., & Tordo, N. (2001). Host switching in Lyssavirus history from the Chiroptera to the Carnivora orders. *Journal of Virology, 75,* 8096–8104.

Baer, G. M. (1991). Vampire bat and bovine paralytic rabies. In G. M. Baer (Ed.), *The natural history of rabies (pp. 389–403).* Boca Raton: CRC Press.

Baer, G. M., Abelseth, M. K., & Debbie, J. G. (1971). Oral vaccination of foxes against rabies. *American Journal of Epidemiology, 93,* 487–490.

Baer, G. M., Broderson, J. R., & Yager, P. A. (1975). Determination of the site of oral rabies vaccination. *American Journal of Epidemiology, 101,* 160–164.

Baltazard, M., & Bahmanyar, M. (1955). Field trials with rabies vaccine on persons bitten by rabid wolves. *Bulletin of the World Health Organization, 13,* 747–772.

Barbour, A. D., & Pugliese, A. (2004). Convergence of a structured metapopulation model to Levins's model. *Journal of Mathematical Biology, 49,* 468–500.

Batista-da-Costa, M., Bonito, R. F., & Nishioka, S. A. (1993). An outbreak of vampire bat bite in a Brazilian village. *Tropical Medicine and Parasitology, 44,* 219–220.

Beck, A. M. (1975). The public health implications of urban dogs. *American Journal of Public Health, 65,* 1315–1318.

Beck, A. M. (2000). The Human-dog relationship: A tale of two species. In C. N. Macpherson, F. X. Meslin, & A. I. Wandeler (Eds.), *Dogs, zoonoses and public health (pp. 1–16).* New York: CABI Publishing.

Belotto, A., Leanes, L. F., Schneider, M. C., Tamayo, H., & Correa, E. (2005). Overview of rabies in the Americas. *Virus Research, 111,* 5–12.

Belotto, A. J. (2004). The Pan American Health Organization (PAHO) role in the control of rabies in Latin America. *Developments in Biologicals (Basel), 119,* 213–216.

Beran, G. W. (1982). Ecology of dogs in the Central Philippines in relation to rabies control efforts. *Comparative Immunology and Microbiology of Infectious Diseases, 5,* 265–270.

Bingham, J. (2005). Canine rabies ecology in southern Africa. *Emerging Infectious Diseases, 11,* 1337–1342.

Blanton, J. D., Krebs, J. W., Hanlon, C. A., & Rupprecht, C. E. (2006). Rabies surveillance in the United States during 2005. *Journal of the American Veterinary Medical Association, 229,* 1897–1911.

Blanton, J. D., Palmer, D., Dyer, J., & Rupprecht, C. E. (2011). Rabies surveillance in the United States during 2010. *Journal of the American Veterinary Medical Association, 239,* 773–783.

Blanton, J. D., Palmer, D., & Rupprecht, C. E. (2010). Rabies surveillance in the United States during 2009. *Journal of the American Veterinary Medical Association, 237,* 646–657.

Bogel, K., Arata, A. A., Moegle, H., & Knorpp, F. (1974). Recovery of reduced fox populations in rabies control. *Zentralblatt fuer Veterinary Medizin B, 21,* 401–412.

Bogel, K., & Meslin, F. X. (1990). Economics of human and canine rabies elimination: Guidelines for programme orientation. *Bulletin of the World Health Organization, 68,* 281–291.

Borchert, M., Mutyaba, I., Van Kerkhove, M. D., Lutwama, J., Luwaga, H., Bisoborwa, G., et al. (2011). Ebola haemorrhagic fever outbreak in Masindi District, Uganda: Outbreak description and lessons learned. *BMC Infectious Diseases, 11,* 357.

Bordignon, J., Brasil-Dos-Anjos, G., Bueno, C. R., Salvatiera-Oporto, J., Davila, A. M., Grisard, E. C., et al. (2005). Detection and characterization of rabies virus in Southern Brazil by PCR amplification and sequencing of the nucleoprotein gene. *Archives of Virology, 150,* 695–708.

Botvinkin, A. D., Poleschuk, E. M., Kuzmin, I. V., Borisova, T. I., Gazaryan, S. V., Yager, P., et al. (2003). Novel lyssaviruses isolated from bats in Russia. *Emerging Infectious Diseases, 9,* 1623–1625.

Bourhy, H., Dacheux, L., Strady, C., & Mailles, A. (2005). Rabies in Europe in 2005. *European Surveillance, 10,* 213–216.

Bourhy, H., Kissi, B., Audry, L., Smreczak, M., Sadkowska-Todys, M., Kulonen, K., et al. (1999). Ecology and evolution of rabies virus in Europe. *Journal of General Virology, 80* (Pt 10), 2545–2557.

Bourhy, H., Kissi, B., & Tordo, N. (1993). Molecular diversity of the Lyssavirus genus. *Virology, 194,* 70–81.

Boyer, J. P., Canac-Marquis, P., Guerin, D., Mainguy, J., & Pelletier, F. (2011). Oral vaccination against raccoon rabies: Landscape heterogeneity and timing of distribution influence wildlife contact rates with the ONRAB vaccine bait. *Journal of Wildlife Diseases, 47,* 593–602.

Brass, D. A. (1994). *Rabies in bats.* Ridgefield: Livia Press.

Brito, M. G., Chamone, T. L., da Silva, F. J., Wada, M. Y., Miranda, A. B., Castilho, J. G., et al. (2011). Antemortem diagnosis of human rabies in a veterinarian infected when handling a herbivore in Minas Gerais, Brazil. *Revista do Instituto de Medicina Tropical de Sao Paulo, 53,* 39–44.

Brooks, R. (1990). Survey of the dog population of Zimbabwe and its level of rabies vaccination. *Veterinary Record, 127,* 592–596.

Caraballo, A. J. (1996). Outbreak of vampire bat biting in a Venezuelan village. *Revista de Saude Publica (Sao Paulo), 30,* 483–484.

Carini, A. (1911). Sur une grande epizootie de rage. *Annals of the Institute Pasteur (Paris), 25,* 843–846.

Castrodale, L., Westcott, M., Dobson, J., & Rupprecht, C. (2008). Rabies in a three-month-old puppy in south-western Alaska. *Veterinary Record, 163,* 92.

Centers for Disease Control and Prevention, (1999). Mass treatment of humans who drank unpasteurized milk from rabid cows--Massachusetts, 1996–1998. *Journal of the American Medical Association, 281,* 1371–1372.

Centers for Disease Control and Prevention, (2003). First human death associated with raccoon rabies--Virginia, 2003. *MMWR Morbidity Mortality Weekly Report, 52,* 1102–1103.

Centers for Disease Control and Prevention, (2004). Investigation of rabies infections in organ donor and transplant recipients--Alabama, Arkansas, Oklahoma, and Texas, 2004. *MMWR Morbidity Mortality Weekly Report, 53,* 586–589.

Centers for Disease Control and Prevention, (2006). Human rabies--Mississippi, 2005. *MMWR Morbidity Mortality Weekly Report, 55,* 207–208.

Centers for Disease Control and Prevention, (2010). Presumptive abortive human rabies - Texas, 2009. *MMWR Morbidity Mortality Weekly Report, 59,* 185–190.

Centers for Disease Control and Prevention, (2012). Recovery of a patient from clinical rabies—California, 2011. *MMWR Morbidity Mortality Weekly Report, 61,* 61–65.

Chang, H. G., Eidson, M., Noonan-Toly, C., Trimarchi, C. V., Rudd, R., Wallace, B. J., et al. (2002). Public health impact of reemergence of rabies, New York. *Emerging Infectious Diseases, 8,* 909–913.

Charlton, K. M., Artois, M., Prevec, L., Campbell, J. B., Casey, G. A., Wandeler, A. I., et al. (1992). Oral rabies vaccination of skunks and foxes with a recombinant human adenovirus vaccine. *Archives of Virology, 123,* 169–179.

Childs, J. E., Curns, A. T., Dey, M. E., Real, L. A., Feinstein, L., Bjornstad, O. N., et al. (2000). Predicting the local dynamics of epizootic rabies among raccoons in the United States. *Proceedings of the National Academy of Sciences, U.S.A, 97,* 13666–13671.

Childs, J. E., Krebs, J. W., & Smith, J. S. (2002). Public health surveillance and the molecular epidemiology of rabies. In T. Leitner (Ed.), *The molecular epidemiology of human viruses (pp. 273–312).* Dordrecht: Kluwer Academic.

Childs, J. E., Robinson, L. E., Sadek, R., Madden, A., Miranda, M. E., & Miranda, N. L. (1998). Density estimates of rural dog populations and an assessment of marking methods during a rabies vaccination campaign in the Philippines. *Preventive Veterinary Medicine, 33,* 207–218.

Chua, K. B., Goh, K. J., et al. (1999). Fatal encephalitis due to Nipah virus among pig-farmers in Malaysia. *Lancet, 354*(9186), 1257–1259.

Cisterna, D., Bonaventura, R., Caillou, S., Pozo, O., Andreau, M. L., Fontana, L. D., et al. (2005). Antigenic and molecular characterization of rabies virus in Argentina. *Virus Research, 109,* 139–147.

Cleaveland, S., Fevre, E. M., Kaare, M., & Coleman, P. G. (2002). Estimating human rabies mortality in the United Republic of Tanzania from dog bite injuries. *Bulletin of the World Health Organization, 80,* 304–310.

Cleaveland, S., Kaare, M., Tiringa, P., Mlengeya, T., & Barrat, J. (2003). A dog rabies vaccination campaign in rural Africa: Impact on the incidence of dog rabies and human dog-bite injuries. *Vaccine, 21,* 1965–1973.

Cliquet, F., Robardet, E., Must, K., Laine, M., Peik, K., Picard-Meyer, E., et al. (2012). Eliminating rabies in Estonia. *PLoS Neglected Tropical Diseases, 6,* e1535.

Coleman, P. G., & Dye, C. (1996). Immunization coverage required to prevent outbreaks of dog rabies. *Vaccine, 14,* 185–186.

Compendium of animal rabies prevention and control, (2004). Compendium of animal rabies prevention and control, 2004: National Association of State Public Health Veterinarians, Inc. (NASPHV). *Morbidity Mortality Weekly Report Recommendations and Reports, 53,* 1–8.

Constantine, D. G. (1962). Rabies transmission by nonbite route. *Public Health Reports, 77,* 287–289.

Coyne, M. J., Smith, G., & McAllister, F. E. (1989). Mathematic model for the population biology of rabies in raccoons in the mid-Atlantic states. *American Journal of Veterinary Research, 50,* 2148–2154.

Crawford-Miksza, L. K., Wadford, D. A., & Schnurr, D. P. (1999). Molecular epidemiology of enzootic rabies in California. *Journal of Clinical Virology, 14,* 207–219.

Daoust, P. Y., Wandeler, A. I., & Casey, G. A. (1996). Cluster of rabies cases of probable bat origin among red foxes in Prince Edward Island, Canada. *Journal of Wildlife Diseases, 32,* 403–406.

de Mattos, C. A., Favi, M., Yung, V., Pavletic, C., & de Mattos, C. C. (2000). Bat rabies in urban centers in Chile. *Journal of Wildlife Diseases, 36,* 231–240.

de Mattos, C. C., de Mattos, C. A., Loza-Rubio, E., Aguilar-Setien, A., Orciari, L. A., & Smith, J. S. (1999). Molecular characterization of rabies virus isolates from Mexico: Implications for transmission dynamics and human risk. *American Journal of Tropical Medicine and Hygiene, 61,* 587–597.

DEFRA, (2011). Defra seeks views on controlling an outbreak of rabies. *Veterinary Record, 169*, 511.

Delpietro, H. A., & Russo, R. G. (1996). Ecological and epidemiologic aspects of the attacks by vampire bats and paralytic rabies in Argentina and analysis of the proposals carried out for their control. *Revue Scientifique et Technique (Paris), 15*, 971–984.

Denduangboripant, J., Wacharapluesadee, S., Lumlertdacha, B., Ruankaew, N., Hoonsuwan, W., Puanghat, A., et al. (2005). Transmission dynamics of rabies virus in Thailand: Implications for disease control. *BMC Infectious Diseases, 5*, 52.

Dietzschold, B., Schnell, M., & Koprowski, H. (2005). Pathogenesis of rabies. *Current Topics in Microbiology and Immunology, 292*, 45–56.

Drosten, C., Gunther, S., et al. (2003). Identification of a novel coronavirus in patients with severe acute respiratory syndrome. *New England Journal of Medicine, 348*(20), 1967–1976.

Eng, T. R., Fishbein, D. B., Talamante, H. E., Hall, D. B., Chavez, G. F., Dobbins, J. G., et al. (1993). Urban epizootic of rabies in Mexico: Epidemiology and impact of animal bite injuries. *Bulletin of the World Health Organization, 71*, 615–624.

Everard, C. O., & Everard, J. D. (1992). Mongoose rabies in the Caribbean. *Annals of the New York Academy of Sciences, 653*, 356–366.

Fagbami, A. H., Anosa, V. O., & Ezebuiro, E. O. (1981). Hospital records of human rabies and antirabies prophylaxis in Nigeria 1969–78. *Transactions of the Royal Society of Tropical Medicine and Hygiene, 75*, 872–876.

Favi, C. M., Bassaletti, C. A., Lopez, D. J., Rodriguez, A. L., & Yung, P. V. (2011). Epidemiological description of rabies reservoir in bats in the Metropolitan Region: Chile. 2000–2009. *Revista Chilena d'Infectologie, 28*, 223–228.

Favi, M., de Mattos, C. A., Yung, V., Chala, E., Lopez, L. R., & de Mattos, C. C. (2002). First case of human rabies in chile caused by an insectivorous bat virus variant. *Emerging Infectious Diseases, 8*, 79–81.

Favi, M., Nina, A., Yung, V., & Fernandez, J. (2003). Characterization of rabies virus isolates in Bolivia. *Virus Research, 97*, 135–140.

Favoretto, S. R., de Mattos, C. C., Morais, N. B., Alves Araujo, F. A., & de Mattos, C. A. (2001). Rabies in marmosets (*Callithrix jacchus*), Ceara, Brazil. *Emerging Infectious Diseases, 7*, 1062–1065.

Feder, H. M., Jr., Petersen, B. W., Robertson, K. L., & Rupprecht, C. E. (2012). Rabies: Still a uniformly fatal disease? Historical occurrence, epidemiological trends, and paradigm shifts. *Current Infectious Diseases Report, 14*, 408–422.

Fehlner-Gardiner, C., Rudd, R., Donovan, D., Slate, D., Kempf, L., & Badcock, J. (2012). Comparing ONRAB(R) and RABORAL V-RG(R) oral rabies vaccine field performance in raccoons and striped skunks, New Brunswick, Canada, and Maine, USA. *Journal of Wildlife Diseases, 48*, 157–167.

Fekadu, M., Endeshaw, T., Alemu, W., Bogale, Y., Teshager, T., & Olson, J. G. (1996). Possible human-to-human transmission of rabies in Ethiopia. *Ethiopean Medical Journal, 34*, 123–127.

Fevre, E. M., Kaboyo, R. W., Persson, V., Edelsten, M., Coleman, P. G., & Cleaveland, S. (2005). The epidemiology of animal bite injuries in Uganda and projections of the burden of rabies. *Tropical Medicine & International Health (Oxford)h, 10*, 790–798.

Fielding, J. E., & Nayda, C. L. (2005). Postexposure prophylaxis for Australian bat lyssavirus in South Australia, 1996 to 2003. *Australian Veterinary Journal (New South Wales), 83*, 233–234.

Fishbein, D. B., Corboy, J. M., & Sasaki, D. M. (1990). Rabies prevention in Hawaii. *Hawaii Medical Journal, 49*, 98–101.

Flores, C. R., Said, F. S., De Anda, L. D., Ibarra, V. F., & Anaya, R. M. (1979). [A new technic for the control of vampire bats: Intramuscular inoculation of cattle with warfarin]. *Boletin de la Oficina Sanitaria Panamericana (Washington, DC), 87*, 283–299.

Fooks, A. R., McElhinney, L. M., Pounder, D. J., Finnegan, C. J., Mansfield, K., Johnson, N., et al. (2003). Case report: Isolation of a European bat lyssavirus type 2a from a fatal human case of rabies encephalitis. *Journal of Medical Virology, 71*, 281–289.

Freuling, C. M., Beer, M., Conraths, F. J., Finke, S., Hoffmann, B., Keller, B., et al. (2011). Novel lyssavirus in Natterer's bat, Germany. *Emerging Infectious Diseases, 17*, 1519–1522.

Frontini, M. G., Fishbein, D. B., Garza, R. J., Flores, C. E., Balderas Torres, J. M., Quiroz, H. G., et al. (1992). A field evaluation in Mexico of four baits for oral rabies vaccination of dogs. *American Journal of Tropical Medicine and Hygiene, 47*, 310–316.

Gacouin, A., Bourhy, H., Renaud, J. C., Camus, C., Suprin, E., & Thomas, R. (1999). Human rabies despite postexposure vaccination. *European Journal of Clinical Microbiology and Infectious Diseases, 18*, 233–235.

Gibbons, R. V., Holman, R. C., Mosberg, S. R., & Rupprecht, C. E. (2002). Knowledge of bat rabies and human exposure among United States cavers. *Emerging Infectious Diseases, 8*, 532–534.

Goddard, A., Donaldson, N., Kosmider, R., Kelly, L., Adkin, A., Horton, D. et al. (2010). *Qualitative risk assessment on the change in likelihood of rabies introduction into the United Kingdom as a consequence of adopting the existing harmonised Community rules for the non-commerical movement of pet animals; Final Report.*

Godin, A. J. (2012). *Wild mammals of New England.* Baltimore: The Johns Hopkins University Press.

Gordon, E. R., Curns, A. T., Krebs, J. W., Rupprecht, C. E., Real, L. A., & Childs, J. E. (2004). Temporal dynamics of rabies in a wildlife host and the risk of cross-species transmission. *Epidemiology and Infections, 132*, 515–524.

Gordon, E. R., Krebs, J. W., Rupprecht, C. R., Real, L. A., & Childs, J. E. (2005). Persistence of elevated rabies prevention costs following post-epizootic declines in rates of rabies among raccoons (*Procyon lotor*). *Preventive Veterinary Medicine, 68*, 195–222.

Greenwood, R. J., Newton, W. E., Pearson, G. L., & Schamber, G. J. (1997). Population and movement characteristics of radio-collared striped skunks in North Dakota during an epizootic of rabies. *Journal of Wildlife Diseases, 33*, 226–241.

Gruzdev, K. N. (2008). The rabies situation in Central Asia. *Developments in Biologicals (Basel), 131*, 37–42.

Guerra, M. A., Curns, A. T., Rupprecht, C. E., Hanlon, C. A., Krebs, J. W., & Childs, J. E. (2003). Skunk and raccoon rabies in the eastern United States: Temporal and spatial analysis. *Emerging Infectious Diseases, 9*, 1143–1150.

Hanlon, C. A., Kuzmin, I. V., Blanton, J. D., Weldon, W. C., Manangan, J. S., & Rupprecht, C. E. (2005). Efficacy of rabies biologics against new lyssaviruses from Eurasia. *Virus Research, 111*, 44–54.

Hanlon, C. A., Niezgoda, M., Hamir, A. N., Schumacher, C., Koprowski, H., & Rupprecht, C. E. (1998). First North American field release of a vaccinia-rabies glycoprotein recombinant virus. *Journal of Wildlife Diseases, 34*, 228–239.

Hanlon, C. L., Hayes, D. E., Hamir, A. N., Snyder, D. E., Jenkins, S., Hable, C. P., et al. (1989). Proposed field evaluation of a rabies recombinant vaccine for raccoons (Procyon lotor): Site selection, target species characteristics, and placebo baiting trials. *Journal of Wildlife Diseases, 25*, 555–567.

Hanna, J. N., Carney, I. K., Smith, G. A., Tannenberg, A. E., Deverill, J. E., Botha, J. A., et al. (2000). Australian bat lyssavirus infection: A second human case, with a long incubation period. *Medical Journal of Australia, 172*, 597–599.

Hattwick, M. A., Rubin, R. H., Music, S., Sikes, R. K., Smith, J. S., & Gregg, M. B. (1974). Postexposure rabies prophylaxis with human rabies immune globulin. *Journal of the American Medical Association, 227*, 407–410.

Hayman, D. T., Johnson, N., Horton, D. L., Hedge, J., Wakeley, P. R., Banyard, A. C., et al. (2011). Evolutionary history of rabies in Ghana. *PLoS Neglected Tropical Diseases, 5*, e1001.

Held, J. R., & Adaros, H. L. (1972). Neurological disease in man following administration of suckling mouse brain antirabies vaccine. *Bulletin of the World Health Organization, 46,* 321–327.

Held, J. R., Tierkel, E. S., & Steele, J. H. (1967). Rabies in man and animals in the United States, 1946–65. *Public Health Reports, 82,* 1009–1018.

Helmick, C. G., Tauxe, R. V., & Vernon, A. A. (1987). Is there a risk to contacts of patients with rabies? *Reviews of Infectious Diseases, 9,* 511–518.

Hemachudha, T., Mitrabhakdi, E., Wilde, H., Vejabhuti, A., Siripataravanit, S., & Kingnate, D. (1999). Additional reports of failure to respond to treatment after rabies exposure in Thailand. *Clinical Infectious Diseases, 28,* 143–144.

Houff, S. A., Burton, R. C., Wilson, R. W., Henson, T. E., London, W. T., Baer, G. M., et al. (1979). Human-to-human transmission of rabies virus by corneal transplant. *New England Journal of Medicine, 300,* 603–604.

Hughes, G. J., Orciari, L. A., & Rupprecht, C. E. (2005). Evolutionary timescale of rabies virus adaptation to North American bats inferred from the substitution rate of the nucleoprotein gene. *Journal of General Virology, 86,* 1467–1474.

Humphrey, G. L., Kemp, G. E., & Wood, E. G. (1960). A fatal case of rabies in a woman bitten by an insectivorous bat. *Public Health Reports, 75,* 317–326.

Hurst, E. W., & Pawan, J. L. (1959). An outbreak of rabies in Trinidad without history of bites, and with the symptoms of acute ascending myelitis. *Carbbean Medical Journal, 21,* 11–24.

Irons, J. V., Eads, R. B., Grimes, J. E., & Conklin, A. (1957). The public health importance of bats. *Texas Reports in Biology and Medicine, 15,* 292–298.

Ito, M., Arai, Y. T., Itou, T., Sakai, T., Ito, F. H., Takasaki, T., et al. (2001). Genetic characterization and geographic distribution of rabies virus isolates in Brazil: Identification of two reservoirs, dogs and vampire bats. *Virology, 284,* 214–222.

Ito, M., Itou, T., Shoji, Y., Sakai, T., Ito, F. H., Arai, Y. T., et al. (2003). Discrimination between dog-related and vampire bat-related rabies viruses in Brazil by strain-specific reverse transcriptase-polymerase chain reaction and restriction fragment length polymorphism analysis. *Journal of Clinical Virology, 26,* 317–330.

Javadi, M. A., Fayaz, A., Mirdehghan, S. A., & Ainollahi, B. (1996). Transmission of rabies by corneal graft. *Cornea, 15,* 431–433.

Johnson, N., Freuling, C., Horton, D., Muller, T., & Fooks, A. R. (2011). Imported rabies, European Union and Switzerland, 2001–2010. *Emerging Infectious Diseases, 17,* 753–754.

Johnson, N., Phillpotts, R., & Fooks, A. R. (2006). Airborne transmission of lyssaviruses. *Journal of Medical Microbiology, 55,* 785–790.

Johnston, W. B., & Walden, M. B. (1996). Results of a national survey of rabies control procedures. *Journal of the American Veterinary Medical Association, 208,* 1667–1672.

Jones, R. D., Kelly, L., Fooks, A. R., & Wooldridge, M. (2005). Quantitative risk assessment of rabies entering Great Britain from North America via cats and dogs. *Risk Analysis, 25,* 533–542.

Kayali, U., Mindekem, R., Hutton, G., Ndoutamia, A. G., & Zinsstag, J. (2006). Cost-description of a pilot parenteral vaccination campaign against rabies in dogs in N'Djamena, Chad. *Tropical Medicine & International Health (Oxford)h, 11,* 1058–1065.

Keesing, F., Belden, L. K., et al. (2010). Impacts of biodiversity on the emergence and transmission of infectious diseases. *Nature, 468*(7324), 647–652.

Kayali, U., Mindekem, R., Yemadji, N., Vounatsou, P., Kaninga, Y., Ndoutamia, A. G., et al. (2003). Coverage of pilot parenteral vaccination campaign against canine rabies in N'Djamena, Chad. *Bulletin of the World Health Organization, 81,* 739–744.

Knobel, D. L., Cleaveland, S., Coleman, P. G., Fevre, E. M., Meltzer, M. I., Miranda, M. E., et al. (2005). Re-evaluating the burden of rabies in Africa and Asia. *Bulletin of the World Health Organization, 83,* 360–368.

Kobayashi, Y., Sato, G., Kato, M., Itou, T., Cunha, E. M., Silva, M. V., et al. (2007). Genetic diversity of bat rabies viruses in Brazil. *Archives of Virology, 152,* 1995–2004.

Kobayashi, Y., Sato, G., Mochizuki, N., Hirano, S., Itou, T., Carvalho, A. A., et al. (2008). Molecular and geographic analyses of vampire bat-transmitted cattle rabies in central Brazil. *BMC Veterinary Research, 4,* 44.

Kobayashi, Y., Sato, G., Shoji, Y., Sato, T., Itou, T., Cunha, E. M., et al. (2005). Molecular epidemiological analysis of bat rabies viruses in Brazil. *Journal of Veterinary Medical Science, 67,* 647–652.

Krebs, J. W., Long-Marin, S. C., & Childs, J. E. (1998). Causes, costs, and estimates of rabies postexposure prophylaxis treatments in the United States. *Journal of Public Health Management Practices, 4,* 56–62.

Krebs, J. W., Mandel, E. J., Swerdlow, D. L., & Rupprecht, C. E. (2005). Rabies surveillance in the United States during 2004. *Journal of the American Veterinary Medical Association, 227,* 1912–1925.

Kuzmin, I. V. (1999). An arctic fox rabies virus strain as the cause of human rabies in Russian Siberia. *Archives of Virology, 144,* 627–629.

Kuzmin, I. V., Hughes, G. J., Botvinkin, A. D., Gribencha, S. G., & Rupprecht, C. E. (2008). Arctic and Arctic-like rabies viruses: Distribution, phylogeny and evolutionary history. *Epidemiology and Infections, 136,* 509–519.

Kuzmin, I. V., Hughes, G. J., Botvinkin, A. D., Orciari, L. A., & Rupprecht, C. E. (2005). Phylogenetic relationships of Irkut and West Caucasian bat viruses within the Lyssavirus genus and suggested quantitative criteria based on the N gene sequence for lyssavirus genotype definition. *Virus Research, 111,* 28–43.

Kuzmin, I. V., Hughes, G. J., & Rupprecht, C. E. (2006). Phylogenetic relationships of seven previously unclassified viruses within the family Rhabdoviridae using partial nucleoprotein gene sequences. *Journal of General Virology, 87,* 2323–2331.

Kuzmin, I. V., Mayer, A. E., Niezgoda, M., Markotter, W., Agwanda, B., Breiman, R. F., et al. (2010). Shimoni bat virus, a new representative of the Lyssavirus genus. *Virus Research, 149,* 197–210.

Kuzmin, I. V., Orciari, L. A., Arai, Y. T., Smith, J. S., Hanlon, C. A., Kameoka, Y., et al. (2003). Bat lyssaviruses (Aravan and Khujand) from Central Asia: Phylogenetic relationships according to N, P and G gene sequences. *Virus Research, 97,* 65–79.

Kuzmin, I. V., Shi, M., Orciari, L. A., Yager, P. A., Velasco-Villa, A., Kuzmina, N. A., et al. (2012). Molecular Inferences Suggest Multiple Host Shifts of Rabies Viruses from Bats to Mesocarnivores in Arizona during 2001–2009. *PLoS Pathogens, 8,* e1002786.

Leach, C. N., & Johnson, H. N. (1940). Human rabies, with special reference to virus distribution and titer. *American Journal of Tropical Medicine and Hygiene, 20,* 335–340.

Lembo, T., Attlan, M., Bourhy, H., Cleaveland, S., Costa, P., de, B. K., et al. (2011). Renewed global partnerships and redesigned roadmaps for rabies prevention and control. *Veterinary Medicine International, 2011,* 923149.

Leslie, M. J., Messenger, S., Rohde, R. E., Smith, J., Cheshier, R., Hanlon, C., et al. (2006). Bat-associated rabies virus in Skunks. *Emerging Infectious Diseases, 12,* 1274–1277.

Lewis, R. L. (2007). A 10-year-old boy evacuated from the Mississippi Gulf Coast after Hurricane Katrina presents with agitation, hallucinations, and fever. *Journal of Emergency Nursing, 33,* 42–44.

Linhart, S. B., Flores, C. R., & Mitchell, G. C. (1972). Control of vampire bats by means of an anticoagulant. *Boletin de la Oficina Sanitaria Panamericana (Washington, DC), 73,* 100–109.

Lopez, A., Miranda, P., Tejada, E., & Fishbein, D. B. (1992). Outbreak of human rabies in the Peruvian jungle. *Lancet, 339,* 408–411.

Lucey, B. T., Russell, C. A., Smith, D., Wilson, M. L., Long, A., Waller, L. A., et al. (2002). Spatiotemporal analysis of epizootic raccoon rabies propagation in Connecticut, 1991–1995. *Vector-borne and Zoonotic Diseases, 2,* 77–86.

Lumbiganon, P., & Wasi, C. (1990). Survival after rabies immunisation in newborn infant of affected mother. *Lancet, 336,* 319.

Macdonald, D. W., & Bacon, P. J. (1982). Fox society, contact rate and rabies epizootiology. *Comparative Immunology and Microbiology of Infectious Diseases, 5,* 247–256.

MacInnes, C. D., Smith, S. M., Tinline, R. R., Ayers, N. R., Bachmann, P., Ball, D. G., et al. (2001). Elimination of rabies from red foxes in eastern Ontario. *Journal of Wildlife Diseases, 37,* 119–132.

MacInnes, C. D., Tinline, R. R., Voigt, D. R., Broekhoven, L. H., & Rosatte, R. R. (1988). Planning for rabies control in Ontario. *Reviews of Infectious Diseases, 10(Suppl 4),* S665–S669.

Maier, T., Schwarting, A., Mauer, D., Ross, R. S., Martens, A., Kliem, V., et al. (2010). Management and outcomes after multiple corneal and solid organ transplantations from a donor infected with rabies virus. *Clinical Infectious Diseases, 50,* 1112–1119.

Malerczyk, C., Detora, L., & Gniel, D. (2011). Imported human rabies cases in europe, the United States, and Japan, 1990 to 2010. *Journal of Travel Medicine, 18,* 402–407.

Manning, S. E., Rupprecht, C. E., Fishbein, D., Hanlon, C. A., Lumlertdacha, B., Guerra, M., et al. (2008). Human rabies prevention–United States, 2008: Recommendations of the advisory committee on immunization practices. *Morbidity Mortality Weekly Report Recommendations and Reports, 57,* 1–28.

Markotter, W., Van, E. C., Kuzmin, I. V., Rupprecht, C. E., Paweska, J. T., Swanepoel, R., et al. (2008). Epidemiology and pathogenicity of African bat lyssaviruses. *Developments in Biologicals (Basel), 131,* 317–325.

Marston, D. A., Horton, D. L., Ngeleja, C., Hampson, K., McElhinney, L. M., Banyard, A. C., et al. (2012). Ikoma lyssavirus, highly divergent novel lyssavirus in an African civet. *Emerging Infectious Diseases, 18,* 664–667.

Martinez-Burnes, J., Lopez, A., Medellin, J., Haines, D., Loza, E., & Martinez, M. (1997). An outbreak of vampire bat-transmitted rabies in cattle in northeastern Mexico. *Candadian Veterinary Journal, 38,* 175–177.

Mayen, F. (2003). Haematophagous bats in Brazil, their role in rabies transmission, impact on public health, livestock industry and alternatives to an indiscriminate reduction of bat population. *Journal of Veterinary Medicine. B Infectious Diseases and Veterinary Public Health, 50,* 469–472.

McQuiston, J. H., Yager, P. A., Smith, J. S., & Rupprecht, C. E. (2001). Epidemiologic characteristics of rabies virus variants in dogs and cats in the United States, 1999. *Journal of the American Veterinary Medical Association, 218,* 1939–1942.

Meslin, F. X. (2005). Rabies as a traveler's risk, especially in high-endemicity areas. *Journal of Travel Medicine, 12(Suppl 1),* S30–S40.

Messenger, S. L., Smith, J. S., Orciari, L. A., Yager, P. A., & Rupprecht, C. E. (2003). Emerging pattern of rabies deaths and increased viral infectivity. *Emerging Infectious Diseases, 9,* 151–154.

Messenger, S. L., Smith, J. S., & Rupprecht, C. E. (2002). Emerging epidemiology of bat-associated cryptic cases of rabies in humans in the United States. *Clinical Infectious Diseases, 35,* 738–747.

Moran, G. J., Talan, D. A., Mower, W., Newdow, M., Ong, S., Nakase, J. Y., et al. (2000). Appropriateness of rabies postexposure prophylaxis treatment for animal exposures. Emergency ID Net Study Group. *Journal of the American Medical Association, 284,* 1001–1007.

Morimoto, K., Patel, M., Corisdeo, S., Hooper, D. C., Fu, Z. F., Rupprecht, C. E., et al. (1996). Characterization of a unique variant of bat rabies virus responsible for newly emerging human cases in North America. *Proceedings of the National Academy of Sciences, U.S.A, 93,* 5653–5658.

Muller, W., Cox, J., & Muller, T. (2004). Rabies in Germany, Denmark, and Austria. In A. A. King, A. R. Fooks, M. Aubert, & A. I. Wandeler (Eds.), *Historical perspective of rabies in europe and the mediteranean basin (pp. 79–92).* Paris: OIE.

Mutinelli, F. (2010). Rabies and feral cat colonies in Italy. *Veterinary Record*, *166*, 537–538.

Mutinelli, F., Stankov, M., Hristovski, V., Theoharakou, H., & Vodopija, I. (2004). Rabies in Italy. In A. A. King, A. R. Fooks, M. Aubert, & A. Wandeler (Eds.), *Historical perspective of rabies in europe and the mediterranean basin (pp. 91–188)*. Paris: OIE.

Nadin-Davis, S., & Bingham, J. (2004). Europe as a source of rabies for the rest of the World. In A. A. King, A. R. Fooks, M. Aubert, & A. I. Wanderler (Eds.), *Historical perspective of rabies in europe and the mediteranean basin (pp. 259–292)*. Paris: OIE.

Nadin-Davis, S. A., Casey, G. A., & Wandeler, A. I. (1994). A molecular epidemiological study of rabies virus in central Ontario and western Quebec. *Journal of General Virology*, *75(Pt 10)*, 2575–2583.

Nadin-Davis, S. A., Muldoon, F., & Wandeler, A. I. (2006). A molecular epidemiological analysis of the incursion of the raccoon strain of rabies virus into Canada. *Epidemiology and Infections*, *134*, 534–547.

Nathwani, D., McIntyre, P. G., White, K., Shearer, A. J., Reynolds, N., Walker, D., et al. (2003). Fatal human rabies caused by European bat Lyssavirus type 2a infection in Scotland. *Clinical Infectious Diseases*, *37*, 598–601.

Nel, L., Jacobs, J., Jaftha, J., von, T. B., Bingham, J., & Olivier, M. (2000). New cases of Mokola virus infection in South Africa: A genotypic comparison of Southern African virus isolates. *Virus Genes*, *20*, 103–106.

Nel, L. H. (2005). Vaccines for lyssaviruses other than rabies. *Expert Review of Vaccines*, *4*, 533–540.

Nel, L. H., & Rupprecht, C. E. (2007). Emergence of lyssaviruses in the Old World: The case of Africa. *Current Topics in Microbiology and Immunology*, *315*, 161–193.

Nel, L. H., Sabeta, C. T., von, T. B., Jaftha, J. B., Rupprecht, C. E., & Bingham, J. (2005). Mongoose rabies in southern Africa: A re-evaluation based on molecular epidemiology. *Virus Research*, *109*, 165–173.

Nelson, R. S., Mshar, P. A., Cartter, M. L., Adams, M. L., & Hadler, J. L. (1998). Public awareness of rabies and compliance with pet vaccination laws in Connecticut, 1993. *Journal of the American Veterinary Medical Association*, *212*, 1552–1555.

Nettles, V. F., Shaddock, J. H., Sikes, R. K., & Reyes, C. R. (1979). Rabies in translocated raccoons. *American Journal of Public Health*, *69*, 601–602.

Noah, D. L., Drenzek, C. L., Smith, J. S., Krebs, J. W., Orciari, I., Shaddock, J., et al. (1998). Epidemiology of human rabies in the United States, 1980 to 1996. *Annals of Internal Medicine*, *128*, 922–930.

Noah, D. L., Smith, M. G., Gotthardt, J. C., Krebs, J. W., Green, D., & Childs, J. E. (1996). Mass human exposure to rabies in New Hampshire: Exposures, treatment, and cost. *American Journal of Public Health*, *86*, 1149–1151.

Organización Panamericana de la Salud (OPS), (1983). *Estragia y Plan de Acción para le Eliminación de la Rabia Urbana en América Latina para et final de la década de 1980* Guayaquil: Ecuador.

Paez, A., Nuncz, C., Garcia, C., & Boshell, J. (2003). Molecular epidemiology of rabies epizootics in Colombia: Evidence for human and dog rabies associated with bats. *Journal of General Virology*, *84*, 795–802.

Paez, A., Saad, C., Nunez, C., & Boshell, J. (2005). Molecular epidemiology of rabies in northern Colombia 1994–2003. Evidence for human and fox rabies associated with dogs. *Epidemiology and Infections*, *133*, 529–536.

Pappaioanou, M., Fishbein, D. B., Dreesen, D. W., Schwartz, I. K., Campbell, G. H., Sumner, J. W., et al. (1986). Antibody response to preexposure human diploid-cell rabies vaccine given concurrently with chloroquine. *New England Journal of Medicine*, *314*, 280–284.

Para, M. (1965). An outbreak of post-vaccinal rabies (rage de laboratoire) in Fortaleza, Brazil, in 1960. Residual fixed virus as the etiological agent. *Bulletin of the World Health Organization*, *33*, 177–182.

Parker, R. L. (1975). Rabies in Skunks. In G. M. Baer (Ed.), *The natural history of rabies* (pp. 41–51). New York: Academic Press.

Perry, B. D. (1993). Dog ecology in eastern and southern Africa: Implications for rabies control. *Onderstepoort Journal of Veterinary Research, 60,* 429–436.

Pool, G. E., & Hacker, C. S. (1982). Geographic and seasonal distribution of rabies in skunks, foxes and bats in Texas. *Journal of Wildlife Diseases, 18,* 405–418.

Potzsch, C. J., Kliemt, A., Kloss, D., Schroder, R., & Muller, W. (2006). Rabies in Europe--trends and developments. *Developments in Biologicals (Basel), 125,* 59–68.

Pybus, M. J. (1988). Rabies and rabies control in striped skunks (Mephitis mephitis) in three prairie regions of western North America. *Journal of Wildlife Diseases, 24,* 434–449.

Robbins, A., Eidson, M., Keegan, M., Sackett, D., & Laniewicz, B. (2005). Bat incidents at children's camps, New York State, 1998–2002. *Emerging Infectious Diseases, 11,* 302–305.

Robinson, L. E., Miranda, M. E., Miranda, N. L., & Childs, J. E. (1996). Evaluation of a canine rabies vaccination campaign and characterization of owned-dog populations in the Philippines. *Southeast Asian Journal of Tropical Medicine and Public Health, 27,* 250–256.

Rohde, R. E., Neill, S. U., Clark, K. A., & Smith, J. S. (1997). Molecular epidemiology of rabies epizootics in Texas. *Clinical and Diagnostic Virology, 8,* 209–217.

Rosatte, R. C., Donovan, D., Allan, M., Bruce, L., Buchanan, T., Sobey, K., et al. (2009). The control of raccoon rabies in Ontario Canada: Proactive and reactive tactics, 1994–2007. *Journal of Wildlife Diseases, 45,* 772–784.

Rotz, L. D., Hensley, J. A., Rupprecht, C. E., & Childs, J. E. (1998). Large-scale human exposures to rabid or presumed rabid animals in the United States: 22 cases (1990–1996). *Journal of the American Veterinary Medical Association, 212,* 1198–1200.

Ruiz, M., & Chavez, C. B. (2010). Rabies in Latin America. *Neurological Research, 32,* 272–277.

Rupprecht, C. E., Briggs, D., Brown, C. M., Franka, R., Katz, S. L., Kerr, H. D., et al. (2009). Evidence for a 4-dose vaccine schedule for human rabies post-exposure prophylaxis in previously non-vaccinated individuals. *Vaccine, 27,* 7141–7148.

Rupprecht, C. E., Glickman, L. T., Spencer, P. A., & Wiktor, T. J. (1987). Epidemiology of rabies virus variants. Differentiation using monoclonal antibodies and discriminant analysis. *American Journal of Epidemiology, 126,* 298–309.

Rupprecht, C. E., Hanlon, C. A., & Slate, D. (2004). Oral vaccination of wildlife against rabies: Opportunities and challenges in prevention and control. *Developments in Biologicals (Basel), 119,* 173–184.

Russell, C. A., Smith, D. L., Childs, J. E., & Real, L. A. (2005). Predictive spatial dynamics and strategic planning for raccoon rabies emergence in Ohio. *PLoS Biology, 3,* e88.

Russell, C. A., Smith, D. L., Waller, L. A., Childs, J. E., & Real, L. A. (2004). A priori prediction of disease invasion dynamics in a novel environment. *Proceedings of the Royal Society B: Biological Sciences, 271,* 21–25.

Sabeta, C. T., Markotter, W., Mohale, D. K., Shumba, W., Wandeler, A. I., & Nel, L. H. (2007). Mokola virus in domestic mammals, South Africa. *Emerging Infectious Diseases, 13,* 1371–1373.

Sacramento, D., Bourhy, H., & Tordo, N. (1991). PCR technique as an alternative method for diagnosis and molecular epidemiology of rabies virus. *Molecular and Cellular Probes, 5,* 229–240.

Samaratunga, H., Searle, J. W., & Hudson, N. (1998). Non-rabies Lyssavirus human encephalitis from fruit bats: Australian bat Lyssavirus (pteropid Lyssavirus) infection. *Neuropathology and Applied Neurobiology, 24,* 331–335.

Schaefer, R., Batista, H. B., Franco, A. C., Rijsewijk, F. A., & Roehe, P. M. (2005). Studies on antigenic and genomic properties of Brazilian rabies virus isolates. *Veterinary Microbiology, 107,* 161–170.

Schneider, M. C., Aguilera, X. P., Barbosa da Silva, J. J., Ault, S. K., Najera, P., Martinez, J., et al. (2011). Elimination of neglected diseases in latin america and the Caribbean: A mapping of selected diseases. *PLoS Neglected Tropical Diseases, 5,* e964.

Scientific Committee on Animal Health and Animal Welfare (2006). *Scientific Opinion of the Scientific Panel on Animal Health and Welfare on a request from the Commission regarding an* Assessment of the risk of rabies introduction into the UK, Ireland, Sweden, Malta, as a consequence of abandoning the serological test measuring protective antibodies to rabies (Rep. No. 436). The European Food Safety Authority Journal.

Shao, X. Q., Yan, X. J., Luo, G. L., Zhang, H. L., Chai, X. L., Wang, F. X., et al. (2011). Genetic evidence for domestic raccoon dog rabies caused by Arctic-like rabies virus in Inner Mongolia, China. *Epidemiology and Infections, 139,* 629–635.

Shoji, Y., Kobayashi, Y., Sato, G., Gomes, A. A., Itou, T., Ito, F. H., et al. (2006). Genetic and phylogenetic characterization of rabies virus isolates from wildlife and livestock in Paraiba, Brazil. *Acta Virologica, 50,* 33–37.

Sidwa, T. J., Wilson, P. J., Moore, G. M., Oertli, E. H., Hicks, B. N., Rohde, R. E., et al. (2005). Evaluation of oral rabies vaccination programs for control of rabies epizootics in coyotes and gray foxes: 1995–2003. *Journal of the American Veterinary Medical Association, 227,* 785–792.

Sipahioglu, U., & Alpaut, S. (1985). Transplacental rabies in humans. *Mikrobiyoloji Bülteni, 19,* 95–99.

Smith, D. L., Lucey, B., Waller, L. A., Childs, J. E., & Real, L. A. (2002). Predicting the spatial dynamics of rabies epidemics on heterogeneous landscapes. *Proceedings of the National Academy of Sciences, U.S.A, 99,* 3668–3672.

Smith, J. S. (1988). Monoclonal antibody studies of rabies in insectivorous bats of the United States. *Reviews of Infectious Diseases, 10*(Suppl 4), S637–S643.

Smith, J. S. (1989). Rabies virus epitopic variation: Use in ecologic studies. *Advances in Virus Research, 36,* 215–253.

Smith, J. S., Sumner, J. W., Roumillat, L. F., Baer, G. M., & Winkler, W. G. (1984). Antigenic characteristics of isolates associated with a new epizootic of raccoon rabies in the United States. *Journal of Infectious Diseases, 149,* 769–774.

Smith, J. S., Orciari, L. A., Yager, P. A., Seidel, H. D., & Warner, C. K. (1992). Epidemiologic and historical relationships among 87 rabies virus isolates as determined by limited sequence analysis. *Journal of Infectious Diseases, 166,* 296–307.

Speare, R., Skerratt, L., Foster, R., Berger, L., Hooper, P., Lunt, R., et al. (1997). Australian bat lyssavirus infection in three fruit bats from north Queensland. *Communicable Diseases Intelligence, 21,* 117–120.

Srinivasan, A., Burton, E. C., Kuehnert, M. J., Rupprecht, C., Sutker, W. L., Ksiazek, T. G., et al. (2005). Transmission of rabies virus from an organ donor to four transplant recipients. *New England Journal of Medicine, 352,* 1103–1111.

Steck, F., & Wandeler, A. (1980). The epidemiology of fox rabies in Europe. *Epidemiology and Infection, 2,* 71–96.

Sterner, R. T., Meltzer, M. I., Shwiff, S. A., & Slate, D. (2009). Tactics and economics of wildlife oral rabies vaccination, Canada and the United States. *Emerging Infectious Diseases, 15,* 1176–1184.

Swaddiwuthipong, W., Weniger, B. G., Wattanasri, S., & Warrell, M. J. (1988). A high rate of neurological complications following Semple anti-rabies vaccine. *Transactions of the Royal Society of Tropical Medicine and Hygiene, 82,* 472–475.

Tabel, H., Corner, A. H., Webster, W. A., & Casey, C. A. (1974). History and epizootiology of rabies in Canada. *Candadian Veterinary Journal, 15,* 271–281.

Talbi, C., Holmes, E. C., de, B. P., Faye, O., Nakoune, E., Gamatie, D., et al. (2009). Evolutionary history and dynamics of dog rabies virus in western and central Africa. *Journal of General Virology, 90,* 783–791.

Talbi, C., Lemey, P., Suchard, M. A., Abdelatif, E., Elharrak, M., Nourlil, J., et al. (2010). Phylodynamics and human-mediated dispersal of a zoonotic virus. *PLoS Pathogens, 6,* e1001166.

Tariq, W. U., Shafi, M. S., Jamal, S., & Ahmad, M. (1991). Rabies in man handling infected calf. *Lancet*, *337*, 1224.

Tierkel, E. S., Arbona, G., Rivera, A., & De Juan, A. (1952). Mongoose rabies in Puerto Rico. *Public Health Reports*, *67*, 274–278.

Tierkel, E. S., Graves, L. M., Tuggle, H. G., & Wadley, S. L. (1950). Effective control of an outbreak of rabies in Memphis and Shelby County, Tennessee. *American Journal of Public Health Nations.Health*, *40*, 1084–1088.

van Thiel, P. P., de Bie, R. M., Eftimov, F., Tepaske, R., Zaaijer, H. L., van Doornum, G. J., et al. (2009). Fatal human rabies due to Duvenhage virus from a bat in Kenya: Failure of treatment with coma-induction, ketamine, and antiviral drugs. *PLoS Neglected Tropical Diseases*, *3*, e428.

Velasco-Villa, A., Gomez-Sierra, M., Hernandez-Rodriguez, G., Juarez-Islas, V., Melendez-Felix, A., Vargas-Pino, F., et al. (2002). Antigenic diversity and distribution of rabies virus in Mexico. *Journal of Clinical Microbiology*, *40*, 951–958.

Velasco-Villa, A., Messenger, S. L., Orciari, L. A., Niezgoda, M., Blanton, J. D., Fukagawa, C., et al. (2008). New rabies virus variant in Mexican immigrant. *Emerging Infectious Diseases*, *14*, 1906–1908.

Velasco-Villa, A., Orciari, L. A., Souza, V., Juarez-Islas, V., Gomez-Sierra, M., Castillo, A., et al. (2005). Molecular epizootiology of rabies associated with terrestrial carnivores in Mexico. *Virus Research*, *111*, 13–27.

Vetter, J. M., Frisch, L., Drosten, C., Ross, R. S., Roggendorf, M., Wolters, B., et al. (2011). Survival after transplantation of corneas from a rabies-infected donor. *Cornea*, *30*, 241–244.

Wandeler, A. (2004). Epidemiology and ecology of fox rabies in Europe. In A. A. King, A. R. Fooks, M. Aubert, & A. I. Wandeler (Eds.), *Historical perspective of rabies in europe and the mediterranean basin (pp. 201–214)*. Paris: OIE.

Wandeler, A., Wachendorfer, G., Forster, U., Krekel, H., Schale, W., Muller, J., et al. (1974). Rabies in wild carnivores in central Europe. I. Epidemiological studies. *Zentralblatt fuer Veterinary Medizin B*, *21*, 735–756.

Wandeler, A. I. (2008). The rabies situation in Western Europe. *Developments in Biologicals (Basel)*, *131*, 19–25.

Wandeler, A. I., Capt, S., Kappeler, A., & Hauser, R. (1988). Oral immunization of wild-life against rabies: Concept and first field experiments. *Reviews of Infectious Diseases*, *10(Suppl 4)*, S649–S653.

Wandeler, A. I., & Salsberg, E. B. (1999). ONTARIO. Raccoon rabies in eastern Ontario. *Canadian Veterinary Journal*, *40*, 731.

Warrilow, D. (2005). Australian bat lyssavirus: A recently discovered new rhabdovirus. *Current Topics in Microbiology and Immunology*, *292*, 25–44.

Warrilow, D., Smith, I. L., Harrower, B., & Smith, G. A. (2002). Sequence analysis of an isolate from a fatal human infection of Australian bat lyssavirus. *Virology*, *297*, 109–119.

Wilde, H., Khawplod, P., Khamoltham, T., Hemachudha, T., Tepsumethanon, V., Lumlerdacha, B., et al. (2005). Rabies control in South and Southeast Asia. *Vaccine*, *23*, 2284–2289.

Wilde, H., Sirikawin, S., Sabcharoen, A., Kingnate, D., Tantawichien, T., Harischandra, P. A., et al. (1996). Failure of postexposure treatment of rabies in children. *Clinical Infectious Diseases*, *22*, 228–232.

Willoughby, R. E., Jr., et al., Tieves, K. S., Hoffman, G. M., Ghanayem, N. S., Amlie-Lefond, C. M., Schwabe, M. J., et al. (2005). Survival after treatment of rabies with induction of coma. *New England Journal of Medicine*, *352*, 2508–2514.

Winkler, W. G. (1968). Airborne rabies virus isolation. *Wildlife Diseases*, *4*, 37–40.

Winkler, W. G., Fashinell, T. R., Leffingwell, L., Howard, P., & Conomy, P. (1973). Airborne rabies transmission in a laboratory worker. *Journal of the American Medical Association, 226*, 1219–1221.

World Health Organization, (2005). WHO Expert Consultation on rabies. *World Health Organization Technical Report Series, 931*, 1–88. (back).

Wright, A., Rampersad, J., Ryan, J., & Ammons, D. (2002). Molecular characterization of rabies virus isolates from Trinidad. *Veterinary Microbiology, 87*, 95–102.

Wunner, W. H., & Briggs, D. J. (2010). Rabies in the 21 century. *PLoS Neglected Tropical Diseases, 4*, e591.

Yung, V., Favi, M., & Fernandez, J. (2012). Typing of the rabies virus in Chile, 2002–2008. *Epidemiology and Infections*, 1–6.

Molecular Epidemiology

Susan A. Nadin-Davis

Animal Health Microbiology Research, Ottawa Laboratory (Fallowfield),
Canadian Food Inspection Agency, Ottawa, ON K2J 4S1, Canada

1 INTRODUCTION

The discipline of molecular epidemiology, in which patterns of disease transmission are followed using selected markers that distinguish different populations of the disease-causing agent, is now a well-established central theme in the study of infectious diseases. Knowledge generated by such an approach improves understanding of viral emergence and spread and contributes to better disease control. Over the last few decades, many technological innovations, particularly in the tools used for genetic characterization of viruses, have enabled the development and application of increasingly sophisticated and sensitive viral typing methods. This chapter will summarize the knowledge gained by applying such methods to rabies and the rabies-related viruses that constitute the *Lyssavirus* genus.

All negative strand RNA viruses, including the lyssaviruses, are especially amenable to molecular epidemiological methods of analysis due to their use of error prone RNA polymerases for replication; the resulting infidelity in genome copying ensures that any population of these viruses does not exist as a discrete entity but as a collection of genetic variants that are distributed around a central consensus sequence, a phenomenon frequently referred to as a quasispecies (Lauring & Andino, 2010). However, given the limitations inherent in the term *quasispecies*, it has been suggested that the more general term *population sequence heterogeneity* is a more accurate measure of the genetic structure of viral populations (Nadin-Davis & Real, 2011). Several studies have documented the heterogeneity of individual rabies virus isolates (Benmansour et al., 1992; Kissi et al., 1999; Nadin-Davis, Muldoon & Wandeler, 2006a), and it was suggested that this characteristic could help the virus overcome barriers to its spread in the

Rabies: Scientific Basis of the Disease and its Management
DOI: http://dx.doi.org/10.1016/B978-0-12-396547-9.00004-3

individual host and between species (Morimoto et al., 1998). However, as frequently observed for RNA viruses (Pybus et al., 2007), genetic diversity of lyssavirus field strains appears to be limited by the process of purifying selection which removes transient deleterious mutations and effectively places substantial limitations on viral mutant fixation (Holmes, Woelk, Kassis & Bourhy, 2002; Bourhy et al., 2008). Accordingly, the observed diversity of a particular viral strain or variant is often very limited and changes slowly with time due to acquisition of many neutral mutations. Sometimes a virus population undergoes a severe bottleneck and the sub-population that successfully emerges establishes a founder lineage. If the consensus sequence of this founder lineage differs consistently from that of its precursor, even by just a small number of mutations, these differences can be the basis of molecular epidemiological investigation.

Lyssaviruses have several specific but important biological properties to be considered when applying molecular epidemiological principles. First, it is generally established that a given viral lineage is always associated with a particular animal species, known as the reservoir host (Hanlon, Niezgoda & Rupprecht, 2007); this is due to the relative efficiency with which virus is propagated by individuals of this animal species and transmitted between conspecifics. It thus follows that the geographical spread of a particular viral population is determined both by the overall range of the animal species itself as well as aspects of the host animal's biology that determines the extent of interaction between different sub-populations of the host. The maintenance of this virus–host relationship is assumed to involve significant co-adaptation by both parties but the mechanisms responsible, at the level of either the host or the virus, are not yet understood. Long-term maintenance of such a virus–host relationship will isolate this virus population from others and lead to the emergence (e.g., via genetic drift) of a virus with distinct distinguishing markers.

A "spillover event" occurs when a rabies virus that is normally maintained in a reservoir host successfully infects another animal species. It is assumed that due to virus–host co-adaptation, sustained transmission of virus within the second host is unlikely and spillover transmission is thus regarded as a "dead-end" infection in the vast majority of situations. However, rarely such an event can initiate a new virus–host relationship in which sustained propagation and intra-species transmission of the virus within the new host occurs. Such a "species jump" will usually be associated with the emergence of a distinct viral population and hence a new lineage.

2 METHODS AND DEFINITION OF TERMS

In this chapter, a viral variant is defined as a viral population that is maintained within a particular host reservoir in a geographically defined area and which can be distinguished from other sympatric or allopatric viral populations. The two main methods for typing of lyssaviruses employ antigenic and genetic methods and the advantages and drawbacks of both techniques will be discussed.

2.1 Antigenic Typing

Antigenic typing is conducted using panels of monoclonal antibodies (MAbs). The discriminatory ability of this method relies on the ability of each MAb of the panel to react with a specific epitope of a viral protein; the epitope will either be present or absent and MAb reactivity will be scored as positive or negative accordingly. Epitope structure depends on the protein's primary sequence as well as its secondary and tertiary structure in many cases and occasionally on the presence of post-translational modifications. Since each MAb theoretically binds to just one epitope, a structure that usually comprises just a small number of amino acids, it follows that the greater the number of independent epitopes that can be targeted with a MAb panel, the greater is the likelihood of improving the discriminatory capability of the panel. In practice this requires the generation and screening of a large number of MAbs. Development of the most appropriate MAb panel must consider the viral populations to be targeted within a particular geographical region so that a panel developed for use in one country or region will not necessarily be useful for strain typing in another geographic area.

Many of our current founding principles of rabies epidemiology were formulated based on antigenic typing studies (Rupprecht, Dietzschold, Wunner & Koprowski, 1991). The nucleoprotein (N), product of the N gene, is the viral protein traditionally targeted by this method since it is produced in large quantities in infected brain tissue and thereby provides an abundant target. MAb binding to this protein is usually assayed by an indirect fluorescent antibody test that can readily be performed in well-established rabies diagnostic facilities. Other proteins that exhibit greater variation in their primary structure, particularly the glycoprotein (G) and phosphoprotein (P), represent other potential targets for antigenic typing. Several groups of anti-P MAbs have been shown to exhibit a wide variety of specificities of utility for variant discrimination (Nadin-Davis et al., 2000; Nadin-Davis, Fehlner-Gardiner, Sheen & Wandeler, 2010b) whereas a group of anti-G MAbs, applied to formalin-fixed tissues in a histochemical approach, was reported to differentiate between several strains of rabies and other lyssaviruses (Warner, Fekadu, Whitfield & Shaddock, 1999a). However, to date, anti-N MAbs have been applied most extensively to viral typing schemes.

To illustrate the utility of antigenic typing, Tables 4.1 and 4.2 summarize the differential reactivity of selected MAbs against particular

TABLE 4.1 Staining Patterns of Ontario Rabies Virus Variants in Indirect Immunofluorescence with Nucleoprotein-Specific Monoclonal Antibodies (MAbs)

MAb	Immunizing Virus	Rabies Virus Variant Staining[a]									
		ERA	EON/Arctic	MAC	BBB1	BBB2	BBB3	LBB	HRB	RDB	SHB
5DF12[b]	SAD	+	+	+	+	+	+	+	+	+	+
11DD1	SAD	+	-	-	-	-	-	-	+	+	-
M992	Duvenhage	+	+	+	+	+	-	+	+	+	-
M1341	MAC	+	+	+	+	+	+	+	+	+	+
26BH11	SL dog	+	+	-	+	-	-	+	+	+	+
20CB11	SL dog	+	+	+	-	+	-	+	+	+	+
24FF1	SL dog	+	-	+	-	-	-	-	-	-	-
M993	Mokola	+	+	+	-	+	-	+	-	-	-
26AF11	SL dog	+	+	-	+	-	-	+	+/-	+	+
26BD6	SL dog	-	+	+	+	+	+	-	-	-	+
32FE10	Lima dog	+	+	+	+	+	+	+	-	+	+
32FF1	Lima dog	+	+	+	-	+	-	+	+	+	+
38FG5	EBLV-1	+	+	+	-	+	-	+	+	+	+
M1347	MAC	-	-	+	-	+	-	+	+	+	+
7D2-7-4	BBB1	-	-	+	+	+	+	+	-	-	-

Reproduced with permission from Fehlner-Gardiner et al. (2008).

[a]Definitions: EON/Arctic = Eastern Ontario/Arctic fox; MAC = mid-Atlantic raccoon; BBB = big brown bat; LBB = little brown bat; HRB = hoary bat; RDB = red bat; SHB = silver-haired bat; SL = Sri Lankan; EBLV = European Bat Lyssavirus; SAD = Street Alabama Dufferin strain; Gaps indicate missing data.

[b]Broadly cross-reactive MAb used as a positive control.

TABLE 4.2 Use of the CDC Panel of 8 Monoclonal Antibodies to Discriminate Rabies Virus Types

RABV Antigenic Variant	Reactivity Pattern With The Following MAbs:								Reservoir Host
	C1	C4	C9	C10	C12	C15	C18	C19	
1	+	+	+	+	+	+	–	+	Dog/mongoose
2	+	+	–	+	+	+	–	+	Dog
3	–	+	+	+	+	–	–	+	D. rotundus
4	–	+	+	+	+	–	–	–	T. brasiliensis
5	–	+	Var	+	+	Var	–	Var	D. rotundus
6	Var	+	+	+	+	–	–	–	Lasiurus sp.
7	+	+	+	–	+	+	–	+	Gray fox
8	–	+	+	+	–	+	+	+	Skunk (south central USA)
9	+	+	+	+	+	–	–	–	T. brasiliensis
10	+	+	+	+	–	+	–	+	Skunk (Baja California)
11	–	+	+	+	–	–	–	+	D. rotundus
CVS/SAD	+	+	+	+	+	+	+	+	Laboratory strain

All profiles are as reported by Diaz, Papo, Rodriguez & Smith, 1994 and Delpietro, Gury-Dhomen, Larghi, Mena-Segura & Abramo, 1997. Var = variable reactivity.

rabies virus variants that circulate in the Americas. In Table 4.1, the differential binding of 15 MAbs to 10 rabies virus variants is shown. Two of these variants associate with terrestrial hosts; the Ontario/Arctic fox (EON/Arctic) variant circulates in southeastern Ontario, albeit at low levels now due to provincial control efforts, and persists across northern Canada; the mid-Atlantic raccoon (MAC) variant circulates throughout the eastern seaboard of the USA, but continues to threaten southern Canada with sporadic incursions. Another seven viral variants circulate in chiropteran hosts, including three types associated with the big brown bat. Also included is the reaction of the panel to the live attenuated vaccine strain, Evelyn-Rokitnicki-Abelseth (ERA), which occasionally elicits clinical disease in wildlife. Based on determination of the reactivity profile of the MAb panel with representative reference specimens and comparison with the profile of an unknown isolate, the variant type of the unknown can usually be unequivocally assigned.

A MAb panel developed by the Centers for Disease Control and Prevention (CDC) has been used to discriminate many strains circulating in the United States (Smith, 2002) and a version of this panel, as depicted in Table 4.2, has been used extensively to investigate the viruses present throughout Latin America. This panel discriminates between eleven recognized rabies virus types, including several that are associated with the two major reservoir hosts of the region, the dog and the vampire bat. It has also significantly contributed to the identification and recognition of additional sylvatic rabies reservoirs (Favoretto, De Mattos, Morais, Alves Araújo & De Mattos, 2001; Velasco-Villa et al., 2002; Yung, Favi & Fernández, 2002; Cisterna et al., 2005). Occasionally, an isolate yields an antigenic pattern previously unrecognized with this panel and then genetic methods provide valuable additional information (see Section 4.4.3). Further isolations of the same viral type are then required to unambiguously identify the host reservoir. Some studies in which both antigenic and genetic methods have been directly compared have noted that genetic analysis can reveal variation not identified antigenically (Albas et al., 2011). Accordingly, as knowledge of sympatric strains circulating in a specific area improves, MAb panel refinement to meet regional and local needs is necessary to preserve the value of antigenic typing regimens. Thus, MAbs useful for discriminating between distinct viral populations circulating in dogs and foxes were identified by a rational combination of genetic and antigenic evaluation of a collection of Brazilian isolates (Bernardi et al., 2005). However, if two genetically distinct but closely related viral variants encode similar viral proteins, antigenic typing will not discriminate between them.

Several studies have explored the precise epitopes recognised by certain MAbs, but in only a few cases have specific amino acid substitutions in the primary sequence of a protein been shown to affect MAb binding (see Smith, 2002) because the analysis is complicated by the fact that epitopes are often conformational in nature and involve two or more portions of the protein.

2.2 Genetic Typing

The development of nucleic acid amplification technologies and improvements in sequencing capability has facilitated the generation of lyssavirus sequences from a large number of specimens, many of which are available in publicly accessible databases. This knowledge has in turn permitted the development of tools and techniques to improve our ability to detect novel lyssaviruses and expand our knowledge of the diversity of this genus. Methods for lyssavirus detection are summarized in Chapter 11 on diagnostic evaluation, whereas this chapter focuses on the methods that facilitate discrimination of lyssaviruses and permit phylogenetic studies of their diversity. In this latter regard, the reverse-transcription polymerase chain reaction (RT-PCR) technique, which allows generation of dsDNA copies of portions of the viral genome, remains a pivotal method for genetic characterization of lyssaviruses (Sacramento, Bourhy & Tordo, 1991).

The genomes of all lyssaviruses studied to date (see Tordo, Poch, Ermine, Keith & Rougeon, 1986; Le Mercier, Jacob & Tordo, 1997; Warrilow, Smith, Harrower & Smith, 2002; Kuzmin, Hughes, Botvinkin, Orciari & Rupprecht, 2005; Marston et al., 2007; Delmas et al., 2008; Markotter, Kuzmin, Rupprecht & Nel, 2008) have an extremely similar organization in which the genome encodes five coding regions: N, P, M, G and L in 3' to 5' order along its length. Negative strand RNA viruses are believed to exhibit limited recombination capability (Chare, Gould & Holmes, 2003), probably due to the nature of their genome replication mechanism; indeed, a recent report citing evidence of rabies virus recombination (Liu, Liu, Liu, Zhai & Xie, 2011) requires further confirmation. Consequently, lyssavirus mutation almost always occurs via single base substitutions, necessary to preserve the reading frame of encoded proteins, or through small insertions/deletions in non-coding regions. Within protein coding regions, the majority of mutations are third base synonymous changes, although first and second base changes are also observed; the less common second base changes are always non-synonymous. Observation of a mutation requires first that a base change be incorporated into a newly synthesized viral genome, but also that the change becomes predominant within the viral population, even if only within a single individual. Although it is assumed that the initial mutation event occurs by chance at a constant rate along the genome, as dictated by the error rate of the viral RNA polymerase, only those mutations that confer a fitness advantage to the virus or those which are neutral in nature and are retained by chance become fixed in the population long enough to be observed. Consequently, levels of nucleotide similarity, as observed between two divergent members of the lyssaviruses (rabies virus (PV strain) and Mokola virus), are highly variable along the length of the genome (Le Mercier et al., 1997). To different degrees, functional constraints operate to limit mutations that are retained throughout the genome; N and L genes are the most conserved, the P and G genes are more variable, while the M gene exhibits intermediate variability. Variation in similarity values within each gene further identifies specific protein coding regions or domains that are more or less conserved; for example, highly variable and conserved domains of the P gene have been identified (Nadin-Davis, Abdel-Malik, Armstrong & Wandeler, 2002) and the central region of the N gene is known to be less variable than either the N- or C-terminus of N (Kissi, Tordo & Bourhy, 1995). The G gene region that encodes the transmembrane and cytoplasmic domains of the G is the least conserved coding region of the lyssavirus genome (Le Mercier et al., 1997).

The non-coding G-L intergenic region is the most variable region of the genome (Ravkov, Smith & Nichol, 1995), and it was presumed that this was due to a lack of function. However, the presence of mutational hot spots along the length of this region suggests that some mutational constraints operate on parts of this sequence; this may indicate that some G-L region sequences have one or more functions (Szanto, Nadin-Davis, Rosatte & White, 2011).

Although direct nucleotide sequencing of the PCR product (amplicon) is currently the method most commonly used for viral characterization and typing, a number of other methods have been used to provide useful epidemiological information in the past and some of these may remain viable low cost options for some studies. The principal methods employed for typing of rabies virus amplicons, typically generated from the N gene, include the following:

2.2.1 Restriction Fragment Length Polymorphism Analysis

This method employs collections of commercially available restriction endonucleases, each of which cleaves dsDNA at specific sequences usually comprising a 4–6bp palindromic motif. Due to their absolute sequence specificity for DNA cleavage they can discriminate between single nucleotide polymorphisms (SNPs) if such a change results in the acquisition or loss of a cutting site, hence the term Restriction Fragment Length Polymorphism (RFLP). Accordingly, treatment of a PCR product with an appropriate panel of restriction endonucleases will generate a reproducible pattern of digestion products. Once a reference collection of certain viral types having specific digestion patterns is established, unidentified samples can be assigned to a particular type by comparison to these reference patterns. Such a strategy has been applied in several studies to identify the variants responsible for cases of unexplained human rabies in the USA (Smith, Fishbein, Rupprecht & Clark, 1991), to identify variants of the arctic fox virus that circulate in geographically restricted areas of Canada (Nadin-Davis, Casey & Wandeler, 1994), to map distinct rabies virus variants in Europe (Bourhy et al., 1999), to distinguish street and vaccine strains in Estonia (Kulonen & Boldina, 1993), to discriminate between several viral types circulating in Mexico (Loza-Rubio et al., 1999) and Iran (Nadin-Davis, Simani, Armstrong, Fayaz & Wandeler, 2003b), to compare viruses associated with cats and dogs in Bangkok, Thailand (Kasempimolporn, Saengseesom, Tirawatnapong, Puempumpanich & Sitprija, 2004), and to distinguish various bat-associated variants in Brazil (Schaefer, Batista, Franco, Rijsewijk & Roehe, 2005).

2.2.2 Heteroduplex Mobility Assay

To discriminate between PCR amplicons generated from a collection of rabies viruses recovered in Turkey, Johnson et al., (2003a) employed a heteroduplex mobility assay (HMA) in which two 606 bp PCR products, generated from the N genes of two different specimens, were compared by annealing them together and then assessing the change in mobility of the resulting products by standard gel electrophoresis. The greater the genetic difference between the two products, the greater the degree of retardation of the heteroduplex compared to the homoduplex. By this approach, which the authors considered a cost effective method of comparing large numbers of PCR products, three main viral lineages were identified and subsequently confirmed by sequence analysis.

2.2.3 Strain-specific PCRs and probes

Another strategy has employed PCR primers that target variable sequences; by adjusting annealing conditions so that these primers hybridize selectively only to their matched target sequence, a PCR product that is specific for a particular viral type can be generated. Such an approach was used to develop a multiplex RT-PCR that generated amplicons of different sizes to discriminate between the mid-Atlantic raccoon variant and other rabies virus variants maintained in wildlife in Ontario, Canada (Nadin-Davis, Huang & Wandeler, 1996) as partially illustrated in Figure 4.1. Similar approaches have been used to discriminate between rabies virus variants circulating in Texas, USA (Rohde, Neill, Clark & Smith, 1997), to distinguish between two distinct biotypes of rabies viruses that circulate in South Africa (Nel, Bingham, Jacobs & Jaftha, 1998) and to discriminate several distinct viral types circulating in Brazil (Sato et al., 2005). Ito et al., (2003) used a combination of multiplex RT-PCR and RFLP to discriminate dog and vampire bat associated viral variants. Methods employing genotype-specific PCR primers used in combination with oligonucleotide probes and Southern blotting or ELISA

(A)

```
91N1578T1   GAAGACTGGACCAGCTATGGGATCCTGATTGCACGGAAAGGAGATAAGATCACCCCAGATTCTCTGGTGGAGATAAAGCGTACCGGTGTAGAAGGGAATTGGGCT
91N2756T2   ...............................................T.......................................................
91N783T3    .......................................................................................................
90N9196T4   ........................................................................................T..............
93N1090CAN  ........................................................................................T..............

516NYRAC    ..................A....T........A..........C.....A..T...A.C.....C..A..C.....A.....A.A...............C......
V125FLRAC   ..........T.......A....T........A..........C.....A..T...A.C.....C..A..C.....A.....A.A...............C......
99N3306RAC  ..........T.......A....T........A..........C.....A..T...A.C.....C..A..C.....A.....A.A...............C......

89L1461EF   ..T...............A.............G.....C......T..A.......T.....C..C..A.....AA...G.......C.C......
92L415ML    ..T...............A.............G.....C..........A.......T...........A.....TCA..G...............
88L1053LAN  ..T......GTT.....C..A.......A..G...GG..C..C...........A.......T.....T....GA....TAA...G............
92L2108LC   ..T......GTT.....C.......T..A.......G..C..C...........GCA....T.....T....GA.......AA...G.......G.......
```

<pre>
            └────────────────────────────────┘   └────────────────────┘
                   BAT-specific primer site   AFX-specific primer site
</pre>

(B)

```
91N1578T1   E  D  W  T  S  Y  G  I  L  I  A  R  K  G  D  K  I  T  P  D  S  L  V  E  I  K  R  T  G  V  E  G  N  W  A
91N2756T2   .  .  .  .  .  .  .  .  .  .  .  .  .  .  .  .  .  .  .  .  .  .  .  .  .  .  .  .  .  .  .  .  .  .  .
91N783T3    .  .  .  .  .  .  .  .  .  .  .  .  .  .  .  .  .  .  .  .  .  .  .  .  .  .  .  .  .  .  .  .  .  .  .
90N9196T4   .  .  .  .  .  .  .  .  .  .  .  .  .  .  .  .  .  .  .  .  .  .  .  .  .  .  .  .  .  .  .  .  .  .  .
93N1090CAN  .  .  .  .  .  .  .  .  .  .  .  .  .  .  .  .  .  .  .  .  .  .  .  .  .  .  .  .  .  .  .  .  .  .  .

516NYRAC    .  .  .  .  .  .  .  .  .  .  .  .  .  .  .  .  .  .  N  .  .  .  .  D  .  .  .  .  D  .  .  .  .  .  .
V125FLRAC   .  .  .  .  .  .  .  .  .  .  .  .  .  .  .  .  .  .  N  .  .  .  .  D  .  .  .  .  D  .  .  .  .  .  .
99N3306RAC  .  .  .  .  .  .  .  .  .  .  .  .  .  .  .  .  .  .  N  .  .  .  .  D  .  .  .  .  D  .  .  .  .  .  .

89L1461EF   D  .  .  .  .  .  .  .  .  .  .  .  .  .  .  .  .  .  N  .  .  .  .  D  .  .  .  .  N  .  .  .  H  .  .
92L415ML    D  .  .  .  .  .  .  .  .  .  .  .  .  .  .  .  .  .  N  .  .  .  .  .  .  .  .  .  H  .  .  .  .  .  .
88L1053LAN  D  .  .  V  .  .  .  .  .  .  .  R  .  .  .  .  .  .  N  .  .  .  D  .  R  .  .  N  .  .  .  .  .  .
92L2108LC   D  .  .  V  .  .  .  .  .  .  .  .  .  .  .  .  .  .  G  T  .  .  D  .  R  .  .  N  .  .  .  S  .  .
```

FIGURE 4.1 **Alignments of 105 bases of internal coding region of the N gene (A) and corresponding amino acid sequences of the encoded nucleoprotein (B) for representative specimens of several viral groups circulating in Canada.** The upper group of five sequences represents four distinct variants of the arctic fox (AFX) strain that circulate in Ontario (T1 to T4) and a variant from northern Canada (Genbank Accession #s L20673-20676 and U03769 respectively). The second group of three sequences comprises the North American raccoon strain with representation from New York (516NYRAC), Florida (V125FLRAC) and Ontario (99N3306RAC), Genbank Accession #s U27218, U27220 and AF351826 respectively. The third group consists of four distinct bat strains that are harbored by different species of insectivorous bats in Canada thus: big brown bat (89L1461EF), little brown bat (92L415ML), silver-haired bat (88L1053LAN) and hoary bat (92L2108LC), with Genbank Accession #s AF351831, AF351839, AF351840 and AF351845 respectively. Amino acid sequences were predicted from the nucleotide sequences of 1A using the DNAsis package (Hitachi). Alignments were performed using CLUSTALX v1.8 (Thompson, Gibson, Plewniak, Jeanmougin & Higgins, 1997) using the Type 1 Ontario variant of the AFX strain (specimen 91N1578T1) as the reference. For all other specimens, dots represent identity to that reference sequence and only base substitutions/amino acid differences are indicated. Two of the sequences targeted by strain-specific primers, as detailed in Section 4.2.2.3, are identified below the nucleotide alignment. Since these primers were designed in the negative-sense orientation, their 3′ termini critical for dictating primer specificity would correspond to the 5′ nucleotide in the figure.

techniques discriminated rabies viruses from other rabies-related lyssaviruses (Black et al., 2000; Picard-Meyer et al., 2004b).

A method that addressed the need to type viruses present in formalin-fixed brain tissues employed several strain-specific probes, designed to target a highly variable region of the P gene corresponding to sequences within bases 1600–2200 of the PV reference genome (GenBank Accession no. M13215), in an *in situ* hybridization format; this procedure can discriminate most of the terrestrial and chiropteran rabies virus types that circulate in Canada (Nadin-Davis, Sheen & Wandeler, 2003a).

2.2.4 *Quantitative PCR*

While strain-discriminatory probes were potentially a valuable tool for viral typing, the labor-intensive nature of performing nucleic acid hybridizations limited their application. This changed with the development of real-time PCR technology, and in particular the TaqMan (5′ nuclease) assay format, which enabled development of rapid highly automated assays that use fluorophore-labeled probes for lyssavirus detection and discrimination, an approach described in detail in Chapter 11 on diagnostic evaluation.

2.2.5 *Nucleotide Sequencing and Principles of Phylogenetic Analysis*

Direct nucleotide sequence determination applied to RT-PCR products over a predetermined sequence window is the most commonly applied method for genetically comparing a collection of lyssaviruses. Alignments of such sequence data allow direct comparison of isolates to reference sequences and hence allocation to a particular type. The sequence window targeted for a phylogenetic study will be determined in part by its goal. In general, it has been observed that similar conclusions on the overall epidemiological relationships of particular lyssaviruses are obtained regardless of the specific sequence window targeted. Thus, studies on representatives of all the recognized lyssavirus genotypes do generally exhibit similar branching patterns irrespective of the use of a specific gene, a portion of a gene or the whole viral genome (Wu, Franka, Velasco-Villa & Rupprecht., 2007), an observation consistent with the notion that members of this genus do not undergo recombination events. However, the analysis requires enough information to generate phylogenies having sufficiently robust statistical support. While short target sequences (200–300 bases) can yield phylogenetic predictions, the longer a sequence window the greater the likelihood of finding differences between samples; as genetic variation increases, the number of informative characters available to a phylogenetic analysis increases, and this improves the chances of obtaining well supported phylogenetic trees (Kissi et al., 1995; Nadin-Davis et al., 2001). Thus, for studies that seek to monitor variation within a closely related viral population, determination of a longer sequence window (>500 bases) or a relatively variable region of the genome, or both, is the most effective approach to show relationships that are strongly supported by statistical analyses. Because comparison of an unidentified viral sample with viruses representative of a region require that each be characterized over the same portion of the genome, the choice of the genome target may be strongly influenced by the availability of sequence data generated by prior studies. The majority of comparative rabies virus studies, which include viruses recovered in many parts of the world, have employed the N gene, or parts thereof, and most of these data are now widely accessible through the GenBank sequence database maintained by the National Center for Biotechnology Information (NCBI, n.d.), National Institutes of Health in the USA. While databases for other genes (G and P) and whole genomes are now increasing in scope, the N gene repository is currently by far the most extensive.

Figure 4.1 shows examples of sequence alignments using data from three groups of North American rabies virus variants. From visual inspection of the aligned nucleotide sequences (Figure 4.1A), it is apparent that within the arctic fox (AFX) and raccoon rabies virus variants there is very limited intra-group variation, but significant inter-group differences. The isolates within the bat group exhibit a significant degree of nucletotide substitutions, and, indeed, the four isolates examined here represent different bat-associated variants. As shown in Figure 4.1B, only some of the nucleotide variation found within this data set results in coding differences in the translation products, but consistent inter-group differences are evident. Such a comparison can easily be performed manually for a small sample set. However, where large numbers of isolates are being compared with respect to long sequences of nucleotides, the complexity of the dataset requires that the analytical output be in the form of a phylogenetic tree, a diagram of hierarchical branches that depicts the evolutionary relationships between samples according to their nucleotide substitution patterns. Figure 4.2 illustrates several examples of such trees that depict the diversity of the *lyssavirus* genus, as described in more detail in Section 4.3. A tree generated by the commonly employed neighbour joining (NJ) method is shown in Figure 4.2A. Within this diagram, samples that form a discrete cluster on one branch of the tree are said to form a clade; where there is strong support for this cluster (see below) the samples are said to form a monophyletic clade, indicating that all members originated from a common precursor. A taxon is normally defined as a species or group of species that clearly identifies a specific group of organisms; this term is sometimes used to refer to a particular group of specimens that form a monophyletic clade. Terms that describe the groupings of a phylogenetic tree, such as *clade, cluster, group, type* and *lineage,* can be, and often are, used interchangeably. However, for clarity when describing a phylogenetic study, it is helpful to assign specific designations to particular levels of sample association and apply these designations consistently.

Phylogenetic trees are generated from sequence data by computer programs that employ a variety of different algorithms and methods for tree reconstruction. Some of the most commonly used software packages include MEGA (Molecular Evolutionary Genetic Analysis) (Tamura, Dudley, Nei & Kumar, 2007), PAUP (Phylogenetic Analysis Using Parsimony) available for purchase from Sinauer Associates, and the older PHYLIP (*PHYL*ogeny *I*nference *P*ackage) software developed

(A)

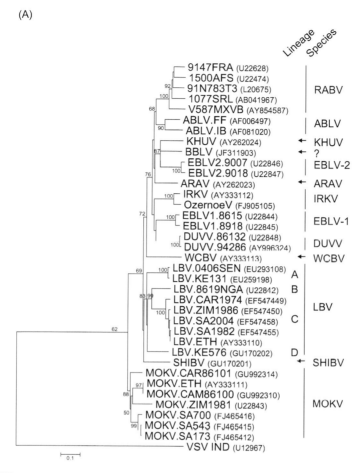

FIGURE 4.2   Phylogenetic analysis of complete N gene sequences of 35 representative members of the lyssavirus genus using the N gene of Vesicular Stomatitis Virus (Indiana strain), *Vesiculovirus* genus, as the outgroup. Sequences were aligned using CLUSTALX and imported into the MEGA4 (Tamura et al., 2007) or PhyML (Guindon et al., 2010) programmes. Several different analyses were implemented and the resulting trees generated as follows: A. Phylogram of a NJ analysis employing 1000 bootstrap replications of the data; a distance scale is shown at bottom left. Clades representing each of the 12 established species are indicated while a question mark identifies the unclassified isolate. Members of the three proposed LBV lineages are also indicated; the five RABV samples each represent a different RABV lineage as shown in more detail in Figure 4. 3. Following each specimen designation its GenBank Accession number is indicated in brackets. B. Circular representation of the tree generated by a MP analysis (with 100 bootstrap replications of the data). C. Unrooted tree of a ML analysis using 100 bootstrap replicates of the data generated using the PhyML programme with the GTR + I+G model identified as that best fitting the data using Modeltest (Posada & Crandall, 1998); a radial depiction of this same tree, shown in the inset, clearly illustrates division of the genus into three phylogroups, I to III. In panels A and C, bootstrap values (shown as a percentage) are indicated for many major branches of the trees. A full listing of all samples used for phylogenetic analysis is available on-line (Supplementary Table 4.5).

(B)

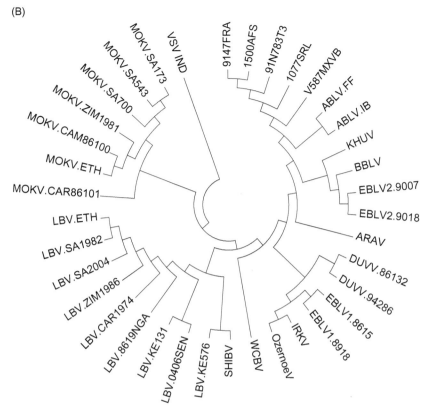

FIGURE 4.2   (Continued)

and supported online by Felsenstein (1993), together with links to many other sources of software for sequence alignments, phylogenetic analysis, tree illustration, Bayesian methods, and other analyses. Most phylogenetic analysis packages incorporate a variety of algorithms and methods for phylogenetic tree construction, the principles of which are explained in detail by Graur and Li (2000). The popular distance-based methods, such as NJ or unweighted pair-group method with arithmetic means (UPGMA), consider the overall genetic distance between all pairs of sequences rather than the actual sequences themselves. Alternative algorithms include the character-based maximum parsimony (MP) and maximum likelihood (ML) methods, which use individual substitutions to determine all possible tree constructions supported by the data and then identify the optimal tree by a comparative process. MP identifies the optimal tree by selecting the minimal number of evolutionary steps required to explain the data; Figure 4.2B illustrates a tree generated by MP analysis of the lyssavirus dataset drawn in circular format. ML, which can be implemented

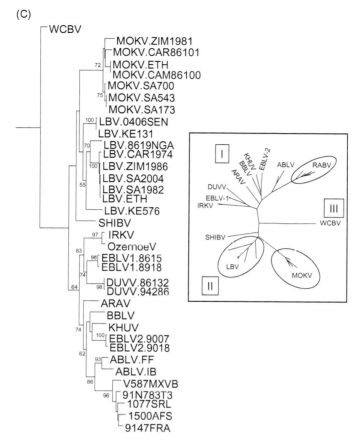

FIGURE 4.2 (Continued)

using the relatively rapid PhyML program (Guindon et al., 2010), identifies the optimal tree as that most likely to have occurred according to the assumed evolutionary model; this package is often used in conjunction with Modeltest (Posada & Crandall, 1998), a program for identifying the evolutionary model best supported by a particular database. Figure 4.2C illustrates a ML analysis of the lyssavirus dataset using this approach.

The choice of method to be employed for a particular analysis will often depend on the study's purpose. Distance methods are frequently preferred from a practical standpoint due to their relatively rapid execution and their ability to identify groups or clades as efficiently as other more computationally intensive methods. For the purpose of epidemiological studies, the association of an isolate within a clade, rather than its precise position within this clade, normally identifies the strain or variant responsible for that case thereby providing the key information sought. Thus, for many analyses, distance methods are sufficiently predictive. However, where the data are to be analyzed for the purpose of exploring mechanisms of mutation, use of the other algorithms is advantageous. When independent analysis of a certain data set by several different algorithms consistently predicts a particular branching pattern and the existence of specific clades, the overall predictive strength of the study is significantly enhanced.

An alternative strategy to explore the robustness of phylogenetic predictions is the use of nonparametric bootstrap analysis. Most phylogeny software can incorporate this statistical method into their analyses, and all molecular epidemiological studies should be encouraged to include this statistic. The method is valuable because even relatively small datasets generate multiple possible branching patterns or trees and statistical methods must be employed to predict the most likely branching pattern, often referred to as the consensus tree. In nonparametric bootstrap analysis, the nucleotide sequence data are resampled randomly with replacement thereby generating pseudoreplicates of the original data; the number of replicates is set by the operator, usually between 100 and 1,000. A smaller number of replicates is generally employed when applying computationally intensive programmes such as MP, while a NJ analysis can normally readily incorporate 1,000 replicates. Upon analysis by one of these algorithms, the proportion of times that each clade occurs within all trees is calculated and this value is considered as a measure of support for that grouping. In the case of RNA viruses, bootstrap values >90% are generally regarded as providing strong support for a clade, while values >70% are often considered significant (Bauldauf, 2003).

Another useful tool for tree construction is the inclusion of outgroups. An outgroup is a representative of a taxon that is known or assumed to be less closely related to all other taxa of the analysis than these taxa are to each other. Thus, the outgroup essentially provides a root to the tree and determines the direction of character change. An unexpected association of an unknown sample with the outgroup implies that the original assumptions regarding the origins of the unknown were incorrect and a different sample set should be employed for an alternate phylogenetic prediction.

Phylogenetic trees can also be constructed with protein translation products of coding regions using approaches similar to those applied to nucleotide sequence data. Comparison of trees generated using nucleotide and amino acid sequence data of the lyssavirus N locus show rather similar topologies, although the former tend to have slightly higher bootstrap values (Kissi et al., 1995). This observation may simply reflect the fact that nucleotide sequences, with three times more datapoints than amino acid sequences, can include synonymous mutations and thus contain greater informative diversity.

Another important aspect of any molecular epidemiological study is the selection of samples to be included for comparison. In general, an appropriately representative selection of samples, as required to answer the questions posed by the phylogenetic study, should be included. Compliance with this criterion may not be simple. Most collections of rabies viruses are drawn from passive surveillance systems that can introduce significant bias. Areas of low human population density, lack of human contact with rabies reservoir species, or an inadequate infrastructure for laboratory confirmation of suspect cases all limit the numbers of samples available for study. The first two limitations have precluded collection of many samples from northern temperate and arctic regions, while the latter issue has severely hampered studies of the nature of the lyssaviruses circulating in many parts of Africa.

While generation of a consensus sequence of an isolate by direct sequencing of an amplicon is the norm, occasionally the extent of viral diversity within an isolate is of interest (Benmansour et al., 1992; Kissi et al., 1999; Nadin-Davis et al., 2006a). To explore the latter scenario, the amplicon has traditionally been inserted into a plasmid vector by standard molecular cloning techniques and several resulting clones sequenced. However, future studies on isolate diversity will undoubtedly benefit from rapidly evolving cost-effective Next Generation Sequencing (NGS) technologies (Wright et al., 2011).

The relatively recent development of coalescent theory, and the availability of computer programs that apply mathematical tools employing this approach, also permit inference of the time-frame of evolution of viral lineages. The BEAST (Bayesian Evolutionary Analysis Sampling Trees) package, (Drummond & Rambaut, 2007), is often used for this purpose. BEAST can apply a number of evolutionary models with either a strict or relaxed molecular clock to estimate the viral nucleotide substitution rate from which the time frame of emergence of specific lineages can be predicted. Such analyses have been applied to the entire rabies virus species and its seven main lineages (Bourhy et al., 2008). Moreover, by exploring patterns of viral population dynamics, coalescent approaches can provide insight into the demographics of viral outbreaks (Biek, Henderson, Waller, Rupprecht & Real, 2007). The application of coalescent methods to studies on lyssaviruses has been reviewed recently (Nadin-Davis & Real, 2011).

2.3 Typing Methods: Antigenic versus Genetic

Traditionally, the less technically demanding and less costly methods of antigenic typing were considered the best for routine application to large numbers of cases, and in many diagnostic laboratories this is still the case. Antigenic analysis is usually sufficient to assign a specimen to a particular viral type, and often this epidemiological information is all that is required. Genetic methods that target the viral genome by employing techniques such as RFLP, HMA or strain-specific PCRs, including

TABLE 4.3   Advantages and Disadvantages of Antigenic and Genetic Methods of Virus Typing

Method Type:	Reagent Costs	Sample Throughput	Technical Complexity	Typing Precision
Antigenic	Low	High	Low	Moderate
Genetic:				
RFLP	Moderate	Medium	Moderate	Moderate
HMA	Low	Medium	Moderate	Low – Moderate
Strain-specific/qPCR	Moderate	Medium	Moderate	Moderate – High
Sequencing	High	Low	High	High

real-time PCR, may potentially be more discriminatory due to the targeting of RNA sequences rather than protein primary structure. However, only nucleotide sequence analysis followed by phylogenetic investigation provides information about the evolution and emergence of specific viral strains, and this approach thereby provides additional information—albeit at a greater cost. Thus, the choice of typing method(s) to apply to a situation will depend on the information sought and the resources available. The differences between these methods in terms of cost, degree of sample throughput, technical complexity, and likely level of typing precision are summarized in Table 4.3. Ideally, a combination of these methods, in which a wide selection of isolates is characterized antigenically or by other simple genetic typing regimens, supplemented with nucleotide sequence determination on representative specimens, is the preferred and most cost-effective means of generating comprehensive epidemiological data. Such a strategy is recommended when the viruses of a previously unstudied area are characterized for the first time. Subsequent surveillance activities in areas that continue to report significant case numbers may be best undertaken by routine antigenic typing with additional investigation by phylogenetic methods on selected isolates that are antigenically unusual, unexpected, or ambiguous. In areas where positive sample numbers are low, exclusive use of the more precise genetic methods might be preferred.

3 LYSSAVIRUS TAXONOMY

3.1 Lyssavirus Species

As detailed in Table 4.4, all lyssaviruses are currently divided into 12 different species, with rabies virus assigned as the type species of the genus (International Committee on Taxonomy of Viruses [ICTV],

TABLE 4.4 Species Currently Assigned to the *Lyssavirus* Genus

Species	Distribution	Reservoir
Rabies virus (RABV)	Worldwide, with the exception of Antarctica and some islands. Some regions traditionally affected (e.g., western Europe) have recently eradicated the disease and attained rabies-free status.	*Canivora*: many species of domestic and wild canids, foxes, mongooses, skunks, and the raccoon, *Chiroptera*: (Americas only) several species of insectivorous, vampire, and frugivorous bats.
Aravan virus (ARAV)	Only a single case reported from Kyrgizstan	*Microchiroptera*: *Myotis blythi*[*]
Australian bat lyssavirus (ABLV)	Australia and possibly areas of SE Asia	*Megachiroptera*: Pteropid species
		Microchiroptera: the insectivorous yellow-bellied sheathtail bat (*Saccolaimus flavicentris*)
Duvenhage virus (DUVV)	African nations, including Guinea, South Africa, Zimbabwe,	*Microchiroptera*: Single cases assigned to *Miniopterus schreibersii*, *Nycteris gambiensis* and *N. thebaica*
European bat lyssavirus 1 (EBLV-1)	Europe, including Denmark, France, Germany, The Netherlands, Poland, Russia, Spain, Ukraine.	*Microchiroptera*: *Eptesicus serotinus*
European bat lyssavirus 2 (EBLV-2)	Several countries of western Europe, particularly The Netherlands, Switzerland, Finland, and the UK.	*Microchiroptera*: *Myotis* species, especially *M. dasycneme* and *M. daubentonii*
Irkut virus (IRKV)	One case from Irkutsk province, Russia. A closely related virus (Ozernoe Virus) was recovered in Primorye Territory, Russia	*Microchiroptera*: Irkut virus was recovered from a bat (*Murina leucogaster*)[*], while Ozernoe virus was from a human case after exposure to an unidentified bat.
Khujand virus (KHUV)	Single case from Tajikistan	*Microchiroptera*: single case from a bat (*Myotis daubentonii*)[*]
Lagos bat virus (LBV)	Several African countries, including Central African Republic, Ethiopia, Nigeria, Senegal, South Africa, Zimbabwe	*Megachiroptera*: Fruit bats, including *Eidolon helvum*, *Rousettus aegyptiacus*, and *Epomophorus wahlbergi*
Mokola virus (MOKV)	Several African countries, including Cameroon, Central African Republic, Ethiopia, Nigeria, South Africa, Zimbabwe	Reservoir unknown—single cases in shrews (*Crocidura sp.*) and a small rodent (*Lopyhromys sikapusi*), with most reported cases in domestic cats and dogs

(Continued)

RABIES: SCIENTIFIC BASIS OF THE DISEASE AND ITS MANAGEMENT

TABLE 4.4   (Continued)

Species	Distribution	Reservoir
Shimoni bat virus (SHIBV)	Single case from a bat in Kenya	Commerson's leaf-nosed bat (*Hipposideros commersoni*)[*]
West Caucasian bat virus (WCBV)	Single case from the Krasnodar region of Russia	*Microchiroptera*: *Miniopteris schreibersi*[*]
Proposed species to be confirmed:		
Bokeloh bat lyssavirus (BBLV)	Two cases in Germany and France	Natterer's bat (*Myotis nattererii*)[*]
Ikoma virus (IKOV)	Single case in Tanzania	Isolated from an African civet (*Civettictis civetta*) but bat reservoir possible[*]
Lleida bat lyssavirus (LLEBV)	Single case in Spain	Isolated from *Miniopterus schreibersii*[*]

Reservoir host to be confirmed.

2011). Three other recently described lyssaviruses, Bokeloh bat lyssavirus (BBLV) (Freuling et al., 2011), Ikoma virus (IKOV) (Marston et al., 2012) and Lleida bat lyssavirus (LLEBV) (Aréchiga et al., 2012) may represent yet other species. This classification is based on serological, immunological, and epidemiological studies of these viruses together with substantial nucleotide sequence information on many representative specimens (Bourhy, Kissi & Tordo, 1993; Kuzmin et al., 2005). Calculations of pairwise genetic distances between several of these viruses at the N gene locus have suggested that 80–82% nucleotide identity is an appropriate cut-off value for discriminating between species (Kuzmin et al., 2005). At the protein level, a distance value of 0.065 for the nucleoprotein appears to provide a convenient cut-off, whereby viruses that differ by more than this value are consistently described as different species. Within a species, viruses exhibit a range of nucleotide identities between 82–100%, with all members of a particular variant usually exhibiting >95% identity, while values within the 82–95% range often define distinct lineages within the species.

3.2 Molecular Epidemiology of Rabies-related viruses

With the exception of the type species, rabies virus, the numbers of isolates collected for many lyssavirus species are limited. Hence, our knowledge of their diversity and range is incomplete. Current knowledge, as illustrated in Figure 4.2 and summarized below, will be expanded upon as more isolates are recovered.

3.2.1 Australian Bat Lyssavirus

The presence of a rabies-like virus that circulates in fruit bats in Australia was first recognized in 1996 (Fraser et al., 1996) and subsequently designated as Australian bat lyssavirus (ABLV) (Hooper et al., 1997). Sequence analysis of several isolates of this virus readily classified them as a distinct monophyletic group that can be differentiated into two biotypes associated with *Pteropus* (flying fox) species and the insectivorous bat (*Saccolaimus flaviventris*), respectively (Gould et al., 1998; Gould, Kattenbelt, Gumley & Lunt, 2002; Guyatt et al., 2003; Warrilow et al., 2002). Based on substitution rates at synonymous and non-synonymous sites respectively, Warrilow et al., (2002) estimated that these two biotypes diverged between 950 to 1,700 years ago; studies using Bayesian methods have not been described but based on studies of other lyssaviruses may yield rather different age estimates. The public health implications of the discovery of these viruses were soon realized with the identification of two human cases of rabies caused by ABLVs (reviewed in Moore, Jansen, Graham, Smith & Craig, 2010). The pteropid-associated virus has been recovered from all four species of flying foxes that inhabit the Australian mainland; indeed, this biotype's geographical range appears to correlate with that of the host although prevalence rates are low (Warrilow, 2005; Moore et al., 2010). ABLV range may extend beyond Australia given some serological evidence for the circulation of these, or closely related viruses, in bat populations of the Philippines (Arguin et al., 2002).

3.2.2 Duvenhage Virus

Beginning in 1970, there have to date been only five reports of the isolation of Duvenhage virus (DUVV) (Paweska et al., 2006; Van Eeden, Markotter & Nel, 2011; van Thiel et al., 2009; Weyer et al., 2011). The first isolate was in a human from South Africa, the second from an insectivorous bat in South Africa, initially identified as *Miniopterus schreibersii*, but now designated as *Miniopterus natalensis*, the third from another species of insectivorous bat, *Nycteris thebaica*, captured in Zimbabwe and the fourth was again from a human in South Africa (Paweska et al., 2006). The last report was of a human case diagnosed in The Netherlands after exposure of the patient to a bat in Kenya (van Thiel et al., 2009). Although a review of all DUVV sequence data might suggest a larger number of isolations, the assignment of individual isolates with different designations by different groups has confused the issue significantly. The four DUVV viruses recovered from southern Africa are highly homogeneous, possibly reflecting the limited geographical range over which they were retrieved (Amengual, Whitby, King, Serra-Cobo & Bourhy, 1997;

Badrane, Bahloul, Perrin & Tordo, 2001; Delmas et al., 2008; Johnson, McElhinney, Smith, Lowings & Fooks, 2002; Nadin-Davis et al., 2002; Paweska et al., 2006). The isolate from Kenya, which is not represented in Figure 4.2 because only partial N gene sequence is currently available, is genetically the most diverse (van Thiel et al., 2009) and may represent a second lineage of this species (Van Eeden et al., 2011). This specimen clearly indicates that our knowledge of the extent of DUVV diversity and geographical range will increase significantly with additional isolations across Africa.

3.2.3 European Bat Lyssaviruses

Our knowledge of the association of lyssaviruses with European bats has increased enormously over the last 50 years and two separate viral species, designated European bat lyssavirus type 1 (EBLV-1) and type 2 (EBLV-2), are now recognized to be responsible for the vast majority of reported cases (Bourhy et al., 1993; Fooks, Brookes, Johnson, McElhinney & Hutson, 2003a; Kuzmin & Rupprecht, 2007). The host reservoirs for these viruses were identified as *Eptesicus serotinus* (EBLV-1) and species of *Myotis* (EBLV-2), particularly *M. daubentonii* and *M. dasycneme* (Bourhy, Kissi, Lafon, Sacramento & Tordo, 1992), with occasional isolations from other bat species thought to represent spillover infections from these reservoirs. A more complex situation may exist in Spain, where significant numbers of EBLV-1 infections were identified in several other bat species, particularly members of the *Myotis* genus (Amengual, Bourhy, López-Roig & Serra-Cobo, 2007; Serra-Cobo, Amengual, Abellán & Bourhy, 2002).

EBLV-1, by far the most frequently isolated lyssavirus from European bats, is widely distributed throughout many European countries (Amengual et al., 1997; Fooks et al., 2003a; Vázquez-Morón et al., 2008). EBLV-1 can be divided into two genetic lineages, 1a and 1b, that exhibit differences in their evolution and dispersal patterns and which may represent two independent introductions of this biotype into Europe (Amengual et al., 1997; Davis et al., 2005). EBLV-1a occurs predominantly in northern countries of Germany, Denmark, The Netherlands, and Poland in areas comprised of flat, low-lying landscape with a mix of urban and rural areas and some forest; this type of landscape, which is ideal habitat for *Eptesicus serotinus*, may have facilitated rapid east-west spread of the relatively homogeneous EBLV-1a sub-type across Northern Europe (Müller et al., 2007). In contrast, the more divergent EBLV-1b specimens appear to exhibit a north-south axis of spread and occur in France, where spatial clustering of the 1a and 1b types was observed (Picard-Meyer et al., 2004a, b), in The Netherlands (van der Poel et al., 2005) and in southern Germany (Johnson et al., 2007b). In Spain, EBLV-1 isolates form a distinct clade separate from other EBLV-1 samples, likely as a result of their association with a different host species, *Eptesicus isabellinus*, (Vázquez-Morón, Juste, Ibáñez, Berciano & Echevarría, 2011); this study suggested a more complex evolutionary history for EBLV-1b viruses by dividing them into four distinct clades and cast doubt on the hypothesis that EBLV-1 was widely introduced into European bats after incursion of a virus from North Africa into Spain (Davis et al., 2005). Surveillance for the presence of related lyssaviruses in bats of North Africa may shed new light on the origins of these European viruses. Using a strict molecular clock to estimate the nucleotide substitution rate for EBLV-1 as 5×10^{-5} substitutions per site per year, it was proposed that EBLV-1 emerged between 500 and 750 years ago (Davis et al., 2005). However, a later study employing a relaxed molecular clock predicted a much higher substitution rate (1.1×10^{-4} per site per year), a value in line with estimates for rabies viruses, and this suggested that the species is of more recent origin (70–300 years) (Hughes, 2008).

EBLV-2, first isolated in 1985 from a Swiss bat biologist working on bats in Finland, has now been identified in many European countries, including The Netherlands, Switzerland, Denmark (Fooks et al., 2003a; Harris et al., 2007), Germany (Freuling et al., 2008), and the United Kingdom (Banyard et al., 2009; Harris et al., 2007; Johnson et al., 2003b) where enhanced surveillance followed a case of human rabies in Scotland caused by EBLV-2 acquired after exposure to a bat (Fooks et al., 2003b). Genetic studies on bat populations suggest that entry of EBLV-2 into the UK has been facilitated by regular movement of *M. daubentonii* between mainland Europe and the UK; in contrast, *E. serotinus* populations are more fragmented, thereby limiting opportunities for incursion of EBLV-1 into the country (Smith et al., 2011). The small number of EBLV-2 specimens available for genetic analysis limits our current understanding of the phylogeny of this species, although it appears to segregate into two distinct lineages, 2a and 2b (Amengual et al., 1997).

While bats are the established reservoirs for EBLVs, spillover infection of EBLV-1 to terrestrial species has occasionally been reported: in a group of Danish sheep (Rønsholt, 2002; Tjørnehøj, Fooks, Agerholm & Rønsholt, 2006), a stone marten in Germany (Müller et al., 2001), and in two cats in France (Dacheux et al., 2009). Experimental studies have shown that EBLVs can cause clinical features of rabies in red and silver foxes (Picard-Meyer et al., 2008; Vos et al., 2004), but the likelihood that these viruses could adapt and become capable of sustained transmission in a terrestrial host appears to be small.

A novel and currently unclassified lyssavirus, designated as Bokeloh bat lyssavirus (BBLV), recovered from a rabid Natterer's bat (*M. nattererii*) in Germany, clustered most closely to KHUV with 80% nucleotide identity at the N gene locus (Freuling et al., 2011). A second isolation of this virus was recently reported in eastern France in 2012, again from *M. nattererii*

(Picard-Meyer et al., 2012). In addition, another novel lyssavirus recovered in Spain from a *Miniopterus schreibersii* bat and designated Lleida bat lyssavirus (LLEBV) appears by N gene comparison to be more closely related to IKOV and WCBV than to the EBLVs (Aréchiga et al., 2012). The full characterization of these viruses is awaited with interest as they appear to extend the known diversity of bat lyssaviruses circulating in western Europe.

3.2.4 Eurasian Bat Lyssaviruses

The identification of four additional lyssavirus species across Eurasia, designated as Aravan virus (ARAV), Irkut virus (IRKV), Khujand virus (KHUV), and West Caucasian bat lyssavirus (WCBV), was supported by extensive phylogenetic analysis of single isolates of each type from various bat species (Kuzmin et al., 2003, 2005). The recently described Ozernoe virus (Leonova et al., 2009), recovered from a human after exposure to a bat, appears to cluster with IRKV (see Table 4.4, Figure 4.2) and thus represents the same species. Further isolates are needed to establish the true reservoirs of these lyssaviruses and their geographical ranges.

Serological evidence has suggested that these or closely related viruses may circulate in certain bat populations of Thailand (Lumlertdacha et al., 2005), Cambodia (Reynes et al., 2004), and Bangladesh (Kuzmin et al., 2006). Moreover, the identification of neutralizing antibodies against WCBV in *Miniopterus* bats of Kenya has fueled the suggestion that this species also circulates in Africa (Kuzmin et al., 2008b).

3.2.5 Lagos Bat Virus

Since the first Lagos Bat Virus (LBV) isolate was recovered from the brain of a Nigerian fruit bat (*Eidolon helvum*) in 1956, and later identified as a rabies-related virus, a number of LBV infections have been reported, but only 16 LBV isolations have been documented (Markotter et al., 2008). LBV has a wide range across much of the African continent, having been identified in Nigeria, the Central African Republic, Guinea and Senegal, South Africa, Ethiopia, Zimbabwe, and Kenya (Kuzmin et al., 2008c; Markotter et al., 2006a,b, 2008), while a single isolation made in Europe originated from a bat imported from Africa (Picard-Meyer et al., 2004b). Although one LBV isolate was from an insectivorous bat, *Nycteris gambiensis*, these viruses are usually recovered from several species of frugivorous bats. Based on viral isolations, *Epomophorus wahlbergi* is the likely LBV reservoir species in South Africa (Markotter et al., 2008), while serological studies suggest reservoir roles for *Eidolon helvum* in Ghana and Nigeria (Dzikwi et al., 2010; Hayman et al., 2008; Wright et al., 2010), for both *Eidolon elvum* and *Rousettus aegyptiacus* in Kenya (Kuzmin et al., 2008c), and for *Eidolon dupreanum* in Madagascar (Reynes et al., 2011). The virus has also been recovered from dogs, several cats, and a mongoose (*Atilax paludinosus*), presumably due to spillover infections, but it has never been reported in humans. Despite the limited sampling, significant antigenic and genetic variation occurs within this species (Badrane & Tordo, 2001; Bourhy et al., 1993; Johnson et al., 2002; Kuzmin et al., 2008c; Nadin-Davis et al., 2002; Mebatsion, Cox & Conzelmann, 1993; Markotter et al., 2008). It appears that the greater the geographical separation between samples, the greater their genetic diversity and four distinct lineages (A-D) of LBV have been described (Kuzmin et al., 2008c). Due to its high divergence from the other lineages, it was suggested that LBV lineage A be considered a separate species, designated Dakar bat lyssavirus (DBLV), for which the isolate from Dakar, Senegal, would represent the type virus (Markotter et al., 2008).

3.2.6 Shimoni Bat Virus

The recently described Shimoni Bat Virus (SHIBV) isolate, recovered from the brain of a Commerson's leaf-nosed bat (*Hipposideros commersoni*) in Kenya (Kuzmin et al., 2010), is an outlier to the LBV clade, but is now recognized as a distinct species. Serological studies support the role of the Commerson's leaf-nosed bat as a reservoir for this virus (Kuzmin et al., 2011).

3.2.7 Mokola Virus

Since the original isolation of Mokola Virus (MOKV) from shrews in Nigeria, this viral species has been recovered from human rabies cases in Nigeria and from other species in several African countries, including Cameroon, Zimbabwe, the Central African Republic, Ethiopia, and South Africa (reviewed by Nel, Jacobs, Jaftha, von Teichman & Bingham, 2000). Several recent isolates have been recovered in South Africa with the help of enhanced antigenic tools for screening all rabies-positive specimens (Nel et al., 2000; Sabeta et al., 2007b, 2010). The host species from which this virus has been recovered include shrews, cats and dogs, and a single isolation from a rodent. Although the high proportion of cases in cats suggests a bat or rodent reservoir for MOKV, the actual reservoir host is currently unknown.

As for the LBVs, the relatively few MOKVs that have been characterized exhibit significant genetic variation according to their geographical source. Isolates from southern Africa define three distinct viral groups, two from geographically distant areas of South Africa and a third represented by samples from Zimbabwe, while isolates from more northern countries lie on distinct branches of the MOKV clade (Nel et al., 2000; Nadin-Davis et al., 2002; Sabeta et al., 2010).

3.2.8 Ikoma Lyssavirus

Another isolate, Ikoma lyssavirus (IKOV), recovered from an African civet in Tanzania, is most closely related to WCBV, but it is sufficiently genetically distant, based on partial N gene sequence, from any other species yet described as to potentially represent another species (Marston et al., 2012). Although this virus promises to further expand known lyssavirus diversity, further studies of its complete genome and likely reservoir host are needed for future classification. This sample is not included in Figure 4.2 due to unavailability of a complete N gene sequence.

3.2.9 Lyssavirus Phylogroups

Based upon their genetic, immunologic and pathologic characteristics, members of the *Lyssavirus* genus have been proposed to form three distinct phylogroups, as illustrated in Figure 4.2C: phylogroup 1 is comprised of RABV, ABLV, ARAV, DUVV, IRKV, KHUV and both EBLVs, phylogroup 2 constitutes the distinctive African viruses LBV, SHIBV and MOKV, and phylogroup 3 is represented by the single divergent WCBV isolate (Badrane et al., 2001; Kuzmin et al., 2005). As more information on the newly identified bat lyssaviruses and IKOV becomes available this division will be refined. Moreover, as surveillance using current identification tools intensifies, particularly in many developing countries, it is anticipated that the geographical distribution of these lyssavirus species will be better defined, many new viruses may be discovered and our knowledge of the diversity of this genus may yet be greatly expanded.

3.3 Molecular Epidemiology of Rabies viruses

All currently known rabies viruses can be divided into seven major lineages which are estimated to have emerged within the last 1,500 years (Bourhy et al., 2008), an estimate that would exclude anecdotal rabies cases described in ancient times (see Section 4.4). The tree illustrated in Figure 4.3 depicts all seven lineages and provides a framework for the following description of the global status of rabies virus molecular epidemiology for which only major themes can be described.

3.3.1 Cosmopolitan Lineage

The geographically widely distributed cosmopolitan lineage is thought to have originated in Europe, perhaps as recently as 1750 (David, Dveres, Yakobson & Davidson, 2009), but was distributed to many other parts of the world by colonial activities that resulted in human-assisted movement of infected animals; in these new environments viral progenitors survived and adapted to persist in several different indigenous mammalian hosts (Kissi et al., 1995; Nadin-Davis & Bingham, 2004; Smith, Orciari, Yager, Seidel & Warner, 1992). Within this lineage several distinct clades are recognized. The viruses that emerged in red foxes in the early 20th century in Europe form one monophyletic clade that in turn is subdivided into several geographically restricted variants circulating in western (WE), central (CE), eastern (EE) and north-eastern (NEE) regions of Europe (Bourhy et al., 1999;

(A)

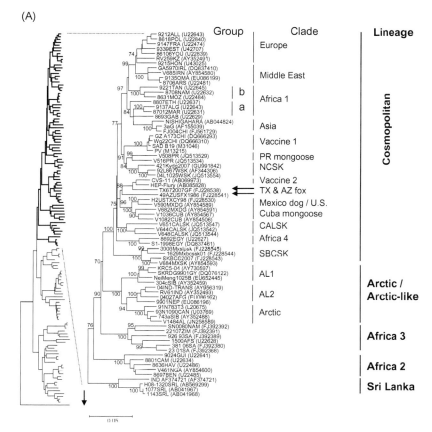

FIGURE 4.3 Phylogenetic tree of complete N gene coding sequences for 130 rabies virus isolates and a single ABLV isolate used as an outgroup. Sequences were analyzed in MEGA4 by the NJ method and the Kimura two-parameter nucleotide substitution model. For clarity the complete tree, illustrated on the left of each panel (A, B), has been divided into two parts, the upper half is shown in panel A and the lower half in panel B. Bootstrap values are indicated as described for Figure 4.2 and the groupings and lineages discussed in the text are identified to the right of the tree. Each specimen designation is followed by its GenBank Accession number in brackets; a full listing of all samples used for phylogenetic analysis is available on-line (Supplementary Table 4.5). Abbreviations are: FB, ferret badger variant; AZ skunk, bat variant maintained by skunks in Arizona, USA; TX & AZ fox, two distinct variants maintained by foxes in Texas and Arizona, USA, respectively.

Metlin, Rybakov, Gruzdev, Neuvonen & Huovilainen, 2007). The raccoon dog is a wild canid native to east Asia often hunted for its fur and now widespread in several European countries after introduction into western Russia (Kauhala & Kowalczyk, 2011). This canid has become a major rabies reservoir in Baltic countries as a result of adaptation of the NEE variant to this host (Bourhy et al., 1999). Closely related viruses also spread to southeast Europe and currently circulate in dogs and wildlife throughout

(B)

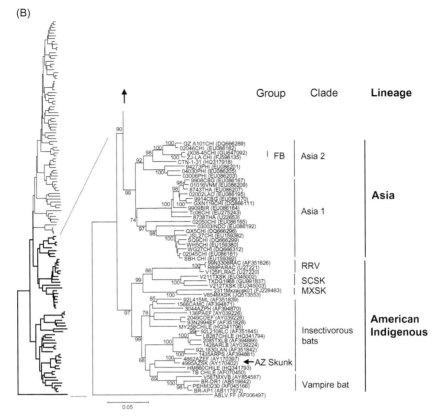

FIGURE 4.3   (Continued)

many parts of the Balkans (Johnson, Fooks, Valtchovski & Müller, 2007a; McElhinney et al., 2011; Turcitu et al., 2010) and Turkey (Johnson et al., 2003a; Johnson, Un, Vos, Aylan & Fooks, 2006). Closely related viruses also circulate in the Middle East. In Israel, jackals and foxes served as reservoirs for several rabies virus variants of this Middle Eastern strain (David, Yakobson, Smith & Stram, 2000; David et al., 2007) but in recent years a new dog-associated variant (Israel VII) that is phylogenetically closely related to Turkish isolates has emerged in the north of the country (David et al., 2009). In Iran, two distinct variants of the cosmopolitan lineage differentially distributed across the country may represent two independent incursions into the country (Nadin-Davis et al., 2003b). An outlying clade of the cosmopolitan lineage, designated Africa 4 and represented by a small number of isolates recovered from Egypt and Israel, appears to be maintained by dog populations of the region (David et al., 2009).

In Asia, the cosmopolitan lineage has been recorded in much of the former Soviet Union (Kuzmin et al., 2004), where red and steppe foxes are host reservoirs, and in Mongolia (Boldbaatar et al., 2010) and China (Zhang et al., 2006; Meng et al., 2007, 2011), where the dog is the most important maintenance host.

Another clade contained within the cosmopolitan lineage, Africa 1, is widely distributed throughout the continent, and studies in the 1990s clearly divided it into two geographically segregated sub-clades, 1a and 1b (Kissi et al., 1995). Africa 1a was identified in northern and eastern Africa including Algeria, Ethiopia and Sudan (Johnson, McElhinney, Ali, Saeed & Fooks, 2004b; Johnson et al., 2010a; Marston et al., 2009), Gabon, Madagascar, Morocco and Tunisia (Amouri et al., 2011). Canid solates from the Central African Republic, Kenya, Mozambique, Namibia, Tanzania (Lembo et al., 2007), Zaire, Zambia (Muleya et al., 2012), Botswana (Johnson, Letshwenyo, Baipoledi, Thobokwe & Fooks, 2004a), Zimbabwe, and South Africa (Sabeta, Bingham & Nel, 2003) constituted Africa 1b (Kissi et al., 1995). Members of both sub-clades were identified in Uganda, which may define a border zone of both types (Hirano, Itou, Shibuya, Kashiwazaki & Sakai, 2010a). However, this clear geographical separation of the Africa 1 sub-types has been complicated with the recent identification of these viruses in west African countries such as Nigeria (Talbi et al., 2009) and Ghana (Hayman et al., 2011), where the Africa 2 lineage was thought to circulate exclusively (see below); these observations may reflect the recent spread of Africa 1 viruses into the region (Hayman et al., 2011). Geopolitical boundaries seem to limit rabies virus spread, but human-mediated dispersal of the disease is proposed, based on phylodynamic studies that suggest that dog-to-dog transmissions alone could not have led to the current extent of disease spread (Talbi et al., 2010); indeed, a single Africa 1b isolate recovered in Ghana may be the result of a long-distance translocation of a diseased animal (Hayman et al., 2011). Although dogs are the major viral reservoir, in some jurisdictions certain wildlife species can maintain particular variants. In southern Africa, epizootics associated with bat-eared foxes (Sabeta, Mansfield, McElhinney Fooks & Nel, 2007a), jackals (Coetzee & Nel, 2007; Zulu, Sabeta & Nel, 2009), and kudu antelope (Mansfield et al., 2006a) have been described.

The cosmopolitan lineage is also represented by many variants on the American continent. The viruses associated with mongoose populations on some Caribbean islands clearly cluster within the cosmopolitan lineage and not with variants associated with mongooses in Africa (Africa 3) and Asia. The distinctness of the viruses present in the islands of Puerto Rico and Cuba suggests independent introductions (Nadin-Davis et al., 2006b; Nadin-Davis, Velez, Malaga & Wandeler, 2008). In continental North America, skunks are associated with several distinct cosmopolitan variants, including the north central skunk strain (NCSK) that circulates in north central USA and western Canada (Nadin-Davis, Huang & Wandeler, 1997; Nadin-Davis et al., 2002; Smith,

Orciari & Yager, 1995), the California skunk (CALSK) variant (Crawford-Miksza, Wadford & Schnurr, 1999; Velasco-Villa et al., 2008b), and the Mexican South Baja California skunk (SBCSK) variant (de Mattos et al., 1999; Nadin-Davis & Loza-Rubio, 2006; Velasco-Villa et al., 2002, 2005). Two distinct viral variants are associated with foxes in Texas (Rohde et al., 1997) and Arizona, with the latter extending into northern Mexico (Velasco-Villa et al., 2002, 2005) with recorded spillover into bobcats (de Mattos et al., 1999; Nadin-Davis & Loza-Rubio, 2006). The Mexican dog type can be divided into a number of regional variants (Velasco-Villa et al., 2005), from one of which emerged a coyote epizootic in Texas in the 1990s (Clark et al., 1994; Velasco-Villa et al., 2005). Several laboratory and vaccine strains, such as PV, SADB19 and CVS, also cluster closely with these field variant viruses (Nadin-Davis et al., 2002). The genetic relatedness of many of these variants, despite their geographical separation and association with different wildlife species, provides strong evidence for their historical emergence from dog-related enzootics introduced during colonization of North America (Velasco-Villa et al., 2008b). Indeed, cross-species transmissions from dogs to wildlife species is a recurring theme around the world (Nadin-Davis & Bingham, 2004).

Throughout Central and South America, the cosmopolitan lineage is represented by dog associated variants known to circulate in the region since the early 19th century, following colonization by Spanish and English explorers (Baer, 2007). Detailed antigenic and genetic studies have been described for samples from Argentina (Cisterna et al., 2005), Bolivia (Favi, Nina, Yung & Fernández, 2003), Venezuela (De Mattos et al., 1996), and Colombia (Hughes, Páez, Bóshell & Rupprecht, 2004), where dog variants appear to have been introduced into fox populations (Páez, Saad, Núñez & Bóshell, 2005). In Brazil, dog rabies has historically been the most prevalent biotype with respect to human disease (Ito et al., 2001), although recently foxes have been identified as a rabies reservoir in northeastern Brazil, where they harbor variants closely related to those circulating in dogs (Bernardi et al., 2005; Carnieli et al., 2008, 2009a; Favoretto et al., 2006; Shoji et al., 2006). It has been estimated that the introduction of the currently circulating viral variants into Brazil occurred in the late 19th to early 20th centuries, after the colonization period but during a time of high levels of immigration from Europe (Kobayashi et al., 2011).

3.3.2 Africa 2

The Africa 2 lineage, first described by Kissi et al., (1995), comprises a group of dog-associated viruses that now circulates widely across sub-Saharan western and central Africa, despite its probable introduction within the last 200 years (Talbi et al., 2009). Multiple regionally restricted variants exist, and there is evidence for a slow east-west spread across the continent (Talbi et al., 2009), but detailed studies on the viruses circulating within specific countries are limited, for example, Burkina Faso (de Benedictis et al., 2009b). A study of Africa 2 viruses recovered from a restricted area in Ghana identified three distinct variants indicative of some incursions from neighboring countries (Hayman et al., 2011). Talbi & Bourhy (2011) have recently reviewed the phylogeny and phylogeography of dog-associated Africa 1 and Africa 2 rabies virus groups across the continent.

3.3.3 *Africa 3*

The distinct Africa 3 lineage (Kissi et al., 1995) consists of a group of viruses known to circulate in mongooses in South Africa, Botswana and Zimbabwe (Johnson et al., 2004a; Nel et al., 2005; von Teichman, Thomson, Meredith & Nel, 1995), although its true range may extend to other parts of Africa (Nel et al., 2005). The yellow mongoose is the principal host for these viruses in South Africa although these viruses are often also reported in various viverrid species (e.g. civets in Zimbabwe) that may contribute to virus maintenance (Sabeta et al., 2008). High levels of genetic heterogeneity and strong geographical partitioning of variants of this lineage have been described (Nel et al., 2005). The lineage's age has been estimated at 159 or 229 years using sequence datasets based on G and N genes respectively (van Zyl, Markotter & Nel, 2010) or 73 years based on G-L intergenic sequences (Davis, Rambaut, Bourhy & Holmes, 2007). In accord with anecdotal records, both studies suggest that mongoose rabies is a more ancient lineage than canid rabies in the region. Davis et al., (2007) have suggested that the mongoose and canid biotypes exhibit distinct evolutionary dynamics.

3.3.4 *Arctic/Arctic-Like*

This lineage, originally designated simply as the Arctic lineage, was historically considered to be limited to red and arctic fox populations of arctic and temperate regions of North America (Canada and Alaska, USA), Greenland and Russia, especially Siberia, with the occasional incursion into the remote Svalbard archipelago of Norway, probably by animals venturing over pack ice from Russia (Kuzmin et al., 2004; Mansfield et al., 2006b; Mørk, Bohlin, Fuglei, Åsbakk & Tryland, 2011; Nadin-Davis et al., 1994). The Arctic-like (AL) designation was added in recognition of its extensive geographical range into many areas south of the temperate zone including Afghanistan, Pakistan, Korea, Iran, Nepal, Bhutan, Mongolia and most of India (Boldbaatar et al., 2010; Bourhy et al., 2008; Hyun et al., 2005; Kuzmin, Hughes, Botvinkin, Gribencha & Rupprecht, 2008a; Nadin-Davis, Turner, Paul, Madhusudana & Wandeler, 2007; Nagarajan et al., 2006, 2009; Park, Shin & Kwon, 2005; Reddy et al., 2011; Shao et al., 2011; Tenzin et al., 2011; Yang et al., 2011) where other species, particularly dogs, other wild canids (e.g., steppe fox, wolf) and raccoon dogs (Korea), act as viral reservoirs. This lineage was also present in Japan in the 1950s prior to rabies eradication from this country (Hatakeyama, Sadamasu & Kai, 2011). It now appears that this lineage emerged somewhere within Asia, perhaps within the last 200 years (Nadin-Davis, Sheen & Wandeler, 2012a), and spread rapidly to yield the more recent Arctic clade which is divided into four groups, A-1 to A-4, that exhibit distinct temporal and spatial distributions (Kuzmin et al., 2008a; Nadin-Davis et al., 2012a).

3.3.5 Sri Lanka

This distinctive lineage is found only in Sri Lanka (Arai et al., 2001; Nanayakkara, Smith & Rupprecht, 2003) and limited parts of southern India (Jayakumar, Tirumurugaan, Ganga, Kumanan & Mahalinga Nainar, 2004) including Tamil Nadu and Kerala (AravindhBabu, Manoharan, Ramadass & Chandran, 2011; Nagarajan et al., 2009), suggesting human-mediated transfer of the strain from Sri Lanka to India. The dog is the main host with frequent spill-over to other domestic animals but wildlife reservoirs may also maintain distinct variants of this lineage in Sri Lanka (Matsumoto et al., 2011).

3.3.6 Asia

This lineage, which circulates across much of mainland and southeast Asia, is divided into two main branches (Bourhy et al., 2008); viruses from China and Indonesia make up the Asia 1 sub-lineage while samples from most other countries of southeast Asia including Thailand, Malaysia, Vietnam, Cambodia, Myanmar (Burma), Laos and the Philippines as well as a small number of samples from China (Meng et al., 2011) comprise the Asia 2 sub-lineage. The representation of Chinese samples in both sub-lineages (Zhang et al., 2009) supports the idea that rabies spread from China to several other countries throughout the region during periods of extensive movement of Chinese migrants (Gong et al., 2010; Meng et al., 2011). Within each sub-lineage several distinct viral clades can be identified, and these are often localized to specific countries or regions. Various estimates of the year of emergence of the Asia lineage range between mean values of 1412 AD to 1742 AD (Gong et al., 2010; Meng et al., 2011; Ming et al., 2010).

In China, as in the rest of Asia, dogs are the most important vector host and frequently transmit the disease to humans and other domestic animals (Tang et al., 2005). However, in the Chinese province of Jiangxi, the ferret badger may constitute a wildlife reservoir responsible for independent maintenance of a rabies virus variant phylogenetically related to dog-associated variants (Zhang et al., 2010). Molecular analyses, targeting different regions of the viral genome, have allowed construction of detailed phylogenies of the viruses circulating in several Chinese provinces (Hirano et al., 2010b; Liu et al., 2007; Meng et al., 2007; Tao et al., 2009; Zhang et al., 2006) and have also identified a small number of viruses of the cosmopolitan lineage. Particular viral variants are often regionally localized thereby illustrating local spread of specific viral types; conversely provinces can harbour multiple variants, an observation that supports frequent long-distance movement of viruses as a result of human-mediated transportation of diseased animals.

Isolates from Indonesia, which formed a monophyletic clade clustering closely with the main Chinese Asia 1 clade, could be divided into three main clusters and several sub-clusters according to their geographical distribution across this island nation (Susetya et al., 2008), a pattern consistent with human-mediated introduction of rabies into Indonesia from China with subsequent spread of the virus between islands.

Detailed epidemiological studies of the viruses circulating in Thailand and southern Vietnam showed that they comprised a reasonably homogeneous monophyletic group within the Asia 2 sub-lineage (Ito et al., 1999; Khawplod et al., 2006; Yamagata et al., 2007) although genetic variation that discriminated between variants circulating within and around Bangkok was described (Kasempimolporn et al., 2004). A focus of another variant more closely associated with typical Chinese isolates, and which may represent a recent incursion of Chinese variants, was recently identified in northern Vietnam (Nguyen et al., 2011). Isolates from the Philippines have defined two sub-types that appear to be sequestered differentially within the Philippine archipelago (Nishizono et al., 2002).

3.3.7 American Indigenous

The American indigenous lineage, found only on the American continent, is comprised of viruses with a very distinct evolutionary origin compared to other rabies viruses. These viruses are found predominantly in many species of *Chiroptera* while a few distinctive strains are harbored by members of the *Carnivora*. The latter include the south central skunk (SCSK) strain associated with skunks of the south central USA, another viral variant associated with skunks of the Sinaloa-Durango region of Mexico (De Mattos et al., 1999; Nadin-Davis & Loza-Rubio, 2006; Smith et al., 1995), and the North American raccoon variant (RRV). A rabies virus variant believed to be associated with the white-tufted ear marmoset, also known as the Sagai monkey (*Callithrix jacchus*), a small diurnal primate, in Ceará state Brazil, is not closely related to any other American rabies virus (Favoretto et al., 2001) but falls within the American indigenous lineage.

The RRV, first reported in Florida in the 1940s, was inadvertently introduced, through transportation of infected animals, into the mid-Atlantic region in the 1970s from where it spread rapidly throughout the eastern seaboard of the USA (Winkler & Jenkins, 1991); indeed incursions across the border into Canada have occurred periodically (Wandeler & Salsberg, 1999). The virus's limited genetic variation suggests its relatively recent emergence (Nadin-Davis et al., 2006a) and indeed using nucleotide sequences the estimated date for TMRCA of this clade is 1946 AD (Szanto et al., 2011) while the year, 1973 AD, inferred for its introduction into Virginia is in accord with surveillance reports (Biek et al., 2007). Detailed genetic studies have revealed the emergence and spread of distinct clades of this variant (Biek et al., 2007; Szanto et al., 2011). Patterns of viral population dynamics can be explained by the effects of landscape topography on host movements and are in agreement with the extensive surveillance records available for this epizootic (Biek et al., 2007). The emergence of the RRV after a spillover event of the SCSK virus into raccoon populations is consistent with their close phylogenetic relationship (Nadin-Davis et al., 2002), although there is little anecdotal evidence to suggest the presence of the SCSK virus in Florida in the 1940s; alternatively, the raccoon variant may have emerged after spillover from a bat variant.

The SCSK variant circulates across many central U.S. states, from Texas in the south as far north as South Dakota. A comparison of the genetic diversity of this variant with that of the NCSK variant, with which its range overlaps in the states of Kansas, Nebraska, and South Dakota, has revealed some interesting differences in their molecular epidemiology and evolution (Barton, Gregory, Davis, Hanlon & Wisely, 2010). Some of the viruses from other skunk specimens from Mexico (Velasco-Villa et al., 2005) also fall into the American indigenous lineage but lie on branches distinct from those of the SCSK variant, suggesting that rabies virus has been introduced into this host on multiple occasions. Indeed, recent events in Flagstaff, Arizona, where a big brown bat variant is being maintained in skunks, demonstrate the opportunity for emergence of new virus-host associations in this species (Leslie et al., 2006).

The existence of enzootic rabies in bats of the Americas was recognized after large outbreaks of rabies in cattle in Trinidad and Latin America early in the 20th century were traced back to vampire bat exposures (see Baer, 1991). Indeed, the introduction of large-scale cattle ranching activities over extensive areas of South America has facilitated large increases in populations of the common vampire bat (*Desmodus rotundus*) and increased spread of the viral variant it harbors. Currently, vampire bat rabies is distributed throughout much of the vampire bat's range in Latin America and the Caribbean, although epidemiological studies of this variant have relied to a large extent on the isolates recovered from exposed domestic livestock. The CDC MAb panel identifies three antigenic variants, AgV3, AgV5 and AgV11, that are associated with vampire bats; all isolates of these variants form a monophyletic clade within the American indigenous lineage (Carnieli, Castilho, Fahl, Véras & Timenetsky, 2009b; Cisterna et al., 2005; De Mattos et al., 1996; Favi et al., 2003; Macedo et al., 2010; Nadin-Davis & Loza-Rubio, 2006; Ito et al., 2001; Velasco-Villa et al., 2002; Wright, Rampersad, Ryan & Ammons, 2002). Despite limited genetic diversity within this clade, distinct localized viral variants exist possibly as a consequence of the impact of topographical features such as mountain ranges that restrict movement of the bat host (Kobayashi et al., 2006, 2008). Viruses of AgV3, the variant typically found in common vampire bats of Brazil, have also been detected in hairy-legged vampire bats in this country possibly as a result of interspecific transmission from *Desmodus rotundus* (Castilho et al., 2010b). A small group of rabies viruses isolated from frugivorous bats (*Artibeus* spp.) in Brazil are closely related to the vampire bat variant. Currently, it is unclear whether these frugivorous bat viruses represent an independent transmission cycle (Shoji et al., 2004). Several cases of human rabies have been recorded following vampire bat attacks. Investigations of human disease outbreaks in Amazonian regions of Peru (Warner et al., 1999b), Brazil (da Rosa et al., 2006; Barbosa et al., 2008), and Ecuador (Castilho et al., 2010a) have demonstrated the serious public health consequences of vampire bat exposure.

The association of rabies viruses with insectivorous bats of North America was first recorded in the 1950s (reviewed by Kuzmin and Rupprecht, 2007); subsequent surveillance and viral typing of isolates from across the continent have identified many distinct variants that are associated with particular bat species. The species most frequently recorded as rabies positive include the big brown bat (*Eptesicus fuscus*), which harbors several distinct variants (Nadin-Davis et al., 2001; Nadin-Davis, Feng, Mousse, Wandeler & Aris-Brosou, 2010a; Neubaum et al., 2008; Shankar et al., 2005); members of the *Lasiurus* genus, including hoary (*L. cinereus*), red (*L. borealis*), and northern yellow (*L. intermedius*) bats, each of which harbor distinct but closely related variants; the silver-haired bat (*Lasionycteris noctivagans*), which together with the tri-colored bat (*Perimyotis subflavus*), formerly known as the eastern pipistrelle, harbors a variant frequently implicated in human rabies cases (Messenger, Smith, Orciari, Yager & Rupprecht, 2003); members of the genus *Myotis*; and, in the USA and Mexico, the Brazilian free-tailed bat (*Tadarida brasiliensis*) (Nadin-Davis et al., 2001; Shankar et al., 2005; Smith et al., 1995; Streicker et al., 2010; Velasco-Villa et al., 2002, 2006; Rohde, Mayes, Smith & Neill, 2004). Additional viral variants that are associated with more rarely reported species, for example, the western pipistrelle (*Pipistrellus hesperus*) in the USA (Franka et al., 2006), have been documented. However, the reservoirs for other variants recovered variously from bats, wild and domestic animals, and humans are yet to be determined (Aréchiga-Ceballos et al., 2010; Nadin-Davis & Loza-Rubio, 2006; Streicker et al., 2010; Velasco-Villa et al., 2008a). Although control of bat rabies through human-mediated efforts remains problematic due to the aerial lifestyle and feeding habits of these hosts, the decline in populations of certain bat species, attributable to the recent white-nose syndrome outbreak (Frick et al., 2010), may have a significant impact on bat rabies epidemiology in eastern North America.

As information on the viruses associated with insectivorous bats of North America grew, increased surveillance of such bats in South America identified several rabies reservoir species. In the most southern parts of the continent, in Chile and Argentina, variants associated with small numbers of specimens of the *Myotis, Histiotus,* and *Lasiurus* genera have been recovered, but the Brazilian free-tailed bat, *Tadarida brasiliensis*, is the principal bat rabies reservoir (De Mattos, Favi, Yung, Pavletic & De Mattos, 2000; Cisterna et al., 2005; Yung et al., 2002). Interestingly, distinct viral variants are associated with Brazilian free-tailed bats in the southern and northern hemispheres (Oliveira et al., 2010; Velasco-Villa et al., 2006).

In Brazil, viral variants associated with *Eptesicus furinalis, Molossus molossus,* and species of the genera *Histiotus, Nyctinomops, Myotis,* and *Lasiurus* have been identified (Albas et al., 2011; Bernardi et al., 2005; Carnieli et al., 2010; Kobayashi et al., 2005, 2007; Oliveira et al., 2010; Queiroz et al., 2012). In addition, single reports of rabies in other species have appeared (Castilho et al., 2008; da Rosa et al., 2011). To date, rabies has been diagnosed in 41 bat species in this country (Sodré, da Gama & de Almeida, 2010); however, the role of many of these species as rabies reservoirs requires further confirmation because interspecific transmission (see Kobayashi et al., 2007) or difficulties in bat speciation can confound the identification of true reservoir hosts. Interestingly, G gene sequencing of some bat-associated rabies viruses from Brazil identified a number of amino acid substitutions at position 333 (Sato et al., 2009) other than the normal Arg or Lys residues important for the virulent phenotype of wild-type viruses (Tuffereau et al., 1989). This raises the question of whether these rabies viruses are less pathogenic than other wild-type viruses or if amino acid substitutions in other parts of the glycoprotein compensate for these mutations.

The presence of a viral variant typical of those associated with lasiurine bats has been reported in Paraguay (Sheeler-Gordon & Smith, 2001). In Colombia, rabies specimens recovered from *Eptesicus* and *Molossus* species may correspond to the variants found in these hosts further south (Páez, Núñez, Garcia & Bóshell, 2003), but this requires direct comparison.

Attempts have been made to reconstruct the evolutionary history of all the viruses associated with American bats. Hughes, Orciari & Rupprecht (2005) estimated dates of TMRCA of all bat viral variants using Bayesian (1660 AD, 95% HPD range 1267–1782) and ML (1651 AD, 95% HPD range 1254–1773) methods. This study suggested that variants associated with the principal chiropteran hosts of Latin America emerged first, followed by the lineages associated with the most common North American hosts. Nadin-Davis et al., (2010a) used Bayesian methods to estimate TMRCA of the viruses associated with the North American big brown bat host to 1573 AD, 95% HPD range 1338–1763. The estimates generated by these two studies appear to be somewhat discordant, but their relatively large ranges clearly overlap. It should also be noted that there is evidence that the big brown bat clade is not monophyletic but paraphyletic and that some of these variants are more similar to viruses associated with solitary species, such as the lasiurine and silver-haired bats.

4 FUTURE TRENDS

Although development of novel MAbs can facilitate improvements in lyssavirus antigenic typing methods (Nadin-Davis, Elmgren, Sheen, Sabeta & Wandeler, 2011), genetic typing methods are likely to be favored in the future because of ongoing advances in this technology. As nucleotide sequencing methods rapidly evolve, determination of complete genes and even complete genomes (see Delmas et al., 2008; NCBI, n.d.) is becoming routine. Complete genome sequencing is now regarded as a prerequisite for confirmation of new species (Markotter et al., 2008) and has been invaluable for defining small differences between closely related vaccine strains (Geue et al., 2008). Such complete information on the genome of many viral isolates will help to elucidate viral features that facilitate adaptation to specific hosts. Moreover, the rapid development of next generation sequencing technology (Wright et al., 2011) promises to further enhance our ability to explore the genetic heterogeneity of the virus at an individual case level and may provide further insights into the evolutionary factors at play in this genus. With the availability of more complex sequence databases, their analysis will increasingly rely on more sophisticated computer-based tools, including BEAST, MigraPhyla (see Meng et al., 2011) and other spatial analysis programs (Carnieli, Oliveira, Macedo & Castilho, 2011), to help us understand the phylodynamics of viral disease outbreaks (Holmes & Grenfell, 2009).

In contrast, when all that is required in support of rabies control programs is the viral type of a specimen, accurate, rapid, and cost effective methods of genetic examination such as pyrosequencing (De Benedictis et al., 2011) and reverse transcription loop-mediated isothermal amplification assay (RT-LAMP) (Saitou et al., 2010) may offer increasingly useful alternative tools. A resequencing microarray technique that uses a broadly reactive set of probes for lyssavirus detection (Dacheux et al., 2010) may enable the identification of diverse lyssaviruses and even new species. As knowledge of the genetic and antigenic diversity of the members of this genus is expanded, concerns regarding the efficacy of current biologicals to protect humans and animals against infection by some lyssaviruses have increased (Hanlon et al., 2005), and this may spur efforts to develop more broadly effective vaccines (Nel, 2005).

Increasingly, particularly in support of lyssavirus studies in bats, genetic profiling of the infected host, often by sequencing of mitochondrial loci, is being used to confirm the species involved (Harris et al., 2008; Streicker et al., 2010). Such an approach removes the uncertainties often inherent in the use of morphological keys for species identification (Nadin-Davis, Guerrero, Knowles & Feng, 2012b) and helps to confirm newly identified wildlife reservoir hosts (Carnieli et al., 2008). In addition, exploration of host sub-population structure, as determined through examination of microsatellites or variable mitochondrial loci, can help explain viral variant range and patterns of disease spread (Nadin-Davis et al., 2010a; Neubaum et al., 2008; Smith et al., 2011). To better understand the dispersal of the raccoon rabies virus, the population structure of the raccoon host over many parts of North America has been examined using microsatellite and mitochondrial loci. Such analyses have suggested that in the host's northern range, rivers impede host movement and disease spread, while further south other topographical features such as valleys and ridges appeared to have minimal effect (Cullingham et al., 2008: Cullingham, Kyle, Pond, Rees & White, 2009; Root et al., 2009). Analysis of skunk microsatellite loci showed limited host population sub-structure in the USA, where both NCSK and SCSK variants circulate, thus strongly implicating differences in the biology of these two viruses as being responsible for the different disease patterns they produce (Barton et al., 2010).

5 CONCLUSIONS

At a practical level, molecular epidemiology of lyssaviruses provides important information on the source of human rabies infections, especially in countries where the disease is rare and may be acquired either from indigenous sources (Messenger et al., 2003; Johnson et al., 2010b) or as a result of travel to countries with endemic rabies (Malerczyk, DeTora & Gniel, 2011). On occasion, such cases have illustrated the long incubation periods sometimes observed in humans (Johnson, Fooks & McColl, 2008) and identified new foci of disease (van Thiel et al., 2009). Molecular epidemiology has also been critical in the identification of sources of animal disease incursion into countries previously considered rabies free, as in a recent outbreak in Italy (De Benedictis et al., 2009a), or to confirm the spread of a variant into a new area, for example, incursion of RRV into Canada (Szanto et al., 2011). Such information helps support appropriate rabies control efforts. Despite success in rabies elimination in western Europe (Brochier, Boulanger, Costy & Pastoret, 1994) and major reductions in rabies in parts of the Americas (Belotto, Leanes, Schneider, Tamayo & Correa, 2005; Rosatte et al., 2007), ongoing vigilance is needed to guard against re-introduction of the disease through illegal movement of animals (Johnson, Freuling, Horton, Müller & Fooks, 2011); unsustainable domestic animal vaccination programs, especially in the developing world; and the emergence of novel viral variants that result in new foci of disease in wildlife (Leslie et al., 2006).

On a more fundamental level, molecular phylogenetic investigation has in recent years significantly extended the known diversity of the *Lyssavirus* genus and highlighted the role of the *Chiroptera* as their principal hosts (Badrane & Tordo, 2001). However, the emergence of one lyssavirus species, rabies virus, probably by a chance spillover event from a bat, has had by far the most impact with respect to human health due to its association in most parts of the world with the dog, an animal adopted by humans for its value as a social companion, protector, and hunter. As modern human populations have spread around the world, their accompanying companion animals have frequently been responsible for spreading rabies to many parts of the globe (Bourhy et al., 2008; Gong et al., 2010; Meng et al., 2011; Nadin-Davis & Bingham, 2004; Talbi et al., 2010). The ability of dog-associated rabies viruses to occasionally cross the species barrier and persist in other members of the *Carnivora* order has allowed the emergence of new enzootics in many countries, thereby complicating enormously the prospects for disease eradication.

Although spillover infections are not an especially rare occurrence, only a small fraction of these events result in new virus–host associations, a phenomenon studied in some detail in American bats (Streicker et al., 2010). That study demonstrated that close physical contact between bat species influenced the frequency of interspecific spillover events, while the extent of genetic similarity between the donor and recipient species significantly impacted the likelihood that the new virus–host association could be permanently maintained. A corollary of this observation may be that natural transmission of bat associated rabies-like viruses such as the EBLVs to non-flying terrestrial hosts is very uncommon (see Section 4.3.2.3) and has limited potential to lead to persistent transmission within canid hosts (Picard-Meyer et al., 2008; Vos et al., 2004). However, the possibility that such rare events have occurred, and may occur again, cannot be ignored.

Indeed, it has been estimated that all current rabies virus lineages have evolved within the last 1,500 years (Bourhy et al., 2008), a value that is at odds with historical records suggesting that humans of ancient times were aware of a rabies-like disease in dogs (Baer, 2007). One explanation of these observations is that over the last several millennia dog-associated rabies has emerged on several occasions with new viral species replacing older ones over time (Nel & Markotter, 2007). Indeed, recent studies of the Arctic/AL lineage (Nadin-Davis et al., 2012a) provide evidence for such a phenomenon. Accordingly, ongoing rabies surveillance coupled with viral typing will be essential to the timely identification of the emergence of new lyssavirus variants and species.

ACKNOWLEDGEMENTS

I thank D. Ogunremi and M. Lin for critical review and suggestions for improvement to this work.

References

Albas, A., Campos, A. C., Araujo, D. B., Rodrigues, C. S., Sodré, M. M., Durigon, E. L., et al. (2011). Molecular characterization of rabies virus isolated from non-haematophagous bats in Brazil. *Revista da Sociedade de Medicina Tropical, 44*, 678–683.

Amengual, B., Whitby, J. E., King, A., Serra-Cobo, J., & Bourhy, H. (1997). Evolution of European bat lyssaviruses. *Journal of General Virology, 78*, 2319–2328.

Amengual, B., Bourhy, H., López-Roig, M., & Serra-Cobo, J. (2007). Temporal dynamics of European bat lyssavirus type 1 and survival of Myotis myotis bats in natural colonies. *PLoS One, 2*(6), e566.

Amouri, I. K., Kharmachi, H., Djebbi, A., Saadi, M., Hogga, N., Zakour, L. B., et al. (2011). Molecular characterization of rabies virus isolated from dogs in Tunisia: Evidence of two phylogenetic variants. *Virus Research, 158*, 246–250.

Arai, Y. T., Takahashi, H., Kameoka, Y., Shiino, T., Wimalaratne, O., & Lodmell, D. L. (2001). Characterization of Sri Lanka rabies virus isolates using nucleotide sequence analysis of nucleoprotein gene. *Acta Virologica, 45*, 321–333.

AravindhBabu, R. P., Manoharan, S., Ramadass, P., & Chandran, N. D. J. (2011). Rabies in South Indian cows: An evidence of Sri Lankan rabies virus variant infection based on the analysis of partial nucleoprotein gene. *Indian Journal of Virology, 22*, 138–141.

Aréchiga, N., Vázquez-Morón, S., Berciano, J., Nicolás, O., Aznar, C., Juste, J., et al. (2012). Novel lyssavirus from a *Minioterus schreibersii* bat in Spain. 23rd*Rabies in the Americas meeting held in São Paulo*, Brazil, October 14–19.

Aréchiga-Ceballos, N., Velasco-Villa, A., Shi, M., Flores-Chávez, S., Barrón, B., Cuevas-Domínguez, E., et al. (2010). New rabies virus variant found during an epizootic in white-nosed coatis from the Yucatán peninsula. *Epidemiology and Infection, 138*, 1586–1589.

Arguin, P. M., Murray-Lillibridge, K., Miranda, M. E. G., Smith, J. S., Calaor, A. B., & Rupprecht, C. E. (2002). Serologic evidence of lyssavirus infections among bats, the Philippines. *Emerging Infectious Diseases, 8*, 258–262.

Badrane, H., & Tordo, N. (2001). Host switching in *Lyssavirus* history from the Chiroptera to the Carnivora orders. *Journal of Virology, 75*, 8096–8104.

Badrane, H., Bahloul, C., Perrin, P., & Tordo, N. (2001). Evidence of two *Lyssavirus* phylogroups with distinct pathogenicity and immunogenicity. *Journal of Virology, 75*, 3268–3276.

Baer, G. M. (1991). Vampire bat and bovine paralytic rabies. In G. M. Baer (Ed.), *The natural history of rabies (pp. 390–406)* (2nd ed.). Boca Raton, USA: CRC Press.

Baer, G. M. (2007). The history of rabies. In A. C. Jackson & W. H. Wunner (Eds.), *Rabies (pp. 1–19)* (2nd ed.). Oxford, UK: Elsevier.

Banyard, A. C., Johnson, N., Voller, K., Hicks, D., Nunez, A., Hartley, M., et al. (2009). Repeated detection of European bat lyssavirus type 2 in dead bats found at a single roost site in the UK. *Archives of Virology, 154*, 1847–1850.

Barbosa, T. F. S., Medeiros, D. B. A., da Rosa, E. S. T., Casseb, L. M. N., Medeiros, R., Pereira, A. S., et al. (2008). Molecular epidemiology of rabies virus isolated from different sources during a bat-transmitted human outbreak occurring in Augusto Correa municipality, Brazilian Amazon. *Virology, 370*, 228–236.

Barton, H. D., Gregory, A. J., Davis, R., Hanlon, C. A., & Wisely, S. M. (2010). Contrasting landscape epidemiology of two sympatric rabies virus strains. *Molecular Ecology, 19*, 2725–2738.

Bauldauf, S. L. (2003). Phylogeny for the faint of heart: A tutorial. *Trends in Genetics, 19*, 345–351.

Belotto, A., Leanes, L. F., Schneider, M. C., Tamayo, H., & Correa, E. (2005). Overview of rabies in the Americas. *Virus Research, 111*, 5–12.

Benmansour, A., Brahimi, M., Tuffereau, C., Coulon, P., Lafay, F., & Flamand, A. (1992). Rapid sequence evolution of street rabies glycoprotein is related to the highly heterogeneous nature of the viral population. *Virology, 187*, 33–45.

Bernardi, F., Nadin-Davis, S. A., Wandeler, A. I., Armstrong, J., Gomes, A. A. B., Lima, F. S., et al. (2005). Antigenic and genetic characterization of rabies viruses isolated from domestic and wild animals of Brazil identifies the hoary fox as a rabies reservoir. *Journal of General Virology, 86*, 3153–3162.

Biek, R., Henderson, J. C., Waller, L. A., Rupprecht, C. E., & Real, L. A. (2007). A high-resolution genetic signature of demographic and spatial expansion in epizootic rabies virus. *Proceedings of the National Academy of Sciences USA, 104*, 7993–7998.

Black, E. M., McElhinney, L. M., Lowings, J. P., Smith, J., Johnstone, P., & Heaton, P. R. (2000). Molecular methods to distinguish between classical rabies and the rabies-related European bat lyssaviruses. *Journal of Virological Methods, 87*, 123–131.

Boldbaatar, B., Inoue, S., Tuya, N., Dulam, P., Batchuluun, D., Sugiura, N., et al. (2010). Molecular epidemiology of rabies virus in Mongolia, 2005–2008. *Japanese Journal of Infectious Diseases, 63*, 358–363.

Bourhy, H., Kissi, B., Lafon, M., Sacramento, D., & Tordo, N. (1992). Antigenic and molecular characterization of bat rabies virus in Europe. *Journal of Clinical Microbiology, 30*, 2419–2426.

Bourhy, H., Kissi, B., & Tordo, N. (1993). Molecular diversity of the *Lyssvirus* genus. *Virology, 194*, 70–81.

Bourhy, H., Kissi, B., Audry, L., Smreczak, M., Sadkowska-Todys, M., Kulonen, K., et al. (1999). Ecology and evolution of rabies in Europe. *Journal of General Virology, 80*, 2545–2557.

Bourhy, H., Reynes, J. -M., Dunham, E. J., Dacheux, L., Larrous, F., Huong, V. T. Q., et al. (2008). The origin and phylogeography of dog rabies virus. *Journal of General Virology, 89*, 2673–2681.

Brochier, B., Boulanger, D., Costy, F., & Pastoret, P. -P. (1994). Towards rabies elimination in Belgium by fox vaccination using a vaccinia-rabies glycoprotein recombinant virus. *Vaccine, 12*, 1368–1371.

Carnieli, P., Jr., Fahl, W. O., Castilho, J. G., Oliveira, R. N., Macedo, C. I., Durymanova, E., et al. (2008). Characterization of rabies virus isolated from canids and identification of the main wild canid host in northeastern Brazil. *Virus Research, 131*, 33–46.

Carnieli, P., Jr., Castilho, J. G., Fahl, W. O., Véras, N. M. C., Carrieri, M. L., & Kotait, I. (2009a). Molecular characterization of rabies virus isolates from dogs and crab-eating foxes in northeastern Brazil. *Virus Research, 141*, 81–89.

Carnieli, P., Jr., Castilho, J. G., Fahl, W. O., Véras, N. M. C., & Timenetsky, M. C. S. T. (2009b). Genetic characterization of rabies virus isolates from cattle between 1997 and 2002 in an epizootic area in the state of São Paulo, Brazil. *Virus Research, 144,* 215–224.

Carnieli, P., Jr., Fahl, W. O., Brandão, P. E., Oliveira, R. N., Macedo, C. I., Durymanova, E., et al. (2010). Comparative analysis of rabies virus isolates from Brazilian canids and bats based on the G gene and G-L intergenic region. *Archives of Virology, 155,* 941–948.

Carnieli, P., Jr., Oliveira, R. N., Macedo, C. I., & Castilho, J. G. (2011). Phylogeography of rabies virus isolated from dogs in Brazil between 1985 and 2006. *Archives of Virology, 156,* 1007–1012.

Castilho, J. G., Canello, F. M., Scheffer, K. C., Achkar, S. M., Carrieri, M. L., & Kotait, I. (2008). Antigenic and genetic characterization of the first rabies virus isolated from the bat *Eumops perotis* in Brazil. *Revista do Instituto de Medicina Tropical de São Paulo, 50,* 95–99.

Castilho, J. G., Carnieli, P., Jr., Durymanova, E. A., de Oliveira Fahl, W., de Novaes Oliveira, R., Macedo, C. I., et al. (2010a). Human rabies transmitted by vampire bats: Antigenic and genetic characterization of rabies virus isolates from the Amazon region (Brazil and Ecuador). *Virus Research, 153,* 100–105.

Castilho, J. G., Carnieli, P., Jr., Oliveira, R. N., Fahl, W. O., Cavalcante, R., Santana, A. A., et al. (2010b). A comparative study of rabies virus isolates from hematophagous bats in Brazil. *Journal of Wildlife Diseases, 46,* 1335–1339.

Chare, E. R., Gould, E. A., & Holmes, E. C. (2003). Phylogenetic analysis reveals a low rate of homologous recombination in negative-sense RNA viruses. *Journal of General Virology, 84,* 2691–2703.

Cisterna, D., Bonaventura, R., Caillou, S., Pozo, O., Andreau, M. L., Fontana, L. D., et al. (2005). Antigenic and molecular characterization of rabies virus in Argentina. *Virus Research, 109,* 139–147.

Clark, K. A., Neill, S. U., Smith, J. S., Wilson, P. J., Whadford, V. W., & Mckirahan, G. W. (1994). Epizootic canine rabies transmitted by coyotes in south Texas. *Journal of the American Veterinary Medical Association, 204,* 536–540.

Coetzee, P., & Nel, L. H. (2007). Emerging epidemic dog rabies in coastal South Africa: A molecular epidemiological analysis. *Virus Research, 126,* 186–195.

Crawford-Miksza, L. K., Wadford, D. A., & Schnurr, D. P. (1999). Molecular epidemiology of enzootic rabies in California. *Journal of Clinical Microbiology, 14,* 207–219.

Cullingham, C. I., Pond, B. A., Kyle, C. J., Rees, E. E., Rosatte, R. C., & White, B. N. (2008). Combining direct and indirect genetic methods to estimate dispersal for informing wildlife disease management decisions. *Molecular Ecology, 17,* 4874–4886.

Cullingham, C. I., Kyle, C. J., Pond, B. A., Rees, E. E., & White, B. N. (2009). Differential permeability of rivers to raccoon gene flow corresponds to rabies incidence in Ontario, Canada. *Molecular Ecology, 18,* 43–53.

Da Rosa, E. S. T., Kotait, I., Barbosa, T. F. S., Carrieri, M. L., Brandão, P. E., Pinheiro, A. S., et al. (2006). Bat-transmitted human rabies outbreaks, Brazilian Amazon. *Emerging Infectious Diseases, 12,* 1197–1202.

Da Rosa, A. R., Kataoka, A. P. A. G., Favoretto, S. R., Sodré, M. M., Netto, J. T., Campos, A. C. A., et al. (2011). First report of rabies infection in bats, *Molossus molossus, Molossops neglectus* and *Myotis riparius* in the city of São Paulo, state of São Paulo, southeastern Brazil. *Revista de Sociedade Brasileira de Medicina Tropical, 44,* 146–149.

Dacheux, L., Larrous, F., Mailles, A., Boisseleau, D., Delmas, O., Biron, C., et al. (2009). European bat lyssavirus transmission among cats, Europe. *Emerging Infectious Diseases, 15,* 280–284.

Dacheux, L., Berthet, N., Dissard, G., Holmes, E. C., Delmas, O., Larrous, F., et al. (2010). Application of broad-spectrum resequencing microarray for genotyping rhabdoviruses. *Journal of Virology, 84,* 9557–9574.

David, D., Yakobson, B., Smith, J. S., & Stram, Y. (2000). Molecular epidemiology of rabies virus isolates from Israel and other middle- and near-Eastern Countries. *Journal of Clinical Microbiology, 38,* 755–762.

David, D., Hughes, G. J., Yakobson, B. A., Davidson, I., Un, H., Aylan, O., et al. (2007). Identification of novel canine rabies virus clades in the Middle East and North Africa. *Journal of General Virology, 88,* 967–980.

David, D., Dveres, N., Yakobson, B. A., & Davidson, I. (2009). Emergence of dog rabies in the Northern region of Israel. *Epidemiology and Infection, 137,* 544–548.

Davis, P. L., Holmes, E. C., Larrous, F., Van der Poel, W. H. M., Tjørnehøj, K., Alonso, W. J., et al. (2005). Phylogeography, population dynamics, and molecular evolution of European bat lyssaviruses. *Journal of Virology, 79,* 10487–10497.

Davis, P. L., Rambaut, A., Bourhy, H., & Holmes, E. C. (2007). The evolutionary dynamics of canid and mongoose rabies virus in southern Africa. *Archives of Virology, 152,* 1251–1258.

De Benedictis, P., Capua, I., Mutinelli, F., Wernig, J. M., Arič, T., & Hostnik, P. (2009a). Update on fox rabies in Italy and Slovenia. *WHO Rabies Bulletin Europe, 33*(1), 5–7.

De Benedictis, P., Sow, A., Fusaro, A., Veggiato, C., Talbi, C., Kaboré, A., et al. (2009b). Phylogenetic analysis of rabies virus from Burkina Faso, 2007. *Zoonoses and Public Health, 57*(7-8), e42–46.

De Benedictis, P., De Battisti, C., Dacheux, L., Marciano, S., Ormelli, S., Salomoni, A., et al. (2011). Lyssavirus detection and typing using pyrosequencing. *Journal of Clinical Microbiology, 49,* 1932–1938.

De Mattos, C. A., De Mattos, C. C., Smith, J. S., Miller, E. T., Papo, S., Utrera, A., et al. (1996). Genetic characterization of rabies field isolates from Venezuela. *Journal of Clinical Microbiology, 34,* 1553–1558.

De Mattos, C. C., De Mattos, C. A., Loza-Rubio, E., Aguilar-Setién, A., Orciari, L. A., & Smith, J. S. (1999). Molecular characterization of rabies virus isolates from Mexico: Implications for transmission dynamics and human risk. *American Journal of Tropical Medicine and Hygiene, 61,* 587–597.

De Mattos, C. A., Favi, M., Yung, V., Pavletic, C., & De Mattos, C. C. (2000). Bat rabies in urban centers in Chile. *Journal of Wildlife Diseases, 36,* 231–240.

Delmas, O., Holmes, E. C., Talbi, C., Larrous, F., Dacheux, L., Bouchier, C., et al. (2008). Genomic diversity and evolution of the lyssaviruses. *PLoS ONE, 3*(4), e2057.

Delpietro, H. A., Gury-Dhomen, F., Larghi, O. P., Mena-Segura, C., & Abramo, L. (1997). Monoclonal antibody characterization of rabies virus strains isolated in the river plate basin. *Journal of Veterinary Medicine, 44,* 477–483.

Diaz, A-M., Papo, S., Rodriguez, A., & Smith, J. S. (1994). Antigenic analysis of rabies-virus isolates from Latin America and the Caribbean. *Journal of Veterinary Medicine, 41,* 153–160.

Drummond, A. J., & Rambaut, A. (2007). BEAST: Bayesian evolutionary analysis by sampling trees. *BMC Evolutionary Biology, 7,* 214. Freely available from http://beast.bio.ed.ac.uk/Main_Page.

Dzikwi, A. A., Kuzmin, I. V., Umoh, J. U., Kwaga, J. K. P., Ahmad, A. A., & Rupprecht, C. E. (2010). Evidence of Lagos bat virus circulation among Nigerian fruit bats. *Journal of Wildlife Diseases, 46,* 267–271.

Favi, M., Nina, A., Yung, V., & Fernández, J. (2003). Characterization of rabies virus isolates in Bolivia. *Virus Research, 97,* 135–140.

Favoretto, S. R., de Mattos, C. C., Morais, N. B., Alves Araújo, F. A., & de Mattos, C. A. (2001). Rabies in marmosets (Callithrix jacchus), Ceará, Brazil. *Emerging and Infectious Diseases, 7,* 1062–1065.

Favoretto, S. R., de Mattos, C. C., de Morais, N. B., Carrieri, M. L., Rolim, B. N., Silva, L. M., et al. (2006). Rabies virus maintained by dogs in humans and terrestrial wildlife, Ceará state, Brazil. *Emerging Infectious Diseases, 12,* 1978–1981.

Fehlner-Gardiner, C., Nadin-Davis, S., Armstrong, J., Muldoon, F., Bachmann, P., & Wandeler, A. (2008). ERA vaccine-derived cases of rabies in wildlife and domestic animals in Ontario, Canada, 1989–2004. *Journal of Wildlife Diseases, 44,* 71–85.

Felsenstein, J. (1993). *PHYLIP: Phylogeny inference package (Version 3.52c).* Seattle, Washington: University of Washington. (Freely available from http://evolution.

gs.washington.edu/phylip.html and with links to other software packages at http://evolution.genetics.washington.edu/phylip/software.html).

Fooks, A. R, Brookes, S. M., Johnson, N., McElhinney, L. M., & Hutson, A. M. (2003a). European bat lyssaviruses: An emerging zoonosis. *Epidemiology and Infection, 131,* 1029–1039.

Fooks, A. R., McElhinney, L. M., Pounder, D. J., Finnegan, C. J., Mansfield, K., Johnson, N., et al. (2003b). Case report: isolation of a European bat lyssavirus type 2a from a fatal human case of rabies encephalitis. *Journal of Medical Virology, 71,* 281–289.

Franka, R., Constantine, D. G., Kuzmin, I., Velasco Villa, A., Reeder, S. A., Streicker, D., et al. (2006). A new phylogenetic lineage of rabies virus associated with western pipistrelle bats (*Pipistrellus Hesperus*). *Journal of General Virology, 87,* 2309–2321.

Fraser, G. C., Hooper, P. T., Lunt, R. A., Gould, A. R., Gleeson, L. J., Hyatt, A. D., et al. (1996). Encephalitis caused by a lyssavirus in fruit bats in Australia. *Emerging Infectious Diseases, 2,* 327–331.

Freuling, C., Grossmann, E., Conraths, F. J., Schameitat, A., Kliemt, J., Auer, E., et al. (2008). First isolation of EBLV-2 in Germany. *Veterinary Microbiology, 131,* 26–34.

Freuling, C. M., Beer, M., Conraths, F. J., Finke, S., Hoffmann, B., Keller, B., et al. (2011). Novel lyssavirus in Natterer's bat, Germany. *Emerging Infectious Diseases, 17,* 1519–1522.

Frick, W. F., Pollock, J. F., Hicks, A. C., Langwig, K. E., Reynolds, D. S., Turner, G. G., et al. (2010). An emerging disease causes regional population collapse of a common North American bat species. *Science, 329,* 679–682.

Geue, L., Schares, S., Schnick, C., Kliemt, J., Beckert, A., Freuling, C., et al. (2008). Genetic characterisation of attenuated SAD rabies virus strains used for oral vaccination of wildlife. *Vaccine, 26,* 3227–3235.

Gong, W., Jiang, Y., Za, Y., Zeng, Z., Shao, M., Fan, J., et al. (2010). Temporal and spatial dynamics of rabies viruses in China and southeast Asia. *Virus Research, 150,* 111–118.

Gould, A. R., Hyatt, A. D., Lunt, R., Kattenbelt, J. A., Hengstberger, S., & Blacksell, S. D. (1998). Characterisation of a novel lyssavirus isolated from *Pteropid* bats in Australia. *Virus Research, 54,* 165–187.

Gould, A. R., Kattenbelt, J. A., Gumley, S. G., & Lunt, R. A. (2002). Characterisation of an Australian bat lyssavirus variant isolated from an insectivorous bat. *Virus Research, 89,* 1–28.

Graur, D., & Li, W. -H. (2000). Molecular phylogenetics: *Fundamentals of molecular evolution.* Sunderland, MA: Sinauer Associates, Inc.. pp. 165–247.

Guindon, S., Dufayard, J. F., Lefort, V., Anisimova, M., Hordijk, W., & Gascuel, O. (2010). New Algorithms and methods to estimate maximum-likelihood phylogenies: Assessing the performance of PhyML 3.0. *Systematic Biology, 59,* 307–321. Available from http://www.atgc-montpellier.fr/phyml/binaries.php.

Guyatt, K. J., Twin, J., Davis, P., Holmes, E. C., Smith, G. A., Smith, I. L., et al. (2003). A molecular epidemiological study of Australian bat lyssavirus. *Journal of General Virology, 84,* 485–496.

Hanlon, C. A., Kuzmin, I. V., Blanton, J. D., Weldon, W. C., Manangan, J. S., & Rupprecht, C. E. (2005). Efficacy of rabies biologics against new lyssaviruses from Eurasia. *Virus Research, 111,* 44–54.

Hanlon, C. A., Niezgoda, M., & Rupprecht, C. E. (2007). Rabies in terrestrial animals. In A. C. Jackson & W. H. Wunner (Eds.), *Rabies (pp. 201–258)* (2nd ed.). San Diego: Academic Press.

Harris, S. L., Mansfield, K., Marston, D. A., Johnson, N., Pajamo, K., O'Brien, N., et al. (2007). Isolation of European bat lyssavirus type 2 from a Daubenton's bat (*Myotis daubentonii*) in Shropshire. *Veterinary Record, 161,* 384–386.

Harris, S. L., Johnson, N., Brookes, S. M., Hutson, A. M., Fooks, A. R., & Jones, G. (2008). The application of genetic markers for EBLV surveillance in European bat species. *Developments in Biologicals (Basel), 131,* 347–363.

Hatakeyama, K., Sadamasu, K., & Kai, A. (2011). Phylogenetic analysis of rabies viruses isolated from animals in Tokyo in the 1950s. *Kansenshōgaku zasshi. (The Journal of the Japanese Association for Infectious Diseases)*, *85*, 238–243.

Hayman, D. T. S., Fooks, A. R., Horton, D., Suu-Ire, R., Breed, A. C., Cunningham, A. A., et al. (2008). Antibodies against Lagos Bat Virus in megachiroptera from West Africa. *Emerging Infectious Diseases*, *14*, 926–928.

Hayman, D. T. S., Johnson, N., Horton, D. L., Hedge, J., Wakeley, P. R., Banyard, A. C., et al. (2011). Evolutionary history of rabies in Ghana. *PLoS Neglected Tropical Diseases*, *5*(4), e1001.

Hirano, S., Itou, T., Shibuya, H., Kashiwazaki, Y., & Sakai, T. (2010a). Molecular epidemiology of rabies virus isolates in Uganda. *Virus Research*, *147*, 135–138.

Hirano, S., Sato, G., Kobayashi, Y., Itou, T., Luo, T. R., Liu, Q., et al. (2010b). Analysis of Chinese rabies virus isolates from 2003–2007 based on P and M protein genes. *Acta virologica*, *54*, 91–98.

Holmes, E. C., & Grenfell, B. T. (2009). Discovering the phylodynamics of RNA viruses. *PLoS Computational Biology*, *5*(10), e1000505.

Holmes, E. C., Woelk, C. H., Kassis, R., & Bourhy, H. (2002). Genetic constraints and the adaptive evolution of rabies viruses in nature. *Virology*, *292*, 247–257.

Hooper, P. T., Lunt, R. A., Gould, A. R., Samaratunga, H., Hyatt, A. D., Gleeson, L. J., et al. (1997). A new lyssavirus – the first endemic rabies-related virus recognized in Australia. *Bulletin du Institut Pasteur*, *95*, 209–218.

Hughes, G. J. (2008). A reassessment of the emergence time of the European bat lyssavirus type 1. *Infections, Genetics and Evolution*, *8*, 820–824.

Hughes, G. J., Páez, A., Bóshell, J., & Rupprecht, C. E. (2004). A phylogenetic reconstruction of the epidemiological history of canine rabies virus variants in Colombia. *Infection, Genetics and Evolution*, *4*, 45–51.

Hughes, G. J., Orciari, L. A., & Rupprecht, C. E. (2005). Evolutionary timescale of rabies virus adaptation to North American bats inferred from the substitution rate of the nucleoprotein gene. *Journal of General Virology*, *86*, 1467–1474.

Hyun, B. H., Lee, K. K., Kim, I. J., Lee, K. W., Park, H. J., Lee, O. S., et al. (2005). Molecular epidemiology of rabies virus isolates from South Korea. *Virus Research*, *114*, 113–125.

International Committee on Taxonomy of Viruses. (2011). ICTV Files and Discussions. ICTV 2011 Master Species List – Version 1, February 21, 2012. (http://talk.ictvonline.org/files/ictv_documents/m/msl/4090.aspx). Accessed March 7, 2012.

Ito, N., Sugiyama, M., Oraveerakul, K., Piyaviriyakul, P., Lumlertdacha, B., Arai, Y. T., et al. (1999). Molecular epidemiology of rabies in Thailand. *Microbiology and Immunology*, *43*, 551–559.

Ito, M., Arai, Y. T., Itou, T., Sakei, T., Ito, F. H., Takasaki, T., et al. (2001). Genetic characterization and geographic distribution of rabies virus isolates in Brazil: identification of two reservoirs, dog and vampire bats. *Virology*, *284*, 214–222.

Ito, M., Itou, T., Shoji, Y., Sakei, T., Ito, F. H., Arai, Y. T., et al. (2003). Discrimination between dog-related and vampire bat-related rabies viruses in Brazil by strain-specific reverse transcriptase-polymerase chain reaction and restriction fragment length polymorphism analysis. *Journal of Clinical Virology*, *26*, 317–330.

Jayakumar, R., Tirumurugaan, K. G., Ganga, G., Kumanan, K., & Mahalinga Nainar, A. (2004). Characterization of nucleoprotein gene sequence of an Indian isolate of rabies virus. *Acta Virologica*, *48*, 47–50.

Johnson, N., McElhinney, L. M., Smith, J., Lowings, P., & Fooks, A. R. (2002). Phylogenetic comparison of the genus *Lyssavirus* using distal coding sequences of the glycoprotein and nucleoprotein genes. *Archives of Virology*, *147*, 2111–2123.

Johnson, N., Black, C., Smith, J., Un, H., McElhinney, L. M., Aylan, O., et al. (2003a). Rabies emergence among foxes in Turkey. *Journal of Wildlife Diseases*, *39*, 262–270.

Johnson, N., Selden, D., Parsons, G., Healy, D., Brookes, S. M., McElhinney, L. M., et al. (2003b). Isolation of a European bat lyssavirus type 2 from a Daubenton's bat in the United Kingdom. *The Veterinary Record, 152,* 383–387.

Johnson, N., Letshwenyo, M., Baipoledi, E. K., Thobokwe, G., & Fooks, A. R. (2004a). Molecular epidemiology of rabies in Botswana: A comparison between antibody typing and nucleotide sequence phylogeny. *Veterinary Microbiology, 101,* 31–38.

Johnson, N., McElhinney, L. M., Ali, Y. H., Saeed, I. K., & Fooks, A. R. (2004b). Molecular epidemiology of canid rabies in Sudan: evidence for a common origin of rabies with Ethiopia. *Virus Research, 104,* 201–205.

Johnson, N., Un, H., Vos, A., Aylan, O., & Fooks, A. R. (2006). Wildlife rabies in Western Turkey: The spread of rabies through the western provinces of Turkey. *Epidemiology and Infection, 134,* 369–375.

Johnson, N., Fooks, A. R., Valtchovski, R., & Müller, T. (2007a). Evidence for trans-border movement of rabies by wildlife reservoirs between countries in the Balkan Peninsular. *Veterinary Microbiology, 120,* 71–76.

Johnson, N., Freuling, C., Marston, D. A., Tordo, N., Fooks, A. R., & Müller, T. (2007b). Identification of European bat lyssavirus isolates with short genomic insertions. *Virus Research, 128,* 140–143.

Johnson, N., Fooks, A., & McColl, K. (2008). Reexamination of human rabies case with long incubation, Australia. *Emerging Infectious Diseases, 14,* 1950–1951.

Johnson, N., Mansfield, K. L., Marston, D. A., Wilson, C., Goddard, T., Selden, D., et al. (2010a). A new outbreak of rabies in rare Ethiopian wolves (*Canis simensis*). *Archives of Virology, 155,* 1175–1177.

Johnson, N., Vos, A., Freuling, C., Tordo, N., Fooks, A. R., & Müller, T. (2010b). Human rabies due to lyssavirus infection of bat origin. *Veterinary Microbiology, 142,* 151–159.

Johnson, N., Freuling, C., Horton, D., Müller, T., & Fooks, A. R. (2011). Imported rabies, European Union and Switzerland, 2001–2010. *Emerging Infectious Diseases, 17,* 753–754.

Kasempimolporn, S., Saengseesom, W., Tirawatnapong, T., Puempumpanich, S., & Sitprija, V. (2004). Genetic typing of feline rabies virus isolated in greater Bangkok, Thailand. *Microbiology and Immunology, 48,* 307–311.

Kauhala, K., & Kowalczyk, R. (2011). Invasion of the raccoon dog Nyctereutes procyonoides in Europe: history of colonization, features behind its success, and threats to native fauna. *Current Zoology, 57,* 584–598.

Khawplod, P., Shoji, Y., Ubol, S., Mitmoonpitak, C., Wilde, H., Nishizono, A., et al. (2006). Genetic analysis of dog rabies viruses circulating in Bangkok. *Infection, Genetics and Evolution, 6,* 235–240.

Kissi, B., Tordo, N., & Bourhy, H. (1995). Genetic polymorphism in the rabies virus nucleoprotein gene. *Virology, 209,* 526–537.

Kissi, B., Badrane, H., Audry, L., Lavenu, A., Tordo, N., Brahimi, M., et al. (1999). Dynamics of rabies virus quasispecies during serial passages in heterologous hosts. *Journal of General Virology, 80,* 2041–2050.

Kobayashi, Y., Sato, G., Shoji, Y., Sato, T., Itou, T., Cunha, E. M. S., et al. (2005). Molecular epidemiological analysis of bat rabies viruses in Brazil. *Journal of Veterinary Medical Science, 67,* 647–652.

Kobayashi, Y., Ogawa, A., Sato, G., Sato, T., Itou, T., Samara, S. I., et al. (2006). Geographical distribution of vampire bat-related cattle rabies in Brazil. *Journal of the Veterinary Medical Sciences, 68,* 1097–1100.

Kobayashi, Y., Sato, G., Kato, M., Itou, T., Cunha, E. M. S., Silva, M. V., et al. (2007). Genetic diversity of bat rabies virus in Brazil. *Archives of Virology, 152,* 1995–2004.

Kobayashi, Y., Sato, G., Mochizuki, N., Hirano, S., Itou, T., Carvalho, A. A. B., et al. (2008). Molecular and geographic analyses of vampire bat-transmitted cattle rabies in central Brazil. *BMC Veterinary Research, 4,* 44.

Kobayashi, Y., Suzuki, Y., Itou, T., Ito, F. H., Sakai, T., & Gojobori, T. (2011). Evolutionary history of dog rabies in Brazil. *Journal of General Virology, 92*, 85–90.

Kulonen, K., & Boldina, I. (1993). Differentiation of two rabies strains in Estonia with reference to recent Finnish isolates. *Journal of Wildlife Diseases, 29*, 209–213.

Kuzmin, I. V., & Rupprecht, C. E. (2007). Bat rabies. In A. C. Jackson & W. H. Wunner (Eds.), *Rabies (pp. 259–307)* (2nd ed.). London: Academic Press, Elsevier.

Kuzmin, I. V., Orciari, L. A., Arai, Y. T., Smith, J. S., Hanlon, C. A., Kameoka, Y., et al. (2003). Bat lyssaviruses (Aravan and Khujand) from Central Asia: phylogenetic relationships according to N, P and G gene sequences. *Virus Research, 97*, 65–79.

Kuzmin, I. V., Botvinkin, A. D., McElhinney, L. M., Smith, J. S., Orciari, L. A., Hughes, G. J., et al. (2004). Molecular epidemiology of terrestrial rabies in the former Soviet Union. *Journal of Wildlife Diseases, 40*, 617–631.

Kuzmin, I. V., Hughes, G. J., Botvinkin, A. D., Orciari, L. A., & Rupprecht, C. E. (2005). Phylogenetic relationshsips of Irkut and West Caucasian bat viruses within the Lyssavirus genus and suggested quantitative criteria based on the N gene sequence for lyssavirus genotype definition. *Virus Research, 111*, 28–43.

Kuzmin, I. V., Niezgoda, M., Carroll, D. S., Keeler, N., Hossain, M. J., Breiman, R. F., et al. (2006). Lyssavirus surveillance in bats, Bangladesh. *Emerging Infectious Diseases, 12*, 486–488.

Kuzmin, I. V., Hughes, G. J., Botvinkin, A. D., Gribencha, S. G., & Rupprecht, C. E. (2008a). Arctic and Arctic-like rabies viruses: distribution, phylogeny and evolutionary history. *Epidemiology and Infection, 136*, 509–519.

Kuzmin, I. V., Niezgoda, M., Franka, R., Agwanda, B., Markotter, W., Beagley, J. C., et al. (2008b). Possible emergence of West Caucasian Bat Virus in Africa. *Emerging Infectious Diseases, 14*, 1887–1889.

Kuzmin, I. V., Niezgoda, M., Franka, R., Agwanda, B., Markotter, W., Beagley, J. C., et al. (2008c). Lagos bat virus in Kenya. *Journal of Clinical Microbiology, 46*, 1451–1461.

Kuzmin, I. V., Mayer, A. E., Niezgoda, M., Markotter, W., Agwanda, B., Breiman, R. F., et al. (2010). Shimoni bat virus, a new representative of the *Lyssavirus* genus. *Virus Research, 149*, 197–210.

Kuzmin, I. V., Turmelle, A. S., Agwanda, B., Markotter, W., Niezgoda, M., Breiman, R. F., et al. (2011). Commerson's leaf-nosed bat (Hipposideros commersoni) is the likely reservoir of Shimoni Bat Virus. *Vector Borne Zoonotic Diseases, 11*, 1465–1470.

Lauring, A. S., & Andino, R. (2010). Quasispecies theory and the behaviour of RNA viruses. *PLoS Pathogens, 6(7)*, e1001005.

Le Mercier, P., Jacob, Y., & Tordo, N. (1997). The complete Mokola virus genome sequence: structure of the RNA dependent RNA polymerase. *Journal of General Virology, 78*, 1571–1576.

Lembo, T., Haydon, D. T., Velasco-Villa, A., Rupprecht, C. E., Packer, C., Brandão, P. E., et al. (2007). Molecular epidemiology identified only a single rabies virus variant circulating in complex carnivore communities of the Serengeti. *Proceedings of the Royal Society B, 274*, 2123–2130.

Leonova, G. N., Belikov, S. I., Kondratov, I. G., Krylova, N. V., Pavlenko, E. V., Romanova, E. V., et al. (2009). A fatal case of bat lyssavirus infection in Primorye Territory of the Russian Far East. *Rabies Bulletin Europe, 33(4)*, 5–7.

Leslie, M. J., Messenger, S., Rohde, R. E., Smith, J., Cheshier, R., Hanlon, C., et al. (2006). Bat-associated rabies virus in skunks. *Emerging Infectious Diseases, 12*, 1274–1277.

Liu, Q., Xiong, Y., Luo, T. R., Wei, Y. C., Nan, S. J., Liu, F., et al. (2007). Molecular epidemiology of rabies in Guangxi province, south of China. *Journal of Clinical Virology, 39*, 295–303.

Liu, W., Liu, Y., Liu, J., Zhai, J., & Xie, Y. (2011). Evidence for inter- and intra-clade recombination in rabies virus. *Infection, Genetics and Evolution, 11*, 906–912.

Loza-Rubio, E., Aguilar-Setien, A., Bahloul, C., Brochier, B., Pastoret, P-P., & Tordo, N. (1999). Discrimination between epidemiological cycles of rabies in Mexico. *Archives of Medical Research, 30*, 144–149.

Lumlertdacha, B., Boongird, K., Wanghongsa, S., Wacharapluesadee, S., Chanhome, L., Khawplod, P., et al. (2005). Survey for bat lyssaviruses, Thailand. *Emerging and Infectious Diseases, 11,* 232–236.

Macedo, C. I., Carnieli, P., Jr., de Oliveira Fahl, W., Lima, J. Y. O., Oliveira, R. N., Achkar, S. M., et al. (2010). Genetic characterization of rabies virus isolated from bovines and equines between 2007 and 2008, in the states of São Paulo and Minas Gerais. *Revista da Sociedade Brasileira de Medicina Tropical, 43,* 116–120.

Malerczyk, C., DeTora, L., & Gniel., D. (2011). Imported human rabies cases in Europe, the United States, and Japan, 1990 to 2010. *Journal of Travel Medicine, 18,* 402–407.

Mansfield, K. L., McElhinney, L. M., Hübschle, O., Mettler, F., Sabeta, C., Nel, L. H., et al. (2006a). A molecular epidemiological study of rabies epizootics in kudu (*Tragelaphus strepsiceros*) in Namibia. *BMC Veterinary Research, 2,* 2.

Mansfield, K. L., Racloz, V., McElhinney, L. M., Marston, D. A., Johnson, N., Ronsholt, L., et al. (2006b). Molecular epidemiological study of Arctic rabies virus isolates from Greenland and comparison with isolates from throughout the Arctic and Baltic regions. *Virus Research, 116,* 1–10.

Markotter, W., Kuzmin, I., Rupprecht, C. E., Randles, J., Sabeta, C. T., Wandeler, A. I., et al. (2006a). Isolation of Lagos Bat Virus from water mongoose. *Emerging Infectious Diseases, 12,* 1913–1918.

Markotter, W., Randles, J., Rupprecht, C. E., Sabeta, C. T., Taylor, P. J., Wandeler, A. I., et al. (2006b). Lagos bat virus South Africa. *Emerging Infectious Diseases, 12,* 504–506.

Markotter, W., Kuzmin, I., Rupprecht, C. E., & Nel, L. H. (2008). Phylogeny of Lagos Bat Virus: Challenges for lyssavirus taxonomy. *Virus Research, 135,* 10–21.

Marston, D. A., McElhinney, L. M., Johnson, N., Müller, T., Conzelmann, K. K., Tordo, N., et al. (2007). Comparative analysis of the full genome sequence of European bat lyssavirus type 1 and type 2 with other lyssaviruses and evidence for a conserved transcription termination and polyadenylation motif in the G-L 3' non-translated region. *Journal of General Virology, 88,* 1302–1314.

Marston, D. A., McElhinney, L. M., Ali, Y. H., Intisar, K. S., Ho, S. M., Freuling, C., et al. (2009). Phylogenetic analysis of rabies viruses from Sudan provides evidence of a viral clade with a unique molecular signature. *Virus Research, 145,* 244–250.

Marston, D. A., Horton, D. L., Ngeleja, C., Hampson, K., McElhinney, L. M., Banyard, A. C., et al. (2012). Ikoma lyssavirus, highly divergent novel lyssavirus in an African civet. *Emerging Infectious Diseases, 18,* 664–667.

Matsumoto, T., Ahmed, K., Wimalaratne, O., Nanayakkara, S., Perera, D., Karunanayake, D., et al. (2011). Novel sylvatic rabies virus variant in endangered golden palm civet, Sri Lanka. *Emerging Infectious Diseases, 17,* 2346–2349.

McElhinney, L. M., Marston, D. A., Freuling, C. M., Cragg, W., Stankov, S., Lalosević, D., et al. (2011). Molecular diversity and evolutionary history of rabies virus strains circulating in the Balkans. *Journal of General Virology, 92,* 2171–2180.

Mebatsion, T., Cox, J. H., & Conzelmann, K. K. (1993). Molecular analysis of rabies-related viruses from Ethiopia. *Onderstepoort Journal of Veterinary Research, 60,* 289–294.

Meng, S. L., Yan, J. X., Xu, G. L., Nadin-Davis, S. A., Ming, P. G., Liu, S. Y., et al. (2007). A molecular epidemiological study targeting the glycoprotein gene of rabies virus isolates from China. *Virus Research, 124,* 125–138.

Meng, S., Sun, Y., Wu, X., Tang, J., Xu, G., Lei, Y., et al. (2011). Evolutionary dynamics of rabies viruses highlights the importance of China rabies transmission in Asia. *Virology, 410,* 403–409.

Messenger, S. L., Smith, J. S., Orciari, L. A., Yager, P. A., & Rupprecht, C. E. (2003). Emerging pattern of rabies deaths and increased viral infectivity. *Emerging Infectious Diseases, 9,* 151–154.

Metlin, A. E., Rybakov, S., Gruzdev, K., Neuvonen, E., & Huovilainen, A. (2007). Genetic heterogeneity of Russian, Estonian and Finnish field rabies viruses. *Archives of Virology, 152,* 1645–1654.

Ming, P., Yan, J., Rayner, S., Meng, S., Xu, G., Tang, Q., et al. (2010). A history estimate and evolutionary analysis of rabies virus variants in China. *Journal of General Virology, 91*, 759–764.

Moore, P. R., Jansen, C. C., Graham, G. C., Smith, I. L., & Craig, S. B. (2010). Emerging tropical diseases in Australia. Part 3. Australian bat lyssavirus. *Annals of Tropical Medicine and Parasitology, 104*, 613–621.

Morimoto, K., Hooper, D. C., Carbaugh, H., Fu, Z. F., Koprowski, H., & Dietzschold, B. (1998). Rabies virus quasispecies: Implications for pathogenesis. *Proceedings of the National Academy of Sciences USA, 95*, 3152–3156.

Mørk, T., Bohlin, J., Fuglei, E., Åsbakk, K., & Tryland, M. (2011). Rabies in the arctic fox population, Svalbard, Norway. *Journal of Wildlife Diseases, 47*, 945–957.

Muleya, W., Namangala, B., Mweene, A., Zulu, L., Fandamu, P., Banda, D., et al. (2012). Molecular epidemiology and a loop-mediated isothermal amplification method for diagnosis of infection with a rabies virus in Zambia. *Virus Research, 163*, 160–168.

Müller, T., Cox, J., Peter, W., Schäfer, R., Bodamer, P., Wulle, U., et al. (2001). Infection of a stone marten with European bat lyssavirus (EBL1). *Rabies Bulletin Europe, 25*, 9–11.

Müller, T., Johnson, N., Freuling, C. M., Fooks, A. R., Selhorst, T., & Vos, A. (2007). Epidemiology of bat rabies in Germany. *Archives of Virology, 152*, 273–288.

Nadin-Davis, S. A., & Bingham, J. (2004). Europe as a source of rabies for the rest of the world. In A. A. King, A. R. Fooks, M. Aubert, & A. I. Wandeler (Eds.), *Historical perspective of rabies in Europe and the Mediterranean basin (pp. 259–280)*. Paris: OIE.

Nadin-Davis, S. A., & Loza-Rubio, E. (2006). The molecular epidemiology of rabies associated with chiropteran hosts in Mexico. *Virus Research, 117*, 215–226.

Nadin-Davis, S. A., & Real, L. A. (2011). Molecular phylogenetics of the lyssaviruses—insights from a coalescent approach. In A. Jackson (Ed.), *Advances in virus research* (Vol. 79, pp. 203–238). Burlington: Academic Press.

Nadin-Davis, S. A., Casey, G. A., & Wandeler, A. I. (1994). A molecular epidemiological study of rabies virus in central Ontario and western Quebec. *Journal of General Virology, 75*, 2575–2583.

Nadin-Davis, S. A., Huang, W., & Wandeler, A. I. (1996). The design of strain-specific polymerase chain reactions for discrimination of the raccoon rabies virus strain from indigenous rabies viruses of Ontario. *Journal of Virological Methods, 57*, 141–156.

Nadin-Davis, S. A., Huang, W., & Wandeler, A. I. (1997). Polymorphism of rabies viruses within the phosphoprotein and matrix protein genes. *Archives of Virology, 142*, 979–992.

Nadin-Davis, S. A., Sheen, M., Abdel Malik, M., Elmgren, L., Armstrong, J., & Wandeler, A. I. (2000). A panel of monoclonal antibodies targeting the rabies virus phosphoprotein identifies a highly variable epitope of value for sensitive strain discrimination. *Journal of Clinical Microbiology, 38*, 1397–1403.

Nadin-Davis, S. A., Huang, W., Armstrong, J., Casey, G. A., Bahloul, C., Tordo, N., et al. (2001). Antigenic and genetic divergence of rabies viruses from bat species indigenous to Canada. *Virus Research, 74*, 139–156.

Nadin-Davis, S. A., Abdel-Malik, M., Armstrong, J., & Wandeler, A. I. (2002). Lyssavirus P gene characterisation provides insights into the phylogeny of the gene and identifies structural similarities and diversity within the encoded phosphoprotein. *Virology, 298*, 286–305.

Nadin-Davis, S. A., Sheen, M., & Wandeler, A. I. (2003a). Use of discriminatory probes for strain typing of formalin-fixed, rabies virus-infected tissues by *in situ* hybridization. *Journal of Clinical Microbiology, 41*, 4343–4352.

Nadin-Davis, S. A., Simani, S., Armstrong, J., Fayaz, A., & Wandeler, A. I. (2003b). Molecular and antigenic characterization of rabies viruses from Iran identifies variants with distinct epidemiological origins. *Epidemiology and Infection, 131*, 777–790.

Nadin-Davis, S. A., Muldoon, F., & Wandeler, A. I. (2006a). A molecular epidemiological analysis of the incursion of the raccoon strain of rabies virus into Canada. *Epidemiology and Infection, 134*, 534–547.

Nadin-Davis, S. A., Torres, G., de los Angeles Ribas, M., Guzman, M., Cruz de la Paz, R., Morales, M., et al. (2006b). A molecular epidemiological study of Rabies in Cuba. *Epidemiology and Infection, 134,* 1313–1324.

Nadin-Davis, S. A., Turner, G., Paul, J. P. V., Madhusudana, S. N., & Wandeler, A. I. (2007). Emergence of Arctic-like rabies lineage in India. *Emerging Infectious Diseases, 13,* 111–116.

Nadin-Davis, S. A., Velez, J., Malaga, C., & Wandeler, A. I. (2008). A molecular epidemiological study of rabies in Puerto Rico. *Virus Research, 131,* 8–15.

Nadin-Davis, S. A., Feng, Y., Mousse, D., Wandeler, A. I., & Aris-Brosou, S. (2010a). Spatial and temporal dynamics of rabies virus variants in big brown bat populations across Canada: Footprints of an emerging zoonosis. *Molecular Ecology, 19,* 2120–2136.

Nadin-Davis, S. A., Fehlner-Gardiner, C., Sheen, M., & Wandeler, A. I. (2010b). Characterization of a panel of anti-phosphoprotein monoclonal antibodies generated against the raccoon strain of rabies virus. *Virus Research, 152,* 126–136.

Nadin-Davis, S. A., Elmgren, L., Sheen, M., Sabeta, C., & Wandeler, A. I. (2011). Generation and characterization of a panel of anti-phosphoprotein monoclonal antibodies directed against Mokola virus. *Virus Research, 160,* 238–245.

Nadin-Davis, S. A., Sheen, M., & Wandeler, A. I. (2012a). Recent emergence of the Arctic rabies virus lineage. *Virus Research, 163,* 352–362.

Nadin-Davis, S. A., Guerrero, E., Knowles, M. K., & Feng, Y. (2012b). DNA barcoding facilitates bat species identification for improved surveillance of bat-associated rabies across Canada. *Open Journal of Zoology, 5* (Suppl 1-M5), 27–37.

Nagarajan, T., Mohanasubramanian, B., Seshagiri, E. V., Nagendrakumar, S. B., Saseendraneth, M. R., Satyanarayana, M. L., et al. (2006). Molecular epidemiology of rabies virus isolates in India. *Journal of Clinical Microbiology, 44,* 3218–3224.

Nagarajan, T., Nagendrakumar, S. B., Mohanasubramanian, B., Rajalakshmi, S., Hanumantha, N. R., Ramya, R., et al. (2009). Phylogenetic analysis of nucleoprotein gene of dog rabies virus isolates from Southern India. *Infection Genetics and Evolution, 9,* 976–982.

Nanayakkara, S., Smith, J. S., & Rupprecht, C. E. (2003). Rabies in Sri Lanka: Splendid isolation. *Emerging Infectious Diseases, 9,* 368–371.

National Center for Biotechnology Information (n.d.). Available at http://www.ncbi.nlm.nih.gov/.

Nel, L. H. (2005). Vaccines for lyssaviruses other than rabies. *Expert Reviews in Vaccines, 4,* 533–540.

Nel, L. H., & Markotter, W. (2007). Lyssaviruses. *Critical Reviews in Microbiology, 33,* 301–324.

Nel, L. H., Bingham, J., Jacobs, J. A., & Jaftha, J. B. (1998). A nucleotide-specific polymerase chain reaction assay to differentiate rabies virus biotypes in South Africa. *Onderstepoort Journal of Veterinary Research, 65,* 297–303.

Nel, L., Jacobs, J., Jaftha, J., von Teichman, B., & Bingham, J. (2000). New cases of Mokola virus infection in South Africa: a genotypic comparison of southern African virus isolates. *Virus Genes, 20,* 103–106.

Nel, L. H., Sabeta, C. T., von Teichman, B., Jaftha, J. B., Rupprecht, C. E., & Bingham, J. (2005). Mongoose rabies in southern Africa: A re-evaluation based on molecular epidemiology. *Virus Research, 109,* 165–173.

Neubaum, M. A., Shankar, V., Douglas, M. R., Douglas, M. E., O'Shea, T. J., & Rupprecht, C. E. (2008). An analysis of correspondence between unique rabies virus variants and divergent big brown bat (*Eptesicus fuscus*) mitochondrial DNA lineages. *Archives of Virology, 153,* 1139–1142.

Nguyen, A. K. T., Nguyen, D. V., Ngo, G. C., Nguyen, T. T., Inoue, S., Yamada, A., et al. (2011). Molecular epidemiology of rabies virus in Vietnam (2006–2009). *Japanese Journal of Infectious Diseases, 64,* 391–396.

Nishizono, A., Mannen, K., Elio-Villa, L. P., Tanaka, S., Li, K-S., Mifune, K., et al. (2002). Genetic analysis of rabies virus isolates in the Philippines. *Microbiology and Immunology, 46*, 413–417.

Oliveira, R. N., de Souza, S. P., Lobo, R. S. V., Castilho, J. G., Macedo, C. I., Carnieli, P., Jr., et al. (2010). Rabies virus in insectivorous bats: Implications of the diversity of the nucleoprotein and glycoprotein genes for molecular epidemiology. *Virology, 405*, 352–360.

Páez, A., Núñez, C., Garcia, C., & Bóshell, J. (2003). Molecular epidemiology of rabies epizootics in Colombia: Evidence for human and dog rabies associated with bats. *Journal of General Virology, 84*, 795–802.

Páez, A., Saad, C., Núñez, C., & Bóshell, J. (2005). Molecular epidemiology of rabies in northern Colombia 1994–2003. Evidence for human and fox rabies associated with dogs. *Epidemiology and Infection, 133*, 529–536.

Park, Y. J., Shin, M. K., & Kwon, H. M. (2005). Genetic characterization of rabies virus isolates in Korea. *Virus Genes, 30*, 341–347.

Paweska, J. T., Blumberg, L. H., Liebenberg, C., Hewlett, R. H., Grobbelaar, A. A., Leman, P. A., et al. (2006). Fatal human infection with rabies-related Duvenhage virus, South Africa. *Emerging Infectious Diseases, 12*, 1965–1967.

Picard-Meyer, E., Barrat, J., Tissot, E., Barrat, M. J., Bruyère, V., & Cliquet, F. (2004a). Genetic analysis of European bat lyssavirus type 1 isolates from France. *Veterinary Record, 154*, 589–595.

Picard-Meyer, E., Barrat, J., Wasniewski, M., Wandeler, A., Nadin-Davis, S., Lowings, J. P., et al. (2004b). Epidemiology of rabid bats in France, 1989 to 2002. *Veterinary Record, 155*, 774–777.

Picard-Meyer, E., Brookes, S. M., Barrat, J., Litaize, E., Patron, C., Biarnais, M., et al. (2008). Experimental infection of foxes with European bat lyssaviruses type 1 and -2. *Developments in Biology (Basel), 131*, 339–345.

Picard-Meyer, E., Borel, C., Jouan, D., Servat, A., Wasniewski, M., Moinet, M., et al. (2012). Monitoring of a French bat colony naturally infected by EBLV-1 from 2009–2012. Discovery of a new infection case in the colony three years after the first positive rabies diagnostic. *23rd Rabies in the Americas meeting held in São Paulo, Brazil, October 14-19, 2012.*

Posada, D., & Crandall, K. A. (1998). Modeltest: Testing the model of DNA substitution. *Bioinformatics, 14*, 817–818. Freely available from http://darwin.uvigo.es/software/modeltest.html.

Pybus, O. G., Rambaut, A., Belshaw, R., Freckleton, R. P., Drummond, A. J., & Holmes, E. C. (2007). Phylogenetic evidence for deleterious mutation load in RNA viruses and its contribution to viral evolution. *Molecular Biology and Evolution, 24*, 845–852.

Queiroz, L. H., Favoretto, S. R., Cunha, E. M. S., Campos, A. C. A., Lopes, M. C., de Carvalho, C., et al. (2012). Rabies in southeast Brazil: A change in the epidemiological pattern. *Archives of Virology, 157*, 93–105.

Ravkov, E. V., Smith, J. S., & Nichol, S. T. (1995). Rabies virus glycoprotein contains a long 3′ noncoding region which lacks pseudogene properties. *Virology, 206*, 718–723.

Reddy, G. B. M., Singh, R., Singh, R. P., Singh, K. P., Gupta, P. K., Desai, A., et al. (2011). Molecular characterization of Indian rabies virus isolates by partial sequencing of nucleoprotein (N) and phosphoprotein (P) genes. *Virus Genes, 43*, 13–17.

Reynes, J. -M., Molia, S., Audry, L., Hout, S., Ngin, S., Walston, J., et al. (2004). Serologic evidence of lyssavirus infection in bats, Cambodia. *Emerging and Infectious Diseases, 10*, 2231–2234.

Reynes, J. -M., Andriamandimby, S. F., Razafitrimo, G. M., Razainirina, J., Jeanmaire, E. M., Bourhy, H., et al. (2011). Laboratory surveillance of rabies in humans, domestic animals, and bats in Madagascar from 2005 to 2010. *Advances in Preventive Medicine, 2011*, 727821.

Rohde, R. E., Neill, S. U., Clark, K. A., & Smith, J. S. (1997). Molecular epidemiology of rabies epizootics in Texas. *Clinical and Diagnostic Virology, 8*, 209–217.

Rohde, R. E., Mayes, B. C., Smith, J. S., & Neill, S. U. (2004). Bat rabies, Texas, 1996–2000. *Emerging Infectious Diseases, 10*, 948–952.

Root, J. J., Puskas, R. B., Fischer, J. W., Swope, C. B., Neubaum, M. A., Reeder, S. A., et al. (2009). Landscape genetics of raccoons (*Procyon lotor*) associated with ridges and valleys of Pennsylvania: Implications for oral rabies vaccination programs. *Vector-Borne and Zoonotic Diseases, 9*, 583–588.

Rønsholt, L. (2002). A new case of European bat lyssavirus (EBL) infection in Danish sheep. *Rabies Bulletin Europe, 2*, 15.

Rosatte, R. C., Power, M. J., Donovan, D., Davies, J. C., Allan, M., Bachmann, P., et al. (2007). Elimination of arctic variant rabies in red foxes, metropolitan Toronto. *Emerging Infectious Diseases, 13*, 25–27.

Rupprecht, C. E., Dietzschold, B., Wunner, W. H., & Koprowski, H. (1991). Antigenic relationships of lyssaviruses. In G. M. Baer (Ed.), *The natural history of rabies (pp. 69–100)*. Boca Raton: CRC Press.

Sabeta, C. T., Bingham, J., & Nel, L. H. (2003). Molecular epidemiology of canid rabies in Zimbabwe and South Africa. *Virus Research, 91*, 203–211.

Sabeta, C. T., Mansfield, K. L., McElhinney, L. M., Fooks, A. R., & Nel, L. H. (2007a). Molecular epidemiology of rabies in bat eared foxes (*Otocyon megalotis*) in South Africa. *Virus Research, 129*, 1–10.

Sabeta, C. T., Markotter, W., Mohale, D. K., Shumba, W., Wandeler, A. I., & Nel, L. H. (2007b). Mokola virus in domestic mammals, South Africa. *Emerging Infectious Diseases, 13*, 1371–1373.

Sabeta, C. T., Shumba, W., Mohale, D. K., Miyen, J. M., Wandeler, A. I., & Nel, L. H. (2008). Mongoose rabies and the African civet in Zimbabwe. *Veterinary Record, 163*, 580.

Sabeta, C., Blumberg, L., Miyen, J., Mohale, D., Shumba, W., & Wandeler, A. (2010). Mokola virus involved in a human contact (South Africa). *FEMS Immunology and Medical Microbiology, 58*, 85–90.

Sacramento, D., Bourhy, H., & Tordo, N. (1991). PCR technique as an alternative method for diagnosis and molecular epidemiology of rabies virus. *Molecular and Cellular Probes, 5*, 229–240.

Saitou, Y., Kobayashi, Y., Hirano, S., Mochizuki, N., Itou, T., Ito, F. H., et al. (2010). A method for simultaneous detection and identification of Brazilian dog- and vampire bat-related rabies virus by reverse transcription loop-mediated isothermal amplification assay. *Journal of Virological Methods, 168*, 13–17.

Sato, G., Tanabe, H., Shoji, Y., Itou, T., Ito, F. H., Sato, T., et al. (2005). Rapid discrimination of rabies viruses isolated from various host species in Brazil by multiplex reverse transcription-polymerase chain reaction. *Journal of Clinical Virology, 33*, 267–273.

Sato, G., Kobayashi, Y., Motizuki, N., Hirano, S., Itou, T., Cunha, E. M. S., et al. (2009). A unique substitution at position 333 on the glycoprotein of rabies virus street strains isolated from non-hematophagous bats in Brazil. *Virus Genes, 38*, 74–79.

Schaefer, R., Batista, H. B. R., Franco, A. C., Rijsewijk, F. A. M., & Roehe, P. M. (2005). Studies on antigenic and genomic properties of Brazilian rabies virus isolates. *Veterinary Microbiology, 107*, 161–170.

Serra-Cobo, J., Amengual, B., Abellán, C., & Bourhy, H. (2002). European bat *Lyssavirus* infection in Spanish bat populations. *Emerging and Infectious Diseases, 8*, 413–420.

Shankar, V., Orciari, L. A., de Mattos, C., Kuzmin, I. V., Pape, W. J., O'Shea, T. J., et al. (2005). Genetic divergence of rabies viruses from bat species of Colorado, USA. *Vector-borne and Zoonotic Diseases, 5*, 330–341.

Shao, X. Q., Yan, X. J., Luo, G. L., Zhang, H. L., Chai, X. L., Wang, F. X., et al. (2011). Genetic evidence for domestic raccoon dog rabies caused by arctic-like rabies virus in Inner Mongolia, China. *Epidemiology and Infection, 139*, 629–635.

Sheeler-Gordon, L. L., & Smith, J. S. (2001). Survey of bat populations from Mexico and Paraguay for rabies. *Journal of Wildlife Diseases, 37*, 582–593.

Shoji, Y., Kobayashi, Y., Sato, G., Itou, T., Miura, Y., Mikami, T., et al. (2004). Genetic characterization of rabies viruses isolated from frugivorous bat (Artibeus spp.) in Brazil. *Journal of Veterinary and Medical Science, 66*, 1271–1273.

Shoji, Y., Kobayashi, Y., Sato, G., Gomes, A. A. B., Itou, T., Ito, F. H., et al. (2006). Genetic and phylogenetic characterization of rabies virus isolates from wildlife and livestock in Paraiba, Brazil. *Acta Virologica, 50*, 33–38.

Smith, J. S. (2002). Molecular epidemiology. In A. C. Jackson & W. H. Wunner (Eds.), *Rabies (pp. 79–111)* (1st ed.). San Diego: Academic Press.

Smith, J. S., Fishbein, D. B., Rupprecht, C. E., & Clark, K. (1991). Unexplained rabies in three immigrants in the United States. A virologic investigation. *New England Journal of Medicine, 324*, 205–211.

Smith, J. S., Orciari, L. A., Yager, P. A., Seidel, H. D., & Warner, C. K. (1992). Epidemiologic and historical relationships among 87 rabies virus isolates as determined by limited sequence analysis. *Journal of Infectious Diseases, 166*, 296–307.

Smith, J. S., Orciari, L. A., & Yager, P. A. (1995). Molecular epidemiology of rabies in the United States. *Seminars in Virology, 6*, 387–400.

Smith, G. C., Aegerter, J. N., Allnutt, T. R., MacNicoll, A. D., Learmount, J., Hutson, A. M., et al. (2011). Bat population genetics and lyssavirus presence in Great Britain. *Epidemiology and Infection, 139*, 1463–1469.

Sodré, M. M., da Gama, A. R., & de Almeida, M. F. (2010). Updated list of bat species positive for rabies in Brazil. *Revista do Instituto de Medicina Tropical de São Paulo, 52*, 75–81.

Streicker, D. G., Turmelle, A. S., Vonhof, M. J., Kuzmin, I. V., McCracken, G. F., & Rupprecht, C. E. (2010). Host phylogeny constrains cross-species emergence and establishment of rabies virus in bats. *Science, 329*, 676–679.

Susetya, H., Sugiyama, M., Inagaki, A., Ito, N., Nudiarto, G., & Minamoto, N. (2008). Molecular epidemiolgy of rabies in Indonesia. *Virus Research, 135*, 144–149.

Szanto, A. G., Nadin-Davis, S. A., Rosatte, R. C., & White, B. N. (2011). Genetic tracking of the raccoon variant of rabies virus in Eastern North America. *Epidemics, 3*, 76–87.

Talbi, C., & Bourhy, H. (2011). La rage canine en Afrique au travers de l'analyse génétique, spatiale et temporelle des isolats. *Virologie, 15*, 307–318.

Talbi, C., Holmes, E. C., de Benedictis, P., Faye, O., Nakouné, E., Gamatié, D., et al. (2009). Evolutionary history and dynamics of dog rabies virus in western and central Africa. *Journal of General Virology, 90*, 783–791.

Talbi, C., Lemey, P., Suchard, M. A., Abdelatif, E., Elharrak, M., Jalal, N., et al. (2010). Phylodynamics and human-mediated dispersal of a zoonotic virus. *PLoS Pathogens, 6(10)*, e1001166.

Tamura, K., Dudley, J., Nei, M., & Kumar, S. (2007). MEGA4: Molecular Evolutionary Genetics Analysis (MEGA) software version 4.0. *Molecular Biology and Evolution, 24*, 1596–1599. Freely available from http://www.megasoftware.net.

Tang, X., Luo, M., Zhang, S., Fooks, A. R., Hu, R., & Tu, C. (2005). Pivotal role of dogs in rabies transmission, China. *Emerging Infectious Diseases, 11*, 1970–1972.

Tao, X. Y., Tang, Q., Li, H., Mo, Z. J., Zhang, H., Wang, D. M., et al. (2009). Molecular epidemiology of rabies in southern People's Republic of China. *Emerging Infectious Diseases, 15*, 1192–1198.

Tenzin, T., Wacharapluesadee, S., Denduangboripant, J., Dhand, N. K., Dorji, R., Tshering, D., et al. (2011). Rabies virus strains circulating in Bhutan: Implications for control. *Epidemiology and Infection, 139*, 1457–1462.

Thompson, J. D., Gibson, T. J., Plewniak, F., Jeanmougin, F., & Higgins, D. G. (1997). The ClustalX windows interface: flexible strategies for multiple sequence alignment aided by quality analysis tools. *Nucleic Acids Research, 25*, 4876–4882.

Tjørnehøj, K., Fooks, A. R., Agerholm, J. S., & Rønsholt, L. (2006). Natural and experimental infection of sheep with European bat lyssavirus type-1 of Danish bat origin. *Journal of Comparative Pathology, 134*, 190–201.

Tordo, N., Poch, O., Ermine, A., Keith, G., & Rougeon, F. (1986). Walking along the rabies genome: Is the large G-L intergenic region a remnant gene? *Proceedings of the National Academy of Sciences USA*, *83*, 3914–3918.

Tuffereau, C., Leblois, H., Bénéjean, J., Coulon, P., Lafay, F., & Flamand, A. (1989). Arginine or lysine in position 333 of ERA and CVS glycoprotein is necessary for rabies virulence in adult mice. *Virology*, *172*, 206–212.

Turcitu, M. A., Barboi, G., Vuta, V., Mihai, I., Boncea, D., Dumitrescu, F., et al. (2010). Molecular epidemiology of rabies virus in Romania provides evidence for a high degree of heterogeneity and virus diversity. *Virus Research*, *150*, 28–33.

Van der Poel, W. H., Van der Heide, R., Verstraten, E. R., Takumi, K., Lina, P. H., & Kramps, J. A. (2005). European bat lyssaviruses, The Netherlands. *Emerging Infectious Diseases*, *11*, 1854–1859.

Van Eeden, C., Markotter, W., & Nel, L. H. (2011). Molcular phylogeny of duvenhage virus. *South African Journal of Science*, *107* (article # 177).

van Thiel, P. P., de Bie, R. M., Eftimov, F., Tepaske, R., Zaaijer, H. L., van Doornum, G. J., et al. (2009). Fatal human rabies due to duvenhage virus from a bat in kenya: Failure of treatment with coma-induction, ketamine, and antiviral drugs. *PLoS Neglected Tropical Diseases*, *3*(7), e428.

Van Zyl, N., Markotter, W., & Nel, L. H. (2010). Evolutionary history of African mongoose rabies. *Virus Research*, *150*, 93–102.

Vázquez-Morón, S., Juste, J., Ibáñez, C., Ruiz-Villamor, E., Avellón, A., Vera, M., et al. (2008). Endemic circulation of European bat lyssavirus type 1 in serotine bats, Spain. *Emerging Infectious Diseases*, *14*, 1263–1266.

Vázquez-Morón, S., Juste, J., Ibáñez, C., Berciano, J. M., & Echevarría, J. E. (2011). Phylogeny of European bat lyssavirus 1 in *Eptesicus isabellinus* bats, Spain. *Emerging Infectious Diseases*, *17*, 520–523.

Velasco-Villa, A., Gómez-Sierra, M., Hernández-Rodríguez, G., Juárez-Islas, V., Meléndez-Félix, A., Vargas-Pino, F., et al. (2002). Antigenic diversity and distribution of rabies virus in Mexico. *Journal of Clinical Microbiology*, *40*, 951–958.

Velasco-Villa, A., Orciari, L. A., Souza, V., Juárez-Islas, V., Gomez-Sierre, M., Castillo, A., et al. (2005). Molecular epizootiology of rabies associated with terrestrial carnivores in Mexico. *Virus Research*, *111*, 13–27.

Velasco-Villa, A., Orciari, L. A., Juárez-Islas, V., Gómez-Sierra, M., Padilla-Medina., I., Flisser, A., et al. (2006). Molecular diversity of rabies virases associated with bats in Mexico and other countries of the Americas. *Journal of Clinical Microbiology*, *44*, 1697–1710.

Velasco-Villa, A., Messenger, S. L., Orciari, L. A., Niezgoda, M., Blanton, J. D., Fukagawa, C., et al. (2008a). Identification of new rabies virus variant in Mexican immigrant. *Emerging Infectious Diseases*, *14*, 1906–1908.

Velasco-Villa, A., Reeder, S. A., Orciari, L. A., Yager, P. A., Franka, R., Blanton, J. D., et al. (2008b). Enzootic rabies elimination from dogs and re-emergence in wild terrestrial carnivores, United States. *Emerging Infectious Diseases*, *14*, 1849–1854.

Von Teichman, B. F., Thomson, G. R., Meredith, C. D., & Nel, L. H. (1995). Molecular epidemiology of rabies virus in South Africa: Evidence for two distinct virus groups. *Journal of General Virology*, *76*, 73–82.

Vos, A., Müller, T., Neubert, L., Zurbriggen, A., Botteron, C., Pöhle, D., et al. (2004). Rabies in red foxes (*Vulpes vulpes*) experimentally infected with European bat lyssavirus type 1. *Journal of Veterinary Medicine Series B*, *51*, 327–332.

Wandeler, A. I., & Salsberg, E. (1999). Raccoon rabies in eastern Ontario. *Canadian Veterinary Journal*, *40*, 731.

Warner, C., Fekadu, M., Whitfield, S., & Shaddock, J. (1999a). Use of anti-glycoprotein monoclonal antibodies to characterize rabies virus in tissues. *Journal of Virological Methods*, *77*, 69–74.

Warner, C. K., Zaki, S. R., Shieh, W. J., Whitfield, S. G., Smith, J. S., Orciari, L. A., et al. (1999b). Laboratory investigation of human deaths from vampire bat rabies in Peru. *American Journal of Tropical Medicine and Hygiene*, 60, 502–507.

Warrilow, D. (2005). Australian bat lyssavirus: a recently discovered new rhabdovirus. *Current Topics in Microbiology and Immunology*, 292, 25–44.

Warrilow, D., Smith, I. L., Harrower, B., & Smith, G. A. (2002). Sequence analysis of an isolate from a fatal human infection of Australian bat lyssavirus. *Virology*, 297, 109–119.

Weyer, J., Szmyd-Potapczuk, A. V., Blumberg, L. H., Leman, P. A., Markotter, W., Swanepoel, R., et al. (2011). Epidemiology of human rabies in South Africa, 1983–2007. *Virus Research*, 155, 283–290.

Winkler, W. G., & Jenkins, S. R. (1991). Raccoon rabies. In G. M. Baer (Ed.), *The natural history of rabies (pp. 325–340)* (2nd ed.). Boca Raton: CRC Press.

Wright, A., Rampersad, J., Ryan, J., & Ammons, D. (2002). Molecular characterization of rabies virus isolates from Trinidad. *Veterinary Microbiology*, 87, 95–102.

Wright, E., Hayman, D. T. S., Vaughan, A., Temperton, N. J., Wood, J. L. N., Cunningham, A. A., et al. (2010). Virus neutralising activity of African fruit bat (*Eidolon helvum*) sera against emerging lyssaviruses. *Virology*, 408, 183–189.

Wright, C. F., Morelli, M. J., Thébaud, G., Knowles, N. J., Herzyk, P., Paton, D. J., et al. (2011). Beyond the consensus: dissecting within-host viral population diversity of foot-and-mouth disease virus by using next-generation genome sequencing. *Journal of Virology*, 85, 2266–2275.

Wu, X., Franka, R., Velasco-Villa, A., & Rupprecht, C. E. (2007). Are all lyssavirus genes equal for phylogenetic analyses? *Virus Research*, 129, 91–103.

Yamagata, J., Ahmed, K., Khawplod, P., Mannen, K., Xuyen, D. K., Loi, H. H., et al. (2007). Molecular epidemiology of rabies in Vietnam. *Microbiology and Immunology*, 51, 833–840.

Yang, D. -K., Shin, E. K., Oh, Y. -I., Kang, H. -K., Lee, K. -W., Cho, S. -D., et al. (2011). Molecular epidemiology of rabies virus circulating in South Korea, 1998–2010. *Journal of Veterinary Medical Science*, 73, 1077–1082.

Yung, V., Favi, M., & Fernández, J. (2002). Genetic and antigenic typing of rabies virus in Chile. *Archives of Virology*, 147, 2197–2205.

Zhang, Y. Z., Xiong, C. L., Zou, Y., Wang, D. M., Jiang, R. J., Xiao, Q. Y., et al. (2006). Molecular characterization of rabies virus isolates in China during 2004. *Virus Research*, 121, 179–188.

Zhang, Y. Z., Xiong, C. L., Lin, X. D., Zhou, D. J., Jiang, R. J., Xiao, Q. Y., et al. (2009). Genetic diversity of Chinese rabies viruses: Evidence for the presence of two distinct clades in China. *Infection, Genetics and Evolution*, 9, 87–96.

Zhang, S., Zhao, J., Liu, Y., Fooks, A. R., Zhang, F., & Hu, R. (2010). Characterization of a rabies virus isolate from a ferret badger (*Melogale moschata*) with unique molecular differences in glycoprotein antigenic site III. *Virus Research*, 149, 143–151.

Zulu, G. C., Sabeta, C. T., & Nel, L. H. (2009). Molecular epidemiology of rabies: focus on domestic dogs (*Canis familiaris*) and black-backed jackals (*Canis mesomelas*) from northern South Africa. *Virus Research*, 140, 71–78.

5

Rabies in Terrestrial Animals

Cathleen A. Hanlon

Kansas State University Rabies Laboratory,
2005 Research Park Circle, Manhattan, KS 66502, USA

1 INTRODUCTION

The word *rabies* elicits different reactions among different groups of people, probably due to the diversity of life experiences, acquired knowledge, and occupation or avocation, as well as geographic location and travel history. These reactions may vary from knowing nothing about the disease and thinking that it is irrelevant to being keenly interested or deathly afraid of this significant source of human and animal mortality. The term rabies describes an acute fatal disease of the nervous system caused by viruses in the genus Lyssavirus, family Rhabdoviridae. The species of animals in which these viruses are maintained are restricted to the orders Chiroptera (bats) and Carnivora. However, all mammals are susceptible to rabies in varying degrees; the disease does not occur in amphibians, reptiles, birds, or invertebrates. The imminent threat of developing rabies in Europe and North America was significantly reduced a number of decades ago through potent biologics effective in the prevention of rabies in humans after an exposure (Manning et al., 2008; World Health Organization, 2005). Equally important over the last half-century, human exposures to rabies were substantially reduced in many regions of the world through the elimination of canine rabies virus variants which are perpetuated by dog-to-dog transmission. Nonetheless, canine rabies has not been eliminated globally, wildlife rabies virus variants are diverse and occur in a number of carnivore and chiropteran species, and humans travel and animals are translocated, both intentionally and inadvertently, making the risk of rabies exposure global.

Rabies: Scientific Basis of the Disease and its Management
DOI: http://dx.doi.org/10.1016/B978-0-12-396547-9.00005-5

2 WHAT IS A SUSCEPTIBLE HOST?

A number of events are required for the establishment of lyssavirus infection. These include direct contact with a receptive host cell and entry into the cell followed by uncoating of virus particles. Then, intracytoplasmic production of nucleic acid and viral proteins must occur. Finally, virions must be assembled and then successfully exit the cells of the infected host, and, of course, in order to perpetuate, a new host must be infected. Innate immune cells of the host detect pathogens via pattern recognition receptions, such as Toll-like receptors (TLE) and others. Pathogen recognition receptors respond to distinct pathogen-associated or danger-associated molecular patterns. Receptor activation induces production of pro-inflammatory cytokines and signals that recruit immune cells and activate an inflammatory response. In addition, they are required to induce adaptive immune responses (Akira, Uematsu, & Takeuchi, 2006). Pro-inflammatory cytokines, such as type I interferon (IFN) acts in either a direct manner through the inhibition of viral replication or in an indirect manner through the recruitment and activation of antigen-presenting cells (e.g., Kupffer cells, dendritic cells). Innate and adaptive immune responses act as two interlocking defense lines against rabies virus infection. It is likely that elimination of rabies virus following an exposure requires initial virus suppression by the innate immune response, and this may then be accompanied by recruitment of effector T cells, leading to activation of an adaptive immune response. The fine-tuning and the interaction of these responses, as well as effective timing, in the prevention of disease following an exposure still need to be defined. The viral strategy of entrance, self-production, and exit of progeny, extrapolated from the biochemical, cellular, and tissue levels of organization to an entire organism, forms the basis of susceptibility, or the innate capability of an animal to become rabid and infectious to other hosts. In theory, a single virion can initiate the process. But in actuality, due in part to innate immunity rather than adaptive immunity, multiple infectious units are required. A very susceptible species or individual host would be one in which only a small number of virus particles are required to result in an infection.

Once an animal is exposed to rabies, the probability of infection depends upon the individual, the species, the rabies virus variant, the amount of virus, and the severity and route of the exposure. Multiple deep bites to the head and face from a proven rabid animal would be more likely to cause infection than a superficial wound at a distant extremity (Turkmen et al., 2012). Virus is introduced into local tissues through a bite from the infectious host. There may be viral replication locally but only in limited amounts. At present, there is no reliable method to determine the risk of a potential exposure and the probability of developing disease. The virus is highly neurotopic; there is no viremia or viral shedding in the urogenitary or gastrointestinal tract. The timing of the progression of viral infection from local neurons at the site of infection to the central nervous system of an infected host remains unpredictable. The incubation period is variable and is generally inversely related to viral dose and severity of exposure (i.e., multiple damaging bites versus pinprick puncture of a single tooth), proximity to the brain (i.e., face, head, and neck versus a peripheral site such as a toe), the species of the animal, and the variant of the virus. For example, raccoons inoculated with 1 ml of a skunk salivary gland suspension with a virus titer of $10^{5.9}$ mouse intracerebral lethal dose 50% ($MICLD_{50}$) remained uninfected and healthy, whereas two skunks inoculated with $10^{0.7}$ and three skunks inoculated with $10^{2.1}$ $MICLD_{50}$ of the same preparation developed clinical signs in 22 and 24 days at the lower dose and 17 or 18 days with the higher dose and succumbed to rabies on days 31 and 35 with the lower dose, and on days 27, 31 and 37 with the higher dose (Hill, Smith, Beran, & Beard, 1993). The majority of infected animals become clinically rabid within several weeks to several months following exposure. On the basis of experimental and field data, a 6-month quarantine (or euthanasia to prevent the development of rabies) is imposed on exposed, unvaccinated (or previously vaccinated but out-of-date) domestic animals, as this is considered a probable maximum incubation period. Once virus reaches the central nervous system, infection may spread quickly through numerous neuronal tracts. Thus, the clinical illness is relatively brief; on the order of days rather than weeks. Once the brain is infected, virus is then capable of spreading centrifugally from the central nervous system through the innervation of major organ systems as long as the host remains alive. Virus is produced in high amounts in the salivary glands, leading to the presence of virus in the saliva (Niezgoda, Briggs, Shaddock, & Rupprecht, 1998; Vaughn, Gerhardt, & Paterson, 1963; Vaughn, Jr., Gerhardt, & Newell, 1965; Tepsumethanon, Wilde, & Sitprija, 2008). Profound neuronal dysfunction occurs and, even with assisted preservation of an airway and ventilation, often manifests as autonomic instability sufficient to result in death.

Host susceptibility is greatly influenced by attributes unique to various species, and, perhaps to a lesser extent, by individual factors, such as immune status and age (Casals, 1940). Some generalities about susceptibility may be inferred from experimental data, and within the limitations of field observations. However, specific species attributes have not been exhaustively characterized, due in part to confounding variables such as virus variant, ecological characteristics of the hosts which affect exposure risk and outcome, and surveillance bias that limits the source and number of specimens examined. From a practical standpoint, high-risk species include certain canids (i.e., dogs, foxes, wolves), herpestidae (i.e., the

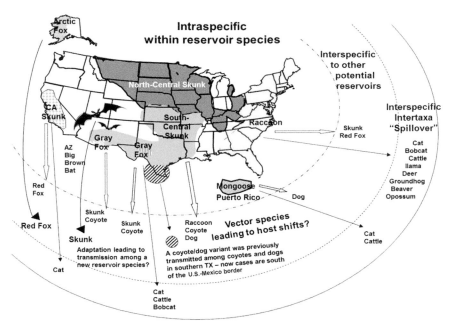

FIGURE 5.1   **Patterns of rabies transmission among reservoir, vector and "spillover" species in the USA.**

mongoose and their allies), mustelids (i.e., skunk, ferrets, mink), procyonids (i.e., the common raccoon and its relatives), and bats. Moderate-risk species consist of felids, ungulates, and primates. Low-risk species comprise monotremes, marsupials, insectivores, rodents, and lagomorphs, among others. Nonetheless, an individual mammal of any species is considered susceptible to rabies.

It is unclear how mammalian reservoirs and associated viral variants may have co-evolved. The greatest genetic differences of rabies virus variants occur between those of terrestrial species and those found in bats. The ability to distinguish among variants has become more precise in the past three decades due to the usefulness of monoclonal antibodies for antigenic characterization (Wiktor & Koprowski, 1978) and, later, the development of the reverse-transcriptase polymerase chain reaction (PCR) assay and genetic sequencing for the most precise characterization of genomic and inferred amino acid differences (Smith, 1996; Tordo, 1996). These tools have led to a clearer understanding of the natural epizootiology of rabies virus variants in their host species, as well as the risk and intensity of spillover into different animals. Reservoir host species are those that are capable of sustained intra-species maintenance of a virus variant within a geographic area (Figure 5.1); they are not "subclinically" infected or chronic shedders of rabies virus but rather the population serves as a source of susceptible individuals within which the virus is exquisitely adapted to achieve transmission to the next host. Critical characteristics of reservoir species include distribution, abundance, population density and contact rate, in order to support the virus–host relationship. In addition, virus maintenance may also take advantage of potentially competent co-existing species, which may serve as an efficient ancillary vector (Bell, 1980; Childs, Trimarchi, & Krebs, 1994). Vector species may closely overlap with the primary reservoir in ecological and behavioral characteristics, as well as in temporal and spatial patterns in a rabies-enzootic area. Examples of this include: 1) a rabies virus variant associated with coyotes in Mexico and Texas which appears to be capable of sustained transmission among susceptible (i.e., unvaccinated) domestic dogs (Clark et al., 1994), 2) red fox rabies virus variants that may be effectively maintained in Europe among raccoon dogs (Singer, Kauhala, Holmala, & Smith, 2009) and in Canada among skunks (Nadin-Davis, Casey, & Wandeler, 1994), and 3) canine rabies virus variants occurring in wild canids, such as jackals, mongooses, and other species, in the Middle East and Africa (Barnard, 1979; Bingham, Foggin, Wandeler, & Hill, 1999; Davis, Rambaut, Bourhy, & Holmes, 2007; Smith, Yager, & Orciari, 1993; Zulu, Sabeta, & Nel, 2009). Within the reservoir or maintenance species, the probability is high that the virus will establish infection; induce behavioral changes, such as aggressive biting behavior; and result in a long enough clinical survival period so that salivary virus shedding takes place coincident with behavior facilitating infection of a new host; all of these factors contribute to optimal virus transmission.

3 PATHOGENESIS

Traditional pathogenesis studies have been conducted in dogs, cats, and ferrets. These studies provide the virological data which, when taken together with epidemiological observations over several decades, support the practice of a 10-day observation period (Figure 5.2). Thus, if a healthy dog, cat, or ferret bites, or otherwise potentially exposes, a person or animal, and rabies needs to be ruled in or out, the animal, irrespective of vaccination status, may simply be observed for 10 days (Niezgoda et al., 1998; Vaughn, Gerhardt, & Paterson, 1963; Vaughn, Jr. et al., 1965; Tepsumethanon et al., 2008). If the animal remains alive and well during this 10-day period, there is no virologic or epidemiologic data suggesting a risk of rabies transmission from the potential exposure to this animal. If there were a risk of viral shedding and transmission, the animal would have manifested signs of clinical rabies (sometimes including acute death) during the observation period. If an animal

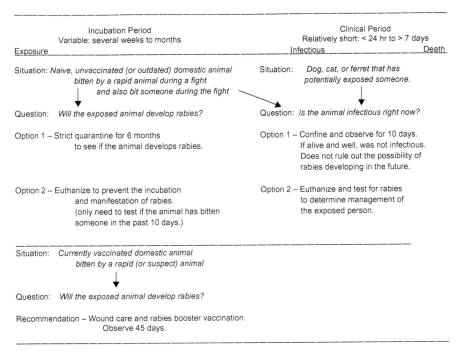

Incubation Period Variable: several weeks to months	Clinical Period Relatively short: < 24 hr to > 7 days
Exposure	Infectious Death

Situation: *Naive, unvaccinated (or outdated) domestic animal bitten by a rapid animal during a fight and also bit someone during the fight*

Situation: *Dog, cat, or ferret that has potentially exposed someone.*

Question: *Will the exposed animal develop rabies?*

Question: *Is the animal infectious right now?*

Option 1 – Strict quarantine for 6 months to see if the animal develops rabies.

Option 1 – Confine and observe for 10 days. If alive and well, was not infectious. Does not rule out the possibility of rabies developing in the future.

Option 2 – Euthanize to prevent the incubation and manifestation of rabies. (only need to test if the animal has bitten someone in the past 10 days.)

Option 2 – Euthanize and test for rabies to determine management of the exposed person.

Situation: *Currently vaccinated domestic animal bitten by a rapid (or suspect) animal*

Question: *Will the exposed animal develop rabies?*

Recommendation – Wound care and rabies booster vaccination. Observe 45 days.

FIGURE 5.2 **Implications of pathogenesis on potential exposure and animal and human management.**

develops clinical illness compatible with rabies during the 10-day observation, then it should be humanely euthanized. If it dies acutely during the 10-day period, or is euthanized due to lack of ownership or concerns about viciousness, the acutely dead, ill, or unwanted animal that was humanely euthanized will need to be tested for rabies as soon as possible to determine the need for post-exposure prophylaxis for the exposed persons, as well as management of exposed animals. If individuals of other species bite or otherwise potentially expose a person, the risk of transmission is assessed based on local transmission patterns, circumstances and severity of exposure, and other related factors. Local and state public health authorities may allow extrapolation of the data on which the 10-day observation is based to domestic livestock and exotic hooved stock and impose a prolonged observation period of up to 30 days. If these animals remain alive and well, the potentially exposed person does not need to undergo post-exposure prophylaxis. The animal may still develop rabies in the future from a prior exposure, but the animal was not infectious at the time of the potential exposure to the human.

As described in Chapter 11, a battery of diagnostic methods on a panel of four or more samples are applied antemortem in humans with encephalitis compatible with rabies. Also, as previously described for animals, a positive correlation was observed between nerve fibers in skin biopsies tested with anti-rabies virus conjugate and traditional brain material testing from ten species of animals with naturally acquired rabies (Blendon et al., 1983). A total of 232 animals, half rabies-positive and half rabies-negative, were evaluated, including dogs, cats, cattle, swine, horses, foxes (gray and red), striped skunks, raccoons, and mongooses. Some animals that ultimately developed rabies were found to have immunofluorescence-positive skin biopsy results two or more days before the onset of clinical signs. The percentage of those with positive skin immunofluorescence results increased as the onset of clinical illness approached. From the mid-course period of illness to death, the correlation between skin and brain approached 100%. Although these 30-year-old data suggest that examination of surgical biopsy specimens by immunofluorescence for rabies virus antigen is a useful and reliable diagnostic tool to evaluate the rabies status of biting dogs or cats, or to confirm a clinical diagnosis of rabies prior to post-mortem evaluation of brain material, these diagnostic tools have not been accepted for early antemortem detection of rabies in clinically suspect animals.

4 PREVENTION

The Compendium of Animal Rabies Control is updated and published regularly (Brown et al., 2011). It is the guiding document on animal rabies control in the USA. Pre-exposure vaccination of domestic animals, with booster vaccination if exposed, is an effective method of disease prevention. Although no vaccine is 100% effective, rabies vaccines are licensed following demonstration of protection of a majority of a vaccinated test group of animals against a lethal rabies virus challenge. The greatest risk of rabies in a vaccinated animal occurs: 1) when the vaccine coverage period has expired (i.e., the animal is outdated on its vaccination and not currently vaccinated); 2) when the animal is young and has received only a single vaccination, to date, in its lifetime; 3) when a severe exposure has occurred; and 4) when the potential exposure is not recognized and no primary wound care or post-exposure booster vaccination is sought by the owner. Rabies is prevented in most exposed animals if they are currently vaccinated and receive wound care consisting of copious flushing of the site (which alone reduces the probability of a productive infection), as well as a booster rabies vaccination. As a precaution, the animal should also be confined and observed for up to 45 days.

Prevention of rabies in wildlife is conceptually achievable through oral vaccination. The greatest successes have been achieved in control of red fox rabies in Europe and Canada. The practicality of the approach is dependent upon: 1) the ease of vaccination of a particular species by the oral route; 2) population densities and ecological characteristics (i.e., family groups with dominant animals so that reaching all equally is not feasible); and 3) accessibility of the target population to efficient bait distribution methods and constraints of bait distribution near high human population densities. Novel biologics are often required for these approaches. Practicing veterinarians may be consulted when domestic animals contact or consume baits distributed free-choice for oral wild-life vaccination. It is ideal for practicing veterinarians to be familiar with wildlife vaccination projects and associated biologics as they may pose a risk for infection and adverse events in uniquely susceptible humans and domestic animals.

5 CLINICAL SIGNS

The initial clinical signs of rabies are often non-specific and may include general lethargy, poor appetite, and diarrhea or vomiting. However, the clinical course is one-way in that there is little waxing and waning of clinical signs but instead daily, if not hourly, progressive deterioration in clinical condition. Changes in behavior may be one of the first clinical signs and may consist of episodes of mild or dramatic abnormalities. For example, an animal may become more reclusive or attention-seeking than normal or it may unpredictably and intermittently attack animate (humans or other animals), inanimate, or unseen or unapparent objects (e.g., humans report hallucinatory episodes; animals may appear to be "snapping at flies"). Rabies is neurotropic and may result in irritation or parasthesias at the site of initial exposure even though the wound inflicted by the rabid animal may have healed. A combination of increased saliva production and a decreased ability to swallow may present as profound contamination of the mouth, chin, and forelegs with potentially infectious saliva (Figure 5.3). Cranial nerve involvement may be focal and unilateral, presenting as unequal pupil size with dysfunction, facial or tongue paresis, and changes in phonation. As the clinical period progresses, unpredictable episodes of attempts to bite may be invoked by auditory, visual, or tactile stimuli with aggression to the point of self-mutilation. In the end stages, most animals become profoundly moribund.

6 DOMESTIC DOGS

The incubation period of rabies in dogs may be as short as 10 days (rarely less) or as long as several months. A commonly imposed

FIGURE 5.3 Salivation in a clinically rabid dog. Photograph courtesy of Ivan Kuzmin, Poxvirus and Rabies Branch, Centers for Disease Control and Prevention, Atlanta, GA 30333.

quarantine period for an exposed dog (and often cat, ferret, horse, cow, and other animals) is 6 months. This period is based on epizootiologic and experimental evidence that if an exposed dog will develop rabies from an exposure, there is a strong likelihood that it will occur within six months of the exposure, and most often, between several weeks to several months following exposure. At the beginning of the clinical course, nonspecific signs may include anorexia, fever, dysphagia, and a change in behavior, such as being more reclusive than normal, or restlessness. The animal may be more easily startled by noise, light, or touch. As each day passes, other clinical abnormalities may appear and the initial signs may become more pronounced. Substantial injury may occur due to aggression towards objects or other animals. Self-mutilation may occur. Salivation may appear profuse due to the inability to swallow. Dehydration and anorexia, sometimes secondary to dysphagia, may result in acute weight loss. Tremors, ataxia, paresis, paralysis, and generalized seizures may develop. The actual physiologic mechanism resulting in death can vary but may result from profound autonomic dysfunction resulting in alternations in respiration, body temperature, blood pressure, cardiac rhythm, and the like.

When a dog has potentially exposed a person to rabies virus through a bite or other exposure route, the dog is commonly subjected to a 10-day confinement and observation period. The 10-day time period is derived from experimental observations of clinical shedding of rabies virus (Vaughn, Jr. et al., 1965), as well as supportive field data. If a dog is capable of shedding virus in its saliva (which can only develop after infection of the brain) at the time that it bit someone, the dog may reliably be expected to develop clinical signs of rabies within the 10-day confinement and observation period. If the dog remains alive and well, there is no epizootiologic or epidemiologic data to contradict the presumption that, at the time of the bite, there was insufficient or a complete lack of virus in the dog's saliva such that there was no risk of infection to the bitten individual.

There are rare reports in the literature of apparent exceptions to the widely practiced public health rabies management protocol in North America of a 10-day confinement and observation period for a biting dog. In one report, four beagles were inoculated with mouse brain passaged saliva from an apparently healthy dog from Ethiopia that had been reported to excrete virus intermittently (Fekadu, 1972; Fekadu, Shaddock, & Baer, 1981; Fekadu, Shaddock, Chandler, & Baer, 1983).Two inoculated dogs remained clinically normal. Two dogs developed signs of rabies but apparently recovered. One later died of a bacterial pneumonia. Salivary swabs from the surviving dog were routinely collected and inoculated into mice. Virus was reported to have been present in samples taken on days 42, 169, and 305 days post-inoculation. The amount of virus was extremely low, fewer than 10 infectious particles. In a related study, 39 dogs were injected intramuscularly with either an Ethiopian strain or a Mexican strain of rabies virus. Virus was recovered from the submaxillary salivary glands of 9 of 17 dogs that died following inoculation with the Ethiopian virus variant. Four of these dogs were reported to have virus in the saliva up to 13 days before overt signs of rabies were observed. Virus was recovered from the submaxillary salivary glands of 16 of 22 dogs that died following inoculation with the Mexican virus variant. Eight of these dogs also excreted virus in the saliva up to seven days before definitive signs of rabies were observed (Fekadu, Shaddock, & Baer, 1982; Fekadu et al., 1983). At least one observation in Asia suggested that, on rare occasions, a dog under a 10-day observation period can still be alive and apparently healthy, while the bitten human may develop signs and symptoms of rabies during that period (Somayajulu & Reddy, 1989). The veracity of these reports is unknown. It is possible that the patient may have been bitten previously by a different dog several weeks to several months prior to the temporally associated situation. However, if these observations have merit, two potential interpretations may be possible. The dogs may be remaining healthy and shedding rabies virus for longer than the expected 10-day period prior to clinical development of rabies. Alternatively, these dogs may reflect the possibility of recovery from rabies with the potential for intermittent shedding. In the past 20 years, virus detection and identification techniques have greatly advanced in sophistication and sensitivity, yet there has been a conspicuous lack of additional reports of similar findings. No such field observations have come from North or South America. Moreover, retrospective and prospective data from Thailand were collected on 1,222 dogs and 303 cats that potentially exposed humans during a 10-day observation period (Tepsumethanon et al., 2008). A total of 644 dogs and 58 cats died within 10 days of observation and were confirmed to be rabid by fluorescent antibody testing of the brain. The other animals remained clinically normal during the 10-day observation and were released to owners or transferred to a municipal animal shelter. This study confirms the practice of observing dogs or cats for 10 days, preferably under veterinary supervision, that have bitten humans and in which there is a need to rule out potential rabies exposure.

7 RACCOONS

The common raccoon (*Procyon lotor*) is widely distributed from Canada through Central America. It was introduced into Russia and Western Europe (Kaufmann, 1982), as well as onto a number of islands, including Japan. Over the past 50 years, raccoons have become recognized as a significant host of wildlife rabies in North America. A number of rabid raccoons were first diagnosed in Florida during the 1940s, but the viral origin of these cases remains speculative. The cases in raccoons gradually spread into the southeastern United States over the next 30 years. Due to their importance in the fur trade and in hunting, it was not unusual for raccoons to be moved between regions. Rabid raccoons were discovered in some shipments of raccoons being moved for hunting purposes (Nettles, Shaddock, Sikes, & Reyes, 1979). In the late 1970s, a new focus of rabid raccoons appeared in the mid-Atlantic region. By 1991, the number of reported rabid raccoons had outnumbered rabid skunks. In 2000, the affected area stretched east of the Appalachian Mountains from eastern Ontario to Florida, and westward into Alabama and Ohio, making the raccoon the single most important rabies reservoir in the United States (Anonymous, 2000; Jenkins & Winkler, 1987). Raccoons occur throughout a large geographical range in North America and due to their high adaptability, often occur in remarkably high density in suburban and urban environments. Together with the existence of a species adapted viral variant (Smith, Sumner, Roumillat, Baer, & Winkler, 1984), raccoon rabies is a particularly salient demonstration of the threat of an old "known" disease that can newly emerge in a different species or geographic area (Hanlon & Rupprecht, 1998).

Despite its relevance as an abundant and capable rabies reservoir species, rabies pathogenesis in raccoons is neither well understood nor well described experimentally. Most published responses of raccoons to rabies virus infection have been conducted using a number of other isolates, including a fox (Sikes & Tierkel, 1961), a bat (Constantine, 1966), a skunk (Hill & Beran, 1992; Hill et al., 1993), and a dog (Rupprecht, Hamir, Johnston, & Koprowski, 1988). Experimental studies investigating the pathogenesis of raccoon rabies virus in raccoons are limited (McLean, 1975; Winkler, Shaddock, & Bowman, 1985; Winkler & Jenkins, 1991), partially due to the inability to definitively characterize such variants prior to the advent of monoclonal antibodies in 1978 (Wiktor & Koprowski, 1978).

In a recent study of rabies in raccoons, animals were inoculated with rabies virus several times over one year, and their responses were monitored (Niezgoda, Diehl, Hanlon, & Rupprecht, 1991). The incubation period was independent of viral dose and was ~50 days (range 23 to 92 days), and the morbidity period was ~4–5 days (range 2 to 10 days). Not all raccoons succumbed or seroconverted. All raccoons euthanized with clinical signs of rabies were positive for rabies virus antigen in CNS, whereas all inoculated animals that did not manifest clinical rabies remained healthy and were free of antigen upon euthanasia.

Following an exposure to a rabid animal, raccoons may or may not develop clinical signs. They may remain uninfected and healthy (McLean, 1975). This may be due to a variety of reasons. Rabies virus may not be shed from the infected animal during the exposure event, or the concentration of excreted virus may be minimal, or the severity of the exposure might be insufficient to cause infection. Despite a lack of infection, raccoons may be sufficiently exposed to antigen or even infectious virus but a productive viral infection never develops. In some individuals, an immune response may be detectable as measured by rabies virus neutralizing antibodies. In contrast, no antibodies may be detected following multiple rabies exposures, but protective immunity may have developed, nonetheless, perhaps consisting of cell-mediated immunity.

Based on both laboratory and field observations, several theoretical groups may exist in free-ranging raccoons. For example, in a study conducted in four different areas of Pennsylvania, the lowest geometric mean titer (GMT) of rabies virus neutralizing antibodies was found in a "rabies-free" area. A higher GMT was found at the presumptive front in two areas, and the highest GMT occurred within a rabies epizootic area (Winkler & Jenkins, 1991). All of these antibody positive raccoons were clinically normal when originally live-trapped. Additionally, in raccoons that were live-trapped during a raccoon rabies epizootic in Philadelphia's Fairmount Park during 1990, only 1 of 14 clinically rabid animals had detectable rabies virus antibody, and all were later confirmed rabid by laboratory diagnosis. Hence, the demonstration of rabies virus neutralizing antibody in a free-ranging raccoon simply indicates that an adaptive immune response from exposure to rabies virus antigen has occurred. The source of immunizing antigen may have been through exposure to whole live street virus or through vaccination. The detection of antibodies may be reflective of acquired immunity, rather than indicative of a viral incubation phase and imminent illness.

Similarly, rabid raccoons that succumb to a productive infection may or may not live long enough after productive infection to develop virus neutralizing antibody. Experimental data suggest that raccoons in terminal stages of disease will probably be seropositive, whereas during earlier stages it would not be unusual for an animal to be antibody negative. Also, if seroconversion does occur at all, the GMT may be higher in raccoons that succumb to infection, in comparison to those that are exposed, and seroconvert, without detectable illness. Under common field scenarios, rabid raccoons may succumb much earlier than in laboratory confinement and thus would most likely remain antibody negative.

In areas believed to be free of terrestrial rabies based on surveillance (such as much of the eastern United States prior to the late 1970s), a majority of raccoons are presumed to be largely naive, non-exposed, antibody-negative animals. Occurrence of minimal seroprevalence for example of up to 3% in low to non-enzootic areas may be coincident with rare rabies exposures, such as a raccoon consuming an infected bat. It must be kept in mind that the limit of detection and quantitation of true rabies virus neutralizing antibody titers may be different for various species, and that method validation for these various purposes is essential so as to evaluate potential nonspecific viral inhibitors within sera of individuals or populations (Hill & Beran, 1992; Hill, Beran, & Clark, 1992; Hill et al., 1993). Alternatively, intentional human intervention through parenteral immunization with inactivated rabies vaccine via wildlife rehabilitation, purposeful trap-vaccinate-release programs, or more recently, programs evaluating the effect of oral rabies vaccination upon wildlife, could also account for virus neutralizing activity.

Categorically, to include complete theoretical outcomes following exposure, a certain small fragment of raccoon population could be exposed to rabies virus, develop neurologic disease, yet recover, with or without obvious sequelae. Nevertheless, documented recovery from clinical rabies is quite rare. A single report exists of an experimentally infected raccoon that showed compatible clinical signs but recovered, and it was previously vaccinated (Rupprecht et al., 1988). Moreover, it is difficult to postulate how an acute encephalomyelitis such as rabies could be successfully cleared spontaneously without some residual neurological damage. The probability for survival in the field for such an obviously debilitated animal would be assuredly low. To date, there is no experimental evidence suggesting a chronic carrier state in raccoons, that is, infected animals that shed virus in the saliva, but remain clinically normal. Rather, a scenario of exposure leading to clinical rabies and death, or the induction of protective immunity, seems a much better fit with current experimental and field data.

8 SKUNKS

The occurrence of rabies in skunks was first described in the New World in the 19th century. Consisting of at least four different species, and several sub-species, skunks are small- to moderate-sized carnivores with specialized anal glands and are widespread in parts of North America. Rabies in skunks was first detailed in the mid-1800s from the American prairies, with northward spread into Canada by the next century. Rabid skunks were noted to be tenacious biters, and at times the disease seemed so prevalent that some proposed the name Rabies mephitica to describe its occurrence in association with spotted skunks (Steele & Fernandez, 1991). Today, in North America, reports of rabies in skunks occur mainly in four geographic regions: (1) the eastern United States; (2) the north central United States and the Canadian provinces of Manitoba, Saskatchewan, and Alberta; (3) California; and (4) the south central United States and Mexico (Aranda & Lopez-de, 1999; Crawford-Miksza, Wadford, & Schnurr, 1999; de Mattos et al., 1999; Gremillion-Smith & Woolf, 1988). Rabies in these areas (in skunks and, to a large extent, in other terrestrial mammals) is caused mainly by four different street virus variants. In the eastern United States, cases are primarily related to spillover from infected raccoons, whereas in the other three, the viruses are adapted to skunks as the primary reservoir. Earlier work suggested that basic differences in host susceptibility would help to explain geographic patterns of maintenance, in that red foxes could succumb relatively quickly to large doses of virus inoculated by rabid skunks (Sikes, 1962). This observation was not found to be universally true, and predated the discovery of antigenic variants of rabies virus. Moreover, the role of several independently maintained rabies virus variants and different skunk species is not precisely understood. Other experimental studies suggest that species specificity of enzootic rabies is due, at least partly, to host immune response, and differences in the route, dose, and pathogenicity of these variants of rabies virus (Charlton, Casey, & Campbell, 1987; Charlton, Webster, Casey, & Rupprecht, 1988; Charlton, Casey, Wandeler, & Nadin-Davis, 1996).

Besides route of inoculation and virus dose, the rabies virus variant may also produce different outcomes in skunks. For example, groups of skunks were inoculated intramuscularly with different dilutions of virus from salivary glands of naturally infected skunks in Canada and the United States collected in areas reflective of either the red fox or raccoon rabies outbreak. While there was no significant difference in basic susceptibility, incubation period, or the spectrum of clinical signs, skunks infected with the rabies isolate from the United States exhibited a morbidity period and shed virus in the saliva over 2–3 days. However, skunks infected with an isolate transmitted among red foxes in Ontario, Canada, were clinically rabid and shed virus in the saliva for an average of 1 week, suggestive of significant differences in the infective potential of the two different variants (Charlton et al., 1988; Charlton, Webster, & Casey, 1991; Hill et al., 1993).

Other experimental investigations of rabies in skunks suggest a major role of local virus replication in muscle tissue during the incubation period (Charlton, Nadin-Davis, Casey, & Wandeler, 1997). Two months post-inoculation, muscle at the inoculation site contained viral RNA, even though other tissues involved in the route of viral migration and early entrance into the CNS were negative. Viral antigen was located within striated muscle fibers and fibrocytes, thus confirming earlier observations, which indicated that virus may replicate in local tissues prior to invasion of the nervous system.

Rabies virus strains may preferentially infect different cell populations in the CNS of skunks, even by the same route of infection. For example, examination of the cerebellum of skunks experimentally infected with either a skunk isolate of street rabies virus or the fixed virus strain, challenge virus standard (CVS), revealed a differential response based on viral inoculum (Jackson, Phelan, & Rossiter, 2000). The skunk rabies virus variant displayed prominent infection of glia cells, with a relatively small amount of antigen in the perikarya of Purkinje cells. In contrast, the highly adapted CVS strain showed many intensely labeled Purkinje cells and relatively few infected glia cells. Previously, it was suggested that the relative accumulations and effects of different rabies viruses in the CNS could account for either furious or paralytic phases of the disease (Smart & Charlton, 1992). Thus, while lyssaviruses are highly neurotropic, rabies virus variants and laboratory strains of attenuated viruses differ in their predilection for various neural elements and locations.

Although experimental studies have somewhat illuminated the pathobiology of rabies in skunks, fewer observations have been made from naturally infected animals. In the end stages of the disease, rabies virus may be detected most consistently in the submandibular salivary glands and sometimes with high concentrations of virus (Charlton, Casey, & Webster, 1984). Serendipitously, during a study of skunk behavior using radio telemetry, the behavior of normal and rabid animals was described in North Dakota (Greenwood, Newton, Pearson, & Schamber, 1997). In 1991, only one of 23 skunks under study was found to be rabid, whereas during 1992, 35 of 50 (70%) were diagnosed as rabid. The estimated survival rate of skunks was 85% during the spring of 1991, but dropped to 17% during the same season in 1992. Nearly a third of rabid skunks were located in dens below ground. No differences were observed between healthy and rabid skunks in estimated mean rate of travel per hour, distance traveled per night, or home range. Among rabid skunks, mean rate of travel, and distance traveled per night, tended to decrease with the onset of illness. Mean home ranges of male skunks were greater than for females before illness, but not after the demonstration of clinical signs. The home range of females did not differ when compared before or after signs of rabies. Rabid animals were more spatially clumped than expected, but no relationship was detected between locations of rabid skunks and dates of death. The finding of this study provided a glimpse into how rabies may affect an animal's use of space, an estimate of the relative impact of rabies upon a population, and, perhaps more importantly, how many cases of rabies in wildlife typically escape detection through routine passive public health surveillance.

In the raccoon rabies enzootic area of the United States, skunks are the most frequently reported species after raccoons (Krebs, Smith, Rupprecht, & Childs, 1999). To date, all skunks examined have been associated with the raccoon rabies virus variant and appear to be related to infection from raccoons, based on spatial and temporal submission patterns. However, on a few occasions (e.g., in Massachusetts, Rhode Island, etc.), rabid skunks have outnumbered rabid raccoons. Whether this variant will evolve towards or is already capable of independent rabies transmission in skunks remains to be seen. Such an emergence would contribute to increased public health implications in regard to human and domestic animal risk of exposure, as well as provide further substantial challenges to wildlife rabies control, perhaps through oral vaccination, due to the absence of an effective oral rabies vaccine packaged in baits for skunks.

9 MONGOOSES AND THEIR ALLIES

Historically, mongoose rabies viruses in Zimbabwe were isolated on several occasions in the 1970s from herpestid species, primarily the slender mongoose (*Galerella sanguinea*). These observations led to the assumption that the slender mongoose was the reservoir host species for this rabies virus (RABV) variant in that country (Foggin, 1988). However, the mongoose variant was further isolated from mustelids (such as the honey badger [*Mellivora capensis*]) and increasingly from the African civet (*Civettictis civetta*) from the 1990s to date (Bingham, Hill, & Matema, 1994; Sabeta et al., 2008). At least three rabies viruses of the mongoose biotype have been identified from the African civet. It has since been realized that the herpestids (mongooses) are morphologically and genetically distinct from the viverrids (genets and civets) (Flynn, Finarelli, Zehr, Hsu, & Nedbal, 2005), and this has led to the recognition of a "mongoose RABV" biotype. Mongooses infected with rabies virus in Zimbabwe are thought to be incidental hosts, whereas in South Africa, the yellow mongoose (*Cynictis penicillata*) was demonstrated to be the reservoir host (Nel et al., 2005; Sabeta, Bingham, & Nel, 2003).

Although there is much in the literature about molecular epidemiology of rabies in mongooses, there is a paucity of published information about the clinical presentation in mongooses. In one study, mongooses were inoculated with either a canine rabies virus variant or a mongoose-associated virus variant (Chaparro & Esterhuysen, 1993). Mongooses infected with the *Cynictis* isolate shed virus in saliva for up to seven days before death. It was also observed that those infected with smaller doses shed more virus in the saliva than those infected with higher doses. The susceptibility of mongooses to the canine isolate was much lower (2 of 18 developed rabies; only one of the two shed virus in saliva) than those inoculated with the mongoose-associated virus (9 of 18 died of rabies). The clinical signs were described as variable and included a loss of fear of humans or conversely reclusive behavior, and vocalization characterized as "barking." The clinical development of disease progressed to incoordination, paresis, paralysis, and coma. Aggression was observed in one animal, whereas two animals were found moribund with no other premonitory signs. Another example of clinical signs of lyssavirus infection in a mongoose involves a case report from August 2004 that describes an animal with disorientation, attacking inanimate objects, and alternating between friendly and aggressive (Markotter et al., 2006). The suspected rabid mongoose was captured in a residential area in a marshy valley due to abnormal behavior. A Lagos Bat virus (LBV) isolate was characterized by antigenic typing with antinucleocapsid monoclonal antibodies, sequencing of the nucleoprotein gene, and evaluation of pathogenicity by the peripheral route in laboratory mice. This was the first identification of LBV in a wild terrestrial species, which was identified as *Atilax paludinosus*, through analysis of cytochrome b sequencing data, and is commonly known as the water or marsh mongoose.

With regard to rabies in mongooses in Cuba and Puerto Rico, it appears that the variant evolved from a progenitor related to canine rabies in Mexico. In contrast, the viruses that circulate in other mongoose reservoirs in Africa and Asia are very different both from each other and from the Cuban viruses. The occurrence of rabies in mongoose in Sri Lanka (Arai et al., 2001) also indicates a close relationship to canine rabies virus variants in the local geographic area.

As previously described rabies in mongoose in Cuba and Sri Lanka appears to originate from adaptation of canine rabies virus variants to mongoose, whereas it appears possible that the mongoose rabies variant isolated from slender mongooses in Zimbabwe (Foggin, 1988) is not a descendent of the cosmopolitan dog variants but is unique unto itself

(Nel et al., 2005). Subsequent isolations of this variant from other species such as the honey badger and African civet demonstrated the involvement of a larger pool of wildlife, similar to the situation occurring in South Africa (Bingham, Javangwe, Sabeta, Wandeler, & Nel, 2001). Considering the large territorial habitat of the slender mongoose, covering around 75% of the African continent, the potential for undiscovered cycles of mongoose rabies north of Zimbabwe may be significant. Adequate primary disease surveillance and advanced diagnostic approaches to elucidate the epizootiology rabies and related virus variants are essential to successful rabies control and prevention. Following successful extinction of rabies virus variants associated with domestic dogs in North America and Western Europe, a marked increase of disease discovery in wildlife occurred (Chang et al., 2002; Krebs, Smith, Rupprecht, & Childs, 2000; Krebs, Williams, Smith, Rupprecht, & Childs, 2003; Rupprecht, Smith, Fekadu, & Childs, 1995). Rabies virus variants specifically adapted to various wildlife species, such as raccoons, skunks, and arctic foxes, and a variety of others were identified (Krebs et al., 2000; Nadin-Davis et al., 2001; Smith, 1988). It is clear that understanding the molecular epidemiology of wildlife virus isolates, in addition to those involved in the canine-to-canine transmission in the same geographical area, will facilitate appropriate animal rabies prevention measures, as well as public and medical education towards the prevention of human rabies. The relatively low case numbers of mongoose rabies and paucity of data from Africa underscores the need for enhanced surveillance of lyssaviruses and accurate identification not only in humans, dogs, and other domestic animals, but also among wildlife species.

10 FOXES

The red fox (Vulpes vulpes) is one of the most widely distributed and abundant wild carnivores in the world, occurring in Eurasia, North America, and northern Africa, and, through intentional introduction, in Australia. Due to the significance of this species as a major reservoir, and the influence of directed research during the past 30 years, coordinated by the World Health Organization (Wandeler et al., 1974a,b,c), more may be known about the rabies virus variants occurring in this species and the ecologic characteristics contributing to their successful maintenance than for any other variant or host species of rabies viruses.

Foxes appear quite susceptible to experimental infection. The amount of virus needed to result in 50% or more infection of red foxes in captivity varied from less than or equal to 1–5 mouse intracerebral lethal dose 50% ($MICLD_{50}$) in the masseter, neck, or cervical muscles, to 16 $MICLD_{50}$ or more in the gluteal muscles (Black & Lawson, 1970; Parker & Wilsnack, 1966; Sikes, 1962; Wandeler et al., 1974b; Winkler, McLean, & Cowart, 1975). Foxes were more resistant to rabies when infected by isolates of canine, bat, or raccoon dog origin (Black & Lawson, 1970; Blancou, Aubert, & Artois, 1991; Parker & Wilsnack, 1966; Sikes, 1962; Wandeler et al., 1974b; Winkler et al., 1975). It took approximately 5 more logs of rabies virus to infect foxes orally than parenterally (Wandeler, 1980). Minimum incubation periods have ranged from as short as 4 days to longer than 15 months, but most are between 2 weeks and 3 months, are inversely related to dose, as is the proportion of foxes that have virus in the salivary glands, and the relative quantity of virus recovered (Blancou et al., 1991; Wandeler, 1980). Thus, it appears that the higher the dose of rabies virus inoculum, the greater the mortality, but fewer of the rabid animals were shedding virus and thus fewer would be a threat for transmission to conspecifics. Conversely, a lower amount of inoculated virus may result in fewer productive infections, but a greater proportion of animals that develop rabies would be capable of shedding virus in the saliva. Considering that most foxes tend to excrete on the average between 10^3 to 10^4 virus particles, this appears to be more than adequate to ensure transmission, given the susceptibility of this species. Short morbidity periods of 2–3 days are the usual rule regardless of dose, but have ranged from <1 to >14 days (Aubert, 1992; Wandeler, 1980). Signs are variable but commonly include anorexia, restlessness, hyperactivity, ataxia, and aggression (Sykes-Andral, 1982). Radio-tracked rabid foxes experience abnormal behaviors, such as spatial-temporal alterations and frequent prostration, and may acquire wounds from normal foxes responding to territorial incursions (Andral, Artois, Aubert, & Blancou, 1982). Virus may be excreted in the saliva concomitant with, or as long as a month before, obvious clinical signs develop (Aubert, 1992). Unlike the case for some other mammals, herd immunity is not believed to be important in foxes. Why enzootic fox rabies substantially diminished in the rate of spread in France (Blancou et al., 1991), or seemingly disappeared in the United States during the 1970s (Winkler, 1975) remains a mystery.

In addition to the red fox, there are approximately 19 other extant "fox" species throughout the world (Nowak, 1991). Although many possess solitary habits and occupy only geographical remnants of a former domain, others are widespread and seem to meet basic social criteria for a successful rabies reservoir. The Arctic and gray foxes (*Urocyon cinereoargenteus*) have distinct associated rabies virus variants and are regionally important in the epidemiology of the polar and western American regions. However, similar information about other species of foxes and potentially associated rabies virus variants does not exist. Some biological limitations appear obvious, most related to direct human depredation or encroachment and subsequent habitat loss. Such a tale holds for many representatives among the genus Vulpes in the Old World, such as the quite social Eurasian corsac fox of the steppe and desert zones, Blanford's fox, the Bengal fox in India, Pakistan and Nepal, and the Tibetan sand fox. Alternatively, the Fennec fox, *Fennecus zerda*, seems confined to the North African desert biome, and the Cape fox has been severely persecuted in its restricted occurrence in dry areas of southern Africa. Little is known of the otherwise rather gregarious pale fox, found throughout Senegal to the Sudan. The bat eared fox, *Otocyon megalotis*, has been reported from Ethiopia to southern Africa, and is considered to be occasionally important in the local epidemiology of rabies (Nel, 1993; Sabeta, Mansfield, McElhinney, Fooks, & Nel, 2007), but it, too, declines near human habitation and suffers destruction from humans and domestic dogs. The New World Channel Islands fox, *Urocyon littoralis*, is restricted to offshore sites of southwestern California, and the kit and swift foxes are endangered species with extremely fragmented distributions, having disappeared over much of their former ranges. The hoary fox, *Lycalopex*, is confined to Brazil, where it is often killed for presumed predation upon fowl. The Culpeo fox, *Pseudalopex*, (and related species) extends over a fairly wide area of South America from the equator to Argentina, but it is a habitat specialist and may be limited in population density, as well as restricted to sanctuaries among park reserves. One possible candidate deserving of further study is the crab-eating fox, *Cerdocyon thous*. Populations of this 3–8 kg canid occur in a variety of savanna to woodland habitats from Columbia to Argentina, are omnivorous and locally abundant, and could prove to be an important species of concern in Latin America, given adequate attention to rabies surveillance.

11 COYOTES

The coyote is an adaptable and abundantly occurring medium-sized carnivore in the United States. Cases of rabies in coyotes were only sporadically reported from 1960 through the mid-1980s. For example, a Sonora canine rabies virus variant would occasionally be detected in animals along the west Texas border with Mexico (Rohde, Neill, Clark, & Smith, 1997). This situation began to change slowly at a focus near the south Texas–Mexico border, associated with another rabies virus variant known at least from the region since 1978, in coyotes and domestic dogs (Clark et al., 1994). During 1988, a south Texas county reported 6 confirmed cases of rabies in coyotes, and 2 cases in dogs. At the same time, an adjacent county reported 9 cases of rabid dogs. During 1989–90, 7 rabid coyotes and 65 dogs were reported in these areas. By 1991, the outbreak expanded approximately 160 km north, with a total of 42 rabid coyotes and 25 dogs over 10 counties. In 1992, it rose to 70 rabid coyotes and 41 dogs from a 12-county area, and by 1993, 71 of the 74 total cases in coyotes, and 42 of 130 total cases in dogs reported from the entire United States were from south Texas. By comparison, no other state reported more than 7 cases in 1993 in dogs. The risk of translocation to other areas was realized when the coyote rabies virus variant was identified from a dog infected on a compound in Alabama, where imported coyotes from Texas were released for hunting purposes (Krebs, Strine, Smith, Rupprecht, & Childs, 1994). Over some 18 counties in 1994, the number of coyote rabies cases reached 77, with 32 cases in dogs, and peaked at 80 rabid coyotes, with 36 cases in dogs, in 20 counties during 1995, when an oral vaccination program and perhaps the effect of drought on population numbers began to dampen the outbreak (Fearneyhough et al., 1998). During November and December 1994, rabies was diagnosed in five dogs from two associated kennels in Florida. In addition, two other dogs at one of the kennels died with suspected, but unconfirmed, rabies. The rabies virus recovered from these dogs was identified, as a rabies virus variant not previously found in Florida, but rather the same virus enzootic among coyotes in south Texas. The suspected source of infection was translocation of infected coyotes from Texas to Florida, also used in hunting enclosures and constituted the second translocation of coyote rabies (Anonymous, 1995a,b). Fortunately, cases of the dog/coyote rabies virus variant continued to decline each year from 1996–99, with 19, 4, 4 and 2 reports, respectively. One case was reported in 2000 and another in 2004; with no further in the US due to this variant since then. At least 2 human cases were associated with the Texas coyote rabies outbreak, in 1991 and 1994, but the historics surrounding the human exposures were unclear (Anonymous, 1991, 1995a).

Coyotes are susceptible to a number of rabies virus variants which is easily demonstrated by the case occurrences in a single year such as in 2010, when four coyotes in Texas were found to be infected with the Texas gray fox rabies virus variant, seven occurring in the raccoon rabies virus variant area were infected with this variant, two cases in the south central skunk rabies virus variant area were infected with this variant, and four in Arizona were infected with the Arizona gray fox virus variant. Despite the clear threat that this species and an associated rabies virus variant present to public health, the relative lack of information about rabies in coyotes may be best summed up by the following past opinion: "...although they are a potential hazard as a reservoir or vector of rabies, they do not appear to be of major epidemiologic significance..." (Sikes & Tierkel, 1966).

12 JACKALS

Jackals consist of three species of small- to medium-sized carnivores, namely, the black-backed jackal (*Canis mesomelas*) and the side-striped jackal (*Canis adustus*) of sub-Saharan Africa, and the golden jackal (*Canis aureus*) of northern Africa and south-central Eurasia. The black-backed and side-striped jackals are closely related to each other, whereas the larger golden jackal is more closely related to wolves and coyotes (Jhala & Moehlman, 2001; Lindblad-Toh et al., 2005). There is virtually nothing in the peer-reviewed published literature about the clinical presentation of rabies in jackals. At the level of the local population, case occurrences support a pattern of maintaining independent epizootics and then being free of rabies for periods of time is essentially similar in domestic dogs, jackals, and other canids. As such, some studies have questioned the ability of jackals to support rabies virus cycles (Cleaveland & Dye, 1995; Rhodes, Atkinson, Anderson, & Macdonald, 1998). Based on the study of discrete local dog populations, domestic dogs appear to support rabies virus infection endemically, whereas it is thought that local populations of jackals do not (Rhodes et al., 1998). At least some authors exert that the jackal populations of Zimbabwe and the rabies viruses they support should not be considered metapopulations because they do not have, as dog populations do, the spatial separation or interpopulation migration that would be necessary for them to be considered metapopulations. If population structure is viewed as a nested hierarchy of subpopulations, many smaller epidemics may occur in different subpopulations, where most transmission occurs at this local level, and broader spreading of a disease is driven by occasional long-range individual transport. An absence of a metapopulation structure may explain the failure of rabies persistence in jackal populations. This absence, rather than their inability to maintain virus, is proposed to distinguish jackals from dogs as hosts of rabies virus variants, particularly in Zimbabwe. But, as stated in the opening sentence of this section, there is an obvious lack of descriptive and research studies about the pathogenesis and clinical presentation of rabies in jackals, irrespective of rabies virus variant.

13 WOLVES

The oral tradition and written folk tales of Eurasia and North America contain many accounts of wolves attacking and killing people. Wolves appear to have always been implicated in transmitting rabies to humans in Europe and Asia. Some of the earliest reports date back to the 13th century (Butzeck, 1987). It is widely accepted that rabies in wolves represents incidental spillover cases from primary reservoir species and that it is unlikely that wolves have a species-specific rabies virus variant that is adapted for wolf-to-wolf transmission (Johnson, 1995). In areas where rabies occurs in red foxes, jackals, and domestic dogs, case intensity in wolves can be high (Chapman, 1978; Randall et al., 2004; Sikes & Tierkel, 1966). Wolves tend to develop the "furious" form of rabies (Turkmen et al., 2012). When the physical size, strength, and easy travel distance of a wolf is considered, a rabid wolf is perhaps the most dangerous rabid animal of all.

14 CATS

Members of the Felidae belong to one of two subfamilies: Pantherinae (which includes the tiger, the lion, the jaguar, and the leopard), and Felinae (which includes the cougar, the cheetah, the lynxes, the ocelot, and the domestic cat). Due to their strict carnivorous characteristics, felids are one of the most specialized groups in the Carnivora. With few exceptions, they exploit a rather solitary existence. Acquisition of a large body size in some groups (e.g., tiger, panther, jaguar, and leopard) has added utility in hunting hoofed stock but tends to limit absolute numbers. Rabid felids can be effective vectors of rabies. For example, after raccoons, skunks, foxes, and coyotes, the bobcat is the most common wild carnivore found with rabies, typically infected with the local terrestrial rabies virus variant such as those found in raccoons or skunks, in the United States, and human cases have resulted from infection by these wild felids. To date, no reservoirs have been described for any feline species.

It is estimated that the number of free-roaming abandoned and feral cats in the United States may be as high as that of owned cats (about 73 million in 2000)(Brown et al., 2005). Given the high rate of sterilization among owned cats, these unowned cats are the primary source of cat overpopulation. Animal shelters nationwide receive several million unwanted cats each year. Due to a shortage of available homes, approximately 75% of these cats are euthanized.

The impact of both owned and unowned free-roaming cats upon the environment is an ongoing subject of debate. Even well-fed cats will hunt and kill prey. These predations cause a significant—and preventable—loss of birds, small mammals, reptiles, and amphibians. Both owned and unowned, free-roaming cats pose small but important threats to human health. Zoonotic agents include rabies virus, Toxoplasma gondii, Bartonella spp., Toxocara cati, Microsporum canis, Cryptosporidium spp., Campylobacter spp., Yersina pestis, Cheyletiella spp., and Francisella tularensis (Brown et al., 2005). Also, human injury can occur if feral cats are handled without proper precautions or experience. If reportable zoonotic diseases are diagnosed, appropriate health officials must be notified.

Surveys indicate that 7 to 22% of U.S. households feed unowned cats, thus increasing their numbers. Few of these cats have been neutered. Public policies for addressing the free-roaming abandoned and feral cat situation should take into account the lack of public awareness about the seriousness of the problem, the bonding of caretakers to unowned cats, and the growing societal opposition to euthanasia. The veterinary profession can play an important role in preventing abandonment, and in providing education about feral cat issues.

Although not ideal, an interim solution to the problem of free-roaming abandoned and feral cats are responsibly managed colonies. The goal of colony management should be the eventual reduction of the colony through attrition. If feral cat colonies are maintained by humans, all cats should receive: a health examination, surgical neutering, vaccinations against rabies, feline panleukopenia, feline herpesvirus -1 , and calicivirus, identification by ear-tipping, testing for feline leukemia virus and feline immunodeficiency virus, adoption of kittens and socialized adult cats if homes are available, and return or removal from the colony of cats that cannot be adopted.

Permanent, enduring solutions to the problem of free-roaming abandoned and feral cats will be achievable when: animal control agencies are adequately funded; population numbers are effectively controlled; environmentally safe and effective contraceptive approaches are developed; appropriate public education campaigns concerning the problems and solutions associated with unowned cats are implemented; and when veterinary, wildlife, public health and other organizations with a focus on humane animal management work cooperatively toward common solutions.

15 CATTLE

The greatest economic and public health impact of rabies in cattle occurs in Latin America (Delpietro & Russo, 1996; Lord, 1992; Martinez-Burnes et al., 1997). The major source of infection is from vampire bats, predominantly the common vampire bat (*Desmodus rotundus*). Cattle may also be infected from dogs or foxes, in areas where such rabies is endemic. In North America, predominant sources of infection are from wild carnivore species, such as from skunks in the Midwest and raccoons in the east. In addition, cattle may infrequently become infected with insectivorous bat rabies virus variants. Among 47 rabies-positive bovine samples provided to the Centers for Disease Control and Prevention from various state health departments within the US, 8% (4/47) were bat-associated variants (J. S. Smith, personal communication).

In one description of rabies in cattle (n = 20) and sheep (n = 5) (Hudson, Weinstock, Jordan, & Bold-Fletcher, 1996a), clinical signs included excessive salivation (100%), behavioral change (100%), muzzle tremors (80%), vocalization (bellowing; 70%), aggression, hyperaesthesia and/or hyperexcitability (70%), and pharyngeal paresis/paralysis (60%). The furious form of rabies was seen in 70% of the cattle and in 80% of sheep (n = 5). In the diseased cattle, the average incubation period was 15 days and the average morbidity period was nearly 4 days. In many cases, rabies may be considered only in retrospect (Stoltenow, Solemsass, Niezgoda, Yager, & Rupprecht, 2000). Often, the animal exhibits signs of choking, and individuals may insert a hand in the mouth to search for a foreign body.

16 HORSES AND DONKEYS

Among 21 experimentally infected horses, the average incubation period was 12 days and average morbidity was nearly 6 days (Hudson, Weinstock, Jordan, & Bold-Fletcher, 1996b). Naive animals had significantly shorter incubation and morbidity periods (P<0.05) than test animals vaccinated with products under development. Tremor of the muzzle was most frequently observed (81%) and the most common initial sign. Other common signs were pharyngeal spasm or pharyngeal paresis (71%), ataxia or paresis (71%), lethargy or somnolence (71%). Although some initial presentations began as the dumb form, ultimately 43% of horses developed furious rabies.

17 SHEEP AND GOATS

The effects of the inoculation of a canine strain of rabies virus in sheep were studied using 10 animals which received different amounts of this virus (Soria, Artois, & Blancou, 1992). Two subjects, inoculated with $10(5.4)$ $MICLD_{50}$, died from rabies after 19 and 40 days of incubation. Clinical signs were anorexia, emaciation, nervous reactions and prostration before death. The virus was recovered from different parts of the central nervous system and salivary glands with high titers. Only three animals showed an antibody response, at very low levels.

18 OPOSSUMS AND OTHER MARSUPIALS

The only North American marsupial, the opossum (*Didelphis virginianus*), appears relatively resistant to experimental infection (Beamer, Mohr, & Barr, 1960), also reflected by consistently low numbers of naturally occurring cases. One of the greatest sources of rabies virus infection for the opossum is the raccoon, most likely due to ecological overlap of the two species in the suburban environment. The virtual absence of substantial reports despite the diversity of marsupials widely and abundantly distributed throughout Australia and South America, argues in part for fundamental taxonomic differences in species susceptibility and viral–host response.

19 RODENTS AND LAGOMORPHS

As demonstrated by historic numbers, the occurrence of rabies virus infection in rodents and lagomorphs is low and mostly related to body size and ecological characteristics factoring into the risk of exposure to locally circulating rabies virus variants and large-bodied species (Anonymous, 2001). For example, in association with the mid-Atlantic raccoon rabies epizootic, there was a >350% increase in rabies in rodents during the period 1985–1994 above the previous years of 1971–1984 (Childs et al., 1997). The majority of rabid rodents reports were in woodchucks (*Marmota monax*) with rare cases in beavers (*Castor Canadensis*) (Anonymous, 2002), porcupines, (*Erethizon dorsatum*) and others, and were, based on temporal and spatial occurrence, associated with the raccoon rabies virus variant. Rabies cases in small domestic rodents and lagomorphs represent a preventable problem where attention to housing to exclude animal contact alone could remove the risk of exposure. Rabbits infected with the virus usually develop paralytic rabies, but some rodents and lagomorphs develop the furious form of rabies. Similar to observations in other species, some rodents have died of rabies without having clinical signs. In one report, a naturally infected rabid guinea pig was euthanized before clear clinical signs associated with rabies were detected (Eidson et al., 2005). In a study of experimental rabies virus infection in wild rodents in 1972, clinical signs were not detected in approximately half of infected squirrels that died of rabies (Winkler, Schneider, & Jennings, 1972).

20 OTHER SPECIES

Rabies has the capacity to occur in unexpected species such as marine animals, as demonstrated by a case report in a ringed seal from Norway (Odegaard & Krogsrud, 1981). This animal was wounded and appeared confused. It deteriorated over the course of 5 days and later became aggressive. Rabies was confirmed by immunofluorescent testing (Dean & Abelseth, 1973) of the brain, and the case was presumed to be due to an epizootic of rabies among arctic foxes in the area. Within the published literature, and surveillance reports from Eurasia, Africa, and the Americas, there is a zoological mix of species diverse enough to be collated according to an alphabetical list. For example, these involve the aardwolf, armadillo, baboon, badger, bison, camel, caracal, chipmunk, civet, duiker, elephant, fox, squirrel, genet, honey badger, hyena, hyrax, Ictonyx, javelina, kudu, llama, lion, marmoset, Nasua, ocelot, opossum, otter, polecat, rabbit, rat, reindeer, roe deer, springbok, suricate, Taurotragus, ursids, vole, warthog, weasel, wildcat, Xerus, yak, zebra, and a host of others (Anonymous, 1990; Batista-Morais, Neilson-Rolim, Matos-Chaves, de Brito-Neto, & Maria-da-Silva, 2000; Berry, 1993; Cappucci, Jr., Emmons, & Sampson, 1972; Dieterich & Ritter, 1982; Dowda & DiSalvo, 1984; Frye & Cucuel, 1968; Karp, Ball, Scott, & Walcoff, 1999; Leffingwell & Neill, 1989; Rausch, 1975; Stoltenow et al., 2000; Swanepoel et al., 1993; Walroth, Brown, Wandeler, Casey, & MacInnes, 1996; Wimalaratne & Kodikara, 1999).These cases support the observation that practically all mammals are susceptible to rabies. Almost all occur as single, incidental observations. However, multiple occurrences in a herd, pack, or the like, can occur, as illustrated by the thousands of cases reported from kudu (Barnard & Hassel, 1981; Hubschle, 1988) and the recent report of cases among white-tailed deer in Pennsylvania (Petersen et al., 2012), probably enhanced by artificial propagation in game ranching. The tally sheet, however, is simply a record of victims, dead-end infections exposed by those individuals of a reservoir host species with a combination of evolutionary and ecological attributes that optimize transmission of one of the oldest known infectious diseases. Perhaps new reservoirs will continue to be identified, as suggested by the bats and their multitude of lyssavirus variants, among other mammals, whether large or small, or terrestrial, aquatic or aerial.

21 CONCLUSIONS

Humans travel, and animals are increasingly moved along with their human companions. The risk of animal movement in regard to disease introduction has been demonstrated repeatedly. Despite the extinction of dog-to-dog types of rabies viruses through enforcement of stray dog control measures and mandatory vaccination, the risk of re-introduction of related variants remains a real and compelling reason for continuing to require vaccination of domestic animals, especially dogs. Even though vaccination of an individual animal simplifies management in the event of an exposure to endogenous wildlife rabies virus variants, required vaccination of the population provides a measure of biosecurity against potential introduction of exogenous canine rabies virus variants. The recent translocation of young dogs from Puerto Rico, Thailand, India, and Iraq (Anonymous, 2008; Castrodale, Walker, Baldwin, Hofmann, & Hanlon, 2008), which developed rabies from their places of origin upon movement into Massachusetts, California, Washington and Alaska, and New Jersey demonstrate the risk due to the relative ease of global transit for persons and animals. The risk of disease translocation can be mitigated through carefully crafted requirements for animal identification, vaccination, serological monitoring, and advance planning essentially equating to a risk-reducing waiting or quarantine period. The most critical component of risk mitigation is education of animal owners and other members of the public as to the importance of these requirements for the prevention of rabies in domestic animals and the prevention of human exposure to rabies from domestic pets and livestock.

Like many zoonoses and other emerging infections, rabies control and prevention requires the cooperation of animal control personnel, law enforcement, and environmental conservation or natural resource agency personnel, veterinarians, diagnosticians, public health professionals, physicians, and others. Responsibilities start locally and on the state level. Rules and regulations pertinent to rabies control may exist under the purview of public health, agriculture, or, less commonly, wildlife agencies. With the majority of rabies cases now in a range of wildlife species, animal control officers in some localities have been trained and authorized to deal with situations involving a variety of domestic and wildlife species, not just dogs, or dogs and cats. At present, the majority of diagnostic, educational, epidemiologic, and rabies control and prevention responsibility is borne by public health agencies. Close coordination between multiple local, state, and federal entities will be necessary for updating current regulations and remaining prepared for disease emergence or re-introduction, be it unintentional or intentional, through the improvement and practice of comprehensive prevention and preparedness strategies. The need for accessible, interactive, real-time, Geographic Information Systems-based tools is critical for timely display of disease occurrence, planning of interactions, and assessment of disease prevention strategies. They will also be powerful tools for spatio-temporal analysis of land-use features and their interaction with intensity and occurrence of rabies and other diseases and environmental conditions, as well. Diligent attention and dedicated effort will be required to maintain and, indeed, even advance emerging and zoonotic disease control, with rabies as a tangible "best-practices" template, beyond the major advances made in the last 50 years.

References

Akira, S., Uematsu, S., & Takeuchi, O. (2006). Pathogen recognition and innate immunity. *Cell, 124*(4), 783–801.

Andral, L., Artois, M., Aubert, M. F., & Blancou, J. (1982). Radio-tracking of rabid foxes. *Comparative Immunology and Microbiology of Infectious Diseases, 5*(1-3), 285–291.

Anonymous, (1990). Rabies in a llama–Oklahoma. *Morbidity and Mortality Weekly Report, 39*(12), 203–204.

Anonymous, (1991). Human rabies–Texas, Arkansas, and Georgia, 1991. *Morbidity Mortality Weekly Report*, 40(44), 765–769.

Anonymous, (1995a). Human rabies–Alabama, Tennessee, and Texas, 1994. *Morbidity Mortality Weekly Report*, 44(14), 269–272.

Anonymous, (1995b). Translocation of coyote rabies–Florida, 1994. *Morbidity and Mortality Weekly Report*, 44(31), 580–581. 587.

Anonymous, (2000). Update: Raccoon rabies epizootic–United States and Canada, 1999. *Morbidity and Mortality Weekly Report*, 49(2), 31–35.

Anonymous, (2002). Rabies in a beaver–Florida, 2001. *Morbidity Mortality Weekly Report*, 51(22), 481–482.

Anonymous, (2008). Rabies in a dog imported from Iraq–New Jersey, June 2008. *Morbidity and Mortality Weekly Report*, 57(39), 1076–1078.

Arai, Y. T., Takahashi, H., Kameoka, Y., Shiino, T., Wimalaratne, O., & Lodmell, D. L. (2001). Characterization of Sri Lanka rabies virus isolates using nucleotide sequence analysis of nucleoprotein gene. *ACTA Virologica*, 45(5-6), 327–333.

Aranda, M., & Lopez-de, B. L. (1999). Rabies in skunks from Mexico. *Journal of Wildlife Diseases*, 35(3), 574–577.

Aubert, M. F. (1992). Epidemiology of fox rabies. In K. Bogel, F. M. Meslin, & M. M. Kaplan (Eds.), *Wildlife Rabies Control* (pp. 9–18). Kent: Wells Medical LTD.

Barnard, B. J. (1979). The role played by wildlife in the epizootiology of rabies in South Africa and South-West Africa. *Onderstepoort Journal of Veterinary Research*, 46(3), 155–163.

Barnard, B. J., & Hassel, R. H. (1981). Rabies in kudus (Tragelaphus strepsiceros) in South West Africa/Namibia. *Journal of the South African Veterinary Association*, 52(4), 309–314.

Batista-Morais, N., Neilson-Rolim, B., Matos-Chaves, H. H., de Brito-Neto, J., & Maria-da-Silva, L. (2000). Rabies in tamarins (Callithrix jacchus) in the state of Ceara, Brazil, a distinct viral variant? *Memorias do Instituto Oswaldo Cruz*, 95(5), 609–610.

Beamer, P. D., MOHR, C. O., & BARR, T. R. (1960). Resistance of the opossum to rabies virus. *American Joural of Veterinary Research*, 21, 507–510.

Bell, G. P. (1980). A possible case of interspecific transmission of rabies in insectivorous bats. *Journal of Mammology*, 61(3), 528–530.

Berry, H. H. (1993). Surveillance and control of anthrax and rabies in wild herbivores and carnivores in Namibia. *Revue Scientifique et Technique*, 12(1), 137–146.

Bingham, J., Foggin, C. M., Wandeler, A. I., & Hill, F. W. (1999). The epidemiology of rabies in Zimbabwe. 2. Rabies in jackals (Canis adustus and Canis mesomelas). *Onderstepoort Journal of Veterinary Research*, 66(1), 11–23.

Bingham, J., Hill, F. W., & Matema, R. (1994). Rabies incubation in an African civet (Civettictis civetta). *Veterinary Record*, 134(20), 528.

Bingham, J., Javangwe, S., Sabeta, C. T., Wandeler, A. I., & Nel, L. H. (2001). Report of isolations of unusual lyssaviruses (rabies and Mokola virus) identified retrospectively from Zimbabwe. *Journal of the South African Veterinary Association*, 72(2), 92–94.

Black, J. G., & Lawson, K. F. (1970). Sylvatic rabies studies in the silver fox (Vulpes vulpes). Susceptibility and immune response. *Canadian Journal of Comparative Medicine*, 34(4), 309–311.

Blancou, J., Aubert, M. F., & Artois, M. (1991). Fox rabies. In G. M. Baer (Ed.), *The natural history of rabies*. Boca Raton: CRC Press.

Brown, C. M., Conti, L., Ettestad, P., Leslie, M. J., Sorhage, F. E., & Sun, B. (2011). Compendium of animal rabies prevention and control, 2011. *Journal of the American Veterinary Medical Association*, 239(5), 609–617.

Brown, R. R., Elston, T. H., Evans, L., Glaser, C., Gulledge, M. L., Jarboe, L., et al. (2005). Feline zoonoses guidelines from the American Association of Feline Practitioners. *Journal of Feline Medicine and Surgery*, 7(4), 243–274.

Butzeck, S. (1987). [The wolf, Canis lupus L., as a rabies vector in the 16th and 17th centuries]. *Zeitschrift fur die Gesamte Hygiene und Ihre Grenzgebiete*, 33(12), 666–669.

Cappucci, D. T., Jr., Emmons, R. W., & Sampson, W. W. (1972). Rabies in an Eastern fox squirrel. *Journal of Wildlife Diseases, 8*(4), 340–342.

Casals, J. (1940). Influence of age factors on susceptibility of mice to rabies virus. *Journal of Experimental Medicine, 72*(4), 445–451.

Castrodale, L., Walker, V., Baldwin, J., Hofmann, J., & Hanlon, C. (2008). Rabies in a puppy imported from India to the USA, March 2007. *Zoonoses and Public Health, 55*(8-10), 427–430.

Chang, H. G., Eidson, M., Noonan-Toly, C., Trimarchi, C. V., Rudd, R., Wallace, B. J., et al. (2002). Public health impact of reemergence of rabies, New York. *Emerging Infectious Diseases, 8*(9), 909–913.

Chaparro, F., & Esterhuysen, J. J. (1993). The role of the yellow mongoose (Cynictis penicillata) in the epidemiology of rabies in South Africa–preliminary results. *Onderstepoort Journal of Veterinary Research, 60*(4), 373–377.

Chapman, R. C. (1978). Rabies: Decimation of a wolf pack in artic Alaska. *Science, 201*(4353), 365–367.

Charlton, K. M., Casey, G. A., & Campbell, J. B. (1987). Experimental rabies in skunks: Immune response and salivary gland infection. *Comparative Immunology and Microbiology of Infectious Diseases, 10*(3-4), 227–235.

Charlton, K. M., Casey, G. A., Wandeler, A. I., & Nadin-Davis, S. (1996). Early events in rabies virus infection of the central nervous system in skunks (Mephitis mephitis). *Acta Neuropathology, 91*(1), 89–98.

Charlton, K. M., Casey, G. A., & Webster, W. A. (1984). Rabies virus in the salivary glands and nasal mucosa of naturally infected skunks. *Canadian Journal of Comparative Medicine, 48*(3), 338–339.

Charlton, K. M., Nadin-Davis, S., Casey, G. A., & Wandeler, A. I. (1997). The long incubation period in rabies: Delayed progression of infection in muscle at the site of exposure. *Acta Neuropathology, 94*(1), 73–77.

Charlton, K. M., Webster, W. A., & Casey, G. A. (1991). Skunk rabies. In G. M. Baer (Ed.), *The Natural History of Rabies (pp. 307–324).* Boca Raton: CRC Press.

Charlton, K. M., Webster, W. A., Casey, G. A., & Rupprecht, C. E. (1988). Skunk rabies. *Reviews of Infectious Diseases, 10*(Suppl. 4), S626–S628.

Childs, J. E., Colby, L., Krebs, J. W., Strine, T., Feller, M., Noah, D., et al. (1997). Surveillance and spatiotemporal associations of rabies in rodents and lagomorphs in the United States, 1985–1994. *Journal of Wildlife Diseases, 33*(1), 20–27.

Childs, J. E., Trimarchi, C. V., & Krebs, J. W. (1994). The epidemiology of bat rabies in New York State, 1988–92. *Epidemiology and Infections, 113*(3), 501–511.

Clark, K. A., Neill, S. U., Smith, J. S., Wilson, P. J., Whadford, V. W., & McKirahan, G. W. (1994). Epizootic canine rabies transmitted by coyotes in south Texas. *Journal of the American Veterinary Medical Association, 204*(4), 536–540.

Cleaveland, S., & Dye, C. (1995). Maintenance of a microparasite infecting several host species: Rabies in the Serengeti. *Parasitology, 111*(Suppl), S33–S47.

Constantine, D. G. (1966). Transmission experiments with bat rabies isolates: Bite transmission of rabies to foxes and coyote by free-tailed bats. *American Joural of Veterinary Research, 27*(116), 20–23.

Crawford-Miksza, L. K., Wadford, D. A., & Schnurr, D. P. (1999). Molecular epidemiology of enzootic rabies in California. *Journal of Clinical Virology, 14*(3), 207–219.

Davis, P. L., Rambaut, A., Bourhy, H., & Holmes, E. C. (2007). The evolutionary dynamics of canid and mongoose rabies virus in Southern Africa. *Archives of Virology, 152*(7), 1251–1258.

de Mattos, C. C., de Mattos, C. A., Loza-Rubio, E., Aguilar-Setien, A., Orciari, L. A., & Smith, J. S. (1999). Molecular characterization of rabies virus isolates from Mexico: Implications for transmission dynamics and human risk. *American Journal of Tropical Medicine and Hygiene, 61*(4), 587–597.

Dean, D. J., & Abelseth, M. K. (1973). Laboratory techniques in rabies: The fluorescent antibody test. *Monograph Series of the World Health Organization, 23*, 73–84.

Delpietro, H. A., & Russo, R. G. (1996). [Ecological and epidemiologic aspects of the attacks by vampire bats and paralytic rabies in Argentina and analysis of the proposals carried out for their control]. *Revue Scientifique et Technique, 15*(3), 971–984.

Dieterich, R. A., & Ritter, D. G. (1982). Rabies in Alaskan reindeer. *Journal of the American Veterinary Medical Association, 181*(11), 1416.

Dowda, H., & DiSalvo, A. F. (1984). Naturally acquired rabies in an eastern chipmunk (Tamias striatus). *Journal of Clinical Microbiology, 19*(2), 281–282.

Eidson, M., Matthews, S. D., Willsey, A. L., Cherry, B., Rudd, R. J., & Trimarchi, C. V. (2005). Rabies virus infection in a pet guinea pig and seven pet rabbits. *Journal of the American Veterinary Medical Association, 227*(6), 932–935, 918.

Fearneyhough, M. G., Wilson, P. J., Clark, K. A., Smith, D. R., Johnston, D. H., Hicks, B. N., et al. (1998). Results of an oral rabies vaccination program for coyotes. *Journal of the American Veterinary Medical Association, 212*(4), 498–502.

Fekadu, M. (1972). Atypical rabies in dogs in Ethiopia. *Ethiopean Medical Journal, 10*(3), 79–86.

Fekadu, M., Shaddock, J. H., & Baer, G. M. (1981). Intermittent excretion of rabies virus in the saliva of a dog two and six months after it had recovered from experimental rabies. *American Journal of Tropical Medicine and Hygiene, 30*(5), 1113–1115.

Fekadu, M., Shaddock, J. H., & Baer, G. M. (1982). Excretion of rabies virus in the saliva of dogs. *Journal of Infectious Diseases, 145*(5), 715–719.

Fekadu, M., Shaddock, J. H., Chandler, F. W., & Baer, G. M. (1983). Rabies virus in the tonsils of a carrier dog. *Archives of Virology, 78*(1-2), 37–47.

Flynn, J. J., Finarelli, J. A., Zehr, S., Hsu, J., & Nedbal, M. A. (2005). Molecular phylogeny of the carnivora (mammalia): Assessing the impact of increased sampling on resolving enigmatic relationships. *Systemative Biology, 54*(2), 317–337.

Foggin, C. M. (1988). *Rabies and rabies-related viruses in Zimbabwe. Historical, virological and ecological aspects.* (Dissertation, University of Zimbabwe, Zimbabwe).

Frye, F. L., & Cucuel, J. P. (1968). Rabies in an ocelot. *Journal of the American Veterinary Medical Association, 153*(7), 789–790.

Greenwood, R. J., Newton, W. E., Pearson, G. L., & Schamber, G. J. (1997). Population and movement characteristics of radio-collared striped skunks in North Dakota during an epizootic of rabies. *Journal of Wildlife Diseases, 33*(2), 226–241.

Gremillion-Smith, C., & Woolf, A. (1988). Epizootiology of skunk rabies in North America. *Journal of Wildlife Diseases, 24*(4), 620–626.

Hanlon, C. A., & Rupprecht, C. E. (1998). The reemergence of rabies: *Emerging infections I*. Washington: ASM Press. (pp. 59–80).

Hill, R. E., & Beran, G. W. (1992). Experimental inoculation of raccoons (*Procyon lotor*) with rabies virus of skunk origin. *Journal of Wildlife Diseases, 28*(1), 51–56.

Hill, R. E., Beran, G. W., & Clark, W. R. (1992). Demonstration of rabies virus-specific antibody in the sera of free-ranging Iowa raccoons (*Procyon lotor*). *Journal of Wildlife Diseases, 28*(3), 377–385.

Hill, R. E., Smith, K. E., Beran, G. W., & Beard, P. D. (1993). Further studies on the susceptibility of raccoons (Procyon lotor) to a rabies virus of skunk origin and comparative susceptibility of striped skunks (Mephitis mephitis). *Journal of Wildlife Diseases, 29*(3), 475–477.

Hubschle, O. J. (1988). Rabies in the kudu antelope (Tragelaphus strepsiceros). *Reviews of Infectious Diseases, 10*(Suppl. 4), S629–S633.

Hudson, L. C., Weinstock, D., Jordan, T., & Bold-Fletcher, N. O. (1996a). Clinical features of experimentally induced rabies in cattle and sheep. *Zentralblatt fur Veterinary Medizin B, 43*(2), 85–95.

Hudson, L. C., Weinstock, D., Jordan, T., & Bold-Fletcher, N. O. (1996b). Clinical presentation of experimentally induced rabies in horses. *Zentralblatt fur Veterinary Medizin B, 43*(5), 277–285.

Jackson, A. C., Phelan, C. C., & Rossiter, J. P. (2000). Infection of Bergmann glia in the cerebellum of a skunk experimentally infected with street rabies virus. *Canadian Journal of Veterinary Research, 64*(4), 226–228.

Jenkins, S. R., & Winkler, W. G. (1987). Descriptive epidemiology from an epizootic of raccoon rabies in the Middle Atlantic States, 1982–1983. *American Journal of Epidemiology, 126*(3), 429–437.

Jhala, Y. V., & Moehlman, P. D. (2001). Golden Jackal *Canus aureus* Linnaeus 1758. In D. Macdonald (Ed.), *The New Encyclopedia of Mammals (pp. 156–161).* Oxford: Oxford University Press.

Johnson, R. H. (1995). Rabies vaccination of wolf-dog hybrids. *Journal of the American Veterinary Medical Association, 206*(4), 426–427.

Karp, B. E., Ball, N. E., Scott, C. R., & Walcoff, J. B. (1999). Rabies in two privately owned domestic rabbits. *Journal of the American Veterinary Medical Association, 215,* 12(1824-7), 1806.

Kaufmann, J. H. (1982). Raccoons and allies. In J. A. Chapman & G. A. Geldhamer (Eds.), *Wild Mammals of North America (pp. 567–785).* Baltimore: The Johns Hopkins University Press.

Krebs, J. W., Smith, J. S., Rupprecht, C. E., & Childs, J. E. (1999). Rabies surveillance in the United States during 1998. *Journal of the American Veterinary Medical Association, 215*(12), 1786–1798.

Krebs, J. W., Smith, J. S., Rupprecht, C. E., & Childs, J. E. (2000). Mammalian reservoirs and epidemiology of rabies diagnosed in human beings in the United States, 1981–1998. *Annals of the New York Academy of Sciences, 916,* 345–353.

Krebs, J. W., Strine, T. W., Smith, J. S., Rupprecht, C. E., & Childs, J. E. (1994). Rabies surveillance in the United States during 1993. *Journal of the American Veterinary Medical Association, 205*(12), 1695–1709.

Krebs, J. W., Williams, S. M., Smith, J. S., Rupprecht, C. E., & Childs, J. E. (2003). Rabies among infrequently reported mammalian carnivores in the United States, 1960–2000. *Journal of Wildlife Diseases, 39*(2), 253–261.

Leffingwell, L. M., & Neill, S. U. (1989). Naturally acquired rabies in an armadillo (Dasypus novemcinctus) in Texas. *Journal of Clinical Microbiology, 27*(1), 174–175.

Lindblad-Toh, K., Wade, C. M., Mikkelsen, T. S., Karlsson, E. K., Jaffe, D. B., Kamal, M., et al. (2005). Genome sequence, comparative analysis and haplotype structure of the domestic dog. *Nature, 438*(7069), 803–819.

Lord, R. D. (1992). Seasonal reproduction of vampire bats and its relation to seasonality of bovine rabies. *Journal of Wildlife Diseases, 28*(2), 292–294.

Manning, S. E., Rupprecht, C. E., Fishbein, D., Hanlon, C. A., Lumlertdacha, B., Guerra, M., et al. (2008). Human rabies prevention–United States, 2008: Recommendations of the Advisory Committee on Immunization Practices. *Morbity and Mortality Weekly Report Recommendations and Reports, 57*(RR-3), 1–28.

Markotter, W., Kuzmin, I., Rupprecht, C. E., Randles, J., Sabeta, C. T., Wandeler, A. I., et al. (2006). Isolation of Lagos bat virus from water mongoose. *Emerging Infectious Diseases, 12*(12), 1913–1918.

Martinez-Burnes, J., Lopez, A., Medellin, J., Haines, D., Loza, E., & Martinez, M. (1997). An outbreak of vampire bat-transmitted rabies in cattle in northeastern Mexico. *Canadian Veterinary Journal, 38*(3), 175–177.

McLean, R. G. (1975). Raccoon rabies. In G. M. Baer (Ed.), *The Natural History of Rabies (pp. 53–76).* New York: Academic Press.

Nadin-Davis, S. A., Casey, G. A., & Wandeler, A. I. (1994). A molecular epidemiological study of rabies virus in central Ontario and western Quebec. *Journal of General Virology, 75*(Pt 10), 2575–2583.

Nadin-Davis, S. A., Huang, W., Armstrong, J., Casey, G. A., Bahloul, C., Tordo, N., et al. (2001). Antigenic and genetic divergence of rabies viruses from bat species indigenous to Canada. *Virus Research, 74*(1-2), 139–156.

Nel, J. A. (1993). The bat-eared fox: A prime candidate for rabies vector? *Onderstepoort Journal of Veterinary Research, 60*(4), 395–397.

Nel, L. H., Sabeta, C. T., von, T. B., Jaftha, J. B., Rupprecht, C. E., & Bingham, J. (2005). Mongoose rabies in southern Africa: A re-evaluation based on molecular epidemiology. *Virus Research, 109*(2), 165–173.

Nettles, V. F., Shaddock, J. H., Sikes, R. K., & Reyes, C. R. (1979). Rabies in translocated raccoons. *American Journal of Public Health, 69*(6), 601–602.

Niezgoda, M., Briggs, D. J., Shaddock, J., & Rupprecht, C. E. (1998). Viral excretion in domestic ferrets (Mustela putorius furo) inoculated with a raccoon rabies isolate. *American Joural of Veterinary Research, 59*(12), 1629–1632.

Niezgoda, M., Diehl, D., Hanlon, C. A., & Rupprecht, C. E. (1991). Pathogenesis of street rabies virus in raccoons. *Wildlife Disease Association, 40th annual conference* (pp. 57–58). Fort Collins, Colorado.

Nowak, A. (1991). *Walkers, mammals of the world*. Baltimore: The Johns Hopkins University Press.

Odegaard, O. A., & Krogsrud, J. (1981). Rabies in Svalbard: Infection diagnosed in arctic fox, reindeer and seal. *Veterinary Record, 109*(7), 141–142.

Parker, R. L., & Wilsnack, R. E. (1966). Pathogenesis of skunk rabies virus: Quantitation in skunks and foxes. *American Journal of Veterinary Research, 27*(116), 33–38.

Petersen, B. W., Tack, D. M., Longenberger, A., Simeone, A., Moll, M. E., Deasy, M. P., et al. (2012). Rabies in captive deer, Pennsylvania, USA, 2007–2010. *Emerging Infectious Diseases, 18*(1), 138 141.

Randall, D. A., Williams, S. D., Kuzmin, I. V., Rupprecht, C. E., Tallents, L. A., Tefera, Z., et al. (2004). Rabies in endangered Ethiopian wolves. *Emerging Infectious Diseases, 10*(12), 2214–2217.

Rausch, R. L. (1975). Rabies in experimentally infected bears, Ursus spp., with epizootiologic notes. *Zentralblatt fur Veterinary Medizin B, 22*(5), 420–437.

Rhodes, C. J., Atkinson, R. P., Anderson, R. M., & Macdonald, D. W. (1998). Rabies in Zimbabwe: Reservoir dogs and the implications for disease control. *Philosophical Transactions of the Royal Society of London B Biological Sciences, 353*(1371).

Rohde, R. E., Neill, S. U., Clark, K. A., & Smith, J. S. (1997). Molecular epidemiology of rabies epizootics in Texas. *Clincal Diagnostic Virology, 8*(3), 209–217.

Rupprecht, C. E., Hamir, A. N., Johnston, D. H., & Koprowski, H. (1988). Efficacy of a vaccinia-rabies glycoprotein recombinant virus vaccine in raccoons (Procyon lotor). *Reviews of Infectious Diseases, 10*(Suppl. 4), S803–S809.

Rupprecht, C. E., Smith, J. S., Fekadu, M., & Childs, J. E. (1995). The ascension of wildlife rabies: A cause for public health concern or intervention? *Emerging Infectious Diseases, 1*(4), 107–114.

Sabeta, C. T., Bingham, J., & Nel, L. H. (2003). Molecular epidemiology of canid rabies in Zimbabwe and South Africa. *Virus Research, 91*(2), 203–211.

Sabeta, C. T., Mansfield, K. L., McElhinney, L. M., Fooks, A. R., & Nel, L. H. (2007). Molecular epidemiology of rabies in bat-eared foxes (Otocyon megalotis) in South Africa. *Virus Research, 129*(1), 1–10.

Sabeta, C. T., Shumba, W., Mohale, D. K., Miyen, J. M., Wandeler, A. I., & Nel, L. H. (2008). Mongoose rabies and the African civet in Zimbabwe. *Veterinary Record, 163*(19), 580.

Sikes, R. K. (1962). Pathogenesis of rabies in wildlife. I. Comparative effect of varying doses of rabies virus inoculated into foxes and skunks. *American Joural of Veterinary Research, 23*, 1041–1047.

Sikes, R. K., & Tierkel, E. S. (1961). Wildlife rabies studies in the southeast. *Proceedings of the sixtyfourth annual meeting of the US livestock sanitation association* (pp. 268–272). Charleston.

Sikes, R. K., & Tierkel, E. S. (1966). Wolf, fox and coyote rabies: *Proceedings of the national rabies symposium.* Atlanta: US Department of Health, Education and Welfare. (pp. 31–33).

Singer, A., Kauhala, K., Holmala, K., & Smith, G. C. (2009). Rabies in northeastern Europe–the threat from invasive raccoon dogs. *Journal of Wildlife Diseases, 45*(4), 1121–1137.

Smart, N. L., & Charlton, K. M. (1992). The distribution of Challenge virus standard rabies virus versus skunk street rabies virus in the brains of experimentally infected rabid skunks. *Acta Neuropathology*, *84*(5), 501–508.

Smith, J. S. (1988). Monoclonal antibody studies of rabies in insectivorous bats of the United States. *Reviews of Infectious Diseases*, *10*(Suppl. 4), S637–S643.

Smith, J. S. (1996). New aspects of rabies with emphasis on epidemiology, diagnosis, and prevention of the disease in the United States. *Clinical Microbiology Reviews*, *9*(2), 166–176.

Smith, J. S., Sumner, J. W., Roumillat, L. F., Baer, G. M., & Winkler, W. G. (1984). Antigenic characteristics of isolates associated with a new epizootic of raccoon rabies in the United States. *Journal of Infectious Diseases*, *149*(5), 769–774.

Smith, J. S., Yager, P. A., & Orciari, L. A. (1993). Rabies in wild and domestic carnivores of Africa: Epidemiological and historical associations determined by limited sequence analysis. *Onderstepoort Journal of Veterinary Research*, *60*(4), 307–314.

Somayajulu, M. V., & Reddy, G. V. (1989). Live dogs and dead men. *Journal of the Association of Physicians India*, *37*(9), 617.

Soria, B. R., Artois, M., & Blancou, J. (1992). Experimental infection of sheep with a rabies virus of canine origin: Study of the pathogenicity for that species. *Revue Scientifique et Technique*, *11*(3), 829–836.

Steele, J. H., & Fernandez, R. (1991). History of rabies and global aspects. In G. M. Baer (Ed.), *The natural history of rabies (pp. 1–24)*. Boca Raton: CRC Press.

Stoltenow, C. L., Solemsass, K., Niezgoda, M., Yager, P., & Rupprecht, C. E. (2000). Rabies in an American bison from North Dakota. *Journal of Wildlife Diseases*, *36*(1), 169–171.

Swanepoel, R., Barnard, B. J., Meredith, C. D., Bishop, G. C., Bruckner, G. K., Foggin, C. M., et al. (1993). Rabies in southern Africa. *Onderstepoort Journal of Veterinary Research*, *60*(4), 325–346.

Sykes-Andral, M. (1982). Behaviour of rabid animals studied at the Centre National d'Etudes sur la Rage. *Comparative Immunology and Microbiology of Infectious Diseases*, *5*(1-3), 337–342.

Tepsumethanon, V., Wilde, H., & Sitprija, V. (2008). Ten-day observation of live rabies suspected dogs. *Developments in Biological Standardization (Basel)*, *131*, 543–546.

Tordo, N. (1996). Characteristics and molecular biology of the rabies virus. In F. X. Meslin, M. M. Kaplan, & H. Koprowski (Eds.), *Laboratory Techniques in Rabies (pp. 28–51)*. Geneva: World Health Organization.

Turkmen, S., Sahin, A., Gunaydin, M., Tatli, O., Karaca, Y., Turedi, S., et al. (2012). A wild wolf attack and its unfortunate outcome: Rabies and death. *Wilderness and Environmental Medicine Journal.Jul*, 13. (Epub ahead of print).

Vaughn, J. B., Jr., Gerhardt, P., & Newell, K. W. (1965). Excretion of street rabies virus in the saliva of dogs. *Journal of the American Medical Association*, *193*, 363–368.

Vaughn, J. B., Gerhardt, P., & Paterson, J. C. (1963). Excretion of street rabies virus in saliva of cats. *Journal of the American Medical Association*, *184*, 705–708.

Walroth, R., Brown, N., Wandeler, A., Casey, A., & MacInnes, C. (1996). Rabid black bears in Ontario. *Canadian Veterinary Journal*, *37*(8), 492.

Wandeler, A., Muller, J., Wachendorfer, G., Schale, W., Forster, U., & Steck, F. (1974a). Rabies in wild carnivores in central Europe. III. Ecology and biology of the fox in relation to control operations. *Zentralblatt fur Veterinary Medizin B*, *21*(10), 765–773.

Wandeler, A., Wachendorfer, G., Forster, U., Krekel, H., Muller, J., & Steck, F. (1974b). Rabies in wild carnivores in central Europe. II. Virological and serological examinations. *Zentralblatt fur Veterinary Medizin B*, *21*(10), 757–764.

Wandeler, A., Wachendorfer, G., Forster, U., Krekel, H., Schale, W., Muller, J., et al. (1974c). Rabies in wild carnivores in central Europe. I. Epidemiological studies. *Zentralblatt fur Veterinary Medizin B*, *21*(10), 735–756.

Wandeler, A. I. (1980). Epdiemiology of fox rabies. In E. Zimen (Ed.), *The Red Fox (pp. 237–249)*. Hingham: Kluwer Boston Inc.

Wiktor, T. J., & Koprowski, H. (1978). Monoclonal antibodies against rabies virus produced by somatic cell hybridization: Detection of antigenic variants. *Proceedings of the National Academy of Sciences USA, 75*(8), 3938–3942.

Wimalaratne, O., & Kodikara, D. S. (1999). First reported case of elephant rabies in Sri Lanka. *Veterinary Record, 144*(4), 98.

Winkler, W. G., Schneider, N. J., & Jennings, W. L. (1972). Experimental rabies infection in wild rodents. *Journal of Wildlife Diseases, 8*(1), 99–103.

Winkler, W. G. (1975). Fox rabies. In G. M. Baer (Ed.), *The Natural History of Rabies (pp. 3–22)*. New York: Academic Press.

Winkler, W. G., & Jenkins, S. R. (1991). Raccoon rabies. In G. M. Baer (Ed.), *The Natural History of Rabies (pp. 325–340)*. Boca Raton: CRC Press.

Winkler, W. G., McLean, R. G., & Cowart, J. C. (1975). Vaccination of foxes against rabies using ingested baits. *Journal of Wildlife Diseases, 11*(3), 382–388.

Winkler, W. G., Shaddock, J. S., & Bowman, C. (1985). Rabies virus in salivary glands of raccoons (Procyon lotor). *Journal of Wildlife Diseases, 21*(3), 297–298.

World Health Organization, (2005). World health organization expert consultation on rabies. *World Health Organ Technical Report Series, 931*, 1–88. (back).

Zulu, G. C., Sabeta, C. T., & Nel, L. H. (2009). Molecular epidemiology of rabies: Focus on domestic dogs (Canis familiaris) and black-backed jackals (Canis mesomelas) from northern South Africa. *Virus Research, 140*(1–2), 71–78.

Bat Rabies

*Ashley C. Banyard[1], David T. S. Hayman[1,2,3],
Conrad M. Freuling[4], Thomas Müller[4], Anthony R.
Fooks[1,5], and Nicholas Johnson[1]*

[1]Wildlife Zoonoses and Vector Borne Diseases Research Group,
Department of Virology, Animal Health and Veterinary Laboratories
Agency, Weybridge, New Haw, Addlestone, Surrey, KT15 3NB, UK,
[2]Cambridge Infectious Diseases Consortium, Department of Veterinary
Medicine, Madingley Road, Cambridge, CB3 0ES, UK, [3]Department
of Biology, Colorado State University, Fort Collins, CO 80523, USA,
[4]Institute of Molecular Biology, Friedrich-Loeffler-Institut, Federal
Research Institute for Animal Health, D-17493 Greifswald - Insel Riems,
Germany, [5]National Consortium for Zoonosis Research, University of
Liverpool, Leahurst, Chester High Road, Neston, Wirral, CH64 7TE, UK

1 INTRODUCTION

The detection of lyssaviruses in bat species during the 20th century
has highlighted the role this complex group of mammals plays as the
principal reservoir for genetically diverse lyssaviruses. There are over
1,100 recognized species of bats classified within the order Chiroptera,
although distinction between species is often technically challenging due
to morphological similarities between different species. The determination of bat species has been aided by the development of genetic tools
that enable differentiation between very closely related species alongside
the more traditional morphological and ecological classification techniques. Such techniques serve to classify groups of mammals as accurately as possible. As a result, bat species continue to be reclassified.

Currently, the order Chiroptera is divided into 19 different families that are further classified into numerous subfamilies and genera

(Table 6.1). Additional classification has subdivided this order into suborders Vespertilioniformes and Pteropodiformes. The Vespertilioniformes includes three superfamilies of bats: the Emballonuroidae (sheath-tailed bats of the Old and the New Worlds that are divided across two subfamilies); the Noctilionoidea (a large highly diverse subfamily with seven families—some of which include omnivorous bat species); and the Vespertilionidea (the largest superfamily of insectivorous bats that includes four distinct families). The Pteropodiformes includes the family Pterpodidae (the Old World fruit bats sometimes referred to as the 'megabats') and five Old World families of insectivorous or carnivorous bats that make up the superfamily Rhinolophoidea, namely: the Craseonycteridae, the Hipposideridae, the Megadermatidae, the Rhinolophidae and the Rhinopomatidae (Hutcheon & Kirsch, 2006; Teeling et al., 2005). Across these suborders, bats can be described according to their feeding habits, and it is this feature of the bat life cycle that we will use to group lyssavirus infections of bats (Altringham, 2011; Giannini & Simmons, 2003) (Table 6.1). Detection of viral pathogens within different bat species, often highlighted through epizootics that have affected both human and animal populations, have fuelled extensive research into bats as hosts of viral pathogens. More recently, the detection of coronaviruses (Shirato et al., 2012), filoviruses (Towner et al., 2009) and henipaviruses (Halpin et al., 2011) have sparked further interest in bats as reservoirs of viral pathogens. However, it is the historical association of rabies virus and related lyssaviruses with bats that establishes this virus genus as the most notable zoonotic pathogen of bat origin.

Within bats, lyssaviruses have been characterized in only a small proportion of recognized species (Table 6.1). However, since the first descriptions of rabies virus in bats (Pawan, 1936), other divergent lyssavirus species have been detected in a wide range of chiropteran hosts. There are currently 12 defined lyssavirus species and a further 2 recently described viruses that all appear to cause, where reported, rabies pathogenesis in humans and terrestrial carnivores (Figure 6.1) (Kuzmin et al., 2010). Importantly, from the perspective of human vaccination, the different lyssavirus species have been divided into three phylogroups according to antigenic divergence and effectiveness of rabies vaccines. Phylogroup I viruses are all effectively neutralized by the current rabies vaccines, based on classical rabies virus strains while viruses categorized in phylogroups II and III are not neutralized by antibodies produced following standard rabies virus vaccination (Figure 6.1). Viruses include: *rabies virus* (RABV), which has been shown to infect a diverse range of species including bats in the New World and carnivores around the globe; *Lagos bat virus* (LBV), which has been detected in frugivorous bats, cats, and dogs but has not been associated with human

TABLE 6.1 The Current Classification of Bat Species and Association with Lyssavirus Infection

Suborder	Superfamily	Family	Genus Associated with Lyssavirus Infection	Lyssavirus Involved	Representative Reference
Pteropodiformes	Pteropodidae (Old world fruit bats)		Cynopterus	unidentified	Smith et al., 1967
			Eidolon	LBV	Kuzmin et al., 2008a
				MOKV	Kemp et al., 1972
			Eonycteris	unidentified	Lumlertdacha et al., 2005
			Epomophorus	LBV	Calisher et al., 2006; Crick et al., 1982
			Epomops	LBV	Calisher et al., 2006; Johnson et al., 2010
			Micropteropus	LBV	Calisher et al., 2006
			Pteropus	ABLV	Wong et al. 2007; Gould et al. 2002
				ARAV	Lumlertdacha et al., 2005
				IRKV	Lumlertdacha et al., 2005
				KHUV	Lumlertdacha et al., 2005
			Rousettus	EBLV-1	Calisher et al., 2006
				LBV	Markotter et al., 2008
				ABLV	Arguin et al., 2002
	Rhinolophoidea	Hipposideridae (Old World leaf-nosed bats)	Hipposideros	SHIBV	Kuzmin et al., 2011a
		Rhinolophidae (Horseshoe bats)	Rhinolophus	EBLV-1	Calisher et al., 2006; Kuzmin and Rupprecht, 1999
				RABV	Jiang et al., 2010

(Continued)

TABLE 6.1 (Continued)

Suborder	Superfamily	Family	Genus Associated with Lyssavirus Infection	Lyssavirus Involved	Representative Reference
Vespertilioniformes	Emballonuroidea	Nycteridae (Slit-faced bats)	Nycteris	LBV	Calisher et al., 2006
		Emballonuridae (Sheath-tailed bats)	Saccolaimus	DUVV	Wong et al., 2007
				ABLV	Calisher et al., 2006; Wong et al., 2007
			Taphozous	ABLV	Arguin et al., 2002
	Noctilionoidea	Phyllostomidae (Leaf-nosed bats)	Anoura	RABV	Sodre et al., 2010
			Artibeus	RABV	Kobayashi et al., 2006; Price and Everard, 1977
			Carollia	RABV	Sodre et al., 2010
			Chrotopterus	RABV	Sodre et al., 2010
			Desmodus	RABV	Favoretto et al., 2002; Lopez et al., 1992
			Diaemus	RABV	Sodre et al., 2010
			Diphylla	RABV	Castilho et al., 2010
			Glossophaga	RABV	Salas-Rojas et al., 2004
			Leptonycteris	RABV	Constantine, 1979
			Lonchorhina	RABV	Sodre et al., 2010
			Lophostoma	RABV	Sodre et al., 2010
			Macrotus	RABV	Constantine, 1979
			Micronycteris	RABV	Salas-Rojas et al., 2004

	Genus	Virus	Reference
Mormoopidae (Ghost-faced bats)	Mormoops	RABV	Constantine, 1979
	Pteronotus	RABV	Salas-Rojas et al., 2004
Phyllostomidae (Leaf-nosed bats)	Phyllostomus	RABV	Sodre et al., 2010
	Platyrrhinus	RABV	Sodre et al., 2010
	Sturnira	RABV	de Alemeida et al., 2011
	Trachops	RABV	Sodre et al., 2010
	Uroderma	RABV	Klug et al., 2011; Nunes et al., 2008
Vespertilionidea — Molossidae (Free-tailed bats)	Cynomops	RABV	Sodre et al., 2010
	Eumops	RABV	Constantine, 1979
	Molossops	RABV	de Rosa et al., 2011
	Molossus	RABV	Scheffer et al., 2007
	Nyctinomops	RABV	Constantine, 1979
	Tadarida	RABV	Constantine, 1979
	Tadarida	EBLV-1	Muller et al., 2007
Miniopteridae (Bent-winged bats)	Miniopterus	ABLV	Arguin et al., 2002
		DUVV	Wong et al., 2007; Markotter et al., 2008
		EBLV-1	Calisher et al., 2006; Kuzmin and Rupprecht, 1999
		WCBV	Calisher et al., 2006

(Continued)

TABLE 6.1 (Continued)

Suborder	Superfamily	Family	Genus Associated with Lyssavirus Infection	Lyssavirus Involved	Representative Reference
		Vespertilionidae (Vesper bats)	Antrozous	RABV	Streicker et al., 2010; Burns et al., 1956
			Barbastella	EBLV-1	Muller et al., 2007
			Corynorhinus	RABV	Streicker et al., 2010; Constantine, 1979
			Eptesicus	RABV	Streicker et al., 2010
				EBLV-1	Calisher et al., 2006
			Euderma	RABV	Constantine, 1979
			Histiotus	RABV	Velasco-Villa et al., 2006
			Lasionycteris	RABV	Streicker et al., 2010
			Lasiurus	RABV	Streicker et al., 2010; Constantine, 1979
			Murina	IRKV	Calisher et al., 2006
			Myotis	ARAV	Arai et al., 2003; Gonzalez et al., 2008
				RABV	Streicker et al., 2010
				EBLV-2	Johnson et al., 2003
				EBLV-1	Muller et al., 2007
				KHUV	Kuzmin et al., 2003
			Nyctalus	EBLV-1	Muller et al., 2007
				RABV	Selimov et al., 1991
			Nycticeius	RABV	Constantine, 1979
			Philetor	ABLV	Arguin et al., 2002

Genus	Virus	Reference
Pipistrellus	RABV	Constantine, 1979
	EBLV-1	Muller et al., 2007; Kuzmin and Botvinkin, 1996
Plecotus	RABV	Constantine, 1979
	EBLV-1	Muller et al., 2004; Bouhry et al., 1992
Scotophilus	ABLV	Arguin et al., 2002
Vespertilio	EBLV-1	Kuzmin et al., 1994; Gonzalez et al., 2008
Vespertilio	RABV	Selimov et al., 1991

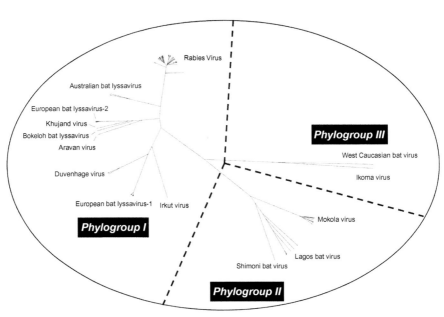

FIGURE 6.1   **Phylogenetic distance between lyssavirus species based on substitutions per site.** One hundred and ten lyssavirus nucleoprotein or full genomes sequences were analyzed in a Bayesian framework (BEAST v1.6.1). Posterior clade probabilities of all major nodes between lyssavirus species was >0.99.

infection (Kuzmin et al., 2008a; Markotter et al., 2006); *Mokola virus* (MOKV), which has caused natural infection of shrews, rats, cats, dogs, and two human cases (Sabeta et al., 2007); *Duvenhage virus* (DUVV), which has been isolated from insectivorous bats and has caused human fatalities following exposure to bat bites (Weyer et al., 2011); *European bat lyssavirus*-1 (EBLV-1), which has been detected primarily in insectivorous bats but has also been associated with cross species transmission (CST) events into what are considered dead-end hosts, including sheep, a stone marten, cats, and humans (Müller et al., 2007); *European bat lyssavirus*-2 (EBLV-2), which has been isolated from insectivorous bats alongside two documented human fatalities (Fooks et al., 2003a, b); *Australian bat lyssavirus* (ABLV) is endemic in frugivorous and insectivorous bats across Australasia and has also 'spilled over' into human populations (Moore et al., 2010); and *Irkut virus* (IRKV) that has caused human infection, *Aravan virus* (ARAV), *Khujand virus* (KHUV) and *West Caucasian bat virus* (WCBV) (Arai et al., 2003; Kuzmin et al., 2003, 2005, 2008b) that are all associated with insectivorous bats, although only single isolates exist for each species (Anonymous, 2009). Finally, in 2009, a virus isolated from a bat in Kenya and named *Shimoni bat lyssavirus* (SHIBV) (Kuzmin et al., 2010, 2011a) was classified by the ICTV into the lyssavirus genus. Alongside these isolates, two currently unclassified lyssaviruses have also been identified that share high levels of genetic and morphological traits with the other lyssaviruses and are considered to be lyssaviruses. These have been reported in both volant and non-volant species and include the following. In 2010, a novel lyssavirus was detected in a Natterer's bat (*Myotis nattererii*) in Germany and was named *Bokeloh bat lyssavirus* (BBLV) (Freuling et al., 2011); and in 2011 a novel lyssavirus was detected in an African civet (*Civettictis civetta*) in Tanzania and was subsequently named *Ikoma Lyssavirus* (IKOV) from the location where the civet was captured (Marston et al., 2012). Each of these isolates has yet to be classified within the lyssavirus genus but ultimately represents new lyssavirus species. Where novel lyssaviruses have been associated with bat infections, they are detailed further in sections that follow.

Along with the ecological evidence that Chiroptera are the original reservoirs of lyssaviruses, there is strong phylogenetic support for this with evidence of bat-derived viruses having evolved long before those RABV of terrestrial carnivore origin (Badrane & Tordo, 2001). Despite this, there remain two distinct lyssaviruses that have not yet been detected in bat species: MOKV and IKOV. In the case of MOKV, the reservoir remains unclear. Numerous virus isolates have been reported from different species, including organ pools from shrews (*Crocidura* sp) in Nigeria (1968) (Kemp et al., 1972; Shope et al., 1970) and Cameroon (1974) (Le-Gonidec, Rickenbach, Robin, & Heme, 1978) (Swanepoel, 1994); domestic cats (*Felis catus*) in South Africa (1970, 1995–1998), Zimbabwe (1981–1982) (Foggin, 1983) and Ethiopia (1989–1990) (Mebatsion, Cox, & Frost, 1992); and the rusty-bellied brush furred rat (*Lophuromys sikapusi*) in the Central African Republic (1983) (Saluzzo et al., 1984). Although human infection has been reported, a fatal (Sabeta et al., 2007) outcome of infection was only recorded in one instance, and so the threat to the human population remains hard to determine (Familusi et al., 1972; Sabeta et al., 2007). Infections with MOKV are rare, with scant surveillance initiatives and limited diagnostic capacity, perhaps perpetuating our lack of knowledge of this lyssavirus (Sabeta et al., 2007). Although MOKV antibodies have been detected in bat species, it is widely accepted that this is due to cross reactivity with antibodies against LBV, rather than evidence of MOKV circulation, although further sampling of different bat species may discover that MOKV also circulates in bats in the Old World (Dzikwi et al., 2010; Kuzmin et al., 2008a). The phylogenetic relationship and current thoughts regarding host susceptibility of MOKV have been reviewed by Markotter and co-workers (2008).

In contrast to MOKV, where multiple isolates have been characterized and CST events have been identified, for IKOV, only a single isolate exists. The discovery of IKOV was the first detection of a non-rabies lyssavirus in

African civets. Furthermore, the detection of this virus within the boundaries of the Serengeti National Park, Tanzania, during routine dog rabies surveillance activities highlights the importance of molecular characterization of virus isolates where surveillance initiatives are in place (Marston et al., 2012; Mebatsion, Cox, & Frost, 1992). Analysis of sequence data generated from infected samples indicates that IKOV is the most divergent lyssavirus reported (Figure 6.1). The Serengeti National Park has been free of classical dog rabies for many years, with rabies cases clustering outside of the park in neighboring villages. Therefore, the isolated detection of this lyssavirus within the park from a terrestrial and largely solitary species strongly suggests that IKOV represents a CST event that could be of bat origin given the viruses close phylogenetic relationship with WCBV (Marston et al., 2012).Without further isolates of this and more closely related lyssaviruses the reservoir for IKOV remains to be elucidated.

The evolutionary history of lyssaviruses within bats remains unresolved. Across the New World, only variants of classical RABV are associated with the infection of bat species, whereas other lyssaviruses remain completely absent across the New World in either volant or non-volant mammalian populations. Indeed, RABV has been detected in more than 50 species of insectivorous bat across the New World alone (see Section 6.2). In direct contrast, across the Old World there have been no detections of classical RABV strains in any bat species despite the presence of classical RABV in terrestrial species. Instead, 12 genetically divergent non-RABV lyssaviruses have been characterized from different bat species in the Old World. The reason for this situation remains unclear, however, tests for neutrality using various evolutionary algorithms on a panel of lyssavirus G protein sequences suggest neutral evolution occurs for lyssaviruses, and therefore niche-partitioning and radiation may explain this genetic pattern (Badrane & Tordo, 2001). Interestingly, where different bat species have been associated with lyssavirus infection, there often appears to be, to some extent, a host restriction to infection with genetically distinct lyssaviruses. RABV itself may be exploiting relatively new niches in previously uninfected New World bats (Streicker et al., 2010). The focus of this chapter will be the association of different lyssavirus species with bats across the globe. Lyssavirus species will be described according to the bat species most frequently associated with infection and to avoid repetition, different species will be described according to their feeding patterns.

2 LYSSAVIRUSES AND INSECTIVOROUS BATS

In this section, we will discuss the association of classical RABV with numerous insectivorous bat species across the New World and comment on recent findings regarding the occurrence of CST events. We will then detail the occurrence of lyssaviruses in insectivorous bats in the Old World, concentrating on what appear to be principal reservoirs of each of the genetically distinct lyssaviruses.

2.1 Rabies Virus in Insectivorous Bats of the New World

2.1.1 Historical perspectives

Prior to the 1950s, insectivorous bats were not considered a source of rabies in the New World. This changed in 1951 with the first documented case of human rabies resulting from a bat bite (Enright et al., 1955). Human infection with insectivorous bat-derived RABV was again reported in the United States in 1954. In the summer of 1953, a seven-year-old boy was attacked and bitten by a rabid Florida yellow bat (*Lasiurus intermedius*) (Venters et al., 1954). Subsequent surveillance for RABV in bats detected 40 cases between 1953 and 1958 (Scatterday, 1954) and the direct transmission of RABV following the bite of a naturally infected insectivorous bat was shown experimentally through transmission to suckling mice (Bell, 1959). This unequivocally highlighted the public health threat of rabid insectivorous bats and brought a new aspect of RABV epidemiology to the interest of the scientific community. Since these early reports of rabies in insectivorous bat populations, surveillance throughout the United States has detected rabies within indigenous insectivorous bat species from almost every state (Mondul, Krebs, & Childs, 2003). The first rabid bats in Canada were reported in 1957 (Avery & Tailyour, 1960). Infected species included the big brown bat (*Eptesicus fuscus*), the little brown bat (*Myotis lucifugus*), and the silver-haired bat (*Lasionycteris noctivagans*), all reported from British Colombia. Detection of RABV infected insectivorous bats has been reported in Mexico since 1953 and South America since the 1960s (Baer & Smith, 1991).

2.1.2 Natural and Experimental Infection of Insectivorous Bats

Insectivorous bats infected with RABV show a variety of disease signs that parallel infection in other mammals. A number of studies have described the susceptibility to infection and the clinical signs of disease in different insectivorous bat species. An early study by Baer and Bales (1967) with the Mexican free-tailed bat (*Tadarida braziliensis*) confirmed the susceptibility to infection with a RABV isolated from the same species. Inoculation by the intracranial, intramuscular, and subcutaneous routes caused clinical disease in bats over a range of virus titers. Infected bats displayed signs of aggression, weakness, and anorexia with relatively short morbidity periods ranging from 4 to 20 days. Importantly, the intermittent excretion of virus in saliva was observed during this period, a feature of virus transmission that remains poorly understood.

No evidence of virus excretion in the absence of disease development was observed. A recent infection study reported RABV infection of the big brown bat (*Eptesicus fuscus*), again using virus derived from the salivary glands of the same species (Jackson et al., 2008). In this study, bats infected intramuscularly exhibited weight loss, ataxia, and paresis. Transmission by bite is considered the most common method of RABV transmission in bats although there is evidence that aerosolized virus may cause infection in roosts where large numbers of bats congregate, such as the Mexican free-tailed bat (Baer & Smith, 1991). Interestingly, experimental studies of North American RABV variants did not demonstrate evidence of disease in bats but showed seroconversion against RABV in response to exposure to aerosolized virus (Davis, Rudd, & Bowen, 2007). This may account for the levels of seropositivity observed in many bat populations around the world.

The striking feature of the ecology of lyssaviruses in the New World in comparison to that of the Old World is the diversity of bat species that appear to act as a virus reservoir. Prior to the introduction of molecular differentiation techniques it was impossible to differentiate between RABV isolates. The application of panels of monoclonal antibodies that targeted the virus nucleoprotein revealed antigenic differences between RABV and other members of the lyssavirus genus (Wiktor & Koprowski, 1980). This approach demonstrated that RABV variants from North America could be differentiated and that different virus isolates were often associated with the reservoir host, including a range of non-flying species such as the red fox (*Vulpes vulpes*), racoons (*Procyon lotor*), and skunks (*Mephitis mephitis*) (Smith, 1988). RABV from particular bat species also showed distinctive monoclonal antibody binding patterns. This was confirmed in both Canada (Nadin-Davis et al., 2001) and Mexico (Velasco-Villa et al., 2002). Importantly, it was also noted that the only lyssavirus species isolated in New World bats, and indeed all non-bat reservoirs, are variants of RABV, with no detection of non-RABV lyssaviruses being reported in the Americas. This has been confirmed by the application of phylogenetic techniques based on the RABV genome, which show that particular bat species are associated with specific RABV variants (Hughes, Orciari, & Rupprecht, 2005; Streicker et al., 2010; Velasco-Villa et al., 2006). This is described further in later sections.

Numerous studies from individual states within the United States (Burnett, 1989; Childs, Trimarchi, & Krebs, 1994; Crawford-Miksza, Wadford, & Schnurr, 1999) and Canada (Nadin-Davis et al., 2001) have reported multiple insectivorous bat species to be rabid, although it remains unclear what properties of bats make them a successful host for rabies virus. Table 6.2 compares the number of rabid bats reported in the United States between 1993 and 2000 (Mondul, Krebs, & Childs, 2003),

TABLE 6.2   A Comparison of Bat Species Testing Positive for Rabies Virus from Two Reports from within the United States

Species	Common Name	1993–2000[a]	2010[b]
		No. (% of species tested)	
Antozous pallidus	Desert pallid bat	21 (21.0)	2 (6.9)
Eptesicus fuscus	Big brown bat	1,216 (5.8)	324 (3.8)
Euderma maculatum	Spotted bat	1 (100)	NR
Lasionycteris noctivagans	Silver-haired bat	73 (12.9)	14 (6.7)
Lasiurus borealis	Red bat	47 (9.0)	40 (23.5)
Lasiurus cinereus	Hoary bat	97 (38.2)	16 (33.3)
Lasiurus ega	Southern yellow bat	7 (21.9)	2 (33.3)
Lasiurus intermedius	Northern yellow bat	3 (100)	16 (18.8)
Lasiurus seminolus	Seminole bat	0 (0.0)	6 (30.0)
Lasiurus xanthinus	Western yellow bat	NR	3 (100.0)
Myotis californicus	California myotis	12 (3.6)	0 (0.0)
Myotis evotis	Long-eared myotis	19 (9.7)	6 (14.6)
Myotis keenii	Keen's myotis	11 (1.9)	NR
Myotis lucifugus	Little brown bat	96 (1.7)	26 (2.9)
Myotis thysanodes	Fringed myotis	0 (0.0)	1 (50.0)
Myotis volans	Long-legged myotis	3 (13.0)	1 (16.7)
Myotis yumanensis	Yuma myotis	4 (1.7)	1 (5.9)
Myotis (unspeciated)	Unspeciated myotis	41 (6.1)	8 (8.4)
Nycticeius humeralis	Evening bat	6 (9.7)	5 (4.8)
Nyctinimops (Tadarida) femorosaccus	Pocketed free-tailed bat	7 (13.2)	NR
Tadarida macrotis	Big free-tailed bat	3 (21.4)	4 (57.1)
Parastrellus Hesperus	Canyon bat	NR	14 (21.5)
Myotis austroriparius	Southeastern myotis	NR	1 (11.1)
Perimyotis subflavus	Tri-colored bat	NR	5 (25.0)
Pipistrellus Hesperus	Western pipistrelle	41 (21.2)	NR
Pipistrellus subflavus	Eastern pipistrelle	20 (17.1)	NR
Plecotus townsendii	Townsend's big-eared bat	3 (10.3)	0 (0.0)

(Continued)

TABLE 6.2   (Continued)

Species	Common Name	1993–2000[a]	2010[b]
		No. (% of species tested)	
Tadarida brasiliensis	Brazilian/Mexican free-tailed bat	214 (31.8)	286 (65.4)
Tadarida unspeciated	Unspeciated free-tailed bats	1 (33.3)	NR
Unspeciated	Unspeciated bats	NR	648 (4.9)

NR Not Reported.
[a]Data obtained from Mondul et al. (2003). A total of 31,380 bats were tested with 1,946 (6.2%) testing positive for rabies virus. Bat species that were negative have been removed from the table.
[b]Data obtained from Blanton et al. (2011). A total of 24,298 bats were tested with 1,430 (5.9%) testing positive for rabies virus.

with the numbers reported during 2010 (Blanton et al., 2011). A number of species are consistently reported with high levels of rabid submissions in both surveys, including the big brown bat (*Eptesicus fuscus*), the silver-haired bat (*Lasionycteris noctivagans*), the red bat (*Lasiurus borealis*), the hoary bat (*Lasiurus cinereus*), the little brown bat (*Myotis lucifugus*), and the Mexican free-tailed bat (*Tadarida braziliensis*). A comparison of four species of North American insectivorous bats that are most commonly reported as rabid (Table 6.3) indicates that traits such as colony size or geographical distribution do not support any particular hypothesis for their association with RABV. Indeed, it appears that the majority of North American bat species have been associated with RABV infection to varying extents. Recent phylogenetic analysis of RABV species isolated from North American bat species suggests that a model of CST followed by virus evolution within a particular host species fits the current diversity of the virus within insectivorous bats (Streicker et al., 2010). This implies that the majority of RABV transmissions within bats are intra-species events. It also suggests that the transmission of bat variants of RABV from a bat to other mammals, including humans, is a rare event. Experiments from the 1960s demonstrated that bat variants of RABV could infect a range of carnivores, although some species showed greater susceptibility than others (Constantine, 1966a,b). In addition to inoculation by intramuscular injection, transmission to conspecifics was also demonstrated by repeated biting, in some cases over one hundred times, by an infected Mexican free-tailed bat (Constantine, 1966c). Examples of bat variant infections in non-bat species have been reported, including from Canada where CST of a bat variant to a squirrel was reported (Webster, Casey, & Charlton, 1988) as well as that of

TABLE 6.3   Comparison of Four Insectivorous Bat Species Associated with
Transmission of Rabies to Humans in North America

Characteristic	Tadarida Brasiliensis	Eptesicus Fuscus	Lasionycteris Noctivagens	Perimyotis Subflavus
Common name	Mexican/ Brazilian free-tail bat	Big brown bat	Silver-haired bat	Tri-colored bat
Distribution	Found throughout the Western hemisphere from northern United States to Eastern Brazil	Found throughout southern Canada and United States to northern South America	Found throughout Canada, United States, and northern Mexico	Found in the eastern regions of Canada and United States
Description	The species grows to about 9 cm in length and weighs around 12 g	The species grows to 12 cm in length and can weigh about 20 g	The species grows to about 10 cm in length, and adults can weigh between 8 and 12 g	The species grows to 9 cm in length, and adults weigh between 4 and 10 g
Colony size	Can reach up to 20 million in size	Maternity colonies have been measured in the hundreds (300–600)	Solitary species.	Generally solitary, although females form small maternity colonies (<20 individuals)
Roost in human structures	Yes	Yes	No	No

CST to two red foxes and a cow (Webster, Casey, & Charlton, 1989). Such CST events are rare, and in the majority of cases there has been no evidence of onward transmission and maintenance of the infecting virus isolate within the new host species. However, two very rare events where CSTs have led to the maintenance of virus in the non-chiropteran host have been documented. One such event occurred on Prince Edward Island, Canada, where a cluster of rabies cases in red foxes (*V. vulpes*) was observed. Screening of the viruses isolated, with a panel of 15 monoclonal antibodies, suggested that they were of bat origin, possibly the little brown bat (*Myotis lucifugus*) (Daoust, Wandeler, & Casey, 1996). A larger epizootic occurred following the emergence of rabies in the skunk (*M. mephitis*) population in the state of Arizona (Leslie et al., 2006). Sequence analysis of the RABV genome recovered from 19 skunks demonstrated that the source of the infection was likely to be from a

RABV variant normally found in the big brown bat (*Eptesicus fuscus*). It is notable that CST events are more commonly reported with RABV than other lyssaviruses, possibly as RABV was more recently introduced into the New World bats. In contrast the bat-associated lyssaviruses in the Old World are considered to be evolutionarily older viruses (Tordo and Badrane, 2001; Noel Tordo, personal communication). These examples of sustained transmission among non-bat hosts are likely examples of how RABV emerged as a pathogen of carnivores, including domestic dogs.

2.1.3 Human Infection

Human infection with RABV of insectivorous bat origin is increasingly being recognized now that rabies has been eliminated from the domestic dog population in North America. At least 61 cases of human rabies between 1950 and 2007 were identified that could be attributed to transmission from bats (De Serres et al., 2008; Messenger, Smith, & Rupprecht, 2002). In one extreme case, transplant material from a human case of rabies of bat origin was transferred to four recipients who then developed the disease (Srinivasan et al., 2005). Table 6.4 shows the number of human cases in the United States between 2002 and 2010. From these data, it is apparent that the transmission to humans, although rare, occurs from a small group of species. Investigation of one of these variants isolated from the eastern pipistrelle bat (*Pipistrellus subflavus*) suggested that this virus possessed certain virological characteristics, such as increased replication in epithelial cells and growth at lower temperatures, which may enhance the ability of the virus to be transmitted by a bite (Kuzmin et al., 2012; Morimoto et al., 1996). Interestingly, one rare survivor of RABV infection was believed to have been infected with a bat-variant of the virus (Willoughby et al., 2005). Due to the recognition that rabies virus transmission from bats represents a high risk following contact, rabies post-exposure prophylaxis is recommended for any encounter with an insectivorous bat.

Until recently, the association of RABV with South American insectivorous bats has been neglected, mainly due to the burden of rabies cases associated with hematophagous species of bat. However, the report of a human rabies case transmitted from a Brazilian free-tailed bat (*Tadarida braziliensis*) (Favi et al., 2002) highlighted the role that insectivorous bats play in the ecology of RABV and the risks to public health. An additional study updated the list of all bat species (insectivorous, frugivorous, and hematophagus) that have been shown to be positive for RABV in Brazil and included 41 species from the families *Molossidae, Phyllostomidae,* and *Vespertilionidae* (Sodre, da Gama, & de Almeida, 2010). The wealth of sequence data available for RABV isolates from New World bat species has led to detailed attempts to create genealogical trees that include all bat reservoir species (Oliveira et al., 2010). This approach confirmed earlier findings on the association between particular virus lineages and

TABLE 6.4   Human Cases of Rabies in the United States of Bat Origin

Year	Total Human Rabies Cases	Cases Due to Bat Variants	Suspected Bat Reservoir
2002	3	3	*Tadarida brasiliensis*
			Perimyotis subflavus
			Lasionycteris noctivagens/ Perimyotis subflavus
2003	3	1	*Lasionycteris noctivagens*
2004	8	6	*Tadarida brasiliensis* (×5)[a]
			Unknown bat
2005	1	1	Unknown bat
2006	3	2	*Tadarida brasiliensis*
			Lasionycteris noctivagens
2007	1	1	Unknown bat
2008	2	2	*Tadarida brasiliensis* related
			Lasionycteris noctivagens
2009	4	3	Unknown bat
			Perimyotis subflavus
			Desmodus rotundus
2010	2	2	*Desmodus rotundus*
			Perimyotis subflavus

(Adapted from Blanton et al., 2011.)
[a]*related to transplantation in four cases.*

bat species separating RABV variants into at least 14 groups. A general observation was the influence of geographical separation of certain species that are located in both North and South America. For example, the RABV variants associated with *Tadarida braziliensis* formed two clusters dependent on the geographical origin of the bat. This was consistently shown for both *Myotis* and *Eptesicus* species. The surprising exceptions to this were those RABV viruses associated with *Lasiurus* species (*L. cinereus*, *L. borelialis* and *L. ega*), which formed a single group (Oliveira et al., 2010). The reasons for this close association between North and South American RABVs are uncertain. However, the hoary bat (*L. cinereus*) is known to make large distance migrations although this is limited to migration from the interior of the United States to over-wintering sites on the coastline and possibly central Mexico

(Cryan & Wolf, 2003). It is possible that there is sufficient overlap of migrating hoary bat populations between Canada and Brazil to have enabled the observed passage of RABV between both continents.

2.2 Rabies in Insectivorous Bats of the Old World

In direct contrast to the situation seen in the New World, where closely related RABV isolates are associated with numerous different insectivorous bat species, in the Old World highly divergent non-RABV isolates have been reported from very few bat species. Due to the paucity of knowledge about other lyssaviruses of the Old World associated with the infection of insectivorous bats, we concentrate on what is known about infection with the EBLVs.

2.2.1 Lyssavirus Infection of Serotine Bats

Serotine bats are considered to play a key role in the epidemiology of bat rabies in Europe. Indeed, the first link to a rabies-like infection was established in 1954 when the first report of a rabid bat in Europe was made in Germany (Mohr, 1957). The species of bat involved in this incident was not reported, although subsequent detections of lyssaviruses in serotine bats may suggest that early bat rabies cases in Europe involved this species. Only with the advent of antigenic typing and molecular tools could the virus isolated in European bats be distinguished from classical RABV and other lyssavirus species (Bourhy et al., 1992; Schneider & Cox, 1994). Serotine bats belong to the genus *Eptesicus* of the subfamily Vespertilioninae, chiropteran family Vespertilionidae of the suborder Vespertilioniformes. Up to 39 insectivorous species of *Eptesicus* are known to exist (Hutson, Mickleburgh, & Racey, 2001; Niethammer & Krapp, 2011). Of the 39 species of *Eptesicus*, *E. serotinus isabellinus* and *E. serotinus* deserve special attention, as they are primarily associated with transmission of EBLV-1. *E. serotinus* is the most widespread and abundant Palaearctic species of *Eptesicus* in Europe. Its present range extends from the Atlantic Ocean and the Mediterranean basin to Northern Denmark and the Eastern parts of the Eurasian continent as far as the Pacific seaboards, with evidence for an expansion into Scandinavia (Figure 6.2A) (Hutson et al., 2008). *E. s. isabellinus* has been reported in North Africa and Southern Iberia and is considered to be genetically divergent from *E. serotinus* (Ibanez, García-Mudarra, Ruedi, Stadelmann, & Juste, 2006).

2.2.1.1 EPIDEMIOLOGY OF LYSSAVIRUS INFECTION OF SEROTINE BATS

Serotine bats are one of the most studied lyssavirus reservoir species, revealing some fundamental insights into virus–host interactions and, hence, possibly the epidemiology of bat lyssavirus infections in

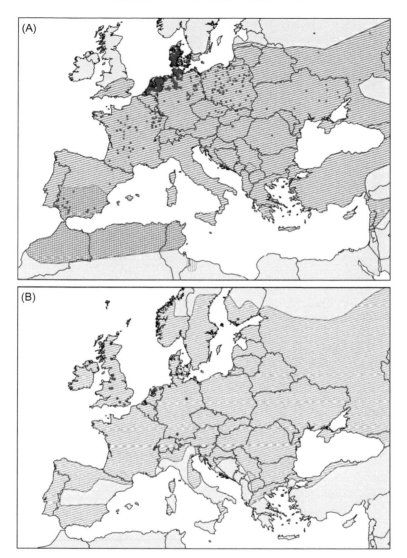

FIGURE 6.2   Distribution of the insectivorous bat species found to be infected with, and reported cases of, EBLV-1 and EBLV-2. A) Range of *E. serotinus* (Red) and *E. isabellinus* (Blue) and reported EBLV-1 cases (dots); B) Range of *M. daubentonii* (Red) and EBLV-2 (dots) cases in Europe 1977–2010. Distribution of bats according to IUCN and Vazquez-Moron, 2011.

insectivorous bat species in the Old World. *E. serotinus* and *E. s. isabellinus* are known to be associated with EBLV-1 and thus, are considered the primary reservoir hosts for this particular bat-associated lyssavirus. Interestingly, of the 959 cases of bat rabies officially reported between 1977 and 2010 to the database of the Rabies Bulletin Europe (RBE) about

95% have been observed in *E. serotinus* (Schatz et al., 2012). The involvement of *E. s. isabellinus* in the epidemiology of EBLV-1 was demonstrated in southern Spain when rabid serotine bats were later genetically identified as Isabelline serotine bats (Perez-Jorda et al., 1995; Vazquez-Moron et al., 2008a; Vazguez-Moron et al., 2008b). During the past 50 years, the vast majority of rabid serotine bats in Europe have been diagnosed in densely populated countries of The North European Plain that covers large areas of Denmark, The Netherlands, and Germany. However, this clustering of reported bat rabies cases suggests a bias in surveillance activities that reflects the location of established networks of bat handlers and their activities (Figure 6.2A). Regardless of this potential bias in sampling, it seems clear that the serotine bat has particular host species-specific ecological traits that may promote the maintenance and transmission of EBLV-1. Indeed *E. serotinus* is a very adaptable species with habitats including semi-desert, temperate, and subtropical dry forest, Mediterranean-type scrubland, farmland, and suburban areas. Furthermore, in contrast to other bat species, it is not experiencing a decline in population size. Hence, a greater abundance of *E. serotinus* might favor virus amplification and may explain the clusters of bat rabies cases detected in densely populated areas. The northern distribution range of serotine bats seems to be limited by a threshold level of mean summer temperatures (Harbusch, 2003). Interestingly, the great majority of EBLV-1–infected *E. serotinus* have been reported from this northern range. This is supported by surveillance for EBLV-1 in Germany, a country with a north-south gradient in elevation where over several decades notably higher numbers of cases have been reported from the northern regions of the country, where the highest density of *E. serotinus* is found (Müller et al., 2007). Taking this into account might explain why in some European countries with existing *E. serotinus* populations, no or only sporadic cases of EBLV-1 infections are observed (Anonymous, 2012; Van der Poel et al., 2005). Despite the limitations of surveillance, based on the current knowledge of EBLV-infections it is likely that serotine bats may be infected with EBLV-1 throughout Eurasia (Schatz et al., 2012). For reasons unknown, however, long-term active and passive surveillance for lyssaviruses has provided no evidence for EBLV-1 circulation in serotine bat populations in southern England, although specific antibodies have been detected in a single serotine bat (Harris et al., 2009). This disparity between mainland Europe and the UK may be due to the limited geographical distribution and population size of *E. serotinus* within the UK preventing EBLV-1 persistence.

The molecular characterization of EBLV-1 isolates from serotine bats has revealed valuable clues to virus–host interactions and help resolve questions surrounding the epidemiological puzzle of this virus–host relationship. With approximately 5×10^5 substitutions per site per year the evolutionary rate of nucleotide substitution of EBLV-1 is one of the lowest recorded for RNA viruses, suggesting that the current genetic diversity of the virus in serotine bats arose between 500 and 750 years ago (Davis et al., 2005) although other estimates predict a more recent evolution (Hughes, 2008). Either way, this is believed to reflect an evolutionary stability between EBLV-1 and its two host species (Davis et al., 2005). Based on unique molecular traits, however, there is evidence that the phylogeny of this virus is more complex than previously thought and clearly associated with geographically different host evolutionary history (Vazquez-Moron et al., 2011). Early genomic sequencing indicated that in European populations of serotine bats two distinct genetic lineages of EBLV-1 evolved, which seemed to geographically cluster (Amengual et al., 1997). Serotine bats throughout the North and East European Plains, for example, located between the North Sea and the Baltic Sea to the north, and the Central European Highlands to the south, are infected by EBLV-1 lineage "a." These isolates have been reported to exhibit an extremely high sequence identity, as confirmed by phylogenetic analysis of the N-gene (Amengual et al., 1997; Davis et al., 2005; Freuling et al., 2008; Van der Poel et al., 2005). Despite its extensive West-East geographic range, the low genetic diversity in connection with a consistent star-like structure of the phylogenetic network suggests a relatively recent introduction, evolution, and rapid expansion of EBLV-1a variants in populations of serotine bats in these parts of Europe (Vazquez-Moron et al., 2011). Although in early studies no, or only a less pronounced, correlation between geographic and genetic distances was observed with lineage 1a using both a hypothetical introduction point or a model based origin of a putative common ancestor (Amengual et al., 1997; Davis et al., 2005), cluster analysis showed a significant positive correlation between genetic and temporal and spatial distance based on N-gene sequence and provided evidence for a geographical segregation (Freuling et al., 2008a, 2012a). EBLV-1a isolates from serotine bats within certain geographical regions of The Netherlands and Germany have almost identical nucleotide sequences. This indicates genomic stability during the transmission cycle of these virus variants, with little geographic spread or intermixing (Freuling et al., 2008a; Van der Poel et al., 2005). Serotine bat colonies usually number 10 to 50, although occasionally up to 300 individuals have been recorded (Hutson et al., 2008; Niethammer & Krapp, 2011). Considering the sedentary nature of the serotine bat (Hutson et al., 2008) transmission of EBLV-1a over long distances may not play a major role in EBLV-1 epidemiology (Freuling et al., 2008a; Van der Poel et al., 2005) and therefore somewhat contradicts a rapid expansion hypothesis of EBLV-1a variants in populations of serotine bats.

In contrast to EBLV-1a, EBLV-1 lineage 'b' appears to follow a more North-South distribution (Figure 6.2A) and is genetically more heterogeneous (Amengual et al., 1997). This heterogeneity is reflected by the presence of diverse sublineages that are predominantly associated with populations of serotine bats on the Iberian Peninsula, in The Netherlands, France, and southwestern Germany (Davis et al., 2005; Müller et al., 2007). A recent phylogenetic analysis segregated *E. s. isabellinus* sequences from the *E. serotinus* sequences and hypothesized that the *E. s. isabellinus* isolates form a further independent lineage, EBLV-1 lineage 'c' (Vazquez-Moron et al., 2011). Considering that serotine bats do not normally migrate over large distances, recent single discoveries of EBLV-1b in Central Poland and Eastern Germany (Müller et al., unpublished data; Smreczak et al., 2007) may suggest that the distribution of this lineage is more widespread in Europe than previously considered. As there is evidence of genetic flow between *E. s. isabellinus* populations on both sides of the Strait of Gibraltar, there is reason to believe that North African conspecifics may also be infected with this lineage (Juste et al., 2009).

It remains unclear, however, exactly how populations of serotine bats in different parts of Europe favor the current geographic distribution and clustering of the two or three different EBLV-1 lineages. Additional genetic data from both the host species and virus isolates are required to fully understand the evolutionary EBLV-1-serotine bat interaction. In any case, serotine bats in the westernmost parts of the North European Plains represent the only geographical location where both EBLV-1a and EBLV-1b are detected (Van der Poel et al., 2005; Picard-Meyer et al., 2006). However, as co-infection of serotine colonies or infection of individuals with both lineages has not been reported, this may suggest independent infectious cycles among *E. serotinus*. Interestingly, results of recapture data and long-term active rabies surveillance in maternity colonies of *E. s. isabellinus* strongly suggest close communities in which variants of EBLV-1 independently circulate (Vazquez-Moron et al., 2008b). Also, provided the hypothesis of an inverse correlation between CST of lyssaviruses and phylogenetic distance between bat species as recently demonstrated for North American bat RABV also applies to European bats (Streicker et al., 2010), then future targeted surveillance should discover the existence of known or new EBLV-1 lineages.

2.2.1.2 CLINICAL INFECTION AND DISEASE PATTERNS

Based on the limited observational studies and experimental data of infection in serotine bats, our understanding of the transmission and pathogenesis of EBLV-1 under natural conditions is incomplete. Hence, clinical implications of EBLV-1 infections in serotine bats are discussed. It is clear that *E. serotinus* is known to succumb to clinical infections and therefore is likely to be submitted for rabies diagnosis (Freuling et al.,

2009a). However, reports on EBLV-1 related clinical signs in serotine bats are rare and stem either from animals kept in captivity in bat sanctuaries or experimental studies. Although no information exists concerning the incubation period in naturally infected *E. serotinus* bats, the incubation periods in experimentally infected bats varies depending on the route of inoculation, with the intracerebral route leading to the shortest incubation period (7–13 days), followed by subcutaneous (17–18 days) and intramuscular (26 days) routes (Freuling et al., 2009b). Because intranasal inoculation was ineffective in experimentally inducing clinical rabies, transmission of EBLV-1 among serotine bats via non-bite exposures in nature seems highly unlikely (Johnson, Phillpotts, & Fooks, 2006a). The subcutaneous route of inoculation was relatively efficient in inducing mortality and concomitant shedding of virus in saliva immediately before the development of disease, while intramuscular inoculation was less efficient. Interestingly, experimental intramuscular infection of RABV in Mexican free-tailed bats (*Tadarida braziliensis*) caused fewer virus-positive salivary glands compared with that resulting from subcutaneous infection (Baer & Bales, 1967). From these experimental studies it has been postulated that this may indicate a host-specific adaptation of EBLV-1 and that the subcutaneous route of infection is highly likely to mimic natural transmission among natural hosts via multiple dose infection (Freuling et al., 2009b). In contrast, inoculation of the North American sibling species *E. fuscus* with EBLV-1 by the intramuscular route led to the development of disease in 50% of challenged animals (Franka et al., 2008). Of note, the detection of EBLV-1 infection in infected *E. serotinus* taste buds was observed, mimicking studies with human rabies (Jackson et al., 1999). The significance of this observation, however, remains unclear with respect to potential excretion of virus (Freuling et al., 2009a).

Only a few detailed descriptions are available of naturally EBLV-infected serotine bats and most of these have involved human contacts (Vos et al., 2007). Where clinical disease in naturally infected serotine bats has been observed, animals demonstrated to a lesser or greater degree the following signs: inability to fly, loss of weight, weakness, hypersensitivity for high frequency sounds, prolonged vocalization and uncontrolled wing beats after high frequency sounds, and a strong tendency to bite and react aggressively to stimuli (Anonymous, 1986, 1989; Bruijn, 2003). Experimentally, EBLV-1 causes disease in serotine bats indistinguishable from that observed with RABV infection of North American bats with clinical signs being consistent with those described above (Freuling et al., 2009b). Whether the different EBLV-1 lineages exhibit differences in pathogenicity and result in different disease patterns in serotine bats is currently unknown. The same applies to the recently discovered mutated forms of lineages 1a and 1b possessing either single or hexameric nucleotide insertions within the 3' untranslated region (UTR) of the nucleoprotein or 35 nucleotide deletions within the UTR between the glycoprotein (G) and polymerase (L) gene (Freuling et al., 2012a; Johnson et al., 2007). The effects of these changes on viral replication remains elusive, as such mutations had not been previously described for lyssaviruses (Harris et al., 2009).

It appears that *E. serotinus* and *E. s. isabellinus* can survive infection with EBLV-1, as evidenced by the repeated captures of individual sero-positive bats over a number of years during serological surveillance initiatives (Harris et al., 2009) and experimental infection (Freuling et al., 2009b). In the latter case, serotine bats surviving in this instance should be considered to have had an aborted infection (Banyard et al., 2011). Although no seroconversion was detected in serotine bats from the Balearic Islands (Spain) and Switzerland (Megali et al., 2010; Serra-Cobo et al., 2002), measurable virus neutralizing antibody (VNA) titers have been detected in free-living con-specifics from several countries with reported prevalences for seroconversion in single colonies in the UK (0.3%) (Brookes et al., 2005; Harris et al., 2009), Germany (5.1%) (Müller et al., unpublished data), in different studies across Spain (4.8%, 7.7%, 19%, and 74%) (Perez-Jorda et al., 1995), and in France (25%) (Picard-Meyer et al., 2011). Similar observations were made in Spanish colonies of *E. s. isabellinus*, where the overall seroprevalence was estimated to be 9% (Echevarria et al., 2001; Vazquez-Moron et al., 2008b). However, in an experimental study, none of the *E. serotinus* infected with EBLV-1 developed VNAs (Freuling et al., 2009b), while in the North American sibling species *E. fuscus* VNAs were detected in survivors and deceased animals (Franka et al., 2008). Although these serological observations in free-living *Eptesicus spp.* have to be interpreted with caution (see the following), this has led some to hypothesize that endemic EBLV-1 is associated with subclinical or mild disease (Vazquez-Moron et al., 2008b). The only experimental study with EBLV-1 in its host reservoir did not provide any evidence for a subclinical carrier state in serotine bats (Freuling et al., 2009b) as repeatedly suggested (Echevarria et al., 2001; Serra-Cobo et al., 2002), and hence the existence of such a state remains to be conclusively demonstrated. The assumption that the clinical expression of EBLV-1 in bats is a non-fatal, extra-neurological infection without important consequences for the health of the *E. serotinus* populations (Echevarria et al., 2001; Vazquez-Moron et al., 2011) is disproved by experimental data and routine rabies diagnostics. Further studies, however, are needed to investigate the circumstances under which neurological manifestations develop in these bats (Vazquez-Moron, et al., 2008a).

2.2.1.3 SEROTINE BATS AS VECTORS FOR EBLV-1

Limited experimental data has shown that other insectivorous bat species, for example, Daubenton's bats (*Myotis daubentonii*), Brandt's bats

(*Myotis brandtii*), common pipistrelle bats (*Pipistrellus pipistrellus*), Egyptian flying foxes (*Rousettus aegyptiacus*), and big brown bats (*E. fuscus*), are susceptible to infection with EBLV-1 (Botvinkin, Kuzmin, & Chernov, 1992; Franka et al., 2008; Kuzmin, Botvinkin, & Shaimardanan, 1994; Kuzmin & Botvinkin, 1996; van der Poel et al., 2000). Because European *Eptesicus* bats are known to cohabit in both mating and wintering roosts with other insectivorous bat species it is not surprising that other bat species have been infected with EBLV-1. Reports of sporadic detection of EBLV-1 in the noctule Bat (*Nyctalus noctula*), the Nathusius' pipistrelle bat (*Pipistrellus nathusii*), the common pipistrelle bat (*Pipistrellus pipistrellus*), the greater horseshoe bat (*Rhinolophus ferrumequinum*), and the parti-colored bat (*Verspertilio murinus*) (King, Haagsma, & Kappeler, 2004; Müller et al., 2007; Selimov et al., 1991) have been made. Further to this, EBLV-1 RNA has been detected in the Schreiber's bent-winged bat (*Miniopterus schreibersii*); the Natterer's bat (*Myotis nattereri*); the greater horseshoe bat (*R. ferrumequinum*) and the greater mouse-eared bat (*Myotis myotis*) from Spain (Serra-Cobo et al., 2002), indicating that EBLV-1 may circulate in the absence of clinical disease in a range of other bat species under natural conditions. However, whether high EBLV-1 specific sero-positivity in free-living bat species such as *M. nattereri*, *R. ferrumequinum*, *M. schreibersii*, and *M. myotis*, and even antigen detection in blood clots in *M. myotis* (Amengual et al., 2007; Serra-Cobo et al., 2002) suggest CST, or evidence of independent reservoirs for these viruses remains questionable.

EBLV-1 is rarely transmitted to terrestrial mammals and humans. Incidents of CST infections of EBLV-1 in sheep, a stone marten (*Martes foina*), and domestic cats have been reported (Dacheux et al., 2009; Müller et al., 2004; Ronsholt, 2002). Furthermore, in contrast to infection with RABV of terrestrial wildlife in Europe, peripheral inoculation of sero-tine-associated EBLV-1 in terrestrial mammals, including foxes, ferrets, cats, dogs, and sheep demonstrated limited susceptibility of those species (Picard-Meyer et al., 2008; Vos et al., 2004a). The virulence of EBLV-1 is considered to be low with sheep and foxes surviving intramuscular inoculation with a high dose of virus (Baltazar, 1988; Brookes et al., 2007; Cliquet et al., 2009; Fekadu et al., 1988; Vos et al., 2004b).

Although numerous human contacts with serotine bats, primarily from handling sick or injured animals, have been reported (Brass, 1994), only two EBLV-1 induced human casualties have been conclusively demonstrated: in Ukraine in Voroshilovgrad (1977) and in Belgorod, Russia (1985) (Fooks, 2004). Although both cases are known to have involved transmission from a bat to human, the species of bat was not defined for either case. Furthermore, the Russian case was genetically typed as being EBLV-1 while the Ukraine case is only assumed to be EBLV-1 from antigenic profiling (Botvinkin et al., 2005; Selimov et al., 1989). Two further reports of human deaths have been reported following encounters with bats in continental Europe, although neither has been confirmed as EBLV-1 nor has an involvement of *E. serotinus* been confirmed (Banyard et al., 2011). In conclusion, the probability of serotine bats transmitting EBLV-1 to other species including humans is minimal and there is no evidence that serotine bats will cause EBLV-1 associated epizootics among terrestrial animals or significant human fatalities (Brass, 1994).

2.2.1.4 CONTROL OF SEROTINE BAT-MEDIATED RABIES

In contrast to North Africa and South Asia, where no specific conservation actions are known, across Europe there are internationally recognized legal frameworks in place for the protection of European bats through the Bonn Convention (Eurobats), the Bern Convention, and Annex IV of the EU Habitats and Species Directive. Hence, all bat species are protected by national legislation in member states (Hutson et al., 2008). Therefore, any specific population control measures aimed at reducing disease prevalence in serotine bats is prohibited and from a scientific point of view also unwarranted (Brass, 1994). Rather, in keeping with ongoing conservation measures, the main focus of activities should be on (i) the establishment of adequate surveillance, (ii) objectively increasing public awareness about bat rabies and (iii) targeted pre- and post-exposure prophylaxis (PEP). Although guidelines on passive and active bat lyssavirus surveillance were established and adopted by the United Nations (Anonymous, 2006) and on request from the European Food Safety Authority (Cliquet et al., 2010), it is not clear what affect these guidelines will have on European bat rabies surveillance (Schatz et al., 2012). Vaccination of serotine bats using modified live RABV vaccines as discussed for vampire bats (Aguilar-Setien et al., 2002) is largely unfeasible. Instead, consistent preventive vaccination of all persons working with bats and PEP of all persons with bat exposures should be given as priority as the available human inactivated rabies vaccines stimulate cross neutralizing antibodies and therefore are considered to confer cross protection with EBLV-1, -2 and BBLV (Fekadu et al., 1988; Lafon, Herzog and Sureau, 1986; Malerczyk et al., 2009; Freuling et al., unpublished data). Any media panic and fearmongering that may have a negative impact on serotine bat populations should be avoided.

2.2.2 *Lyssavirus Infection of Myotis Bats*

As observed with *E. serotinus* and EBLV-1, there appears to be a strong association of EBLV-2 with *Myotis* species. The *Myotis* genus comprises more than 100 species in three main phylo-geographic clades, that is, North America, South America, and the Old World (Lack et al., 2010). In the latter, rabies in Daubenton's bats (*Myotis daubentonii*) and pond bats (*Myotis dasycneme*) has been genetically characterized as EBLV-2.

While the pond bat is confined to Central and Eastern continental Europe, the Daubenton's bat is widely distributed throughout Eurasia (Stebbings & Griffith, 1986) (Figure 6.2B).

In fact, only after the tragic death of a Swiss biologist in Finland in 1985 and subsequent characterization of the causative agent did it become evident that bat rabies in Europe at that time was caused by two different lyssavirus species, EBLV-1 and 2 (Bourhy et al., 1992; Lumio et al., 1986). The first isolation of EBLV-2 from a bat was made in The Netherlands from a Pond bat (Nieuwenhuijs, 1987; Nieuwenhuijs, Haagsma & Lina, 1992). From 1987 to 1993, five Pond bats tested rabies positive, of which the majority were characterized as EBLV-2 (Davis et al., 2005; Nieuwenhuijs, Haagsma, & Lina, 1992). Despite intensive surveillance efforts, no further cases of rabies in the Pond bat were detected in The Netherlands (Van der Poel et al., 2005). All of the Dutch EBLV-2 isolates form a separate cluster, away from the other EBLV-2 sequences (Davis et al., 2005), possibly reflecting geographical or host species relationships. However, since no other *M. dascyneme* associated isolates have been characterized elsewhere and given the distant relationship between these two bat species (Ruedi & Mayer, 2001), it is also likely that cases in *M. dascyneme* resemble CST events from Daubenton's bats. In fact, rabies was also reported from Pond bats in Denmark and Germany (Kappeler, 1989), and Daubenton's bats in Russia. However, the isolates from Denmark and Germany were not characterized, and, surprisingly, isolates from Russia appeared to be RABV, suggesting possible contamination in the diagnostic laboratory (Kuzmin and Rupprecht, 2007).

Elsewhere in Europe, EBLV-2 has been isolated sporadically from the Daubenton's bat in Switzerland (Megali et al., 2010), the United Kingdom (Banyard et al., 2009; Whitby et al., 2000), Germany (Freuling et al., 2012b) and Finland (Jakava-Viljanen et al., 2010). A Daubenton's bat captured during net trapping in Finland showed abnormal behavior, died, and subsequently tested EBLV-2 positive. Molecular characterization then indicated that the first human victim in 1985 had contracted the infection in Finland (Jakava-Viljanen et al., 2010). Besides this Finnish case, a second human rabies case occurred after EBLV-2 infection of a bat biologist in the UK (*Supplementary Case history box 6.1*) (Fooks et al., 2003b; Nathwani et al., 2003). No further EBLV-2 CST events have been reported.

2.2.2.1 NATURAL AND EXPERIMENTAL INFECTION OF MYOTIS SPECIES

As with disease progression seen in natural infection of bats with EBLV-1, the infection of *Myotis* species with EBLV-2 have involved vocalization, agitation, and aggressiveness. Both naturally and experimentally infected bats exhibited a reduced food and water intake a few days before they died (Freuling et al., 2008b, 2012b; Johnson et al., 2006b).

Occasionally, infection of bats has been observed during the rehabilitation of grounded bats in specialist wildlife centers. In such cases, the apparent 'activation' of virus can come following considerable incubation periods ranging from seven weeks (Johnson et al., 2003) to as long as nine months (Pajamo et al., 2008). In comparison, a captive Daubenton's bat infected by experimental subdermal inoculation with a standard dose developed disease after 32 days (Johnson et al., 2008) although numerous alternative routes of infection failed to cause clinical disease.

Following both experimental infection (Johnson et al., 2008) and natural exposure, EBLV-2 RNA was detected in the brain and to a lesser extent in other organs, including the tongue and salivary glands (Banyard et al., 2009; Freuling et al., 2008b, 2012b; Johnson et al., 2003). The heterogeneous pattern in different non-neuronal organs may be linked to the relative time point of death and the degree of centrifugal viral dissemination rather than a specific effect of EBLV-2 (Johnson et al., 2006b).

How EBLV-2 persists in its natural host is still poorly understood. In the UK, where a passive bat rabies surveillance initiative has been established (Brookes et al., 2005; Harris et al., 2009), EBLV-2 specific antibodies are often detected at low levels within bat populations, suggesting that the virus is endemic within the Daubenton's bat population (Banyard, Hartley and Fooks, 2010). Serosurveillance studies have also indicated a low level seroprevalence in the bat population in the UK (Harris et al., 2009), although the mechanisms of persistence at the population level remain unclear. Virus neutralizing antibodies have also been detected in Daubenton's bats in Switzerland and Sweden (Megali et al., 2010; National Veterinary Institute, 2010). It is postulated that transmission of EBLV-2 is through biting or scratching, as virus was detected in oral swabs of a subdermally infected bat in an experimental setting. Furthermore, oral swabs from individual Daubenton's bats from Scotland (2008) and Switzerland (2009) yielded EBLV-2 specific RNA, but not viable virus (Schatz et al., 2012). The perpetuation of EBLV-2 infection in Daubenton's bat populations is driven by virological and ecological parameters (Vos et al., 2007), and rates may confer a virus–host co-evolutionary process. Models indicate that the behavior of the Daubenton's bats result in high gene flow, which may allow EBLV-2 to become established and rapidly spread throughout the population (Smith et al., 2011).

As well as the infection of Myotis bats with EBLV-2, several other lyssaviruses have been isolated from Myotis species. In 1991, an apparently healthy lesser mouse-eared bat (*Myotis blythi*) captured in the Aravan district, Kyrgyzstan, tested positive for rabies by the mouse inoculation test (Kuzmin et al., 1992). In 2001, near the town of Khujand, Tajikistan, a whiskered bat (*Myotis mystacinus*) that was grounded and collected by hand also tested positive (Botvinkin et al., 2003). Subsequent characterization of the isolated viruses revealed they represent new lyssavirus species Aravan virus (ARAV) and (KHUV) Khujand virus. Little is known about the epidemiology of these lyssaviruses in bats, as only single isolations have been made. Also, as the taxonomy of whiskered bats was modified, it is likely that the bat species from which KHUV was isolated was actually the steppe whiskered bat, *Myotis aurascens* (Benda et al., 2008).

In 2009, a virus isolated from a Natterer's bat (*Myotis nattereri*) from Germany was shown to be antigenically and genetically distant to all previously known lyssaviruses (Supplementary Case history box 6.2), and the virus detected, named Bokeloh bat lyssavirus (BBLV), (Freuling et al., 2011). Natterer's bats are among the species routinely submitted under passive surveillance schemes in Europe, albeit in low numbers (Schatz et al., 2012). It is thus enigmatic why this virus was only discovered recently. Surprisingly, viral RNA (EBLV-1) was previously detected in the brain of a Natterer's bat in Spain, although both fluorescent antibody testing and virus isolation was unsuccessful (Serra-Cobo et al., 2002). Likewise, high levels of VNAs in the greater mouse-eared bat from Spain (Amengual et al., 2007) is only corroborated by viral RNA (EBLV-1) detected in the brain of a few individuals (Serra-Cobo et al., 2002). Elsewhere in Europe, only two cases of bat rabies were confirmed in greater mouse-eared bats, one in Germany and one in Poland (Schatz et al., 2012). Unfortunately, in both instances the viruses involved were not fully characterized. The isolation of closely related lyssavirus species, that is, EBLV-2, ARAV, KHUV, and BBLV from *Myotis* species suggests that this bat genus may play a key role in the transmission and maintenance of this clade of phylogroup I lyssaviruses.

2.2.3 Lyssavirus Infection of Other Insectivorous Bat Species

Alongside the infection of serotine and myotis species there have been a number of individual isolations of lyssaviruses from other bat species. Of these, bats within the *Miniopteridae*, commonly referred to as the bent-winged bats and found across much of Africa, Asia, Australia and southern Europe, have been associated with several highly divergent lyssavirus species. The first lyssavirus to be associated with *Miniopterus spp.* was first isolated in 1970 following the death of a human that had been bitten in South Africa, by an insectivorous bat, while sleeping. The virus was named Duvenhage virus (DUVV) after the individual bitten, and it is thought from the description given that it was most likely a bent-winged bat (Meredith, Prossouw, & Koch, 1971). During the 1980s there were two further reports of DUVV from insectivorous bats. In 1981, the virus was isolated from what was believed to have been a bent-winged bat caught by a domestic cat in Makhado town, Limpopo Province,

South Africa (King & Crick, 1988), although again the species of the infected bat was never confirmed, and in 1986 an Egyptian slit-faced bat (*Nycteris thebaica*) that was trapped during a survey in Zimbabwe also tested positive for DUVV (Foggin, 1988). Another lyssavirus predominantly isolated from African bats, LBV has also been isolated from the insectivorous Gambian slit-faced bat (*Nycteris gambiensis*) (Markotter et al., 2006), although LBV is predominantly associated with the infection of frugivorous bats (see Section 6.4).

Two further human cases involving DUVV have also been reported. The first occurred in 2006 when a 77-year-old man was scratched on the face by an insectivorous bat in North West Province, South Africa. The individual was unknowingly infected and died 14 days following the onset of clinical disease (Paweska et al., 2006). The second case involved the infection of a 34-year-old woman who died 28 days after the onset of clinical disease (van Thiel et al., 2008). The lack of adequate rabies surveillance and characterization of isolates prevents the development of a detailed understanding between lyssaviruses and bats in Africa. Moreover, the number of bat-associated human cases of rabies in Africa remains unknown.

The common bent-wing bat (*M. schreibersii*) became the subject of further interest among rabies scientists when a bat of this species was net-trapped in Russia near the Georgian border and subsequently tested rabies positive for lyssavirus infection (Botvinkin et al., 2003). The virus was only detected conclusively as being a member of the lyssavirus genus following fluorescent antibody testing (FAT) on brain material from mice inoculated with suspect material via the intracranial route that subsequently developed neurological disease. The virus, named West Caucasian bat virus (WCBV), was a genetically divergent bat-derived member of the lyssavirus genus with no serological cross-reactivity to other lyssaviruses (Kuzmin et al., 2005; Franka et al., 2008; Kuzmin et al., 2008b; Horton et al., 2010). Given this observation, neutralizing antibodies against WCBV in *Miniopterus* bats collected in Kenya (Kuzmin et al., 2008b) indicate that WCBV or some other antigenically similar virus also circulates in Africa and in the rest of the Old World where *Miniopterus* bats are abundant (Kuzmin et al., 2011b). The recent discovery of IKOV makes these data even more intriguing, although the serological response to IKOV has not yet been defined with respect to the cross-neutralization of other lyssaviruses.

It is unknown whether the common bent-wing bat is also affected by European lyssavirus species, although EBLV-1 RNA was detected in one *Miniopterus* specimen in Spain (Serra-Cobo et al., 2002). Virus neutralizing antibodies that did not cross-react with WCBV were also found in *M. schreibersii* from France and Spain (Serra-Cobo et al., 2002). A serological survey of bats in the Philippines showed that virus neutralizing antibodies to Australian bat lyssavirus were most prevalent in *M. schreibersii* (4/11, 36%) (Arguin et al., 2002). However, in none of the 422 bats of this species tested was evidence for an ABLV infection detected (Field, 2004). In any case, the association of *M. schreibersii* with numerous lyssaviruses has led to the suggestion that this species may facilitate cross-continental transmission of different lyssavirus species. This is supported by the global distribution of *M. schreibersii* from southern Europe and Africa and across the Middle East and Caucasus mountains.

Another insectivorous species associated with lyssavirus infection are bats in the subfamily *Murininae*, Genus *Murina*, more commonly known as the tube-nosed bats. In 2002, a bat that had entered an apartment in the town of Irkutsk was captured and died after approximately 10 days with signs of general exhaustion, poor appetite, and weakness. The bat was subjected to rabies diagnostics. Following antigenic and genetic typing, the virus was classified as a lyssavirus and named Irkut virus (IRKV) (Botvinkin et al., 2003; Kuzmin et al., 2005). Initially, the bat was classified as a greater tube-nosed bat (*Murina leucogaster*), although the exact species of the bat from which IRKV was derived remains unclear. A single human case of rabies was reported due to infection with IRKV in 2007, although the bat species involved remains unknown (Leonova et al., 2009).

In 2009, another divergent lyssavirus, named Shimoni bat virus (SHIBV), was isolated from a dead Commerson's leaf-nosed bat (*Hipposideros commersoni*) during a search for pathogens in bats in Kenya (Kuzmin et al., 2010). Antigenically, SHIBV demonstrates similarity to MOKV and LBV, but is genetically divergent from these species (Kuzmin et al., 2010). Additional comparative serological surveys in *Rousettus aegyptiacus* bats and *Hipposideros commersoni* bats in roosts in Kenya where these species sympatrically roost demonstrated that the seroprevalence to SHIBV was equivalent in the presence or absence of *R. aegyptiacus* bats. These data supports the suggestion that *H. commersoni* is the host species of SHIBV (Kuzmin et al., 2011a).

Although initially discovered in and predominantly associated with fruit-eating bat species (see Section 6.3), the Australian bat lyssavirus (ABLV) has also been associated with infection of insectivorous bat species. In Australia, 63 microbat species are detected (Richards, 1998), although ABLV infection has only been confirmed by virus isolation in one insectivorous species, the yellow-bellied sheath-tailed bat (*Saccolaimus flaviventris*). This bat species is a member of the sheath-tailed bats in the family Emballonuridae found in Australia and Papua New Guinea. However, while no further direct isolations have been reported, serological evidence of exposure to ABLV has been shown in seven genera, representing five of the six families of Australian insectivorous bats, that is, *Chaerephon* and *Tadarida* (Molossidae), *Chalinolobus* and *Vespadelus* (Vespertilionidae), *Hipposideros* (Hipposideridae), *Macroderma*

(Megadermatidae), and *Saccolaimus* (Emballonuridae) (Animal Health Australia, 2009). In the latter genus, the yellow-bellied sheath-tailed bat had significantly higher antibody prevalence (up to 62.5%) than other species, suggesting that this bat plays an important role in the ecology of ABLV (Field, 2004). Interestingly, viral antigen was not detected in any of the 668 clinically normal wild-caught microchiroptera collected around Australia; but only in sick, injured, or found dead animals, suggesting that the virus causes disease in bats.

Genetic characterization also revealed that the source of infection in the first human case of ABLV in 1996 was likely *S. flaviventris* (Gould et al., 2002). In a second human case in 1998, subsequent sequence analysis of the viral isolate indicated it was of the variant characteristic of flying foxes (Warrilow et al., 2002) (see Section 6.3).

3 LYSSAVIRUSES AND FRUGIVOROUS BATS

Frugivorous bats belong to the order Chiroptera, suborder Pteropodiformes, superfamily Pteropodidae. Frugivorous bats are present throughout most of the world, performing vital ecological roles such as pollinating flowers and dispersing fruit seeds (Simmons, 2005). Given this broad distribution, the detection of lyssaviruses within only a few species is somewhat surprising (Banyard et al., 2011). The *Pteropodidae* is divided into seven subfamilies with 186 total extant species, represented by 44 to 46 genera. Of these, only a few genera have been directly associated with lyssavirus infection (Table 6.1).

Lyssavirus infection of frugivorous bats is largely restricted to the large fruit bats present across Africa and Australia, with numerous highly diverse lyssaviruses detected. Genetic analysis of isolates by a number of research groups has led to the suggestion that African bat populations are the original hosts for lyssaviruses (Nel & Rupprecht, 2007). However, despite this geographical bias in detection of lyssaviruses in African fruit bat populations, other fruit bat populations across the globe have also yielded different lyssaviruses. Here we describe the principal associations of lyssaviruses with frugivorous bats.

3.1 Detection of Lagos Bat Virus in Frugivorous Bats

From an African perspective, RABV itself is completely absent from fruit bat populations, in direct contrast to the situation seen in carnivores. In place, LBV is the lyssavirus most frequently associated with different fruit bat species. Indeed, LBV or antibodies that neutralise LBV have been isolated from several fruit bat species, including the straw-colored fruit bat (*Eidolon helvum*), Wahlberg's epauletted fruit bat

(*Epomorphorus wahlbergi*), and the Egyptian fruit bat (*Rousettus aegyptiacus*) (Dzkiwi et al., 2010; Hayman et al., 2008, 2010). Studies with African fruit bat species have detected a high seroprevalence of antibodies against LBV in these three different colonial fruit bat species, with antibody levels ranging from 14–67% and 29–46%, for the former two species, respectively. Initial studies showed that older *R. aegyptiacus* and *E. helvum* were shown to have a higher level of seroprevelance within populations of both species (Kuzmin et al., 2008a; Dzikwi et al., 2010; Hayman et al., 2010). Data from *E. helvum* populations derived over a four-year study period demonstrated that seroprevalence in mature bats was significantly greater than that found in juvenile or sexually immature bats. Seroprevalence in sexually mature adults fluctuated between 23% (15–33%) and 49% (39–52%), with no significance seen between the proportion of seropositive sexually mature adults between sampling occasions over the study period (Hayman et al., 2012). Despite significant levels of LBV seropositivity within bat populations, attempts to isolate virus have often been problematic. Indeed, Hayman et al., (2012) tested 796 oral swabs from healthy *E. helvum* using RT-PCR but did not detect viral RNA in any sample. Similarly, Kuzmin et al., (2008a) tested 931 oral swabs and 1,182 brain samples from at least 30 different species of bat by RT-PCR, with LBV RNA only being detected in a single *E. helvum* bat. As expected, high levels of viral RNA were detected in the brain, but also in the salivary gland and tongue, and so salivary excretion was postulated as the means of virus transmission (Kuzmin et al., 2008a). Possible explanations for the seroprevalence levels seen in healthy *E. helvum* bats include: infection with seroconversion and recovery; seroconversion; and latent infection, although no evidence for latent infection has been shown. Further studies are required to enhance our knowledge of how LBV circulates in these bat populations (Hayman et al., 2012).

Both serological and genomic analyses have suggested that LBV isolates form four highly divergent lineages that in some cases are as divergent as some of the other, individually classified lyssaviruses. These have been tentatively termed lineages A to D. Lineage A includes isolates from Senegal (1985) (Swanepoel, 1994), a Kenyan (2007) (Kuzmin et al., 2008a) and a French isolate (either Togolese or Egyptian origin, (1999) (Aubert, 1999); lineage B includes the original Nigerian isolate (1956) (Boulger and Porterfield, 1958); lineage C includes viruses from the Central African Republic (1974) (Sureau et al., 1977), Zimbabwe (1986) (King and Crick, 1988), and South Africa (1980) (King and Crick, 1988) (2003–2005) (Markotter et al., 2006); and lineage D includes a single isolate from Kenya taken from a healthy Egyptian fruit bat (2008) (Kuzmin et al., 2010). The most striking observation from these studies is the close sequence identity of isolates in lineage C that have been reported over a 25-year period. This suggests a high level of genome stability within these viruses (Markotter et al., 2008).

The detection of LBV in a rabid African water mongoose (*Atilax paludinosus*) was the first isolation of LBV from a terrestrial wildlife species (Markotter et al., 2006). Previous isolations in terrestrial mammals were all from domestic animals, including cats (*Felis catus*) (King & Crick, 1988) and dogs (*Canis familiaris*) (Mebatsion, Cox, & Frost, 1992; Markotter et al., 2008). Alongside the rare detection of natural CST events, the potential for CST events has also, as with other lyssaviruses, been assessed experimentally. Early studies with experimental infection with LBV involved infection of guinea pigs (n = 2) and a monkey (*Cercocebus torquatus*) (Boulger & Porterfield, 1958). In these early studies, the virus was reported to be apathogenic and alongside other studies (Badrane & Tordo, 2001) led to the conclusion that phylogroup II lyssaviruses, when inoculated by a peripheral route, were reduced in pathogenicity. Inoculation by the IC route, however, highlighted the ability of both LBV and MOKV to cause encephalitis and death (Tignor et al., 1973). Experimental studies with LBV, MOKV and RABV have suggested comparable mortality following IM inoculation of LBV and RABV, although IM inoculation with MOKV caused reduced mortality (Markotter et al., 2009).

As seen in the epidemiology of RABV and EBLV-1, CST events have also been reported for LBV infection. Indeed, the isolation of LBV from non-volant wildlife species such as mongoose highlights our poor understanding of the incidence and host range of lyssaviruses in Africa. Until molecular tools such as PCR are widely available to assess individual samples, the opportunity to characterize viruses at the molecular level will be missed, a fact highlighted by the detection of a IKOV (Marston et al., 2012). Findings such as this reiterate the need for a thorough molecular analysis of samples to ensure a complete understanding of the epidemiology of these viruses (Fooks et al., 2009).

3.2 Australian Bat Lyssavirus and Flying Foxes

ABLV is predominantly associated with infection of frugivorous flying foxes. Genetic characterization of ABLV detected in *Pteropus* species have shown them to be genetically distinct from those isolated from insectivorous bats, and within each lineage the genetic variation is very narrow (Gould et al., 2002; Guyatt et al., 2003). The initial isolation of ABLV was from a black flying fox (*Pteropus alecto*) in 1996 (Crerar et al., 1996; Fraser et al., 1996). Since then, ABLV has been detected in all four frugivorous bat species present in Australia (Fraser et al., 1996; Gould et al., 1998, 2002; Guyatt et al., 2003). Unlike other lyssaviruses, no correlation has been seen between genetic variation and the geographical distribution of ABLV isolates within Australia (Gould et al., 2002). Although all flying fox species seem to be affected by ABLV, no species-specific genetic associations have been observed (Barrat, 2004; Guyatt et al., 2003). Interestingly, ABLV is the only lyssavirus species which, from evidence gathered to date, has reservoirs in both mega- and microbats species. The correlation between CST and genetic distances of bat species, as demonstrated for North American bat RABV (Streicker et al., 2010), does not seem to fit as a model for the maintenance and transmission of ABLV. In any case, ABLV appears to be maintained in bat species in Australia. Furthermore, serosurveillance of bat populations in the Philippines and surrounding islands has suggested that lyssavirus infection of bats might be more widespread than previously thought (Arguin et al., 2002), although virus isolations are required due to cross-neutralization between phylogroup I viruses.

From a clinical perspective, the gray-headed flying fox (*Pteropus poliocephalus*) has been assessed for susceptibility to infection with ABLV with the outcome of clinical disease being seen in a proportion of the animals inoculated (McColl et al., 2002). Three of ten intramuscularly inoculated animals developed clinical signs consistent with a lyssavirus infection, but virus neutralizing antibodies were only detected in five of the seven survivors (McColl et al., 2002). In contrast, infection of companion animals with ABLV led to survival in all inoculated cats and dogs, although some animals did show transient neurological signs (McColl et al., 2007). Although no viral antigen could be detected at post mortem in any tissues, strong serological profiles were seen in all infected animals.

3.3 Infection of Fruit Bats with Other Lyssaviruses

With the exception of LBV and ABLV, no other non-rabies lyssaviruses have been detected in frugivorous bats. However, from a molecular perspective, there have been reports of EBLV-1 RNA in Egyptian fruit bats (*Rousettus aegyptiacus*). In 1997, following the transport of a group of Egyptian fruit bats from a Dutch zoo to a Danish zoo, EBLV-1a RNA was detected in not only the exported bats but also within the Dutch colony from which the bats had been exported (Rønsholt et al., 1998). Furthermore, during the investigations surrounding these reports, no clinical disease was observed in any of the bats, and it was concluded that EBLV-1a was being maintained in healthy Egyptian fruit bats (Wellenberg et al., 2002). Following these reports, experimental studies with these bats have shown that, in some instances, these bats are able to survive infection with EBLV-1 following intracerebral inoculation (van der Poel et al., 2000). Further studies are required to understand bat immune responses to lyssaviruses.

3.4 Interaction of Rabies Virus with New World Frugivorous Bats

The frugivorous bat populations of the Americas belong to the Phyllostomidae, and are genetically distant from the true Pteropodidae of the Old World. However, while the fruit bats in the Old World have been associated exclusively with infection of non-rabies lyssaviruses, in the New World the fruit bat populations are only associated with infection with RABV. Pawan (1936) was the first to describe rabies virus infection of frugivorous bats in Trinidad in 1931. This initial report was followed by several other reports of RABV infection of frugivorous bats, including the great fruit eating bat (*Artibeus lituratus*) (Price & Everard, 1977; Stouraitis & Salvatierra, 1978). In these cases, bats were reported to be displaying abnormal behavior and were characterized as being infected with RABV, although it is believed that this infection represented spillover from vampire bat species. Delpietro, Gury-Dhomen, Larghi, Mena-Segura and Abramo (1997) reported that RABV isolates circulating in frugivorous species in South America were antigenically closely related to the vampire bat related viruses commonly described. Shoji et al., (2004) genetically typed a vampire bat rabies variant in *Artibeus spp.* Most recently, complete genome sequencing has shown 97% identity across the entire genomes of a vampire bat derived and a frugivorous bat derived isolate (Mochizuki et al., 2011). Numerous reports across South America detail infection of fruit bats with RABV. But compared to the threat to both human and livestock populations from hematophagous bats (Section 6.4) and the infection of numerous insectivorous bats across the Americas (Section 6.2), the significance of rabies in frugivorous bats in South America is small (Carneiro et al., 2009; Cunha et al., 2006).

4 RABIES IN HEMATOPHAGOUS BATS

Evolution has resulted in one group of bats that are highly effective in delivering a bite capable of damaging the skin and, as a consequence, are highly effective at transmitting rabies to other species: the vampire bats. Many of the biological adaptations of the vampire bat justify this claim, from the dentition that has evolved to cut through the skin of vertebrates, active secretion of saliva containing anti-coagulants to maintain blood flow in a wound, a colonial habit that provides the opportunity for transmission to con-specifics, and a wide selection of prey species that includes humans. For these reasons, vampire bats are a significant cause of death of livestock and responsible for increasing reports of rabies in humans.

Vampire bats are those that derive all their nutrition from feeding on the blood of other vertebrates. Only three genera of bats are blood feeders or hematophagus. All three belong to the family Phyllostomidae or New-World leaf-nosed bats, and each genus contains a single species (Koopman, 1988). These are the common vampire bat, *Desmodus rotundus*; the hairy-legged vampire bat, *Diphylla ecaudata*; and the white-winged vampire bat, *Diaemus youngi*. The latter species primarily feed on birds, whereas *D. rotundus* feeds primarily on mammals and is responsible for virtually all cases of rabies virus transmission. Therefore, this species will be considered in this section.

Blood-feeding by bats has not evolved outside of Central and South America, and all of the fossil evidence for vampire bats going back to the Pleistocine period (2.5 million years ago to 12,000 years ago) have been found in the Americas (Arellano-Sota, 1988). The association of rabies with the vampire bat is a relatively recent observation, despite the obvious advantage the species has in transmitting rabies. A number of authors have speculated that rabies was present within vampire bat species prior to the Spanish conquest of the New World in the fifteenth century. This is based mainly on the writings of the Spanish colonists (de Oviedo y Valdes, 1950) and the limited information from the indigenous peoples that has survived and been translated (Vos et al., 2011). It is also speculated that the introduction of livestock and horses by the Spanish settlers increased the numbers of potential prey animals and has led to an increase in the population of vampire bats (Greenhall, 1988). A further factor that may have increased both the numbers of vampire bats and brought them into closer contact with humans has been the introduction of manmade structures, such as mines and bridges, that provide additional roosting sites (Greenhall, 1988).

Historically, early writings report rabies as the cause of disease in cattle in southern Brazil following detection of Negri bodies in the brains of cattle (Carini, 1911; Haupt & Rehaag, 1921). However, the link to vampire bats as the cause was not initially accepted, despite the absence of rabid dogs in the affected areas. Later reports, primarily by Joseph Lennox Pawan (1887–1957) on the island of Trinidad, established clearly the link between rabies in vampire bats and outbreaks in livestock in 1925 and humans in 1929 (Hurst & Pawan, 1932; Pawan, 1938). The close proximity between Trinidad and the South American mainland suggested that migration of vampire bats infected with RABV was the likely explanation for the emergence of disease. One of the striking features of these reports was the observation of vampire bat activity during the day and what appeared to be aggressive attacks on humans and tethered livestock (Pawan, 1936).

4.1 Biology of Vampire Bats

The common vampire bat (*D. rotundus*) is widely distributed from northern Mexico to central Chile and northern Argentina. It occupies a range of ecosystems including rainforests, wetlands, and desert. It is also found at a range of elevations from sea level to 3,500 meters and is not present over much of the Andes and central Mexico (Koopman, 1988).

Common vampire bats roost in colonies of variable size, ranging from less than 10 bats to over 300. Occasionally colonies of 2000 or more individuals have been reported (Arellano-Sota, 1988). Common vampire bats often roost in caves with other species, and both autogrooming and allogrooming are common, potentially providing opportunities for spread of RABV. Furthermore, common vampire bats participate in altruistic feeding, whereby a bat that has fed will regurgitate blood for one that has not. This could provide a further opportunity for transmission of virus.

4.2 Rabies in the Vampire Bat

Pawan (1936) described the appearance of rabies in wild-caught bats. These included both the furious and paralytic forms, as well as sudden death with no apparent disease. Moreno and Baer (1980) conducted the definitive experimental study using RABV prepared from a naturally infected vampire bat. They demonstrated that both intramuscular and subcutaneous inoculation of captive vampire bats resulted in the infection of the bat and that this infection was fatal. Many of the infected bats excreted virus in saliva prior to development of disease but showed no evidence for a carrier status whereby virus could be excreted long-term in a healthy vampire bat. Later studies confirmed the susceptibility of vampire bats to intramuscular inoculation of virus and reported a decrease in blood consumption and dehydration as early signs of infection. Wing and hind limb paralysis was observed (Aguilar-Setien et al., 1998), as well as muscular spasms and tremors (Almeida et al., 2005). Seroconversion was observed in a number of bats that survived inoculation with RABV. A further study observed salivary excretion on three occasions by a small number of bats that survived inoculation with RABV (Aguilar-Setien et al., 2005), although this has not been corroborated by other researchers.

4.3 Epidemiology of Rabies in Vampire Bats

Characterization of RABV in reservoirs in the New World was initially attempted using antigenic differentiation using panels of monoclonal antibodies raised against the RABV nucleocapsid protein (Rupprecht & Dietzschold, 1987; Diaz et al., 1994). This initially classified rabies viruses as those transmitted by terrestrial and non-terrestrial reservoirs and proved useful to identify the source of infection in humans. This has been further developed to distinguish between a variety of antigenic variants, of which there are currently 11 variants that are present across North and South America (Velasco-Villa et al., 2006). This approach to classification has been further refined through the use of nucleic acid sequence data derived from discrete fragments of the RABV genome. This began with short fragments of the nucleoprotein gene (Smith, 1996), which corroborated the separation of reservoirs based on antigenic variation and allowed further discrimination that began to link particular RABV variants with species (Nadin-Davis & Loza-Rubio, 2006; Franka et al., 2006; Mochizuki et al., 2011).

The main purpose of molecular epidemiology is to trace the variation of RABV in reservoir populations in fine detail. This has been achieved for the rabies virus variants associated with vampire bats, particularly when this has affected cattle (Kobayashi et al., 2006, 2008) or humans (da Rosa et al., 2006; Barbosa et al., 2008; Castilho et al., 2010). The major economic impact of vampire bat predation is through the transmission of rabies to livestock causing death. Estimates have suggested that over a 9-year period losses due to livestock death in the Americas were over $50 million (Belotto et al., 2005). Seasonal peaks in the vampire bat population due to births have been associated with an increase in livestock rabies (Lord, 1992).

Attacks on humans are rare; however, one of the most recent cases of rabies in Louisiana occurred in a migrant worker who was likely to have contracted the disease while in his home state of Michoacán, Mexico (Balsamo et al., 2011). Anecdotal evidence suggests that the incidence of vampire bat attacks on people resulting in rabies is on the increase. This could be due to an actual increase in virus prevalence in the reservoir, or an increase in attacks by vampire bats on humans, or an increase in incident reporting. Changes in human activity have been responsible for increasing vampire predation. For example, removal of livestock can result in increased biting attacks on humans. Many of the reported outbreaks occur in remote communities with limited access to health care professionals and facilities (da Rosa et al., 2006; Gonçalves et al., 2002; Schneider et al., 2001).

4.4 Control of Vampire Bats and Future Perspectives

The main approach to the control of RABV transmission has been to control the bat reservoir, mainly through destruction. This takes the form of habitat destruction, that is, destroying roosting caves, or direct trapping and killing of vampire bats (Mayen, 2003). However, this often results in the destruction of other bat species. A more focused approach is the application of anticoagulants, either to cattle (Thompson, Mitchell, & Burns, 1972), or applied directly to captured bats, which are then released. This approach provides a short-term respite but creates an ecological niche that tends to be filled quickly by dispersing vampire bats. Parenteral vaccination of cattle is an effective means of preventing death from rabies infection, although is costly and requires repeated application to every generation. An alternative to this is oral vaccination with plant-derived feed that has been genetically modified to express RABV glycoprotein (McGarvey et al., 1995; Loza-Rubio et al., 2008). Further investigations have considered vaccination of vampire bats (Aguilar-Setien et al., 1998, 2002). These studies demonstrated that anti-rabies vaccination in the common vampire bat was effective but is costly and unlikely to reach the widespread populations found throughout Latin America.

A further consideration is the potential for vampire bats to increase their range both north into the United States and south to larger areas of Argentina. Climate change models that predict modest increases in winter temperatures suggest that the range of the common vampire bat could spread north into the United States (Mistry & Moreno-Valdez, 2008). Despite their effectiveness at transmitting RABV, there is no evidence for transmission of vampire bat variants to other reservoirs (bats or dogs) with establishment of the variant within the new host and sustained transmission.

5 CONCLUSIONS

In summary, bats have played and continue to play a key role in lyssavirus ecology and evolution. Clearly, many different bat species from different geographical locations and with different feeding habits and ecologies have been found to be infected with different lyssaviruses. However, the unusual distribution of lyssaviruses around the world and their association with particular bat species remains unexplained. Despite the genetic divergence of lyssaviruses detected in different bat species, the disease is similar in all species observed, causing neurological disease where clinical signs are observed. With the exception of vampire bats, the public health risk of rabies in bats remains low, although the consequences of infection are severe. In all species, the mechanisms by which abortive infections occur are unclear. However, there remains no convincing evidence for the existence of a carrier status in bats. While it appears that CST events of lyssaviruses to other mammals, including humans, are rare, the potential for sustained transmission in non-bat hosts exists, although further studies are needed in order to define the many gaps in our understanding of bat lyssaviruses.

ACKNOWLEDGEMENTS

ACB, ARF and NJ are supported in part by the European Commission Seventh Framework Programme under ANTIGONE (project number 278976), as well as DEFRA grants SE0426, SV3500 and SEO421. DTSH is funded by the Wellcome Trust, and ARF and DTSH are funded by the Research and Policy for Infectious Disease Dynamics (RAPIDD) program of the Science and Technology Directorate, Department of Homeland Security, Fogarty International Centre, National Institutes of Health. CMF and TM received funding from the German Federal Ministry for Education and Research (BMBF, grant 01KI1016A).

References

Aguilar-Setien, A., Brochier, B., Tordo, N., De Paz, O., Desmettre, P., Peharpre, D., et al. (1998). Experimental rabies infection and oral vaccination in vampire bats (*Desmodus rotundus*). *Vaccine*, 16(11-12), 1122–1126.

Aguilar-Setien, A., Campos, Y. L., Cruz, E. T., Kretschmer, R., Brochier, B., & Pastoret, P. P. (2002). Vaccination of vampire bats using recombinant vaccinia-rabies virus. *Journal of Wildlife Diseases*, 38(3), 539–544.

Aguilar-Setien, A., Loza-Rubio, E., Salas-Rojas, M., Brisseau, N., Cliquet, F., Pastoret, P. P., et al. (2005). Salivary excretion of rabies virus by healthy vampire bats. *Epidemiology and Infection*, 133(3), 517–522.

Almeida, M. F., Martorelli, L. F., Aires, C. C., Sallum, P. C., Durigon, E. L., & Massad, E. (2005). Experimental rabies infection in haematophagous bats Desmodus rotundus. *Epidemiology and Infection*, 133(3), 523–527.

Altringham, J. D. (2011). *Bats: From Evolution to Conservation* (2nd ed.). Oxford, New york: Oxford University Press.

Amengual, B., Bourhy, H., Lopez-Roig, M., & Serra-Cobo, J. (2007). Temporal dynamics of European bat Lyssavirus type 1 and survival of Myotis myotis bats in natural colonies. *PLoS One*, 2(6), 1–7.

Amengual, B., Whitby, J. E., King, A., Cobo, J. S., & Bourhy, H. (1997). Evolution of European bat lyssaviruses. *Journal of General Virology*, 78(9), 2319–2328.

Animal Health Australia. (2009). Disease strategy Australian bat lyssavirus version 3.0, 2009. AUSVETPLAN–Australian veterinary emergency plan. <http://www.animal-healthaustralia.com.au/programs/emergency-animal-disease-preparedness/ausvet-plan/disease-strategies/>. Accessed 20.12.12.

Anonymous. (1986). Bat rabies in the Union of Soviet Socialist Republics. *Rabies Bulletin Europe*, 10, 12–14.

Anonymous. (1989). Observations on the course of a bat rabies case in the Federal Republic in Germany. *Rabies Bulletin Europe*, 13, 11.

Anonymous. (2006). Agreement on the conservation of populations of bats in Europe (EUROBATS) Annex 5: bat rabies.

Anonymous. (2009). ICTV Official taxonomy updates since the 8th report, 2009.

Anonymous. (2012). Rabies Bulletin Europe, <www.who-rabies-bulletin.org>. Accessed 20.12.12.

Arai, Y. T., Kuzmin, I., Kameoka, Y., & Botvinkin, A. D. (2003). New Lyssavirus genotype from the lesser mouse-eared bat (Myotis blythi), Kyrghyzstan. *Emerging Infectious Diseases*, 9(3), 333–337.

Arellano-Sota, C. (1988). Biology, ecology, and control of the vampire bat. *Reviews of Infectious Diseases*, 10(Suppl 4), S615–619.

Arguin, P. M., Murray-Lillibridge, K., Miranda, M. E. G., Smith, J. S., Calaor, A. B., & Rupprecht, C. E. (2002). Serologic evidence of Lyssavirus infections among bats, the Philippines. *Emerging Infectious Diseases*, 8(3), 258–262.

Aubert, M. F. (1999). Rabies in individual countries, France. *Rabies Bulletin Europe*, 23, 6.

Avery, R. J., & Tailyour, J. M. (1960). The isolation of rabies virus from insectivorous bats in British Columbia. *Canadian Journal Comparative Medicine*, 24, 143–146.

Badrane, H., & Tordo, N. (2001). Host switching in Lyssavirus history from the Chiroptera to the Carnivora orders. *Journal of Virology*, 75(17), 8096–8104.

Baer, G. M., & Bales, G. L. (1967). Experimental rabies infection in the Mexican freetail bat. *Journal of Infectious Diseases, 117*(1), 82–90.

Baer, G. M., & Smith, J. S. (1991). Rabies in nonhematophagous bats. In G. M. Baer (Ed.), *The Natural History of Rabies* (pp. 341–366) (2nd ed.). Boca Raton, Florida, USA: CRC Press.

Balsamo, G., Ratard, R. C., Thoppil, D. R., Thoppil, M., Pino, F. V., Rupprecht, C. E., et al. (2011). Human rabies from exposure to a vampire bat in Mexico – Louisiana, 2010. *Morbidity and Mortality Weekly, 60*, 1050–1052.

Baltazar, R. S. (1988). Study of rabies virus isolated from a European bat "Eptesicus serotinus"–Virulence for sheep and fox. *Revue Medicine Veterinaire, 139*(7), 615–621.

Banyard, A. C., Hartley, M., & Fooks, A. R. (2010). Reassessing the risk from rabies: A continuing threat to the UK? *Virus Research, 152*(1-2), 79–84.

Banyard, A. C., Hayman, D. T. S., Johnson, N., McElhinney, L., & Fooks, A. R. (2011). Bats and lyssaviruses. *Advances in Virus Research, 79*, 239–289.

Banyard, A. C., Johnson, N., Voller, K., Hicks, D., Nunez, A., Hartley, M., et al. (2009). Repeated detection of European bat lyssavirus type 2 in dead bats found at a single roost site in the UK. *Archives of Virology, 154*(11), 1847–1850.

Barbosa, T. F., Medeiros, D. B., Travassos da Rosa, E. S., Casseb, L. M., Medeiros, R., Pereira, A., et al. (2008). Molecular epidemiology of rabies virus isolated from different sources during a bat-transmitted human outbreak occurring in Augusto Correa municipality, Brazilian Amazon. *Virology, 370*(2), 228–236.

Barrat, J. (2004). *Australian Bat Lyssavirus*. St. Lucia: The University of Queensland.

Bell, J. F. (1959). Transmission of rabies to laboratory animals by bite of a naturally infected bat. *Science, 129*(3361), 1490–1491.

Belotto, A., Leanes, L. F., Schneider, M. C., Tamayo, H., & Correa, E. (2005). Overview of rabies in the Americas. *Virus Research, 111*(1), 5–12.

Benda, P., Aulagnier, S., Hutson, A. M., Tsytsulina, K., Karataş, A., Palmeirim, J., et al. (2008). Myotis aurascens IUCN 2011. <www.iucnredlist.org/> IUCN Red List of Threatened Species. Version 2011.2. Accessed 20.12.12.

Blanton, J. D., Palmer, D., Dyer, J., & Rupprecht, C. E. (2011). Rabies surveillance in the United States during 2010. *Journal of the American Veterinary Medical Association, 239*(6), 773–783.

Botvinkin, A. D., Kuzmin, I., & Chernov, S. M. (1992). Experimental infection of bats with Lyssaviruses serotype-1 and serotype-4. *Voprosy Virusologii, 37*, 215–218.

Botvinkin, A. D., Poleschuk, E. M., Kuzmin, I., Borisova, T. I., Gazaryan, S. V., Yager, P., et al. (2003). Novel lyssaviruses isolated from bats in Russia. *Emerging Infectious Diseases, 9*, 1623–1625.

Botvinkin, A. D., Selnikova, O. P., Antonova, L. A., Moiseeva, A. B., & Nesterenko, E. Y. (2005). Human rabies case caused from a bat bite in Ukraine. *Rabies Bulletin Europe, 29*, 5–7.

Boulger, L. R., & Porterfield, J. S. (1958). Isolation of a virus from Nigerian fruit bats. *Transactions of the Royal Society of Tropical Medicine and Hygiene, 52*(5), 421–424.

Bourhy, H., Kissi, B., Lafon, M., Sacramento, D., & Tordo, N. (1992). Antigenic and Molecular Characterization of Bat Rabies Virus in Europe. *Journal of Clinical Microbiology, 30*(9), 2419–2426.

Brass, D. A. (1994). *Rabies in Bats: Natural History and Public Health Implications*. Ridgefield CN: Livia Press.

Brookes, S. M., Klopfleisch, R., Muller, T., Healy, D. M., Teifke, J. P., Lange, E., et al. (2007). Susceptibility of sheep to European bat lyssavirus type-1 and -2 infection: A clinical pathogenesis study. *Veterinary Microbiology, 125*(3–4), 210–223.

Bruijn, Z. (2003). Het gedrag van hondsdolle vleermuizen. *Zoogdier, 14*(3), 27–28.

Burnett, C. D. (1989). Bat rabies in Illinois: 1965 to 1986. *Journal of Wildlife Diseases, 25*(1), 10–19.

Burns, K. F., Farinacci, C. F., & Murnane, T. G. (1956). Insectivorous bats naturally infected with rabies in Southwestern United States. *American Journal of Public Health and the Nations Health, 46*(9), 1089–1097.

Calisher, C. H., Childs, J. E., Field, H. E., Holmes, K. V., & Schountz, T. (2006). Bats: important reservoir hosts of emerging viruses. *Clinical Microbiology Reviews, 19*(3), 531–545.

Carini, (1911). About one large epizootie of rabies. *Annals Institute Pasteur, 25*, 843–846.

Carneiro, N. F., Caldeira, A. P., Antunes, L. A., Carneiro, V. F., & Carneiro, G. F. (2009). Rabies in Artibeus lituratus bats in Montes Claros, State of Minas Gerais. *Revista da Sociedade Brasileira de Medicina Tropical, 42*(4), 449–451.

Castilho, J. G., Carnieli, P., Jr., et al., Durymanova, E. A., Fahl Wde, O., Oliveira Rde, N., Macedo, C. I., et al. (2010). Human rabies transmitted by vampire bats: Antigenic and genetic characterization of rabies virus isolates from the Amazon region (Brazil and Ecuador). *Virus Research, 153*(1), 100–105.

Childs, J. E., Trimarchi, C. V., & Krebs, J. W. (1994). The epidemiology of bat rabies in New York State, 1988–1992. *Epidemiology and Infection, 113*(3), 501–511.

Cliquet, F., Freuling, C., Smreczak, M., van der Poel, W. H. M., Horton, D. L., Fooks, A. R., et al. (2010). Development of harmonised schemes for monitoring and reporting of rabies in animals in the European Union. Scientific Report Submitted to EFSA (Q-2010-00078).

Cliquet, F., Picard-Meyer, E., Barrat, J., Brookes, S. M., Healy, D. M., Wasniewski, M., et al. (2009). Experimental infection of foxes with European Bat Lyssaviruses type-1 and 2. *BMC Veterinary Research, 5*, 19.

Constantine, D. G. (1966a). Transmission experiments with bat rabies isolates: responses of certain Carnivora to rabies virus isolated from animals infected by non-bite route. *American Journal of Veterinary Research, 27*(116), 13–15.

Constantine, D. G. (1966b). Transmission experiments with bat rabies isolates: reaction of certain Carnivora, opossum, and bats to intramuscular inoculations of rabies virus isolated from free-tailed bats. *American Journal of Veterinary Research, 27*(116), 16–19.

Constantine, D. G. (1966c). Transmission experiments with bat rabies isolates: bite transmission of rabies to foxes and coyote by free-tailed bats. American. *Journal of Veterinary Research, 27*(116), 20–23.

Constantine, D. G. (1979). An updated list of rabies-infected bats in North America. *Journal of Wildlife Diseases, 15*(2), 347–349.

Crawford-Miksza, L. K., Wadford, D. A., & Schnurr, D. P. (1999). Molecular epidemiology of enzootic rabies in California. *Journal of Clinical Virology, 14*(3), 207–219.

Crerar, S., Longbottom, H., Rooney, J., & Thornber, P. (1996). Human health aspects of a possible lyssavirus in a flying fox. *Communicable Disease Intelligence, 20*, 325.

Crick, J., Tignor, G. H., & Moreno, K. (1982). A new isolate of Lagos bat virus from the Republic of South Africa. *Transactions of the Royal Society of Tropical Medicine and Hygiene, 76*(2), 211–213.

Cryan, P. M., & Wolf, B. O. (2003). Sex differences in the thermoregulation and evaporative water loss of a heterothermic bat, *Lasiurus cinereus*, during its spring migration. *Journal of Experimental Biology, 206*(Pt 19), 3381–3390.

Cunha, E. M., Silva, L. H., Lara Mdo, C., Nassar, A. F., Albas, A., Sodre, M. M., et al. (2006). Bat rabies in the north-northwestern regions of the state of Sao Paulo, Brazil: 1997–2002. *Revista de Saude Publica, 40*(6), 1082–1086.

da Rosa, E. S., Kotait, I., Barbosa, T. F., Carrieri, M. L., Brandao, P. E., Pinheiro, A. S., et al. (2006). Bat-transmitted human rabies outbreaks, Brazilian Amazon. *Emerging Infectious Diseases, 12*(8), 1197–1202.

Dacheux, L., Larrous, F., Mailles, A., Boisseleau, D., Delmas, O., Biron, C., et al. (2009). European bat lyssavirus transmission among cats, Europe. *Emerging Infectious Diseases, 15*(2), 280–284.

Daoust, P. Y., Wandeler, A. I., & Casey, G. A. (1996). Cluster of rabies cases of probable bat origin among red foxes in Prince Edward Island, Canada. *Journal of Wildlife Diseases*, 32(2), 403–406.

Davis, A. D., Rudd, R. J., & Bowen, R. A. (2007). Effects of aerosolized rabies virus exposure on bats and mice. *Journal of Infectious Diseases*, 195(8), 1144–1150.

Davis, P. L., Holmes, E. C., Larrous, F., van der Poel, W. H., Tjornehoj, K., Alonso, W., et al. (2005). Phylogeography, population dynamics, and molecular evolution of European bat lyssaviruses. *Journal of Virology*, 79(16), 10487–10497.

de Almeida, M. F., Favoretto, S. R., Martorelli, L. A., Trezza-Netto, J., Campos, A., Ozahata, C. H., et al. (2011). Characterization of rabies isolates from a colony of Eptesicus furinalis bats in Brazil. *Revista do Instituto de Medicina Tropical de São Paulo*, 53(1), 31–37.

De Oviedo y Valdes, F. (1950). Sumario de la Natural Historia de las Indias,1526. Reprint Fondo de Cultura Economia.

de Rosa, A. R., Kataoka, A. P. A. G., Favoretto, S. R., Sodre, M. M., Trezza Netto, J., Campos, A. C. A., et al. (2011). First report of rabies infection in bats, Molossus molossus, Molossops neglectus and Myotis riparius in the city of Sao Paulo, State of Sao Paulo, southeastern Brazil. *Revista da Sociedade Brasileira de Medicina Tropical*, 44(2), 146–149.

De Serres, G., Dallaire, F., Cote, M., & Skowronski, D. M. (2008). Bat rabies in the United States and Canada from 1950 through 2007: Human cases with and without bat contact. *Clinical Infectious Diseases*, 46(9), 1329–1337.

Delpietro, H. A., Gury-Dhomen, F., Larghi, O. P., Mena-Segura, C., & Abramo, L. (1997). Monoclonal antibody characterization of rabies virus strains isolated in the River Plate Basin. *Zentralbl Veterinarmed B*, 44(8), 477–483.

Diaz, A. M., Papo, S., Rodriguez, A., & Smith, J. S. (1994). Antigenic analysis of rabies-virus isolates from Latin America and the Caribbean. *Zentralbl Veterinarmed B*, 41(3), 153–160.

Dzikwi, A. A., Kuzmin, I. I., Umoh, J. U., Kwaga, J. K., Ahmad, A. A., & Rupprecht, C. E. (2010). Evidence of Lagos bat virus circulation among Nigerian fruit bats. *Journal of Wildlife Diseases*, 46(1), 267–271.

Echevarria, J. E., Avellon, A., Juste, J., Vera, M., & Ibanez, C. (2001). Screening of active lyssavirus infection in wild bat populations by viral RNA detection on oropharyngeal swabs. *Journal of Clinical Microbiology*, 39(10), 3678–3683.

Enright, J. B., Sadler, W. W., Moulton, J. E., & Constantine, D. (1955). Isolation of rabies virus from an insectivorous bat (Tadarida mexicana) in California. *Proceedings of the Society for Experimental Biology and Medicine*, 89(1), 94–96.

Familusi, J. B., Osunkoya, B. O., Moore, D. L., Kemp, G. E., & Fabiyi, A. (1972). A fatal human infection with Mokola virus. *American Journal of Tropical Medicine and Hygiene*, 21(6), 959–963.

Favi, M., de Mattos, C. A., Yung, V., Chala, E., Lopez, L. R., & de Mattos, C. C. (2002). First case of human rabies in chile caused by an insectivorous bat virus variant. *Emerging Infectious Diseases*, 8(1), 79–81.

Favoretto, S. R., Carrieri, M. L., Cunha, E. M., Aguiar, E. A., Silva, L. H., Sodre, M. M., et al. (2002). Antigenic typing of Brazilian rabies virus samples isolated from animals and humans, 1989–2000. *Revista do Instituto de Medicina Tropical de Sao Paulo*, 44(2), 91–95.

Fekadu, M., Shaddock, J. H., Chandler, F. W., & Sanderlin, D. W. (1988). Pathogenesis of rabies virus from a Danish bat (Eptesicus serotinus): neuronal changes suggestive of spongiosis. *Archives of Virology*, 99(3-4), 187–203.

Field, H. (2004). The ecology of Hendra virus and Australian bat lyssavirus. Brisbane, Australia: University of Queensland.

Foggin, C. M. (1983). Mokola virus infection in cats and a dog in Zimbabwe. *Veterinary Record*, 113(5), 115.

Foggin, C. M. (1988). *Rabies and Rabies Related Viruses in zimbabwe: Historical, Virological and Ecological Aspects* (Vol. PhD). Harare: University of Zimbabwe.

Fooks, A. (2004). The challenge of new and emerging lyssaviruses. *Expert Review of Vaccines*, 3(4), 333–336.

Fooks, A. R., Brookes, S. M., Johnson, N., McElhinney, L. M., & Hutson, A. M. (2003a). European bat lyssaviruses: an emerging zoonosis. *Epidemiology and Infection*, 131(3), 1029–1039.

Fooks, A. R., Johnson, N., Freuling, C. M., Wakeley, P. R., Banyard, A. C., McElhinney, L. M., et al. (2009). Emerging technologies for the detection of rabies virus: challenges and hopes in the 21st century. *PLoS Neglected Tropical Diseases*, 3(9), e530.

Fooks, A. R., McElhinney, L. M., Pounder, D. J., Finnegan, C. J., Mansfield, K., Johnson, N., et al. (2003b). Case report: Isolation of a European bat lyssavirus type 2a from a fatal human case of rabies encephalitis. *Journal of Medical Virology*, 71(2), 281–289.

Franka, R., Constantine, D. G., Kuzmin, I., Velasco-Villa, A., Reeder, S. A., Streicker, D., et al. (2006). A new phylogenetic lineage of rabies virus associated with western pipistrelle bats (*Pipistrellus hesperus*). *Journal of General Virology*, 87(Pt 8), 2309–2321.

Franka, R., Johnson, N., Muller, T., Vos, A., Neubert, L., Freuling, C., et al. (2008). Susceptibility of North American big brown bats (*Eptesicus fuscus*) to infection with European bat lyssavirus type 1. *Journal of General Virology*, 89, 1998–2010.

Fraser, G. C., Hooper, P. T., Lunt, R. A., Gould, A. R., Gleeson, L. J., Hyatt, A. D., et al. (1996). Encephalitis caused by a Lyssavirus in fruit bats in Australia. *Emerging Infectious Diseases*, 2(4), 327–331.

Freuling, C., Johnson, N., Marston, D. A., Selhorst, T., Geue, L., Fooks, A. R., et al. (2008a). A random grid based molecular epidemiological study on EBLV isolates from Germany. *Developmental Biology (Basel)*, 131, 301–309.

Freuling, C., Grossmann, E., Conraths, F. J., Schameitat, A., Kliemt, J., Auer, E., et al. (2008b). First isolation of EBLV-2 in Germany. *Veterinary Microbiology*, 131(1–2), 26–34.

Freuling, C., Vos, A., Johnson, N., Fooks, A. R., & Muller, T. (2009a). Bat rabies–a Gordian knot? *Berliner und Munchener Tierarztliche Wochenschrift*, 122(11–12), 425–433.

Freuling, C., Vos, A., Johnson, N., Kaipf, I., Denzinger, A., Neubert, L., et al. (2009b). Experimental infection of serotine bats (Eptesicus serotinus) with European bat lyssavirus type 1a. *Journal of General Virology*, 90(10), 2493–2502.

Freuling, C. M., Beer, M., Conraths, F. J., Finke, S., Hoffmann, B., Keller, B., et al. (2011). Novel Lyssavirus in Natterer's Bat, Germany. *Emerging Infectious Diseases*, 17(8), 1519–1522.

Freuling, C. M., Hoffmann, B., Selhorst, T., Conraths, F. J., Kliemt, J., Schatz, J., et al. (2012a). New insights into the genetics of EBLV-1 from Germany. *Berl Munch Tierarztl Wochenschr*, 125(5–6), 259–263.

Freuling, C., Kliemt, J., Schares, S., Heidecke, D., Driechciarz, R., Schatz, J., et al. (2012b). Detection of European bat lyssavirus 2 (EBLV-2) in a Daubenton's bat from Magdeburg, Germany. *Berliner und Munchener Tierarztliche Wochenschrift*, 122(11–12), 425–433.

Giannini, N. P., & Simmons, N. B. (2003). A phylogeny of megachiropteran bats (Mammalia: Chiroptera: Pteropodidae) based on direct optimization analysis of one nuclear and four mitochondrial genes. *Cladistics*, 19(6), 496–511.

Goncalves, M. A., Sa-Neto, R. J., & Brazil, T. K. (2002). Outbreak of aggressions and transmission of rabies in human beings by vampire bats in northeastern Brazil. *Revista da Sociedade Brasileira de Medicina Tropical*, 35(5), 461–464.

Gonzalez, J. P., Gouilh, A. R., Reyes, J. M., & Leroy, E. (2008). Bat-borne viral diseases. In C. J. P. Colfer (Ed.), *Human Health and Forests: A Global Overview of Issues, Practice, and Policy (pp. 161–196)*. UK and USA: Earthscan.

Gould, A. R., Hyatt, A. D., Lunt, R., Kattenbelt, J. A., Hengstberger, S., & Blacksell, S. D. (1998). Characterisation of a novel lyssavirus isolated from Pteropid bats in Australia. *Virus Research*, 54(2), 165–187.

Gould, A. R., Kattenbelt, J. A., Gumley, S. G., & Lunt, R. A. (2002). Characterisation of an Australian bat lyssavirus variant isolated from an insectivorous bat. *Virus Research*, 89(1), 1–28.

Greenhall, A. M. (1988). Feeding behaviour. In A. M. Greenhall & U. Schmidt (Eds.), *Natural History of Vampire Bats (pp. 111–131)*. Florida: CRC Press Inc.

Guyatt, K. J., Twin, J., Davis, P., Holmes, E. C., Smith, G. A., Smith, I. L., et al. (2003). A molecular epidemiological study of Australian bat lyssavirus. *Journal of General Virology, 84*, 485–496.

Halpin, K., Hyatt, A. D., Fogarty, R., Middleton, D., Bingham, J., Epstein, J. H., et al. (2011). Pteropid bats are confirmed as the reservoir hosts of henipaviruses: a comprehensive experimental study of virus transmission. *American Journal of Tropical Medicine and Hygiene, 85*(5), 946–951.

Harbusch, C. (2003). *Aspects of the ecology of Serotine bats (Eptesicus serotinus, Schreber 1774) in contrasting landscapes in southwest Germany and Luxembourg*. (PhD thesis), University of Aberdeen, UK.

Harris, S. L., Aegerter, J. N., Brookes, S. M., McElhinney, L. M., Jones, G., Smith, G. C., et al. (2009). Targeted surveillance for European bat lyssaviruses in English bats (2003–2006). *Journal of Wildlife Diseases, 45*(4), 1030–1041.

Haupt, H., & Rehaag, H. (1921). Epizootic rabies in a herd from Santa Carina (south-east Brazil) transmitted by bats. *Zeitschr Infektions Hyg Haustiere, 76*–90 *(XXII)*, 104–107.

Hayman, D. T., Emmerich, P., Yu, M., Wang, L. F., Suu-Ire, R., Fooks, A. R., et al. (2010). Long-term survival of an urban fruit bat seropositive for Ebola and Lagos bat viruses. *PLoS One, 5*, 8.

Hayman, D. T., Fooks, A. R., Horton, D., Suu-Ire, R., Breed, A. C., Cunningham, A. A., et al. (2008). Antibodies against Lagos bat virus in megachiroptera from West Africa. *Emerging Infectious Diseases, 14*(6), 926–928.

Hayman, D. T. S., Fooks, A. R., Rowcliffe, J. W., McCrea, R., Restif, O., Baker, K. S., et al. (2012). Endemic Lagos bat virus infection in Eidolon helvum. *Epidemiology and Infection*. (In press).

Hughes, G. J., Orciari, L. A., & Rupprecht, C. E. (2005). Evolutionary timescale of rabies virus adaptation to North American bats inferred from the substitution rate of the nucleoprotein gene. *Journal of General Virology, 86*(Pt 5), 1467–1474.

Hughes, G. J., (2008). A reassessment of the emergence time of European bat lyssavirus type-1. *Infections, Genetics and Evolution, 8*(6), 820–824.

Hurst, E. W., & Pawan, J. L. (1932). A further account of the Trinidad outbreak of acute rabies myelitis. *Journal of Pathological Bacteriology, 35*, 301–321.

Hutcheon, J. M., & Kirsch, J. A. W. (2006). A moveable face: deconstructing the Microchiroptera and a new classification of extant bats. *Acta Chiropteralogica, 8*(1), 1–10.

Hutson, A. M., Mickleburgh, S. P., & Racey, P. A. (2001). *Microchiropteran Bats: Global Status Survey and Conservation Action Plan*. Oxford: Information Press.

Hutson, A. M., Spitzenberger, F., Aulagnier, S. Alcald, J. T., Csorba, G., Bumrungsri, S., et al., (2008, 2009/01/20/). Eptesicus serotinus, from <http://www.iucnredlist.org/>. Accessed 20.12.12.

Ibanez, C., García-Mudarra, J. L., Ruedi, M., Stadelmann, B., & Juste, J. (2006). The Iberian contribution to cryptic diversity in European bats. *Acta Chiropterologica, 8*(2), 277–297.

Jackson, A. C., Ye, H., Phelan, C. C., Ridaura-Sanz, C., Zheng, Q., Li, Z., et al. (1999). Extraneural organ involvement in human rabies. *Laboratory Investigation, 79*(8), 945–951.

Jackson, F. R., Turmelle, A. S., Farino, D. M., Franka, R., McCracken, G. F., & Rupprecht, C. E. (2008). Experimental rabies virus infection of big brown bats (Eptesicus fuscus). *Journal of Wildlife Diseases, 44*(3), 612–621.

Jakava-Viljanen, M., Lilley, T., Kyheroinen, E. M., & Huovilainen, A. (2010). First encounter of European bat lyssavirus type 2 (EBLV-2) in a bat in Finland. *Epidemiology and Infection, 138*(11), 1581–1585.

Jiang, Y., Wang, L., Lu, Z., Xuan, H., Han, X., Xia, X., et al. (2010). Seroprevalence of rabies virus antibodies in bats from southern China. *Vector Borne Zoonotic Diseases, 10*(2), 177–181.

Johnson, N., Freuling, C. M., Marston, D. A., Tordo, N., Fooks, A. R., & Muller, T. (2007). Identification of European bat lyssavirus isolates with short genomic insertions. *Virus Research, 128*(1–2), 140–143.

Johnson, N., Phillpotts, R., & Fooks, A. R. (2006a). Airborne transmission of lyssaviruses. *Journal of Medical Microbiology, 55*(Pt 6), 785–790.

Johnson, N., Selden, D., Parsons, G., Healy, D., Brookes, S. M., McElhinney, L. M., et al. (2003). Isolation of a European bat lyssavirus type 2 from a Daubenton's bat in the United Kingdom. *Veterinary Record, 152*(13), 383–387.

Johnson, N., Vos, A., Neubert, L., Freuling, C., Mansfield, K., Kaipf, I., et al. (2008). Experimental study of European bat lyssavirus type-2 infection in Daubenton's bats (Myotis daubentonii). *Journal of General Virology, 89*(11), 2662–2672.

Johnson, N., Wakeley, P. R., Brookes, S. M., & Fooks, A. R. (2006b). European Bat Lyssavirus Type 2 RNA in Myotis daubentonii. *Emerging Infectious Diseases, 12*(7), 1142–1144.

Juste, J., Bilgin, R., Munoz, J., & Ibanez, C. (2009). Mitochondrial DNA signature at different spatial scales: from the effects of the Straits of Gibraltar to population structure in the meridional serotine bat (*Eptesicus isabellinur*). *Heredity, 103*, 178–187.

Kappeler, A. (1989). Bat rabies surveillance in Europe. *Rabies Bulletin Europe, 13*, 12–13.

Kemp, G. E., Causey, O. R., Moore, D. L., Odelola, A., & Fabiyi, A. (1972). Mokola virus. Further studies on IbAn 27377, a new rabies-related etiologic agent of zoonosis in Nigeria. *American Journal of Tropical Medicine and Hygiene, 21*(3), 356–359.

King, A., & Crick, J. (1988). Rabies-related viruses. In K. M. Campbell & J. B. Campbell (Eds.), *Rabies* (pp. 177–200). Boston: Kluwer Academic Publishers.

King, A. A., Haagsma, J., & Kappeler, A. (2004). Lyssavirus infections in European bats. In A. A. King, A. R. Fooks, M. Aubert, & A. I. Wandeler (Eds.), *Historical Perspective of Rabies in Europe and the Mediterranean Basin* (pp. 221–241). Paris: OIE (World Organisation for Animal Health).

Klug, B. J., Turmelle, A. S., Ellison, J. A., Baerwald, E. F., & Barclay, R. M. (2011). Rabies prevalence in migratory tree-bats in Alberta and the influence of roosting ecology and sampling method on reported prevalence of rabies in bats. *Journal of Wildlife Diseases, 47*(1), 64–77.

Kobayashi, Y., Ogawa, A., Sato, G., Sato, T., Itou, T., Samara, S. I., et al. (2006). Geographical distribution of vampire bat-related cattle rabies in Brazil. *Journal of Veterinary Medical Science, 68*(10), 1097–1100.

Kobayashi, Y., Sato, G., Mochizuki, N., Hirano, S., Itou, T., Carvalho, A. A., et al. (2008). Molecular and geographic analyses of vampire bat-transmitted cattle rabies in central Brazil. *BMC Veterinary Research, 4*, 44.

Koopman, K. F. (1988). Systematics and distribution. In A. M. Greenhall & U. Schmidt (Eds.), *Natural History of Vampire Bats* (pp. 7–17). Boca Ratón: CRC Press.

Kuzmin, I. V., & Botvinkin, A. D. (1996). The behaviour of bats Pipistrellus pipistrellus after experimental inoculation with rabies and rabies-like viruses and some aspects of pathogenesis. *Myotis, 34*, 93–99.

Kuzmin, I. V., Botvinkin, A. D., Rybin, S. N., & Bayaliev, A. B. (1992). A lyssavirus with an unusual antigenic structure isolated from a bat in southern Kyrghyzstan. *Voprosy Virusologii, 37*(5-6), 256–259.

Kuzmin, I. V., Botvinkin, A. D., & Shaimardanov, R. T. (1994). Experimental Rabies and Rabies-related infections in bats. *Voprosy Virusologii, 39*(1), 17–21.

Kuzmin, I. V., Bozick, B., Guagliardo, S. A., Kunkel, R., Shak, J. R., Tong, S., et al. (2011b). Bats, emerging infectious diseases, and the rabies paradigm revisited. *Emerging Health Threats Journal, 4*, 7159.

Kuzmin, I. V., Hughes, G. J., Botvinkin, A. D., Orciari, L. A., & Rupprecht, C. E. (2005). Phylogenetic relationships of Irkut and West Caucasian bat viruses within the Lyssavirus genus and suggested quantitative criteria based on the N gene sequence for lyssavirus genotype definition. *Virus Research, 111*(1), 28–43.

Kuzmin, I. V., Shi, M., Orciari, L. A., Yager, P. A., Vellasco-Villa, A., Kuzmina, N. A., et al. (2012). Molecular inferences suggest multiple host shifts of rabies viruses from bats to mesocarnivores in Arizona during 2001-2009. *PLOS Pathogens, 8*(6), 1–11.

Kuzmin, I. V., Mayer, A. E., Niezgoda, M., Markotter, W., Agwanda, B., Breiman, R. F., et al. (2010). Shimoni bat virus, a new representative of the Lyssavirus genus. *Virus Research, 149*(2), 197–210.

Kuzmin, I. V., Niezgoda, M., Franka, R., Agwanda, B., Markotter, W., Beagley, J. C., et al. (2008a). Lagos bat virus in Kenya. *Journal of Clinical Microbiology, 46*(4), 1451–1461.

Kuzmin, I. V., Niezgoda, M., Franka, R., Agwanda, B., Markotter, W., Beagley, J. C., et al. (2008b). Possible emergence of West Caucasian bat virus in Africa. *Emerging Infectious Diseases, 14*(12), 1887–1889.

Kuzmin, I. V., Orciari, L. A., Arai, Y. T., Smith, J. S., Hanlon, C. A., Kameoka, Y., et al. (2003). Bat lyssaviruses (Aravan and Khujand) from Central Asia: phylogenetic relationships according to N, P and G gene sequences. *Virus Research, 97*(2), 65–79.

Kuzmin, I. V., & Rupprecht, C. E. (1999). *Bat Rabies Handbook of Animal Models of Infection: Experimental Models in Antimicrobial Chemotherapy*. Academic Press. pp. 259.

Kuzmin, I. V., Turmelle, A. S., Agwanda, B., Markotter, W., Niezgoda, M., Breiman, R. F., et al. (2011a). Commerson's leaf-nosed bat (Hipposideros commersoni) is the likely reservoir of Shimoni Bat Virus. *Vector-Borne and Zoonotic Diseases, 11*(11), 1465–1470.

Lack, J. B., Roehrs, Z. P., Stanley, C. E., Jr., Ruedi, M., & Van den Bussche, R. A. (2010). Molecular phylogenetics of Myotis indicate familial-level divergence for the genus Cistugo (Chiroptera). *Journal of Mammalogy, 91*(4), 976–992.

Lafon, M., Herzog, M., & Sureau, P. (1986). Human rabies vaccines induce neutralizing antibodies against the European bat rabies virus (Duvenhage). *Lancet, 2*(8505), 515.

Le-Gonidec, G., Rickenbach, A., Robin, Y., & Heme, G. (1978). Isolation of a strain of Mokola virus in Cameroon. *Annals Microbiologie (Paris), 129*(2), 245–249.

Leonova, G. N., Belikov, S. I., Kondratov, I. G., Krylova, N. V., Pavlenko, E. V., Romanova, E. V., et al. (2009). A fatal case of bat lyssavirus infection in Primorye Territory of the Russian Far East. *Rabies Bulletin Europe, 33*(4), 5–8.

Leslie, M. J., Messenger, S., Rohde, R. E., Smith, J., Cheshier, R., Hanlon, C., et al. (2006). Bat-associated rabies virus in skunks. *Emerging Infectious Diseases, 12*(8), 1274–1277.

Lopez, A., Miranda, P., Tejada, E., & Fishbein, D. B. (1992). Outbreak of human rabies in the Peruvian jungle. *Lancet, 339*(8790), 408–411.

Lord, R. D. (1992). Seasonal reproduction of vampire bats and its relation to seasonality of bovine rabies. *Journal of Wildlife Diseases, 28*(2), 292–294.

Loza-Rubio, E., Rojas, E., Gomez, L., Olivera, M. T., & Gomez-Lim, M. A. (2008). Development of an edible rabies vaccine in maize using the Vnukovo strain. *Developmental Biology (Basel), 131*, 477–482.

Lumio, J., Hillbom, M., Roine, R., Ketonen, L., Haltia, M., Valle, M., et al. (1986). Human rabies of bat origin in Europe. *Lancet, 15*, 378.

Lumlertdacha, B., Wacharapluesadee, S., Chanhome, L., & Hemachudha, T. (2005). Bat lyssavirus in Thailand. *Journal of the Medical Association of Thailand, 88*(7), 1011–1014.

Malerczyk, C., Selhorst, T., Tordo, N., Moore, S. A., & Müller, T. (2009). Antibodies induced by vaccination with purified chick embryo cell culture vaccine (PCECV) cross-neutralize non-classical bat lyssavirus strains. *Vaccine, 27*(39), 5320–5325.

Markotter, W., Kuzmin, I. V., Rupprecht, C. E., & Nel, L. H. (2009). Lagos bat virus virulence in mice inoculated by the peripheral route. *Epidemiology and Infection, 137*(8), 1155–1162.

Markotter, W., Randles, J., Rupprecht, C. E., Sabeta, C. T., Taylor, P. J., Wandeler, A., et al. (2006). Lagos bat virus, South Africa. *Emerging Infectious Diseases, 12*(3), 504–506.

Markotter, W., Van Eeden, C., Kuzmin, I. V., Rupprecht, C. E., Paweska, J. T., Swanepoel, R., et al. (2008). Epidemiology and pathogenicity of African bat lyssaviruses. *Developments in Biologicals (Basel), 131*, 317–325.

Marston, D. A., Horton, D. L., Ngeleja, C., Hampson, K., McElhinney, L. M., Banyard, A. C., et al. (2012). Ikoma Lyssavirus, evidence for a highly divergent novel lyssavirus in an African civet (*Civettictis civetta*). *Emerging infectious Diseases, 18*, 4.

Mayen, F. (2003). Haematophagous bats in Brazil, their role in rabies transmission, impact on public health, livestock industry and alternatives to an indiscriminate reduction of bat population. *Journal Veterinary Medicine B Infectious Diseases Veterinary Public Health, 50*(10), 469–472.

McColl, K. A., Chamberlain, T., Lunt, R. A., Newberry, K. M., Middleton, D., & Westbury, H. A. (2002). Pathogenesis studies with Australian bat lyssavirus in grey-headed flying foxes (*Pteropus poliocephalus*). *Australian Veterinary Journal, 80*(10), 636–641.

McColl, K. A., Chamberlain, T., Lunt, R. A., Newberry, K. M., & Westbury, H. A. (2007). Susceptibility of domestic dogs and cats to Australian bat lyssavirus (ABLV). *Veterinary Microbiology, 123*(1–3), 15–25.

McGarvey, P. B., Hammond, J., Dienelt, M. M., Hooper, D. C., Fu, Z. F., Dietzschold, B., et al. (1995). Expression of the rabies virus glycoprotein in transgenic tomatoes. *Bio-Technology, 13*(13), 1484–1487.

Mebatsion, T., Cox, J. H., & Frost, J. W. (1992). Isolation and characterization of 115 street rabies virus isolates from Ethiopia by using monoclonal antibodies: identification of 2 isolates as Mokola and Lagos bat viruses. *Journal of Infectious Diseases, 166*(5), 972–977.

Megali, A., Yannic, G., Zahno, M. L., Brugger, D., Bertoni, G., Christe, P., et al. (2010). Surveillance for European bat lyssavirus in Swiss bats. *Archives of Virology, 155*(10), 1655–1662.

Meredith, C. D., Prossouw, A. P., & Koch, H. P. (1971). An unusual case of human rabies thought to be of chiropteran origin. *South African Medical Journal, 45*(28), 767–769.

Messenger, S. L., Smith, J. S., & Rupprecht, C. E. (2002). Emerging epidemiology of bat-associated cryptic cases of rabies in humans in the United States. *Clinical Infectious Diseases, 35*(6), 738–747.

Mistry, S., & Moreno-Valdez, A. (2008). Climate change and bats: Vampire bats offer clues to the future. *Bats, 26*, 8–11.

Mochizuki, N., Kobayashi, Y., Sato, G., Hirano, S., Itou, T., Ito, F. H., et al. (2011). Determination and molecular analysis of the complete genome sequence of two wild-type rabies viruses isolated from a haematophagous bat and a frugivorous bat in Brazil. *Journal of Veterinary Medical Science, 73*(6), 759–766.

Mohr, W. (1957). Die Tollwut. *Medizinische Klinik, 52*(24), 1057–1060.

Mondul, A. M., Krebs, J. W., & Childs, J. E. (2003). Trends in national surveillance for rabies among bats in the United States (1993–2000). *Journal of the American Veterinary Medical Association, 222*(5), 633–639.

Moore, P. R., Jansen, C. C., Graham, G. C., Smith, I. L., & Craig, S. B. (2010). Emerging tropical diseases in Australia. Part 3. Australian bat lyssavirus. *Annals of Tropical Medicine and Parasitology, 104*(8), 613–621.

Moreno, J. A., & Baer, G. M. (1980). Experimental rabies in the vampire bat. *American Journal of Tropical Medicine and Hygiene, 29*(2), 254–259.

Morimoto, K., Patel, M., Corisdeo, S., Hooper, D. C., Fu, Z. F., Rupprecht, C. E., et al. (1996). Characterization of a unique variant of bat rabies virus responsible for newly emerging human cases in North America. *Proceedings of the National Academy of Sciences of the United States of America, 93*(11), 5653–5658.

Müller, T., Cox, J., Peter, W., Schäfer, R., Johnson, N., McElhinney, L. M., et al. (2004). Spillover of European bat lyssavirus type 1 into a stone marten (Martes foina) in Germany. *Journal of Veterinary Medicine Series B, 51*(2), 49–54.

Muller, T., Johnson, N., Freuling, C. M., Fooks, A. R., Selhorst, T., & Vos, A. (2007). Epidemiology of bat rabies in Germany. *Archives of Virology, 152*(2), 273–288.

Oliveira, R. N., Souza, S. P., Lobo, R. S., Castilho, J. G., Macedo, C. I., Carnieli, P., Jr., et al. (2010). Rabies virus in insectivorous bats: implications of the diversity of the nucleoprotein and glycoprotein genes for molecular epidemiology. *Virology, 405*(2), 352–360.

Nadin-Davis, S. A., Huang, W., Armstrong, J., Casey, G. A., Bahloul, C., Tordo, N., et al. (2001). Antigenic and genetic divergence of rabies viruses from bat species indigenous to Canada. *Virus Research, 74*(1-2), 139–156.

Nadin-Davis, S. A., & Loza-Rubio, E. (2006). The molecular epidemiology of rabies associated with chiropteran hosts in Mexico. *Virus Research, 117*(2), 215–226.

Nathwani, D., McIntyre, P. G., White, K., Shearer, A. J., Reynolds, N., Walker, D., et al. (2003). Fatal human rabies caused by European bat Lyssavirus type 2a infection in Scotland. *Clinical Infectious Diseases, 37*(4), 598–601.

National Veterinary Institute. (2010). Surveillance of zoonotic and other animal disease agents in Sweden 2009.

Nel, L. H., & Rupprecht, C. E. (2007). Emergence of lyssaviruses in the Old World: the case of Africa. *Current Topics in Microbiology and Immunology, 315*, 161–193.

Niethammer, J., & Krapp, F. (2011). *Die Fledermäuse Europas.* Wiebelsheim, Germany: AULA-Verlag.

Nieuwenhuijs, J., Haagsma, J., & Lina, P. (1992). Epidemiology and control of rabies in bats in the Netherlands. *Revue Scientifique Technologique Office International Epizootic, 11*(4), 1155–1161.

Nieuwenhuijs, J. H. (1987). Veterinary chief inspection for public health. Rabies in bats. *Tijdschrift Voor Diergeneeskunde, 112*(20), 1193–1197.

Nunes, K. N., da Rosa, E. S. T., Barbosa, T. F., Pereira, A. S., Medeiros, D. B., Casseb, L. M., et al. (2008). Genetic characterization of the rabies virus strain QR 18867 (Rhabdoviridae, Lyssavirus) isolated from the *Uroderma bilobatum* bat in Portel Municipality, Para State, 2004. *American Journal of Tropical Medical and Hygiene, 79*, 21–23.

Pajamo, K., Harkess, G., Goddard, T., Marston, D., McElhinney, L., Johnson, N., et al. (2008). Isolation of European bat lyssavirus type 2 (EBLV-2) in a Daubenton's bat in the UK with a minimum incubation period of 9 months. *Rabies Bulletin Europe, 32*(2), 6–7.

Pawan, J. L. (1936). Rabies in the vampire bat of Trinidad, with special reference to the clinical course and the latency of infection. *Annals of Tropical Medicine and Parasitology, 30*, 410–422.

Pawan, J. L. (1938). The transmission of paralytic rabies in Trinidad by the vampire bat. *Annals of Tropical Medicine and Parasitology, 30*, 101–128.

Paweska, J. T., Blumberg, L. H., Liebenberg, C., Hewlett, R. H., Grobbelaar, A. A., Leman, P. A., et al. (2006). Fatal human infection with rabies-related Duvenhage virus, South Africa. *Emerging Infectious Diseases, 12*(12), 1965–1967.

Perez-Jorda, J. L., Ibanez, C., Munozcervera, M., & Tellez, A. (1995). Lyssavirus in *Eptesicus serotinus* (Chiroptera, Vespertilionidae). *Journal of Wildlife Diseases, 31*(3), 372–377.

Picard-Meyer, E., Barrat, J., Tissot, E., Verdot, A., Patron, C., Barrat, M. J., et al. (2006). Bat rabies surveillance in France, from 1989 through May 2005. *Development in Biological Standardization, 125*, 283–288.

Picard-Meyer, E., Brookes, S. M., Barrat, J., Litaize, E., Patron, C., Biarnais, M., et al. (2008). Experimental infection of foxes with European bat lyssaviruses type-1 and -2. *Developments in Biologicals (Basel), 131*, 339–345.

Picard-Meyer, E., Dubourg-Savage, M. J., Arthur, L., Barataud, M., Becu, D., Bracco, S., et al. (2011). Active surveillance of bat rabies in France: a 5-year study (2004–2009). *Veterinary Microbiology, 151*(3-4), 390–395.

Price, J. L., & Everard, C. O. (1977). Rabies virus and antibody in bats in Grenada and Trinidad. *Journal of Wildlife Diseases, 13*(2), 131–134.

Richards, G. C. (1998). Order chiroptera: Bats. In R. Strahan & S. van Dyck (Eds.), *The Mammals of Australia.* Sydney: New Holland Publishers.

Rønsholt, L., Soerensen, K. J., Bruschke, C. J. M., Wellenberg, G. J., van Orischot, J. T., Johnstone, P., et al. (1998). Clinically silent rabies infection in (zoo) bats. *Veterinary Record, 142*, 519–520.

Ronsholt, L. (2002). A new case of European Bat Lyssavirus (EBL) infection in Danish sheep. *Rabies Bulletin Europe, 26*(2), 15.

Ruedi, M., & Mayer, F. (2001). Molecular systematics of bats of the genus Myotis (Vespertilionidae) suggests deterministic ecomorphological convergences. *Molecular Phylogenetics and Evolution, 21*(3), 436–448.

Rupprecht, C. E., & Dietzschold, B. (1987). Perspectives on rabies virus pathogenesis. *Laboratory Investigation, 57*(6), 603–606.

Sabeta, C. T., Markotter, W., Mohale, D. K., Shumba, W., Wandeler, A. I., & Nel, L. H. (2007). Mokola virus in domestic mammals, South Africa. *Emerging Infectious Diseases, 13*(9), 1371–1373.

Salas-Rojas, M., Sanchez-Hernandez, C., Romero-Almaraz Md Mde, L., Schnell, G. D., Schmid, R. K., & Aguilar-Setien, A. (2004). Prevalence of rabies and LPM paramyxovirus antibody in non-hematophagous bats captured in the Central Pacific coast of Mexico. *Transactions of the Royal Society of Tropical Medicine and Hygiene, 98*(10), 577–584.

Saluzzo, J. F., Rollin, P. E., Daugard, C., Digoutte, J. P., Georges, A. J., & Sureau, P. (1984). Premier isolement du virus Mokola a patir d'une rongeur (*Lophuromys sikapusi*). *Annales de Institut Pasteur Virologie, 135E*, 57–66.

Scatterday, J. E. (1954). Bat rabies in Florida. *Journal of the American Veterinary Medical Association, 124*(923), 125.

Schatz, J., Fooks, A. R., McElhinney, L., Horton, D., Echevarria, J. E., Vazquez-Moron, S., et al. (2012). Bat rabies Surveillance in Europe. *Zoonosis and Public Health*. Sept 11.

Scheffer, K. C., Carrieri, M. L., Albas, A., Santos, H. C., Kotait, I., & Ito, F. H. (2007). Rabies virus in naturally infected bats in the State of Sao Paulo, Southeastern Brazil. *Revista de Saude Publica, 41*(3), 389–395.

Schneider, L. G., & Cox, J. H. (1994). Bat Lyssavirus in Europe. *Current Topics in Microbiology and Immunolology, 187*, 207–218.

Schneider, M. C., Aron, J., Santos-Burgoa, C., Uieda, W., & Ruiz-Velazco, S. (2001). Common vampire bat attacks on humans in a village of the Amazon region of Brazil. *Cadernos de Saude Publica, 17*(6), 1531–1536.

Selimov, M. A., Smekhov, A. M., Antonova, L. A., Shablovskaya, E. A., King, A. A., & Kulikova, L. G. (1991). New strains of rabies-related viruses isolated from bats in the Ukraine. *Acta Virologica, 35*(3), 226–231.

Selimov, M. A., Tatarov, A. G., Botvinkin, A. D., Klueva, E. V., Kulikova, L. G., & Khismatullina, N. A. (1989). Rabies-related Yuli virus; identification with a panel of monoclonal antibodies. *Acta Virologica, 33*(6), 542–546.

Serra-Cobo, J., Amengual, B., Abellan, C., & Bourhy, H. (2002). European bat Lyssavirus infection in Spanish bat populations. *Emerging Infectious Diseases, 8*(4), 413–420.

Shirato, K., Maeda, K., Tsuda, S., Suzuki, K., Watanabe, S., Shimoda, H., et al. (2012). Detection of bat coronaviruses from *Miniopterus fuliginosus* in Japan. *Virus Genes, 44*(1), 40–44.

Shoji, Y., Kobayashi, Y., Sato, G., Itou, T., Miura, Y., Mikami, T., et al. (2004). Genetic characterization of rabies viruses isolated from frugivorous bat (Artibeus spp.) in Brazil. *Journal of Veterinary Medical Science, 66*(10), 1271–1273.

Shope, R. E., Murphy, F. A., Harrison, A. K., Causey, O. R., Kemp, G. E., Simpson, D. I., et al. (1970). Two African viruses serologically and morphologically related to rabies virus. *Journal of Virology, 6*(5), 690–692.

Simmons, N. B. (2005). Order Chiroptera. In D. E. Wilson & D. M. Reeder (Eds.), *Mammal species of the world* (Vol. 1, pp. 312–529). Baltimore: Baltimore University Press.

Smith, G. C., Aegerter, J. N., Allnutt, T. R., MacNicoll, A. D., Learmount, J., Hutson, A. M., et al. (2011). Bat population genetics and Lyssavirus presence in Great Britain. *Epidemiology and Infection, 139*(10), 1463–1469.

Smith, J. S. (1988). Monoclonal antibody studies of rabies in insectivorous bats of the United States. *Reviews of Infectious Diseases, 10*(Suppl 4), S637–643.

Smith, J. S. (1996). New aspects of rabies with emphasis on epidemiology, diagnosis, and prevention of the disease in the United States. *Clinical Microbiology Reviews, 9*(2), 166–176.

Smith, P. C., Lawhaswasdi, K., Vick, W. E., & Stanton, J. S. (1967). Isolation of rabies virus from fruit bats in Thailand. *Nature, 216*(5113), 384.

Smreczak, M., Orlowska, A., Trebas, P., & Zmudzinski, J. F. (2007, 27–30 May 2007). *The first case of European bat lyssavirus type 1b infection in E. serotinus in Poland.* Paper presented at "Towards the Elimination of Rabies in Eurasia". Paris, France.

Sodre, M. M., da Gama, A. R., & de Almeida, M. F. (2010). Updated list of bat species positive for rabies in Brazil. *Revista do Instituto de Medicina Tropical de Sao Paulo, 52*(2), 75–81.

Srinivasan, A., Burton, E. C., Kuehnert, M. J., Rupprecht, C., Sutker, W. L., Ksiazek, T. G., et al. (2005). Transmission of rabies virus from an organ donor to four transplant recipients. *New England Journal of Medicine, 352*(11), 1103–1111.

Stebbings, R. E., & Griffith, F. (1986). *Distribution and status of bats in Europe.* Huntington: Institue of Terrestrial Ecology.

Stouraitis, P., & Salvatierra, J. (1978). Isolation of rabies virus from bats in Bolivia. *Tropical Animal Health and Production, 10*(2), 101–102.

Streicker, D., Turmelle, A. S., Vonhof, M. J., Kuzmin, I., McCracken, G. F., & Rupprecht, C. E. (2010). Host Phylogeny Constrains Cross-Species Emergence and Establishment of Rabies Virus in bats. *Science, 329,* 676–679.

Sureau, P., Germain, M., Herve, J. P., Geoffroy, B., Cornet, J. P., Heme, G., et al. (1977). Isolation of the Lagos-bat virus in the Central African Republic. *Bulletin de la Societe de Pathologie Exotique et de Ses Filiales, 70*(5), 467–470.

Swanepoel, R. (1994). Rabies. In J. A. W. Coetzer, G. R. Thompson, & R. C. Tustin (Eds.), *Infectious Diseases of Livestock with Secial Reference to Southern Africa (pp. 493–553).* Cape Town: Oxford University Press/NECC.

Teeling, E. C., Springer, M. S., Madsen, O., Bates, P., O'Brien, S. J., & Murphy, W. J. (2005). A molecular phylogeny for bats illuminates biogeography and the fossil record. *Science, 307*(5709), 580–584.

Thompson, R. D., Mitchell, G. C., & Burns, R. J. (1972). Vampire bat control by systemic treatment of livestock with an anticoagulant. *Science, 177*(51), 806–808.

Tignor, G. H., Shope, R. E., Bhatt, P. N., & Percy, D. H. (1973). Experimental infection of dogs and monkeys with two rabies serogroup viruses, Lagos bat and Mokola (IbAn 27377): clinical, serologic, virologic, and fluorescent-antibody studies. *Journal of Infectious Diseases, 128*(4), 471–478.

Towner, J. S., Amman, B. R., Sealy, T. K., Carroll, S. A., Comer, J. A., Kemp, A., et al. (2009). Isolation of genetically diverse Marburg viruses from Egyptian fruit bats. *PLoS Pathogens, 5*(7), e1000536.

van der Poel, W. H. M., Van der Heide, R., Van Amerongen, G., Van Keulen, L. J., Wellenberg, G. J., Bourhy, H., et al. (2000). Characterisation of a recently isolated lyssavirus in frugivorous zoo bats. *Archives of Virology, 145*(9), 1919–1931.

Van der Poel, W. H. M., Van der Heide, R., Verstraten, E. R. A. M., Takumi, K., Lina, P. H. C., & Kramps, J. A. (2005). European bat lyssaviruses, the Netherlands. *Emerging Infectious Diseases, 11*(12), 1854–1859.

van Thiel, P. P., van den Hoek, J. A., Eftimov, F., Tepaske, R., Zaaijer, H. J., Spanjaard, L., et al. (2008). Fatal case of human rabies (Duvenhage virus) from a bat in Kenya: The Netherlands, December 2007. *Euro Surveillance, 13,* 2.

Vazquez-Moron, S., Juste, J., Ibanez, C., Aznar, C., Ruiz-Villamor, E., & Echevarria, J. E. (2008a). Asymptomatic rhabdovirus infection in meridional serotine bats (*Eptesicus isabellinus*) from Spain. *Developmental Biology (Basel)*, *131*, 311–316.

Vazquez-Moron, S., Juste, J., Ibanez, C., Berciano, J. M., & Echevarria, J. E. (2011). Phylogeny of European bat Lyssavirus 1 in *Eptesicus isabellinus* bats, Spain. *Emerging Infectious Diseases*, *17*(3), 520–523.

Vazquez-Moron, S., Juste, J., Ibanez, C., Ruiz-Villamor, E., Avellon, A., Vera, M., et al. (2008b). Endemic circulation of European bat lyssavirus type 1 in serotine bats, Spain. *Emerging Infectious Diseases*, *14*(8), 1263–1266.

Velasco-Villa, A., Gomez-Sierra, M., Hernandez-Rodriguez, G., Juarez-Islas, V., Melendez-Felix, A., Vargas-Pino, F., et al. (2002). Antigenic diversity and distribution of rabies virus in Mexico. *Journal of Clinical Microbiology*, *40*(3), 951–958.

Velasco-Villa, A., Orciari, L. A., Juarez-Islas, V., Gomez-Sierra, M., Padilla-Medina, I., Flisser, A., et al. (2006). Molecular diversity of rabies viruses associated with bats in Mexico and other countries of the Americas. *Journal of Clinical Microbiology*, *44*(5), 1697–1710.

Venters, H. D., Hoffert, W. R., Schatterday, J. E., & Hardy, A. V. (1954). Rabies in bats in Florida. *American Journal Public Health and the Nations Health*, *44*(2), 182–185.

Vos, A., Kaipf, I., Denzinger, A., Fooks, A. R., Johnson, N., & Müller, T. (2007). European bat lyssaviruses–an ecological enigma. *Acta chiropterologica*, *9*(1), 283–296.

Vos, A., Muller, T., Cox, J., Neubert, L., & Fooks, A. R. (2004a). Susceptibility of ferrets (Mustela putorius furo) to experimentally induced rabies with European Bat Lyssaviruses (EBLV). *Journal of Veterinary Medicine Series B-Infectious Diseases and Veterinary Public Health*, *51*(2), 55–60.

Vos, A., Muller, T., Neubert, L., Zurbriggen, A., Botteron, C., Pohle, D., et al. (2004b). Rabies in Red Foxes (Vulpes vulpes) Experimentally Infected with European Bat Lyssavirus Type 1. *Journal of Veterinary Medicine Series B*, *51*(7), 327–332.

Vos, A., Nunan, C., Bolles, D., Muller, T., Fooks, A. R., Tordo, N., et al. (2011). The occurrence of rabies in pre-Columbian Central America: an historical search. *Epidemiology and Infection*, *139*(10), 1445–1452.

Warrilow, D., Smith, I. L., Harrower, B., & Smith, G. A. (2002). Sequence analysis of an isolate from a fatal human infection of Australian bat lyssavirus. *Virology*, *297*(1), 109–119.

Webster, W. A., Casey, G. A., & Charlton, K. M. (1988). Ontario. Rabies in a squirrel. *Canadian Veterinary Journal*, *29*(12), 1015.

Webster, W. A., Casey, G. A., & Charlton, K. M. (1989). Atlantic Canada. Bat-induced rabies in terrestrial mammals in Nova Scotia and Newfoundland. *Canadian Veterinary Journal*, *30*(8), 679.

Wellenberg, G. J., Audry, L., Ronsholt, L., van der Poel, W. H., Bruschke, C. J., & Bourhy, H. (2002). Presence of European bat lyssavirus RNAs in apparently healthy *Rousettus aegyptiacus* bats. *Archives of Virology*, *147*(2), 349–361.

Weyer, J., Szmyd Potapczuk, A. V., Blumberg, L. H., Leman, P. A., Markotter, W., Swanepoel, R., et al. (2011). Epidemiology of human rabies in South Africa, 1983–2007. *Virus Research*, *155*(1), 283–290.

Whitby, J. E., Heaton, P. R., Black, E. M., Wooldridge, M., McElhinney, L. M., & Johnstone, P. (2000). First isolation of a rabies-related virus from a Daubenton's bat in the United Kingdom. *Veterinary Record*, *147*(14), 385–388.

Wiktor, T. J., & Koprowski, H. (1980). Antigenic variants of rabies virus. *Journal of Experimental Medicine*, *152*(1), 99–112.

Willoughby, R. E., Jr., et al., Tieves, K. S., Hoffman, G. M., Ghanayem, N. S., Amlie-Lefond, C. M., Schwabe, M. J., et al. (2005). Survival after treatment of rabies with induction of coma. *New England Journal of Medicine*, *352*(24), 2508–2514.

Wong, S., Lau, S., Woo, P., & Yuen, K. Y. (2007). Bats as a continuing source of emerging infections in humans. *Reviews in Medical Virology*, *17*(2), 67–91.

Human Disease

Alan C. Jackson

Departments of Internal Medicine (Neurology) and Medical Microbiology
University of Manitoba, Winnipeg, Manitoba R3A 1R9, Canada

1 INTRODUCTION

Since antiquity, rabies has been one of the most feared diseases. Human rabies remains an important public health problem in many developing countries where dog rabies is endemic. Worldwide, there are at least 55,000 (World Health Organization, 2005) and perhaps more than 75,000 human deaths each year due to rabies. Beginning in the 1990s, up to six human cases of rabies were diagnosed per year in the USA, and many of these infections were acquired indigenously from unrecognized exposures to insectivorous bats (Noah et al., 1998). A significant number of additional rabies cases probably go unrecognized in the USA and Canada because undiagnosed acute and fatal neurologic illnesses are common and there may be no history of an animal exposure.

2 EXPOSURES, INCUBATION PERIOD, AND PRODROMAL SYMPTOMS

The infectious cycle of rabies virus is perpetuated mainly through animal bites and the deposition of rabies virus-laden saliva into subcutaneous tissues and muscles. With respect to human rabies, worldwide, dogs are by far the most common and important rabies vector; bats are most important in the USA and Canada, although there are reservoirs in various terrestrial animals. Other types of non-bite exposures, including contamination of an open wound, scratch, abrasion, or mucous membrane by saliva or central nervous system (CNS) tissue from an infected animal, are quite common, although they are rarely responsible

Rabies: Scientific Basis of the Disease and its Management
DOI: http://dx.doi.org/10.1016/B978-0-12-396547-9.00007-9

TABLE 7.1   Human Rabies Cases Transmitted by Corneal Transplantation

Location	Year	Age of Patient (recipient)	Time to Death (days)	Reference
United States	1978	37	50	(Houff et al., 1979)
France	1979	36	41	(Galian et al., 1980)
Thailand	1981	41	22	(Thongcharoen et al., 1981)
Thailand	1981	25	33	(Thongcharoen et al., 1981)
India	1987	62	15	(Gode & Bhide, 1988)
India	1988	48	264[a]	(Gode & Bhide, 1988)
Iran	1994	40	27	(Javadi, Fayaz, Mirdehghan, & Ainollahi, 1996)
Iran	1994	35	41	(Javadi, Fayaz, Mirdehghan, & Ainollahi, 1996)

[a]*Patient received two doses of rabies vaccine about one month after the transplant.*

for transmission of rabies virus. Handling and skinning of infected carcasses and perhaps consumption of raw infected meat have resulted in transmission of rabies virus (Kureishi, Xu, Wu, & Stiver, 1992; Tariq, Shafi, Jamal, & Ahmad, 1991; Wallerstein, 1999). Rarely, but notably, inhalation of aerosolized rabies virus in caves containing millions of bats (Constantine, 1962) or in laboratories (Tillotson, Axelrod, & Lyman, 1977; Winkler, Fashinell, Leffingwell, Howard, & Conomy, 1973) has resulted in human rabies. At least eight cases of rabies have resulted from transplantation (human-to-human) of rabies virus-infected corneas (Table 7.1). Rabies also developed in a patient from India 16 days after corneal transplantation, but the source of the infection was unknown in this case, (Masthi et al., 2012) and it is unlikely that it was due to the transplanted cornea. In other reports, transmission did not occur after corneal transplantation from a donor with rabies in France (Sureau, Portnoi, Rollin, Lapresle, & Chaouni-Berbich, 1981) and from another donor in Germany, in which there were two cornea transplant recipients (Johnson, Brookes, Fooks, & Ross, 2005). In 2004, transplantations of organs and a vascular artery segment in Texas were associated with transmission of rabies virus and the development of fatal rabies in four recipients (Srinivasan et al., 2005) (Table 7.2). The donor for these cases presented with gastrointestinal symptoms, throat pain, intermittent periods of confusion and agitation, and he had mild fever and ballistic trunk movements (Burton et al., 2005). The initial CT head scan showed a small subarachnoid hemorrhage. In retrospect, it is highly doubtful that this clinical presentation

TABLE 7.2   Cases of Human Rabies Associated with Organ Transplantation in the USA and Germany

	Sex/Age	Organ Transplanted	Onset of Clinical Rabies Post-Transplantation (days)
Donor in USA	Male/20	–	–
Recipient 1	Male/53	Liver	21
Recipient 2	Female/50	Kidney	27
Recipient 3	Male/18	Kidney	27
Recipient 4	Female/55	Iliac artery segment	27
Donor in Germany	Female/26	–	–
Recipient 1	Female/46	Lung	6 weeks
Recipient 2	Male/72	Kidney	5 weeks
Recipient 3	Male/47	Kidney/pancreas	5 weeks

From the USA (Burton et al., 2005; Srinivasan et al., 2005) and Germany (Maier et al., 2010)

could be explained by a small subarachnoid hemorrhage. Subsequently, there was neurologic deterioration and a repeat CT head scan showed a large subarachnoid hemorrhage with evidence of herniation. He progressed to brain death, and his organs (lungs, kidneys, and liver) and iliac vessels were harvested (Burton et al., 2005). The four recipients of the liver, kidneys, and an iliac artery segment (for a liver transplant) developed clinical rabies within a month and died. The donor had anti-rabies virus antibodies in serum at the time of death and three of the four recipients had antibodies on postoperative days 35 and 36 (Srinivasan et al., 2005). Immunosuppression of the recipients in order to prevent organ rejection results in a favorable environment for viral replication and spread. Only later, it was determined that the donor had been bitten by a bat, and antigenic typing indicated that the rabies virus variant was associated with Brazilian (Mexican) free-tail bats (Krebs, Mandel, Swerdlow, & Rupprecht, 2005). All four transplantation recipients had histopathologic features of encephalitis with cytoplasmic inclusions characteristic of Negri bodies and rabies virus antigen was detected in neurons with immunohistochemical staining from multiple areas of the CNS. Rabies virus antigen was also observed in peripheral nerves of the transplanted kidneys, liver, and arterial graft (Figure 7.1). Transmission occurred again from a donor to organ transplant recipients in Germany that resulted in three fatal cases in 2005 (Maier et al., 2010) (Table 7.2). It is clear, and it should be emphasized, that tissues or organs should not be transplanted from a donor who dies from an undiagnosed neurologic

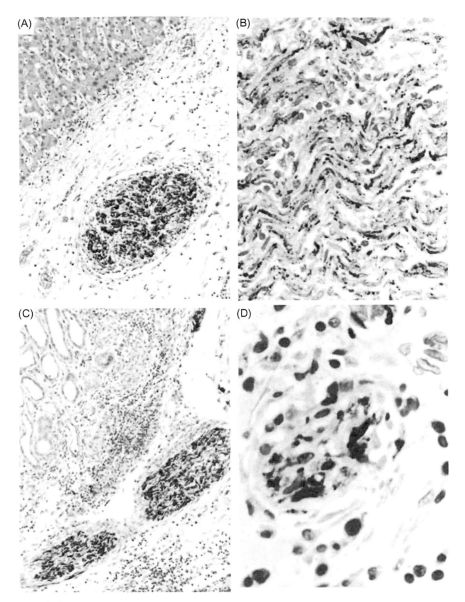

FIGURE 7.1 Immunohistochemical staining for rabies virus antigen (red) in peripheral nerves of the liver (A and B), kidney (C), and arterial graft (D) transplants. *(Reproduced with permission from Srinivasan et al. Transmission of rabies virus from an organ donor to four transplant recipients. New England Journal of Medicine 352, 1103–1111, 2005 Copyright © 2005, Massachusetts Medical Society. All rights reserved.)*

disease because the risk of transmitting unsuspected infectious agents, including rabies virus, is unacceptably high. Most other reported cases of human-to-human transmission have not been well documented. Two patients with rabies from Ethiopia were described, and their only known exposure was contact with family members who died of rabies (Fekadu et al., 1996). In this report a 41-year-old female died of rabies 33 days after her 5-year-old son died of rabies; he had bitten his mother on her little finger. A 5-year-old boy presented with rabies 36 days after his mother died of rabies; he had repeatedly received kisses from his mother on his mouth during her illness. Sexual transmission of rabies virus has not been documented. Although natural human-to-human transmission of rabies likely occurs very rarely, anyone in direct contact with rabies patients, including family members and health care workers, should employ barrier nursing techniques in order to minimize the risk of transmission of the virus via saliva or other secretions (Remington, Shope, & Andrews, 1985). There is evidence of transplacental transmission of rabies virus in a single report from Turkey (Sipahioglu & Alpaut, 1985).

The incubation period for human rabies is usually 20–90 days after exposure, although occasionally disease develops after only a few days (Anderson, Nicholson, Tauxe, & Winkler, 1984) and rare cases have occurred a year or more following exposure. Three immigrants from Laos, the Philippines, and Mexico developed rabies in the USA due to rabies virus strains from their countries of origin with incubation periods of at least 11 months and 4 and 6 years, which were based on the time of their immigration (Smith, Fishbein, Rupprecht, & Clark, 1991). A case of rabies in a 10-year-old Vietnamese girl in Australia in 1990 was also likely acquired at least 5 years earlier (Bek, Smith, Levy, Sullivan, & Rubin, 1992; McColl et al., 1993). The incubation period (from exposure to onset of disease) in rabies is longer and more variable than for most other infectious diseases, which may cause considerable emotional stress to the patient. Very long incubation periods raise the possibility of another unrecognized or forgotten exposure in rabies endemic areas. Severe multiple bites and facial bites are associated with shorter incubation periods (Warrell & Warrell, 1991), although there is a lack of a correlation between the site of the bite and the incubation period (Dupont & Earle, 1965). There may be no history of a bite exposure because it was unrecognized, particularly with insectivorous bat bites because they may be very small (Jackson & Fenton, 2001) (Figure 7.2), or because a bite was either forgotten or no inquiry was made while the patient was still lucid. In the United States and Canada, 20% of patients who acquire rabies from bats have no history of bat contact, and only 38% have a history of a bite or scratch (Jackson, 2011). With known bite exposures from rabid animals, the following has been observed in untreated persons who develop rabies: 50–80% occurrence after head bites, 15–40% after hand

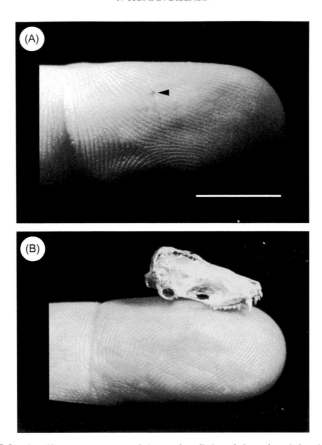

FIGURE 7.2   Small puncture wound (arrowhead) involving the right ring finger of
a bat biologist (A) caused by a defensive bite from a canine tooth of a silver-haired bat
(*Lasionycteris noctivagans*) (Bar = 10 mm). Skull of a silver-haired bat (B) (length of
17.1 mm) is resting on a distal phalanx, which demonstrates the small size of the bat and its
teeth. (*Reproduced from Jackson and Fenton in Lancet* **357**, *1714, 2001 Copyright © 2001, Elsevier.*)

or arm bites, and 3–10% after leg bites. The risk is about 0.1% for contamination of minor wounds with saliva, including scratches (Hattwick,
1974). The biologic bases for these observations are unclear, but a number of factors may be responsible, including the density of rabies virus
receptors in affected tissues, the degree of innervation in tissues in different anatomical locations, the quantity of virus inoculated, and the
properties of the rabies virus variant. Some individuals with rabies virus
exposures may have inapparent rabies virus infection and develop naturally acquired immunity (Doege & Northrop, 1974). Low titers of rabies
virus neutralizing antibodies (VNA) have been found in Canadian Inuit
hunters (7 of 20) and their wives (2 of 11) (Orr, Rubin, & Aoki, 1988).
A survey in fox trappers in northern Alaska identified a 68-year-old

aboriginal male who had trapped for about 47 years and had a VNA concentration of 2.30 IU/mL; he did not recall ever being bitten by a fox. Black and Wiktor (Black & Wiktor, 1986) also observed low titers of rabies VNA in 17% (5 of 30) of Florida raccoon hunters, but not in a control group of hunters. VNA were detected in five humans in Amazonian communities in Peru (at high risk of vampire bat depredation), who had not previously received rabies vaccine; four had a history of a bat bite (Gilbert et al., 2012). In four, the VNA titers ranged from 0.4 to 0.6 IU/mL, whereas one had a high titer of 2.8 IU/mL. Surprisingly, VNA were also detected in 6.6% (15 of 226) of unimmunized students and faculty members of a veterinary medical school at the inception of a rabies vaccine trial (Ruegsegger, Black, & Sharpless, 1961). These studies suggest that exposure of humans to rabies virus under natural conditions can rarely result in result in immunization (natural vaccination) without the development of clinical disease.

Non-specific prodromal symptoms of rabies, including fever, chills, malaise, fatigue, insomnia, anorexia, headache, anxiety, and irritability may last for up to 10 days prior to the onset of neurologic symptoms (Warrell, 1976). About 30–70% of patients develop pain, paresthesias, and/or pruritus at or close to the site of the bite, and the bite wound has often healed by the time these symptoms develop (Dupont & Earle, 1965; Hattwick, 1974). The pruritus may result in severe excoriations from scratching. Retro-orbital pain also occurred as an early symptom in some patients with transmission by corneal transplantation. Local neurologic symptoms may reflect infection involving local peripheral sensory ganglia (dorsal root or trigeminal ganglia) (Mitrabhakdi et al., 2005). The initial neurologic symptoms may occasionally occur at a site distant from the bite, although the pathogenetic basis for this phenomenon is not clear. Two patients bitten on their toes developed rabies with early severe itching of their ears (Hemachudha, 1994). Tremor has also been described involving the bitten extremity (Warrell, 1976).

3 CLINICAL FORMS OF DISEASE

3.1 Encephalitic Rabies

About 80% of patients develop an encephalitic or classical (also called *furious*) form of rabies, and about 20% have a paralytic form of disease. In encephalitic rabies, patients have episodes of generalized arousal or hyperexcitability, which are separated by lucid periods (Warrell, 1976) and these features reflect brain involvement with the infection. Intermittent episodes may occur with confusion, hallucinations, agitation, and aggressive behavior, which typically last for periods of 1–5 minutes (Hattwick, 1974; Hemachudha, 1997; Warrell & Warrell, 1991). The episodes may occur spontaneously or be precipitated by a variety of sensory stimuli (tactile, auditory, visual or olfactory). Biting behavior of patients with rabies has been described (Dupont & Earle, 1965; Emmons et al., 1973; Warrell, 1976), but it is unusual. Fever is common and may be quite high (over 42° C/107°F), and there may be signs of autonomic dysfunction, including hypersalivation, lacrimation, sweating, piloerection (gooseflesh), and dilated pupils. The autonomic dysfunction may result from the infection directly involving the autonomic nervous system centers or pathways in the hypothalamus, spinal cord, and/or autonomic ganglia. Parasympathetic stimulation may increase the production of saliva above the normal volume of about 1 liter per 24 hours. Often patients appear frightened with wide palpebral fissures, dilated pupils, and an open mouth (Nicholson, 1994). Movement disorders have been noted (Warrell, 1976). Seizures, including convulsions, may occur, but they are not common and they usually occur late in the illness. Cranial nerve signs may be present, including ophthalmoplegia, facial weakness, impaired swallowing, and tongue weakness. There may also be nuchal rigidity, reflecting leptomeningeal inflammation.

About 50–80% of patients develop hydrophobia, which is a characteristic and the most specific manifestation of rabies. Hydrophobia is not a feature of any other diseases. The term *hydrophobia* is derived from the Greek word meaning "fear of water." Patients may initially experience pain in the throat or difficulty swallowing. On attempts to swallow, they experience contractions of the diaphragm, sternocleidomastoids, scalenes, and other accessory muscles of inspiration, which last for about 5 to 15 seconds and may be associated with epigastric pain (Figure 7.3). These symptoms may be followed by contraction of neck muscles, resulting in flexion or extension of the neck and rarely with opisthotonic posturing. There may be associated retching, vomiting, coughing, aspiration into the trachea, grimacing, convulsions, and hypoxia (Editorial, 1975). Patients may die during severe spasms with the development of cardiorespiratory arrest if supportive care measures are not initiated (Warrell & Warrell, 1991). During the spasms, there is an associated feeling of terror, often without associated pain. Patients avoid drinking for long periods of time, even despite intense thirst, resulting in dehydration. Subsequently, the sight, sound, or even mention of water (or liquids) may trigger these spasms, indicating that hydrophobia is reinforced by conditioning (Warrell et al., 1976). Hydrophobic spasms may also occur spontaneously, particularly later in the course of the illness. A draft of air on the skin or the breath of an examiner may have the same effect, which has been termed *aerophobia*, and a variety of other stimuli, including water splashed on the skin, attempts by the patient to speak, and stimulation from bright lights or loud sounds, also may precipitate spasms

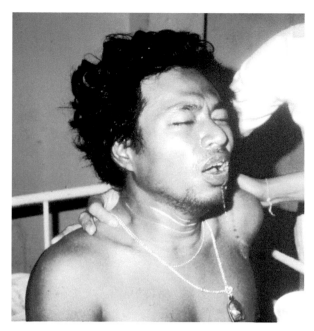

FIGURE 7.3 Hydrophobic spasm of inspiratory muscles associated with terror in a patient with furious rabies encephalitis attempting to swallow water *(Copyright D.A. Warrell, Oxford, UK).*

(Warrell, 1976). Patients may wear heavy clothing in order to avoid drafts. The fan test, elicited by fanning a current of air across the face and observing the patient for spasms of the pharyngeal and neck muscles, has been used as a bedside diagnostic test for the presence of aerophobia (Wilson, Hettiarachchi, & Wijesuriya, 1975). Sobbing respiration (like a child who has been crying) with a two-stage (sniff-sniff) inspiration followed by a slow, full expiration has been described (Pearson, 1976). Later these spasms merge with the development of periodic, apneustic or ataxic breathing as the patient's level of consciousness deteriorates (Warrell et al., 1976). The hydrophobia of rabies is likely due to selective infection of neurons that inhibit the inspiratory motor neurons in the region of the nucleus ambiguus in the brainstem (Warrell et al., 1976; Warrell, 1976). This results in exaggeration of defensive reflexes that protect the respiratory tract. Vocal cord weakness may result in a change in the voice, and patients may make bark-like sounds. Increased libido, priapism (painful spontaneous erections) and spontaneous ejaculations occasionally occur in rabies, and they may be early manifestations of the disease (Bhandari & Kumar, 1986; Dutta, 1996; Gardner, 1970; Talaulicar, 1977; Udwadia, Udwadia, Rao, & Kapadia, 1988). Some clinical features in rabies have a greater association with particular rabies virus variants.

For example, bat-acquired rabies more frequently has tremor and myoclonus, whereas dog-acquired rabies more frequently has hydrophobia and aerophobia (Udow, Marrie, & Jackson, 2012). In encephalitic rabies there is often progression to severe flaccid paralysis, coma, and multiple organ failure. The paralysis that develops either in association with or after the development of coma should not be confused with paralytic rabies in which the muscle weakness develops, in contrast, early in the course of the illness (see Section 3.2). Rabies is almost always fatal, and death often occurs within 14 days of the onset of clinical manifestations, although the time of death may be influenced by critical care measures.

A wide variety of medical complications can develop in patients with rabies. Many of these complications may also occur in critically ill patients with other acute neurological disorders, but some are likely related to the widespread infection in the CNS with systemic (extraneural) organ involvement due to infection of autonomic or sensory neurons (Jackson et al., 1999). Cardiopulmonary complications are the most common and important. Respiratory complications include hyperventilation, hypoxemia, respiratory depression with apnea, atelectasis, and aspiration with secondary pneumonia (Hattwick, 1974). Sinus tachycardia is a common cardiac feature, and the degree of the tachycardia is often greater than that expected for the degree of fever (Warrell et al., 1976). Cardiac arrhythmias (including wandering atrial/nodal pacemaker, sinus bradycardia and supraventricular or ventricular ectopic beats), hypotension, heart failure, and cardiac arrest may occur (Hattwick, 1974; Warrell et al., 1976). Cardiac arrhythmias may account for the sudden death of patients who are alert and do not have advanced neurologic signs of rabies. Cardiac manifestations may reflect infection involving the autonomic nervous system or myocardium (Cheetham, Hart, Coghill, & Fox, 1970; Jackson et al., 1999; Metze & Feiden, 1991; Raman, Prosser, Spreadbury, Cockcroft, & Okubadejo, 1988; Ross & Armentrout, 1962). Either hyperthermia or hypothermia may be present, which may reflect hypothalamic involvement of the infection. Gastrointestinal hemorrhage, especially hematemesis, is a common complication (Kureishi et al., 1992). Endocrine complications include both inappropriate secretion of antidiuretic hormone and diabetes insipidus (Bhatt, Hattwick, Gerdsen, Emmons, & Johnson, 1974; Hattwick, 1974).

3.2 Paralytic Rabies

In paralytic rabies, flaccid muscle weakness develops early in the course of the disease, and the weakness is prominent. Patients are frequently misdiagnosed with this clinical form of the disease, especially if a history of an animal bite is not obtained. The earliest description of paralytic rabies was recorded in 1887 (Gamaleia, 1887). Paralytic rabies has also been called *dumb rabies*. Patients may be literally dumb or mute due to laryngeal muscle weakness, but the term *dumb rabies* usually refers to the quieter clinical features and prominent weakness rather than specifically to the presence of anarthria (Editorial, 1978; Mills, Swanepoel, Hayes, & Gelfand, 1978). The development of paralytic rabies is not related to the anatomical site of the bite (Tirawatnpong et al., 1989), and the incubation period is similar to that in encephalitic rabies. Patients are usually alert, with a normal mental status, at the onset of this clinical form of rabies. The weakness often begins in the bitten extremity and spreads to involve the other extremities, sometimes in an ascending pattern. Muscle fasciculations may be present (Phuapradit, Manatsathit, Warrell, & Warrell, 1985). The facial muscles are frequently weak bilaterally. Associated bilateral deafness has been reported (Phuapradit et al., 1985). Although patients may have local pain, paresthesias or pruritus at the site of the bite, the sensory examination is usually normal in patients with paralytic rabies. The clinical picture may be confused with the Guillain-Barré syndrome, including both the acute inflammatory demyelinating polyradiculopathy and the more severe motor-sensory neuropathy of acute onset with predominant axonal involvement (called the *axonal* Guillain-Barré syndrome) (Feasby et al., 1986; Griffin et al., 1996; Sheikh et al., 2005). Sphincter involvement, especially with urinary incontinence, is common in paralytic rabies, but this is not a feature of the Guillain-Barré syndrome (Asbury & Cornblath, 1990). In addition, pain and sensory disturbances may occur in paralytic rabies. Myoedema has been reported as a sign observed in paralytic rabies, but not in encephalitic rabies (Hemachudha, Phanthumchinda, Phanuphak, & Manutsathit, 1987). However, myoedema has not been confirmed as an important sign of paralytic rabies in other reports. In myoedema, percussion of a muscle (e.g., deltoid or thigh muscle) with a tendon hammer results in local mounding of the muscle without propagated contractions and with electrical silence; the mounding disappears over a few seconds. Myoedema is thought to be a normal physiological phenomenon and its presence does not indicate neuromuscular pathology (Hornung & Nix, 1992). Hence, the importance of this sign in rabies requires future clarification. Bulbar and respiratory muscles eventually become weak in paralytic rabies, resulting in death. Hydrophobia is more unusual in the paralytic form of the disease, although mild inspiratory spasms are commonly observed (Hemachudha et al., 1988). Survival in paralytic rabies is usually longer (up to 30 days) than in encephalitic rabies (Hemachudha, Wacharapluesadee, Mitrabhakdi, Morimoto, & Lewis, 2005). It is unclear if the hydrophobic spasms *per se* lead to death in the first few days of illness in encephalitic disease (Editorial, 1978), or if they reflect a more life-threatening distribution of the brain infection.

An unusual human outbreak of rabies affecting over 70 people occurred in Trinidad between 1929 and 1937 with transmission of the virus from vampire bats (Hurst & Pawan, 1931; Pawan, 1939; Waterman, 1959). All patients in this outbreak had the paralytic form of the disease. This led to diagnostic uncertainty and initially poliomyelitis and botulism were suspected. Nine miners died of paralytic rabies transmitted by vampire bats in British Guiana (presently Guyana) in 1953 (Nehaul, 1955). Similarly, seven children died of paralytic rabies in Surinam in 1973–1974, and vampire bats were probably also the responsible vector (Verlinde, Li-Fo-Sjoe, Versteeg, & Dekker, 1975). However, rabies virus transmitted by vampire bats does not always produce paralytic rabies. A 1990 outbreak of human rabies in Peru with transmission from vampire bats exclusively produced cases of encephalitic rabies (Lopez, Miranda, Tejada, & Fishbein, 1992). Recent outbreaks of human rabies with transmission from vampire bats have occurred in the Amazon region of northern Brazil in 2004. Apparently, all of the 21 cases had the paralytic form of rabies, but only limited clinical and pathological information has been reported (da Rosa et al., 2006; Fernandes et al., 2011). Furthermore, it has been observed that a dog may bite two individuals, and one develops encephalitic rabies and the other develops paralytic rabies (Hemachudha et al., 1988; Wilde & Chutivongse, 1988).

The pathogenetic basis for the two different clinical forms of rabies has not been determined (Hemachudha et al., 2005). In a small series, there were no marked differences in the regional distribution of rabies virus antigen or in the inflammatory changes (Tirawatnpong et al., 1989). However, at the time of death, the distribution of the viral infection may be much more widespread and not closely reflect the distribution at the time of the patient's presentation with paralytic rabies. Electrophysiologic studies have indicated that peripheral nerve involvement, including demyelination, likely contributes to the weakness in paralytic rabies (Mitrabhakdi et al., 2005). It is curious that an earlier serum neutralizing antibody response was observed in patients with encephalitic rabies than in those with paralytic rabies (Hemachudha, 1994). There is evidence that patients with paralytic rabies have defects in immune responsiveness, including lack of lymphocyte proliferative responses to rabies virus antigen (Hemachudha et al., 1988) and lower levels of serum cytokines, including interleukin-6 and the soluble interleukin-2 receptor, than patients with encephalitic rabies (Hemachudha, Panpanich, Phanuphak, Manatsathit, & Wilde, 1993). In contrast, a Chinese case that was misdiagnosed as axonal Guillain-Barré syndrome (Griffin et al., 1996) had pathologic changes that were most marked in the ventral spinal nerve roots without prominent inflammation or motor neuron degeneration (Sheikh et al., 2005). This may have occurred either because axonal degeneration can be an early morphologic consequence of rabies virus-infected motor neurons or because axonal degeneration may be caused by immune injury. In support of the latter hypothesis, a case of encephalitic rabies was treated with high-dose intravenous rabies immune globulin and developed severe paralysis (Hemachudha et al., 2003). However, the lack of anti-rabies virus antibodies in some paralytic rabies cases argues that antibody-mediated injury to nerves is not the only mechanism resulting in paralysis in rabies.

4 INVESTIGATIONS

4.1 Imaging Studies

Computed tomographic (CT) studies of the brain usually are normal in rabies (Faoagali, De Buse, Strutton, & Samaratunga, 1988; Mrak & Young, 1993; White et al., 1994), although hypodense cortical lesions (Sow et al., 1996) and nonenhancing basal ganglia hypodensities (Awasthi, Parmar, Patankar, & Castillo, 2001) have been described. There have been reports of magnetic resonance imaging (MRI) studies of the brain with normal findings (Mrak & Young, 1993; Sing & Soo, 1996), but imaging may show lesions that are usually located in gray matter areas of the brain parenchyma, including the brainstem (Awasthi et al., 2001; Hantson et al., 1993; Laothamatas, Sungkarat, & Hemachudha, 2011; Pleasure & Fischbein, 2000). For example, increased signals were observed on T_2-weighted images in the medulla (Figure 7.4) and pons

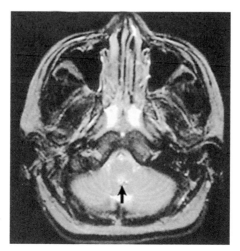

FIGURE 7.4   An axial T2-weighted magnetic resonance image through the medulla demonstrating focal increased signal in the dorsal midline (arrow) (General Electric 1.5-T Signa system; TR, 2500ms, TE, 80ms). *(Reproduced with permission from Pleasure and Fischbein, Correlation of clinical and neuroimaging findings in a case of rabies encephalitis. Archives of Neurology 57, 1765–1769, 2000. Copyright © 2000, American Medical Association. All rights reserved.)*

with only minimal gadolinium enhancement in these areas in a patient from California infected by a rabies virus strain associated with Brazilian (Mexican) free-tailed bats (Pleasure & Fischbein, 2000). Gadolinium enhancement of cervical nerve roots was described in a patient with paralytic rabies (Laothamatas, Hemachudha, Tulyadechanont, & Mitrabhakdi, 1997). Gadolinium enhancement involving the medulla and hypothalamus was also described in the same report in another patient with paralytic rabies. This indicates imaging evidence of brain infection, which has been shown in histopathologic studies at the time of death (Chopra, Banerjee, Murthy, & Pal, 1980). MRI findings in both the brain and spinal cord were found to be similar in a small number of patients with encephalitic and paralytic rabies (Laothamatas, Hemachudha, Mitrabhakdi, Wannakrairot, & Tulayadaechanont, 2003).

4.2 Laboratory Studies

The electroencephalogram may be normal or show non-specific abnormalities in human rabies. Slow wave activity has been observed, as well as periodic (Komsuoglu, Dora, & Kalabay, 1981) and epileptiform activity. Electrophysiological studies showed evidence of peripheral nerve and/or anterior horn cell involvement in a small series, and features of peripheral nerve demyelination were observed in the paralytic cases (Mitrabhakdi et al., 2005). Electrophysiologic evidence of a primary axonal neuropathy was found in two patients with paralytic rabies in another report (Prier et al., 1979; Prier, Gibert, Bodros, & Krymolieres, 1979). Hematologic and biochemical tests are usually normal, although hyponatremia may occur secondary to inappropriate secretion of antidiuretic hormone. Cerebrospinal fluid (CSF) analysis often becomes abnormal in human rabies. A CSF pleocytosis (elevated number of white cells) was found in 59% of cases in the first week of illness and in 87% after the first week (Anderson et al., 1984). The white cell count is usually less than 100 cells/µL, and the leukocytes are predominantly mononuclear cells. The CSF protein concentration may be mildly elevated and glucose is usually in the normal range, although low CSF glucose levels have occasionally been reported (Chotmongkol, Vuttivirojana, & Cheepblangchai, 1991; Roine et al., 1988). Serum neutralizing antibodies against rabies virus are not usually present in unimmunized patients until the second week of the illness, and patients may die of rabies without developing a detectable serum antibody level (Anderson et al., 1984; Hattwick, 1974; Kasempimolporn, Hemachudha, Khawplod, & Manatsathit, 1991). Antibody had not developed in serum by 10 days after the onset of clinical symptoms in five of 18 (28%) patients with rabies in the USA (Noah et al., 1998). One patient, who had received interferon therapy, had not developed antibodies by

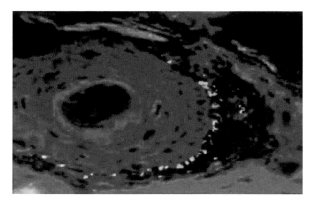

FIGURE 7.5   **Human hair follicle from nuchal skin biopsy.** Nerve fibers surrounding the follicle are stained by specific fluorescence indicating the presence of rabies virus antigen (direct fluorescence antibody method on a frozen section; ×250 magnification). *(From Trimarchi and Nadin-Davis in Diagnostic evaluation, in Rabies, 2 ed., Alan C. Jackson and William H. Wunner (eds.), pp. 411–469. London: Elsevier, 2007 Copyright © 2007, Elsevier.)*

the time of death 24 days after the onset of symptoms (Sibley et al., 1981). Rabies virus antibodies develop in the CSF later than in the serum and the CSF titer is lower. Very high titers of rabies virus antibodies in the CSF have been interpreted as evidence of rabies encephalitis in vaccinated patients (Alvarez et al., 1994; Hattwick, 1974; Madhusudana, Nagaraj, Uday, Ratnavalli, & Kumar, 2002; Porras et al., 1976; Tillotson et al., 1977). Rabies virus may occasionally be isolated from saliva and rarely from the CSF or urine sediment (Anderson et al., 1984). Virus isolation is more likely during early disease before neutralizing antibodies appear, because they produce "autosterilization" of tissues. Rabies virus antigen may be demonstrated antemortem by using the fluorescent antibody technique in frozen sections from skin biopsies (Figure 7.5). A skin biopsy should be obtained containing hair follicles (minimum of 10) by using a full-thickness punch biopsy (5–6 mm in diameter), which is typically taken from the posterior region of the neck at the hairline. Many sections of the biopsy specimen, which should include several hair follicles, are examined with fluorescent antibody staining for rabies virus antigen that is found in adjacent small sensory nerves (Bryceson et al., 1975; Warrell et al., 1988). Antigen detection has also been performed on corneal impression smears, but the sensitivity of the method is low and false positive results may occur (Anderson et al., 1984; Koch, Sagartz, Davidson, & Lawhaswasdi, 1975; Mathuranayagam & Rao, 1984; Noah et al., 1998; Warrell et al., 1988).

4.2.1 Detection of Rabies Virus RNA

Small amounts of rabies virus RNA from saliva, brain tissue, or CSF can be amplified using the reverse transcriptase polymerase chain reaction (RT-PCR), and this technique has proven to be a valuable diagnostic tool for rabies. RT-PCR was initially used on CSF specimens and subsequently on saliva to confirm a diagnosis of rabies (Crepin et al., 1998; Kamolvarin et al., 1993; McColl et al., 1993). Saliva for RT-PCR analysis can be collected with a sterile eyedropper pipette. In a study on both saliva and CSF samples from nine patients with confirmed rabies, the premortem diagnosis of rabies was confirmed by positive RT-PCR in five of nine patients (56%) in saliva and in only two of nine patients (22%) in CSF (Crepin et al., 1998). In comparison, skin biopsies were positive for rabies virus antigen using the fluorescent antibody technique in six of seven patients (86%). These findings led to a recommendation that both skin biopsy and saliva specimens be obtained for testing with immunofluorescence and RT-PCR, respectively. Rabies virus RNA can also be detected in skin biopsies using RT-PCR (Dacheux et al., 2008). Of 20 human rabies cases diagnosed before death in the USA between 1980 and 1996, rabies virus RNA was detected in saliva from all 10 patients who had the test performed, including three who had negative viral isolation from saliva (Noah et al., 1998). It should be emphasized that negative tests for the detection of antigens or RNA never excludes a diagnosis of rabies, and, if clinical suspicion is high, then these tests should be repeated.

4.2.2 Brain Tissues

The presence of Negri bodies in neurons is a pathologic hallmark of rabies observed on routine histologic staining, but these characteristic inclusion bodies in the cytoplasm of infected neurons (see Chapter 9) may be absent. The diagnosis of rabies in humans using brain biopsies has not been assessed adequately, but rabies virus antigen was detected in brain tissues obtained by biopsy from three of three cases in the USA from 1980 to 1996 (Noah et al., 1998). Post-mortem brain tissue may be obtained by a needle (e.g., Vim-Silverman or trucut needle) aspiration technique through either the orbit or foramen magnum (Sow et al., 1996; Tong, Leung, & Lam, 1999; Warrell, 1996) and assessed for viral isolation, rabies virus antigen or rabies virus RNA, although false negative results may occur. Hence, a full autopsy may not be required to confirm a diagnosis when rabies is clinically suspected but unconfirmed antemortem. Rabies may not be diagnosed until postmortem neuropathologic examination of the brain is performed because this diagnosis was not considered by the patient's physicians (Geyer et al., 1997; King et al., 1978; Munoz et al., 1996; Parker et al., 2003; Silverstein et al., 2003). A range of diagnostic investigations may be performed on postmortem human tissues, including virus isolation, the fluorescent antibody test (on fresh or formalin-fixed, paraffin-embedded (Whitfield et al., 2001) specimens), immunoperoxidase staining for rabies virus antigen or *in situ* hybridization for rabies virus RNA (Jackson & Wunner, 1991) or detection of rabies virus RNA by using reverse transcriptase polymerase chain reaction amplification (see Section 4.2.1).

5 DIFFERENTIAL DIAGNOSIS

For patients and their relatives who are unable to recall an animal exposure, even when questioned directly, it may prove more difficult to make a diagnosis of rabies. Most cases without a clear history of rabies exposure are due to bat rabies viruses. There may be a history of recent travel in a rabies endemic area; dog bites are usually recognized, but may not receive appropriate medical attention. Rabies is most commonly misdiagnosed as either a psychiatric or laryngopharyngeal disorder. The disease may also present with bizarre neuropsychiatric symptoms mimicking conditions such as schizophrenic psychosis or acute mania (Goswami, Shankar, Channabasavanna, & Chattopadhyay, 1984).

Patients often become quite fearful about the possibility of developing rabies after an animal bite or exposure. Rabies hysteria is a conversion disorder (classified as a somatoform disorder) in which patients exhibit clinical features similar to rabies due to unconscious motivation that involves poorly understood neural networks (Ron, 2001; Wilson et al., 1975), which should not be confused with malingering (feigning) in which there is deception by the patient. Rabies hysteria is probably the most difficult differential diagnosis. In general, it is characterized by a shorter incubation period (often a few hours or a day or two) than rabies, an early onset of inability of the patient to communicate, bizarre spasms, spitting out of water taken in the mouth with no actual attempt at swallowing, barking, biting, aggressive behavior directed toward health care workers, lack of fever and neurologic signs, and a long clinical course with recovery. Village practitioners in endemic areas may establish a reputation that they can cure rabies due to recovery of patients with rabies hysteria (Wilson et al., 1975). However, it should be emphasized that the clinical picture may be so bizarre in patients with rabies that they may be misdiagnosed as having hysteria (Bisseru, 1972).

Other viral encephalitides may show behavioral disturbances with fluctuations in the level of consciousness. However, hydrophobic spasms are not observed in these conditions, and it is unusual for a conscious patient to have prominent brainstem signs in other encephalitides. Herpes *simiae* (B virus) encephalomyelitis, which is transmitted by monkey bites, is often associated with a shorter incubation period than rabies (e.g., 3–5 days); vesicles may be present at the site of the bite (also in the monkey's oral cavity), and recovery may occur (Whitley, 2004). Anti-N-methyl-D-aspartate receptor (anti-NMDA) encephalitis occurs in young patients (especially females) and is characterized by behavioral changes, autonomic instability, hypoventilation, and seizures, and it has recently been recognized that this autoimmune disease rivals viral etiologies as a cause of encephalitis (Gable, Sheriff, Dalmau, Tilley, & Glaser, 2012). Two cases of rabies in the USA were misdiagnosed as Creutzfeldt-Jakob disease (Geyer et al., 1997) and both of these patients had a rapidly progressive neurological illness with prominent myoclonus.

Tetanus, a disease caused by the neurotoxin from the bacteria *Clostridium tetani*, may develop in association with a dirty wound caused by an animal bite. Tetanus has a shorter incubation period (usually 3–21 days) than rabies and, unlike rabies, it is characterized by sustained muscle rigidity involving axial muscles, including paraspinal, abdominal, masseter (trismus), laryngeal, and respiratory muscles, with superimposed brief recurrent muscle spasms (Bleck & Brauner, 2004). In tetanus, the mental state is not affected, there is no CSF pleocytosis, and the prognosis is much better than in rabies.

Post-vaccinal encephalomyelitis is another important differential diagnosis, particularly in patients who have been immunized with a vaccine derived from neural tissues (e.g., Semple vaccine). Post-vaccinal encephalomyelitis usually develops within 2 weeks of initiation of vaccination, which is helpful in the differential diagnosis. Local sensory symptoms (paresthesias, pain, and pruritus), alternating intervals of agitation and lucidity, and hydrophobia are clinical features that strongly suggest a diagnosis of rabies rather than post-vaccinal encephalomyelits.

Paralytic rabies resembles the Guillain-Barré syndrome, including both acute inflammatory demyelinating polyradiculopathy and acute motor-sensory axonal neuropathy. In a recent pathologic series of the latter (Griffin et al., 1996), one case (case number 1 in the report) was subsequently demonstrated to have paralytic rabies (Sheikh et al., 2005). Local symptoms at the site of the bite, piloerection, early or persistent bladder dysfunction, and fever are more suggestive of paralytic rabies. The Guillain-Barré syndrome may occasionally occur as a post-vaccinal complication from rabies vaccines derived from neural tissues, particularly the suckling mouse brain vaccine (Toro, Vergara, & Roman, 1977).

6 RABIES DUE TO OTHER LYSSAVIRUS GENOTYPES

In addition to rabies virus, which is *Lyssavirus* genotype 1, there are six other Lyssavirus genotypes and five have been associated with cases of human rabies: Mokola virus (genotype 3), Duvenhage virus (genotype 4), European bat lyssavirus 1 (genotype 5), European bat lyssavirus 2 (genotype 6), and Australian bat lyssavirus (genotype 7) (Table 7.3). In addition, one human case was reported due to Irkut virus infection (Leonova et al., 2009), which has not yet been designated in a genotype. They are commonly called *rabies-like* or *rabies-related viruses*.

TABLE 7.3 Reported Human Rabies Cases Due to Other *Lyssavirus* Genotypes

Virus (Genotype)	Year	Location	Age of Patient	Reference
Mokola (3)[a]	1968	Nigeria	3.5	(Familusi & Moore, 1972)
Mokola (3)	1971	Nigeria	6	(Familusi et al., 1972)
Duvenhage (4)	1970	South Africa	31	(Meredith et al., 1971)
Duvenhage (4)	2006	South Africa	77	(Paweska et al., 2006)
Duvenhage (4)	2007	Kenya	34	(van Thiel et al., 2009)
European bat lyssavirus 1 (5)	1985	Russia	11	(Selimov et al., 1989)
European bat lyssavirus 2 (6)	1985	Finland	30	(Roine et al., 1988)
European bat lyssavirus 2 (6)	2002	Scotland	55	(Nathwani et al., 2003)
Australian bat lyssavirus (7)	1996	Australia	39	(Samaratunga et al., 1998)
Australian bat lyssavirus (7)	1998	Australia	37	(Hanna et al., 2000)
Irkut (pending)	2007	Russia	20	(Leonova et al., 2009)

[a]*It is doubtful that this patient's clinical picture was actually caused by Mokola virus infection.*

Lagos bat virus (genotype 2), which was first isolated from fruit-eating bats in Nigeria, is the only genotype that has not been associated with human disease. Although not yet designated genotypes, Aravan virus and Khujand virus are lyssaviruses isolated from bats in Central Asia, whereas Irkut virus and West Caucasian bat virus are lyssaviruses isolated from bats in Russia (Botvinkin et al., 2003; Kuzmin et al., 2003; Kuzmin, Hughes, Botvinkin, Orciari, & Rupprecht, 2005) and to date, of these four lyssaviruses, only Irkut virus has been reported to be associated with human disease (see Chapter 6).

6.1 Duvenhage Virus

In 1970, a 31-year-old man from rural South Africa developed an illness with fever, excessive sweating, hydrophobia, and spasms of his face, arms, and torso that were precipitated by being touched (Meredith, Rossouw, & Koch, 1971). He also exhibited confusion, irritability, and marked aggressiveness. He died after an illness lasting about 5 days. He lived outside the recognized enzootic and epizootic areas for rabies. He had been bitten on the lip by a bat while sleeping about 4 weeks earlier. The virus isolated from his brain was a new virus and was characterized and named *Duvenhage virus* (genotype 4). This patient's clinical illness was indistinguishable from that caused by rabies virus (genotype 1).

In February, 2006 a 77-year-old male was scratched on the cheek by an insectivorous bat in North West Province, South Africa (Paweska et al., 2006). He did not receive post-exposure treatment and he became ill 27 days later with an influenza-like illness and hallucinations. On his third day of illness, he was admitted to hospital with fever, generalized rigidity, and involuntary grimacing, and the next day he had generalized seizures. He died on day 14 of his illness. Duvenhage virus was identified by RT-PCR on antemortem saliva and postmortem brain tissue and confirmed with sequencing of the nucleoprotein amplicons.

In 2007, a bat flew in the face of a 34-year-old Dutch female in Kenya; she sustained two small superficial wounds on the right side of her nose, and she did not receive postexposure rabies prophylaxis (van Thiel et al., 2009). Twenty-three days later, she experienced malaise, dizziness, muscle aches, and headache, and two days later she had difficulty with speech and swallowing. She later had fever, hypersalivation, limb weakness, and a seizure. She died 45 days after the bat exposure (on day 20 of her hospital admission), after receiving aggressive medical interventions. Duvenhage virus was confirmed antemortem by PCR studies on a nuchal skin biopsy.

6.2 Mokola Virus

Mokola virus was first isolated from shrews in Nigeria (Shope et al., 1970). In 1968, a 3½-year-old girl from Nigeria presented with a sudden onset of fever and convulsions (Familusi & Moore, 1972). She rapidly made a complete recovery. There were no cells in her CSF, and the CSF protein and glucose were normal. Mokola virus was isolated from her CSF, although the shrew isolate of Mokola virus was handled in the same laboratory during the same time period. Cross-contamination of specimens in the laboratory remains a possible explanation for this viral isolation. The girl's neutralizing antibody titers were very low and disappeared within several months. The febrile convulsion was unlikely related to Mokola virus infection.

A 6-year-old girl died in Nigeria in 1971 after a six-day illness (Familusi, Osunkoya, Moore, Kemp, & Fabiyi, 1972). She presented with drowsiness, confusion, and weakness involving her extremities and trunk and progressed to coma. Her CSF was normal without a pleocytosis. At autopsy, there were large eosinophilic inclusion bodies in the cytoplasm of neurons, and Mokola virus was isolated from her brain. Shrews were known to be plentiful around the house where she lived, although there was no documented evidence that she had actually been bitten. Mokola virus infection was associated with meningoencephalitis in this case without the typical features of brainstem involvement seen in encephalitic rabies.

6.3 European Bat Lyssavirus 1

In 1985, an 11-year-old girl from Belgorod, Russia, was bitten on the lower lip by an unidentified bat and died with signs of rabies (Selimov et al., 1989). The viral isolate was called *Yuli virus* and classified as European bat lyssavirus type 1 (genotype 5) (Bourhy, Kissi, Lafon, Sacramento, & Tordo, 1992). There was an earlier fatal case in a 15-year-old female in Voroshilovgrad (now Lugansk), Ukraine, in 1977 that developed after a bat bite (Anonymous, 1986). However, no viral isolate is available for molecular characterization, and it is uncertain if the infection was actually caused by European bat lyssavirus type 1 or another lyssavirus. A similar situation applies in the case of a 34-year-old male who died with clinical rabies (with hypersalivation and hydrophobia) 45 days after a bat bite in the Lugansk province, Ukraine, in 2002, which is only 50 km from the site of the 1977 case (Botvinkin et al., 2005).

6.4 European Bat Lyssavirus 2

In 1985, a 30-year-old zoologist from Finland developed numbness in his right arm and neck with leg weakness (Roine et al., 1988). His CSF was normal without a pleocytosis. Subsequently, he developed myoclonus of his legs, agitation, hyperexcitability, inspiratory spasms, dysarthria, dysphagia, and hypersalivation. He had a delirium that progressed to coma. Diabetes insipidus occurred, and he died 23 days after the onset of the illness. He had never been vaccinated against rabies and had been bitten by bats in several countries over the prior 5-year period, including an exposure in southern Finland 51 days prior to the onset of his symptoms. A virus was isolated that resembled the enzootic European bat rabies virus isolates, and it was classified as European bat lyssavirus type 2 (genotype 6) (Bourhy et al., 1992). This patient also had a clinical illness that was indistinguishable from rabies associated with genotype 1.

In 2002 a 55-year-old bat conservationist presented with hematemesis and a 5-day history of arm paresthesias, left arm and shoulder pain and difficulty swallowing (Fooks et al., 2003; Nathwani et al., 2003). About 19 weeks prior, he had been bitten on his left finger by a Daubenton's bat in Angus, Scotland, although he also had a remote history of other bat bites. On admission to hospital in Dundee, Scotland, he was febrile and had gaze-evoked nystagmus, dysarthria, truncal, limb, and gait ataxia, areflexia in the arms, and hyperreflexia in the legs, and his behavior was inappropriately familiar. CT and MRI scans did not show significant abnormalities. CSF showed a normal cell count and a mildly elevated

CSF protein (58 mg/dL). He was treated with intravenous immuno-globulin and also subsequently with high-dose methylprednisolone and cyclophosphamide (Fooks et al., 2003). On day 5 of hospitalization he became acutely confused, agitated, and aggressive. He was sedated and a repeat CSF examination showed a pleocytosis and CSF protein elevated at 1.09 mg/dL. His mental state subsequently deteriorated, and his limbs became flaccid; he died on day 14 of hospitalization. Hemi-nested RT-PCR was positive for lyssavirus RNA in saliva obtained on day 9 of hospitalization, with high homology with previous EBLV type 2a isolates obtained from bats in the UK. No rabies virus antibodies were detected in sera or CSF, and virus could not be isolated from saliva, skin biopsies, or CSF. The lyssavirus was cultured from postmortem brain tissue. This was the first case of indigenous human rabies in the United Kingdom in 100 years and has had important public health implications for bat exposures in the region.

6.5 Australian Bat Lyssavirus

In 1996, a 39-year-old female from Australia died after a 20-day illness (Samaratunga, Searle, & Hudson, 1998). She cared for fruit bats and had sustained numerous scratches to her left arm over 4 weeks prior to the onset of her illness, and she was likely bitten by a yellow-bellied sheathtail bat, *Saccolaimus flaviventris* (an insectivorous bat) in her care (Hanna et al., 2000). She developed progressive left arm weakness. Her CSF showed 100 white cells/μL (80% mononuclear cells and 20% polymorphonuclear leukocytes). She deteriorated with diplopia, dysarthria, dysphagia, and ataxia. She later developed progressive limb and facial weakness with reduced deep tendon reflexes and fluctuations in her level of consciousness prior to her death. Small eosinophilic cytoplasmic inclusions were observed in neurons in gray matter areas. RT-PCR amplification of RNA extracted from brain tissue and CSF indicated that she was infected with a virus identical to Australian bat lyssavirus (genotype 7) that had been identified previously in flying foxes, which are fruit-eating bats. This patient had typical brainstem involvement of rabies, which quickly progressed to diffuse brain involvement.

In 1998, a 37-year-old woman from Mackay, Queensland, was admitted to hospital with a 5-day history of fever, paresthesias around the dorsum of her left hand, pain about the left shoulder girdle, and sore throat with difficulty swallowing (Hanna et al., 2000). There were pharyngeal spasms, evidence of autonomic instability, and progressive neurologic deterioration. She died 19 days after the onset of the illness. Twenty-seven months prior to the onset of her illness (during 1996), she was bitten at the base of her left little finger by a flying fox (fruit bat) in the course of removing the bat from the back of a young child. She did not receive rabies post-exposure prophylaxis. Hemi-nested PCR analyses on multiple tissues and saliva were positive for the flying fox (*Pteropus* spp.) variant of Australian bat lyssavirus (Hanna et al., 2000). Although *Lyssavirus* infections of flying foxes have only recently been recognized in Australia, rabies virus infection was recognized in a gray-head flying fox (*Pteropus poliocephalus*) that died in India in 1978 (Pal et al., 1980) and also in two dog-faced fruit bats (*Cyanopterus brachyotis*) from Thailand (Smith, Lawhaswasdi, Vick, & Stanton, 1967). It is unclear exactly when and how Australian bat lyssavirus obtained its foothold in Australian frugivorous and insectivorous bats, but it is clear that the virus poses a threat to human health in this region.

6.6 Irkut Virus

In 2007, a 20-year-old female died of rabies in the Primorye Territory, which is in the Russian Far East (Leonova et al., 2009). On August 10, 2007, this female was bitten on the lip by a bat of unknown species, and she did not receive any post-exposure therapy. A month later, on September 10th, she developed fever, headache, vomiting, diplopia, and head and hand tremor. She was admitted to hospital on the next day with bulbar symptoms. She progressed into a deep stupor with flaccid paresis and died after 11 days. Molecular characterization of a virus isolated from her brain identified it as a bat lyssavirus called Irkut virus, which had previously been isolated from a dead greater tubenosed bat (*Murina leucogaster*) in Irkutsk (Botvinkin et al., 2003).

References

Alvarez, L., Fajardo, R., Lopez, E., Pedroza, R., Hemachudha, T., Kamolvarin, N., et al. (1994). Partial recovery from rabies in a nine-year-old boy. *The Pediatric Infectious Disease Journal, 13*, 1154–1155.

Anderson, L. J., Nicholson, K. G., Tauxe, R. V., & Winkler, W. G. (1984). Human rabies in the United States, 1960 to 1979: Epidemiology, diagnosis, and prevention. *Annals of Internal Medicine, 100*, 728–735.

Anonymous, (1986). Bat rabies in the Union of Soviet Socialist Republics. *Rabies Bulletin Europe, 10*, 12–14.

Asbury, A. K., & Cornblath, D. R. (1990). Assessment of current diagnostic criteria for Guillain-Barre syndrome. *Annals of Neurology, 27*(Suppl), S21–S24.

Awasthi, M., Parmar, H., Patankar, T., & Castillo, M. (2001). Imaging findings in rabies encephalitis. *American Journal of Neuroradiology, 22*(4), 677–680.

Bek, M. D., Smith, W. T., Levy, M. H., Sullivan, E., & Rubin, G. L. (1992). Rabies case in New South Wales, 1990: Public health aspects. *Medical Journal of Australia, 156*, 596–600.

Bhandari, M., & Kumar, S. (1986). Penile hyperexcitability as the presenting symptom of rabies. *British Journal of Urology, 58*, 224–233.

Bhatt, D. R., Hattwick, M. A. W., Gerdsen, R., Emmons, R. W., & Johnson, H. N. (1974). Human rabies: Diagnosis, complications, and management. *American Journal of Diseases of Children, 127*, 862–869.

Bisseru, B. (1972). Human rabies. In B. Bisseru (Ed.), *Rabies (pp. 385–453)*. London: William Heinemann Medical Books Ltd.

Black, D., & Wiktor, T. J. (1986). Survey of raccoon hunters for rabies antibody titers: Pilot study. *Journal of the Florida Medical Association, 73*, 517–520.

Bleck, T. P., & Brauner, J. S. (2004). Tetanus. In W. M. Scheld, R. J. Whitley, & C. M. Marra (Eds.), *Infections of the central nervous system (pp. 625–648)* (3 ed.). Philadelphia: Lippincott Williams and Wilkins.

Botvinkin, A. D., Poleschuk, E. M., Kuzmin, I. V., Borisova, T. I., Gazaryan, S. V., Yager, P., et al. (2003). Novel lyssaviruses isolated from bats in Russia. *Emerging Infectious Diseases, 9*(12), 1623–1625.

Botvinkin, A. D., Selnikova, O. P., Antonova, L. A., Moiseeva, A. B., Nesterenko, E. Y., & Gromashevsky, L. V. (2005). Human rabies case caused from a bat bite in Ukraine. *Rabies Bulletin Europe, 29*(3), 5–7.

Bourhy, H., Kissi, B., Lafon, M., Sacramento, D., & Tordo, N. (1992). Antigenic and molecular characterization of bat rabies virus in Europe. *Journal of Clinical Microbiology, 30*(9), 2419–2426.

Bryceson, A. D. M., Greenwood, B. M., Warrell, D. A., Davidson, N. M., Pope, H. M., Lawrie, J. H., et al. (1975). Demonstration during life of rabies antigen in humans. *Journal of Infectious Diseases, 131*(1), 71–74.

Burton, E. C., Burns, D. K., Opatowsky, M. J., El-Feky, W. H., Fischbach, B., Melton, L., et al. (2005). Rabies encephalomyelitis: Clinical, neuroradiological, and pathological findings in 4 transplant recipients. *Archives of Neurology, 62*(6), 873–882.

Cheetham, H. D., Hart, J., Coghill, N. F., & Fox, B. (1970). Rabies with myocarditis: Two cases in England. *Lancet, 1*, 921–922.

Chopra, J. S., Banerjee, A. K., Murthy, J. M. K., & Pal, S. R. (1980). Paralytic rabies: A clinicopathological study. *Brain, 103*, 789–802.

Chotmongkol, V., Vuttivirojana, A., & Cheepblangchai, M. (1991). Unusual manifestation in paralytic rabies. *Southeast Asian Journal of Tropical Medicine and Public Health, 22*, 279–280.

Constantine, D. G. (1962). Rabies transmission by nonbite route. *Public Health Reports, 77*, 287–289.

Crepin, P., Audry, L., Rotivel, Y., Gacoin, A., Caroff, C., & Bourhy, H. (1998). Intravitam diagnosis of human rabies by PCR using saliva and cerebrospinal fluid. *Journal of Clinical Microbiology, 36*(4), 1117–1121.

da Rosa, E. S. T., Kotait, I., Barbosa, T. F. S., Carrier, M. L., Brandão, P. E., Pinheiro, A. S., et al. (2006). Bat-transmitted human rabies outbreaks, Brazilian Amazon. *Emerging Infectious Diseases, 12*(8), 1197–1202.

Dacheux, L., Reynes, J. M., Buchy, P., Sivuth, O., Diop, B. M., Rousset, D., et al. (2008). A reliable diagnosis of human rabies based on analysis of skin biopsy specimens. *Clinical Infectious Diseases, 47*(11), 1410–1417.

Doege, T. C., & Northrop, R. L. (1974). Evidence for inapparent rabies infection. *Lancet, 2*, 826–829.

Dupont, J. R., & Earle, K. M. (1965). Human rabies encephalitis. A study of forty-nine fatal cases with a review of the literature. *Neurology, 15*, 1023–1034.

Dutta, J. K. (1996). Excessive libido in a woman with rabies. *Postgraduate Medical Journal, 72*, 554.

Editorial, (1975). Diagnosis and management of human rabies. *British Medical Journal, 3*(5986), 721–722.

Editorial, (1978). Dumb rabies. *Lancet, 2*(8098), 1031–1032.

Emmons, R. W., Leonard, L. L., DeGenaro, F., Jr., et al., Protas, E. S., Bazeley, P. L., Giammona, S. T., et al. (1973). A case of human rabies with prolonged survival. *Intervirology, 1*(1), 60–72.

Familusi, J. B., & Moore, D. L. (1972). Isolation of a rabies related virus from the cerebrospinal fluid of a child with "aseptic meningitis." *African Journal of Medical Science, 3*, 93–96.

Familusi, J. B., Osunkoya, B. O., Moore, D. L., Kemp, G. E., & Fabiyi, A. (1972). A fatal human infection with Mokola virus. *American Journal of Tropical Medicine and Hygiene*, 21, 959–963.

Faoagali, J. L., De Buse, P., Strutton, G. M., & Samaratunga, H. (1988). A case of rabies. *Medical Journal of Australia*, 149, 702–707.

Feasby, T. E., Gilbert, J. J., Brown, W. F., Bolton, C. F., Hahn, A. F., Koopman, W. F., et al. (1986). An acute axonal form of Guillain-Barre polyneuropathy. *Brain*, 109, 1115–1126.

Fekadu, M., Endeshaw, T., Alemu, W., Bogale, Y., Teshager, T., & Olson, J. G. (1996). Possible human-to-human transmission of rabies in Ethiopia. *Ethiopian Medical Journal*, 34(2), 123–127.

Fernandes, E. R., de Andrade, H. F. J., Lancellotti, C. L., Quaresma, J. A., Demachki, S., da Costa Vasconcelos, P. F., et al. (2011). In situ apoptosis of adaptive immune cells and the cellular escape of rabies virus in CNS from patients with human rabies transmitted by *Desmodus rotundus*. *Virus Research*, 156(1-2), 121–126.

Fooks, A. R., McElhinney, L. M., Pounder, D. J., Finnegan, C. J., Mansfield, K., Johnson, N., et al. (2003). Case report: Isolation of a European bat lyssavirus type 2a from a fatal human case of rabies encephalitis. *Journal of Medical Virology*, 71(2), 281–289.

Gable, M. S., Sheriff, H., Dalmau, J., Tilley, D. H., & Glaser, C. A. (2012). The frequency of autoimmune N-methyl-D-aspartate receptor encephalitis surpasses that of individual viral etiologies in young individuals enrolled in the California encephalitis project. *Clinical Infectious Diseases*, 54(7), 899–904.

Galian, A., Guerin, J. M., Lamotte, M., Le Charpentier, Y., Mikol, J., Dureaux, J. B., et al. (1980). Human-to-human transmission of rabies via a corneal transplant - France. *Morbidity and Mortality Weekly Report*, 29, 25–26.

Gamaleia, N. (1887). Etude sur la rage paralytique chez l'homme. *Annales de L'Institut Pasteur (Paris)*, 1, 63–83.

Gardner, A. M. N. (1970). An unusual case of rabies (Letter). *Lancet*, 2, 523.

Geyer, R., Van Leuven, M., Murphy, J., Damrow, T., Sastry, L., Miller, S., et al. (1997). Human rabies – Montana and Washington, 1997. *Morbidity and Mortality Weekly Report*, 46(33), 770–774.

Gilbert, A. T., Peterson, B. W., Recuenco, S., Niezgoda, M., Gómez, J., Laguna-Torres, V. A., et al. (2012). Evidence of rabies virus exposure among humans in the Peruvian Amazon. *American Journal of Tropical Medicine and Hygiene*, 87(2), 206–215.

Gode, G. R., & Bhide, N. K. (1988). Two rabies deaths after corneal grafts from one donor (Letter). *Lancet*, 2, 791.

Goswami, U., Shankar, S. K., Channabasavanna, S. M., & Chattopadhyay, A. (1984). Psychiatric presentations in rabies: A clinico-pathologic report from south India with a review of literature. *Tropical and Geographical Medicine*, 36(1), 77–81.

Griffin, J. W., Li, C. Y., Ho, T. W., Tian, M., Gao, C. Y., Xue, P., et al. (1996). Pathology of the motor-sensory axonal Guillain-Barre syndrome. *Annals of Neurology*, 39, 17–28.

Hanna, J. N., Carney, I. K., Smith, G. A., Tannenberg, A. E. G., Deverill, J. E., Botha, J. A., et al. (2000). Australian bat lyssavius infection: A second human case, with a long incubation period. *Medical Journal of Australia*, 172(12), 597–599.

Hantson, P., Guerit, J. M., de Tourtchaninoff, M., Deconinck, B., Mahieu, P., Dooms, G., et al. (1993). Rabies encephalitis mimicking the electrophysiological pattern of brain death. A case report. *European Neurology*, 33, 212–217.

Hattwick, M. A. W. (1974). Human rabies. *Public Health Review*, 3, 229–274.

Hemachudha, T. (1994). Human rabies: Clinical aspects, pathogenesis, and potential therapy. In C. E. Rupprecht, B. Dietzschold, & H. Koprowski (Eds.), *Lyssaviruses (pp. 121–143)*. Berlin: Springer-Verlag. Current Topics in Microbiology and Immunology.

Hemachudha, T. (1997). Rabies. In K. L. Roos (Ed.), *Central nervous system infectious diseases and therapy (pp. 573–600)*. New York: Marcel Dekker.

Hemachudha, T., Panpanich, T., Phanuphak, P., Manatsathit, S., & Wilde, H. (1993). Immune activation in human rabies. *Transactions of the Royal Society of Tropical Medicine and Hygiene, 87*, 106–108.

Hemachudha, T., Phanthumchinda, K., Phanuphak, P., & Manutsathit, S. (1987). Myoedema as a clinical sign in paralytic rabies (Letter). *Lancet, 1*, 1210.

Hemachudha, T., Phanuphak, P., Sriwanthana, B., Manutsathit, S., Phanthumchinda, K., Siriprasomsup, W., et al. (1988). Immunologic study of human encephalitic and paralytic rabies: Preliminary report of 16 patients. *American Journal of Medicine, 84*, 673–677.

Hemachudha, T., Sunsaneewitayakul, B., Mitrabhakdi, E., Suankratay, C., Laothamathas, J., Wacharapluesadee, S., et al. (2003). Paralytic complications following intravenous rabies immune globulin treatment in a patient with furious rabies (Letter). *International Journal of Infectious Diseases, 7*(1), 76–77.

Hemachudha, T., Wacharapluesadee, S., Mitrabhakdi, E., Morimoto, K., & Lewis, R. A. (2005). Pathophysiology of human paralytic rabies. *Journal of Neurovirology, 11*(1), 93–100.

Hornung, K., & Nix, W. A. (1992). Myoedema. A clinical and electrophysiological evaluation. *European Neurology, 32*(3), 130–133.

Houff, S. A., Burton, R. C., Wilson, R. W., Henson, T. E., London, W. T., Baer, G. M., et al. (1979). Human-to-human transmission of rabies virus by corneal transplant. *New England Journal of Medicine, 300*, 603–604.

Hurst, E. W., & Pawan, J. L. (1931). An outbreak of rabies in Trinidad without history of bites, and with the symptoms of acute ascending myelitis. *Lancet, 2*, 622–628.

Jackson, A. C. (2011). Update on rabies. *Research and Reports in Tropical Medicine, 2*, 31–43.

Jackson, A. C., & Fenton, M. B. (2001). Human rabies and bat bites (Letter). *Lancet, 357*(9269), 1714.

Jackson, A. C., & Wunner, W. H. (1991). Detection of rabies virus genomic RNA and mRNA in mouse and human brains by using in situ hybridization. *Journal of Virology, 65*(6), 2839–2844.

Jackson, A. C., Ye, H., Phelan, C. C., Ridaura-Sanz, C., Zheng, Q., Li, Z., et al. (1999). Extraneural organ involvement in human rabies. *Laboratory Investigation, 79*(8), 945–951.

Javadi, M. A., Fayaz, A., Mirdehghan, S. A., & Ainollahi, B. (1996). Transmission of rabies by corneal graft. *Cornea, 15*(4), 431–433.

Johnson, N., Brookes, S. M., Fooks, A. R., & Ross, R. S. (2005). Review of human rabies cases in the UK and in Germany. *Veterinary Record, 157*(22), 715.

Kamolvarin, N., Tirawatnpong, T., Rattanasiwamoke, R., Tirawatnpong, S., Panpanich, T., & Hemachudha, T. (1993). Diagnosis of rabies by polymerase chain reaction with nested primers. *Journal of Infectious Diseases, 167*, 207–210.

Kasempimolporn, S., Hemachudha, T., Khawplod, P., & Manatsathit, S. (1991). Human immune response to rabies nucleocapsid and glycoprotein antigens. *Clinical and Experimental Immunology, 84*(2), 195–199.

King, D. B., Sangalang, V. E., Manuel, R., Marrie, T., Pointer, A. E., & Thomson, A. D. (1978). A suspected case of human rabies - Nova Scotia. *Canadian Diseases Weekly Report, 4*, 49–51.

Koch, F. J., Sagartz, J. W., Davidson, D. E., & Lawhaswasdi, K. (1975). Diagnosis of human rabies by the cornea test. *American Journal of Clinical Pathology, 63*, 509–515.

Komsuoglu, S. S., Dora, F., & Kalabay, O. (1981). Periodic EEG activity in human rabies encephalitis (Letter). *Journal of Neurology, Neurosurgery and Psychiatry, 44*, 264–265.

Krebs, J. W., Mandel, E. J., Swerdlow, D. L., & Rupprecht, C. E. (2005). Rabies surveillance in the United States during 2004. *Journal of the American Veterinary Medical Association, 227*(12), 1912–1925.

Kureishi, A., Xu, L. Z., Wu, H., & Stiver, H. G. (1992). Rabies in China: Recommendations for control. *Bulletin of the World Health Organization, 70*, 443–450.

Kuzmin, I. V., Hughes, G. J., Botvinkin, A. D., Orciari, L. A., & Rupprecht, C. E. (2005). Phylogenetic relationships of Irkut and West Caucasian bat viruses within the

Lyssavirus genus and suggested quantitative criteria based on the N gene sequence for lyssavirus genotype definition. *Virus Research*, *111*(1), 28–43.

Kuzmin, I. V., Orciari, L. A., Arai, Y. T., Smith, J. S., Hanlon, C. A., Kameoka, Y., et al. (2003). Bat lyssaviruses (Aravan and Khujand) from Central Asia: Phylogenetic relationships according to N, P and G gene sequences. *Virus Research*, *97*(2), 65–79.

Laothamatas, J., Hemachudha, T., Mitrabhakdi, E., Wannakrairot, P., & Tulayadaechanont, S. (2003). MR imaging in human rabies. *American Journal of Neuroradiology*, *24*(6), 1102–1109.

Laothamatas, J., Hemachudha, T., Tulyadechanont, S., & Mitrabhakdi, E. (1997). Neuroimaging in paralytic rabies. *Ramathibodi Medical Journal*, *20*(3), 149–156.

Laothamatas, J., Sungkarat, W., & Hemachudha, T. (2011). Neuroimaging in rabies. *Advances in Virus Research*, *79*, 309–327.

Leonova, G. N., Belikov, S. I., Kondratov, I. G., Krylova, N. V., Pavlenko, E. V., Romanova, E. V., et al. (2009). A fatal case of bat lyssavirus infection in Primorye Territory of the Russian Far East. *Rabies Bulletin Europe*, *33*(4), 5–8.

Lopez, R. A., Miranda, P. P., Tejada, V. E., & Fishbein, D. B. (1992). Outbreak of human rabies in the Peruvian jungle. *Lancet*, *339*(8790), 408–411.

Madhusudana, S. N., Nagaraj, D., Uday, M., Ratnavalli, E., & Kumar, M. V. (2002). Partial recovery from rabies in a six-year-old girl (Letter). *International Journal of Infectious Diseases*, *6*(1), 85–86.

Maier, T., Schwarting, A., Mauer, D., Ross, R. S., Martens, A., Kliem, V., et al. (2010). Management and outcomes after multiple corneal and solid organ transplantations from a donor infected with rabies virus. *Clinical Infectious Diseases*, *50*(8), 1112–1119.

Masthi, N. R., Raviprakash, D., Gangasagara, S. B., Sriprakash, K. S., Ashwin, B. Y., Ullas, P. T., et al. (2012). Rabies in a blind patient: Confusion after corneal transplantation. *National Medical Journal of India*, *25*(2), 83–84.

Mathuranayagam, D., & Rao, P. V. (1984). Antemortem diagnosis of human rabies by corneal impression smears using immunofluorescent technique. *Indian Journal of Medical Research*, *79*, 463–467.

McColl, K. A., Gould, A. R., Selleck, P. W., Hooper, P. T., Westbury, H. A., & Smith, J. S. (1993). Polymerase chain reaction and other laboratory techniques in the diagnosis of long incubation rabies in Australia. *Australian Veterinary Journal*, *70*(3), 84–89.

Meredith, C. D., Rossouw, A. P., & Koch, H. v. P. (1971). An unusual case of human rabies thought to be of chiropteran origin. *South African Medical Journal*, *45*(28), 767–769.

Metze, K., & Feiden, W. (1991). Rabies virus ribonucleoprotein in the heart (Letter). *New England Journal of Medicine*, *325*, 1814–1815.

Mills, R. P., Swanepoel, R., Hayes, M. M., & Gelfand, M. (1978) Dumb rabies: Its development following vaccination in a subject with rabies. *Central African Journal of Medicine*, *24*(6), 115–117.

Mitrabhakdi, E., Shuangshoti, S., Wannakrairot, P., Lewis, R. A., Susuki, K., Laothamatas, J., et al. (2005). Difference in neuropathogenetic mechanisms in human furious and paralytic rabies. *Journal of the Neurological Sciences*, *238*(1–2), 3–10.

Mrak, R. E., & Young, L. (1993). Rabies encephalitis in a patient with no history of exposure. *Human Pathology*, *24*, 109–110.

Munoz, J. L., Wolff, R., Jain, A., Sabino, J., Jacquette, G., Rapoport, M., et al. (1996). Human rabies - Connecticut, 1995. *Morbidity and Mortality Weekly Report*, *45*(10), 207–209.

Nathwani, D., McIntyre, P. G., White, K., Shearer, A. J., Reynolds, N., Walker, D., et al. (2003). Fatal human rabies caused by European bat lyssavirus type 2a infection in Scotland. *Clinical Infectious Diseases*, *37*(4), 598–601.

Nehaul, B. B. G. (1955). Rabies transmitted by bats in British Guiana. *American Journal of Tropical Medicine and Hygiene*, *4*, 550–553.

Nicholson, K. G. (1994). Human rabies. In R. R. McKendall & W. G. Stroop (Eds.), *Handbook of neurovirology (pp. 463–480)*. New York: Marcel Dekker. Neurological disease and therapy.

Noah, D. L., Drenzek, C. L., Smith, J. S., Krebs, J. W., Orciari, L., Shaddock, J., et al. (1998). Epidemiology of human rabies in the United States, 1980 to 1996. *Annals of Internal Medicine, 128*(11), 922–930.

Orr, P. H., Rubin, M. R., & Aoki, F. Y. (1988). Naturally acquired serum rabies neutralizing antibody in a Canadian Inuit population. *Arctic Medical Research, 47*(*Suppl 1*), 699–700.

Pal, S. R., Arora, B., Chhuttani, P. N., Broor, S., Choudhury, S., Joshi, R. M., et al. (1980). Rabies virus infection of a flying fox bat, *Pteropus poliocephalus* in Chandigarh, Northern India. *Tropical and Geographical Medicine, 32*(3), 265–267.

Parker, R., McKay, D., Hawes, C., Daly, P., Bryce, E., Doyle, P., et al. (2003). Human rabies, British Columbia – January 2003. *Canada Communicable Disease Report, 29*(16), 137–138.

Pawan, J. L. (1939). Paralysis as a clinical manifestation in human rabies. *Annals of Tropical Medicine and Parasitology, 33*, 21–29.

Paweska, J. T., Blumberg, L. H., Liebenberg, C., Hewlett, R. H., Grobbelaar, A. A., Leman, P. A., et al. (2006). Fatal human infection with rabies-related Duvenhage virus, South Africa. *Emerging Infectious Diseases, 12*(12), 1965–1967.

Pearson, C. A. (1976). Rabies (Letter). *Lancet, 1*, 206.

Phuapradit, P., Manatsathit, S., Warrell, M. J., & Warrell, D. A. (1985). Paralytic rabies: Some unusual clinical presentations. *Journal of the Medical Association of Thailand, 68*, 106–110.

Pleasure, S. J., & Fischbein, N. J. (2000). Correlation of clinical and neuroimaging findings in a case of rabies encephalitis. *Archives of Neurology, 57*(12), 1765–1769.

Porras, C., Barboza, J. J., Fuenzalida, E., Adaros, H. L., Oviedo, A. M., & Furst, J. (1976). Recovery from rabies in man. *Annals of Internal Medicine, 85*, 44–48.

Prier, S., Gibert, C., Bodros, A., & Krymolieres, F. (1979). Neurophysiological changes in non-vaccinated rabies patients (Letter). *Lancet, 1*, 620.

Prier, S., Gibert, C., Bodros, A., Vachon, F., Atanasiu, P., & Masson, M. (1979). Les neuropathies de la rage humaine: Etude clinique et electrophysiologique de deux cas [Human rabies neuropathies: Clinical and electrophysiological study in two cases]. *Revue Neurologique, 135*, 161–168.

Raman, G. V., Prosser, A., Spreadbury, P. L., Cockcroft, P. M., & Okubadejo, O. A. (1988). Rabies presenting with myocarditis and encephalitis. *Journal of Infection, 17*, 155–158.

Remington, P. L., Shope, T., & Andrews, J. (1985). A recommended approach to the evaluation of human rabies exposure in an acute-care hospital. *Journal of the American Medical Association, 254*, 67–69.

Roine, R. O., Hillbom, M., Valle, M., Haltia, M., Ketonen, L., Neuvonen, E., et al. (1988). Fatal encephalitis caused by a bat-borne rabies-related virus: Clinical findings. *Brain, 111*, 1505–1516.

Ron, M. (2001). Explaining the unexplained: Understanding hysteria (Editorial). *Brain, 124*(Pt 6), 1065–1066.

Ross, E., & Armentrout, S. A. (1962). Myocarditis associated with rabies: Report of a case. *New England Journal of Medicine, 266*, 1087–1089.

Ruegsegger, J. M., Black, J., & Sharpless, G. R. (1961). Primary antirabies immunization of man with HEP Flury virus vaccine. *American Journal of Public Health, 51*, 706–716.

Samaratunga, H., Searle, J. W., & Hudson, N. (1998). Non-rabies lyssavirus human encephalitis from fruit bats: Australian bat lyssavirus (pteropid lyssavirus) infection. *Neuropathology and Applied Neurobiology, 24*(4), 331–335.

Selimov, M. A., Tatarov, A. G., Botvinkin, A. D., Klueva, E. V., Kulikova, L. G., & Khismatullina, N. A. (1989). Rabies-related Yuli virus; Identification with a panel of monoclonal antibodies. *Acta Virologica, 33*, 542–546.

Sheikh, K. A., Ramos-Alvarez, M., Jackson, A. C., Li, C. Y., Asbury, A. K., & Griffin, J. W. (2005). Overlap of pathology in paralytic rabies and axonal Guillain-Barré syndrome. *Annals of Neurology, 57*(5), 768–777.

Shope, R. E., Murphy, F. A., Harrison, A. K., Causey, O. R., Kemp, G. E., Simpson, D. I. H., et al. (1970). Two African viruses serologically and morphologically related to rabies virus. *Journal of Virology, 6,* 690–692.

Sibley, W. A., Ray, C. G., Petersen, E., Ryan, K., Graham, A. R., Gibbs, M. A., et al. (1981). Human rabies acquired outside the United States from a dog bite. *Morbidity and Mortality Weekly Report, 43,* 537–540.

Silverstein, M. A., Salgado, C. D., Bassin, S., Bleck, T. P., Lopes, M. B., Farr, B. M., et al. (2003). First human death associated with raccoon rabies -- Virginia, 2003. *Morbidity and Mortality Weekly Report, 52*(45), 1102–1103.

Sing, T. M., & Soo, M. Y. (1996). Imaging findings in rabies. *Australasian Radiology, 40*(3), 338–341.

Sipahioglu, U., & Alpaut, S. (1985). Transplacental rabies in a human [Turkish]. *Mikrobiyoloji Bulteni, 19,* 95–99.

Smith, J. S., Fishbein, D. B., Rupprecht, C. E., & Clark, K. (1991). Unexplained rabies in three immigrants in the United States: A virologic investigation. *New England Journal of Medicine, 324*(4), 205–211.

Smith, P. C., Lawhaswasdi, K., Vick, W. E., & Stanton, J. S. (1967). Isolation of rabies virus from fruit bats in Thailand. *Nature, 216*(113), 384.

Sow, P. S., Diop, B. M., Ndour, C. T. Y., Soumare, M., Ndoye, B., Faye, M. A., et al. (1996). Occipital cerebral aspiration ponction: Technical procedure to take a brain specimen for postmortem virological diagnosis of human rabies in Dakar. *Medecine Et Maladies Infectieuses, 26*(5), 534–536.

Srinivasan, A., Burton, E. C., Kuehnert, M. J., Rupprecht, C., Sutker, W. L., Ksiazek, T. G., et al. (2005). Transmission of rabies virus from an organ donor to four transplant recipients. *New England Journal of Medicine, 352*(11), 1103–1111.

Sureau, P., Portnoi, D., Rollin, P., Lapresle, C., & Chaouni-Berbich, A. (1981). Prevention of inter-human rabies transmission after corneal graft (French). *Comptes Rendus de l'Académie des Sciences - Series III, Sciences de la Vie, 293*(13), 689–692.

Talaulicar, P. M. S. (1977). Persistent priapism in rabies. *British Journal of Urology, 49,* 462.

Tariq, W. U. Z., Shafi, M. S., Jamal, S., & Ahmad, M. (1991). Rabies in man handling infected calf (Letter). *Lancet, 337,* 1224.

Thongcharoen, P., Wasi, C., Sirikavin, S., Boonthai, P., Bedavanij, A., Dumavibhat, P., et al. (1981). Human-to-human transmission of rabies via corneal transplant - Thailand. *Morbidity and Mortality Weekly Report, 30,* 473–474.

Tillotson, J. R., Axelrod, D., & Lyman, D. O. (1977). Rabies in a laboratory worker - New York. *Morbidity and Mortality Weekly Report, 26,* 183–184.

Tirawatnpong, S., Hemachudha, T., Manutsathit, S., Shuangshoti, S., Phanthumchinda, K., & Phanuphak, P. (1989). Regional distribution of rabies viral antigen in central nervous system of human encephalitic and paralytic rabies. *Journal of the Neurological Sciences, 92,* 91–99.

Tong, T. R., Leung, K. M., & Lam, A. W. S. (1999). Trucut needle biopsy through superior orbital fissure for diagnosis of rabies. *Lancet, 354*(9196), 2137–2138.

Toro, G., Vergara, I., & Roman, G. (1977). Neuroparalytic accidents of antirabies vaccination with suckling mouse brain vaccine: Clinical and pathologic study of 21 cases. *Archives of Neurology, 34,* 694–700.

Udow, S. J., Marrie, R. A., & Jackson, A. C. (2012). Clinical features of dog- and bat-acquired rabies in humans. Platform presentation at the XXIIIrd International Meeting on Research Advances and Rabies Control in the Americas in Sao Paulo, Brazil on October 15, 2012. *Research Advances and Rabies Control in the Americas.*

Udwadia, Z. F., Udwadia, F. E., Rao, P. P., & Kapadia, F. (1988). Penile hyperexcitability with recurrent ejaculations as the presenting manifestation of a case of rabies. *Postgraduate Medical Journal, 64,* 85–86.

van Thiel, P. P., de Bie, R. M., Eftimov, F., Tepaske, R., Zaaijer, H. L., van Doornum, G. J., et al. (2009). Fatal human rabies due to Duvenhage virus from a bat in Kenya: Failure of treatment with coma-induction, ketamine, and antiviral drugs. *PLoS Neglected Tropical Diseases, 3*(7), e428.

Verlinde, J. D., Li-Fo-Sjoe, E., Versteeg, J., & Dekker, S. M. (1975). A local outbreak of paralytic rabies in Surinam children. *Tropical and Geographical Medicine, 27*(2), 137–142.

Wallerstein, C. (1999). Rabies cases increase in the Philippines. *British Medical Journal, 318*(7194), 1306.

Warrell, D. A. (1976). The clinical picture of rabies in man. *Transactions of the Royal Society of Tropical Medicine and Hygiene, 70*, 188–195.

Warrell, D. A., Davidson, N. M., Pope, H. M., Bailie, W. E., Lawrie, J. H., Ormerod, L. D., et al. (1976). Pathophysiologic studies in human rabies. *American Journal of Medicine, 60*(2), 180–190.

Warrell, D. A., & Warrell, M. J. (1991). Rabies. In H. P. Lambert (Ed.), *Infections of the central nervous system (pp. 317–328)*. Philadelphia: B.C. Decker Inc.

Warrell, M. J. (1996). Rabies. In G. C. Cook (Ed.), *Manson's tropical diseases (pp. 700–720)* (20 ed.). London: W.B. Saunders.

Warrell, M. J., Looareesuwan, S., Manatsathit, S., White, N. J., Phuapradit, P., Vejjajiva, A., et al. (1988). Rapid diagnosis of rabies and post-vaccinal encephalitides. *Clinical and Experimental Immunology, 71*, 229–234.

Waterman, J. A. (1959). Acute ascending rabic myelitis. Rabies - transmitted by bats to human beings and animals. *Caribbean Medical Journal, 21*, 46–74.

White, M., Davis, A., Rawlings, J., Neill, S., Hendricks, K., Simpson, D., et al. (1994). Human rabies – Texas and California, 1993. *Morbidity and Mortality Weekly Report, 43*(6), 93–96.

Whitfield, S. G., Fekadu, M., Shaddock, J. H., Niezgoda, M., Warner, C. K., & Messenger, S. L. (2001). A comparative study of the fluorescent antibody test for rabies diagnosis in fresh and formalin-fixed brain tissue specimens. *Journal of Virological Methods, 95*(1–2), 145–151.

Whitley, R. J. (2004). B virus. In W. M. Scheld, R. J. Whitley, & C. M. Marra (Eds.), *Infections of the central nervous system (pp. 197–203)* (3 ed.). Philadelphia: Lippincott Williams and Wilkins.

Wilde, H., & Chutivongse, S. (1988). Rabies: Current management in Southeast Asia. *Medical Progress, 15*, 14–23.

Wilson, J. M., Hettiarachchi, J., & Wijesuriya, L. M. (1975). Presenting features and diagnosis of rabies. *Lancet, 2*, 1139–1140.

Winkler, W. G., Fashinell, T. R., Leffingwell, L., Howard, P., & Conomy, J. P. (1973). Airborne rabies transmission in a laboratory worker. *Journal of the American Medical Association, 226*, 1219–1221.

World Health Organization, (2005). *WHO expert consultation on rabies: First report*. Geneva: World Health Organization.

8

Pathogenesis

Alan C. Jackson[1] and Zhen F. Fu[2,3]

[1]Departments of Internal Medicine (Neurology) and Medical Microbiology, University of Manitoba, Winnipeg, MB R3A 1R9, Canada, [2]Department of Pathology, University of Georgia, 501 D. W. Brooks Dr., Athens, GA 30602, USA, [3]State-Key Laboratory of Agricultural Microbiology, Huazhong Agricultural University, 1 Shizishan Road, Wuhan, Hubei 430070, China

1 INTRODUCTION

Rabies virus is a highly neurotropic virus that spreads along neural pathways and invades the central nervous system (CNS), where it causes an acute infection. Most of what we know about the events that take place during rabies infection has been learned from experimental models using animals. Fixed laboratory strains of rabies virus and rodent models have commonly been used, because they are easier to handle and less expensive, although the events in these models may not closely mimic the disease under natural conditions either in humans or in rabies vectors. There are a number of sequential steps that occur after peripheral inoculation of rabies virus from an animal bite, which is the most common mechanism of transmission (Figure 8.1). The steps include replication in peripheral tissues, spread along peripheral nerves and the spinal cord to the brain, dissemination within the CNS, and centrifugal spread from the CNS along nerves to various organs, including the salivary glands. Each of the pathogenetic steps will be discussed in this chapter. In addition, mechanisms of immune-mediated pathology and neuronal dysfunction in rabies will be addressed.

Rabies: Scientific Basis of the Disease and its Management
DOI: http://dx.doi.org/10.1016/B978-0-12-396547-9.00008-0

299

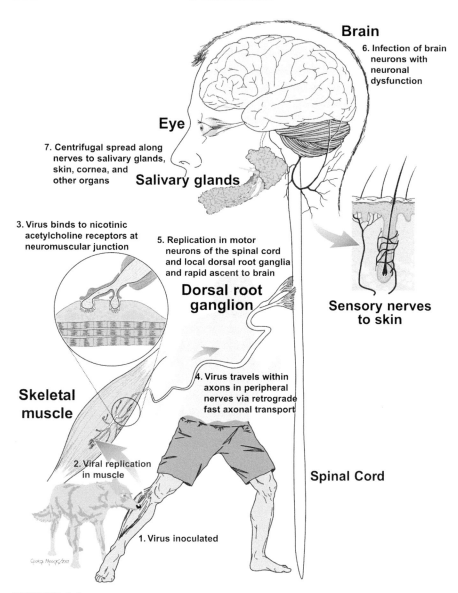

FIGURE 8.1   Schematic diagram showing the sequential steps in the pathogenesis of rabies after an animal bite.

2 VIRUS ENTRY INTO THE NERVOUS SYSTEM

2.1 Earliest Events

Early studies in rabies pathogenesis, which were performed in order to establish the pathways and rate of viral spread, involved amputation of the tail or leg of an animal proximal to the site of inoculation with a "fixed" or "street" (wild-type) strain of rabies virus. The development of rabies could be prevented with amputation, and the timing of the procedure was found to be critical. In later studies, neurectomy of the sciatic nerve was performed instead of amputation, and similar results were observed (Baer, Shantha, & Bourne, 1968; Baer, Shanthaveerappa, & Bourne, 1965). These experiments clearly demonstrated that there was an incubation period in rabies during which there was time-dependent movement of virus along peripheral nerves from the site of inoculation to the CNS, and models using street rabies virus supported the idea that the virus remains at or near the site of entry for most of the long incubation period (Baer & Cleary, 1972). However, the time periods in which the procedures were life-saving in rodents infected with fixed rabies virus strains were relatively short (Baer et al., 1965; Dean, Evans, & McClure, 1963), suggesting a different mechanism of viral entry for fixed viruses than in natural rabies (due to street viruses).

Under natural conditions, humans and animals may experience long and variable incubation periods following a bite exposure. This may play a role in maintaining enzootic rabies, especially in high density, high contact populations where there is a tendency for the disease to "burn" itself out by rapidly reducing the number of susceptible animals. In humans, the incubation period is usually between 20 and 90 days, although incubation periods rarely may be as short as a few days or longer than a year (Smith, Fishbein, Rupprecht, & Clark, 1991). There is uncertainty about the events that occur during this incubation period. There has been speculation that macrophages may sequester rabies virus *in vivo* because persistent *in vitro* infections of human and murine monocytic cell lines and of primary murine bone marrow macrophages have been demonstrated with different rabies virus strains (Ray, Ewalt, & Lodmell, 1995). However, this has not yet been demonstrated in animal models. The best experimental animal studies to date examining the events that take place during the incubation period were performed in striped skunks using a Canadian isolate of street rabies virus obtained from skunk salivary glands (Charlton, Nadin-Davis, Casey, & Wandeler, 1997). These studies, which used reverse transcriptase-polymerase chain reaction (RT-PCR) amplification, showed that viral genomic RNA was frequently present in the inoculated muscle (found in four of nine skunks), but not in either spinal ganglia or the spinal cord when skunks were sacrificed 62–64 days post-inoculation. Immunohistochemical studies performed prior to the development of clinical disease showed evidence of infection of extrafusal muscle fibers and occasional fibrocytes at the site of inoculation. Although it is unclear, the infection of muscle fibers may be a critical pathogenetic step for the virus to gain access to the peripheral nervous system. In a highly susceptible host after intramuscular inoculation, rabies virus-infected suckling hamsters showed early infection of striated muscle cells near the site of inoculation, and, shortly afterward, neuro-muscular and neurotendinal spindles became infected near the site of inoculation, which was followed by evidence of infection of small nerves within muscles, tendons, and adjoining connective tissues (Murphy, Bauer, Harrison, & Winn, 1973). However, these events occurred within a few days of inoculation and do not mimic the situation with the longer incubation periods seen in natural infections.

In mouse models, early infection of muscle or other extraneural tis-sues was not observed following inoculation of fixed rabies virus strains (Coulon et al., 1989; Johnson, 1965). Virus-specific RNA was not detected with RT-PCR amplification in the masseter muscle of adult mice between 6 and 30 hours after inoculation of the challenge virus standard (CVS) strain of fixed rabies virus in the muscle, although viral RNA was identi-fied in trigeminal ganglia at 18 hours and in the brainstem at 24 hours after inoculation (Shankar, Dietzschold, & Koprowski, 1991). These studies strongly suggest that rabies virus is capable of direct entry into peripheral nerves without a replicative cycle in extraneural cells during the short incubation period. This is likely the mechanism of viral entry in rodent models using fixed strains of rabies virus, accounting for the short period of time during which amputation or neurectomy is protec-tive after peripheral inoculation of fixed rabies virus (Baer et al., 1965; Dean et al., 1963). Unfortunately, these models provide little information about events that take place during the long incubation period of natural rabies.

2.2 Superficial and Non-Bite Exposures

The vast majority of human rabies cases that occur without a history of an exposure are thought to be due to unrecognized or forgotten bites. Molecular characterization of the rabies virus strains has indicated that they are most frequently from the strain found in silver-haired bats and tricolored bats (formerly called eastern pipistrelle bats) in the USA (Noah et al., 1998), which are small bats. Experimental studies on the silver-haired bat rabies virus (SHBRV) indicate that the virus replicates well at lower than normal body temperatures (34°C) and is associated with higher infectivity in cell types present in the dermis, including fibro-blasts and epithelial cells, than with coyote street virus (Morimoto et al., 1996). Hence, the SHBRV was likely selected for efficient local replication in the dermis, which could explain the success of this strain. However, after superficial exposures, it is unclear how or at precisely what sites the virus invades peripheral nerves in the skin or subcutaneous tissues because this has not yet been studied in pathogenesis studies.

Humans have rarely been infected by bat viruses via the airborne route, either in caves where millions of bats roost (Constantine, 1962) or in laboratory accidents by aerosolized rabies virus (Tillotson, Axelrod, & Lyman, 1977; Winkler, Fashinell, Leffingwell, Howard, & Conomy, 1973). Viral entry by the olfactory and oral routes is much less common than by bites. Relatively little experimental work has been done with routes of viral entry other than one simulating a bite exposure (using inoculation techniques). The nasal mucosa has been shown to act as a site of viral entry by suckling guinea pigs that have inhaled street rabies virus (Hronovsky & Benda, 1969). Rabies virus antigen was initially found in nasal mucosa cells 6 days later. Early brain infection was prominent in the olfactory bulbs, suggesting that rabies virus spread into the brain by an olfactory pathway. Similar results were obtained using a variety of rabies virus strains in mice and hamsters (Fischman & Schaeffer, 1971). Rabies virus antigen has been observed in olfactory receptor cells of naturally infected Brazilian free-tailed bats obtained from a cave, suggesting that the nasal mucosa is a portal of entry in natural infection of bats by airborne rabies virus in caves (Constantine, Emmons, & Woodie, 1972). Experimental studies showing transmission of rabies to a variety of species of carnivorous animals caged in a cave containing millions of bats supported infection by the airborne route (Constantine, 1962). However, it is thought that the presence of a very large number (millions) of bats in an unventilated area is necessary for airborne transmission of rabies virus.

Oral transmission of rabies virus might occur naturally by consumption of carcasses of rabid animals by wildlife and may also be important when humans eat raw dog meat (Wallerstein, 1999). Low susceptibility was observed when mice (Charlton & Casey, 1979a) and skunks (Charlton & Casey, 1979c) were given CVS or street rabies virus either by the oral route or intestinal instillation. Mice, hamsters, guinea pigs, and rabbits of different ages were infected with CVS either orally or by gastric tube administration (Fischman & Ward, 1968). In CVS-infected weanling mice and hamsters that were infected by this route, rabies virus antigen was not observed in intestinal mucosal cells, but was found in neurons in Auerbach's and Meissner's plexuses of the stomach and intestine (Fischman & Schaeffer, 1971). These findings suggest that viral entry by the oral route likely occurs via breaks in the integrity of the gastrointestinal mucosa. However, the importance of oral transmission in natural rabies of animals remains uncertain.

3 RABIES VIRUS RECEPTORS

The most striking feature of rabies virus is its almost exclusive neurotropism. The rabies virus glycoprotein is thought to be of prime importance in this process by binding to neurospecific receptors (see Chapter 2). These receptors have a role in normal cell function and are hijacked by viruses to gain entry into cells. At least three rabies virus receptors have been proposed, and it is very likely that additional ones will be identified in the future.

3.1 Nicotinic Acetylcholine Receptor

The nicotinic acetylcholine receptor (nAChR) was the first identified receptor for rabies virus (Lentz, Burrage, Smith, Crick, & Tignor, 1982). Rabies virus antigen was detected at sites coincident with the nAChR in infected cultured chick myotubes from chicken embryos and also shortly after immersion of mouse diaphragms in a suspension of rabies virus. It was evident from these studies that the distribution of viral antigen detected by fluorescent antibody staining at sites in neuromuscular junctions corresponded to the distribution of nAChRs. The receptors were stained with the rhodamine-conjugated antagonist α-bungarotoxin. Pretreatment of myotubes with either the irreversible binding nicotinic cholinergic antagonist α-bungarotoxin or the reversible binding d-tubo-curarine reduced the number of myotubes that became infected with rabies virus. Studies in other laboratories showed that pretreatment of cultured rat myotubes with α-bungarotoxin had an inhibitory effect on infection (Tsiang, de la Porte, Ambroise, Derer, & Koenig, 1986). Binding of radiolabeled rabies virus to purified *Torpedo* acetylcholine receptor was also inhibited by nicotinic antagonists, but not by atropine (a muscarinic antagonist) (Lentz, Benson, Klimowicz, Wilson, & Hawrot, 1986). Monoclonal antibodies raised against a peptide containing residues 190–203 of the rabies virus glycoprotein also inhibited binding of the rabies virus glycoprotein and α-bungarotoxin to the AChR (Bracci et al., 1988). Both rabies virus and neurotoxins bind to residues 173–204 of the $α_1$-subunit of the AchR and the highest-affinity virus-binding determinants are located within residues 179–192 (Lentz, 1990). These studies have provided strong evidence that rabies virus binds to nicotinic acetylcholine receptors in neuromuscular junctions.

Snake venom neurotoxins are polypeptides that bind with high affinity to nAChRs and competitively block the depolarizing action of acetylcholine. When the amino acid sequence of the rabies virus glycoprotein was compared with that of snake venom neurotoxins, a significant sequence similarity was found between a segment (residues 151–238) of the rabies virus glycoprotein and the entire long neurotoxin sequence (71–74 residues) (Lentz, Wilson, Hawrot, & Speicher, 1984). The glycoprotein showed identity with residues at the end of loop 2 of the long neurotoxin (the "toxic loop"), which is a long central loop projecting from the molecule that is highly conserved among all of the neurotoxins (Figure 8.2). This suggests that this region of the rabies virus glycoprotein is likely a recognition site for the acetylcholine receptor (Lentz, 1985).

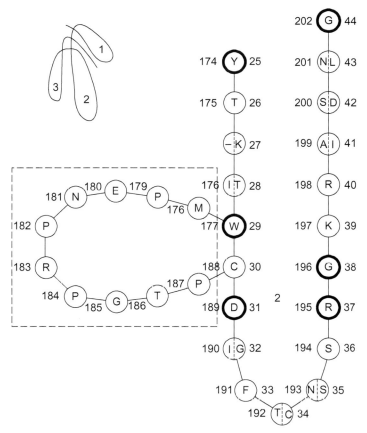

FIGURE 8.2 **A model showing the similarity of the rabies virus glycoprotein with the "toxic" loop of the neurotoxins.** The segment of the glycoprotein (residues 174–202) corresponding to loop 2 of the long neurotoxins (Karlsson positions 25–44) is positioned in relationship to a schematic representation of loop 2. Within circles, residues or gaps in the glycoprotein are shown on the left and those in the neurotoxin on the right. One letter is shown where the glycoprotein and toxin are identical. Bold circles are residues highly conserved or invariant among all the neurotoxins. A 10-residue insertion in the glycoprotein is enclosed in the box. The rabies virus sequence is of the CVS strain, and the neurotoxin sequence is *Ophiophagus hannah*, toxin b. Inset: schematic of neurotoxin structure showing positions of loops 1, 2, and 3. *(Reproduced with permission from Lentz et al., Amino acid sequence similarity between rabies virus glycoprotein and snake venom curaremimetic neurotoxins. Science **226**:847–848, 1984. Copyright 1984 by American Association for the Advancement of Science.)*

Lentz and co-workers indicated that binding of rabies virus to AChRs would localize and concentrate the virus on post-synaptic cells, which would facilitate subsequent uptake and transfer of virus to peripheral motor nerves (Lentz et al., 1982). Studies performed in chick spinal cord-muscle co-cultures showed that the CVS strain of rabies virus and AChR tracers co-localized at neuromuscular junctions and nerve terminals, which provided evidence that the neuromuscular junction is the major site of entry into neurons (Lewis, Fu, & Lentz, 2000). Subsequently, co-localization with endosome tracers indicated that the virus resides in an early endosome compartment. There is also supporting ultrastructural evidence that rabies virus particles enter nerve terminals by endocytosis (Charlton & Casey, 1979b; Iwasaki & Clark, 1975). The acidic interior of the endosome triggers fusion of the viral membrane with the endosome membrane, which allows the viral nucleocapsid to escape into the cytoplasm. However, it has not yet been resolved whether the viral uncoating actually takes place in nerve terminals or in the cell body (perikaryon) after transport in the axon.

Although rabies virus infection with fixed strains is restricted to a small number of cell types *in vivo*, fixed viruses can infect a much larger variety of cell types *in vitro* (Reagan & Wunner, 1985). There is evidence that carbohydrate moieties, phospholipids, highly sialylated gangliosides, and other membrane-associated proteins might contribute to the cellular membrane receptor structure for rabies virus (Broughan & Wunner, 1995; Conti, Superti, & Tsiang, 1986; Superti et al., 1986; Superti, Seganti, Tsiang, & Orsi, 1984).

Variations in animal susceptibility to rabies virus infection have been recognized for many years. When infected by intramuscular inoculation, foxes are highly sensitive to rabies virus infection, dogs are less sensitive, and opossums are highly resistant (Baer, Bellini, & Fishbein, 1990). The difference in susceptibility between the red fox and the opossum could reflect the quantity of acetylcholine receptors in muscle (Baer, Shaddock, Quirion, Dam, & Lentz, 1990). A striking difference in the muscle content, B_{max} of nicotinic acetylcholine receptors was found with 180.5 fmol/mg protein present in red foxes and only 11.4 fmol/mg protein present in opossums, which was a highly significant difference (p < 0.001). No difference was observed in the binding affinity, K_d. In addition, radiolabeled rabies virus bound much better to fox muscles than opossum muscles. Hence, the susceptibility of different animal species to rabies virus may, at least in part, be related to the quantity of nicotinic acetylcholine receptors in their muscles.

Although important in the neuromuscular junction, nAChR has not yet been determined whether it is also an important rabies virus receptor in the CNS. Binding of rabies virus to nicotinic acetylcholine receptors in the brain could cause neuronal dysfunction. An anti-rabies virus glycoprotein monoclonal antibody was used to generate (by immunization) an anti-idiotypic antibody, B9, that selectively binds to nAChRs (Hanham, Zhao, & Tignor, 1993). Immunostaining of neuronal elements in the brains of rabies virus-infected mice with the B9 antibody was greatly reduced. This suggests that rabies virus binds to nAChRs in the brain, but the pathogenetic significance of this binding in producing neuronal dysfunction in rabies has not yet been established.

3.2 Neural Cell Adhesion Molecule (NCAM) Receptor

The NCAM receptor, which is a cell adhesion glycoprotein of the immunoglobulin superfamily on their cell surface, has been identified as rabies virus receptor since it was not found on the surface of resistant cell lines (Thoulouze et al., 1998). Incubation of susceptible cells with rabies virus decreased surface expression of NCAM and had no effect on other integral proteins of the cell membrane, whereas another virus, vaccinia virus, did not affect surface NCAM expression. This is consistent with internalization of rabies virus-NCAM receptor complexes during viral entry by adsorptive endocytosis. Rabies virus infection was also inhibited when NCAM receptor was blocked with heparan sulfate, which is a natural ligand physiologically, and by either polyclonal or monoclonal antibodies directed against NCAM receptor (Thoulouze et al., 1998). Furthermore, soluble NCAM neutralized rabies virus infection, indicating that occupation of the receptor site on virus particles prevented binding to the rabies virus receptors on target cells. When resistant L cells were transfected with NCAM cDNA, the cells became susceptible to rabies virus infection. Hence, there is very strong *in vitro* evidence that NCAM is a rabies virus receptor.

When primary cortex cultures were prepared from NCAM receptor-deficient ("knockout") and their wild-type littermate mice (Thoulouze et al., 1998) and infected with CVS, a significantly lower mean number of cells became infected in NCAM receptor-deficient cultures (7.8 +/− 3.9%) than in wild-type cultures (18.6 +/− 8.9%) (p < 0.005). *In vivo*, after inoculation of CVS into the masseter muscle of NCAM receptor-deficient and wild-type mice, significantly less rabies virus antigen was found in the brainstem/cerebellum, diencephalon and cerebral cortex in NCAM receptor-deficient than in wild-type mice, indicating that viral spread was less efficient without NCAM receptor. After inoculation of CVS into hindlimb muscles, the mean survival of NCAM receptor-deficient mice was 13.6 days compared to 10.2 days in wild-type mice ($P = 0.002$), indicating that the disease progressed slower without NCAM receptor. The absence of NCAM receptor *in vivo* only mildly delayed the death of mice. Interestingly, this suggests that there must be other functionally important rabies virus receptors in the CNS in addition to NCAM receptor. The NCAM receptor is localized in presynaptic membranes and, hence, it is well-positioned for internalization of rabies virus by receptor-mediated endocytosis into vesicles (Lafon, 2005). Subsequently, there is retrograde transport of either these vesicles carrying dissociated rabies virions or of viral nucleocapsids, which are released after uncoating of the virus with fusion of the viral envelope.

3.3 Low-Affinity p75 Neurotrophin Receptor

A report that the low-affinity p75 neurotrophin receptor (p75[NTR]) is a receptor for street rabies virus further suggests multiple candidates for the rabies virus receptor (Tuffereau, Benejean, Blondel, Kieffer, & Flamand, 1998). When a random-primed cDNA library from the mRNA of neuroblastoma cells (NG108) was used to transfect COS7 cells, a single plasmid was identified after subcloning, which, when transfected into BSR cells, bound soluble rabies virus glycoprotein. The 1.3 kb insert of this plasmid showed high amino acid sequence homology with both rat and human p75[NTR]. Most cell lines of non-neuronal cell origin, including BSR cells, are not permissive for street rabies virus infection. However, the BSR cells with stable expression of p75[NTR], were able to bind soluble rabies virus glycoprotein. A fox street rabies virus isolate was also able to infect p75[NTR]-expressing BSR cells, but relatively few untransfected control BSR cells. BSR cells expressing p75[NTR] were only slightly more susceptible to infection with CVS and p75[NTR]-expressing BSR cells were 3 to 10 times more susceptible to CVS infection than control BSR cells in the presence of 10% serum. Subsequently, Tuffereau et al., performed further studies in cultured adult mouse dorsal root ganglion neurons and reported that, although p75[NTR] is a receptor for soluble rabies virus glyco-protein in transfected cells of heterologous systems, a rabies virus glycoprotein-interaction is not necessary for rabies virus infection of primary neurons (Tuffereau et al., 2007).

Since CVS, like street rabies virus strains, is highly neuronotropic *in vivo*, one would expect that CVS would also use the same receptors as street rabies viruses *in vivo*. In addition, evaluation of non-adapted street rabies virus infection of mice would be difficult because, for example, a high incidence of spontaneous recovery with neurologic sequelae has been observed after peripheral inoculation of mice with a fox isolate of street virus (Jackson, Reimer, & Ludwin, 1989). When p75[NTR]-deficient mice were infected intracerebrally with CVS, similar clinical features of disease and pathologic changes were observed in the brain as in mice expressing p75[NTR] (Jackson & Park, 1999). p75[NTR] is not present at the neuromuscular junction, and it is mainly present in the dorsal horn of the spinal cord, suggesting that it could be involved in trafficking of rabies virus by a sensory pathway (Lafon, 2005). Ligand – p75[NTR] complexes are normally internalized by clathrin-coated pits into endosomes (Butowt & Von Bartheld, 2003). Lafon (2005) has speculated that p75[NTR] may play an important role in the retrograde transport of rabies virus by forming a rabies virus – p75[NTR] complex that is transported into the cell in endocytic compartments, possibly following caveolae transcytosis.

4 SPREAD TO THE CNS

Centripetal spread of rabies virus to the CNS occurs within motor and perhaps also sensory axons of peripheral nerves. Colchicine, a microtubule-disrupting agent active for tubulin-containing cytoskeletal structures, is an effective inhibitor of fast axonal transport in the sciatic nerve of rats (Tsiang, 1979). When colchicine was applied locally to the sciatic nerve using elastomer cuffs to obtain high local concentrations of the drug, adverse systemic effects were avoided. Propagation of rabies virus was prevented, providing strong evidence that rabies virus spreads from sites of peripheral inoculation to the CNS by retrograde fast axonal transport. Human dorsal root ganglia neurons in a compartmentalized cell culture system were used to show that viral retrograde transport occurs at a rate of between 50 and 100 mm/day (Tsiang, Ceccaldi, & Lycke, 1991). There is evidence that the rabies virus phosphoprotein, which is a member of the ribonucleocapsid complex (see Chapter 2), interacts with dynein light chain 8 (LC8). Dynein LC8 is a component of both cytoplasmic myosin V and dynein that are involved in actin-based transport (important in early steps of viral entry) and microtubule-based transport (for fast axonal transport) in neurons, respectively (Jacob, Badrane, Ceccaldi, & Tordo, 2000; Raux, Flamand, & Blondel, 2000). This led to speculation that rabies virus phosphoprotein-dynein interaction may be of fundamental importance in axonal transport of rabies virus. However, studies performed in young mice have shown that deletions of the dynein light chain binding region of recombinant SAD-L16, which contained the genetic sequence of the Street Alabama Dufferin (SAD)-B19 strain, resulted in mutant viruses that demonstrated only minor effects on viral spread after peripheral inoculation, and they remained neuroinvasive and neurovirulent (Mebatsion, 2001; Rasalingam, Rossiter, Mebatsion, & Jackson, 2005). Mazarakis et al., (2001) have demonstrated that rabies virus glycoprotein-pseudotyped lentivirus (equine infectious anemia virus)-based vectors enhance gene transfer to neurons by facilitating retrograde axonal transport. Hence, the rabies virus glycoprotein may play a more important role than the phosphoprotein.

Other investigators used mouse and hamster models to demonstrate early and at least near-simultaneous involvement of motor neurons in the spinal cord and primary sensory neurons in dorsal root ganglia (Coulon et al., 1989; Jackson & Reimer, 1989; Johnson, 1965; Murphy et al., 1973). After inoculation of mice in the masseter muscle with CVS, early infection was found in trigeminal ganglia (Jackson, 1991b; Shankar et al., 1991). Studies using RT-PCR amplification showed that infection was detectable in trigeminal ganglia (18 hours post-inoculation) before the brainstem (24 hours post-inoculation) (Shankar et al., 1991). However, elegant transneuronal tracer methods using CVS in rats

(Tang, Rampin, Giuliano, & Ugolini, 1999) and studies in rhesus monkeys (Kelly & Strick, 2000) have not shown early infection of primary sensory neurons. Two days after inoculation of CVS into the bulbospongiosus muscle of rats, the distribution of rabies virus antigen was limited to ipsilateral bulbospongiosus motor neurons in the spinal cord (Tang et al., 1999). One day later (3 days post-inoculation) there was evidence of transfer of antigen to interneurons in the dorsal gray commissure, intermediate zone and sacral parasympathetic nucleus and also to external urethral sphincter motor neurons; at this time there was no labeling of primary sensory neurons in local dorsal root ganglia. This study indicates that a motor pathway rather than a sensory pathway is important in the spread of rabies virus to the CNS. It is unclear if the different results obtained in earlier studies are due to differences in the animal models, including the species of the host and the route of inoculation.

5 SPREAD WITHIN THE CNS

Once CNS neurons (often in the spinal cord) become infected in rodent models, there is rapid dissemination of rabies virus infection along neuroanatomical pathways. Rabies virus also spreads within the CNS, as in the peripheral nervous system, by fast axonal transport. Evidence was provided for axonal transport using stereotaxic brain inoculation in rats (Gillet, Derer, & Tsiang, 1986) and by the administration of colchicine, which inhibited virus transport within the CNS (Ceccaldi, Ermine, & Tsiang, 1990; Ceccaldi, Gillet, & Tsiang, 1989). Studies on cultured rat dorsal root ganglia neurons showed that anterograde fast axonal transport of rabies virus is in the range of 100–400 mm/day (Tsiang, Lycke, Ceccaldi, Ermine, & Hirardot, 1989). However, the importance of this is unclear because transneuronal tracing studies with CVS in rhesus monkeys have indicated that the spread of rabies virus occurs exclusively by retrograde axonal transport with trans-synaptic transport of rabies virus also occurring exclusively in the retrograde direction (Kelly & Strick, 2000). Studies performed with rabies virus glycoprotein gene-deficient recombinant rabies virus showed limited spread in the brains of mice after intracerebral inoculation (Etessami et al., 2000). After stereotaxic inoculation of the recombinant virus into the rat striatum, infection remained restricted to initially infected neurons and there was no evidence of trans-synaptic spread to secondary neurons. Hence, the rabies virus glycoprotein is necessary for trans-synaptic spread of rabies virus from one neuron to another.

Ultrastructural studies in a skunk model indicated that most viral budding occurs on synaptic or adjacent plasma membranes of dendrites, with less prominent budding from the plasma membrane of the perikaryon (Charlton & Casey, 1979b). Most virions were found partially engulfed by an invaginated membrane of an adjacent axon terminal, indicating transneuronal dendroaxonal transfer of virus. Virions were also occasionally observed budding freely into the intercellular space.

After footpad inoculation of mice with CVS, there was early involvement of neurons in the brainstem tegmentum and deep cerebellar nuclei (Jackson & Reimer, 1989). Subsequently, the infection spread to involve cerebellar Purkinje cells and neurons in the diencephalon, basal ganglia and cerebral cortex. Rabies virus, like Borna disease virus (Carbone, Duchala, Griffin, Kincaid, & Narayan, 1987), spread to the hippocampus relatively late after peripheral inoculation. Rabies virus predominantly infected pyramidal neurons of the hippocampus, with relative sparing of neurons in the dentate gyrus in adult mice (Jackson & Reimer, 1989). The basis for cell selectivity is uncertain, although Gosztonyi and Ludwig (Gosztonyi & Ludwig, 2001) speculated that if N-methyl-D-aspartate (NMDA) NR1 receptors are involved as rabies virus receptors, then cell selectivity can be explained by the fact that rabies virus spreads only by retrograde (not by anterograde) fast axonal transport. Therefore, the virus cannot infect dentate granule cells by the perforant path and mossy fibers from CA3 that predominantly have α-amino-3-hydroxy-5-methyl-4-isoxazole propionate (AMPA) and kainate receptors rather than NMDA receptors. Although rabies virus is highly neuronotropic, skunk rabies virus has been observed to infect Bergmann glia in the cerebellum more prominently than Purkinje cells in experimentally infected skunks (Jackson, Phelan, & Rossiter, 2000). In street virus-infected skunks, initial infection was present in the lumbar spinal cord and transit to the brain occurred via a variety of long ascending and descending fiber tracts, including rubrospinal, corticospinal, spinothalamic, spino-olivary, vestibulospinal/spinovestibular, reticulospinal/spinoreticular, cerebellospinal/spinocerebellar, and dorsal column pathways (Charlton, Casey, Wandeler, & Nadin-Davis, 1996).

6 SPREAD FROM THE CNS

Centrifugal spread or viral spread from the CNS to peripheral sites along neuronal routes is essential for transmission of rabies virus to its natural hosts. Salivary gland infection is necessary for the transfer of infectious oral fluids by rabid vectors. The salivary glands receive parasympathetic innervation by the facial (via the submandibular ganglion or Langley's ganglion in some animals) and glossopharyngeal (via the otic ganglion) nerves, sympathetic innervation via the superior (or cranial) cervical ganglion, and afferent (sensory) innervation (Emmelin, 1967). Unilateral excision of a portion of the lingual nerve and the cranial cervical ganglion of dogs and foxes resulted in very low viral titers in denervated salivary glands compared with contralateral salivary glands after street rabies virus infection (Dean et al., 1963). Evidence of widespread infection of salivary gland epithelial cells is a result of viral spread along multiple terminal axons rather than spread between epithelial cells (Charlton, Casey, & Campbell, 1983). Rabies virus antigen was found concentrated in the apical region of mucous acinar cells and ultrastructural studies showed that viral matrices were present in the basal region and there was viral budding on the apical plasma membrane into the acinar lumen and into the intercellular canaliculi and, occasionally, onto membranes of secretory granules (Balachandran & Charlton, 1994). Viral titers in salivary glands may be higher than in CNS tissues (Dierks, 1975).

In addition to salivary gland infection, evidence was found in a suckling hamster model of centrifugal spread involving the central, peripheral and autonomic nervous systems in many peripheral sites (Murphy, Harrison, Winn, & Bauer, 1973). Infection was observed in the ganglion cell layer of the retina and in corneal epithelial cells, which are innervated by sensory afferents via the trigeminal nerve. Epithelial cells in both superficial and deep layers of the cornea were found to be infected (Balachandran & Charlton, 1994). Detection of rabies virus antigen in corneal impression smears has been used as a diagnostic test for human rabies (Koch, Sagartz, Davidson, & Lawhaswasdi, 1975), and rabies virus has been transmitted by corneal transplantation in humans (see Chapter 7). Infection may be found in free sensory nerve endings of tactile hair in a skin biopsy, which is one of the best diagnostic methods of confirming an antemortem diagnosis of rabies in humans (see Chapters 7 and 11). Antigen may be demonstrated in small nerves around hair follicles or in epithelial cells of hair follicles in the skin, which is taken from the nape of the neck because it is rich in hair follicles. Widespread infection may be observed in sensory nerve end organs in the oral and nasal cavities, including the olfactory epithelium and taste buds in the tongue.

Studies in both natural and experimental rabies have demonstrated infection involving neurons in a variety of extraneural organs, including the adrenal medulla, cardiac ganglia, and plexuses in the luminal gastrointestinal tract, major salivary glands, liver and exocrine pancreas (Balachandran & Charlton, 1994; Debbie & Trimarchi, 1970; Jackson et al., 1999). In addition, there is infection involving a variety of non-neuronal cells, including acini in major salivary glands in rabies vectors, epithelium of the tongue, cardiac and skeletal muscle, hair follicles, and even pancreatic islets (Balachandran & Charlton, 1994; Debbie & Trimarchi, 1970; Jackson et al., 1999; Murphy et al., 1973). There are a few reports of myocarditis in human cases of rabies (Araujo, de Brito, & Machado, 1971; Cheetham, Hart, Coghill, & Fox, 1970; Ross & Armentrout, 1962).

7 ANIMAL MODELS OF RABIES VIRUS NEUROVIRULENCE

Viral neurovirulence can be defined as the capacity of a virus to cause disease of the nervous system, especially the CNS. Analysis of neurovirulence has frequently been approached in experimental models by comparing infections in a host with closely related viruses (e.g., different rabies virus strains or a parent rabies virus and a variant) (Jackson, 1991a). The ability of a virus to spread to the CNS from a peripheral site, or *neuroinvasiveness*, is an important component of neurovirulence after natural routes of viral entry. The route of inoculation is often very important in evaluating neurovirulence experimentally. Intracerebral inoculation is commonly used for convenience and a number of peripheral sites also have been used in different models, including footpad, intramuscular, intraperitoneal and intraocular inoculation. Species, age, and the immune status of the host have also proved to be important factors in neurovirulence (Flamand et al., 1984). Monoclonal antibody-resistant (MAR) variant viruses were selected *in vitro* from CVS and ERA laboratory strains of rabies virus with neutralizing anti-glycoprotein antibodies (Dietzschold et al., 1983; Seif, Coulon, Rollin, & Flamand, 1985). Mutations involving antigenic site III are located between amino acid residues 330 and 338 of the CVS and ERA glycoprotein. Variants with a single amino acid change at position 333, with loss of either arginine (Dietzschold et al., 1983; Seif et al., 1985) or lysine (Tuffereau et al., 1989), have been found to have diminished virulence in mice after intracerebral inoculation, whereas variants with amino acid changes at other positions remain neurovirulent. Comparisons of avirulent variants with their parent viruses in mouse and rat models using different routes of inoculation have been a useful approach in understanding the biological bases of rabies virus neurovirulence. Both MAR variants RV194-2 (Dietzschold et al., 1983) and Av01 (Coulon, Rollin, Aubert, & Flamand, 1982) have substitution of a glutamine for the arginine of CVS at position 333 of the glycoprotein.

Avirulent rabies virus variants, but not the parent CVS strain, have been shown to cause infection in extraneural sites close to the site of inoculation in different models. For example, Av01 infected the anterior epithelium of the lens after inoculation into the anterior chamber of the eye in rats (Kucera, Dolivo, Coulon, & Flamand, 1985) (Figure 8.3). Similarly, RV194-2 inoculated into the tongue of mice and rats produced local infection involving epithelial tissues, glandular cells, and muscles (Torres-Anjel, Montano-Hirose, Cazabon, Oakman, & Wiktor, 1984). In these models, the variant viruses demonstrated less restricted cellular tropism than the highly neuronotropic parental CVS strain.

Two independent studies in mice showed no impairment in neuroinvasiveness after peripheral inoculation of AvO1 or RV194-2 (Coulon et al.,

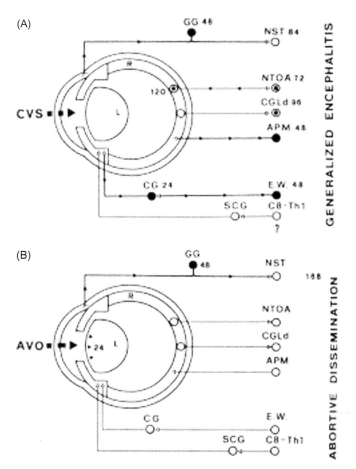

FIGURE 8.3 **Propagation of CVS (A) and avirulent variant rabies virus strain AVO (B) through the trigeminal (top), visual (center) and autonomic (bottom) interconnections between the eye and brain.** Symbols: Open arrows, direction of neurotransmission; closed arrows, direction of propagation of the virus; circles, peripheral and central neuronal somata infected primarily (closed), secondarily (dots), and not infected (open) at each interval of time, indicated in hours after inoculation. APM, area praetectalis medialis; C8-Th 1, spinal preganglionic sympathetic neurons; CG, ciliary ganglion; CGLd, lateral geniculate body (dorsal part); EW, Edinger-Westphal nucleus; GG, trigeminal (gasserian) ganglion; L, lens; NST, terminal trigeminal sensory nucleus; NTOA, terminal nuclei of the accessory optic system; R, retina; SCG, superior cervical sympathetic ganglion. *(Reproduced with permission from Kucera et al., Journal of Virology 55:158–162, 1985 Copyright © 1985, American Society for Microbiology.)*

1989; Jackson, 1991b). An excellent model was developed for studying the pathways of viral spread to the brain by inoculating rabies virus into the anterior chamber of the eye in rats (Kucera et al., 1985). There are six potential neural pathways for viral spread to occur between the eye and brain (see Figure 8.3). Rabies virus was localized in tissues using immunofluorescent staining. After inoculation of CVS, viral antigen was initially detected at 24 hours in the ipsilateral ciliary ganglion and later in the Edinger-Westphal nucleus of the oculomotor nerve (parasympathetic pathway). At 48 hours, virus also spread to the ipsilateral ganglion of the trigeminal nerve (an afferent sensory pathway) and to neurons of the contralateral area praetectalis medialis, which projects to the retina via preopticoretinal fibers. In contrast, AvO1 propagated in the trigeminal pathway but not in either parasympathetic or preopticoretinal fibers. Neurons in the trigeminal ganglion also were infected at 48 hours, indicating a similar rate of spread. Thus, avirulent AvO1 spreads to the brain in this model using more limited pathways than its virulent parent virus.

Intracerebral inoculation is a crude technique in which the inoculum spreads throughout the cerebrospinal fluid (CSF) spaces, including the ventricular system and subarachnoid space (Mims, 1960). A stereotaxic apparatus can deliver an inoculum into a precise location in the brain. AvO1 was surprisingly found to be neurovirulent after stereotaxic inoculation into the neostriatum or cerebellum of adult mice (Yang & Jackson, 1992), although AvO1 infected fewer neurons and deaths occurred later than after stereotaxic inoculation with CVS (Jackson, 1994). After inoculation of AvO1 into the striatum, the infection was widespread in the brain, and there were morphologic changes of apoptosis in neurons (A.C. Jackson, unpublished observations) and also infiltration with inflammatory cells. Serum neutralizing antibodies against rabies virus were produced later and at lower levels than after intracerebral inoculation. AvO1 is likely neurovirulent after stereotaxic brain inoculation because this route produces both a direct site of viral entry into the CNS and a low level of immune stimulation.

Since centrifugal spread of CVS is limited, comparisons of the spread of CVS and variants from the CNS have not been as useful as for comparisons of spread to the CNS and within the CNS. In the model of Kucera et al., (1985) using intraocular inoculation of rats, CVS spread from the nuclei of the accessory optic system to ganglionic cells of the retina in both eyes (see Figure 8.3), and AvO1 did not show evidence of centrifugal spread in this model.

8 STRUCTURAL DAMAGE CAUSED BY RABIES VIRUS INFECTION IN THE CNS

Despite the dramatic and severe clinical neurological signs in rabies, the neuropathological findings are usually quite mild, especially under natural conditions (see Chapter 9). Yet, experimental infection with fixed viruses, on the other hand, induces the expression of innate immune molecules, extensive inflammatory cells into the CNS, and apoptosis in

laboratory animals (Sarmento et al., 2005). The infiltration of inflammatory cells and induction of apoptosis correlates with the enhancement of
blood-brain-barrier (BBB) permeability and the attenuation of the virus
(Kuang et al., 2009). It is thus hypothesized that induction of the innate
immune responses is one of the important mechanisms of rabies virus
attenuation (Kuang, Lackay, Zhao, & Fu, 2009; Wang et al., 2005). These
studies are summarized below.

8.1 Innate Immune Responses

The innate immune system, also known as non-specific immune system, provides immediate defense against infections. Although the brain
was thought traditionally as an immune-privileged site, resident CNS
cells, including microglia, astrocytes, and neurons, are now known to
be capable of initiating innate immune responses (Carson, 2002; Reiss,
Chesler, Hodges, Ireland, & Chen, 2002). Within the brain, these innate
responses are critical in establishing protective immunity, and the
defenses mounted by these cell types are the first to engage and counter viruses or other infectious agents. Innate immune responses also
recruit leukocytes into the CNS and establish a microenvironment that
can potentially direct the activity of infiltrating cells. The induction of
innate immune gene expression has been reported in mice infected with
fixed viruses, and it was found by using RT-PCR that IL-6, IFN-γ, and
TNF-α were up-regulated in the CNS of mice at 4–6 days after infection
with CVS-F3 (Phares et al., 2006). Using Affymetrix microarrays, Prehaud
et al (2005) found that infection with fixed rabies virus CVS strain
induced the expression of innate immune response genes in a human
postmitotic neuron-derivative cell line, NT2-N (Prehaud et al., 2005).
These genes include beta interferon (IFN-β), chemokines (CCL-5, CXCL-
10), and inflammatory cytokines (IL-6, TNF-α, IL-α). The same virus
(CVS) induced the expression of Toll-like receptors (TLR) (McKimmie
et al., 2005), IFN-β, IL-6 and Mx1 (Johnson et al., 2006) in the CNS of
mice after either intracerebral or intramuscular inoculation. However,
many of the innate immune genes (TLRs, chemokines, and cytokines)
were up-regulated only in the CNS of mice infected with fixed virus
(CVS-B2c), but not in mice infected with street rabies virus (SHBRV,
a virus derived from silver-haired bats) when the gene expression was
analyzed with Affymetrix microarrays (Wang et al., 2005). RT-PCR
confirmed that indeed many of these genes are up-regulated in mice
infected with fixed virus, but not in mice infected with street rabies
virus. Furthermore, the protein levels for some of the chemokines and
cytokines were also found to be increased in the CNS of mice infected
with fixed virus (B2c), not in mice infected with street virus (DRV, a
virus derived from a Mexican dog) (Kuang et al., 2009). These studies

indicate that fixed rabies virus induces, while street rabies virus evades, the innate immune responses (Wang et al., 2005). Induction of innate immune responses leads to the extensive infiltration of inflammatory cells into the CNS of mice infected with fixed rabies virus, while infiltration of inflammatory cells is scarce in the CNS of mice infected with street rabies virus (Kuang et al., 2009, Wang et al., 2005).

8.2 Apoptosis

Apoptosis is a process by which cells undergo physiologic cell death in response to diverse stimuli. It is a normal process in embryonic development, maturation of the immune system and in normal tissue turnover (Buja, Eigenbrodt, & Eigenbrodt, 1993; Thompson, 1995). Morphologically, apoptosis is characterized by nuclear and cytoplasmic condensation of single parenchymal cells followed by fragmentation of the nuclear chromatin and the subsequent formation of multiple fragments of condensed nuclear material and cytoplasm (Buja et al., 1993). Phagocytosis of this material occurs, although an inflammatory reaction, is normally absent. In contrast, cellular death due to necrosis is characterized by preservation of cell outlines and there is variable swelling of the cell and its organelles. Cellular fragmentation occurs as a late event in necrosis. There are derangements in energy and substrate metabolism in necrosis that result in breaks in the plasma membrane and organellar membranes. Apoptosis, on the other hand, is associated with endonuclease-mediated cleavage of the DNA of nuclear chromatin, resulting in DNA fragments with sizes in multiples of a single nucleosome length (180 base pairs). The internucleosomal cleavage of the DNA in apoptosis results in a "ladder" appearance of the DNA upon agarose gel electrophoresis, whereas in necrosis there is less specific degradation of DNA into a "smear" containing fragments of various sizes following electrophoresis.

Apoptotic cell death likely plays an important pathogenetic role in a wide variety of viral infections, including those produced by a large number of RNA and DNA viruses and apoptosis occurs in the CNS of humans and experimental animals in many of these infections (Allsopp & Fazakerley, 2000; Hardwick, 1997; Roulston, Marcellus, & Branton, 1999). Strong evidence of apoptotic cell death was found in both cultured cells and neurons in experimental mouse rabies models infected by intracerebral inoculation of fixed rabies virus strains (Jackson, 1999; Jackson & Park, 1998; Jackson & Rossiter, 1997). *In vitro* studies using CVS infected cultured rat prostatic adenocarcinoma (AT3) cells showed striking morphologic changes, revealing apoptosis, at the levels of both light and electron microscopy, whereas AT3 cells transfected with the *bcl*-2 gene (an anti-apoptosis gene) did not demonstrate apoptotic changes

(Jackson & Rossiter, 1997). Terminal deoxynucleotidyltransferase-mediated dUTP-digoxigenin nick end labeling (TUNEL) staining was also demonstrated in infected AT3 cells, indicating evidence of oligonucleosomal DNA fragmentation typical of apoptosis. In addition, *in vitro* infection of mouse neuroblastoma (N18) cells with CVS was associated with apoptosis (Theerasurakarn & Ubol, 1998). *In vitro* studies have also shown that the ERA strain of fixed rabies virus replicates and induces apoptosis in mouse spleen lymphocytes and the human T-lymphocyte cell line Jurkat (Thoulouze, Lafage, Montano-Hirose, & Lafon, 1997) and that cell death was concomitant with expression of the viral glycoprotein. Whereas CVS induces apoptosis in mouse embryonic hippocampal neurons, the extent of apoptosis and pathogenicity was studied in primary neuron cultures infected with two stable variants of CVS-24, CVS-B2c and CVS-N2c (Morimoto, Hooper, Spitsin, Koprowski, & Dietzschold, 1999). It was found that the extent of apoptosis in adult mice was actually lower in primary neuron cultures infected with the more pathogenic variant CVS-N2c than with the less pathogenic variant CVS-B2c, indicating an inverse relationship between and pathogenicity and apoptosis. Guigoni and Coulon (Guigoni & Coulon, 2002) observed that primary cultures of CVS-infected purified rat spinal motoneurons did not show major evidence of apoptosis over a period of 7 days, while infected purified hippocampal neurons showed apoptosis in over 90% of neurons within 3 days, indicating that different neuronal cell types respond differently to rabies virus infection. CVS and Pasteur virus (PV) strains induce only limited apoptosis whereas two vaccine strains, Evelyn-Rokitnicki-Abelseth (ERA) and SN-10, induce strong apoptosis in the human neuroblastoma SK-N-SH cell line and in lymphoblastoid Jurkat cells (Baloul & Lafon, 2003; Lay, Prehaud, Dietzschold, & Lafon, 2003; Prehaud, Lay, Dietzschold, & Lafon, 2003; Thoulouze et al., 1997, 2003). Hence, there are both virus-dependent and cell-dependent mechanisms for induction of apoptosis. Furthermore, rabies virus-induced apoptosis is activated by caspase-dependent and caspase-independent pathways (Sarmento, Tseggai, Dhingra, & Fu, 2006; Thoulouze et al., 2003). There is activation of caspase 8 and caspase 3, but not caspase 9, and poly ADP-ribose polymerase (PARP) is cleaved, confirming activation of downstream caspases and involvement of the extrinsic apoptotic pathway (Kassis, Larrous, Estaquier, & Bourhy, 2004; Sarmento et al., 2006; Ubol, Sukwattanapan, & Utaisincharoen, 1998). Apoptosis-inducing factor is a pro-apoptotic signal transducing molecule that was shown in infection to be up-regulated and translocated from the cytoplasm to the nucleus, where it binds to DNA and provokes chromatin condensation, indicating activation of a caspase-independent pathway (Sarmento et al., 2006).

In adult mice infected intracerebrally with CVS, specific morphologic changes associated with apoptosis were observed in neurons, particularly in pyramidal neurons of the hippocampus and cortical neurons, and there was positive TUNEL staining in the same regions (Jackson & Rossiter, 1997) (Figure 8.4). Double-labeling studies indicated that infected neurons actually underwent apoptosis. However, not all infected neurons (e.g., Purkinje cells) demonstrated these morphologic features of apoptosis or positive TUNEL staining. Increased expression of the pro-apoptotic Bax protein was observed in neurons in areas where apoptosis was prominent (Jackson & Rossiter, 1997). Studies in *bax*-deficient mice showed that neuronal apoptosis was less marked with similar clinical disease as in wild-type littermates, indicating that the Bax protein plays an important role in modulating rabies virus-induced apoptosis under specific experimental conditions (Jackson, 1999).

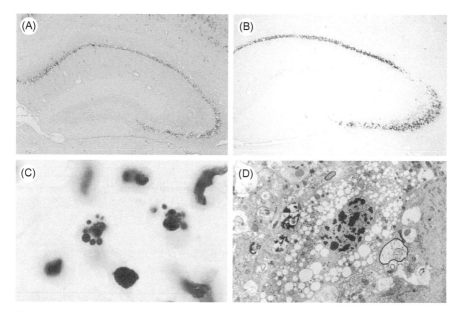

FIGURE 8.4   (A) Immunostaining for rabies virus antigen in the hippocampus of a mouse 7 days after intracerebral inoculation with CVS showing antigen in pyramidal neurons and in cortical neurons; neurons in the dentate gyrus do not demonstrate staining. (B) TUNEL staining in the hippocampus of a mouse 7 days after intracerebral inoculation with CVS showing marked staining is present in pyramidal neurons but not in neurons of the dentate gyrus (B). Note the similarity in the distribution of TUNEL staining in (B) and rabies virus antigen in (A). (C) Neurons in the cerebral cortex 8 days after inoculation with CVS showing multiple condensations of nuclear chromatin in two cells. (D) Hippocampal pyramidal neuron showing a pattern of irregular chromatin condensation and marked cytoplasmic vacuolation. (A: immunoperoxidase-hematoxylin; B: TUNEL staining; C: cresyl violet staining; D: transmission electron microscopy; magnifications: A, B, ×20; C, ×1050; D, ×1450. (*Adapted with permission from Jackson and Rossiter, Journal of Virology* **71**,*5603–5607, 1997 Copyright © 1997, American Society for Microbiology.*)

Both CVS- and SAD-L16 (a vaccine strain based on SAD-B19)-infected suckling mice show widespread and severe morphologic changes of apoptosis with positive TUNEL staining and activation of caspase 3, a downstream caspase, after intracerebral and peripheral routes of inoculation (Jackson & Park, 1998; Rasalingam, Rossiter, & Jackson, 2005; Rasalingam et al., 2005) (Figure 8.5). In suckling mice infected with CVS via intracerebral inoculation, uninfected neurons in the external granular layer of the cerebellum also underwent apoptosis (Figure 8.5A) despite the absence of rabies virus antigen, likely due to indirect mechanisms. The role of the adaptive immune response in producing neuronal apoptosis with intracerebral inoculation was evaluated by comparing the infections in adult C57BL/6J mice with nude mice (T cell deficient) and *Rag1* mice (T and B cell deficient) (Rutherford & Jackson, 2004). Both strains of immunodeficient mice showed very similar clinical disease and neuropathological findings, including marked neuronal apoptosis, indicating that the adaptive immune response is unlikely to be of fundamental importance in producing neuronal apoptosis in this model.

Apoptosis in infected cultured cells, including embryonic cells, does not closely correspond to what is observed in infected animals. Animals peripherally inoculated with CVS strains do not show the prominent apoptosis that is observed in neurons after intracerebral inoculation (Jackson, 2003; Reid & Jackson, 2001). After intracerebral inoculation of mice with SHBRV, an important bat rabies virus variant, significant neuronal apoptosis was not observed in the brain (Sarmento, Li, Howerth, Jackson, & Fu, 2005; Yan et al., 2001), in contrast to observations with fixed (attenuated) strains. Following a low dose of CVS-B2c inoculated intramuscularly into mice, neuronal apoptosis in the spinal cord was associated with failure of the infection to spread to the brain and produce neurological disease, whereas in the infection with the SHBRV, apoptosis was not induced in the spinal cord and spread occurred to the brain (Sarmento et al., 2005). Neonatal mice, on the other hand, peripherally inoculated with SAD-L16 virus, were compared with mice infected with the less virulent SAD-D29 virus, which has an attenuating mutation at position 333 of the glycoprotein. The less virulent SAD-D29 virus actually induced more neuronal apoptosis in the brainstem and cerebellum than SAD-L16 virus (Jackson, Rasalingam, & Weli, 2006), indicating that the inverse relationship between pathogenicity and apoptosis applies *in vivo* in the CNS as well as *in vitro*.

In rabies virus infection, there are complex mechanisms involved in the ensuing cell death or survival of neurons, both *in vitro* and in animal models using different viral strains and routes of inoculation. Both *in vitro* and *in vivo* observations demonstrate that apoptosis may be a protective rather than a pathogenic mechanism in rabies virus infections because the less pathogenic viruses induce more apoptosis than the more pathogenic viruses *in vitro* and also *in vivo* using peripheral routes

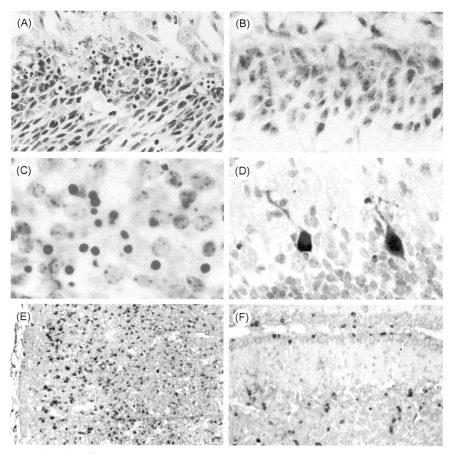

FIGURE 8.5   **Brain sections after intracerebral inoculation with CVS-11 of 6-day-old mice (A–C) and of 7-day-old mice with L16 (D–F).** (A) Nuclear chromatin condensations in multiple cells in the external granular layer of the cerebellum in a CVS-infected suckling mouse. (B) External granular layer of the cerebellum of an uninfected mouse of the same age showing the absence of typical apoptotic morphology. (C) Multiple neurons in the dentate gyrus of the hippocampus of a CVS-infected suckling mouse showing chromatin condensations involving entire nuclei. (D) Immunoperoxidase staining for activated caspase-3 and (E, F) TUNEL staining in L16–infected mouse brains. (D) Activated caspase 3 staining is present in Purkinje cells of the cerebellum 4 days p.i. and (E) TUNEL staining is present in in many neurons in the cerebral cortex and (F) in neurons in the cerebellar external and internal granular layers 4 and 6 days p.i., respectively. A–C: cresyl violet staining; D: caspase 3 immunostaining; E, F: TUNEL staining – methyl green; magnifications: A, ×85; B, ×95; C, ×235; D, ×510; E, ×195; F, ×125.). *A–C, Adapted from Jackson and Park, in Acta Neuropathologica **95**, 159–164, 1998 with kind permission of Springer Science and Business Media, D–F, Adapted with permission from Rasalingam, Rossiter and Jackson, Canadian Journal of Veterinary Research **69**, 100–105, 2005) Copyright © 2005, Canadian Veterinary Medical Association.*

of inoculation (Jackson et al., 2006; Morimoto et al., 1999; Prehaud et al., 2003; Sarmento et al., 2005; Yan et al., 2001).

There is a report demonstrating apoptosis in a single human rabies case (Adle-Biassette et al., 1996), but morphologic evidence of neuronal apoptosis has generally not been prominent in natural rabies in humans or animals. Juntrakul, Ruangvejvorachai, Shuangshoti, Wacharapluesadee, & Hemachudha (2005) reported that TUNEL positive cells were found throughout the neuroaxis in seven cases of human rabies, but this may have been due to non-specific staining. Also, morphologic evidence of neuronal apoptosis was not assessed or illustrated in this report. Jackson et al. did not find evidence of neuronal apoptosis in 12 human rabies cases using histological analysis, TUNEL staining, and staining for cleaved caspase-3 (Jackson, Randle, Lawrance, & Rossiter, 2008) (see Chapter 9).

8.3 Degeneration of Neuronal Processes

Scott et al. have recently comprehensively evaluated CVS-infection in adult transgenic mice expressing the yellow fluorescent protein (YFP; H clone) using hindlimb foot pad inoculation of CVS (Scott, Rossiter, Andrew, & Jackson, 2008). In these mice, YFP expression is driven in a subpopulation of neurons using the *thy1* vector, and there are strong fluorescent signals in dendrites, axons, and presynaptic nerve terminals (Feng et al., 2000). Conventional histopathology showed mild inflammatory changes without significant degenerative neuronal changes, but at late clinical time points with the development of severe clinical neurological disease, fluorescence microscopy showed marked abnormalities, especially beading and/or swelling, in dendrites and axons of layer V cortical pyramidal neurons, severe involvement of axons in the brainstem and the inferior cerebellar peduncle, and severe abnormalities affecting axons of cerebellar mossy fibers (Figure 8.6). The structural changes take a few days to develop, likely because they are mediated by abnormal axoplasmic structural protein function. Toluidine blue-stained resin sections and electron microscopy showed vacuolation in cortical neurons that corresponded to swollen mitochondria, and vacuolation in the neuropil of the cerebral cortex. Axonal swellings, key markers of axonal degeneration, were observed. Vacuolation was also observed in axons and in pre-synaptic nerve endings. These morphological changes are sufficient to explain the severe clinical disease with a fatal outcome.

The morphologic changes in axons have a striking similarity to the neurodegenerative changes that occur in diabetic sensory and autonomic neuropathy, in which a key feature is the presence of axonal swellings that are composed of accumulations of mitochondria and cytoskeletal proteins (e.g., neurofilaments) (Lauria et al., 2003; Schmidt et al., 1997). Diabetes-induced oxidative stress in sensory neurons and peripheral

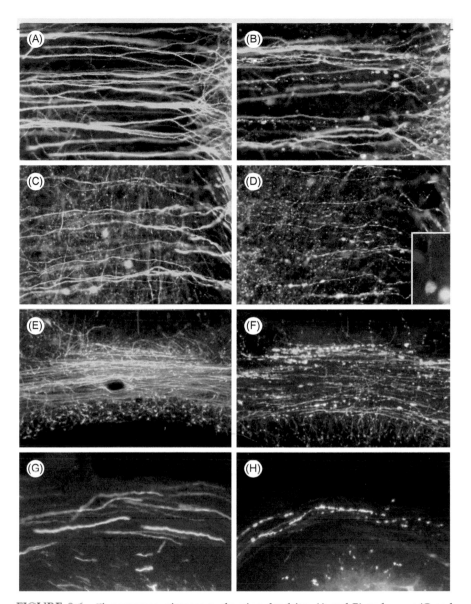

FIGURE 8.6 Fluorescence microscopy showing dendrites (A and B) and axons (C and D) of layer V pyramidal neurons in the cerebral cortex of mock-infected (A and C) and moribund CVS-infected (B, D, and D inset) YFP mice. In infected mice, beading is observed in a minority of dendrites (B), while more axons are involved (D). There are no abnormalities in the dendrites (A) or axons (C) of mock-infected mice. Axons in mock-infected mice are slightly varicose (C), which is characteristic of these fibers. Fluorescence microscopy shows rabies virus antigen (red) in the perikaryon and dendrite of a YFP-expressing neuron (D inset). Morphology of the cerebellar mossy fibers of mock-infected (E) and moribund CVS-infected YFP mice (F). Mossy fiber axons in the cerebellar commissure of moribund mice show severe beading (F), whereas no abnormalities were observed in mock-infected mice (E). Axons in the inferior cerebellar peduncles are normal in mock-infected mice (G) and show marked beading in CVS-infected moribund mice (H). A–D, ×280; D inset, ×265; E, F ×95; G, H, ×230. *Adapted with permission from Scott et al. in Journal of Virology 82:513–521, 2008; doi: 10.1128/JVI.01677-07 Copyright © 2008, American Society for Microbiology.*

nerves is demonstrated by increased production of reactive oxygen species (Nishikawa et al., 2000; Russell et al., 2002), lipid peroxidation (Obrosova et al., 2002), and protein nitrosylation (Obrosova et al., 2005).

8.4 Blood-Brain Barrier (BBB)

The BBB is a separation of circulating blood from the brain extracellular fluid in the CNS. It occurs along all capillaries and consists of tight junctions around the capillaries (Hamilton, Foss, & Leach, 2007). Endothelial cells restrict the diffusion of microscopic objects and large or hydrophilic molecules into the cerebrospinal fluid (CSF), while allowing the diffusion of small hydrophobic molecules (O_2, CO_2, hormones) (Gloor et al., 2001). Cells of the barrier actively transport metabolic products such as glucose across the barrier with specific proteins. The loss of BBB integrity during CNS infection and autoimmunity has generally been associated with the development of neurological signs. However, it was found that infection of attenuated rabies virus CVS-F3 increased BBB permeability and CNS inflammation in the absence of neurological sequelae, leading to virus clearance from the CNS (Phares, Kean, Mikheeva, & Hooper, 2006). The loss of BBB integrity is associated with the expression of several chemokines/cytokines and the accumulation of CD4- and CD19-positive cells in the CNS, particularly in the cerebellum. It was further demonstrated that street rabies virus (for example, SHBRV) can induce a strong virus-specific immune response in the periphery, but unable to enhance the BBB permeability. As a consequence, immune effectors cannot be delivered into the CNS, leading to the death of the infected animals (Roy & Hooper, 2007). In this study, the authors used PLSJL mice that are less susceptible to SHBRV infection than the 129/SvEv mice, largely due to the elevated capacity of PLSJL mice to mediate BBB permeability changes in response to the infection. Treatment of the SHBRV-infected mice with the steroid hormone dehydroepiandrosterone (DHEA) reduced the BBB permeability, resulting in increased mortality. On the other hand, immunization of SHBRV-infected mice with myelin basic protein (MBP) that induces extensive BBB permeability and CNS inflammation results in greater virus clearance and improved survival (Roy & Hooper, 2007; Spitsin et al., 2008). Subsequent studies have demonstrated that only infection with fixed viruses CVS-F3, CVS-B2c, HEP) leads to the enhancement of BBB permeability while infection with street rabies viruses (SHBRV, DRV) does not (Kuang et al., 2009; Roy & Hooper, 2008; Zhao, Toriumi, Kuang, Chen, & Fu, 2009). Fixed rabies viruses not only can lead to clearance of fixed viruses in the CNS but also can clear street rabies viruses from the CNS (Faber et al., 2009; Li et al., 2012; Wang et al., 2011). It has been demonstrated that these fixed viruses enhance the BBB permeability, allowing primed B cells to enter into the CNS and producing virus-neutralizing Ab *in situ* to clear rabies virus from the CNS (Hooper, Phares, Fabis, & Roy, 2009; Roy & Hooper, 2007).

Overall, these studies demonstrate that induction of innate immunity is one of the important mechanisms for rabies virus attenuation. Infection with small doses of fixed rabies virus can induce innate immune responses including the expression of innate immune genes, infiltration of inflammatory cells, induction of apoptosis, and enhancement of BBB permeability. It also indicates that fixed rabies virus can induce neurological diseases via immune-mediated pathogenesis particularly when infected with large doses (Sarmento et al., 2005). On the other hand, evasion of innate immunity by street rabies virus is one of the pathogeneic mechanisms for rabies. However, what leads to the death of the infected individuals is not entirely clear at the moment although neuronal dysfunction has been proposed as the cause of rabies (Tsiang 1982).

9 BRAIN DYSFUNCTION IN RABIES

The dramatic and severe clinical neurological signs in rabies with only mild neuropathological findings under natural conditions (see Chapter 9) led to the hypothesis that rabies results from neuronal dysfunction rather than structural damage (Tsiang 1982). Many experimental studies have been performed to gain an understanding of the bases of this neuronal dysfunction. Although no fundamental underlying defect has been identified to explain this dysfunction, major areas of research in this area will be summarized.

9.1 Neuropeptide Synthesis during Rabies

Although studies performed *in vitro* have shown that rabies virus has little or no inhibitory effect on cellular RNA and protein synthesis (Ermine & Flamand, 1977; Madore & England, 1977; Tuffereau & Martinet-Edelist, 1985), *in vivo* studies using CVS-24–infected rats showed that there was progressive reduction in the expression of the noninducible housekeeping gene that encodes glyceraldehyde-3-phosphate dehydrogenase and the late response gene that encodes pro-enkephalin, possibly due to the global suppression of cellular protein synthesis related to extensive synthesis of rabies virus mRNA (Fu et al., 1993). This occurred in association with induction of immediate-early-response genes (*erg-1*, *junB*, and *c-fos*) in the hippocampus and cerebral cortex, where there was colocalization of expression of these genes with viral mRNA expression. In another study, infection of mice with CVS-N2c resulted in down-regulation of about 90% of genes in the normal brain at more than fourfold lower levels by using subtraction hybridization (Prosniak, Hooper, Dietzschold, & Koprowski, 2001). Only about 1.4% of genes became up-regulated, including genes involved in regulation of cell metabolism, protein synthesis and growth and differentiation. However, Weihe et al., (2008) reported that rabies virus infection of mice caused a strong induction of calcitonin gene-related peptide (CGRP), vasoactive intestinal peptide (VIP), and somatostatin. Surprisingly, the induction of these peptides even occurred in neurons that are not infected with rabies virus, for example, neurons in the dentate gyrus. It has been proposed that the strong RABV-induced up-regulation of CGRP in the brain may be associated with immune evasion because CGRP can have an inhibitory effect on antigen presentation. It is unknown whether the expression of neuropeptides is affected in natural rabies.

9.2 Defective Neurotransmission

9.2.1 Acetylcholine

A hypothesis that defective cholinergic neurotransmission might be the basis for neuronal dysfunction in rabies led to the investigation of specific binding to muscarinic acetylcholine receptors in CVS peripherally infected rat brains. ^{3}H-labeled antagonist, quinuclidinyl benzylate (QNB) was used as an indication of defective neurotransmission (Tsiang, 1982). Binding of ^{3}H-labeled QNB to AChRs in infected brain homogenates was decreased by 96 hours after infection compared with controls and the binding was markedly decreased at 120 hours, 10–20 hours before death was expected to occur. The greatest reduction in binding was found in the hippocampus and smaller reductions were observed in the cerebral cortex and in the caudate nucleus.

When cholinergic neurotransmission was examined in mice infected intracerebrally with CVS and compared with mock-infected control mice, the enzymatic activities of choline acetyltransferase and acetylcholinesterase, which are required for the synthesis and degradation of acetylcholine, respectively, were similar in the cerebral cortex and hippocampus of moribund CVS-infected and control mice (Jackson, 1993). In contrast to the findings in infected rats, QNB binding to muscarinic acetylcholine receptors, which was assessed with ^{3}H-labeled QNB using Scatchard plots, was not significantly different in the cerebral cortex or hippocampus of CVS-infected and uninfected control mice. These findings cast doubt on the importance of rabies virus binding to muscarinic acetylcholine receptors in the brain. However, it is possible that differences in the species (mouse versus rat) or in the route of inoculation (peripheral versus intracerebral) account for the differences in the results of the two studies.

In naturally infected rabid dogs, specific binding of ^{3}H-labeled QNB was reduced in the hippocampus (35%) and in the brainstem (27%), but not in other brain regions, compared with uninfected control dogs (Dumrongphol, Srikiatkhachorn, Hemachudha, Kotchabhakdi, & Govitrapong, 1996). The results were similar whether the clinical disease was of the furious or dumb form. K_d values were increased, indicating a decrease in receptor affinity, and B_{max} values, reflecting receptor content, were unchanged in rabid dogs. Curiously, increased K_d values were found to be similar in the hippocampus whether or not rabies virus antigen was detectable at that site. These findings argue against alteration of muscarinic receptor binding as a specific consequence of rabies virus infection of neurons. They suggest an unknown indirect mechanism for altered receptor affinity that is not related to clinical manifestations of disease or the local viral load.

9.2.2 Serotonin

Insofar as defective neurotransmission involving other neurotransmitters could be important in the pathogenesis of rabies, the role of serotonin has been examined with great interest. Serotonin has a wide distribution in the brain, and it is important in the control of sleep and wakefulness, pain perception, memory, and a variety of behaviors (Julius, 1991). Alterations of sleep stages have been recognized in experimental rabies in mice (see Section 8.3). Again, ligand binding to serotonin (5-HT) receptor subtypes was studied in the brains of CVS-infected rats (Ceccaldi, Fillion, Ermine, Tsiang, & Fillion, 1993). In this case, binding to 5-HT$_1$ receptor sites using [^{3}H] 5 HT was not affected in the hippocampus, but there was a marked decrease in B_{max} in the cerebral cortex 5 days after inoculation of CVS into the masseter muscles. In the presence of drugs that mask 5-HT$_{1A}$, 5-HT$_{1B}$, and 5-HT$_{1C}$ receptors, [^{3}H] 5-HT binding was reduced by 50% in the cerebral cortex 3 days after inoculation, whereas binding of ligands specific for 5-HT$_{1A}$ and 5-HT$_{1B}$ receptor sites was not affected. These results indicate that rabies virus infection must affect other 5-HT receptors in the cerebral cortex. Furthermore, the reduced binding was demonstrated before rabies virus antigen was detected in the cerebral cortex. Hence, the effect of rabies virus on receptor binding is unlikely, due to either direct or indirect effects of viral replication in cortical neurons. There are important serotonergic projections from the dorsal raphe nuclei in the brainstem to the cerebral cortex, and early infection of the midbrain raphe nuclei in experimental rabies in skunks has been documented (Smart & Charlton, 1992). Is it possible that the reduced binding of serotonin to the 5-HT receptors is an indirect effect of the infection at non-cortical sites by unknown mechanisms? Or is it part of a physiological response to the stress produced by the infection? In support of impaired serotonergic neurotransmission in rabies, potassium-evoked release of [^{3}H] 5-HT labeled synaptosomes from the cerebral cortex of CVS-infected rats was decreased 31% compared with controls (Bouzamondo, Ladogana, & Tsiang, 1993). Hence, there is evidence of both impaired release and impaired binding of serotonin, possibly playing an important role in producing the neuronal dysfunction in rabies.

9.2.3 γ-Amino-n-Butyric Acid

Impairments of both release and uptake of γ-amino-n-butyric acid (GABA) have been found in CVS-infected primary rat cortical neuronal cultures (Ladogana, Bouzamondo, Pocchiari, & Tsiang, 1994). A 45% reduction of [^{3}H] GABA uptake was found 3 days after infection, which coincided with the time of peak viral growth in the cultures. Kinetic analysis revealed major reductions in V_{max}, indicating a decrease in the number of fully active GABA transport sites. There were no significant changes in K_m in infected cultures in comparison to controls, reflecting the affinity of the GABA transport system for its substrate. Potassium- and veratridine-induced [^{3}H] GABA release was increased in infected cultures by 98 and 35%, respectively, compared with controls. The importance of these abnormalities in both the uptake and release of GABA on rabies pathogenesis *in vivo* has yet to be determined.

9.3 Electrophysiological Alterations

In addition to effects on neurotransmission, viruses may have important effects on the electrophysiological properties of neurons. Electroencephalographic (EEG) recordings of mice infected with CVS showed that the initial changes were alterations of sleep stages, including the disappearance of rapid-eye-movement (REM) sleep and the development of pseudoperiodic facial myoclonus (Gourmelon, Briet, Court, & Tsiang, 1986). Later, there was a generalized slowing of the EEG recordings (at 2–4 cycles per second). Terminally, there was an extinction of hippocampal slow activity with flattening of cortical activity. Brain electrical activity terminated about 30 minutes before cardiac arrest, indicating that cerebral death in experimental rabies occurs prior to failure of vegetative functions. Street virus-infected mice showed progressive disappearance of all sleep stages with a concomitant increase in the duration of waking stages (indicating insomnia), and these changes occurred before the development of clinical signs of rabies (Gourmelon, Briet, Clarencon, Court, & Tsiang, 1991). There was an absence of EEG abnormalities in street virus-infected mice that lasted through the preagonal phase of the disease. Since pathologic changes are more marked in neurons infected with fixed rabies viruses than street rabies virus strains, these observations are consistent with the idea that functional impairment of brain neurons is much more important in street rabies virus infection than in infection with fixed rabies virus strains.

9.4 Ion Channels

Defective neurotransmission is not the only potential explanation for functional impairment of neurons in rabies. Viral infections might also have important effects on ion channels of neurons. Studies were performed *in vitro* using rabies virus (RC-HL strain) infection of mouse neuroblastoma NA cells and the whole-cell patch clamp technique (Iwata, Komori, Unno, Minamoto, & Ohashi, 1999). The infection reduced the functional expression of voltage-dependent sodium channels and inward rectifier potassium channels, and there was a decreased resting membrane potential reflecting membrane depolarization. There was no change in the expression of delayed rectifier potassium channels, indicating that nonselective dysfunction of ion channels had not occurred. The reduction in the number of sodium channels and inward rectifier potassium channels could prevent infected neurons from firing action potentials and generating synaptic potentials, resulting in functional impairment.

Rabies virus (RC-HL strain) infection of NG108-15 cells *in vitro* was not found to alter the functional expression of voltage-dependent calcium ion channels (Iwata, Unno, Minamoto, Ohashi, & Komori, 2000). NG108-15 cells express both α_2-adrenoreceptors and muscarinic receptors. Induced voltage-dependent calcium ion channel current inhibition with noradrenaline (for α_2-adrenoreceptors) was decreased significantly in rabies virus infection, whereas carbachol (for muscarinic receptors) inhibition remained unchanged. Since α_2-adrenoreceptor-mediated inhibition of voltage-dependent calcium ion current serves as a brake mechanism to keep neurons from releasing their neurotransmitters beyond physiological requirements, the impaired modulation by α_2-adrenoreceptors could possibly contribute to clinical features of rabies, including hyperexcitability and aggressive behavior (Iwata et al., 2000).

9.5 Nitric Oxide

Nitric oxide (NO) is a short-lived gaseous radical that acts as a biologic mediator for diverse cell types. It is produced by many different cells and mediates a variety of functions, including vasodilation, neurotransmission, immune cytotoxicity, production of synaptic plasticity in the brain and neurotoxicity (Lowenstein, Dinerman, & Snyder, 1994; Nathan, 1992). NO is released by the enzyme nitric oxide synthase (NOS), which also produces other reactive oxides of nitrogen (Nathan, 1992). There are three isoforms of NOS: neuronal NOS (nNOS, also NOS-1), inducible NOS (iNOS, also NOS-2) and endothelial NOS (eNOS, also NOS-3). nNOS is constitutively expressed and inducible by cytokines, including IFN-χ, TNF-α and IL-12, whereas iNOS is inducible with lipopolysaccharides, IFN-χ and TNF-α.

NO plays a variety of roles in different viral infections (Reiss & Komatsu, 1998). In some viral infections (e.g., with Sindbis virus), inhibition of NOS results in increased mortality of infected mice, suggesting that NO plays a protective role in the pathogenesis of the viral infection (Tucker, Griffin, Choi, Bui, & Wesselingh, 1996). During infection with vesicular stomatitis virus (a rhabdovirus), NO has been shown to inhibit viral replication and promote viral clearance and recovery of infected mice (Komatsu, Bi, & Reiss, 1996).

Induction of iNOS mRNA occurred in mice infected experimentally with street rabies virus (Koprowski et al., 1993). iNOS mRNA was detected using RT-PCR amplification in the brains of three of six paralyzed mice, 9–14 days after inoculation of rabies virus in the masseter muscle. iNOS mRNA expression was induced rapidly in the brains of the rabid mice. It was speculated that NO and/or other endogenous neurotoxins may mediate the neuronal dysfunction in rabies and other infectious diseases (Koprowski et al., 1993; Zheng et al., 1993). The onset of clinical signs in rabies virus-infected rats and the clinical progression of the disease correlated with increasing quantities of NO in the brain to levels up to 30-fold more than in controls, which was determined using spin trapping of NO and electron paramagnetic resonance spectroscopy (Hooper et al., 1995). iNOS was detected by immunostaining in CVS-infected rats in many cells throughout the brain near blood vessels, which were identified as microglia and macrophages (Van Dam et al., 1995). CVS-24-infected rats developed a reduction in nNOS activity with reductions in nNOS mRNA and nNOS immunoreactivity and an increase in iNOS activity in the brain in a time-dependent manner (Akaike et al., 1995). Choline acetyltransferase activity in the brain remained unchanged, indicating that the decrease in nNOS activity did not reflect generalized neuronal loss. The NO produced by macrophages may be neurotoxic because its reaction with the superoxide anion O_2^- leads to the formation of peroxynitrate, which is a reactive oxidizing agent capable of causing tissue damage (Akaike et al., 1995). Ubol, Sukwattanapan, & Maneerat (2001) found that mice treated with the iNOS inhibitor, aminoguanidine (AG), delayed the death of CVS-11-infected mice by 1.0 to 1.6 days (depending on the dose). A delay in rabies virus replication was observed in the AG-treated mice. The role of NO in rabies pathogenesis clearly needs further study because it exerts both beneficial and detrimental effects and because complex mechanisms are likely involved.

9.6 Excitotoxicity

Excitatory amino acids (e.g., glutamate) have been recognized to play a role in neuronal injury in a variety of neurological diseases, including stroke, epilepsy, and neurodegenerative disorders. Recently, there is evidence that neurotropic viruses, including human immunodeficiency virus (Kaul & Lipton, 2004; Nath et al., 2000) and Sindbis virus (Darman et al., 2004; Nargi-Aizenman & Griffin, 2001; Nargi-Aizenman et al., 2004), induce neuronal injury through excitotoxic mechanisms. There has been recent speculation that the N-methyl-D-aspartate (NMDA) receptor may be one of the rabies virus receptors (Gosztonyi & Ludwig, 2001). Tsiang and co-workers reported that the noncompetitive NMDA antagonists ketamine and/ or MK-801 inhibited rabies virus infection in primary neuron cultures, inhibited rabies virus genome transcription and restricted viral spread in an experimental model of rabies in rats (Lockhart, Tordo, & Tsiang, 1992; Lockhart, Tsiang, Ceccaldi, & Guillemer, 1991; Tsiang, Ceccaldi, Ermine, Lockhart, & Guillemer, 1991). The *in vitro* doses of ketamine and MK-801 used were much higher than required for stimulation of glutamate receptors, indicating that other mechanisms of actions, including antiviral effects, were likely of primary importance. In more recent studies, Weli, Scott, Ward, & Jackson (2006) observed that CVS-infected cortical and hippocampal mouse embryonic neurons showed loss of trypan blue exclusion, morphologic apoptotic features and activated caspase 3 expression, indicating apoptosis and that the NMDA antagonists, ketamine (125 μM) and MK-801 (60 μM), had no significant neuroprotective effect. Glutamate-stimulated increases of intracellular calcium were reduced in CVS-infected hippocampal neurons compared with mock-infected neurons. Ketamine (120 mg/kg/d intraperitoneally) given to adult ICR mice infected with CVS via the hindlimb foot pad produced no beneficial effects. Hence, there was no supportive evidence that excitotoxicity plays an important role in rabies virus infection or that ketamine is a useful therapeutic agent in this experimental model of rabies.

9.7 Oxidative Stress

Oxidative stress plays a role in neurodegeneration in a variety of diseases, including Parkinson's disease and Alzheimer's disease, and amyotrophic lateral sclerosis (Andersen, 2004; Dexter et al., 1989; Giasson et al., 2000; Pedersen et al., 1998; Sayre et al., 1997). Oxidative stress in viral infections has also been recognized (Schwarz, 1996). Reactive oxygen species (ROS), which can be generated by mitochondria or the family of NADPH oxidases, modulate the permissiveness of cells to viral replication, regulate host inflammatory and immune responses, and cause oxidative damage to both host tissues and progeny virus (Valyi-Nagy & Dermody, 2005). Oxidative injury is observed in experimental acute encephalitis caused by herpes simplex virus 1 in mice (Milatovic et al., 2002; Schachtele, Hu, Little, & Lokensgard, 2010; Valyi-Nagy, Olson, Valyi-Nagy, Montine, & Dermody, 2000). Rabies virus infection of DRG neurons can persist for more than 20 days without any cytopathic effects observed in the cultures (Tsiang et al., 1991), so they are a good cell type with which to evaluate the effects of rabies virus infection on neuronal processes (axons). Rabies virus infection of many other primary neurons results in apoptotic cell death (Morimoto et al., 1999; Weli et al., 2006).

Jackson, Kammouni, Zherebitskaya, & Fernyhough (2010) have evaluated immunostaining in CVS- and mock-infected cultures of DRG neurons derived from adult mice for neuron specific β-tubulin, rabies virus antigen, and for amino acid adducts of 4-hydroxy-2-nonenal (4-HNE), which is a marker of lipid peroxidation and, hence, oxidative stress (Figure 8.7). Neuronal viability (by trypan blue exclusion), TUNEL staining, and axonal growth were also assessed in the cultures. CVS infected 33–54% of cultured DRG neurons, similar to the findings of other investigators. Neuronal viability and TUNEL staining were similar in CVS- and mock-infected DRG neurons. There were significantly more 4-HNE-labeled puncta at 2 and 3 days post-infection (p.i.) in CVS-infected cultures than in mock-infection. Axonal outgrowth was reduced at these time points in CVS infection versus mock-infected cultures. Axonal swellings with 4-HNE-labeled puncta were also associated with aggregations of actively respiring mitochondria, and recently it has been shown that 4-HNE directly impairs mitochondrial function in cultured DRG neurons (Akude, Zherebitskaya, Chowdhury, Girling, & Fernyhough, 2010).

Kammouni et al., (2012) have recently evaluated whether the inducible transcription factor nuclear factor (NF)-κB acts as a critical bridge linking CVS infection and oxidative stress. CVS infection induced expression of NF-κB p50 subunit versus mock infection on Western immunoblotting. Ciliary neurotrophic factor, a potent activator of NF-κB, had no effect on mock-infected rat DRG neurons and reduced the number of 4-HNE-labeled puncta. SN50, a peptide inhibitor of NF-κB, and CVS infection had an additive effect in producing axonal swellings, indicating that NF-κB is neuroprotective. The fluorescent signal for subunit p50 was quantitatively evaluated in the nucleus and cytoplasm of mock- and CVS-infected rat DRG neurons. At 24 hrs post-infection (p.i.) there was a significant increase in the nucleus:cytoplasm ratio, indicating increased transcriptional activity of NF-κB, perhaps as a response to stress. However, at both 48 and 72 hrs p.i., there was significantly reduced nuclear localization of NF-κB. CVS infection may induce oxidative stress by inhibiting nuclear activation of NF-κB. A rabies virus protein may directly inhibit NF-κB activity, which occurs in infections with hepatitis C virus, cowpox, raccoon pox, some strains of vaccinia virus, and African swine fever virus. Further investigations are needed to gain a better understanding of the basic mechanisms involved in the oxidative damage associated with rabies virus infection.

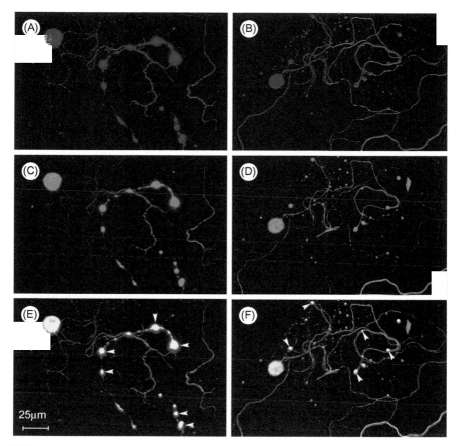

FIGURE 8.7 **CVS-infection causes formation of axonal swellings in DRG cultures.** Fluorescence microscopy showing CVS-infected mouse DRG neurons at 72 hrs p.i. β-tubulin is a marker of DRG neuronal cell bodies and axons (red) and expression of β-tubulin in CVS-infected neurons showed multiple axonal swellings (A, B). Staining for β-tubulin III (A) shows one (large spherical body) at 72 h p.i. (A). Axonal swellings are well established at 72 h p.i. (A–F; indicated by arrowheads in E, F). Rabies virus antigen is strongly expressed in the neuronal cell bodies, axons, and axonal swellings at 72 h p.i. (C, E). Staining for 4-hydroxy-2-nonenal (4-HNE) (green) showed expression in the axons of CVS-infected neurons and showed accumulation in regions with axonal swellings (D, F). In CVS-infected neurons, merging of signals (yellow) for β-tubulin (E) and 4-HNE (F) showed there was strong expression of these elements in axons and in axonal swellings (arrowheads). *Adapted with permission from Jackson et al. in Journal of Virology 84:4697–4705, 2010; doi: 10.1128/JVI.02654-09 Copyright © 2010, American Society for Microbiology.*

9.8 Bases for Behavioral Changes

The neuroanatomical bases for the behavioral changes in animals with rabies have not yet been well characterized. Limbic system infection and dysfunction are suspected to play an important role in the behavioral changes, including alertness, loss of natural timidity, aberrant sexual behavior, and aggressiveness (Johnson, 1971). However, experimental rabies studies in these models have not been particularly helpful in giving insights into the neuroanatomical substrate for behavioral changes because these changes are not normally observed in rodent models and hippocampal infection actually occurs relatively late after peripheral routes of inoculation (Jackson & Reimer, 1989). The neural mechanisms of aggressive behavior are not well understood. Aggressive behavior is associated with lesions in a variety of locations in the brain, including the posterior olfactory bulbs, the ventromedial nucleus of the hypothalamus, and the septal area (Isaacson, 1989). Offensive aggression, which is often impulsive and seemingly unprovoked, has been associated with low CNS serotonergic activity and also increased testosterone in humans and animal studies (Kalin, 1999). Aggressive behavior is essential in most rabies vectors for horizontal transmission of the virus to other hosts by biting. Early and selective brainstem infection in rabies would allow centrifugal spread of the virus to salivary glands as well as involvement of the serotonergic system in the raphe nuclei, resulting in aggressive behavior of animals with adequate cognitive and motor function in order to execute successful viral transmission by biting. Few studies have been performed in natural models of rabies in which aggressive behavior is exhibited. In the best available study, striped skunks inoculated peripherally with a skunk rabies virus isolate were compared with skunks infected with CVS (Smart & Charlton, 1992). The street virus-infected skunks exhibited aggressive responses to presentation of a stick in their cages, whereas this behavior was not observed in CVS-infected skunks. Heavy accumulations of viral antigen were found in the midbrain raphe nuclei, red nucleus, dorsal motor nucleus of the vagus and hypoglossal nucleus in street virus-infected skunks, but not in CVS-infected skunks. Impaired serotonin neurotransmission from the raphe nuclei in the brainstem, however, may account for the development of aggressive behavior in natural vectors of rabies.

10 RECOVERY FROM RABIES AND CHRONIC RABIES VIRUS INFECTION

Although rabies is usually considered a uniformly fatal disease, it has been recognized that animals sometimes may recover from rabies. Recovery from rabies has also been called *abortive rabies*, which can occur either with or without neurologic sequelae (Bell, 1975). There have been a large number of reports of survival after the development of neurologic illness, particularly in experimental animals (Jackson, 1997). Because of limitations on laboratory diagnostic tests performed during life, a conclusive diagnosis of rabies is only rarely made in natural cases that recover. Animals clinically suspected of having rabies are usually killed, and they do not have an opportunity to recover. In a series of five reports from the Pasteur Institute of Southern India, the unusual case of a chronically infected dog has been described. A 14-year-old boy died with hydrophobia 48 days after he stepped on a dog and was bitten in November 1965 (Veeraraghavan, Gajanana, & Rangasami, 1967; Veeraraghavan et al., 1967, 1969, 1970, 1968). The dog was observed at the Pasteur Institute until it died in February 1969 (Veeraraghavan et al., 1970). During that period, rabies virus was isolated from daily saliva samples taken from the dog on 13 occasions between January and May 1966 (Veeraraghavan et al., 1967) and once in January 1967 after the dog was given a course of prednisolone (Veeraraghavan et al., 1968). Rabies virus was not isolated postmortem from the dog's brain, spinal cord, or salivary glands, although fluorescent antibody staining showed rabies virus antigen in its brain and spinal cord (Veeraraghavan et al., 1970). No anti-rabies virus antibodies were found in the dog's blood at any time (Veeraraghavan et al., 1967, 1969, 1970, 1968). Although this is an extremely interesting and unusual series of reports, it is unlikely that this seronegative dog excreted a virulent rabies virus that was responsible for the boy's death. The boy may have become infected from an undocumented rabies exposure months or even years earlier (Smith et al., 1991). There was a poor correlation of this laboratory's results with viral isolation and antigen detection in saliva samples and in CNS tissues from the dog. This might be explained by the presence of neutralizing antibodies in tissues, but this dog was seronegative. The viral isolations from saliva samples could be explained by cross-contamination of specimens in the laboratory. Because of a number of inconsistencies in these reports, the validity of this series of reports remains uncertain.

In another report, five dogs in Ethiopia are described that remained healthy for up to 72 months after the first isolation of rabies virus from their saliva (Fekadu, 1972, 1975). However, exposures from these dogs did not result in any human cases of rabies. In a follow-up study, secretion of rabies virus was documented in the saliva of a dog experimentally infected with an Ethiopian strain of dog rabies virus for up to 6 months after its recovery from rabies (Fekadu, Shaddock, & Baer, 1981).

Finally, remarkable cases of experimental rabies in two cats have been reported (Murphy et al., 1980). Cat 1 developed paralysis, most marked in its hindlimbs, 17 days after inoculation of a rabies virus strain isolated from a big brown bat. The cat showed slow progressive recovery until 100 weeks after inoculation, when it developed progressive neurologic deterioration with aggressive behavior and weakness and atrophy; it was killed for further study 136 weeks after inoculation.

Cat 2 remained well for 120 weeks after inoculation before it developed progressive neurologic deterioration; it was killed at 136 weeks after inoculation. There were high titers of neutralizing antibody in the serum and CSF of both cats. Rabies virus was not isolated from the saliva or tissue suspensions from these cats, but was isolated from the brain of cat 2 by explant culture techniques. Rabies virus antigen was detected at multiple sites in the CNS, and viral inclusions were found in neurons at four sites in the brain of cat 2. Degenerative neuronal changes were noted, and there were extensive inflammatory changes in both cats (Figure 8.8). Perl, Bell, Moore, & Stewart (1977) also reported a similar recrudescent form of rabies in a cat experimentally infected with a bat rabies virus isolate. At necropsy, there were features of chronic encephalitis. Rabies virus could not be isolated from CNS tissues, probably because of the presence of neutralizing antibodies. These well-documented, extraordinary cases indicate that chronic rabies virus infection may occur rarely, at least under experimental conditions. However, it is unclear if chronic rabies infections have any significance in the natural history of rabies, including a role in perpetuation of rabies in natural reservoirs. If animals with chronic rabies are unable to transmit the virus and are incompetent vectors, then this chronic state may not have any biological importance in nature.

Early studies on rabies pathogenesis in vampire bats, which were performed in Trinidad, suggested that bats might be chronically infected with rabies virus and secrete infectious rabies virus over periods lasting up to several months (Pawan, 1936). These early studies were performed before modern virological methods became available and suffered from inadequate diagnostic evaluations, which was largely limited to examination of tissues for Negri bodies. Infections with a variety of other bat

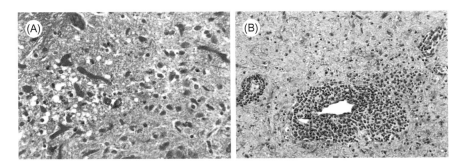

FIGURE 8.8   Medial geniculate body of the thalamus of cat 1, which was killed at 136 weeks post-infection, showing degenerative neuronal changes with vacuolation (A) and massive perivascular lymphocytic and plasmacytic infiltration (B), which were seen throughout the brain. Hematoxylin and eosin; magnifications: A, ×115; B, ×60. (*Courtesy of Dr. Frederick A. Murphy, University of Texas Medical Branch, Galveston, TX.*)

viruses, including Rio Bravo virus, may have been misdiagnosed as rabies virus (Constantine, 1988; Moreno & Baer, 1980). More recent experimental studies have shown that vampire bats have variable incubation periods lasting up to 4 weeks and then develop an acute disease with excretion of virus in the saliva that is not prolonged (Moreno & Baer, 1980). A study of Brazilian free-tailed bats from a dense cave population in New Mexico revealed that 69% of the bats had neutralizing rabies virus antibodies, but only 0.5% had active infection as assessed by direct fluorescent antibody testing of the brain (Steece & Altenbach, 1989). Hence, seroconversion likely occurs in many naturally infected bats, although it is unknown whether any central nervous system involvement normally occurs in this setting, and fatal infections may be relatively infrequent. In Spain, serotine bats (*Eptesicus serotinus*) in which EBLV-1 infection has been recognized were recently studied with RT-PCR amplification of oropharyngeal swabs and simultaneous brain samples. Of 33 bats, a positive RT-PCR result was found in 13 (39%) oropharyngeal swabs and 5 (15%) brains and the positive brains were usually associated with clinical disease (Echevarria, Avellon, Juste, Vera, & Ibanez, 2001). Unless the infection was associated with a previously unrecognized pattern of viral spread in the host, viral RNA was cleared from the brain but not from extraneural tissues in many of these bats. Of course, a positive RT-PCR result does not indicate the presence of infectious virus at a site, rather, it may be a marker of remote infection in some cases.

A study of rabies virus infection in spotted hyenas in the Serengeti changes our perspective about naturally occurring variations in rabies pathogenesis (East et al., 2001). In this study, spotted hyenas were monitored in three social groups for periods of 9 to 13 years. Clinical rabies was never observed. On the basis of rabies virus neutralization antibody (VNA) titers, 37% (37 of 100) were found to be seropositive and repeat studies in six of them indicated that half of the seropositive animals became seronegative. High-ranking hyenas had high VNA titers. They also had high oral (open mouths licked by clan members at rates of over twice an hour) and bite contact rates, and they lived to a mature age of more than 4 years. Although infectious rabies virus was not isolated from saliva, almost half of the seropositive hyenas demonstrated saliva that was positive for rabies virus RNA by RT-PCR. Rabies virus RNA was also detected in three of 23 hyena brain samples in which the hyenas were killed by motor vehicles or other causes. RNA sequence analysis of the isolated viruses showed sequence divergence with strains found in the Serengeti in African wild dogs, bat-eared foxes, and the white-tailed mongoose, and the sequences in the spotted hyenas more closely resembled those found in dogs in the Middle East and Europe. This interesting report really changes our perspective on the ecology of less virulent viral variants and is an exception to the old dogma that rabies virus kills the great majority of exposed individuals. It is likely that, in the future, we will learn that under some circumstances the situation is similar in other species, including bats. Much more research is needed in order to gain a better understanding of the full spectrum of the ecology of rabies virus infection.

11 SUMMARY

Rabies is a normally fatal viral infection of the nervous system in humans and animals with characteristic clinical manifestations. Considerable progress has been made in understanding the pathogenesis of rabies. Rabies virus is highly neurotropic. It binds to the nAChR at the neuromuscular junction and it spreads by axonal transport via peripheral nerves to the CNS, where it causes widespread infection in neurons within the CNS. The combination of virus-induced behavioral changes in rabies vectors and centrifugal spread of the virus to salivary glands allows efficient transmission of the infection. An understanding of rabies virus neurovirulence is emerging from basic studies of virus variants in a variety of animal models. A single amino acid change in the rabies virus glycoprotein at position 333 has dramatic effects on the outcome of infection, and it affects both the efficiency of viral spread involving different afferent and efferent pathways and cellular tropisms. The precise events at the site of viral entry during the long incubation period of rabies remain poorly understood. The fundamental basis for neuronal dysfunction in rabies has not yet been determined, although there are several hypotheses under active study at the present time. A better understanding of rabies pathogenesis will, hopefully, lead to advances in the treatment of rabies and other viral diseases.

References

Adle-Biassette, H., Bourhy, H., Gisselbrecht, M., Chretien, F., Wingertsmann, L., Baudrimont, M., et al. (1996). Rabies encephalitis in a patient with AIDS: A clinicopathological study. *Acta Neuropathologica, 92*(4), 415–420.

Akaike, T., Weihe, E., Schaefer, M., Fu, Z. F., Zheng, Y. M., Vogel, W., et al. (1995). Effect of neurotropic virus infection on neuronal and inducible nitric oxide synthase activity in rat brain. *Journal of Neurovirology, 1*(1), 118–125.

Akude, E., Zherebitskaya, E., Chowdhury, S. K. R., Girling, K., & Fernyhough, P. (2010). 4-Hydroxy-2-nonenal induces mitochondrial dysfunction and aberrant axonal outgrowth in adult sensory neurons that mimics features of diabetic neuropathy. *Neurotoxicology Research, 17*(1), 28–38.

Allsopp, T. E., & Fazakerley, J. K. (2000). Altruistic cell suicide and the specialized case of the virus-infected nervous system. *Trends in Neurological Sciences, 23*(7), 284–290.

Andersen, J. K. (2004). Oxidative stress in neurodegeneration: Cause or consequence? *Nature Medicine, 10*(Suppl.), S18–25.

Araujo, M. D. F., de Brito, T., & Machado, C. G. (1971). Myocarditis in human rabies. *Revista Do Instituto de Medicina Tropical de Sao Paulo, 13*, 99–102.

Baer, G. M., Bellini, W. J., & Fishbein, D. B. (1990). Rhabdoviruses. In B. N. Fields, D. M. Knipe, R. M. Chanock, M. S. Hirsch, J. L. Melnick, T. P. Monath, & B. Roizman (Eds.), *Virology:Volume 1 (pp. 883–930)* (2nd ed.). New York: Raven Press.

Baer, G. M., & Cleary, W. F. (1972). A model in mice for the pathogenesis and treatment of rabies. *Journal of Infectious Diseases, 125*, 520–527.

Baer, G. M., Shaddock, J. H., Quirion, R., Dam, T. V., & Lentz, T. L. (1990). Rabies suscepti- bility and acetylcholine receptor (Letter). *Lancet, 335*, 664–665.

Baer, G. M., Shantha, T. R., & Bourne, G. H. (1968). The pathogenesis of street rabies virus in rats. *Bulletin of the World Health Organization, 38*, 119–125.

Baer, G. M., Shanthaveerappa, T. R., & Bourne, G. H. (1965). Studies on the pathogenesis of fixed rabies virus in rats. *Bulletin of the World Health Organization, 33*, 783–794.

Balachandran, A., & Charlton, K. (1994). Experimental rabies infection of non-nervous tissues in skunks (*Mephitis mephitis*) and foxes (*Vulpes vulpes*). *Veterinary Pathology, 31*, 93–102.

Baloul, L., & Lafon, M. (2003). Apoptosis and rabies virus neuroinvasion. *Biochimie, 85*(8), 777–788.

Bell, J. F. (1975). Latency and abortive rabies. In G. M. Baer (Ed.), *The natural history of rabies (pp. 331–354)*. New York: Academic Press.

Bouzamondo, E., Ladogana, A., & Tsiang, H. (1993). Alteration of potassium-evoked 5-HT release from virus-infected rat cortical synaptosomes. *NeuroReport, 4*(5), 555–558.

Bracci, L., Antoni, G., Cusi, M. G., Lozzi, L., Niccolai, N., Petreni, S., et al. (1988). Antipeptide monoclonal antibodies inhibit the binding of rabies virus glycoprotein and alpha-bungarotoxin to the nicotinic acetylcholine receptor. *Molecular Immunology, 25*, 881–888.

Broughan, J. H., & Wunner, W. H. (1995). Characterization of protein involvement in rabies virus binding to BHK-21 cells. *Archives of Virology, 140*, 75–93.

Buja, L. M., Eigenbrodt, M. L., & Eigenbrodt, E. H. (1993). Apoptosis and necrosis: Basic types and mechanisms of cell death. *Archives of Pathology and Laboratory Medicine, 117*, 1208–1214.

Butowt, R., & Von Bartheld, C. S. (2003). Connecting the dots: Trafficking of neurotro- phins, lectins and diverse pathogens by binding to the neurotrophin receptor p75NTR. *European Journal of Neuroscience, 17*(4), 673–680.

Carbone, K. M., Duchala, C. S., Griffin, J. W., Kincaid, A. L., & Narayan, O. (1987). Pathogenesis of Borna disease in rats: Evidence that intra-axonal spread is the major route for virus dissemination and the determinant for disease incubation. *Journal of Virology, 61*, 3431–3440.

Carson, M. J. (2002). Microglia as liaisons between the immune and central nervous sys- tems: Functional implications for multiple sclerosis. *Glia, 40*(2), 218–231.

Ceccaldi, P. -E., Ermine, A., & Tsiang, H. (1990). Continuous delivery of colchicine in the rat brain with osmotic pumps for inhibition of rabies virus transport. *Journal of Virological Methods, 28*, 79–84.

Ceccaldi, P. -E., Fillion, M. -P., Ermine, A., Tsiang, H., & Fillion, G. (1993). Rabies virus selectively alters 5-HT$_1$ receptor subtypes in rat brain. *European Journal of Pharmacology, 245*(2), 129–138.

Ceccaldi, P. E., Gillet, J. P., & Tsiang, H. (1989). Inhibition of the transport of rabies virus in the central nervous system. *Journal of Neuropathology and Experimental Neurology, 48*, 620–630.

Charlton, K. M., & Casey, G. A. (1979a). Experimental oral and nasal transmission of rabies virus in mice. *Canadian Journal of Comparative Medicine, 43*, 10–15.

Charlton, K. M., & Casey, G. A. (1979b). Experimental rabies in skunks: Immunofluorescence light and electron microscopic studies. *Laboratory Investigation, 41*, 36–44.

Charlton, K. M., & Casey, G. A. (1979c). Experimental rabies in skunks: Oral, nasal, tracheal and intestinal exposure. *Canadian Journal of Comparative Medicine, 43*, 168–172.

Charlton, K. M., Casey, G. A., & Campbell, J. B. (1983). Experimental rabies in skunks: Mechanisms of infection of the salivary glands. *Canadian Journal of Comparative Medicine, 47*, 363–369.

Charlton, K. M., Casey, G. A., Wandeler, A. I., & Nadin-Davis, S. (1996). Early events in rabies virus infection of the central nervous system in skunks (*Mephitis mephitis*). *Acta Neuropathologica, 91*, 89–98.

Charlton, K. M., Nadin-Davis, S., Casey, G. A., & Wandeler, A. I. (1997). The long incubation period in rabies: Delayed progression of infection in muscle at the site of exposure. *Acta Neuropathologica, 94*(1), 73–77.

Cheetham, H. D., Hart, J., Coghill, N. F., & Fox, B. (1970). Rabies with myocarditis: Two cases in England. *Lancet, 1*, 921–922.

Constantine, D. G. (1962). Rabies transmission by nonbite route. *Public Health Reports, 77*, 287–289.

Constantine, D. G. (1988). Transmission of pathogenic organisms by vampire bats. In A. M. Greenhall & U. Schmidt (Eds.), *Natural history of vampire bats (pp. 167–189)*. Boca Raton, Florida: CRC Press.

Constantine, D. G., Emmons, R. W., & Woodie, J. D. (1972). Rabies virus in nasal mucosa of naturally infected bats. *Science, 175*, 1255–1256.

Conti, C., Superti, F., & Tsiang, H. (1986). Membrane carbohydrate requirement for rabies virus binding to chicken embryo related cells. *Intervirology, 26*(3), 164–168.

Coulon, P., Derbin, C., Kucera, P., Lafay, F., Prehaud, C., & Flamand, A. (1989). Invasion of the peripheral nervous systems of adult mice by the CVS strain of rabies virus and its avirulent derivative AvO1. *Journal of Virology, 63*, 3550–3554.

Coulon, P., Rollin, P., Aubert, M., & Flamand, A. (1982). Molecular basis of rabies virus virulence. I. Selection of avirulent mutants of the CVS strain with anti-G monoclonal antibodies. *Journal of General Virology, 61*, 97–100.

Darman, J., Backovic, S., Dike, S., Maragakis, N. J., Krishnan, C., Rothstein, J. D., et al. (2004). Viral-induced spinal motor neuron death is non-cell-autonomous and involves glutamate excitotoxicity. *Journal of Neuroscience, 24*(34), 7566–7575.

Dean, D. J., Evans, W. M., & McClure, R. C. (1963). Pathogenesis of rabies. *Bulletin of the World Health Organization, 29*, 803–811.

Debbie, J. G., & Trimarchi, C. V. (1970). Pantropism of rabies virus in free-ranging rabid red fox *Vulpes fulva. Journal of Wildlife Diseases, 6*, 500–506.

Dexter, D. T., Carter, C. J., Wells, F. R., Javoy-Agid, F., Agid, Y., Lees, A., et al. (1989). Basal lipid peroxidation in substantia nigra is increased in Parkinson's disease. *Journal of Neurochemistry, 52*(2), 381–389.

Dierks, R. E. (1975). Electron microscopy of extraneural rabies infection. In G. M. Baer (Ed.), *The natural history of rabies (pp. 303–318)*. New York: Academic Press.

Dietzschold, B., Wunner, W. H., Wiktor, T. J., Lopes, A. D., Lafon, M., Smith, C. L., et al. (1983). Characterization of an antigenic determinant of the glycoprotein that correlates with pathogenicity of rabies virus. *Proceedings of the National Academy of Sciences of the United States of America, 80*, 70–74.

Dumrongphol, H., Srikiatkhachorn, A., Hemachudha, T., Kotchabhakdi, N., & Govitrapong, P. (1996). Alteration of muscarinic acetylcholine receptors in rabies viral-infected dog brains. *Journal of the Neurological Sciences, 137*(1), 1–6.

East, M. L., Hofer, H., Cox, J. H., Wulle, U., Wiik, H., & Pitra, C. (2001). Regular exposure to rabies virus and lack of symptomatic disease in Serengeti spotted hyenas. *Proceedings of the National Academy of Sciences of the United States of America, 98*(26), 15026–15031.

Echevarria, J. E., Avellon, A., Juste, J., Vera, M., & Ibanez, C. (2001). Screening of active lyssavirus infection in wild bat populations by viral RNA detection on oropharyngeal swabs. *Journal of Clinical Microbiology, 39*(10), 3678–3683.

Emmelin, N. (1967). Nervous control of salivary glands. In C. F. Code (Ed.), *Handbook of physiology, Section 6, Volume II (pp. 595–632)*. Washington, D.C.: American Physiological Society.

Ermine, A., & Flamand, A. (1977). RNA syntheses in BHK_{21} cells infected by rabies virus. *Annals of Microbiology, 128*, 477–488.

Etessami, R., Conzelmann, K. K., Fadai-Ghotbi, B., Natelson, B., Tsiang, H., & Ceccaldi, P. E. (2000). Spread and pathogenic characteristics of a G-deficient rabies virus recombinant: An *in vitro* and *in vivo* study. *Journal of General Virology, 81*, 2147–2153.

Faber, M., Li, J., Kean, R. B., Hooper, D. C., Alugupalli, K. R., & Dietzschold, B. (2009). Effective preexposure and postexposure prophylaxis of rabies with a highly attenuated recombinant rabies virus. *Proceedings of the National Academy of Sciences of the United States of America, 106*(27), 11300–11305.

Fekadu, M. (1972). Atypical rabies in dogs in Ethiopia. *Ethiopian Medical Journal, 10*, 79–86.

Fekadu, M. (1975). Asymptomatic non-fatal canine rabies (Letter). *Lancet, 1*, 569.

Fekadu, M., Shaddock, J. H., & Baer, G. M. (1981). Intermittent excretion of rabies virus in the saliva of a dog two and six months after it had recovered from experimental rabies. *American Journal of Tropical Medicine and Hygiene, 30*, 1113–1115.

Feng, G., Mellor, R. H., Bernstein, M., Keller-Peck, C., Nguyen, Q. T., Wallace, M., et al. (2000). Imaging neuronal subsets in transgenic mice expressing multiple spectral variants of GFP. *Neuron, 28*(1), 41–51.

Fischman, H. R., & Schaeffer, M. (1971). Pathogenesis of experimental rabies as revealed by immunofluorescence. *Annals of the New York Academy of Sciences, 177*, 78–97.

Fischman, H. R., & Ward, F. E. (1968). Oral transmission of rabies virus in experimental animals. *American Journal of Epidemiology, 88*(1), 132–138.

Flamand, A., Coulon, P., Pepin, M., Blancou, J., Rollin, P., & Portnoi, D. (1984). Immunogenic and protective power of avirulent mutants of rabies virus selected with neutralizing monoclonal antibodies. In R. M. Chanock & R. A. Lerner (Eds.), *Modern approaches to vaccines: Molecular and chemical basis of virus virulence and immunogenicity (pp. 289–294)*. Cold Spring Harbor, New York: Cold Spring Harbor Laboratory.

Fu, Z. F., Weihe, E., Zheng, Y. M., Schafer, M. -H., Sheng, H., Corisdeo, S., et al. (1993). Differential effects of rabies and Borna disease viruses on immediate-early- and late-response gene expression in brain tissues. *Journal of Virology, 67*, 6674–6681.

Giasson, B. I., Duda, J. E., Murray, I. V., Chen, Q., Souza, J. M., Hurtig, H. I., et al. (2000). Oxidative damage linked to neurodegeneration by selective alpha-synuclein nitration in synucleinopathy lesions. *Science, 290*(5493), 985–989.

Gillet, J. P., Derer, P., & Tsiang, H. (1986). Axonal transport of rabies virus in the central nervous system of the rat. *Journal of Neuropathology and Experimental Neurology, 45*, 619–634.

Gloor, S. M., Wachtel, M., Bolliger, M. F., Ishihara, H., Landmann, R., & Frei, K. (2001). Molecular and cellular permeability control at the blood-brain barrier. *Brain Research Reviews, 36*(2-3), 258–264.

Gosztonyi, G., & Ludwig, H. (2001). Interactions of viral proteins with neurotransmitter receptors may protect or destroy neurons. *Current Topics in Microbiology and Immunology, 253*, 121–144.

Gourmelon, P., Briet, D., Clarencon, D., Court, L., & Tsiang, H. (1991). Sleep alterations in experimental street rabies virus infection occur in the absence of major EEG abnormalities. *Brain Research, 554*, 159–165.

Gourmelon, P., Briet, D., Court, L., & Tsiang, H. (1986). Electrophysiological and sleep alterations in experimental mouse rabies. *Brain Research, 398*, 128–140.

Guigoni, C., & Coulon, P. (2002). Rabies virus is not cytolytic for rat spinal motoneurons *in vitro. Journal of Neurovirology, 8*(4), 306–317.

Hamilton, R. D., Foss, A. J., & Leach, L. (2007). Establishment of a human in vitro model of the outer blood-retinal barrier. *Journal of Anatomy, 211*(6), 707–716.

Hanham, C. A., Zhao, F., & Tignor, G. H. (1993). Evidence from the anti-idiotypic network that the acetylcholine receptor is a rabies virus receptor. *Journal of Virology, 67*, 530–542.

Hardwick, J. M. (1997). Virus-induced apoptosis. *Advances in Pharmacology, 41*, 295–336.

Hooper, D. C., Ohnishi, S. T., Kean, R., Numagami, Y., Dietzschold, B., & Koprowski, H. (1995). Local nitric oxide production in viral and autoimmune diseases of the central nervous system. *Proceedings of the National Academy of Sciences of the United States of America, 92*(12), 5312–5316.

Hooper, D. C., Phares, T. W., Fabis, M. J., & Roy, A. (2009). The production of antibody by invading B cells is required for the clearance of rabies virus from the central nervous system. *PLoS Neglected Tropical Diseases, 3*(10), e535.

Hronovsky, V., & Benda, R. (1969). Development of inhalation rabies infection in suckling guinea pigs. *Acta Virologica, 13*, 198–202.

Isaacson, R. L. (1989). The neural and behavioural mechanisms of aggression and their alteration by rabies and other viral infections. In O. Thraenhart, H. Koprowski, K. Bögel, & P. Sureau (Eds.), *Progress in rabies control: Proceedings of the second international IMVI ESSEN/WHO symposium on "New Developments in Rabies Control", Essen, 5-7 July 1988; and, report of the WHO consultation on rabies, Essen, 8 July 1988 WHO consultation on rabies (pp. 17–23)*. Royal Tunbridge Wells, Kent: Wells Medical.

Iwasaki, Y., & Clark, H. F. (1975). Cell to cell transmission of virus in the central nervous system. II. Experimental rabies in mouse. *Laboratory Investigation, 33*, 391–399.

Iwata, M., Komori, S., Unno, T., Minamoto, N., & Ohashi, H. (1999). Modification of membrane currents in mouse neuroblastoma cells following infection with rabies virus. *British Journal of Pharmacology, 126*(8), 1691–1698.

Iwata, M., Unno, T., Minamoto, N., Ohashi, H., & Komori, S. (2000). Rabies virus infection prevents the modulation by a_2-adrenoceptors, but not muscarinic receptors, of Ca^{2+} channels in NG108-15 cells. *European Journal of Pharmacology, 404*(1-2), 79–88.

Jackson, A. C. (1991a). Analysis of viral neurovirulence. In J. Brosius & R. T. Fremeau (Eds.), *Molecular genetic approaches to neuropsychiatric diseases (pp. 259–277)*. San Diego: Academic Press.

Jackson, A. C. (1991b). Biological basis of rabies virus neurovirulence in mice: Comparative pathogenesis study using the immunoperoxidase technique. *Journal of Virology, 65*(1), 537–540.

Jackson, A. C. (1993). Cholinergic system in experimental rabies in mice. *Acta Virologica, 37*(6), 502–508.

Jackson, A. C. (1994). Animal models of rabies virus neurovirulence. In C. E. Rupprecht, B. Dietzschold, & H. Koprowski (Eds.), *Current topics in microbiology and immunology, Volume 187: Lyssaviruses (pp. 85–93)*. Berlin: Springer-Verlag.

Jackson, A. C. (1997). Rabies. In N. Nathanson, R. Ahmed, F. Gonzalez-Scarano, D. E. Griffin, K. Holmes, F. A. Murphy, & H. L. Robinson (Eds.), *Viral pathogenesis (pp. 575–591)*. Philadelphia: Lippincott - Raven.

Jackson, A. C. (1999). Apoptosis in experimental rabies in *bax*-deficient mice. *Acta Neuropathologica, 98*(3), 288–294.

Jackson, A. C. (2003). Neuronal apoptosis in experimental rabies: Role of the route of viral entry. *Neurology, 60*(Suppl 1), A102.

Jackson, A. C., Kammouni, W., Zherebitskaya, E., & Fernyhough, P. (2010). Role of oxidative stress in rabies virus infection of adult mouse dorsal root ganglion neurons. *Journal of Virology, 84*(9), 4697–4705.

Jackson, A. C., & Park, H. (1998). Apoptotic cell death in experimental rabies in suckling mice. *Acta Neuropathologica, 95*(2), 159–164.

Jackson, A. C., & Park, H. (1999). Experimental rabies virus infection of p75 neurotrophin receptor - deficient mice. *Acta Neuropathologica, 98*(6), 641–644.

Jackson, A. C., Phelan, C. C., & Rossiter, J. P. (2000). Infection of Bergmann glia in the cerebellum of a skunk experimentally infected with street rabies virus. *Canadian Journal of Veterinary Research, 64*(4), 226–228.

Jackson, A. C., Randle, E., Lawrance, G., & Rossiter, J. P. (2008). Neuronal apoptosis does not play an important role in human rabies encephalitis. *Journal of Neurovirology, 14*(5), 368–375.

Jackson, A. C., Rasalingam, P., & Weli, S. C. (2006). Comparative pathogenesis of recombinant rabies vaccine strain SAD-L16 and SAD-D29 with replacement of Arg333 in the glycoprotein after peripheral inoculation of neonatal mice: Less neurovirulent strain is a stronger inducer of neuronal apoptosis. *Acta Neuropathologica, 111*(4), 372–378.

Jackson, A. C., & Reimer, D. L. (1989). Pathogenesis of experimental rabies in mice: An immunohistochemical study. *Acta Neuropathologica, 78*(2), 159–165.

Jackson, A. C., Reimer, D. L., & Ludwin, S. K. (1989). Spontaneous recovery from the encephalomyelitis in mice caused by street rabies virus. *Neuropathology and Applied Neurobiology, 15*(5), 459–475.

Jackson, A. C., & Rossiter, J. P. (1997). Apoptosis plays an important role in experimental rabies virus infection. *Journal of Virology, 71*(7), 5603–5607.

Jackson, A. C., Ye, H., Phelan, C. C., Ridaura-Sanz, C., Zheng, Q., Li, Z., et al. (1999). Extraneural organ involvement in human rabies. *Laboratory Investigation, 79*(8), 945–951.

Jacob, Y., Badrane, H., Ceccaldi, P. E., & Tordo, N. (2000). Cytoplasmic dynein LC8 interacts with lyssavirus phosphoprotein. *Journal of Virology, 74*(21), 10217–10222.

Johnson, N., McKimmie, C. S., Mansfield, K. L., Wakeley, P. R., Brookes, S. M., Fazakerley, J. K., et al. (2006). Lyssavirus infection activates interferon gene expression in the brain. *Journal of General Virology, 87*(Pt 9), 2663–2667.

Johnson, R. T. (1965). Experimental rabies: Studies of cellular vulnerability and pathogenesis using fluorescent antibody staining. *Journal of Neuropathology and Experimental Neurology, 24*, 662–674.

Johnson, R. T. (1971). The pathogenesis of experimental rabies. In Y. Nagano & F. M. Davenport (Eds.), *Rabies (pp. 59–75)*. Baltimore: University Park Press.

Julius, D. (1991). Molecular biology of serotonin receptors. *Annual Review of Neuroscience, 14*, 335–360.

Juntrakul, S., Ruangvejvorachai, P., Shuangshoti, S., Wacharapluesadee, S., & Hemachudha, T. (2005). Mechanisms of escape phenomenon of spinal cord and brainstem in human rabies. *BMC Infectious Diseases, 5*(1), 104.

Kalin, N. H. (1999). Primate models to understand human aggression. *Journal of Clinical Psychiatry, 60*(Suppl. 15), 29–32.

Kammouni, W., Hasan, L., Saleh, A., Wood, H., Fernyhough, P., & Jackson, A. C. (2012). Role of nuclear factor-kB in oxidative stress associated with rabies virus infection of adult rat dorsal root ganglion neurons. *Journal of Virology, 86*(15), 8139–8146.

Kassis, R., Larrous, F., Estaquier, J., & Bourhy, H. (2004). Lyssavirus matrix protein induces apoptosis by a TRAIL-dependent mechanism involving caspase-8 activation. *Journal of Virology, 78*(12), 6543–6555.

Kaul, M., & Lipton, S. A. (2004). Signaling pathways to neuronal damage and apoptosis in human immunodeficiency virus type 1-associated dementia: Chemokine receptors, excitotoxicity, and beyond. *Journal of Neurovirology, 10*(Suppl. 1), 97–101.

Kelly, R. M., & Strick, P. L. (2000). Rabies as a transneuronal tracer of circuits in the central nervous system. *Journal of Neuroscience Methods, 103*(1), 63–71.

Koch, F. J., Sagartz, J. W., Davidson, D. E., & Lawhaswasdi, K. (1975). Diagnosis of human rabies by the cornea test. *American Journal of Clinical Pathology, 63*, 509–515.

Komatsu, T., Bi, Z., & Reiss, C. S. (1996). Interferon-g induced type I nitric oxide synthase activity inhibits viral replication in neurons. *Journal of Neuroimmunology, 68*(1–2), 101–108.

Koprowski, H., Zheng, Y. M., Heber-Katz, E., Fraser, N., Rorke, L., Fu, Z. F., et al. (1993). *In vivo* expression of inducible nitric oxide synthase in experimentally induced neurologic disease. *Proceedings of the National Academy of Sciences of the United States of America, 90*(7), 3024–3027.

Kuang, Y., Lackay, S. N., Zhao, L., & Fu, Z. F. (2009). Role of chemokines in the enhancement of BBB permeability and inflammatory infiltration after rabies virus infection. *Virus Research, 144*(1–2), 18–26.

Kucera, P., Dolivo, M., Coulon, P., & Flamand, A. (1985). Pathways of the early propagation of virulent and avirulent rabies strains from the eye to the brain. *Journal of Virology, 55*(1), 158–162.

Ladogana, A., Bouzamondo, E., Pocchiari, M., & Tsiang, H. (1994). Modification of tritiated g-amino-*n*-butyric acid transport in rabies virus-infected primary cortical cultures. *Journal of General Virology, 75*(3), 623–627.

Lafon, M. (2005). Rabies virus receptors. *Journal of Neurovirology, 11*(1), 82–87.

Lauria, G., Morbin, M., Lombardi, R., Borgna, M., Mazzoleni, G., Sghirlanzoni, A., et al. (2003). Axonal swellings predict the degeneration of epidermal nerve fibers in painful neuropathies. *Neurology, 61*(5), 631–636.

Lay, S., Prehaud, C., Dietzschold, B., & Lafon, M. (2003). Glycoprotein of nonpathogenic rabies viruses is a major inducer of apoptosis in human Jurkat T cells. *Annals of the New York Academy of Sciences, 1010,* 577–581.

Lentz, T. L. (1985). Rabies virus receptors. *Trends in Neurological Sciences, 8,* 360–364.

Lentz, T. L. (1990). Rabies virus binding to an acetylcholine receptor a-subunit peptide. *Journal of Molecular Recognition, 3,* 82–88.

Lentz, T. L., Benson, R. J. J., Klimowicz, D., Wilson, P. T., & Hawrot, E. (1986). Binding of rabies virus to purified *Torpedo* acetylcholine receptor. *Molecular Brain Research, 387,* 211–219.

Lentz, T. L., Burrage, T. G., Smith, A. L., Crick, J., & Tignor, G. H. (1982). Is the acetylcholine receptor a rabies virus receptor? *Science, 215*(4529), 182–184.

Lentz, T. L., Wilson, P. T., Hawrot, E., & Speicher, D. W. (1984). Amino acid sequence similarity between rabies virus glycoprotein and snake venom curaremimetic neurotoxins. *Science, 226,* 847–848.

Lewis, P., Fu, Y., & Lentz, T. L. (2000). Rabies virus entry at the neuromuscular junction in nerve-muscle cocultures. *Muscle and Nerve, 23*(5), 720–730.

Li, J., Ertel, A., Portocarrero, C., Barkhouse, D. A., Dietzschold, B., Hooper, D. C., et al. (2012). Postexposure treatment with the live-attenuated rabies virus (RV) vaccine TriGAS triggers the clearance of wild-type RV from the Central Nervous System (CNS) through the rapid induction of genes relevant to adaptive immunity in CNS tissues. *Journal of Virology, 86*(6), 3200–3210.

Lockhart, B. P., Tordo, N., & Tsiang, H. (1992). Inhibition of rabies virus transcription in rat cortical neurons with the dissociative anesthetic ketamine. *Antimicrobial Agents and Chemotherapy, 36,* 1750–1755.

Lockhart, B. P., Tsiang, H., Ceccaldi, P. E., & Guillemer, S. (1991). Ketamine-mediated inhibition of rabies virus infection *in vitro* and in rat brain. *Antiviral Chemistry and Chemotherapy, 2,* 9–15.

Lowenstein, C. J., Dinerman, J. L., & Snyder, S. H. (1994). Nitric oxide: A physiologic messenger. *Annals of Internal Medicine, 120,* 227–237.

Madore, H. P., & England, J. M. (1977). Rabies virus protein synthesis in infected BHK-21 cells. *Journal of Virology, 22,* 102–112.

Mazarakis, N. D., Azzouz, M., Rohll, J. B., Ellard, F. M., Wilkes, F. J., Olsen, A. L., et al. (2001). Rabies virus glycoprotein pseudotyping of lentiviral vectors enables retrograde axonal transport and access to the nervous system after peripheral delivery. *Human Molecular Genetics, 10*(19), 2109–2121.

Mebatsion, T. (2001). Extensive attenuation of rabies virus by simultaneously modifying the dynein light chain binding site in the P protein and replacing Arg333 in the G protein. *Journal of Virology, 75*(23), 11496–11502.

Milatovic, D., Zhang, Y., Olson, S. J., Montine, K. S., Roberts, L. J., Morrow, J. D., et al. (2002). Herpes simplex virus type 1 encephalitis is associated with elevated levels of F2-isoprostanes and F4-neuroprostanes. *Journal of Neurovirology, 8*(4), 295–305.

Mims, C. A. (1960). Intracerebral injections and the growth of viruses in the mouse brain. *British Journal of Experimental Pathology, 41,* 52–59.

Moreno, J. A., & Baer, G. M. (1980). Experimental rabies in the vampire bat. *American Journal of Tropical Medicine and Hygiene, 29*(2), 254–259.

Morimoto, K., Hooper, D. C., Spitsin, S., Koprowski, H., & Dietzschold, B. (1999). Pathogenicity of different rabies virus variants inversely correlates with apoptosis and rabies virus glycoprotein expression in infected primary neuron cultures. *Journal of Virology, 73*(1), 510–518.

Morimoto, K., Patel, M., Corisdeo, S., Hooper, D. C., Fu, Z. F., Rupprecht, C. E., et al. (1996). Characterization of a unique variant of bat rabies virus responsible for newly emerging human cases in North America. *Proceedings of the National Academy of Sciences of the United States of America, 93*(11), 5653–5658.

Murphy, F. A., Bauer, S. P., Harrison, A. K., & Winn, W. C. (1973). Comparative pathogenesis of rabies and rabies-like viruses: Viral infection and transit from inoculation site to the central nervous system. *Laboratory Investigation, 28*, 361–376.

Murphy, F. A., Bell, J. F., Bauer, S. P., Gardner, J. J., Moore, G. J., Harrison, A. K., et al. (1980). Experimental chronic rabies in the cat. *Laboratory Investigation, 43*, 231–241.

Murphy, F. A., Harrison, A. K., Winn, W. C., & Bauer, S. P. (1973). Comparative pathogenesis of rabies and rabies-like viruses: Infection of the central nervous system and centrifugal spread of virus to peripheral tissues. *Laboratory Investigation, 29*, 1–16.

Nargi-Aizenman, J. L., & Griffin, D. E. (2001). Sindbis virus-induced neuronal death is both necrotic and apoptotic and is ameliorated by N-methyl-D-aspartate receptor antagonists. *Journal of Virology, 75*(15), 7114–7121.

Nargi-Aizenman, J. L., Havert, M. B., Zhang, M., Irani, D. N., Rothstein, J. D., & Griffin, D. E. (2004). Glutamate receptor antagonists protect from virus-induced neural degeneration. *Annals of Neurology, 55*(4), 541–549.

Nath, A., Haughey, N. J., Jones, M., Anderson, C., Bell, J. E., & Geiger, J. D. (2000). Synergistic neurotoxicity by human immunodeficiency virus proteins Tat and gp120: Protection by memantine. *Annals of Neurology, 47*(2), 186–194.

Nathan, C. (1992). Nitric oxide as a secretory product of mammalian cells. *Federation of American Societies for Experimental Biology Journal, 6*, 3051–3064.

Nishikawa, T., Edelstein, D., Du, X. L., Yamagishi, S., Matsumura, T., Kaneda, Y., et al. (2000). Normalizing mitochondrial superoxide production blocks three pathways of hyperglycaemic damage. *Nature, 404*(6779), 787–790.

Noah, D. L., Drenzek, C. L., Smith, J. S., Krebs, J. W., Orciari, L., Shaddock, J., Sanderlin, D., Whitfield, S., Fekadu, M., Olson, J. G., Rupprecht, C. E., & Childs, J. E. (1998). Epidemiology of human rabies in the United States, 1980 to 1996. *Annals of Internal Medicine, 128*(11), 922–930.

Obrosova, I. G., Pacher, P., Szabo, C., Zsengeller, Z., Hirooka, H., Stevens, M. J., et al. (2005). Aldose reductase inhibition counteracts oxidative-nitrosative stress and poly(ADP-ribose) polymerase activation in tissue sites for diabetes complications. *Diabetes, 54*(1), 234–242.

Obrosova, I. G., Van, H. C., Fathallah, L., Cao, X. C., Greene, D. A., & Stevens, M. J. (2002). An aldose reductase inhibitor reverses early diabetes-induced changes in peripheral nerve function, metabolism, and antioxidative defense. *FASEB Journal, 16*(1), 123–125.

Pawan, J. L. (1936). Rabies in the vampire bat of Trinidad, with special reference to the clinical course and the latency of infection. *Annals of Tropical Medicine and Parasitology, 30*, 401–422.

Pedersen, W. A., Fu, W., Keller, J. N., Markesbery, W. R., Appel, S., Smith, R. G., et al. (1998). Protein modification by the lipid peroxidation product 4-hydroxynonenal in the spinal cords of amyotrophic lateral sclerosis patients. *Annals of Neurology, 44*(5), 819–824.

Perl, D. P., Bell, J. F., Moore, G. J., & Stewart, S. J. (1977). Chronic recrudescent rabies in a cat. *Proceedings of the Society for Experimental Biology and Medicine, 155*, 540–548.

Phares, T. W., Kean, R. B., Mikheeva, T., & Hooper, D. C. (2006). Regional differences in blood-brain barrier permeability changes and inflammation in the apathogenic clearance of virus from the central nervous system. *Journal of Immunology*, *176*(12), 7666–7675.

Prehaud, C., Lay, S., Dietzschold, B., & Lafon, M. (2003). Glycoprotein of nonpathogenic rabies viruses is a key determinant of human cell apoptosis. *Journal of Virology*, *77*(19), 10537–10547.

Prosniak, M., Hooper, D. C., Dietzschold, B., & Koprowski, H. (2001). Effect of rabies virus infection on gene expression in mouse brain. *Proceedings of the National Academy of Sciences of the United States of America*, *98*(5), 2758–2763.

Rasalingam, P., Rossiter, J. P., & Jackson, A. C. (2005). Recombinant rabies virus vaccine strain SAD-L16 inoculated intracerebrally in young mice produces a severe encephalitis with extensive neuronal apoptosis. *Canadian Journal of Veterinary Research*, *69*(2), 100–105.

Rasalingam, P., Rossiter, J. P., Mebatsion, T., & Jackson, A. C. (2005). Comparative pathogenesis of the SAD-L16 strain of rabies virus and a mutant modifying the dynein light chain binding site of the rabies virus phosphoprotein in young mice. *Virus Research*, *111*(1), 55–60.

Raux, H., Flamand, A., & Blondel, D. (2000). Interaction of the rabies virus P protein with the LC8 dynein light chain. *Journal of Virology*, *74*(21), 10212–10216.

Ray, N. B., Ewalt, L. C., & Lodmell, D. L. (1995). Rabies virus replication in primary murine bone marrow macrophages and in human and murine macrophage-like cell lines: Implications for viral persistence. *Journal of Virology*, *69*, 764–772.

Reagan, K. J., & Wunner, W. H. (1985). Rabies virus interaction with various cell lines is independent of the acetylcholine receptor: Brief report. *Archives of Virology*, *84*, 277–282.

Reid, J. E., & Jackson, A. C. (2001). Experimental rabies virus infection in *Artibeus jamaicensis* bats with CVS-24 variants. *Journal of Neurovirology*, *7*(6), 511–517.

Reiss, C. S., Chesler, D. A., Hodges, J., Ireland, D. D., & Chen, N. (2002). Innate immune responses in viral encephalitis. *Current Topics in Microbiology and Immunology*, *265*, 63–94.

Reiss, C. S., & Komatsu, T. (1998). Does nitric oxide play a critical role in viral infections? *Journal of Virology*, *72*(6), 4547–4551.

Ross, E., & Armentrout, S. A. (1962). Myocarditis associated with rabies: Report of a case. *New England Journal of Medicine*, *266*, 1087–1089.

Roulston, A., Marcellus, R. C., & Branton, P. E. (1999). Viruses and apoptosis. *Annual Review of Microbiology*, *53*, 577–628.

Roy, A., & Hooper, D. C. (2007). Lethal silver-haired bat rabies virus infection can be prevented by opening the blood-brain barrier. *Journal of Virology*, *81*(15), 7993–7998.

Roy, A., & Hooper, D. C. (2008). Immune evasion by rabies viruses through the maintenance of blood-brain barrier integrity. *Journal of Neurovirology*, *14*(5), 401–411.

Russell, J. W., Golovoy, D., Vincent, A. M., Mahendru, P., Olzmann, J. A., Mentzer, A., et al. (2002). High glucose-induced oxidative stress and mitochondrial dysfunction in neurons. *FASEB Journal*, *16*(13), 1738–1748.

Rutherford, M., & Jackson, A. C. (2004). Neuronal apoptosis in immunodeficient mice infected with the challenge virus standard strain of rabies virus by intracerebral inoculation. *Journal of Neurovirology*, *10*(6), 409–413.

Sarmento, L., Li, X., Howerth, E., Jackson, A. C., & Fu, Z. F. (2005). Glycoprotein-mediated induction of apoptosis limits the spread of attenuated rabies viruses in the central nervous system of mice. *Journal of Neurovirology*, *11*, 571–581.

Sarmento, L., Tseggai, T., Dhingra, V., & Fu, Z. F. (2006). Rabies virus-induced apoptosis involves caspase-dependent and caspase-independent pathways. *Virus Research*, *121*, 144–151.

Sayre, L. M., Zelasko, D. A., Harris, P. L., Perry, G., Salomon, R. G., & Smith, M. A. (1997). 4-Hydroxynonenal-derived advanced lipid peroxidation end products are increased in Alzheimer's disease. *Journal of Neurochemistry*, *68*(5), 2092–2097.

Schachtele, S. J., Hu, S., Little, M. R., & Lokensgard, J. R. (2010). Herpes simplex virus induces neural oxidative damage via microglial cell Toll-like receptor-2. *Journal of Neuroinflammation*, *7*(1), 35.

Schmidt, R. E., Dorsey, D., Parvin, C. A., Beaudet, L. N., Plurad, S. B., & Roth, K. A. (1997). Dystrophic axonal swellings develop as a function of age and diabetes in human dorsal root ganglia. *Journal of Neuropathology and Experimental Neurology*, *56*(9), 1028–1043.

Schwarz, K. B. (1996). Oxidative stress during viral infection: A review. *Free Radical Biology and Medicine*, *21*(5), 641–649.

Scott, C. A., Rossiter, J. P., Andrew, R. D., & Jackson, A. C. (2008). Structural abnormalities in neurons are sufficient to explain the clinical disease and fatal outcome in experimental rabies in yellow fluorescent protein-expressing transgenic mice. *Journal of Virology*, *82*(1), 513–521.

Seif, I., Coulon, P., Rollin, P. E., & Flamand, A. (1985). Rabies virulence: Effect on pathogenicity and sequence characterization of rabies virus mutations affecting antigenic site III of the glycoprotein. *Journal of Virology*, *53*, 926–935.

Shankar, V., Dietzschold, B., & Koprowski, H. (1991). Direct entry of rabies virus into the central nervous system without prior local replication. *Journal of Virology*, *65*, 2736–2738.

Smart, N. L., & Charlton, K. M. (1992). The distribution of challenge virus standard rabies virus versus skunk street rabies virus in the brains of experimentally infected rabid skunks. *Acta Neuropathologica*, *84*, 501–508.

Smith, J. S., Fishbein, D. B., Rupprecht, C. E., & Clark, K. (1991). Unexplained rabies in three immigrants in the United States: A virologic investigation. *New England Journal of Medicine*, *324*(4), 205–211.

Spitsin, S., Portocarrero, C., Phares, T. W., Kean, R. B., Brimer, C. M., Koprowski, H., et al. (2008). Early blood-brain barrier permeability in cerebella of PLSJL mice immunized with myelin basic protein. *Journal of Neuroimmunology*, *196*(1–2), 8–15.

Steece, R., & Altenbach, J. S. (1989). Prevalence of rabies specific antibodies in the Mexican free-tailed bat (*Tadarida brasiliensis mexicana*) at Lava Cave, New Mexico. *Journal of Wildlife Diseases*, *25*(4), 490–496.

Superti, F., Hauttecoeur, B., Morelec, M. J., Goldoni, P., Bizzini, B., & Tsiang, H. (1986). Involvement of gangliosides in rabies virus infection. *Journal of General Virology*, *67*, 47–56.

Superti, F., Seganti, L., Tsiang, H., & Orsi, N. (1984). Role of phospholipids in rhabdovirus attachment to CER cells. Brief report. *Archives of Virology*, *81*(3–4), 321–328.

Tang, Y., Rampin, O., Giuliano, F., & Ugolini, G. (1999). Spinal and brain circuits to motoneurons of the bulbospongiosus muscle: Retrograde transneuronal tracing with rabies virus. *Journal of Comparative Neurology*, *414*(2), 167–192.

Theerasurakarn, S., & Ubol, S. (1998). Apoptosis induction in brain during the fixed strain of rabies virus infection correlates with onset and severity of illness. *Journal of Neurovirology*, *4*(4), 407–414.

Thompson, C. B. (1995). Apoptosis in the pathogenesis and treatment of disease. *Science*, *267*(5203), 1456–1462.

Thoulouze, M. I., Lafage, M., Montano-Hirose, J. A., & Lafon, M. (1997). Rabies virus infects mouse and human lymphocytes and induces apoptosis. *Journal of Virology*, *71*(10), 7372–7380.

Thoulouze, M. I., Lafage, M., Schachner, M., Hartmann, U., Cremer, H., & Lafon, M. (1998). The neural cell adhesion molecule is a receptor for rabies virus. *Journal of Virology*, *72*(9), 7181–7190.

Thoulouze, M. I., Lafage, M., Yuste, V. J., Baloul, L., Edelman, L., Kroemer, G., et al. (2003). High level of Bcl-2 counteracts apoptosis mediated by a live rabies virus vaccine strain and induces long-term infection. *Virology*, *314*(2), 549–561.

Tillotson, J. R., Axelrod, D., & Lyman, D. O. (1977). Rabies in a laboratory worker – New York. *Morbidity and Mortality Weekly Report*, *26*, 183–184.

Torres-Anjel, M. J., Montano-Hirose, J., Cazabon, E. P. I., Oakman, J. K., & Wiktor, T. J. (1984). A new approach to the pathobiology of rabies virus as aided by immunoperoxidase staining. *American Association of Veterinary Laboratory Diagnosticians, 27th Annual Proceedings*, 1–26.

Tsiang, H. (1979). Evidence for an intraaxonal transport of fixed and street rabies virus. *Journal of Neuropathology and Experimental Neurology, 38,* 286–296.

Tsiang, H. (1982). Neuronal function impairment in rabies-infected rat brain. *Journal of General Virology, 61*(2), 277–281.

Tsiang, H., Ceccaldi, P. -E., Ermine, A., Lockhart, B., & Guillemer, S. (1991). Inhibition of rabies virus infection in cultured rat cortical neurons by an N-methyl-D-aspartate noncompetitive antagonist, MK-801. *Antimicrobial Agents and Chemotherapy, 35*(3), 572–574.

Tsiang, H., Ceccaldi, P. E., & Lycke, E. (1991). Rabies virus infection and transport in human sensory dorsal root ganglia neurons. *Journal of General Virology, 72*(5), 1191–1194.

Tsiang, H., de la Porte, S., Ambroise, D. J., Derer, M., & Koenig, J. (1986). Infection of cultured rat myotubes and neurons from the spinal cord by rabies virus. *Journal of Neuropathology and Experimental Neurology, 45,* 28–42.

Tsiang, H., Lycke, E., Ceccaldi, P. -E., Ermine, A., & Hirardot, X. (1989). The anterograde transport of rabies virus in rat sensory dorsal root ganglia neurons. *Journal of General Virology, 70,* 2075–2085.

Tucker, P. C., Griffin, D. E., Choi, S., Bui, N., & Wesselingh, S. (1996). Inhibition of nitric oxide synthesis increases mortality in Sindbis virus encephalitis. *Journal of Virology, 70,* 3972–3977.

Tuffereau, C., Benejean, J., Blondel, D., Kieffer, B., & Flamand, A. (1998). Low-affinity nerve-growth factor receptor (P75NTR) can serve as a receptor for rabies virus. *European Molecular Biology Organization Journal, 17*(24), 7250–7259.

Tuffereau, C., Leblois, H., Benejean, J., Coulon, P., Lafay, F., & Flamand, A. (1989). Arginine or lysine in position 333 of ERA and CVS glycoprotein is necessary for rabies virulence in adult mice. *Virology, 172,* 206–212.

Tuffereau, C., & Martinet-Edelist, C. (1985). Shut-off of cellular RNA after infection with rabies virus. *Comptes Rendus de l'Académie des Sciences - Series III, Sciences de la Vie, 300,* 597–600.

Tuffereau, C., Schmidt, K., Langevin, C., Lafay, F., Dechant, G., & Koltzenburg, M. (2007). The rabies virus glycoprotein receptor p75NTR is not essential for rabies virus infection. *Journal of Virology, 81*(24), 13622–13630.

Ubol, S., Sukwattanapan, C., & Maneerat, Y. (2001). Inducible nitric oxide synthase inhibition delays death of rabies virus-infected mice. *Journal of Medical Microbiology, 50*(3), 238–242.

Ubol, S., Sukwattanapan, C., & Utaisincharoen, P. (1998). Rabies virus replication induces Bax-related, caspase dependent apoptosis in mouse neuroblastoma cells. *Virus Research, 56*(2), 207–215.

Valyi-Nagy, T., & Dermody, T. S. (2005). Role of oxidative damage in the pathogenesis of viral infections of the nervous system. *Histology and Histopathology, 20*(3), 957–967.

Valyi-Nagy, T., Olson, S. J., Valyi-Nagy, K., Montine, T. J., & Dermody, T. S. (2000). Herpes simplex virus type 1 latency in the murine nervous system is associated with oxidative damage to neurons. *Virology, 278*(2), 309–321.

Van Dam, A. M., Bauer, J., Manahing, W. K. H., Marquette, C., Tilders, F. J. H., & Berkenbosch, F. (1995). Appearance of inducible nitric oxide synthase in the rat central nervous system after rabies virus infection and during experimental allergic encephalomyelitis but not after peripheral administration of endotoxin. *Journal of Neuroscience Research, 40,* 251–260.

Veeraraghavan, N., Gajanana, A., & Rangasami, R. (1967). Hydrophobia among persons bitten by apparently healthy animals: *The pasteur institute of Southern India, Coonoor: Annual report of the director 1965 and scientific report 1966.* Madras: Diocesan Press. (pp. 90–91).

Veeraraghavan, N., Gajanana, A., Rangasami, R., Kumari, C., Saraswathi, K. C., Devaraj, R., et al. (1967). Studies on the salivary excretion of rabies virus by the dog from Surandai: *The pasteur institute of Southern India, Coonoor: Annual report of the director 1965 and scientific report 1966.* Madras: Diocesan Press. (pp. 91–97).

Veeraraghavan, N., Gajanana, A., Rangasami, R., Oonnunni, P. T., Saraswathi, K. C., Devaraj, R., et al. (1969). Studies on the salivary excretion of rabies virus by the dog from Surandai: *The pasteur institute of Southern India, Coonoor: Annual report of the director 1967 and scientific report 1968*. Madras: Diocesan Press. (pp. 68–70).

Veeraraghavan, N., Gajanana, A., Rangasami, R., Oonnunni, P. T., Saraswathi, K. C., Devaraj, R., et al. (1970). Studies on the salivary excretion of rabies virus by the dog from Surandai: *The pasteur institute of Southern India, Coonoor: Annual report of the director 1968 and scientific report 1969*. Madras: Diocesan Press. (pp. 66).

Veeraraghavan, N., Gajanana, A., Rangasami, R., Saraswathi, K. C., Devaraj, R., & Hallan, K. M. (1968). Studies on the salivary excretion of rabies virus by the dog from Surandai: *The pasteur institute of Southern Indian, Coonoor: Annual report of the director 1966 and scientific report 1967*. Madras: Diocesan Press. (pp. 71–78).

Wallerstein, C. (1999). Rabies cases increase in the Philippines. *British Medical Journal, 318*(7194), 1306.

Wang, H., Zhang, G., Wen, Y., Yang, S., Xia, X., & Fu, Z. F. (2011). Intracerebral administration of recombinant rabies virus expressing GM-CSF prevents the development of rabies after infection with street virus. *PLoS ONE, 6*(9), e25414.

Wang, Z. W., Sarmento, L., Wang, Y., Li, X. Q., Dhingra, V., Tseggai, T., et al. (2005). Attenuated rabies virus activates, while pathogenic rabies virus evades, the host innate immune responses in the central nervous system. *Journal of Virology, 79*(19), 12554–12565.

Weihe, E., Bette, M., Preuss, M. A., Faber, M., Schafer, M. K., Rehnelt, J., et al. (2008). Role of virus-induced neuropeptides in the brain in the pathogenesis of rabies. In B. Dodet, A. R. Fooks, T. Müller, & N. Tordo (Eds.), *Towards the elimination of rabies in Eurasia (pp. 73–81)* (131 ed.). Basel: Kager. (Developments in Biologicals, Volume 131).

Weli, S. C., Scott, C. A., Ward, C. A., & Jackson, A. C. (2006). Rabies virus infection of primary neuronal cultures and adult mice: Failure to demonstrate evidence of excitotoxicity. *Journal of Virology, 80*(20), 10270–10273.

Winkler, W. G., Fashinell, T. R., Leffingwell, L., Howard, P., & Conomy, J. P. (1973). Airborne rabies transmission in a laboratory worker. *Journal of the American Medical Association, 226*, 1219–1221.

Yan, X., Prosniak, M., Curtis, M. T., Weiss, M. L., Faber, M., Dietzschold, B., et al. (2001). Silver-haired bat rabies virus variant does not induce apoptosis in the brain of experimentally infected mice. *Journal of Neurovirology, 7*(6), 518–527.

Yang, C., & Jackson, A. C. (1992). Basis of neurovirulence of avirulent rabies virus variant Av01 with stereotaxic brain inoculation in mice. *Journal of General Virology, 73*(4), 895–900.

Zhao, L., Toriumi, H., Kuang, Y., Chen, H., & Fu, Z. F. (2009). The roles of chemokines in rabies virus infection: Overexpression may not always be beneficial. *Journal of Virology, 83*(22), 11808–11818.

Zheng, Y. M., Schafer, M. K. -H., Weihe, E., Sheng, H., Corisdeo, S., Fu, Z. F., et al. (1993). Severity of neurological signs and degree of inflammatory lesions in the brains of rats with Borna disease correlate with the induction of nitric oxide synthase. *Journal of Virology, 67*, 5786–5791.

Pathology

John P. Rossiter[1] and Alan C. Jackson[2]

[1]Department of Pathology and Molecular Medicine, Queen's University
and Kingston General Hospital, Kingston, Ontario K7L 3N6, Canada
[2]Departments of Internal Medicine (Neurology) and Medical Microbiology,
University of Manitoba, Winnipeg, Manitoba R3A 1R9, Canada

1 INTRODUCTION

Investigation of the pathological changes in the central and peripheral nervous systems and extraneural organs of human rabies cases, as well as naturally and experimentally infected animals, has provided an important foundation for ongoing study of the pathogenesis of rabies (see Chapter 8). Following an exposure with deposition of rabies virus at or near the site of a bite from a rabid animal, the virus spreads centripetally towards the central nervous system (CNS) by retrograde fast axonal transport through the peripheral nervous system, typically to the spinal cord. Rapid neuron-to-neuron transsynaptic viral dissemination within the spinal cord and brain results in a polioencephalomyelitis (i.e., an inflammatory disease predominantly involving the gray matter of the brain and spinal cord (Love & Wiley, 2002). This is followed by centrifugal spread away from the CNS along peripheral nerve pathways, with resulting infection of salivary glands, skin, heart, and other viscera.

Many of the cardinal pathological features of rabies virus infection were first described over about a 40-year period extending from the early 1870s to the early 1900s (Abba & Bormans, 1905; Babes, 1892; Benedikt, 1878; Gowers, 1877; Negri, 1903a; Negri, 1909; Nepveu, 1872; Pasteur, Chamberland, Roux, & Thuillier, 1881; Ramón y Cajal & Garcia, 1904; Schaffer, 1888; Van Gehuchten & Nelis, 1900). However, throughout the subsequent century, especially following the introduction of electron microscopy and immunohistochemistry, pathological studies have continued to provide key insights into our understanding of this dreaded disease.

2 MACROSCOPIC FINDINGS

Macroscopic examination of the brain in rabies victims is frequently unremarkable, or it shows a spectrum of relatively mild and non-specific changes (Love & Wiley, 2002; Nieberg & Blumberg, 1972; Perl & Good, 1991; Sükrü-Aksel, 1958). There is often mild cerebral edema, but severe cerebral swelling and associated brain herniation are not features of rabies. There may be congestion of leptomeningeal and paren-chymal blood vessels, sometimes associated with multiple petechiae (Lowenberg, 1928; Tangchai, Yenbutr, & Vejjajiva, 1970). Frank subarach-noid or parenchymal hemorrhage is not a recognized feature of rabies. Thickening of the basal leptomeninges due to prominent inflamma-tory cell infiltration has been reported in a few cases of rabies in chil-dren (Tangchai et al., 1970). Although the brain parenchyma is typically grossly unremarkable, focal changes are seen in some cases, perhaps related to prolongation of survival with critical care measures (Rubin, Sullivan, Summers, Gregg, & Sikes, 1970). Multifocal gray and white matter tissue softening and discoloration were found in an immuno-suppressed male with prolonged survival (Mackenzie, Medvedev, & Thiessen, 2003). A variety of macroscopic abnormalities, including dis-coloration of the cortical mantle, generalized softening of deep gray mat-ter and infarction of the insular cortex, were described in four transplant recipients infected with rabies virus from a common donor (Burton et al., 2005). Softening and congestion of the amygdalae and extensive lami-nar necrosis of the cerebral cortex were described in a case of bat-trans-mitted human rabies, where the patient received human rabies immune globulin and antiviral therapy and died 33 days after the onset of symp-toms. Death was attributed to a direct viral effect rather than anoxic brain injury or autolysis of 'respirator brain' (Dolman & Charlton, 1987). Leptomeningeal and vascular congestion may also be seen in the spinal cord and may be intense (Gowers, 1877; Lowenberg, 1928; Tangchai et al., 1970). In their classic study of the 1929–30 Trinidad outbreak of paralytic rabies, Hurst & Pawan (1932) described the victims' spinal cords as hav-ing the "consistency of butter," likely reflecting extensive tissue injury.

3 PATHOLOGY IN THE CENTRAL NERVOUS SYSTEM

3.1 Overview

Despite the catastrophic clinical outcome of rabies virus encepha-lomyelitis, the histopathological changes observed in the central ner-vous system (CNS) are typically relatively mild, with varying degrees of mononuclear inflammatory cell infiltration of the leptomeninges, perivascular cuffing, microglial activation with formation of 'Babes nodules' and neuronophagia. Moreover, this combination of features is not unique to rabies and can be seen in a variety of other viral encephalitides (Love & Wiley, 2002). However, Negri bodies, eosinophilic cytoplasmic viral inclusions that are unique to rabies, are found in infected neurons in many cases. The extent of the infection of the CNS by rabies virus is best highlighted by immunostaining for rabies virus antigen.

3.2 Inflammation

Some degree of inflammatory cell infiltration of the leptomeninges is usually seen in human rabies cases, although the extent and intensity can vary greatly (Dupont & Earle, 1965; Perl & Good, 1991; Tangchai et al., 1970). The infiltrates are typically composed predominantly of lymphocytes and monocytes, with smaller numbers of plasma cells, but neutrophils can predominate when inflammation is intense, especially in fulminant childhood cases (Perl & Good, 1991; Tangchai et al., 1970). In the classic study by Dupont & Earle (1965) of 49 cases of human rabies encephalitis, frank leptomeningitis was found in three cases, all of them children. Tangchai et al. (1970) observed meningitis in four of 24 cases, again all children, in whom the clinical course was more fulminant by comparison with their other cases. The meninges of the brainstem are frequently involved, especially in cases where leptomeningeal inflammation is sparse overall (Perl & Good, 1991). Paradoxically, in paralytic rabies cases inflammatory cell infiltration of the spinal meninges is typically relatively sparse, despite intense inflammation within the adjacent spinal cord in many of the reported cases (Chopra, Banerjee, Murthy, & Pal, 1980; Hurst & Pawan, 1932).

Perivascular mononuclear inflammatory cell infiltrates (Figure 9.1), consisting predominantly of lymphocytes and monocytes, are seen in the great majority of human rabies cases, but their density and distribution can vary greatly between cases. This perivascular 'cuffing' is seen predominantly in gray matter, especially in the brainstem and spinal cord, with relative sparing of white matter. Dupont & Earle (1965) observed perivascular cuffing in 48 of their 49 cases, in the following locations and approximate proportions of cases: medulla and pons (38%), spinal cord (35%), cerebral cortex (26%), hippocampus (14%), thalamus (29%), basal ganglia (26%), and cerebellum (14%). In the paralytic rabies series of Hurst & Pawan (1932) (3 cases) and Chopra et al. (1980) (11 cases), dense perivascular infiltrates of lymphocytes and some neutrophils were seen in the anterior and posterior horn gray matter of most cases. This was especially prominent in the lumbar and lower thoracic segments in the cases of Chopra et al. (1980), and was associated with extension of inflammation along perivascular spaces into the adjacent white matter.

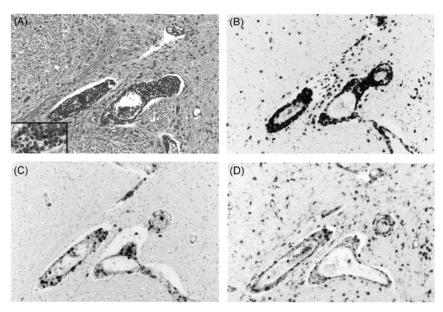

FIGURE 9.1   **Perivascular mononuclear inflammatory cell infiltrates in the brainstem (medulla oblongata) of a human rabies case.** (A) Hematoxylin-Phloxine-Saffron (HPS) stained section (magnification ×80). Inset shows a higher power view of an infiltrate that is composed predominantly of lymphocytes (magnification ×240). (B,C,D) Adjacent immunoperoxidase stained sections showing: (B) CD3-positive T lymphocytes, (C) CD20-positive B lymphocytes and (D) CD68-positive monocyte/macrophages in a perivascular distribution and microglia/macrophages in the neuropil of the brainstem.

By contrast, inflammation was considerably less severe in the brain and predominantly involved the medulla (both series) and dorsal half of the pons (Hurst & Pawan, 1932).

In the case of a patient who died 17 days into the clinical course of encephalitic rabies, Iwasaki et al. (Iwasaki, Sako, Tsunoda, & Ohara, 1993; Iwasaki & Tobita, 2002) found that 50 to 70% of perivascular mononuclear cells were CD3 immunopositive T lymphocytes. Approximately one-third of these were CD4-positive helper T cells. Only occasional CD20-positive B cells were observed and the remaining perivascular cells were CD68-positive monocyte/macrophage lineage cells. More than half of the T lymphocytes were found in the CNS parenchyma surrounding the perivascular spaces. However, in a rabies patient who only survived for 9 days, there was virtually no inflammation or tissue injury, despite the finding of numerous neurons containing Negri bodies (see below), which emphasizes an important point that fatal encephalitic rabies may not necessarily be accompanied by significant inflammation (Iwasaki et al., 1993). The degree of inflammation may at least be partially influenced by the strain of rabies virus. In dogs experimentally inoculated

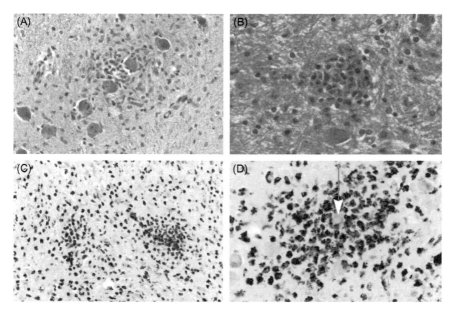

FIGURE 9.2   Microglial nodules ('Babes nodules') in the medulla oblongata of a human rabies case. (A,B) HPS stained sections (magnifications A×160, B×260). (C, D) CD68-immunoperoxidase stained sections showing (C) diffuse and nodular microglial proliferation (magnification ×85) and (D) a microglial nodule containing a neuron (arrow) undergoing neuronophagia (magnification ×170).

with an Ethiopian dog rabies virus strain, widespread inflammation, neuronal degeneration and neuronophagia (see below) were seen, whereas such lesions were generally much less severe in animals infected with a Mexican dog virus strain (Fekadu, Chandler, & Harrison, 1982). An especially florid and widespread encephalitic pattern was reported in a patient presenting with paralytic rabies caused by canine virus transmitted by a fox bite (Suja et al., 2004).

In 1892, Babes described microscopic accumulations of cells surrounding chromatolytic and degenerating neurons in a series of human rabies cases. He called these foci "les ilots inflammatoires pericellulaires de la rage" (pericellular inflammatory islets of rabies) and "nodules rabiques" (rabidic nodules) (Babes, 1892). Subsequently referred to in the literature as "Babes nodules," these microglial nodules (Figure 9.2) are composed predominantly of activated microglia/monocytes and are seen in other viral encephalitides and other infectious disorders (Love & Wiley, 2002). Dupont & Earle (1965) observed activated microglia in many of their cases, especially in the medulla, and in several cases the microglia had a predominantly "rod cell' morphology. Classic Babes nodules were seen in 42% of their cases. In paralytic rabies cases, microglial proliferation, both diffuse and nodular, was found throughout the spinal gray matter and focally in the adjacent white matter in most cases. Marked microglial activation was seen in the medulla and dorsal half of the pons in several of these cases (Chopra et al., 1980; Hurst & Pawan, 1932).

3.3 Cell Injury and Cell Death

Neuronophagia (Figure 9.2D), a microscopic pattern characterized by accumulations of activated microglia/macrophages in the process of phagocytosing degenerating and/or dying neurons, is seen in many rabies cases. Once again, however, the severity and anatomical extent of neuronophagia and resulting neuronal loss can vary greatly between cases. Dupont & Earle (1965) observed neuronophagia in 57% of their cases. The neurons within these foci often have a shrunken appearance, with condensed cytoplasm and pyknotic nuclei (Perl & Good, 1991). Central chromatolysis is a cytological pattern of swelling of the neuronal cell body, disruption and dispersal of Nissl granules from the central part of the perikaryon and peripheral displacement of the nucleus that is classically seen in response to axonal injury and may also be seen in rabies. In some paralytic rabies cases, there was extensive neuronal degeneration and loss in the anterior and posterior horns of the spinal cord and to a lesser extent in the medulla (Chopra et al., 1980), whereas in others there was marked neuronal central chromatolysis but only occasional (Hurst & Pawan, 1932) or absent (Sheikh et al., 2005) neuronophagia. Central chromatolysis of spinal motor neurons has also been described in encephalitic rabies cases (Mitrabhakdi et al., 2005). In addition to chromatolysis, vacuolation of neuronal cytoplasm and degenerative changes in nuclear chromatin have been reported (Lowenberg, 1928; Reisman, Alpers, & Cooper, 1933).

In rodents experimentally infected with street virus, neurons typically remain relatively intact, with little alteration in the structure of organelles. However, with fixed virus infection using intracerebral inoculation in adult animals or using any route of inoculation in immature animals, widespread neuronal injury with cytoplasmic condensation, multivesiculation, increase in lysosome content, intercellular edema, and cell death are frequently seen (Miyamoto & Matsumoto, 1967; Murphy, 1977). Prominent micro-vacuolation of the gray matter neuropil, especially cerebral cortex and thalamus, has been documented in experimentally infected skunks and foxes with the Arctic fox variant, as well as in naturally occurring infection in these species and also in cow, horse, and cat (Bundza & Charlton, 1988; Charlton, 1984; Charlton, Casey, Webster, & Bundza, 1987). This spongiform change closely resembled that of traditional spongiform encephalopathies, although the vacuolation was less extensive than that found in skunks experimentally inoculated with scrapie agent (Bundza & Charlton, 1988). Prominent vacuolation of the neuronal soma has been reported in mice following intracerebral or intramuscular inoculation with the CVS strain of fixed rabies virus (Greenwood, Newton, Pearson, & Schamber, 1997). Spongiform change is not a feature of human rabies cases.

The role of apoptotic cell death in the pathogenesis of experimentally induced and naturally occurring rabies infection has been investigated in considerable detail. The data strongly indicate that neuronal apoptosis does not play an important pathogenetic role in human rabies encephalitis or in canine infection with "street' rabies virus (Jackson, Randle, Lawrance, & Rossiter, 2008; Suja, Mahadevan, Madhusudana, & Shankar, 2011).

Juntrakul, Ruangvejvorachai, Shuangshoti, Wacharapluesadee, & Hemachudha (2005) observed cytoplasmic cytochrome-c immunoreactivity in neurons and numerous TUNEL-labeled (terminal deoxynucleotidyltransferase-mediated DNA nick-end labeling) cells in many regions of the CNS in a series of 10 human rabies cases. Foci of TUNEL-positive neurons were also observed in the brainstem and hippocampus of a patient who developed rabies encephalitis on a background of AIDS (Adle-Biassette et al., 1996). However, neither of these two reports showed any cytological evidence of apoptosis, such as cell shrinkage, nuclear karyorrhexis, and formation of apoptotic bodies, and apoptotic cell death cannot be reliably diagnosed in the absence of such morphological features. Jackson et al. (2008) evaluated brain tissue sections from the cerebral cortex, hippocampus and brainstem in 12 human rabies cases for evidence of neuronal apoptosis, using morphological assessment, TUNEL staining, and immunohistochemical staining for activated (cleaved) caspase-3, a downstream effector of apoptosis. There was a complete lack of morphological evidence of apoptosis in neurons, including neurons with strong immunohistochemical staining for rabies virus antigen. There was TUNEL staining of scattered non-neuronal cells within the neuropil of these cases and focally of apoptotic perivascular inflammatory cells, but there was no evidence of TUNEL staining of neurons. Likewise, in all of these cases, activated caspase-3 was not seen in neurons, although there was multifocal immunostaining of the processes of activated microglia in 9 of 12 cases (Jackson et al., 2008). Suja et al., in their studies of canine (Suja et al., 2011; Suja, Mahadevan, Madhusudhana, Vijayasarathi, & Shankar, 2009) and human (Suja et al., 2011) street virus infected brains, found no morphological, TUNEL, or DNA laddering evidence of neuronal apoptosis in multiple neuroanatomical areas. In canine brains, a few TUNEL labeled perivascular microglial cells, vascular endothelial cells, and white matter glial cells were seen, whereas in human brains occasional TUNEL-positive inflammatory cells were seen in the hippocampus and medulla oblongata (Suja et al., 2011).

In animal models, prominent neuronal apoptosis has been found in the brains of adult and immature mice following intracerebral inoculation with the CVS strain of fixed rabies virus (Fu & Jackson, 2005; Jackson & Park, 1998; Jackson & Rossiter, 1997; Theerasurakarn & Ubol, 1998; Yan et al., 2001). However, following peripheral inoculation of a fruit-eating bat species with rabies challenge virus standard (CVS) strain, apoptosis was not observed (Reid & Jackson, 2001). In six-week old mice inoculated in the hindlimb footpad with CVS strain TUNEL labeling showed no definite neuronal staining and only scattered positive non-neuronal cells (predominantly inflammatory) in the cerebral cortex, while immunostaining for activated caspase-3 showed only occasional non-neuronal cells in the cerebral cortex of moribund animals (Scott, Rossiter, Andrew, & Jackson, 2008). Extensive TUNEL-positivity was found in the brains of mice intracerebrally inoculated with street (wild-type) rabies virus strains in one study (Ubol & Kasisith, 2000), but other investigators found little or none with bat rabies virus infections (Sarmento, Li, Howerth, Jackson, & Fu, 2005; Yan et al., 2001). A compelling experimentally based model for a "subversive neuroinvasive strategy of rabies virus" in naturally occurring infection has been advanced. Preservation of the integrity of the neuronal network by avoidance of neuronal apoptosis, together with induction of apoptosis in potentially protective T lymphocytes, permits dissemination of the virus, its excretion in saliva, and transmission by bite to another host (Baloul & Lafon, 2003; Lafon, 2011).

3.4 Negri and Lyssa Bodies

In the early 1900s, Adelchi Negri undertook a series of detailed studies of rabies virus-infected animal brains and described the characteristic neuronal intracytoplasmic inclusions that now bear his name (Negri, 1903a,b) (see Chapter 1). Although he mistakenly interpreted these intracytoplasmic bodies as a protozoan species, he established their detection as being specifically diagnostic for rabies virus infection. Following his premature death from tuberculosis at age 37, his work was summarized by his wife (Negri-Luzzani, 1913). For many decades before the introduction of electron microscopy and immunofluorescence staining of viral material, light microscopic identification of Negri bodies remained the predominant pathological method for diagnosing rabies encephalomyelitis. Furthermore, Negri's work stimulated numerous studies, extending well into the second half of the 20th century, on the nature of Negri bodies and their significance in the pathogenesis of rabies (see Kristensson, Dastur, Manghani, Tsiang, & Bentivoglio (1996) and Perl & Good (1991) for detailed reviews). In recent years, there has been increasing experimental evidence that Negri bodies may be sites of viral transcription and replication, with exploitation of cellular biology for this purpose (Lahaye et al., 2009; Ménager et al., 2009). However, it is noteworthy that while Negri bodies are a pathognomic finding in most cases of "street" rabies viral infection, they are almost never detected with "fixed" virus strains.

On hematoxylin and eosin stained sections, Negri bodies appear as dense, well-defined, oval or round, eosinophilic cytoplasmic inclusions (Figure 9.3). They are typically 2 to 10 μm in diameter, but may range from 0.5 to 27 μm in size (Negri, 1903a; Nieberg & Blumberg, 1972; Perl & Good, 1991). There is considerable inter-species variation in the average size of Negri bodies, ranging from small in rabbits and raccoons, to large in dogs, guinea pigs, and skunks, and very large in cows (Perl & Good, 1991). Within an individual neuron, Negri bodies may be single or multiple, and they are typically located in the perikaryon but may occasionally be found in dendrites and axons.

Using the Mann methylene blue and eosin staining method (Lepine & Atanasiu, 1996), Negri observed a small, basophilic and granular

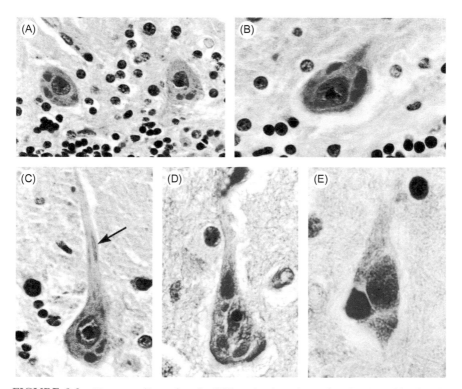

FIGURE 9.3 **Hematoxylin and eosin (HE) stained sections showing Negri bodies in the perikarya of (A–C) cerebellar Purkinje cells and (D,E) pyramidal neurons in the cerebral cortex of human rabies cases.** The arrow in (D) indicates a Negri body in an apical dendrite. (magnifications, A ×315, B ×460, C ×550, D ×730, E ×865).

"Innerkörperchen" (inner body) within the Negri body inclusions (Negri, 1903a, 1909). Negri and subsequent investigators emphasized the presence of this inner body/basophilic granule/basophilic stippling, to the extent that in 1925 Goodpasture (1925) coined the term "lyssa bodies" to distinguish other inclusions that lacked these staining features from classic Negri bodies. The impetus to make this distinction was to recognize the presence of the other small eosinophilic cytoplasmic inclusions that lacked an internal structure, found in healthy neurons in a variety of animal species. The concern was that the other inclusion bodies could be misinterpreted as rabies virus inclusions (Goodpasture, 1925; Szlachta & Habel, 1953; Tierkel, 1973b). However, this is less of an issue with human tissue, as lyssa bodies are typically more numerous than Negri bodies (Mrak & Young, 1994; Sung, Hayano, Mastri, & Okagaki, 1976). They are frequently more irregularly shaped and less clearly demarcated from the surrounding cell cytoplasm than Negri bodies, although they share ultrastructural features with Negri bodies (Iwasaki & Tobita, 2002; Mrak & Young, 1994; Perl & Good, 1991) and are immunoreactive for rabies virus antigen (see below).

Negri bodies have been found in 50 to 90% of street rabies infections in different series of human rabies cases (Dupont & Earle, 1965; Herzog, 1945; Jogai, Radotra, & Banerjee, 2000; Negri-Luzzani, 1913), influenced in part by the species, the extent of tissue sampling, and possibly by the duration of clinical disease. Dupont & Earle (1965) observed Negri bodies in 71% of their cases, and Negri bodies unassociated with inflammation in 15% of cases. Although Negri bodies may be found in virtually any neuronal population in the CNS or peripheral nerve ganglia, they tend to be most numerous and largest in larger neurons (Tangchai et al., 1970), especially in hippocampal pyramidal neurons and cerebellar Purkinje cells. Dupont & Earle (1965) observed Negri bodies in the following locations and approximate proportions of their series of 59 cases: cerebellum (60%), hippocampus (43%), medulla (14%), pontine nuclei (12%), spinal cord (10%), cerebral cortex (7%), midbrain (7%), basal ganglia (5%), thalamus (2%), and peripheral nerve ganglia (5%), and Tangchai et al. (1970) in their series of 24 human rabies cases: cerebellum (54%), hippocampus (50%), brainstem (50%), hypothalamus (42%), thalamus and hypothalamus (42%), cerebral cortex (21%), and spinal cord (21%). In their monumental study of more than one thousand biologically positive rabies cases from a diverse range of species, Tustin & Smit (1962) found that Negri bodies were present in 71% of cases when the hippocampus alone was examined histologically. Remarkably, numerous and diffusely distributed Negri bodies have been reported in two immunocompromised patients, one of whom had AIDS (Adle-Biassette et al., 1996) and the other a renal transplant recipient with prolonged survival (Mackenzie et al., 2003). In a series of experimentally infected dogs, an increasing proportion of neurons contained Negri bodies as the disease progressed (Marinesco & Storesco, 1931). More numerous and widely distributed Negri bodies have also been observed with increasing duration of survival in some human case series (Sandhyamani, Roy, Gode, & Kalla, 1981). Negri bodies are more likely to be found in areas of the CNS where there is little inflammation and are less frequently seen in degenerating neurons and/or in association with inflammatory foci (Dupont & Earle, 1965; Iwasaki & Tobita, 2002; Marinesco & Storesco, 1931; Sükrü-Aksel, 1958).

The ultrastructure of Negri bodies and rabies virions has been investigated in human rabies cases (Adle-Biassette et al., 1996; de Brito, Araujo, & Tiriba, 1973; Gonzalez-Angulo, Marquez-Monter, Feria-Velasco, & Zavala, 1970; Iwasaki, Liu, Yamamoto, & Konno, 1985; Leech, 1971; Manghani, Dastur, Nanavaty, & Patel, 1986; Morecki & Zimmerman, 1969; Mrak & Young, 1993; Sandhyamani et al., 1981) and in experimentally inoculated animals (Charlton & Casey, 1979; Fekadu et al., 1982; Hottle, Morgan, Peers, & Wyckoff, 1951; Iwasaki & Clark, 1975; Iwasaki, Ohtani, & Clark, 1975; Murphy, Bauer, Harrison, & Winn, 1973; Miyamoto & Matsumoto, 1965; Perl, Callaway, & Hicklin, 1972). In addition, many aspects of rabies virus replication and virus/ host cell interaction have been studied ultrastructurally *in vitro* (Davies, Englert, Sharpless, & Cabasso, 1963; Hummeler, Koprowski, & Wiktor, 1967; Iwasaki & Clark, 1977; Iwasaki & Minamoto, 1982; Iwasaki, Wiktor, & Koprowski, 1973; Lewis & Lentz, 1998; Matsumoto & Kawai, 1969; Matsumoto, Schneider, Kawai, & Yonezawa, 1974). Negri bodies show a similar spectrum of ultrastructural features in both human and animal material, being composed of large aggregates of granulo-filamentous matrix material (matrix) and varying numbers of viral particles (Figure 9.4). The matrix consists of randomly oriented viral nucleocapsids (Hummeler, Tomassini, Sokol, Kuwert, & Koprowski, 1968; Schneider et al., 1973). In "immature" inclusions, it has a filamentous appearance with visible substructural coiling of the nucleocapsid strands, but with maturation and increasing density, the individual strands become increasingly difficult to resolve, such that the matrix has a more granular and electron-dense appearance (Matsumoto et al., 1974; Miyamoto & Matsumoto, 1967; Murphy et al., 1973). Bullet-shaped or tubular virions are typically associated with the matrix accumulations (for more detailed descriptions, see Perl & Good (1991), Iwasaki & Tobita (2002), and Chapter 2), but may be absent in some Negri bodies (Morecki & Zimmerman, 1969; Perl & Good, 1991; Sandhyamani et al., 1981). In a study of experimentally infected rhesus monkey tissue, Perl and colleagues (Perl et al., 1972; Perl & Good, 1991) described three basic ultrastructural configurations of Negri bodies. In the first type, seen most often in the thalamus and caudate nuclei, many bullet-shaped rabies virions were found around the periphery of the matrix, where they were often attached to dilations in the endoplasmic

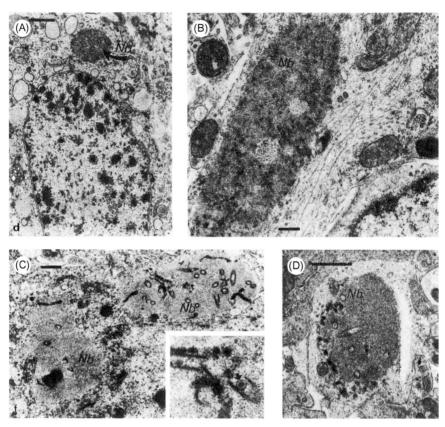

FIGURE 9.4   **Ultrastructural features of Negri bodies.** (A) A circular Negri body (Nb) in the perikaryon of a hippocampal neuron from a human patient. (B) A large elongated Negri body in a dendrite of a cortical neuron from a mouse. (C, D) Negri bodies in infected mouse brains containing bullet-shaped virions. The inset in C shows a virion with its core. Scale bars correspond in A, C, D to 1 μm, in B to 0.5 μm. *(Reproduced from Kristensson et al. (1996) with permission.)*

reticulum. Bullet-shaped virions were also frequently located within deep invaginations of the matrix contour, accounting for the inner bodies (Innerkörperchen) seen by light microscopy. However, entrapment of ribosomes and endoplasmic reticulum has been proposed as an alternative explanation (Iwasaki & Tobita, 2002). In the second type of inclusion, found predominantly in brainstem, cerebellar, and spinal neurons, tubular virions were dispersed throughout the matrix, whereas in the third type (in hippocampal neurons) virions were typically not seen within aggregates of matrix.

Lyssa bodies, without the inner body, exhibit essentially the same ultrastructural features as Negri bodies (Matsumoto, 1963; Miyamoto & Matsumoto, 1965). Furthermore, a much more extensive distribution of lyssa bodies is observed in viral infection by electron microscopy than can be resolved by routine light microscopy. Lyssa bodies, which form small aggregates of virions and matrix, are undetectable by light microscopy in neuronal perikarya and also in dendrites and axons (Charlton & Casey, 1979; Fekadu et al., 1982; Iwasaki et al., 1985; Jenson, Rabin, Wende, & Melnick, 1967; Murphy et al., 1973; Murphy, Harrison, Winn, & Bauer, 1973; Perl & Good, 1991). Small aggregates of matrix and virions have also been observed in astrocytes (Fekadu et al., 1982; Iwasaki & Clark, 1975; Matsumoto, 1963; Perl & Good, 1991) and oligodendrocytes (Perl & Good, 1991).

In the later stages of rabies virus assembly, the virions acquire a lipid bilayer envelope by budding through host cell plasma membranes into the extracellular space (see Chapter 2). Virions also may bud intracytoplasmically through membranes of the endoplasmic reticulum or, less frequently, Golgi apparatus or outer lamella of the nuclear envelope (Charlton & Casey, 1979; Gosztonyi, 1994; Iwasaki & Clark, 1975; Iwasaki et al., 1985; Iwasaki et al., 1975; Murphy et al., 1973). Importantly, cell surface viral budding may not be associated with adjacent Negri bodies or nucleocapsid matrix (Iwasaki & Tobita, 2002). Also, neuron-to-neuron transmission of rabies virus, especially at synaptic junctions, has been clearly established ultrastructurally (Burrage, Tignor, & Smith, 1983; Charlton & Casey, 1979; Iwasaki et al., 1985; Iwasaki et al., 1975).

Recent experimental investigations employing *in vitro* infection of neuronal and non-neuronal cell lines with CVS strain of fixed rabies virus indicate that Negri body-like structures that develop in the infected cells are likely sites of viral transcription and replication (Lahaye et al., 2009) with viral exploitation of cellular compartmentalization and cellular proteins, especially the innate immune response receptor Toll-like receptor 3 (TLR3), to promote viral replication and potentially evade apoptosis (Ménager et al., 2009). However, it is noteworthy that, although Negri bodies are a characteristic cytological feature in many cases of infection with "street" rabies virus strains, they are almost never observed with "fixed" virus strains (Kristensson et al., 1996). This raises the question of the exact correlation between the Negri body-like structures characterized in the above *in vitro* studies using fixed virus strains and the classic Negri bodies seen in wild-type infection. The fact that Negri bodies are almost never found following *in vivo* infection with fixed strains indicates that, whereas their presence is not essential for a fatal outcome, they may be a reflection of a more efficient "subversive neuroinvasive strategy" (Lafon, 2011) in naturally occurring infection but not in the context of experimental infection with attenuated fixed virus strains.

3.5 Degeneration of Neuronal Processes

The fact that the pathological abnormalities described above, that is, perivascular inflammation, microglial activation, neuronophagia, and Negri bodies, may be minimal or even absent in some fatal cases (Jogai et al., 2000), indicates that these are not essential neuropathological accompaniments of neurological dysfunction in rabies virus infection. This suggests that morphological correlates of neuronal dysfunction in rabies, should they exist, are likely to be most apparent at a fine structural scale. The neuropil microvacuolation observed by Charlton and colleagues (Charlton, 1984; Charlton et al., 1987) in experimentally infected skunks and foxes consisted of membrane-bound vacuoles in neuronal processes, predominantly dendrites, and has some similarities to excitotoxic amino acid-induced dendritic swelling (Charlton et al., 1987). Li, Sarmento, & Fu (2005) found severe disorganization and destruction of axons and dendrites, with relative preservation of the neuronal cell bodies, in mice infected with a pathogenic rabies virus strain (N2C virus derived from CVS-24), but not with an attenuated strain (SN-10 derived from the SAD B19 vaccine strain). There was complete loss of neurofilament and MAP-2 (microtubule-associated protein 2) immunoreactivity in neurons infected with the pathogenic strain. These investigators consequently proposed that pathogenic rabies virus infection may induce degeneration of neuronal processes by disrupting cytoskeletal integrity and that this may form the basis for neuronal dysfunction in rabies (Li et al., 2005). The potential pathogenetic role of structural changes in neuronal processes was assessed by Scott et al. (2008) in transgenic mice expressing yellow fluorescent protein (YFP) in subpopulations of CNS neurons following peripheral inoculation with the challenge virus standard (CVS-11) fixed rabies virus strain. Although histopathological changes were minimal in paraffin embedded brain sections from moribund CVS-infected animals in this study, fluorescence microscopy showed marked beading and fragmentation of the dendrites and axons of pyramidal neurons in layer V of the cerebral cortex, cerebellar mossy fibers and of axons in brainstem tracts (see Chapter 8). On resin embedded sections, numerous vacuoles were seen in the perikarya and proximal dendrites of pyramidal neurons in the cerebral cortex and CA1 hippocampal sector and also throughout the neuropil of the cerebral cortex. Ultrastructurally, there were swollen mitochondria within the perkarya, dendrites and axons of many cortical pyramidal neurons, and also frequently at sites of dendritic and axonal beading (Figure 9.5). However, affected cortical neurons, which also typically contained nucleocapsid material, did not otherwise show overt degenerative features, such as perikaryal swelling, loss of plasma or nuclear membrane integrity, or abnormal chromatin condensation. The neuropil vacuolation seen by light microscopy on resin sections corresponded ultrastructurally with swollen neuronal processes and very distended presynaptic nerve endings, with many of the neuropil vacuoles containing swollen mitochondria and membranous-type debris (Figure 9.5). It was concluded that the preceding spectrum of structural changes was sufficient to explain

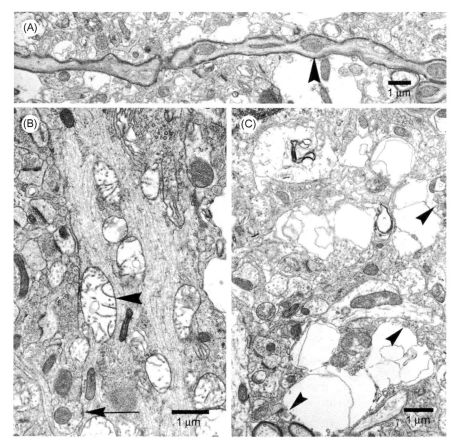

FIGURE 9.5   Electron micrographs of the cerebral cortex of a moribund CVS infected mouse showing (A) an axon of a pyramidal neurons containing swollen mitochondria (arrowhead) corresponding with areas of beading; (B) a dendrite containing swollen mitochondria (arrowhead) and (C) vacuolated presynaptic nerve endings containing synaptic vesicles (arrowheads). *(Reproduced from Scott et al. (2008) with permission; doi: 10.1128/ JVI.01677-07 Copyright © 2008, American Society for Microbiology.)*

the severe clinical disease and fatal outcome in this experimental rabies model (Scott et al., 2008). There is recent evidence that this neuronal process degeneration is a result of oxidative stress (Kammouni et al., 2012; Jackson, Kammouni, Zherebitskaya, & Fernyhough, 2010) (see Chapter 8).

3.6 Distribution of Rabies Virus Antigen

The development and use of immunofluorescence (Bingham & van der Merwe, 2002; Charlton & Casey, 1979; Goldwasser & Kissling, 1958; Johnson, Swoveland, & Emmons, 1980; Murphy et al., 1973) and immunoperoxidase techniques (Fekadu, Greer, Chandler, & Sanderlin, 1988; Iwasaki et al., 1985; Jackson & Reimer, 1989; Last, Jardine, Smit, & van der Lugt, 1994; Tirawatnpong et al., 1989) for the detection of rabies virus antigen represented important methodological advances for both the diagnosis and investigation of rabies virus infection. In 1958, Goldwasser & Kissling (1958) used immunofluorescence microscopy to demonstrate rabies virus antigen in infected brain tissue, and they established that Negri bodies contain viral antigen. The fluorescent antibody test (FAT) subsequently became established as an important routine laboratory method for the rapid diagnosis of rabies (Bingham & van der Merwe, 2002).

Comprehensive immunohistochemical studies of the distribution of rabies virus antigen in human CNS material have been relatively few in number but have shown a fairly consistent pattern. Polyclonal or monoclonal anti-ribonucleoprotein/nucleocapsid antibodies have typically been used in these studies (Jogai et al., 2000; Iwasaki et al., 1985; Johnson et al., 1980; Tirawatnpong et al., 1989). Rabies virus antigen (RVAg) was found throughout the brain and spinal cord in most cases. It was consistently present in far more neurons than those containing Negri bodies and also in cases in which no Negri bodies were found, despite a meticulous search. In a series of 20 cases with a clinical diagnosis of rabies, a histopathological diagnosis could be made in only 17 cases, whereas all of the cases exhibited positive RVAg-immunohistochemical staining (Jogai et al., 2000). RVAg was typically seen in the cytoplasm of neuronal perikarya and in dendrites and axons and appeared as blob-like masses (10–20 μm) and granules (1–3 μm) (Figure 9.6), with the larger masses corresponding with Negri bodies seen on hematoxylin and eosin-stained sections (Jogai et al., 2000). The intensity of staining may vary from cell to cell and some neurons showed diffuse staining of their cytoplasm and processes (Feiden, Feiden, Gerhard, Reinhardt, & Wandeler, 1985; Iwasaki et al., 1985; Johnson et al., 1980). RVAg was also found in processes in the neuropil remote from cell bodies as oval or spindle-shaped masses (Iwasaki et al., 1985) and was inconsistently present in some astrocytes and oligodendrocytes (Feiden et al., 1985; Jogai et al., 2000; Tirawatnpong et al., 1989).

In a quantitative study of neuronal infection in a human case, neurons with Negri bodies contained larger mean amounts of RVAg than those that did not. Of a variety of neuronal populations, cerebellar Purkinje cells and periaqueductal gray matter neurons showed the largest percentage area for both Negri bodies and RVAg signal, whereas neurons in the trochlear nucleus had a much smaller area of Negri bodies, despite a similar RVAg signal (Jackson, Ye, Ridaura-Sanz, & Lopez-Corella, 2001). Rabies virus genomic RNA and mRNA has been detected by *in situ* hybridization in brain tissue from human rabies cases and in

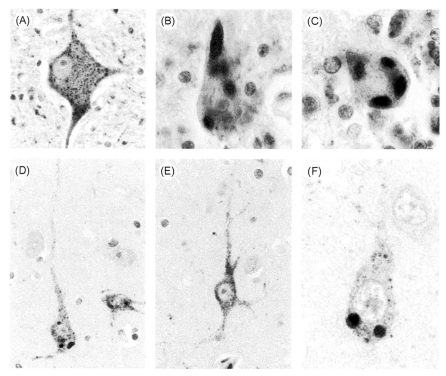

FIGURE 9.6   **Immunoperoxidase staining for rabies virus antigen (mouse monoclonal antirabies virus nucleocapsid protein IgG) in human rabies cases.** (A) Motor neuron in anterior horn of spinal cord; (B,C) cerebellar Purkinje cells; (D–F) pyramidal neurons in cerebral cortex. The larger immunolabeled masses correspond with Negri bodies. (magnifications, A ×256, B ×535, C ×567, D ×300, E ×290, F ×516).

both CVS- and street virus-infected mouse brains. The distribution of virus RNA was similar to that of viral antigen, although the amount of RNA signal was generally lower than that of RVAg, especially in dendrites (Jackson, Reimer, & Wunner, 1989; Jackson & Wunner, 1991; Jackson, 1992).

In a study of three encephalitic rabies cases, Feiden et al. (1985) observed RVAg in all brain regions, the highest amounts being found in the hippocampus, hypothalamus, and tegmental region of the lower brainstem, with many positive neurons also being present in the ventral thalamus and basal portion of the lower brainstem. In the cerebral cortex, basal ganglia and gray matter of the spinal cord, there was a more patchy distribution of fewer virus-containing neurons. In the cerebellar cortex, Purkinje cells as well as neurons in the molecular and internal granule cell layers contained numerous immunopositive inclusions (Feiden et al., 1985). Tirawatnpong et al. (1989), in a study of four encephalitic and three paralytic rabies cases, did not find any correlation between the distribution of RVAg and the presenting clinical manifestations. In patients who survived 7 days or less, there were a greater number of antigen-positive neurons in the brainstem and spinal cord. In those who survived longer than 7 days, a similar degree of widespread neuronal involvement was seen in the spinal cord and in supratentorial and infratentorial structures. RVAg was found in neurons of the dorsal and ventral horns of the spinal cord, regardless of the clinical pattern, and the site of the infecting bite was not associated with a particular antigen distribution. RVAg-positive neurons were typically found in all layers of the cerebral cortex and cortical involvement did not clearly correlate with the degree of disturbance of consciousness (Tirawatnpong et al., 1989). In a study of twelve paralytic and eight encephalitic rabies cases by Jogai et al. (2000), the maximum amount of RVAg was found in the hippocampus, followed by the pons, medulla, and cerebellum, whereas antigen was relatively minimal in the cerebral cortex in most cases. In three cases in which Negri bodies were absent, RVAg was present in all regions examined in one case and was restricted to the pons and medulla in the two other cases. Jogai et al. (2000) found a positive correlation between the degree of inflammation and intensity of RVAg-immunopositivity, whereas Tirawatnpong et al. (1989) did not observe a correlation between either the anatomical distribution or amount of RVAg and inflammation in their cases. In a series of nine human rabies cases (and also in six naturally infected "furious" canines) Suja et al. (2011) observed RVAg in almost all brain areas, but with striking involvement of lateral and ventral group of thalamic nuclei, basal ganglia and limbic structures. There was labeling of multiple Negri bodies and antigen was also observed in oligodendrocytes and the processes of fibrous astrocytes (Suja et al., 2011).

Animal models have been especially valuable for investigating the centripetal and centrifugal spread of rabies virus to and from the CNS and for establishing the anatomic sequence of spread within the CNS (Baer, Shanthaveerappa, & Bourne, 1965; Charlton & Casey, 1979; Charlton, Casey, Wandeler, & Nadin-Davis, 1996; Coulon et al., 1989; Dean, Evans, & McClure, 1963; Huygelen, 1960; Huygelen & Mortelmans, 1959; Jackson & Reimer, 1989; Kliger & Bernkopf, 1943; Murphy, Bauer, et al., 1973; Murphy, Harrison, et al., 1973; Schneider, 1969a,b; Schneider & Hamann, 1969; Shankar, Dietzschold, & Koprowski, 1991; Smart & Charlton, 1992). Schneider (1969a) established an ascending pattern of infection in the mouse by sequential immunofluorescence and infectivity titration studies of parts of the CNS. In experimentally infected mice, rats, and hamsters, spread of virus from the lumbar spinal cord to the brainstem happens within hours (Baer, Shantha, & Bourne, 1968; Baer, Shanthaverrappa, et al., 1965; Murphy et al., 1973).

Murphy (1977) described an "ascending wave of rabies infection in the brain," as demonstrated by a decreasing immunofluorescent gradient from brainstem to forebrain at earlier stages of infection, and massive accumulations of RVAg throughout the brain and spinal cord at terminal stages (Murphy et al., 1973). In skunks inoculated into hindlimb foot muscle with street rabies virus, Charlton & Casey (1979) observed granular RVAg immunoflourescence in scattered neurons in the spinal cord, medulla oblongata, pons, and cerebellum in one animal killed at 10 days and another at 14 days post-inoculation. In all other skunks killed on day 14 or later, there was intense immunofluorescence in gray matter throughout the brain and spinal cord. Analysis of the early events in this skunk model indicated viral entrance and replication at L2 and L3 levels of the spinal cord, local spread by propriospinal neurons, and early and rapid spread to the brain via long ascending and descending fiber tracts (Charlton et al., 1996). RVAg-positive fine particles, aggregates, and filaments were found in most neurons and fine granules aligned within axons were seen in the white matter (Charlton & Casey, 1979; Charlton et al., 1996; Smart & Charlton, 1992).

Following hindlimb footpad inoculation of mice with fixed rabies virus (Jackson & Reimer, 1989), RVAg was initially detected by immunoperoxidase staining on day 4 in lumbar dorsal root ganglia and gray matter of the lumbar spinal cord, with much greater involvement on the side of inoculation. A few positive neurons were also seen in the brainstem tegmentum at this time. On day 5, there were more infected neurons in the lumbar and sacral cord segments, including ventral horn neurons contralateral to the inoculation side. There was heavy involvement of the brainstem tegmentum, prominent infection of the deep cerebellar nuclei, but only a few positive Purkinje cells and no RVAg in the cerebral cortex or hippocampus. On day 6, there was again more signal in the spinal cord and brainstem tegmentum. Numerous Purkinje cells and a few cerebellar internal granule cells were immunopositive. Many neurons in the cerebral cortex were also now infected, but there was still no evidence of hippocampal infection. Hippocampal RVAg was initially seen in the CA3 region on day 8, with spread to CA1 and CA4 and sparse involvement of the dentate gyrus by day 10. The quantity of RVAg in the CNS decreased progressively in surviving mice after day 10, with death occurring between day 10 and 18 (Jackson & Reimer, 1989). One conclusion from this study was that the hippocampus is not a good location for the detection of early CNS infection following peripheral inoculation. A similar conclusion was reached on the basis of fluorescent antibody testing (FAT) of 252 rabies-positive brains from a diverse range of naturally infected species by Bingham & van der Merwe (2002). Whereas Negri bodies, when present, are typically most readily found in the hippocampus, cerebellar Purkinje cells and pyramidal neurons of the cerebral cortex (Tierkel, 1973a), the hippocampus, cerebellum, and different parts of the cerebrum were negative by FAT in 4.9, 4.5 and 3.9–11.1% respectively of Bingham and van der Merwes' series. By contrast, the only structures that were FAT-positive in all 252 animal brains were the thalamus, pons, and medulla oblongata, with a consistent abundance of RVAg in the thalamus (Bingham & van der Merwe, 2002).

Suja et al. (2011) reported significant differences in the patterns of RVAg immunohistochemical staining between mice experimentally inoculated with street virus versus CVS strain. Following intramuscular inoculation with street virus there was widespread infection, with a caudocranial gradient and intraneuronal aggregation of viral antigen into multiple Negri body-like globular masses and extensive dendritic spread. By contrast, in CVS infected animals there was diffuse cytoplasmic staining of neurons and minimal dendritic spread.

4 PATHOLOGY IN THE PERIPHERAL NERVOUS SYSTEM

Spread of rabies virus through the peripheral nervous system (PNS) plays an essential role in the centripetal and centrifugal phases of rabies infection. There are associated inflammatory, reactive, and degenerative changes in many of the structural components of the PNS, including neuronal cell bodies in sensory and autonomic ganglia and their capsular/satellite cells, as well as in sensory and motor axons and their enveloping Schwann cells. This results in varying degrees of degeneration and loss of sensory and autonomic neurons (neuronopathy), reactive proliferation of their satellite cells and demyelination, and Wallerian degeneration of nerve fibers in spinal nerve roots and peripheral nerves. Historically, there has been particular interest in the diagnostic utility of histological changes in the peripheral nerve ganglia.

4.1 Changes in Sensory and Autonomic Ganglia

The proliferation of capsule (satellite) cells was briefly described in 1872 by Nepveau in the Gasserian ganglion (sensory ganglion of the trigeminal nerve) of a human rabies case, but it was Van Gehuchten and Nelis in 1900 who first recognized the importance of ganglionic lesions as a diagnostic marker of rabies. They described marked proliferation of capsular cells surrounding chromatolytic neurons in spinal and cranial nerve ganglia of animal and human rabies cases (Figure 9.7). This capsular cell reaction, together with varying degrees of interstitial lymphocytic infiltration, resulted in grossly apparent enlargement and increased firmness of involved spinal ganglia (Van Gehuchten & Nelis, 1900;

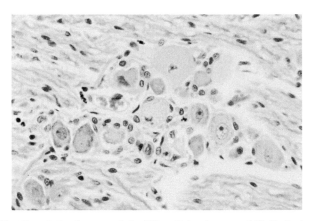

FIGURE 9.7   Van Gehuchten nodule ('Van Gehuchten and Nelis lesion') in rat tri-geminal ganglion following inoculation of street virus into the ipsilateral mental nerve. There is a group of chromatolytic neurons accompanied by proliferation of satellite cells and a sparse lymphocytic infiltrate. (HE stain, magnification ×285). *(Reproduced from Iwasaki and Tobita, 2002, with permission.)*

Perl & Good, 1991; Iwasaki & Tobita, 2002). These changes, often referred to in the literature as "Van Gehuchten and Nelis lesions" or "Van Gehuchten nodules" were reconfirmed in a number of later studies (Hardenbergh, 1916; Marinesco & Storesco, 1931; Tangchai & Vejjajiva, 1971; Sung et al., 1976; Mitrabhakdi et al., 2005).

In a series of 52 human rabies cases, Herzog (1945) could not find Negri bodies in the hippocampus in almost 50% of the cases, whereas he observed capsular cell proliferation and neuronal degenerative changes in the ganglion nodosum of the vagus nerve in all 52 cases, with or without accompanying focal or diffuse inflammatory cell infiltration. In a study of 9 human rabies cases, Tangchai & Vejjajiva (1971) reported leu-kocyte infiltration in 3, hypertrophy and proliferation of capsular cells in 5, neuronal degeneration in 5, and extreme vascular congestion in 2 cases. There was also increased stromal and epineurial collagen in the ganglia. Many of the neurons were moderately swollen, with pale finely vacuolated cytoplasm, and some fragmented neurons undergoing neuro-nophagia by histiocytes were seen. Some neurons also contained round acidophilic cytoplasmic inclusions (5–10 μm), although these lacked inner bodies (Tangchai & Vejjajiva, 1971). Numerous Negri bodies have been found in the Gasserian (Garcia-Tamayo, Avila-Mayor, & Anzola-Perez, 1972) and dorsal root ganglia (Sung et al., 1976) in some cases. Severe lymphocytic inflammation and necrosis has been observed in the infe-rior cervical sympathetic ganglion of an encephalitic rabies case with prolonged survival (Sandhyamani et al., 1981). Mitrabhakdi et al. (2005) reported electrophysiological abnormalities consistent with dorsal root ganglionopathy in one encephalitic rabies patient and two paralytic patients who had severe prodromal paresthesiae. Postmortem analysis showed severe ganglionitis, with infiltration predominantly by CD3-positive T lymphocytes.

In experimentally infected immature hamsters, Murphy et al. (1973) first observed RVAg immunofluorescence in small numbers of ipsilateral lumbar dorsal root ganglion neurons 60 to 70 hours following hindlimb inoculation and in lumbar autonomic ganglia by 72 hours. The RVAg initially had a "dust-like" distribution, with rapid subsequent progression to involvement of nearly all dorsal root ganglia by brilliant aggregate fluorescence. This was most dense at the peripheral margins of individual ganglion cells, with myelinated axon hillocks being especially heavily infected. Satellite cells did not contain RVAg. Ultrastructurally, large masses of nucleocapsid material were seen within ganglion cells, particularly at their margins, with comparatively small numbers of virus particles budding from intracytoplasmic membranes deeper within the cytoplasm (Murphy et al., 1973). In skunks, Charlton & Casey (1979) found RVAg immunofluorescence in scattered lumbar dorsal root ganglia 10 days following hindlimb inoculation, at which time no fluorescence was seen in peripheral nerve fibers. By 14 days post-inoculation, many dorsal root neurons contained antigen and linear arrays of granular fluorescent RVAg were seen in axons of hindlimb and forelimb peripheral nerves. After 14 days, variable but increasing degrees of neuronal chromatolysis, neuronophagia and inflammatory cell infiltration were seen in the trigeminal and dorsal root ganglia (Charlton & Casey, 1979).

In adult mice inoculated in the right hindlimb footpad with the CVS strain, Rossiter, Hsu, & Jackson (2009) first observed RVAg in a right lumbo-sacral dorsal root ganglion of one of three infected animals at 3 days post-inoculation (p.i.) and by 3.5 p.i. antigen could be traced within the dorsal roots and spinal cord. At day 4.5, p.i. antigen was first seen in left-sided dorsal root ganglia. At later time points, there was bilateral infection of ganglia. Multifocal mononuclear inflammatory cells infiltrates of dorsal ganglia were first observed at 4 days p.i., and these infiltrates subsequently became more prominent. Degenerating gangliocytes became increasingly frequent after 4 days p.i. There was a spectrum of light microscopic changes, with varying degrees of nuclear eccentricity and irregularity, cytoplasmic chromatolysis and vacuolation, to more advanced degeneration with neuronophagia. The neuronal nuclei tended to be paler and their nuclei smaller and less intensely stained than those in mock-infected mice. However, there was no evidence of karyorrhectic chromatin condensation characteristic of apoptosis and TUNEL staining or immunolabeling for activated caspase-3 was not observed in gangliocytes of CVS- or mock-infected mice. The above degenerative changes preferentially involved gangliocytes that had a medium to large soma size and that were intermixed with morphologically normal neurons. The satellite (capsule) cells surrounding degenerating gangliocytes tended to be larger and more numerous than those around morphologically intact neurons. Ultrastructurally. there was a spectrum of changes in gangliocytes that included features characteristic of "the axotomy response" (nuclear eccentricity, central accumulation of mitochondria, and Golgi apparatus and preferential localization of rough endoplasmic reticulum towards the periphery of perikarya) and the appearance of numerous autophagic compartments and aggregation of intermediate filaments, while the neurons retained relatively intact mitochondria and plasma membranes. At later stages the neuronal cytoplasm was highly vacuolated, but some cells still contained relatively intact mitochondria (Rossiter et al., 2009).

4.2 Changes in Spinal Nerve Roots and Peripheral Nerves

Pathological studies of spinal nerve roots and peripheral nerves in human rabies cases have been fewer and have tended to be less comprehensive than those of other parts of the nervous system. Knutti (1929) observed from slight focal to extensive necrosis in dorsal nerve roots in a case of paralytic rabies, with only minor changes in the ventral roots. Hurst & Pawan (1932) noted lymphocytic infiltrates in the connective tissue sheath of the sciatic nerve in one of their paralytic cases. In peripheral nerves from 9 encephalitic rabies cases, including material from the facial region, upper and lower limbs, Tangchai & Vejjajiva (1971) observed diffuse perivenous, sub-epineurial and sub-perineurial mononuclear inflammatory cell infiltration (3 cases), degeneration of nerve fibers (7 cases), proliferation and hypertrophy of Schwann cells (5 cases) and sub-epineurial and -perineurial edema (3 cases). Similar changes were seen in dorsal spinal nerve roots, but these were milder. Chopra et al. (1980) studied a total of 4 spinal nerves and 17 peripheral nerves in their series of 11 paralytic rabies cases. In all of the spinal nerves there was both Wallerian degeneration and segmental demyelination, but inflammatory cell infiltration was seen in only one of the spinal nerve specimens. The peripheral nerves showed variable degrees of segmental demyelination and remyelination, loss of myelinated fibers, and Wallerian degeneration. In 9 of 17 nerves, segmental demyelination was the primary lesion. There was no apparent relationship between the degree of spinal or peripheral nerve pathology and the duration of incubation or clinical illness (Chopra et al., 1980). Mitrabhakdi et al. (2005) observed heavy lymphocytic infiltration of dorsal and ventral roots in two paralytic cases, but, by contrast, they observed only a mild degree of inflammation in dorsal and ventral roots in encephalitic rabies cases.

A predominant pattern of acute motor axonal neuropathy involving ventral spinal roots and peripheral nerves, in the absence of prominent inflammation or motor neuron degeneration, has been found in a paralytic rabies case, which was initially diagnosed and reported as an axonal Guillain-Barré syndrome (Sheikh et al., 2005). RVAg was observed in multiple lumbar anterior horn cells and in their dendrites in this case, as well as in a large proportion of ventral root myelinated axons. Ultrastructurally, mature viral particles were seen in some axons in the ventral root exit zone. Double-label immunostaining showed colocalization of human IgG and C3d complement activation marker with RVAg on ventral root axons, supporting the possibility that the pathogenesis of paralytic rabies may include immune-mediated axonal degeneration (Sheikh et al., 2005).

In experimentally infected hamsters, Murphy (1977) first observed RVAg in peripheral nerves proximal to the inoculation site only concomitant with or later than RVAg detection in ipsilateral spinal ganglia or spinal cord, with individual axons showing a fine linear dust-like pattern of immunofluorescence. With disease progression, the majority of axons in peripheral nerves of the inoculated hindlimb, and subsequently throughout the body, showed this pattern. Ultrastructurally, there was concentration of virus particles at nodes of Ranvier, reflecting budding from the high density of membranous organelles at these sites. By contrast, in internodal regions, virus particles only budded individually or in small groups from plasma membranes, resulting in the presence of virions between the axonal and adjacent Schwann cell plasma membranes (Jenson, Rabin, Bentinck, & Melnick, 1969; Murphy et al., 1973). Murphy et al. (1973a) never found viral particles in Schwann cells, whereas Atanasiu & Sisman (1967) did observe Schwann cell infection. In sciatic nerves of mice experimentally inoculated with canine rabies virus, degeneration of approximately 40% of myelinated axons was found, while only occasional degenerating unmyelinated axons were observed (Teixeira et al., 1986). Axonal degeneration and severe demyelination, possibly immunologically mediated, was found in the trigeminal and facial nerves of experimentally infected rats with street rabies virus (Minguetti, Hofmeister, Hayashi, & Montano, 1997).

5 PATHOLOGY INVOLVING THE INOCULATION SITE, EYE, AND EXTRANEURAL ORGANS

Localized replication of rabies virus in extraneural tissue at the inoculation site, including within skeletal muscle, may be an important feature of rabies virus pathogenesis preceding centripetal spread through peripheral nerves to the CNS. Later in the course of infection, centrifugal spread through both somatic sensory and autonomic divisions of the peripheral nervous system results in involvement of a broad range of extraneural tissues and organs, including lacrimal and salivary glands, cornea, skin, heart, gastrointestinal tract, and adrenal glands.

5.1 Changes at Inoculation Site

Rabies infection usually results from inoculation of saliva into deep soft tissues through a bite from a rabid animal. Direct uptake of virus by sensory nerve endings has been documented in several studies (Baer et al., 1968; Coulon et al., 1989; Dean et al., 1963; Kucera, Dolivo, Coulon, & Flamand, 1985; Shankar et al., 1991). However, in many cases initial local replication within skeletal muscle at the inoculation site likely precedes entry into peripheral nerves at motor end plates and neuromuscular and neurotendinal spindles (Charlton & Casey, 1979; Murphy et al., 1973). Furthermore, viral sequestration and delayed progression of infection in skeletal muscle fibers at the site of exposure appears to be a key factor in rabies cases with a long incubation period (Baer & Cleary, 1972; Charlton, Nadin-Davis, Casey, & Wandeler, 1997). Following experimental intramuscular inoculation of immature hamsters (Murphy et al., 1973), the first evidence of viral replication was seen as RVAg in individual nearby striated muscle fibers, rapidly followed by involvement of groups of fibers. Ultrastructurally, moderate numbers of virus particles were found budding from the membranes of the sarcolemma and sarcoplasmic reticulum, in the absence of any associated inflammatory response or significant cytopathological changes. This was rapidly followed by detection of RVAg in neuromuscular and neurotendinal spindles (the sensory stretch receptors found deep within skeletal muscles and tendons respectively), with the unmyelinated nerve endings wrapping these proprioceptors also frequently containing RVAg. Later in the course of infection, following centrifugal spread, infected neuromuscular spindles were found in other parts of the body and were especially numerous in the subcutaneous musculature of the nose (Murphy et al., 1973).

In the case of bat-transmitted rabies, it appears that the particular ability of bat rabies strains to infect and replicate within epithelial cells and fibroblasts accounts for the observation that even superficial bat-inflicted wounds have a high probability of causing clinical disease (Love & Wiley, 2002; Morimoto et al., 1996).

5.2 Ocular Pathology

Given that the retina is a direct extension of the CNS, it is perhaps surprising how few detailed accounts of ocular pathology exist in the literature. Haltia, Tarkkanen, & Kivela (1989) described the ocular pathology in the case of a 30-year-old man who developed rabies after receiving several bites from a bat. These authors observed lymphocytic and plasma cell infiltrates in the ciliary body and focally in the choroid, focal loss of retinal pigment epithelium, perivascular inflammation in the retinal nerve fiber layer, focal endothelial destruction and occlusion of retinal veins, destruction of many retinal ganglion cells, and partial loss of bipolar cells. RVAg was seen in the cytoplasm of many of the surviving ganglion cells. In experimentally infected rabbits, Dejean (1937) observed corneal sensory loss and clouding, retinal venous congestion, choroidal hemorrhages, vitreous clouding, and the presence of Negri-like bodies in retinal ganglion cells. Murphy et al. (1973) observed large aggregates of RVAg in the retinal ganglion cell layer of nearly every terminally infected hamster and focal dust-like antigen in the inner and outer retinal nuclear layers and corneal epithelium in some animals.

Centrifugal spread of virus via sensory fibers to corneal epithelium underlies the use of immunofluorescent staining of corneal impressions for antemortem diagnosis of rabies (Schneider, 1969c) and also explains the rare instances of rabies transmission through infected corneal transplants (Gode & Bhide, 1988; Houff et al., 1979).

5.3 Changes in Extraneural Organs

Centrifugal viral spread to cutaneous nerve endings surrounding hair follicles (especially in the head region) forms the basis for antemortem diagnosis by means of immunostained nuchal skin biopsies, in a large proportion of animal (Blenden, Bell, Tsao, & Umoh, 1983) and human rabies cases (Blenden, Creech, & Torres-Anjel, 1986; Bryceson et al., 1975). In some cases, RVAg is also found in epidermal cells (Bago, Revilla-Fernandez, Allerberger, & Krause, 2005; Balachandran & Charlton, 1994; Jackson et al., 1999).

Centrifugal spread to the major salivary glands, resulting in production of saliva containing high titers of rabies virus, is a central feature of bite transmission of rabies by natural vectors, such as dog (Goldwasser, Kissling, Carski, & Hosty, 1959), fox (Balachandran & Charlton, 1994; Dierks, Murphy, & Harrison, 1969), and skunk (Balachandran & Charlton, 1994). Ultrastructurally, budding of numerous virions from the apical membranes of mucogenic cells and their release into intercellular canaliculi has been documented (Balachandran & Charlton, 1994; Dierks et al., 1969). In human rabies cases RVAg was found in acini of minor salivary glands of the tongue, as well as in skeletal muscle fibers of the tongue, but there was no significant involvement of acini in the major salivary glands (Jackson et al., 1999; Li, Feng, & Ye, 1995).

Widespread distribution of RVAg has been observed in autonomic nerve plexuses related to multiple organs, including cardiac ganglia and the submucosal plexus of Meissner and myenteric plexus of Auerbach

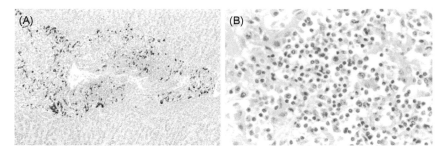

FIGURE 9.8 Adrenal gland showing abundant rabies virus antigen in the adrenal medulla with sparing of the adrenal cortex (A). Mononuclear inflammatory infiltrate in the adrenal medulla (B). A, immunoperoxidase-hematoxylin; B, HE stain. magnifications, A ×80; B ×140.

in the gastrointestinal tract, in both animal (Debbie & Trimarchi, 1970; Fischman & Schaeffer, 1971; Murphy et al., 1973) and human material (Jackson et al., 1999; Jogai, Radotra, & Banerjee, 2002). There is typically an associated mild mononuclear cell inflammatory response (Jackson et al., 1999), although more prominent inflammation and degeneration of enteric ganglion cells has been observed in some cases (Love, 1944). The adrenal medulla, as an extension of the sympathetic nervous system, has been found to be frequently involved, with many cells containing RVAg in both animals (Debbie & Trimarchi, 1970; Fischman & Schaeffer, 1971; Murphy et al., 1973) and humans (Jackson et al., 1999; Jogai et al., 2002), and may be accompanied by moderate to severe inflammation ("medullitis") (Figure 9.8) (Almeida, Teixeira, de Oliveira, Brandao, & Gobbi, 1986; Love, 1944; Lopez-Corella, Ridaura-Sanz, & Samayoa-Palma, 1997; Jackson et al., 1999).

Clinical and/or pathological evidence of cardiac involvement is a recognized feature of some human rabies cases (Burton et al., 2005; Cheetham, Hart, Coghill, & Fox, 1970; Raman, Prosser, Spreadbury, Cockcroft, & Okubadejo, 1988; Ross & Armentrout, 1962; Roux et al., 1976). Myocarditis, characterized by multifocal muscle fiber degeneration/necrosis, has been observed in some cases (Burton et al., 2005; Cheetham et al., 1970; Ross & Armentrout, 1962). RVAg within cardiac myocytes has also been reported in cases with mild or no associated inflammation (Jackson et al., 1999; Metze & Feiden, 1991).

6 SUMMARY AND CONCLUSIONS

As an almost invariably fatal infection of the CNS, rabies shares a number of histopathological features with other viral encephalitides, such as leptomeningeal and perivascular mononuclear inflammatory cell infiltration, microglial activation, and neuronophagia. However, in some cases, inflammatory changes and neuronal cell death are minimal or even absent, indicating that these are not essential contributors to a fatal outcome in rabies. The presence of Negri body viral inclusions in the cytoplasm of neurons is a unique and diagnostic feature in many cases of infection with street rabies virus strains. Recent data indicate that Negri bodies may be sites of viral transcription and replication, with viral "hijacking" of cellular compartmentalization and proteins, especially the innate immune response receptor TLR3, to promote replication and potentially evade the innate immune response and apoptosis.

There is strong evidence that neuronal apoptosis does not play an important pathogenetic role in human rabies encephalitis or in most instances of animal infection with "street" rabies virus. In paralytic rabies, by comparison with encephalitic cases, there tends to be, at least in some cases, more severe involvement of the spinal cord and brainstem and a greater degree of injury in spinal nerve roots and peripheral nerves. Ultrastructural and immunohistochemical studies of human and animal rabies material have contributed greatly to an understanding of rabies pathogenesis. This includes centripetal and centrifugal phases of viral spread through the peripheral nervous system, with resulting involvement of autonomic nerve plexuses in multiple organs and viral spread to extraneural organs. The major salivary glands are of particular importance in the transmission of rabies by natural vectors. Degenerative structural changes in neuronal processes, including swelling of mitochondria in axons and dendrites and vacuolation of presynaptic nerve endings, may be sufficient to explain severe clinical disease and fatal outcome in experimental rabies.

References

Abba, F., & Bormans, A. (1905). Sur le diagnostic histologique de la rage. *Annales de L'Institut Pasteur (Paris)*, *19*, 49.

Adle-Biassette, H., Bourhy, H., Gisselbrecht, M., Chrétien, F., Wingertsmann, L., Baudrimont, M., et al. (1996). Rabies encephalitis in a patient with AIDS: A clinicopathological study. *Acta Neuropathologica*, *92*(4), 415–420.

Almeida, H. d. O., Teixeira, V. d. P., de Oliveira, G., Brandao, M. d. C., & Gobbi, H. (1986). Medulite supra-renalica em casos de raiva humana [Adrenal medullitis in cases of human rabies (Portuguese)]. *Memorias do Instituto Oswaldo Cruz*, *81*(4), 439–442.

Atanasiu, P., & Sisman, J. (1967). Morphological aspects of rabies virus (French). *Bulletin de l' Office International des Epizooties*, *67*(3), 521–533.

Babes, V. M. (1892). Sur certains caractères des lésions histologiques de la rage. *Annales de L'Institut Pasteur*, *6*, 209–223.

Baer, G. M., & Cleary, W. F. (1972). A model in mice for the pathogenesis and treatment of rabies. *Journal of Infectious Diseases*, *125*, 520–527.

Baer, G. M., Shantha, T. R., & Bourne, G. H. (1968). The pathogenesis of street rabies virus in rats. *Bulletin of the World Health Organization*, *38*, 119–125.

Baer, G. M., Shanthaveerappa, T. R., & Bourne, G. H. (1965). Studies on the pathogenesis of fixed rabies virus in rats. *Bulletin of the World Health Organization, 33*, 783–794.

Bago, Z., Revilla-Fernandez, S., Allerberger, F., & Krause, R. (2005). Value of immunohistochemistry for rapid ante mortem rabies diagnosis. *International Journal of Infectious Diseases, 9*(6), 351–352.

Balachandran, A., & Charlton, K. (1994). Experimental rabies infection of non-nervous tissues in skunks (*Mephitis mephitis*) and foxes (*Vulpes vulpes*). *Veterinary Pathology, 31*, 93–102.

Baloul, L., & Lafon, M. (2003). Apoptosis and rabies virus neuroinvasion. *Biochimie, 85*(8), 777–788.

Benedikt, M. (1878). Zur pathologischen Anatomie der Lyssa. *Virchows Archiv für pathologische Anatomie und Physiologie und fur klinische Medizin, 72*, 425–431.

Bingham, J., & van der Merwe, M. (2002). Distribution of rabies antigen in infected brain material: determining the reliability of different regions of the brain for the rabies fluorescent antibody test. *Journal of Virological Methods, 101*(1-2), 85–94.

Blenden, D. C., Bell, J. F., Tsao, A. T., & Umoh, J. U. (1983). Immunofluorescent examination of the skin of rabies-infected animals as a means of early detection of rabies virus antigen. *Journal of Clinical Microbiology, 18*(3), 631–636.

Blenden, D. C., Creech, W., & Torres-Anjel, M. J. (1986). Use of immunofluorescence examination to detect rabies virus antigen in the skin of humans with clinical encephalitis. *Journal of Infectious Diseases, 154*, 698–701.

Bryceson, A. D. M., Greenwood, B. M., Warrell, D. A., Davidson, N. M., Pope, H. M., Lawrie, J. H., et al. (1975). Demonstration during life of rabies antigen in humans. *Journal of Infectious Diseases, 131*(1), 71–74.

Bundza, A., & Charlton, K. M. (1988). Comparison of spongiform lesions in experimental scrapie and rabies in skunks. *Acta Neuropathologica, 76*, 275–280.

Burrage, T. G., Tignor, G. H., & Smith, A. L. (1983). Immunoelectron microscopic localization of rabies virus antigen in central nervous system and peripheral tissue using low- temperature embedding and protein A-gold. *Journal of Virological Methods, 7*(5-6), 337–350.

Burton, E. C., Burns, D. K., Opatowsky, M. J., El-Feky, W. H., Fischbach, B., Melton, L., et al. (2005). Rabies encephalomyelitis: clinical, neuroradiological, and pathological findings in 4 transplant recipients. *Archives of Neurology, 62*(6), 873–882.

Charlton, K. M. (1984). Rabies: spongiform lesions in the brain. *Acta Neuropathologica, 63*, 198–202.

Charlton, K. M., & Casey, G. A. (1979). Experimental rabies in skunks: immunofluorescence light and electron microscopic studies. *Laboratory Investigation, 41*, 36–44.

Charlton, K. M., Casey, G. A., Wandeler, A. I., & Nadin-Davis, S. (1996). Early events in rabies virus infection of the central nervous system in skunks (*Mephitis mephitis*). *Acta Neuropathologica, 91*, 89–98.

Charlton, K. M., Casey, G. A., Webster, W. A., & Bundza, A. (1987). Experimental rabies in skunks and foxes: Pathogenesis of the spongiform lesions. *Laboratory Investigation, 57*, 634–645.

Charlton, K. M., Nadin-Davis, S., Casey, G. A., & Wandeler, A. I. (1997). The long incubation period in rabies: delayed progression of infection in muscle at the site of exposure. *Acta Neuropathologica, 94*(1), 73–77.

Cheetham, H. D., Hart, J., Coghill, N. F., & Fox, B. (1970). Rabies with myocarditis: Two cases in England. *Lancet, 1*, 921–922.

Chopra, J. S., Banerjee, A. K., Murthy, J. M. K., & Pal, S. R. (1980). Paralytic rabies: A clinicopathological study. *Brain, 103*, 789–802.

Coulon, P., Derbin, C., Kucera, P., Lafay, F., Prehaud, C., & Flamand, A. (1989). Invasion of the peripheral nervous systems of adult mice by the CVS strain of rabies virus and its avirulent derivative AvO1. *Journal of Virology, 63*, 3550–3554.

Davies, M. C., Englert, M. E., Sharpless, G. R., & Cabasso, V. J. (1963). The electron microscopy of rabies virus in cultures of chicken embryo tissues. *Virology, 21*, 642–651.

de Brito, T., Araujo, M. D. F., & Tiriba, A. (1973). Ultrastructure of the Negri body in human rabies. *Journal of the Neurological Sciences, 20*, 363–372.

Dean, D. J., Evans, W. M., & McClure, R. C. (1963). Pathogenesis of rabies. *Bulletin of the World Health Organization, 29*, 803–811.

Debbie, J. G., & Trimarchi, C. V. (1970). Pantropism of rabies virus in free-ranging rabid red fox *Vulpes fulva*. *Journal of Wildlife Diseases, 6*, 500–506.

Dejean, C. (1937). Les modifications du fond d'oeil dans la rage chez le lapin. *Bulletin des sociétés d'ophtalmologie de France, 50*, 247–254.

Dierks, R. E., Murphy, F. A., & Harrison, A. K. (1969). Extraneural rabies virus infection. Virus development in fox salivary gland. *American Journal of Pathology, 54*(2), 251–273.

Dolman, C. L., & Charlton, K. M. (1987). Massive necrosis of the brain in rabies. *Canadian Journal of Neurological Sciences, 14*(2), 162–165.

Dupont, J. R., & Earle, K. M. (1965). Human rabies encephalitis. A study of forty-nine fatal cases with a review of the literature. *Neurology, 15*, 1023–1034.

Feiden, W., Feiden, U., Gerhard, L., Reinhardt, V., & Wandeler, A. (1985). Rabies encephalitis: immunohistochemical investigations. *Clinical Neuropathology, 4*, 156–164.

Fekadu, M., Chandler, F. W., & Harrison, A. K. (1982). Pathogenesis of rabies in dogs inoculated with an Ethiopian rabies virus strain. Immunofluorescence, histologic and ultrastructural studies of the central nervous system. *Archives of Virology, 71*, 109–126.

Fekadu, M., Greer, P. W., Chandler, F. W., & Sanderlin, D. W. (1988). Use of the avidin-biotin peroxidase system to detect rabies antigen in formalin-fixed paraffin-embedded tissues. *Journal of Virological Methods, 19*, 91–96.

Fischman, H. R., & Schaeffer, M. (1971). Pathogenesis of experimental rabies as revealed by immunofluorescence. *Annals of the New York Academy of Sciences, 177*, 78–97.

Fu, Z. F., & Jackson, A. C. (2005). Neuronal dysfunction and death in rabies virus infection. *Journal of Neurovirology, 11*(1), 101–106.

Garcia-Tamayo, J., Avila-Mayor, A., & Anzola-Perez, E. (1972). Rabies virus neuronitis in humans. *Archives of Pathology, 94*, 11–15.

Gode, G. R., & Bhide, N. K. (1988). Two rabies deaths after corneal grafts from one donor (Letter). *Lancet, 2*, 791.

Goldwasser, R. A., & Kissling, R. E. (1958). Fluorescent antibody staining of street and fixed rabies virus antigens. *Proceedings of the Society for Experimental Biology and Medicine, 98*, 219–223.

Goldwasser, R. A., Kissling, R. E., Carski, T. R., & Hosty, T. S. (1959). Fluorescent antibody staining of rabies virus antigens in the salivary glands of rabid animals. *Bulletin of the World Health Organization, 20*, 579–588.

Gonzalez-Angulo, A., Marquez-Monter, H., Feria-Velasco, A., & Zavala, B. J. (1970). The ultrastructure of Negri bodies in Purkinje neurons in human rabies. *Neurology, 20*, 323–328.

Goodpasture, E. W. (1925). A study of rabies, with reference to a neural transmission of the virus in rabbits, and the structure and significance of Negri bodies. *American Journal of Pathology, 1*, 547–584.

Gosztonyi, G. (1994). Reproduction of lyssaviruses: ultrastructural composition of lyssavirus and functional aspects of pathogenesis. In C. E. Rupprecht, B. Dietzschold, & H. Koprowski (Eds.), *Current Topics in Microbiology and Immunology, Volume 187: Lyssaviruses (pp. 43–68)*. Berlin: Springer-Verlag.

Gowers, W. R. (1877). The pathological anatomy of hydrophobia. *Transactions of the Pathological Society of London, 28*, 10–23.

Greenwood, R. J., Newton, W. E., Pearson, G. L., & Schamber, G. J. (1997). Population and movement characteristics of radio-collared striped skunks in North Dakota during an epizootic of rabies. *Journal of Wildlife Diseases, 33*(2), 226–241.

Haltia, M., Tarkkanen, A., & Kivela, T. (1989). Rabies: ocular pathology. *British Journal of Ophthalmology, 73,* 61–67.

Hardenbergh, J. B. (1916). The reliability of cell proliferation changes in the diagnosis of rabies. *Journal of the American Veterinary Medical Association, 49,* 663.

Herzog, E. (1945). Histologic diagnosis of rabies. *Archives of Pathology, 39,* 279–280.

Hottle, G. A., Morgan, G., Peers, J. H., & Wyckoff, R. W. G. (1951). The electron microscopy of rabies inclusion (Negri) bodies. *Proceedings of the Society for Experimental Biology and Medicine, 77,* 721–723.

Houff, S. A., Burton, R. C., Wilson, R. W., Henson, T. E., London, W. T., Baer, G. M., et al. (1979). Human-to-human transmission of rabies virus by corneal transplant. *New England Journal of Medicine, 300,* 603–604.

Hummeler, K., Koprowski, H., & Wiktor, T. J. (1967). Structure and development of rabies virus in tissue culture. *Journal of Virology, 1*(1), 152–170.

Hummeler, K., Tomassini, N., Sokol, F., Kuwert, E., & Koprowski, H. (1968). Morphology of the nucleoprotein component of rabies virus. *Journal of Virology, 2*(10), 1191–1199.

Hurst, E. W., & Pawan, J. L. (1932). A further account of the Trinidad outbreak of acute rabic myelitis: histology of the experimental disease. *Journal of Pathology and Bacteriology, 35,* 301–321.

Huygelen, C. (1960). Further observations on the pathogenesis of rabies in guinea-pigs after experimental infection with the Flury strain. *Antonie Van Leeuwenhoek, 26,* 66–72.

Huygelen, C., & Mortelmans, J. (1959). Quantitative determination of the dissemination of Flury rabies virus in the central nervous system of the guinea-pig after intramuscular inoculation in the hind leg. *Antonie Van Leeuwenhoek, 25,* 265–271.

Iwasaki, Y., & Clark, H. F. (1975). Cell to cell transmission of virus in the central nervous system. II. Experimental rabies in mouse. *Laboratory Investigation, 33,* 391–399.

Iwasaki, Y., & Clark, H. F. (1977). Rabies virus infection in mouse neuroblastoma cells. *Laboratory Investigation, 36,* 578–584.

Iwasaki, Y., Liu, D. S., Yamamoto, T., & Konno, H. (1985). On the replication and spread of rabies virus in the human central nervous system. *Journal of Neuropathology and Experimental Neurology, 44,* 185–195.

Iwasaki, Y., & Minamoto, N. (1982). Scanning and freeze-fracture electron microscopy of rabies virus infection in murine neuroblastoma cells. *Comparative Immunology, Microbiology and Infectious Diseases, 5*(1-3), 1–8.

Iwasaki, Y., Ohtani, S., & Clark, H. F. (1975). Maturation of rabies virus by budding from neuronal cell membrane in suckling mouse brain. *Journal of Virology, 15*(4), 1020–1023.

Iwasaki, Y., Sako, K., Tsunoda, I., & Ohara, Y. (1993). Phenotypes of mononuclear cell infiltrates in human central nervous system. *Acta Neuropathologica, 85*(6), 653–657.

Iwasaki, Y., & Tobita, M. (2002). Pathology. In A. C. Jackson & W. H. Wunner (Eds.), *Rabies* (pp. 283–306). San Diego: Academic Press.

Iwasaki, Y., Wiktor, T. J., & Koprowski, H. (1973). Early events of rabies virus replication in tissue cultures: an electron microscopic study. *Laboratory Investigation, 28,* 142–148.

Jackson, A. C. (1992). Detection of rabies virus mRNA in mouse brain by using *in situ* hybridization with digoxigenin-labelled RNA probes. *Molecular and Cellular Probes, 6*(2), 131–136.

Jackson, A. C., Kammouni, W., Zherebitskaya, E., & Fernyhough, P. (2010). Role of oxidative stress in rabies virus infection of adult mouse dorsal root ganglion neurons. *Journal of Virology, 84*(9), 4697–4705.

Jackson, A. C., & Park, H. (1998). Apoptotic cell death in experimental rabies in suckling mice. *Acta Neuropathologica, 95*(2), 159–164.

Jackson, A. C., Randle, E., Lawrance, G., & Rossiter, J. P. (2008). Neuronal apoptosis does not play an important role in human rabies encephalitis. *Journal of Neurovirology, 14*(5), 368–375.

Jackson, A. C., & Reimer, D. L. (1989). Pathogenesis of experimental rabies in mice: an immunohistochemical study. *Acta Neuropathologica, 78*(2), 159–165.

Jackson, A. C., Reimer, D. L., & Wunner, W. H. (1989). Detection of rabies virus RNA in the central nervous system of experimentally infected mice using in situ hybridization with RNA probes. *Journal of Virological Methods, 25*(1), 1–11.

Jackson, A. C., & Rossiter, J. P. (1997). Apoptosis plays an important role in experimental rabies virus infection. *Journal of Virology, 71*(7), 5603–5607.

Jackson, A. C., & Wunner, W. H. (1991). Detection of rabies virus genomic RNA and mRNA in mouse and human brains by using in situ hybridization. *Journal of Virology, 65*(6), 2839–2844.

Jackson, A. C., Ye, H., Phelan, C. C., Ridaura-Sanz, C., Zheng, Q., Li, Z., et al. (1999). Extraneural organ involvement in human rabies. *Laboratory Investigation, 79*(8), 945–951.

Jackson, A. C., Ye, H., Ridaura-Sanz, C., & Lopez-Corella, E. (2001). Quantitative study of the infection in brain neurons in human rabies. *Journal of Medical Virology, 65*(3), 614–618.

Jenson, A. B., Rabin, E. R., Bentinck, D. C., & Melnick, J. L. (1969). Rabiesvirus neuronitis. *Journal of Virology, 3*, 265–269.

Jenson, A. B., Rabin, E. R., Wende, R. D., & Melnick, J. L. (1967). A comparative light and electron microscopic study of rabies and Hart Park virus encephalitis. *Experimental and Molecular Pathology, 7*(1), 1–10.

Jogai, S., Radotra, B. D., & Banerjee, A. K. (2000). Immunohistochemical study of human rabies. *Neuropathology, 20*(3), 197–203.

Jogai, S., Radotra, B. D., & Banerjee, A. K. (2002). Rabies viral antigen in extracranial organs: a post-mortem study. *Neuropathology and Applied Neurobiology, 28*(4), 334–338.

Johnson, K. P., Swoveland, P. T., & Emmons, R. W. (1980). Diagnosis of rabies by immunofluorescence in trypsin-treated histologic sections. *Journal of the American Medicial Association, 244*, 41–43.

Juntrakul, S., Ruangvejvorachai, P., Shuangshoti, S., Wacharapluesadee, S., & Hemachudha, T. (2005). Mechanisms of escape phenomenon of spinal cord and brainstem in human rabies. *BMC Infectious Diseases, 5*(1), 104.

Kammouni, W., Hasan, L., Saleh, A., Wood, H., Fernyhough, P., & Jackson, A. C. (2012). Role of nuclear factor-κB in oxidative stress associated with rabies virus infection of adult rat dorsal root ganglion neurons. *Journal of Virology, 86*(15), 8139–8146.

Kliger, I. J., & Bernkopf, H. (1943). The path of dissemination of rabies virus in the body of normal and immunized mice. *British Journal of Experimental Pathology, 24*, 15–21.

Knutti, R. E. (1929). Acute ascending paralysis and myelitis due to the virus of rabies. *Journal of the American Medical Association, 93*, 754–758.

Kristensson, K., Dastur, D. K., Manghani, D. K., Tsiang, H., & Bentivoglio, M. (1996). Rabies: interactions between neurons and viruses. A review of the history of Negri inclusion bodies. *Neuropathology and Applied Neurobiology, 22*, 179–187.

Kucera, P., Dolivo, M., Coulon, P., & Flamand, A. (1985). Pathways of the early propagation of virulent and avirulent rabies strains from the eye to the brain. *Journal of Virology, 55*(1), 158–162.

Lafon, M. (2011). Evasive strategies in rabies virus infection. *Advances in Virus Research, 79*, 33–53.

Lahaye, X., Vidy, A., Pomier, C., Obiang, L., Harper, F., Gaudin, Y., et al. (2009). Functional characterization of Negri bodies (NBs) in rabies virus infected cells: evidence that NBs are sites of viral transcription and replication. *Journal of Virology, 83*(16), 7948–7958.

Last, R. D., Jardine, J. E., Smit, M. M., & van der Lugt, J. J. (1994). Application of immunoperoxidase techniques to formalin-fixed brain tissue for the diagnosis of rabies in southern Africa. *Onderstepoort Journal of Veterinary Research, 61*(2), 183–187.

Leech, R. W. (1971). Electron-microscopic study of the inclusion body in human rabies. *Neurology, 21*(1), 91–94.

Lepine, P., & Atanasiu, P. (1996). Histopathological diagnosis. In F. -X. Meslin, M. M. Kaplan, & H. Koprowski (Eds.), *Laboratory techniques in rabies (pp. 66–79)* (4th ed.). Geneva: World Health Organization.

Lewis, P., & Lentz, T. L. (1998). Rabies virus entry into cultured rat hippocampal neurons. *Journal of Neurocytology, 27*(8), 559–573.

Li, X. Q., Sarmento, L., & Fu, Z. F. (2005). Degeneration of neuronal processes after infection with pathogenic, but not attenuated, rabies viruses. *Journal of Virology, 79*(15), 10063–10068.

Li, Z., Feng, Z., & Ye, H. (1995). Rabies viral antigen in human tongues and salivary glands. *Journal of Tropical Medicine and Hygiene, 98*(5), 330–332.

Lopez-Corella, E., Ridaura-Sanz, C., & Samayoa-Palma, J. E. (1997). Human rabies. Systemic pathology in 33 autopsies. *Laboratory Investigation, 76,* 140A.

Love, S., & Wiley, C. A. (2002). Viral diseases. In D. I. Graham & P. L. Lantos (Eds.), *Greenfield's neuropathology (pp. 1–105).* London: Arnold.

Love, S. V. (1944). Paralytic rabies: Review of the literature and report of a case. *Journal of Pediatrics, 24,* 312–325.

Lowenberg, K. (1928). Rabies in man. Microscopic observations. *Archives of Neurology and Psychiatry, 19,* 638–646.

Mackenzie, I. R., Medvedev, G., & Thiessen, B. (2003). An unusual case of rabies with prolonged survival and extreme neuropathology. *Canadian Journal of Neurological Sciences, 30,* 408–409.

Manghani, D. K., Dastur, D. K., Nanavaty, A. N., & Patel, R. (1986). Pleomorphism of fine structure of rabies virus in human and experimental brain. *Journal of the Neurological Sciences, 75,* 181–193.

Marinesco, G., & Storesco, G. (1931). Études sur la pathologie de la rage. *Archives Roumaines de Pathologie Experimentales et de Microbiologie, 4,* 243–288.

Matsumoto, S. (1963). Electron microscope studies of rabies virus in mouse brain. *Journal of Cell Biology, 19,* 565–591.

Matsumoto, S., & Kawai, A. (1969). Comparative studies on development of rabies virus in different host cells. *Virology, 39*(3), 449–459.

Matsumoto, S., Schneider, L. G., Kawai, A., & Yonezawa, T. (1974). Further studies on the replication of rabies and rabies-like viruses in organized cultures of mammalian neural tissues. *Journal of Virology, 14*(4), 981–996.

Ménager, P., Roux, P., Mégret, F., Bourgeois, J. P., Le Sourd, A. M., Danckaert, A., et al. (2009). Toll-like receptor 3 (TLR3) plays a major role in the formation of rabies virus Negri bodies. *PLoS Pathogens, 5*(2), e1000315.

Metze, K., & Feiden, W. (1991). Rabies virus ribonucleoprotein in the heart (Letter). *New England Journal of Medicine, 324,* 1814–1815.

Minguetti, G., Hofmeister, R. M., Hayashi, Y., & Montano, J. A. (1997). Ultrastructure of cranial nerves of rats inoculated with rabies virus. *Arquivos de Neuro Psiquiatria, 55*(4), 680–686.

Mitrabhakdi, E., Shuangshoti, S., Wannakrairot, P., Lewis, R. A., Susuki, K., Laothamatas, J., et al. (2005). Difference in neuropathogenetic mechanisms in human furious and paralytic rabies. *Journal of the Neurological Sciences, 238*(1–2), 3–10.

Miyamoto, K., & Matsumoto, S. (1965). The nature of the Negri body. *Journal of Cell Biology, 27,* 677–682.

Miyamoto, K., & Matsumoto, S. (1967). Comparative studies between pathogenesis of street and fixed rabies infection. *Journal of Experimental Medicine, 125,* 447–474.

Morecki, R., & Zimmerman, H. M. (1969). Human rabies encephalitis. Fine structure study of cytoplasmic inclusions. *Archives of Neurology, 20*(6), 599–604.

Morimoto, K., Patel, M., Corisdeo, S., Hooper, D. C., Fu, Z. F., Rupprecht, C. E., et al. (1996). Characterization of a unique variant of bat rabies virus responsible for newly emerging human cases in North America. *Proceedings of the National Academy of Sciences of the United States of America, 93*(11), 5653–5658.

Mrak, R. E., & Young, L. (1993). Rabies encephalitis in a patient with no history of exposure. *Human Pathology, 24,* 109–110.

Mrak, R. E., & Young, L. (1994). Rabies encephalitis in humans: Pathology, pathogenesis and pathophysiology. *Journal of Neuropathology and Experimental Neurology, 53,* 1–10.

Murphy, F. A. (1977). Rabies pathogenesis: brief review. *Archives of Virology, 54,* 279–297.

Murphy, F. A., Bauer, S. P., Harrison, A. K., & Winn, W. C. (1973). Comparative pathogenesis of rabies and rabies-like viruses: Viral infection and transit from inoculation site to the central nervous system. *Laboratory Investigation, 28,* 361–376.

Murphy, F. A., Harrison, A. K., Winn, W. C., & Bauer, S. P. (1973). Comparative pathogenesis of rabies and rabies-like viruses: Infection of the central nervous system and centrifugal spread of virus to peripheral tissues. *Laboratory Investigation, 29,* 1–16.

Negri, A. (1903a). Beitrag zum Studium der Aetiologie der Tollwuth. *Zeitschrift für Hygiene und Infektionskrankheiten, 43,* 507–528.

Negri, A. (1903b). Zur Aetiologie der Tollwuth. Die Diagnose der Tollwuth auf Grund der Neuen Befunde. *Zeitschrift für Hygiene und Infektionskrankheiten, 44,* 519.

Negri, A. (1909). Über die Morphologie und der Entwicklungszyklus des Parasiten der Tollwut (Neurocytes hydrophobiae Calkins). *Zeitschrift für Hygiene und Infektionskrankheiten, 63,* 421–440.

Negri-Luzzani, L. (1913). Le diagnostic de la rage par la demonstration du parasite spécifique. Resultats de dix ans d'expériences. *Annales de L'Institut Pasteur (Paris), 27,* 1039–1064.

Nepveu, M. (1872). Un cas de rage. *Comptes Rendus des Séances et Mémoires de la Société de Biologie, 4,* 133.

Nieberg, K. C., & Blumberg, J. M. (1972). Viral encephalitides. In J. Minckler (Ed.), *Pathology of the nervous system (pp. 2266–2323).* New York: McGraw-Hill Book Company.

Pasteur, L., Chamberland, C. E., Roux, E., & Thuillier, L. (1881). Sur la rage. *Comptes Rendus de l'Académie des Sciences, 92,* 1259–1260.

Perl, D. P., Callaway, C. S., & Hicklin, M. (1972). An ultrastructural study of Negri bodies in experimental rabies following prolonged incubation. *Journal of Neuropathology and Experimental Neurology, 31,* 172.

Perl, D. P., & Good, P. F. (1991). The pathology of rabies in the central nervous system. In G. M. Baer (Ed.), *The natural history of rabies (pp. 163–190)* (2nd ed.). Boca Raton, Florida: CRC Press.

Raman, G. V., Prosser, A., Spreadbury, P. L., Cockcroft, P. M., & Okubadejo, O. A. (1988). Rabies presenting with myocarditis and encephalitis. *Journal of Infection, 17,* 155–158.

Ramón y Cajal, S., & Garcia, D. (1904). Las lesiones del retículo de las células nerviosas en la rabia. *Trabajos del Laboratorio de Investigaciones biológicas de la Universidad de Madrid, 3,* 213.

Reid, J. E., & Jackson, A. C. (2001). Experimental rabies virus infection in *Artibeus jamaicensis* bats with CVS-24 variants. *Journal of Neurovirology, 7*(6), 511–517.

Reisman, D., Alpers, B. J., & Cooper, D. A. (1933). Hydrophobia. Report of two fatal cases with pathologic studies in one. *Archives of Internal Medicine, 51,* 643–655.

Ross, E., & Armentrout, S. A. (1962). Myocarditis associated with rabies: report of a case. *New England Journal of Medicine, 266,* 1087–1089.

Rossiter, J. P., Hsu, L., & Jackson, A. C. (2009). Selective vulnerability of dorsal root ganglia neurons in experimental rabies after peripheral inoculation of CVS-11 in adult mice. *Acta Neuropathologica, 118*(2), 249–259.

Roux, F., Bourgeade, A., Salaun, J. J., Bondurand, A., Ette, M., & Bertrand, E. (1976). L'atteinte cardiaque dans la rage humaine [Cardiac involvement in human rabies]. *Coeur et Medecine Interne, 15*(1), 37–44.

Rubin, R. H., Sullivan, L., Summers, R., Gregg, M. B., & Sikes, R. K. (1970). A case of human rabies in Kansas: epidemiologic, clinical, and laboratory considerations. *Journal of Infectious Diseases, 122,* 318–322.

Sandhyamani, S., Roy, S., Gode, G. R., & Kalla, G. N. (1981). Pathology of rabies: A light- and electron-microscopical study with particular reference to the changes in cases with prolonged survival. *Acta Neuropathologica, 54,* 247–251.

Sarmento, L., Li, X., Howerth, E., Jackson, A. C., & Fu, Z. F. (2005). Glycoprotein-mediated induction of apoptosis limits the spread of attenuated rabies viruses in the central nervous system of mice. *Journal of Neurovirology, 11*, 571–581.

Schaffer, K. (1888). Histologische Untersuchung eines Falles von Lyssa. *Archiv für Psychiatrie und Nervenkrankheiten, 19*, 45–63.

Schneider, L. G. (1969a). Die Pathogenese der Tollwut bei der Maus. I. Die Virusausbreitung vom Infektionsort zum Zentralnervensystem. *Zentralblatt fur Bakteriologie, 211*, 281–308.

Schneider, L. G. (1969b). Die Pathogenese der Tollwut bei der Maus. II. Die Virusausbreitung innerhalb des ZNS. *Zentralblatt fur Bakteriologie, 212*, 1–13.

Schneider, L. G. (1969c). The cornea test; a new method for the intra-vitam diagnosis of rabies. *Zentralblatt Fur Veterinarmedizin - Reihe B, 16*(1), 24–31.

Schneider, L. G., Dietzschold, B., Dierks, R. E., Matthaeus, W., Enzmann, P. J., & Strohmaier, K. (1973). Rabies group-specific ribonucleoprotein antigen and a test system for grouping and typing of rhabdoviruses. *Journal of Virology, 11*(5), 748–755.

Schneider, L. G., & Hamann, I. (1969). Die Pathogenese der Tollwut bei der Maus. III. Die zentrifugale Virusausbreitung und die Virusgeneralisierung im Organismus. *Zentralblatt fur Bakteriologie, 212*, 13–41.

Scott, C. A., Rossiter, J. P., Andrew, R. D., & Jackson, A. C. (2008). Structural abnormalities in neurons are sufficient to explain the clinical disease and fatal outcome in experimental rabies in yellow fluorescent protein-expressing transgenic mice. *Journal of Virology, 82*(1), 513–521.

Shankar, V., Dietzschold, B., & Koprowski, H. (1991). Direct entry of rabies virus into the central nervous system without prior local replication. *Journal of Virology, 65*, 2736–2738.

Sheikh, K. A., Ramos-Alvarez, M., Jackson, A. C., Li, C. Y., Asbury, A. K., & Griffin, J. W. (2005). Overlap of pathology in paralytic rabies and axonal Guillain-Barré syndrome. *Annals of Neurology, 57*(5), 768–772.

Smart, N. L., & Charlton, K. M. (1992). The distribution of challenge virus standard rabies virus versus skunk street rabies virus in the brains of experimentally infected rabid skunks. *Acta Neuropathologica, 84*, 501–508.

Suja, M. S., Mahadevan, A., Madhusudana, S. N., & Shankar, S. K. (2011). Role of apoptosis in rabies viral encephalitis: a comparative study in mice, canine, and human brain with a review of literature. *Patholog. Res. Int, 2011*, 374286.

Suja, M. S., Mahadevan, A., Madhusudhana, S. N., Vijayasarathi, S. K., & Shankar, S. K. (2009). Neuroanatomical mapping of rabies nucleocapsid viral antigen distribution and apoptosis in pathogenesis in street dog rabies--an immunohistochemical study. *Clinical Neuropathology, 28*(2), 113–124.

Suja, M. S., Mahadevan, A., Sundaram, C., Mani, J., Sagar, B. C., Hemachudha, T., et al. (2004). Rabies encephalitis following fox bite–histological and immunohistochemical evaluation of lesions caused by virus. *Clinical Neuropathology, 23*(6), 271–276.

Sükrü-Aksel, I. (1958). Pathologische Anatomie der Lyssa. In O. Lubarsch, F. Henke, & R. Rossle (Eds.), *Handbuch der Speziellen Pathologischen Anatomie und Histologie (pp. 417–435)*. Berlin: Springer-Verlag.

Sung, J. H., Hayano, M., Mastri, A. R., & Okagaki, T. (1976). A case of human rabies and ultrastructure of the Negri body. *Journal of Neuropathology and Experimental Neurology, 35*, 541–559.

Szlachta, H. L., & Habel, R. E. (1953). Inclusions resembling Negri bodies in the brains of nonrabid cats. *Cornell Veterinarian, 43*(2), 207–212.

Tangchai, P., & Vejjajiva, A. (1971). Pathology of the peripheral nervous system in human rabies: a study of nine autopsy cases. *Brain, 94*, 299–306.

Tangchai, P., Yenbutr, D., & Vejjajiva, A. (1970). Central nervous system lesions in human rabies: a study of twenty-four cases. *Journal of the Medical Association of Thailand, 53*, 471–488.

Teixeira, F., Aranda, F. J., Castillo, S., Perez, M., Del Peon, L., & Hernandez, O. (1986). Experimental rabies: ultrastructural quantitative analysis of the changes in the sciatic nerve. *Experimental and Molecular Pathology, 45*(3), 287–293.

Theerasurakarn, S., & Ubol, S. (1998). Apoptosis induction in brain during the fixed strain of rabies virus infection correlates with onset and severity of illness. *Journal of Neurovirology, 4*(4), 407–414.

Tierkel, E. S. (1973a). Laboratory techniques in rabies: rapid microscopic examination for negri bodies and preparation of specimens for biological test. *World Health Organization Monograph Series, 23*, 41–55.

Tierkel, E. S. (1973b). Rapid microscopic examination for Negri bodies and preparation of specimens for biological test. In M. M. Kaplan & H. Koprowski (Eds.), *Laboratory Techniques in Rabies (pp. 41–55)* (3rd ed.). Geneva: World Health Organization.

Tirawatnpong, S., Hemachudha, T., Manutsathit, S., Shuangshoti, S., Phanthumchinda, K., & Phanuphak, P. (1989). Regional distribution of rabies viral antigen in central nervous system of human encephalitic and paralytic rabies. *Journal of the Neurological Sciences, 92*, 91–99.

Tustin, R. C., & Smit, J. D. (1962). Rabies in South Africa. An analysis of histological examination. *Journal of the South African Veterinary Medical Association, 33*, 295–310.

Ubol, S., & Kasisith, J. (2000). Reactivation of Nedd-2, a developmentally down-regulated apoptotic gene, in apoptosis induced by a street strain of rabies virus. *Journal of Medical Microbiology, 49*(11), 1043–1046.

Van Gehuchten, A., & Nelis, C. (1900). Les lésions histologiques de la rage chez les animaux et chez l'homme. *Bulletin de l'Académie royale de médecine de Belgique, 14*, 31–66.

Yan, X., Prosniak, M., Curtis, M. T., Weiss, M. L., Faber, M., Dietzschold, B., et al. (2001). Silver-haired bat rabies virus variant does not induce apoptosis in the brain of experimentally infected mice. *Journal of Neurovirology, 7*(6), 518–527.

10

Immunology

Monique Lafon

Viral Neuroimmunology, Department of Virology, Institut Pasteur,
25 rue du Dr Roux 75724 Paris cedex 15, France

1 INTRODUCTION

Viruses are obligatory parasites. Successful completion of virus cycle and subsequent transmission to a new host relies upon the evolution of strategies that allow the virus to hijack the cellular machinery, modulate host cell survival, and escape host lines of defense. Rabies virus (RABV), a neurotropic virus causing fatal encephalitis, is transmitted in the saliva of an infected animal (mainly dogs but also bats and other animal vectors) after bites or scratches. After entry at the neuromuscular junction or passage through the synaptic cleft, RABV particles propagate towards the cell body by retrograde transport using axonal vesicles (Klingen, Conzelmann, & Finke, 2008). Virus replication occurs in the cell bodies and dendrites (Ugolini, 1995, 2010) from which newly formed viral particles are released. RABV infects neurons almost exclusively and travels from one neuron to the next in the spinal cord to the brainstem, from where it reaches the salivary glands via cranial nerves. Once in the salivary glands, RABV is excreted in saliva and can be then transmitted to a new host.

During its journey, RABV faces host defenses at different steps: At first, RABV particles delivered in the skin or muscle by the bite are rapidly detected by the early line of defense, the innate immune response, which contributes to both eliminate microbes locally and to set up a specific immune response (B and T cells) in the periphery (extraneurally). After its entry into nerves, the virus has to cope with the innate immune response launched by the infected neuron that has the capacity to counter the infection. Once infection is settled in the neurons, the infected neurons are protected from the destruction by infiltrating T cells and by mechanisms limiting the inflammation of neuronal tissue. In addition, the central control of immunological homeostasis operated by the nervous system (NS), resulting in an inappropriate down-regulation of the immune responsiveness in the periphery, might also facilitate the propagation of the virus in the NS. Preservation of the integrity of the neuronal network up to the brainstem gives the opportunity for the virus to reach the salivary glands and be transmitted.

The knowledge we have gained of the interactions of RABV with the immune responses have been learned mainly in models of experimental rabies in mice using laboratory-adapted RABV strains and sometimes street RABV strains injected by intramuscular or intraplantar (foot pad) route to mimic natural transmission by a bite. Fatal rabies encephalitis can be reproduced in this model using challenge virus standard (CVS) in particular. This virus invades the spinal cord and brain regions and causes fatal encephalitis (Camelo, Lafage, & Lafon, 2000; C. H. Park et al., 2006; Xiang, Knowles, McCarrick, & Ertl, 1995). Some mutant strains of RABV with attenuated pathogenicity cause only transient infection of the NS (Galelli, Baloul, & Lafon, 2000; Hooper et al., 1998; Irwin, Wunner, Ertl, & Jackson, 1999; Weiland, Cox, Meyer, Dahme, & Reddehase, 1992; Xiang, et al., 1995). This is the case for Pasteur virus (PV), which results in a non-fatal abortive disease characterized by a transient and restricted infection of the NS followed by irreversible limb paralysis (Galelli, et al., 2000).

2 RABV INNATE IMMUNE RESPONSE

2.1 Innate Immune Response in the Periphery

The innate immune response is the first line of defense against infectious agents. It involves the release of type 1 interferons (IFN-a/b), inflammatory cytokines and chemokines, the activation of complement and the attraction of macrophages, neutrophils, and NK cells into infected tissues. This innate immune response is triggered in the first hours following the entry of pathogens and is not pathogen specific. It contrasts with the adaptive immune response that is tailored to a specific pathogen and requires several days to develop. The innate immune system can sense the presence of micro-organisms through "pattern recognition receptors" (PRRs) that recognize danger signals and "pathogen associated molecular patterns (PAMPs) expressed by microbes. Toll-like receptors (TLRs) or retinoic acid inducible gene (RIG)-like receptors (RLRs) are important PRRs for the recognition of viral dsRNAs and ssRNAs. The TLRs are a family of 13 members. The RLRs family, mainly involved in virus detection, consists of three proteins: RIG protein 1, (RIG-I), melanoma differentiation-associated gene-5 (MDA-5), and laboratory of genetics and physiology 2 (LGP2) proteins. Some of these receptors are at the surface of the cells, detecting the presence of danger signals present in the extra-cellular milieu. This is the case for TLR2 and TLR4. Other receptors are expressed in the cytoplasm (RLRs) or in endosomal vesicles (TLR3, 7–9 and 13), allowing the detection of danger signals produced in the early steps of the entry or replication of intracellular pathogens. Recruitment of particular receptors depends upon the motifs they bind to and the localization of the receptors. For example, TLR3 senses only dsRNA with a length higher than 40–50 base pairs, a constraint allowing the forma-tion of complex homo-dimers gathering two molecules of TLR3 with the dsRNA (Liu et al., 2008). RLRs sense viral ARNs, but only those present in the cytoplasm and encoding a tri-phosphate at the 5'-end (Hornung et al., 2006). Resulting signal transduction cascades involve TRIF, Myd88, or IPS-1 as adaptors of TLR3, of TLRs other than TLR3, and of RLRs, respec-tively. They trigger production of chemokines, inflammatory cytokines, and antiviral molecules such as IFNs.

RABV is inoculated in the skin or in muscle by bites or scratches. The entry of the virus by the host is rapidly detected by host defence mecha-nisms in the periphery. The IFN response triggered at the site of entry has an antiviral effect. Evidence has been indirectly obtained by comparing the viral load in the thigh muscles of two groups of mice, parental mice and mice lacking the type I IFN receptor (IFNAR) after injection of the hind limb with CVS. The viral load measured by the accumulation of viral RNA was increased in mice lacking IFNAR compared to parental strain of mice (Chopy, Detje, Lafage, Kalinke, & Lafon, 2011). This observation suggests that some viral particles might be readily eliminated at this early step of infection. The nature of cells producing type I IFN at the site of the injection is not known with certainty. Candidates are muscle cells, fibroblasts, keratinocytes, dentritic cells (DCs), and macrophages. With the exception of reports in experimental rabies in skunks with a wild-type RABV strain indicating that infected muscle cells can be detected at the site of injection (Charlton & Casey, 1979, 1981), experiments have not yet been performed *in vivo* to determine the nature of cells other than muscle cells that RABV infects *in situ*. Nevertheless, there is experimen-tal evidence that RABV can infect bone marrow derived conventional DCs (cDCs) and macrophages *in vitro*. Despite nonproductive infection, RABV triggers the production of IFNs, cytokines and chemokines in these cells (Faul et al., 2010; Nakamichi, Inoue, Takasaki, Morimoto, & Kurane, 2004). Infection of sentinel cells may be dispensable since macrophages activa-tion was observed *in vitro* when inactivated RABV was added to the cul-ture (Nakamichi, et al., 2004). In cell culture, maturation of cDCs in the presence of RABV is controlled by IFN, which production might rely on the recognition of intra cytoplasmic RABV RNAs through RIG-I and mda-5 receptors and not TLR7 (Faul, et al., 2010), a characteristic of cDCs (Eisenacher, Steinberg, Reindl, & Krug, 2007).

At that time it is unknown whether virulent RABV strains trigger weaker or equal DCs activation versus attenuated strains and whether RABV evades the innate immune response in the periphery. Experiments were performed with highly attenuated recombinant RABV (rRABV) genetically modified to allow the expression of chemokines or multiple copies of G protein in a search for more effective rabies vaccines. These RABVs trigger a stronger activation of cDCs in the periphery than the parental rRABV strains (Li, McGettigan, Faber, Schnell, & Dietzschold, 2008; Pulmanausahakul et al., 2001; Wen et al., 2011), suggesting they could indeed trigger robust vaccine protection.

2.2 Innate Immune Response in the NS

Like most tissues in the organism, the NS expresses different types of receptors capable of sensing danger and pathogen signals (Boivin, Coulombe, & Rivest, 2002; Bottcher et al., 2003; Koedel et al., 2004; McKimmie, Johnson, Fooks, & Fazakerley, 2005; Nguyen, Julien, & Rivest, 2002). Central neurons express TLR1-4 as well as TLR 7 and 8 (Barajon et al., 2009; Kim et al., 2007; Ma, Haynes, Sidman, & Vartanian, 2007; Ma et al., 2006; Prehaud, Megret, Lafage, & Lafon, 2005; Tang et al., 2007). They express the RLRs (RIG-I and Mda-5) (Chopy et al., 2011; Lafon, Megret, Lafage, & Prehaud, 2006; Menager et al., 2009; Peltier, Simms, Farmer, & Miller, 2010), but not the RLR LGP2, which seems to be actively degraded in neurons (Chopy, Pothlichet, et al., 2011). Peripheral nerve plexuses and nerves (dorsal root ganglion sensory neurons and fibers of sciatic nerves) express TLRs (TLR3, 4 and 7) with prominent expression of TLR3 (Barajon, et al., 2009; Cameron et al., 2007; Goethals, Ydens, Timmerman, & Janssens, 2010). Neurons take an active part in the innate immune response in the brain, being both responders to IFN and IFN producers, secreting type I IFN [predominantly IFN-beta in the brain, no IFN-alpha and no Type III IFN lambda] (Delhaye et al., 2006; Prehaud, et al., 2005; Sommereyns, Paul, Staeheli, & Michiels, 2008).

2.2.1 RABV Evasion of the IFN Response in the Infected Neurons

RABV is known to trigger a RIG-I mediated innate immune response in neurons (Hornung, et al., 2006) by detecting the 5'tri phosphate base pairing of the viral genome (Pichlmair et al., 2006). After infection, human neurons can mount a classical primary IFN response (activation of IRF3 and NF-kappa B) , as well as a secondary IFN response, (activation of STATs and IRF7), leading to the production of cytokines (IL-6, TNF-alpha) and chemokines (CXCL10 and CCL5) (Chopy, Detje, et al., 2011; Chopy, Pothlichet, et al., 2011; Prehaud, et al., 2005).

Viruses have evolved sophisticated strategies to escape the innate immune response (Randall & Goodbourn, 2008; Versteeg & Garcia-Sastre, 2010). This is the case for RABV [for review see (Rieder & Conzelmann, 2009)]. The N and the P protein of RABV are multifunctional proteins involved in RNA synthesis and in counteracting the host innate immune response. The N protein limits RIG-I-signaling (Masatani, Ito, Shimizu, Ito, Nakagawa, Abe, et al., 2010; Masatani, Ito, et al., 2010a), whereas the P protein inhibits IRF3 and IRF7 phosphorylation (Brzozka, Finke, & Conzelmann, 2005; Rieder et al., 2011), suppresses STAT1 nuclear translocation (Brzozka, Finke, & Conzelmann, 2006; Vidy, El Bougrini, Chelbi-Alix, & Blondel, 2007) and sequesters an antiviral protein, the promyelocytic leukemia (PML) protein, in the cytoplasm (Blondel, Kheddache, Lahaye, Dianoux, & Chelbi-Alix, 2010). As a result, down-regulation of the IFN response can be observed *in vitro*. For example, in RABV infected human post-mitotic neurons (NT2-N), transcription of *IFN-beta* gene is seen as early as 6 h post infection, and IFN-beta protein is produced during the first 24 h post infection, whereas transcription and production decline thereafter (Prehaud, et al., 2005).

Dampening the IFN response favors RABV infection as demonstrated with the death of mice intracerebrally infected with P protein RABV mutants lacking the capacity to decrease the host IFN response (Ito et al., 2010) and with the earlier death of mice lacking IFNAR specifically in the NS, compared to parental mice after CVS intramuscular injection (Chopy, Detje, et al., 2011). These two experiments indicate that RABV infection is sensible to IFN signaling and that P protein-mediated IFN evasion is efficient. Moreover, it has been shown that virulence, at least for a Japanese vaccine strain (Nishigahara RABV strain), depends upon the capacity of this strain to evade the innate immune response and this process is controlled by the ability of the P and N protein to evade the innate immune response (Masatani, Ito, et al., 2010b; Shimizu, Ito, Sugiyama, & Minamoto, 2006). Thus, evasion of the IFN response in infected neurons may be critical for RABV progression in the NS through the neuronal network allowing the virus to reach the brainstem and the salivary glands.

2.2.2 RABV Limits the Inflammatory Response in the NS

Inflammation is a key component of host responses to cell damage or microbial entry, leading to the production of inflammatory mediators including complement, adhesion molecules, and cyclo-oxygenase enzymes and their products, as well as cytokines or chemokines. Release of these toxic factors has dramatic consequences when the site of inflammation is the NS, where severe dysfunction can lead to significant NS pathology with neuronal death (Brown & Neher, 2010). In the brain, both neurons and glial cells can mount antiviral, inflammatory, and chemokine responses. Astrocytes can respond to the presence of innate immune stimulus in the brain by producing pro-inflammatory cytokines and chemokines (C. Park et al., 2006). Nevertheless, an important role is taken by microglia in the induction of neuroinflammation, a feature which may reflect the density or the subcellular localization of the innate immune receptors (Bsibsi et al., 2006).

Transcriptome and proteomic analysis of the inflammatory response triggered in the NS of mice by various virulent strains of RABV showed that RABV infection stimulates the expression of chemokines (CCL5, CCL2, CCL9 and CXCL9) and inflammatory cytokines (IL-6, IL-12) (Baloul, Camelo, & Lafon, 2004; Camelo, et al., 2000; Chopy, Pothlichet, et al., 2011; Sugiura et al., 2011; Wang et al., 2005). However, the inflammatory reaction in the RABV-infected NS is transient with the expression of a majority of markers being rapidly down-regulated in the spinal cord and with a slight delay in the brain (Chopy, Pothlichet, et al., 2011).

Cells expressing inflammatory markers in the RABV infected NS such as TNF-alpha or IL-1 are not the infected neurons, but neighboring glial or endothelial cells (Marquette et al., 1996; Nuovo, DeFaria, Chanona-Vilchi, & Zhang, 2004; Van Dam et al., 1995).

When compared with other encephalitic virus infections, such as those caused by Borna virus, RABV triggers only limited inflammation (Fu et al., 1993; Shankar et al., 1992). Moreover, comparison of the inflammatory reaction triggered by RABV strains of various degree of pathogenicity indicates that the more pathogenic strains trigger weaker inflammatory responses (Baloul & Lafon, 2003; Hicks et al., 2009; Laothamatas et al., 2008; Wang, et al., 2005). For example, transcriptome analysis performed in the NS of mice infected with wild-type (street) RABV strains, such as a dog RABV strain isolated in China or a bat RABV isolate from North America (silver-haired bat isolate), showed that innate immune response is stimulated but to a limited extent compared with those triggered by laboratory strains (Sugiura, et al., 2011; Wang, et al., 2005; P. Zhao et al., 2011). Dogs infected with RABV causing paralytic rabies showed longer periods of illness and more intense nuclear magnetic resonance (NMR) signals (a marker of inflammation) than dogs infected with strains causing furious rabies. And the pattern of cytokines and chemokines mRNAs expression was greater in paralytic than in furious rabies (Laothamatas, et al., 2008).

With the sole exception of experiments in which mice inoculated with rRABV encoding chemokines died because of the excessive influx of monocytes and T cells into the brain (L. Zhao, Toriumi, Kuang, Chen, & Fu, 2009), most experimental evidence shows that inflammation does not promote RABV infection but instead limits the propagation of the virus through the NS. Immunization of mice with pro-inflammatory myelin basic protein (MBP) prior to RABV infection improved the survival to a challenge with a virulent bat RABV strain and, conversely, treatment

with a steroid hormone decreasing brain inflammation and with minocy-
cline, a tetracycline derivate with anti-inflammatory properties, increased
the mortality rate (Jackson, Scott, Owen, Weli, & Rossiter, 2007; Roy &
Hooper, 2007). Also, over expression of TNF-alpha by a rRABV attenuates RABV replication by inducing a strong T cell infiltration and microglial activation (Faber et al., 2005). It is likely that this low inflammatory
reaction in the infected NS contributes to keeping intact the BBB, a condition that correlates with RABV pathogenicity, with non-pathogenic
RABV strains triggering a transient opening of the BBB, but not pathogenic strains (Phares, Kean, Mikheeva, & Hooper, 2006; Roy, Phares,
Koprowski, & Hooper, 2007).

Altogether, these data indicate that virulent RABV strains trigger a
moderate inflammatory response in the NS and suggest that regulatory
mechanisms are set up in the course of the infection to reduce the RABV-
induced inflammation of the NS.

Limitation of neuroinflammation occurs by several mechanisms. RABV
infection avoids neuronal apoptosis (Lafon, 2011) and rarely infects glial
cells; two intrinsic features of the infection contributing to limit neuroinflammation. In addition, in the course of RABV infection the expression of anti-inflammatory molecules is up-regulated in the NS. This is
the case for the anti-inflammatory soluble proteins TNFR1 and 2, which
can interfere with the binding of TNF to its receptors (Chopy, Pothlichet,
et al., 2011). This is also the case for the suppressors of cytokine signaling (SOCS), a family of proteins that negatively control cytokine signal
transduction, with SOCS-1 being up-regulated in the brains of RABV
infected dogs in noninfected cells in close vicinity of infected neurons
(Nuovo, et al., 2004). More importantly, RABV upregulates the expression of HLA-G, a non classical MHC molecule and B7-H1, the ligand of
PD-1, (programmed death protein-1), in neurons, and also in the infected
NS, in the case of B7-H1 (Laton et al., 2008; Lafon et al., 2005). Besides,
their immune-tolerant properties, which are exploited by RABV, (see
below), HLAG and B7-H1 molecules are now also considered as providing negative feedback that limits tissue inflammation (Carosella, Moreau,
Aractingi, & Rouas-Freiss, 2001; Phares, Stohlman, Hinton, Atkinson, &
Bergmann, 2010). This is the case in particular for B7-H1, which dampens
the expression of pro-inflammatory molecules (such as iNos and TNF-
alpha) during viral encephalitis (Phares, et al., 2010), whereas HLA-G
influences the cytokine balance towards a Th2 pattern by promoting the
secretion of IL-4, IL-3 and IL-10 and down-regulating the production of
IFN-gamma and TNF-alpha (Carosella, et al., 2001).

Limitation of inflammation by RABV infection might be permitted by
1) reducing the entry in the NS of mononuclear leukocytes, monocytes
and macrophages, 2) maintaining the impermeability of the BBB, and
3) minimizing the release of neurotoxic molecules that can compromise

NS function and host survival. These conditions should preserve not only the integrity of the infected neuronal network but also the life of the host, allowing the virus to reach the brainstem and the salivary glands before the premature death of the infected host.

3 RABV ADAPTIVE IMMUNE RESPONSE

Building an adaptive immune response against a microbe, even a neurotropic virus that rapidly enters the NS after its inoculation in muscle, always occurs in the periphery and never in the NS, which is devoid of lymphoid organs (Galea, Bechmann, & Perry, 2007). The triggering of the adaptive immune response takes place in the lymphoid organs such as the lymph nodes or spleen relies on the activation of plasmacytoid DCs, pDCs and of type 1 IFN that they produce in a TLR7 and 9 dependent manner after encountering the microbe (Diebold et al., 2003) (Steinman, 1991).

The CD4$^+$ T lymphocytes recognize foreign antigens that have been processed through the MHC class II exogenous presentation pathway by activated DCs. Once presented by the MHC, the peptides of the digested foreign antigen are recognized by T cells bearing the appropriate T cell receptor (TCR) and CD4 molecule. Signalling via the TCR and CD4 molecule triggers activation and differentiation of T cells into two functional subsets, the T helper 1 (Th1) and T helper 2 (Th2) cells. The distinction of the two subsets, which is clearer in the mouse than in the human immune system, is based on the cytokines they secrete: interferon-gamma is the signature cytokine for Th1 cells, whereas interleukin-4 (IL-4) is the signature cytokine for the Th2 cells. Generation of Th1 cells is under the control of IL-12 produced by macrophages and DCs. Th1 cells limit the proliferation of pathogens via IFN-gamma production and provide help for antibody production by B lymphocytes. CD8$^+$ T cells, in contrast to CD4$^+$ T cells, recognize foreign antigens that have been processed by the endogenous pathway of cells expressing MHC class I molecules. Infected cells export pathogen peptides embedded in the groove of MHC class I molecules to the cell surface. The peptide-charged infected cells activate T cells expressing the CD8 accessory surface molecules and the appropriate TCR. Activated CD8$^+$ T lymphocytes produce IFN-γ and kill the infected cells via cytotoxicity by means of perforin and granzyme release and/or Fas-mediated lysis.

3.1 RABV Specific Immune Response in the Periphery

After the injection of the encephalitic RABV strain, CVS, in the hind limb of mice, the size of the draining popliteal lymph nodes and those of spleen increase. Draining lymph nodes are populated with activated T cells expressing the marker of activation CD69. Activation can also be observed among peripheral blood lymphocytes (Vuaillat et al., 2008).

A strong B cell response is mounted in the spleen. When mice were injected with a less pathogenic virus (the PV strain), similar activation of T cells was observed in lymph nodes and blood suggesting that adaptive immune response is independent of the virulence of the RABV strain. Indeed, when mice were injected with an encephalitic RABV bat strain (silver-haired bat rabies virus, SHBRV) or with a less pathogenic virus (CVS-F3, mutant of CVS encoding a mutation in the G protein), the resulting adaptive immune responses (neutralizing antibodies, CD4$^+$, CD8$^+$ T cells response) were not different (Roy & Hooper, 2007). This suggests that adaptive immune response triggered by RABV strains in the periphery, an event that occurs late after the virus has already entered the NS, is unrelated to RABV pathogenicity. This may explain why patients die of rabies despite having mounted an immune response in the periphery attested by the presence of neutralizing antibodies in the blood (Hunter et al., 2010). Nevertheless, in an experimental model of rabies in skunks, a cyclophosphamide-induced immunosuppression was found to reduce the infection of the salivary glands (Charlton, Casey, & Campbell, 1984), suggesting that the adaptive immune response may control to some extent the final stage of RABV infection. This control may also be exerted on viral particles at the site of entry if entry into the NS has been delayed.

3.2 RABV Provokes the Killing of Migratory T cells

Most of infections of the NS are controlled by infiltrating T cells. This is, for example, observed during the course of West Nile virus brain infection, where CD8$^+$ T cells attracted by the chemokines produced by inflammatory cells in the infected NS are a critical factor for controlling the infection (Klein et al., 2005; Zhang, Chan, Lu, Diamond, & Klein, 2008). In rabies, sterilization of the infection by T cells is inefficient, and is specifically inactivated by the virus (Lafon, 2008). Immunohistochemical studies performed on rabies autopsy cases revealed that the cells undergoing death were leukocytes and not neurons (Hemachudha et al., 2005; Tobiume et al., 2009). This observation was reproduced in mice infected with the encephalitic RABV strain CVS. Immunocytochemistry of brain and spinal cord slices revealed that despite a heavy load of viral antigens, infected neurons do not undergo death. In contrast, the migrating T cells (CD3+) were apoptotic (Baloul, et al., 2004; Baloul & Lafon, 2003; Kojima et al., 2009; Lafon, 2005; Rossiter, Hsu, & Jackson, 2009). Moreover, pathogenicity of the CVS strain was similar in immunocompetent mice Balb/c mice and in Nu/Nu Balb/c mice, indicating that T cells do not control the outcome of encephalitic rabies (Lafon, 2005). In striking contrast, deprivation of T cells transformed an abortive infection into a encephalitic rabies similar to that caused by the encephalitic strain CVS infection, showing that T cells is a critical factor in the restriction of the NS infection caused by an abortive RABV strain. Indeed, when apoptosis was analyzed in the spinal cord of immunocompetent mice infected with the abortive RABV strain PV, killing of T cells was not observed; instead, infected neurons died (Galelli, et al., 2000). Altogether, these observations indicate that T cells have a protective potential to control RABV infection in the NS, nevertheless their capacity to control RABV infection is impeded with the encephalitic RABV strain. The mechanisms by which the encephalitic RABV strain evades the host T cell response was further studied as described below.

3.2.1 Entry of T cells in the RABV-infected NS

Mononuclear leukocytes, monocytes and macrophages are able to be recruited to the NS in pathological conditions, including infections by neurotropic viruses (Davoust, Vuaillat, Androdias, & Nataf, 2008). Once activated, the T and B cells and macrophages from the periphery expressing surface adhesion molecules have the capacity to enter the NS (Engelhardt, 2008). This entry is independent of blood-brain barrier (BBB) integrity that modulates the entry of solutes and not cells (Bechmann, Galea, & Perry, 2007). The absence of T cell protection against an infection by the encephalitic RABV strain might be related to a blockage of T cells into the NS. This is likely not the case because after infection with an encephalitic RABV strain, blood T cells expressed markers of activation (CD69) and were highly positive for collapsing response mediator protein 2 (CRMP2), a marker of T cell polarization and migration. The brain was enriched with this type of cell, indicating that RABV-activated T cells have migratory properties (Vuaillat, et al., 2008). Thus, activation and entry into the NS are not limiting factors for T cell protective function. When infiltration of T cells in the NS was compared in mice infected either with an abortive or an encephalitic RABV strain (Baloul, et al., 2004), the parenchyma became invaded by infiltrating T cells similarly in the two groups of mice. However, this phenomenon was interrupted after a few days of infection by an encephalitic strain, whereas CD3[+] T cells accumulation in PV infected NS was continuous. Disappearance of T cells in the CVS-infected brain and an increase in number of apoptotic cells in the NS were concomitant events. These observations strongly suggest that encephalitic RABV strains, but not abortive strains, trigger unfavorable conditions for T cell survival in the infected NS.

Neutralizing antibodies have been described as a critical factor for protection against RABV (Hooper, et al., 1998; Montano-Hirose et al., 1993; Wiktor et al., 1984; Wunner, Dietzschold, Curtis, & Wiktor, 1983). The entry of B cells into the RABV infected NS and the local secretion of antibody contribute to the clearance of attenuated RABV from NS (Hooper, Phares, Fabis, & Roy, 2009). It is striking to note that during the course of encephalitic RABV infection B cells are almost undetectable in brain (Camelo, et al., 2000; Kojima et al., 2010), suggesting that restricted entry or specific destruction of migratory B cells could also contribute to RABV virulence.

3.2.2 Destruction of T cells in the RABV-infected NS

Tumors evade immune surveillance by multiple mechanisms, including the inhibition of tumour-specific T cell immunity. In order to escape attack from protective T cells, tumor cells up-regulate expression of certain surface molecules such as B7-H1, Fas-L, and HLA-G, which triggers death signalling in activated T cells expressing the corresponding ligands PD-1 for B7-H1, Fas for FasL and CD8 (among others) for HLA-G (Dong et al., 2002; Gratas et al., 1998; Rouas-Freiss, Moreau, Menier, & Carosella, 2003). Studies evaluating whether RABV-infected neurons up-regulate immunosubversive molecules to kill activated T cells following an evasive strategy similar to that selected by tumors cells have been undertaken both *in vivo* and *in vitro*. *In vitro*, RABV infection was found to up-regulate the expression of HLA-G at the surface of human neurons (Lafon, et al., 2005; Megret et al., 2007). *In vivo*, comparison of experimental rabies in mice caused by CVS, which kills T cells, or by PV, which does not kill T cells, leads to the finding that the CVS-infected NS, but not the PV-infected NS, up-regulates the expression of FasL. In mice lacking a functional FasL, there was less T cell apoptosis in the NS than in control mice. Remarkably, RABV morbidity and mortality were reduced in these mice. Destruction of T cells through the Fas/FasL pathway can be enhanced by indoleamine 2, 3 dioxygenase (IDO), which RABV upregulates expression in the infected neurons and brain (Prehaud, et al., 2005; P. Zhao, et al., 2011). The enzyme IDO converts extracellular tryptophan into kynurenine, thereby reducing its concentration in the microenvironment that in turn markedly enhances the sensitivity of any nearby T cell for Fas-ligand induced apoptosis (Kwidzinski et al., 2003).

In addition, RABV-infected brain up-regulates the expression of another immunosubversive molecule, B7-H1 (Lafon, et al., 2008). Whereas non-infected NS was almost devoid of B7-H1 expression, RABV infection triggers neural B7-H1 expression that increases as the infection progresses. Infected neurons and also non-infected neural cells, including astrocyte-like cells, were found positive for B7-H1. RABV infection of B7-H1 deficient mice resulted in a drastic reduction in clinical signs and mortality. Reduction of RABV virulence in $B7H1^{-/-}$ mice was concomitant of a reduction of $CD8^+$ T cell apoptosis among the migratory T cells.

Altogether these experiments indicate that despite the triggering of a classical adaptive immune response in the periphery and the infiltration of the lymphocytes into the infected NS, the protection, which could have been conferred in the NS by this immune response, is drastically impeded by RABV infection.

4 RABV INFECTION TRIGGERS A CNS MEDIATED IMMUNE-UNRESPONSIVENESS

The dampening of immune protection already triggered by RABV is completed by a central immunosupression caused by the neuronal reflex control of immunity triggered by the NS facing an excess of inflammation in an attempt to restore general homeostasis.

RABV infection by a pathogenic strain induces an immune-unresponsiveness (Camelo, Lafage, Galelli, & Lafon, 2001; Hirai et al., 1992; Kasempimolporn, Saengseesom, Mitmoonpitak, Akesowan, & Sitprija, 1997; Kasempimolporn, Tirawatnapong, Saengseesom, Nookhai, & Sitprija, 2001; Perry, Hotchkiss, & Lodmell, 1990; Torres-Anjel, Volz, Torres, Turk, & Tshikuka, 1988; Tshikuka, Torres-Anjel, Blenden, & Elliott, 1992; Wiktor, Doherty, & Koprowski, 1977a, 1977b) characterized by the impairment of T cells functions with an alteration of cytokine pattern, an inhibition of T cells proliferation and the destruction of immune cells without affecting the proportion of immune cells (CD4/CD8 ratio constant) in the lymphoid organs (Perry, et al., 1990). This leads to the atrophy of the spleen and the thymus of RABV infected mammals. TNF-alpha receptor has been found to play a role in RABV immune-unresponsiveness, since immune cells lacking the TNF alpha p55 receptor were less immunosuppressed compared to the wild-type (Camelo, et al., 2000). Most importantly, infection of the brain is required since immune-unresponsivenesss does not occur after the infection of the NS with an abortive RABV strain, which infects the spinal cord only (Camelo, et al., 2001). This suggests that the property of the NS that centrally controls the immune response in the periphery might be triggered (Tracey, 2009). NS modulates the immune functions through two main immune-neuroendocrine pathways: the hypothalamo-pituitary (HPA) axis and the autonomous NS (ANS) composed of sympathic and parasympathic nerves fibres (Johnston & Webster, 2009). The homeostatic reflex is activated after the brain senses the presence of an excess of inflammatory cytokines such as TNF-alpha, IL-1β or IL-6 in the periphery, by neuronal (mainly through local afferent fibres of the vagus nerve) and by humoral pathway (Johnston & Webster, 2009). This input is processed by the NS in frontal, hypothalamic and brainstem centers.

This general immune-unresponsiveness controlled by the NS may be advantageous for RABV propagation since a mouse strain having a less efficient HPA axis is less susceptible to rabies (Roy & Hooper, 2007). This

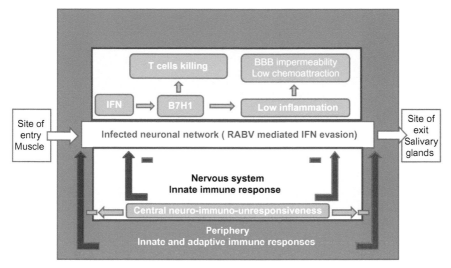

FIGURE 10.1 **Dual role of type I IFN in RABV infection may result from the compartmentalization of the NS.** One can distinguish between the network of the infected neurons, in which RABV evades the IFN response by viral protein-mediated mechanisms (white tube), from the rest of the NS (yellow rectangle) made of glial cells in which IFN can be produced (heterocellular IFN). Production of heterocellular IFN is not controlled because the glial cells are not infected. RABV infection propagates rapidly in the neuronal network in which IFN production is reduced, whereas the heterocellular IFN might promote RABV infection by upregulating the expression of IFN dependent B7-H1 and HLAG proteins functioning both as anti-inflammatory molecules and as immunosubversive molecules allowing the elimination of migratory T cells. In the periphery, innate and adaptive immune responses are triggered by the detection of virus particles at the site of entry (muscle). These host defenses can exert their protective functions at the site of entry and possibly also at the final stage of infection, in the salivary glands. However, they might be down regulated by central immune-unresponsiveness.

central immunosupression may limit peripheral control of infection in the muscle or the salivary glands (see Figure 10.1).

Thus, RABV infection not only actively inhibits the T cell response and inflammation in the NS by upregulating B7-H1 and FasL molecules, but also benefits from the intrinsic capacity of NS to trigger central immunosuppression in order to maintain whole body homeostasis

5 PARADOXICAL ROLE OF IFN IN RABV VIRULENCE

A series of experiments clearly indicate that RABV infection is sensible to IFN signaling and that P protein-mediated IFN evasion is efficient. Nevertheless, in the course of infection, the IFN induction in the whole RABV-infected NS is far from being abrogated (Chopy, Detje, et al.,

2011; Chopy, Pothlichet, et al., 2011; Lafon, et al., 2008; Li et al., 2012; Sugiura, et al., 2011; Wang, et al., 2005). Indeed, after injection of RABV (CVS) into the hindlimbs, a progressive infection within the spinal cord and the brain is accompanied by a robust innate immune response characterized by a type 1 IFN response. This is not a peculiar property of the laboratory strains, since similar observation was made after infection with a highly virulent RABV strain, the DOG-4 strain (Li, et al., 2012).

It may not be surprising that IFN can be produced in the NS during infection because the mechanisms evolved by RABV to escape the IFN response are restricted to infected neurons, the only cell type expressing the P and N proteins. These mechanisms cannot operate in glial cells because they do not express any viral proteins, glial cells being rarely infected *in vivo* (Iwasaki & Clark, 1975). Glia are efficient innate responders (C. Park, et al., 2006) and, in particular, they are IFN responders (Lafon, et al., 2005). They do not need to be infected to mount an innate immune response as shown by treating microglial cultures with inactivated RABV (Nakamichi et al., 2005). Indeed, in the brain of RABV naturally infected dogs or experimentally infected mice, the cells expressing cytokines were not the infected neurons, but non-infected nearby cells with glial or macrophage morphology (Marquette, et al., 1996; Nuovo, et al., 2004; Van Dam, et al., 1995), suggesting that non infected glial cells may be IFN responders and producers of heterocellular IFN, which is produced by neighboring cells and not by infected cells. In this case, we can distinguish the infected part of the NS consisting of the network of infected neurons, in which the IFN response is limited by the evasive mechanisms of the virus, from the non infected part made of the glial cells that are resistant to the RABV evasive mechanisms and in which an heterocellular IFN response can settle. One can wonder what is the function of the heterocellular IFN, if any. It can be speculated the heterocellular IFN makes non neuronal cells refractory to infection. RABV neuronotropism and virus progression could be facilitated by this means as demonstrated for poliovirus, which is another neuronotropic virus (Ida-Hosonuma et al., 2005; Kuss, Etheredge, & Pfeiffer, 2008; Pfeiffer, 2010).

Lessons from the use of recombinant forms of IFN-beta as treatment in relapsing forms of multiple sclerosis highlight that IFN promotes the production of anti-inflammatory molecules and reduces the trafficking of inflammatory cells across the endothelium of brain capillaries (review in Kieseier, 2011). If similar functions were demonstrated in RABV-infected brain, it could be proposed that IFN reduces the trafficking of inflammatory cells, and contributes to the low inflammatory environment triggered by RABV by up-regulating the production of anti-inflammatory molecules.

Beside intrinsic antiviral or anti-inflammatory properties, type I IFN also controls the expression of a large number of genes (ISG, interferon stimulated genes) (Takeuchi & Akira, 2010). Among those genes, are the B7-H1 and the non classical MHC Class I molecule HLA-G, two genes in which expression is up-regulated in RABV infection in an IFN dependent manner (Chopy, Pothlichet, et al., 2011; Lafon, et al., 2008; Lafon, et al., 2005). B7-H1 has been demonstrated to contribute to the killing of migratory T cells in RABV infection (Lafon, et al., 2008). In addition, B7-H1 and HLA-G have anti-inflammatory functions (Carosella, et al., 2001; Phares, et al., 2010).

To explain the dual role of IFN in RABV infection, it can be proposed that heterocellular IFN produced in the NS in the course of RABV infection contributes to the low inflammatory environment set up by RABV infection and to the RABV mediated killing of migratory T cells, whereas IFN antiviral effect takes place in infected neurons (Figure 10.1).

6 CONCLUSIONS

Thus RABV has selected a battery of mechanisms to escape the host immunosurveillance possibly explaining why, in the absence of post-exposure treatment, rabies is one of the very few human infections with a near 100% mortality rate. Despite these well-adapted viral strategies to escape the immune response, RABV infection can be limited if vaccine is injected promptly after exposure suggesting that the viral-mediated paralysis of the host immune response requires some time, which can be exploited for therapy. However, the efficacy of rabies post-exposure treatment requires public education; prompt wound cleansing, vaccination and availability of rabies immunoglobulins. Half of the victims being children, pre-exposure vaccination of young individuals should be considered in an attempt to improve the global health of mankind. In addition, improved knowledge of the immune evasive mechanisms evolved by RABV to infect the NS, may help identify new therapeutical targets such as the central neural immune reflex or neuroinflammation.

References

Baloul, L., Camelo, S., & Lafon, M. (2004). Up-regulation of Fas ligand (FasL) in the central nervous system: A mechanism of immune evasion by rabies virus. *J Neurovirol, 10*(6), 372–382.

Baloul, L., & Lafon, M. (2003). Apoptosis and rabies virus neuroinvasion. *Biochimie, 85*(8), 777–788.

Barajon, I., Serrao, G., Arnaboldi, F., Opizzi, E., Ripamonti, G., Balsari, A., et al. (2009). Toll-like receptors 3, 4, and 7 are expressed in the enteric nervous system and dorsal root ganglia. *J Histochem Cytochem, 57*(11), 1013–1023. doi: jhc.2009.953539 [pii] 10.1369/jhc.2009.953539.

Bechmann, I., Galea, I., & Perry, V. H. (2007). What is the blood-brain barrier (not)? *Trends Immunol, 28*(1), 5–11. doi: S1471-4906(06)00329-2 [pii] 10.1016/j.it.2006.11.007.

Blondel, D., Kheddache, S., Lahaye, X., Dianoux, L., & Chelbi-Alix, M. K. (2010). Resistance to rabies virus infection conferred by the PMLIV isoform. *J Virol*, *84*(20), 10719–10726. doi: JVI.01286-10 [pii] 10.1128/JVI.01286-10.

Boivin, G., Coulombe, Z., & Rivest, S. (2002). Intranasal herpes simplex virus type 2 inoculation causes a profound thymidine kinase dependent cerebral inflammatory response in the mouse hindbrain. *Eur J Neurosci*, *16*(1), 29–43.

Bottcher, T., von Mering, M., Ebert, S., Meyding-Lamade, U., Kuhnt, U., Gerber, J., et al. (2003). Differential regulation of Toll-like receptor mRNAs in experimental murine central nervous system infections. *Neurosci Lett*, *344*(1), 17–20.

Brown, G. C., & Neher, J. J. (2010). Inflammatory neurodegeneration and mechanisms of microglial killing of neurons. *Mol Neurobiol*, *41*(2–3), 242–247. doi:10.1007/s12035-010-8105-9.

Brzozka, K., Finke, S., & Conzelmann, K. K. (2005). Identification of the rabies virus alpha/beta interferon antagonist: Phosphoprotein P interferes with phosphorylation of interferon regulatory factor 3. *J Virol*, *79*(12), 7673–7681.

Brzozka, K., Finke, S., & Conzelmann, K. K. (2006). Inhibition of interferon signaling by rabies virus phosphoprotein P: Activation-dependent binding of STAT1 and STAT2. *J Virol*, *80*(6), 2675–2683.

Bsibsi, M., Persoon-Deen, C., Verwer, R. W., Meeuwsen, S., Ravid, R., & Van Noort, J. M. (2006). Toll-like receptor 3 on adult human astrocytes triggers production of neuroprotective mediators. *Glia*, *53*(7), 688–695. doi:10.1002/glia.20328.

Camelo, S., Lafage, M., Galelli, A., & Lafon, M. (2001). Selective role for the p55 Kd TNF-alpha receptor in immune unresponsiveness induced by an acute viral encephalitis. *J Neuroimmunol*, *113*(1), 95–108.

Camelo, S., Lafage, M., & Lafon, M. (2000). Absence of the p55 Kd TNF-alpha receptor promotes survival in rabies virus acute encephalitis. *J Neurovirol*, *6*(6), 507–518.

Cameron, J. S., Alexopoulou, L., Sloane, J. A., DiBernardo, A. B., Ma, Y., Kosaras, B., et al. (2007). Toll-like receptor 3 is a potent negative regulator of axonal growth in mammals. *J Neurosci*, *27*(47), 13033–13041.

Carosella, E. D., Moreau, P., Aractingi, S., & Rouas-Freiss, N. (2001). HLA-G: A shield against inflammatory aggression. *Trends Immunol*, *22*(10), 553–555.

Charlton, K. M., & Casey, G. A. (1979). Experimental rabies in skunks: Immunofluorescence light and electron microscopic studies. *Lab Invest*, *41*(1), 36–44.

Charlton, K. M., & Casey, G. A. (1981). Experimental rabies in skunks: Persistence of virus in denervated muscle at the inoculation site. *Can J Comp Med*, *45*(4), 357–362.

Charlton, K. M., Casey, G. A., & Campbell, J. B. (1984). Experimental rabies in skunks: Effects of immunosuppression induced by cyclophosphamide. *Can J Comp Med*, *48*(1), 72–77.

Chopy, D., Detje, C. N., Lafage, M., Kalinke, U., & Lafon, M. (2011). The type I interferon response bridles rabies virus infection and reduces pathogenicity. *J Neurovirol*, *17*(4), 353–367. doi:10.1007/s13365-011-0041-6.

Chopy, D., Pothlichet, J., Lafage, M., Megret, F., Fiette, L., Si-Tahar, M., et al. (2011). Ambivalent role of the innate immune response in rabies virus pathogenesis. *J Virol*, *85*(13), 6657–6668. doi: JVI.00302-11 [pii] 10.1128/JVI.00302-11.

Davoust, N., Vuaillat, C., Androdias, G., & Nataf, S. (2008). From bone marrow to microglia: Barriers and avenues. *Trends Immunol*, *29*(5), 227–234. doi: S1471-4906(08)00088-4 [pii] 10.1016/j.it.2008.01.010.

Delhaye, S., Paul, S., Blakqori, G., Minet, M., Weber, F., Staeheli, P., et al. (2006). Neurons produce type I interferon during viral encephalitis. *Proc Natl Acad Sci U S A*, *103*(20), 7835–7840.

Diebold, S. S., Montoya, M., Unger, H., Alexopoulou, L., Roy, P., Haswell, L. E., et al. (2003). Viral infection switches non-plasmacytoid dendritic cells into high interferon producers. *Nature*, *424*(6946), 324–328. doi: 10.1038/nature01783 nature01783 [pii].

Dong, H., Strome, S. E., Salomao, D. R., Tamura, H., Hirano, F., Flies, D. B., et al. (2002). Tumor-associated B7-H1 promotes T-cell apoptosis: A potential mechanism of immune evasion. *Nat Med*, *8*(8), 793–800.

Eisenacher, K., Steinberg, C., Reindl, W., & Krug, A. (2007). The role of viral nucleic acid recognition in dendritic cells for innate and adaptive antiviral immunity. *Immunobiology*, *212*(9–10), 701–714. doi: S0171-2985(07)00112-X [pii] 10.1016/j.imbio.2007.09.007.

Engelhardt, B. (2008). The blood-central nervous system barriers actively control immune cell entry into the central nervous system. *Curr Pharm Des*, *14*(16), 1555–1565.

Faber, M., Bette, M., Preuss, M. A., Pulmanausahakul, R., Rehnelt, J., Schnell, M. J., et al. (2005). Overexpression of tumor necrosis factor alpha by a recombinant rabies virus attenuates replication in neurons and prevents lethal infection in mice. *J Virol*, *79*(24), 15405–15416. doi: 79/24/15405 [pii] 10.1128/JVI.79.24.15405-15416.2005.

Faul, E. J., Wanjalla, C. N., Suthar, M. S., Gale, M., Wirblich, C., & Schnell, M. J. (2010). Rabies virus infection induces type I interferon production in an IPS-1 dependent manner while dendritic cell activation relies on IFNAR signaling. *PLoS Pathog*, *6*(7), e1001016. doi:10.1371/journal.ppat.1001016.

Fu, Z. F., Weihe, E., Zheng, Y. M., Schafer, M. K., Sheng, H., Corisdeo, S., et al. (1993). Differential effects of rabies and borna disease viruses on immediate-early- and late-response gene expression in brain tissues. *J Virol*, *67*(11), 6674–6681.

Galea, I., Bechmann, I., & Perry, V. H. (2007). What is immune privilege (not)? *Trends Immunol*, *28*(1), 12–18. doi: S1471-4906(06)00326-7 [pii] 10.1016/j.it.2006.11.004.

Galelli, A., Baloul, L., & Lafon, M. (2000). Abortive rabies virus central nervous infection is controlled by T lymphocyte local recruitment and induction of apoptosis. *J Neurovirol*, *6*(5), 359–372.

Goethals, S., Ydens, E., Timmerman, V., & Janssens, S. (2010). Toll-like receptor expression in the peripheral nerve. *Glia*, *58*(14), 1701–1709. doi:10.1002/glia.21041.

Gratas, C., Tohma, Y., Barnas, C., Taniere, P., Hainaut, P., & Ohgaki, H. (1998). Up-regulation of Fas (APO-1/CD95) ligand and down-regulation of Fas expression in human esophageal cancer. *Cancer Res*, *58*(10), 2057–2062.

Hemachudha, T., Wacharapluesadee, S., Mitrabhakdi, E., Wilde, H., Morimoto, K., & Lewis, R. A. (2005). Pathophysiology of human paralytic rabies. *J Neurovirol*, *11*(1), 93–100.

Hicks, D. J., Nunez, A., Healy, D. M., Brookes, S. M., Johnson, N., & Fooks, A. R. (2009). Comparative pathological study of the murine brain after experimental infection with classical rabies virus and European bat lyssaviruses. *J Comp Pathol*, *140*(2–3), 113–126.

Hirai, K., Kawano, H., Mifune, K., Fujii, H., Nishizono, A., Shichijo, A., et al. (1992). Suppression of cell-mediated immunity by street rabies virus infection. *Microbiol Immunol*, *36*(12), 1277–1290.

Hooper, D. C., Morimoto, K., Bette, M., Weihe, E., Koprowski, H., & Dietzschold, B. (1998). Collaboration of antibody and inflammation in clearance of rabies virus from the central nervous system. *J Virol*, *72*(5), 3711–3719.

Hooper, D. C., Phares, T. W., Fabis, M. J., & Roy, A. (2009). The production of antibody by invading B cells is required for the clearance of rabies virus from the central nervous system. *PLoS Negl Trop Dis*, *3*(10), e535.

Hornung, V., Ellegast, J., Kim, S., Brzozka, K., Jung, A., Kato, H., et al. (2006). 5'-Triphosphate RNA is the ligand for RIG-I. *Science*, *314*(5801), 994–997.

Hunter, M., Johnson, N., Hedderwick, S., McCaughey, C., Lowry, K., McConville, J., et al. (2010). Immunovirological correlates in human rabies treated with therapeutic coma. *J Med Virol*, *82*(7), 1255–1265. doi:10.1002/jmv.21785.

Ida-Hosonuma, M., Iwasaki, T., Yoshikawa, T., Nagata, N., Sato, Y., Sata, T., et al. (2005). The alpha/beta interferon response controls tissue tropism and pathogenicity of poliovirus. *J Virol*, *79*(7), 4460–4469. doi: 79/7/4460 [pii] 10.1128/JVI.79.7.4460-4469.2005.

Irwin, D. J., Wunner, W. H., Ertl, H. C., & Jackson, A. C. (1999). Basis of rabies virus neuro-virulence in mice: Expression of major histocompatibility complex class I and class II mRNAs. *J Neurovirol, 5*(5), 485–494.

Ito, N., Moseley, G. W., Blondel, D., Shimizu, K., Rowe, C. L., Ito, Y., et al. (2010). Role of interferon antagonist activity of rabies virus phosphoprotein in viral pathogenicity. *J Virol, 84*(13), 6699–6710. doi: JVI.00011-10 [pii]10.1128/JVI.00011-10.

Iwasaki, Y., & Clark, H. F. (1975). Cell to cell transmission of virus in the central nervous system. II. Experimental rabies in mouse. *Lab Invest, 33*(4), 391–399.

Jackson, A. C., Scott, C. A., Owen, J., Weli, S. C., & Rossiter, J. P. (2007). Therapy with mino-cycline aggravates experimental rabies in mice. *J Virol, 81*, 6248–6253.

Johnston, G. R., & Webster, N. R. (2009). Cytokines and the immunomodulatory function of the vagus nerve. *Br J Anaesth, 102*(4), 453–462. doi: aep037 [pii] 10.1093/bja/aep037.

Kasempimolporn, S., Saengseesom, W., Mitmoonpitak, C., Akesowan, S., & Sitprija, V. (1997). Cell-mediated immunosuppression in mice by street rabies virus not restored by calcium ionophore or PMA. *Asian Pac J Allergy Immunol, 15*(3), 127–132.

Kasempimolporn, S., Tirawatnapong, T., Saengseesom, W., Nookhai, S., & Sitprija, V. (2001). Immunosuppression in rabies virus infection mediated by lymphocyte apoptosis. *Jpn J Infect Dis, 54*(4), 144–147.

Kieseier, B. C. (2011). The mechanism of action of interferon-beta in relapsing multiple scle-rosis. *CNS Drugs, 25*(6), 491–502. doi: 10.2165/11591110-000000000-000004 [pii].

Kim, D., Kim, M. A., Cho, I. H., Kim, M. S., Lee, S., Jo, E. K., et al. (2007). A critical role of toll-like receptor 2 in nerve injury-induced spinal cord glial cell activation and pain hypersensitivity. *J Biol Chem, 282*(20), 14975–14983. doi: M607277200 [pii] 10.1074/jbc. M607277200.

Klein, R. S., Lin, E., Zhang, B., Luster, A. D., Tollett, J., Samuel, M. A., et al. (2005). Neuronal CXCL10 directs CD8+ T-cell recruitment and control of West Nile virus encephalitis. *J Virol, 79*(17), 11457–11466. doi: 79/17/11457 [pii] 10.1128/JVI.79.17.11457-11466.2005.

Klingen, Y., Conzelmann, K. K., & Finke, S. (2008). Double-labeled rabies virus: Live track-ing of enveloped virus transport. *J Virol, 82*(1), 237–245. doi: JVI.01342-07 [pii] 10.1128/ JVI.01342-07.

Koedel, U., Rupprecht, T., Angele, B., Heesemann, J., Wagner, H., Pfister, H. W., et al. (2004). MyD88 is required for mounting a robust host immune response to Streptococcus pneu-moniae in the CNS. *Brain, 127*(Pt 6), 1437–1445.

Kojima, D., Park, C. H., Satoh, Y., Inoue, S., Noguchi, A., & Oyamada, T. (2009). Pathology of the spinal cord of C57BL/6J mice infected with rabies virus (CVS-11 strain). *J Vet Med Sci, 71*(3), 319–324.

Kojima, D., Park, C. H., Tsujikawa, S., Kohara, K., Hatai, H., Oyamada, T., et al. (2010). Lesions of the central nervous system induced by intracerebral inoculation of BALB/c mice with rabies virus (CVS-11). *J Vet Med Sci, 72*(8), 1011–1016. doi: JST.JSTAGE/ jvms/09-0550 [pii].

Kuss, S. K., Etheredge, C. A., & Pfeiffer, J. K. (2008). Multiple host barriers restrict polio-virus trafficking in mice. *PLoS Pathog, 4*(6), e1000082. doi:10.1371/journal.ppat.1000082.

Kwidzinski, E., Bunse, J., Kovac, A. D., Ullrich, O., Zipp, F., Nitsch, R., et al. (2003). IDO (indolamine 2,3-dioxygenase) expression and function in the CNS. *Adv Exp Med Biol, 527*, 113–118.

Lafon, M. (2005). Modulation of the immune response in the nervous system by rabies virus. *Curr Top Microbiol Immunol, 289*, 239–258.

Lafon, M. (2008). Immune evasion, a critical strategy for rabies virus. *Dev Biol (Basel), 131*, 413–419.

Lafon, M. (2011). Evasive strategies in rabies virus infection. *Adv Virus Res, 79*, 33–53. doi: B978-0-12-387040-7.00003-2 [pii] 10.1016/B978-0-12-387040-7.00003-2.

Lafon, M., Megret, F., Lafage, M., & Prehaud, C. (2006). The innate immune facet of brain: Human neurons express TLR-3 and sense viral dsRNA. *J Mol Neurosci, 29*(3), 185–194.

Lafon, M., Megret, F., Meuth, S. G., Simon, O., Velandia Romero, M. L., Lafage, M., et al. (2008). Detrimental contribution of the immuno-inhibitor b7-h1 to rabies virus encephalitis. *J Immunol*, *180*(11), 7506–7515.

Lafon, M., Prehaud, C., Megret, F., Lafage, M., Mouillot, G., Roa, M., et al. (2005). Modulation of HLA-G expression in human neural cells after neurotropic viral infections. *J Virol*, *79*(24), 15226–15237.

Laothamatas, J., Wacharapluesadee, S., Lumlertdacha, B., Ampawong, S., Tepsumethanon, V., Shuangshoti, S., et al. (2008). Furious and paralytic rabies of canine origin: Neuroimaging with virological and cytokine studies. *J Neurovirol*, *14*(2), 119–129.

Li, J., Ertel, A., Portocarrero, C., Barkhouse, D. A., Dietzschold, B., Hooper, D. C., et al. (2012). Postexposure treatment with the live-attenuated rabies virus (RV) vaccine TriGAS triggers the clearance of wild-type RV from the Central Nervous System (CNS) through the rapid induction of genes relevant to adaptive immunity in CNS tissues. *J Virol*, *86*(6), 3200–3210. doi: JVI.06699-11 [pii] 10.1128/JVI.06699-11.

Li, J., McGettigan, J. P., Faber, M., Schnell, M. J., & Dietzschold, B. (2008). Infection of monocytes or immature dendritic cells (DCs) with an attenuated rabies virus results in DC maturation and a strong activation of the NFkappaB signaling pathway. *Vaccine*, *26*(3), 419–426. doi: S0264-410X(07)01262-5 [pii] 10.1016/j.vaccine.2007.10.072.

Liu, L., Botos, I., Wang, Y., Leonard, J. N., Shiloach, J., Segal, D. M., et al. (2008). Structural basis of toll-like receptor 3 signaling with double-stranded RNA. *Science*, *320*(5874), 379–381. doi: 320/5874/379 [pii] 10.1126/science.1155406.

Ma, Y., Haynes, R. L., Sidman, R. L., & Vartanian, T. (2007). TLR8: An innate immune receptor in brain, neurons and axons. *Cell Cycle*, *6*(23), 2859–2868.

Ma, Y., Li, J., Chiu, I., Wang, Y., Sloane, J. A., Lu, J., et al. (2006). Toll-like receptor 8 functions as a negative regulator of neurite outgrowth and inducer of neuronal apoptosis. *J Cell Biol*, *175*(2), 209–215. doi: jcb.200606016 [pii] 10.1083/jcb.200606016.

Marquette, C., Van Dam, A. M., Ceccaldi, P. E., Weber, P., Haour, F., & Tsiang, H. (1996). Induction of immunoreactive interleukin-1 beta and tumor necrosis factor-alpha in the brains of rabies virus infected rats. *J Neuroimmunol*, *68*(1–2), 45–51.

Masatani, T., Ito, N., Shimizu, K., Ito, Y., Nakagawa, K., Abe, M., et al. (2010). Amino acids at positions 273 and 394 in rabies virus nucleoprotein are important for both evasion of host RIG-I-mediated antiviral response and pathogenicity. *Virus Res* doi: S0168-1702(10)00352-7 [pii] 10.1016/j.virusres.2010.09.016.

Masatani, T., Ito, N., Shimizu, K., Ito, Y., Nakagawa, K., Sawaki, Y., et al. (2010a). Rabies virus nucleoprotein functions to evade activation of the RIG-I-mediated antiviral response. *J Virol*, *84*(8), 4002–4012. doi: JVI.02220-09 [pii] 10.1128/JVI.02220-09

Masatani, T., Ito, N., Shimizu, K., Ito, Y., Nakagawa, K., Sawaki, Y., et al. (2010b). Rabies Virus Nucleoprotein Functions To Evade Activation of the RIG-I-Mediated Antiviral Response. *Journal of Virology*, *84*(8), 4002–4012. doi:10.1128/Jvi.02220-09.

McKimmie, C. S., Johnson, N., Fooks, A. R., & Fazakerley, J. K. (2005). Viruses selectively upregulate Toll-like receptors in the central nervous system. *Biochem Biophys Res Commun*, *336*(3), 925–933.

Megret, F., Prehaud, C., Lafage, M., Moreau, P., Rouas-Freiss, N., Carosella, E. D., et al. (2007). Modulation of HLA-G and HLA-E expression in human neuronal cells after rabies virus or herpes virus simplex type 1 infections. *Hum Immunol*, *68*(4), 294–302.

Menager, P., Roux, P., Megret, F., Bourgeois, J. P., Le Sourd, A. M., Danckaert, A., et al. (2009). Toll-like receptor 3 (TLR3) plays a major role in the formation of rabies virus Negri Bodies. *PLoS Pathog*, *5*(2), e1000315. doi:10.1371/journal.ppat.1000315.

Montano-Hirose, J. A., Lafage, M., Weber, P., Badrane, H., Tordo, N., & Lafon, M. (1993). Protective activity of a murine monoclonal antibody against European bat lyssavirus 1 (EBL1) infection in mice. *Vaccine*, *11*(12), 1259–1266.

Nakamichi, K., Inoue, S., Takasaki, T., Morimoto, K., & Kurane, I. (2004). Rabies virus stimulates nitric oxide production and CXC chemokine ligand 10 expression in macrophages through activation of extracellular signal-regulated kinases 1 and 2. *J Virol*, *78*(17), 9376–9388.

Nakamichi, K., Saiki, M., Sawada, M., Takayama-Ito, M., Yamamuro, Y., Morimoto, K., et al. (2005). Rabies virus-induced activation of mitogen-activated protein kinase and NF-kappaB signaling pathways regulates expression of CXC and CC chemokine ligands in microglia. *J Virol*, *79*(18), 11801–11812.

Nguyen, M. D., Julien, J. P., & Rivest, S. (2002). Innate immunity: The missing link in neuro-protection and neurodegeneration? *Nat Rev Neurosci*, *3*(3), 216–227.

Nuovo, G. J., DeFaria, D. L., Chanona-Vilchi, J. G., & Zhang, Y. (2004). Molecular detection of rabies encephalitis and correlation with cytokine expression. *Mod Pathol*

Park, C. H., Kondo, M., Inoue, S., Noguchi, A., Oyamada, T., Yoshikawa, H., et al. (2006). The histopathogenesis of paralytic rabies in six-week-old C57BL/6J mice following inoculation of the CVS-11 strain into the right triceps surae muscle. *J Vet Med Sci*, *68*(6), 589–595.

Park, C., Lee, S., Cho, I. H., Lee, H. K., Kim, D., Choi, S. Y., et al. (2006). TLR3-mediated signal induces proinflammatory cytokine and chemokine gene expression in astrocytes: Differential signaling mechanisms of TLR3-induced IP-10 and IL-8 gene expression. *Glia*, *53*(3), 248–256. doi:10.1002/glia.20278.

Peltier, D. C., Simms, A., Farmer, J. R., & Miller, D. J. (2010). Human neuronal cells possess functional cytoplasmic and TLR-mediated innate immune pathways influenced by phosphatidylinositol-3 kinase signaling. *J Immunol*, *184*(12), 7010–7021. doi: jimmunol.0904133 [pii] 10.4049/jimmunol.0904133.

Perry, L. L., Hotchkiss, J. D., & Lodmell, D. L. (1990). Murine susceptibility to street rabies virus is unrelated to induction of host lymphoid depletion. *Journal of immunology*, *144*(9), 3552–3557.

Pfeiffer, J. K. (2010). Innate host barriers to viral trafficking and population diversity: Lessons learned from poliovirus. *Adv Virus Res*, *77*, 85–118. doi: B978-0-12-385034-8.00004-1 [pii] 10.1016/B978-0-12-385034-8.00004-1.

Phares, T. W., Kean, R. B., Mikheeva, T., & Hooper, D. C. (2006). Regional differences in blood-brain barrier permeability changes and inflammation in the apathogenic clearance of virus from the central nervous system. *J Immunol*, *176*(12), 7666–7675.

Phares, T. W., Stohlman, S. A., Hinton, D. R., Atkinson, R., & Bergmann, C. C. (2010). Enhanced antiviral T cell function in the absence of B7-H1 is insufficient to prevent persistence but exacerbates axonal bystander damage during viral encephalomyelitis. *J Immunol*, *185*(9), 5607–5618. doi: jimmunol.1001984 [pii] 10.4049/jimmunol.1001984.

Pichlmair, A., Schulz, O., Tan, C. P., Naslund, T. I., Liljestrom, P., Weber, F., et al. (2006). RIG-I-mediated antiviral responses to single-stranded RNA bearing 5′-phosphates. *Science*, *314*(5801), 997–1001.

Prehaud, C., Megret, F., Lafage, M., & Lafon, M. (2005). Virus infection switches TLR-3-positive human neurons to become strong producers of beta interferon. *J Virol*, *79*(20), 12893–12904.

Pulmanausahakul, R., Faber, M., Morimoto, K., Spitsin, S., Weihe, E., Hooper, D. C., et al. (2001). Overexpression of cytochrome C by a recombinant rabies virus attenuates pathogenicity and enhances antiviral immunity. *J Virol*, *75*(22), 10800–10807. doi:10.1128/JVI.75.22.10800-10807.2001.

Randall, R. E., & Goodbourn, S. (2008). Interferons and viruses: An interplay between induction, signalling, antiviral responses and virus countermeasures. *J Gen Virol*, *89*(Pt 1), 1–47. doi: 89/1/1 [pii] 10.1099/vir.0.83391-0.

Rieder, M., Brzozka, K., Pfaller, C. K., Cox, J. H., Stitz, L., & Conzelmann, K. K. (2011). Genetic dissection of interferon-antagonistic functions of rabies virus phosphoprotein: Inhibition of interferon regulatory factor 3 activation is important for pathogenicity. *J Virol*, *85*(2), 842–852. doi: JVI.01427-10 [pii] 10.1128/JVI.01427-10.

Rieder, M., & Conzelmann, K. K. (2009). Rhabdovirus evasion of the interferon system. *J Interferon Cytokine Res*, *29*(9), 499–509.

Rossiter, J. P., Hsu, L., & Jackson, A. C. (2009). Selective vulnerability of dorsal root ganglia neurons in experimental rabies after peripheral inoculation of CVS-11 in adult mice. *Acta Neuropathol, 118*, 249–259.

Rouas-Freiss, N., Moreau, P., Menier, C., & Carosella, E. D. (2003). HLA-G in cancer: A way to turn off the immune system. *Semin Cancer Biol, 13*(5), 325–336.

Roy, A., & Hooper, D. C. (2007). Lethal silver-haired bat rabies virus infection can be prevented by opening the blood-brain barrier. *J Virol, 81*(15), 7993–7998.

Roy, A., Phares, T. W., Koprowski, H., & Hooper, D. C. (2007). Failure to open the blood-brain barrier and deliver immune effectors to central nervous system tissues leads to the lethal outcome of silver-haired bat rabies virus infection. *J Virol, 81*(3), 1110–1118.

Shankar, V., Kao, M., Hamir, A. N., Sheng, H., Koprowski, H., & Dietzschold, B. (1992). Kinetics of virus spread and changes in levels of several cytokine mRNAs in the brain after intranasal infection of rats with Borna disease virus. *J Virol, 66*(2), 992–998.

Shimizu, K., Ito, N., Sugiyama, M., & Minamoto, N. (2006). Sensitivity of rabies virus to type I interferon is determined by the phosphoprotein gene. *Microbiol Immunol, 50*(12), 975–978.

Sommereyns, C., Paul, S., Staeheli, P., & Michiels, T. (2008). IFN-lambda (IFN-lambda) is expressed in a tissue-dependent fashion and primarily acts on epithelial cells in vivo. *PLoS Pathog, 4*(3), e1000017. doi:10.1371/journal.ppat.1000017.

Steinman, R. M. (1991). The dendritic cell system and its role in immunogenicity. *Annu Rev Immunol, 9*, 271–296. doi:10.1146/annurev.iy.09.040191.001415.

Sugiura, N., Uda, A., Inoue, S., Kojima, D., Hamamoto, N., Kaku, Y., et al. (2011). Gene Expression Analysis of Host Innate Immune Responses in the Central Nervous System following Lethal CVS-11 Infection in Mice. *Jpn J Infect Dis, 64*(6), 463–472.

Takeuchi, O., & Akira, S. (2010). Pattern recognition receptors and inflammation. *Cell, 140*(6), 805–820. doi: S0092-8674(10)00023-1 [pii] 10.1016/j.cell.2010.01.022.

Tang, S. C., Arumugam, T. V., Xu, X., Cheng, A., Mughal, M. R., Jo, D. G., et al. (2007). Pivotal role for neuronal Toll-like receptors in ischemic brain injury and functional deficits. *Proc Natl Acad Sci U S A, 104*(34), 13798–13803. doi: 0702553104 [pii] 10.1073/pnas.0702553104.

Tobiume, M., Sato, Y., Katano, H., Nakajima, N., Tanaka, K., Noguchi, A., et al. (2009). Rabies virus dissemination in neural tissues of autopsy cases due to rabies imported into Japan from the Philippines: Immunohistochemistry. *Pathol Int, 59*(8), 555–566.

Torres-Anjel, M. J., Volz, D., Torres, M. J., Turk, M., & Tshikuka, J. G. (1988). Failure to thrive, wasting syndrome, and immunodeficiency in rabies: A hypophyscal/hypothalamic/thymic axis effect of rabies virus. *Rev Infect Dis, 10*(Suppl 4), S710–725.

Tracey, K. J. (2009). Reflex control of immunity. *Nature reviews. Immunology, 9*(6), 418–428. doi:10.1038/nri2566.

Tshikuka, J. G., Torres-Anjel, M. J., Blenden, D. C., & Elliott, S. C. (1992). The microepidemiology of wasting syndrome, a common link to diarrheal disease, cancer, rabies, animal models of AIDS, and HIV-AIDS YHAIDS). The feline leukemia virus and rabies virus models. *Ann N Y Acad Sci, 653*, 274–296.

Ugolini, G. (1995). Specificity of rabies virus as a transneuronal tracer of motor networks: Transfer from hypoglossal motoneurons to connected second-order and higher order central nervous system cell groups. *J Comp Neurol, 356*(3), 457–480.

Ugolini, G. (2010). Advances in viral transneuronal tracing. *J Neurosci Methods, Epub ahead of print* doi: S0165-0270(09)00623-2 [pii] 10.1016/j.jneumeth.2009.12.001.

Van Dam, A. M., Bauer, J., Man, A., Hing, W. K., Marquette, C., Tilders, F. J., et al. (1995). Appearance of inducible nitric oxide synthase in the rat central nervous system after rabies virus infection and during experimental allergic encephalomyelitis but not after peripheral administration of endotoxin. *J Neurosci Res, 40*(2), 251–260.

Versteeg, G. A., & Garcia-Sastre, A. (2010). Viral tricks to grid-lock the type I interferon system. *Curr Opin Microbiol, 13*(4), 508–516. doi: S1369-5274(10)00070-6 [pii] 10.1016/j.mib.2010.05.009.

Vidy, A., El Bougrini, J., Chelbi-Alix, M. K., & Blondel, D. (2007). The nucleocytoplasmic rabies virus P protein counteracts interferon signaling by inhibiting both nuclear accumulation and DNA binding of STAT1. *J Virol, 81*(8), 4255–4263.

Vuaillat, C., Varrin-Doyer, M., Bernard, A., Sagardoy, I., Cavagna, S., Chounlamountri, I., et al. (2008). High CRMP2 expression in peripheral T lymphocytes is associated with recruitment to the brain during virus-induced neuroinflammation. *J Neuroimmunol, 193*(1-2), 38–51.

Wang, Z. W., Sarmento, L., Wang, Y., Li, X. Q., Dhingra, V., Tseggai, T., et al. (2005). Attenuated rabies virus activates, while pathogenic rabies virus evades, the host innate immune responses in the central nervous system. *J Virol, 79*(19), 12554–12565.

Weiland, F., Cox, J. H., Meyer, S., Dahme, E., & Reddehase, M. J. (1992). Rabies virus neuritic paralysis: Immunopathogenesis of nonfatal paralytic rabies. *J Virol, 66*(8), 5096–5099.

Wen, Y., Wang, H., Wu, H., Yang, F., Tripp, R. A., Hogan, R. J., et al. (2011). Rabies virus expressing dendritic cell-activating molecules enhances the innate and adaptive immune response to vaccination. *J Virol, 85*(4), 1634–1644. doi: JVI.01552-10 [pii] 10.1128/JVI.01552-10.

Wiktor, T. J., Doherty, P. C., & Koprowski, H. (1977a). In vitro evidence of cell-mediated immunity after exposure of mice to both live and inactivated rabies virus. *Proc Natl Acad Sci U S A, 74*(1), 334–338.

Wiktor, T. J., Doherty, P. C., & Koprowski, H. (1977b). Suppression of cell-mediated immunity by street rabies virus. *J Exp Med, 145*(6), 1617–1622.

Wiktor, T. J., Macfarlan, R. I., Reagan, K. J., Dietzschold, B., Curtis, P. J., Wunner, W. H., et al. (1984). Protection from rabies by a vaccinia virus recombinant containing the rabies virus glycoprotein gene. *Proc Natl Acad Sci U S A, 81*(22), 7194–7198.

Wunner, W. H., Dietzschold, B., Curtis, P. J., & Wiktor, T. J. (1983). Rabies subunit vaccines. *J Gen Virol, 64*(Pt 8), 1649–1656.

Xiang, Z. Q., Knowles, B. B., McCarrick, J. W., & Ertl, H. C. (1995). Immune effector mechanisms required for protection to rabies virus. *Virology, 214*(2), 398–404.

Zhang, B., Chan, Y. K., Lu, B., Diamond, M. S., & Klein, R. S. (2008). CXCR3 mediates region-specific antiviral T cell trafficking within the central nervous system during West Nile virus encephalitis. *J Immunol, 180*(4), 2641–2649. doi: 180/4/2641 [pii].

Zhao, L., Toriumi, H., Kuang, Y., Chen, H., & Fu, Z. F. (2009). The Roles of Chemokines in Rabies Virus Infection: Over-Expression May Not Always Be Beneficial. *J Virol.*

Zhao, P., Zhao, L., Zhang, T., Qi, Y., Wang, T., Liu, K., et al. (2011). Innate immune response gene expression profiles in central nervous system of mice infected with rabies virus. *Comp Immunol Microbiol Infect Dis.* doi: S0147-9571(11)00077-4 [pii] 10.1016/j.cimid.2011.09.003.

Laboratory Diagnosis of Rabies

Cathleen A. Hanlon[1] and Susan A. Nadin-Davis[2]

[1]Kansas State University Rabies Laboratory, 2005 Research Park Circle, Manhattan, KS 66502, USA, [2]Animal Health Microbiology Research, Ottawa Laboratory (Fallowfield), Canadian Food Inspection Agency, Ottawa, Ontario, K2J 4S1, Canada

1 LABORATORY-BASED RABIES DIAGNOSTIC TESTING

Rabies diagnosis is one of the most compelling examples of "One Health" in action. The diagnosis of rabies, or, conversely, a negative finding, often has direct impacts on human health by determining the need for medical intervention with expensive post-exposure prophylaxis, as well as directing the appropriate management of potentially exposed animals (Anonymous, 2011; Manning et al., 2008). Reliable diagnosis of animals is critically important for saving human lives. Laboratory-based diagnosis is essential because the signs of rabies may vary considerably and may be clinically indistinguishable from other causes of encephalitis, acute disability such as from trauma, altered mentation due to toxins, or other conditions (Dimaano, Scholand, Alera, & Belandres, 2011). A *positive* diagnosis requires an urgent public health investigation and medical care for all potentially exposed persons, as well as examination and rabies booster vaccination of currently vaccinated pets and livestock. If an animal is exposed to rabies and is overdue on its rabies shots or is unvaccinated, the recommendation is euthanasia or a six-month quarantine, as the animal is at risk of developing rabies from this exposure during this period of time (Anonymous, 2011). A *negative* diagnosis can be as important as a positive diagnosis by ruling out the need for post-exposure prophylaxis (currently in the USA a series of four or five vaccinations and a dose of human rabies immune globulin to prevent rabies) for humans and proper management of domestic animals which may have been exposed to the animal which was tested. If no diagnostic

Rabies: Scientific Basis of the Disease and its Management
DOI: http://dx.doi.org/10.1016/B978-0-12-396547-9.00011-0

testing is available, significant animal control, public health, and health care provider time and effort, as well as essential biologics would need to be applied to every *suspected* case in order to avert human and animal deaths.

In the USA, each year an estimated 100,000 rabies-suspect animals are submitted to one of over 100 laboratories offering primary rabies diagnostic testing (Blanton, Palmer, Dyer, & Rupprecht, 2011). The majority of rabies diagnosis is conducted within state public health laboratories. In some large states, several public health laboratories may provide rabies diagnosis. In some states, state agriculture or animal health laboratories may augment primary diagnosis or conduct testing only on animals where there has been no known human exposure so as to guide appropriate management of potentially exposed domestic animals. In rare cases, university-based laboratories conduct rabies diagnosis for a state or region of the USA. In Canada, rabies diagnosis is available at two laboratories of the Canadian Food Inspection Agency: one located in Ottawa, Ontario, and the other in Lethbridge, Alberta. In Mexico, the national diagnostic reference laboratory conducting rabies diagnosis is the Instituto de Diagnóstico y Referencia Epidemiológicos (InDRE) which is part of Centro Nacional de Vigilancia Epidemiológica y Control de Enfermedados.

Due to the public health importance of rabies diagnosis, the majority of rabies testing is supported by government funding for public health and is often a combination of federal and state sources, although this is not always the case. In some localities in the USA, the potentially exposed person or the owner of an exposed animal may need to pay for the test at a laboratory. The cost of testing typically ranges from $65–145 USD or more per test for the standard direct fluorescent antibody (dFA) test using commercial rabies diagnostic conjugates (Anonymous, 2012). Among the 100,000 various animal samples tested, approximately 6,000 to 10,000 test positive for rabies annually in the USA, with the majority of positive tests occurring among wildlife species (Blanton, Hanlon, & Rupprecht, 2007; Blanton, Krebs, Hanlon, & Rupprecht, 2006; Blanton, Palmer, Christian, & Rupprecht, 2008; Blanton, Palmer, Dyer, & Rupprecht, 2011; Blanton, Palmer, & Rupprecht, 2010; Blanton, Robertson, Palmer, & Rupprecht, 2009). In Canada, too, the majority of positive cases are reported in wildlife species (Fehlner-Gardiner, Muldoon, Nadin-Davis, Wandeler, Kush & Jordan, 2008).

Despite the importance of whether an individual human or animal is rabid, rabies testing by any method is only partially under regulatory jurisdiction. The diagnosis of rabies in an animal by a public health laboratory in the USA using commercial reagents is performed on an "environmental" or non-human sample and thus is not under the direct jurisdiction of the Clinical Laboratory Improvement Amendments (CLIA) of 1988 (Interpretive Guidelines §493.1253(b)(2))(1996). The reagents for primary diagnosis by direct immunofluorescence testing on brain samples, in many cases, may be sold as Analyte Specific Reagents (ASRs), for which, by regulation, the manufacture must specify that the analytical and performance characteristics have not been established. When a laboratory conducts tests using these diagnostic reagents, the laboratory has implicitly accepted the limitations of the use of these products. Each laboratory is expected to establish test method characteristics for each batch of reagents that are used. According to current regulations, these products should only be used by high-complexity laboratories, including those regulated by CLIA and public health laboratories. Execution of the primary method with diagnostic reagents that may be under CLIA regulation is largely the responsibility of each laboratory. Although there are recommendations for promoting quality testing, the implementation of these guidelines is left to the discretion of the laboratory. These include such considerations as: 1) before testing, the arranging for the test and specimen collection, 2) during testing, which consists of control testing, test performance, result interpretation and recording, and 3) after testing, including reporting of the result, documentation, confirmatory testing, and biohazardous waste disposal. And yet, the vast majority of the operating units of the laboratories conducting rabies diagnosis are not under CLIA-regulation per se. The conjugates that they purchase and use every day for primary rabies diagnosis by immunofluorescence on animal tissues may be subject to limited Food and Drug Administration (FDA) regulation, but because they are largely applied to animal brain material, the FDA/CLIA regulation language restricts their specific jurisdiction to testing on *human* samples, even though these tests determine human medical treatment. Implementation of practices reflecting "One Medicine – One Health" would be highly desirable, as these reagents and laboratory procedures are directly relevant to human medical intervention, as well as domestic animal management that may ultimately impact human medical management.

1.1 Historic Approaches

Historic accounts of rabies, based on compatible clinical presentations in humans and animals, date back four millennia, even to a 23rd-century BC document from Babylon. The clinical presentation in humans and the association between human disease and animals, especially dogs, was relatively well-described over numerous centuries by various authors, scientists, and even in art. The first laboratory-based diagnostic method came about through microscopic examination of brain tissue and the identification in 1903 of Negri bodies by an Italian pathologist and microbiologist after which they are named (Negri, 1903). His observation ushered in a flood of novel approaches to staining these bodies (Sellers, 1923). One approach which was quite useful for decades was a special Seller's stain which was performed on wet slide impressions of fresh samples (unfixed) of hippocampus and sometimes other parts of the brain (Young & Sellers, 1927). Although a fundamental laboratory test for nearly half a century, Negri bodies were not always present in an infected animal, but their presence provided relatively good specificity in comparison to a presumptive diagnosis based solely on clinical signs. The next major improvement in sensitivity of a diagnostic method was the practice of virus isolation through intracerebral inoculation of a brain homogenate from the suspected animal into suckling or juvenile mice. This was a fundamental diagnostic technique for many years (Koprowski, 1966, 1973), is still in limited use globally for primary diagnosis, and is of great utility for specific research purposes. The brains of the mice that succumbed to inoculation were historically examined for Negri bodies. A somewhat more technically demanding but similar approach to the isolation of live virus in a sample is cell culture (see Expert committee on rabies; second report, 1954). Fluorescence microscopy and fluorescein-labeled antibodies ushered in the "gold standard" diagnostic technique still in use today (Goldwasser & Kissling, 1958).

1.2 The Direct Fluorescent Antibody Test

The principal laboratory test currently used for rabies diagnosis is the direct fluorescent antibody (dFA) test (Goldwasser & Kissling, 1958). Antibodies to epitopes found on the nucleocapsid protein of rabies viruses are fluorescein-conjugated. When these antibodies bind to proteins in infected neurons, fluorescent foci are observed under fluorescence microscopy. The main limitations of this method are the requirement for a high quality fluorescence microscope and the need for trained and experienced personnel. In addition to outstanding sensitivity and specificity, a compelling strength of this method is the relative brevity of the testing process. Under ideal conditions, a definitive result can be available in a few hours.

In the USA, the need for a national standard protocol for the laboratory diagnosis of rabies was recognized for quite some time (Hanlon, Smith, & Anderson, 1999). Nearly a decade ago, a committee was formed from representatives of national and state public health laboratories to evaluate procedures employed by rabies diagnostic laboratories across the country. The committee identified and described the minimum standards for reliable rabies diagnosis in the USA (Rudd, Smith, Yager, Orciari, & Trimarchi, 2005). This document was a good starting point (see details in the following sections). Now that a minimum standard has been described, it would be ideal to revisit practices and work toward the description of best practices in regard to method validation within each laboratory. Current needs include support, both financial and leadership commitment, for vibrant networking among rabies diagnosticians. At a number of public health, agriculture, and university-based laboratories in the USA, it is not uncommon for there to be a single individual, and a sole designated back-up individual comprising the entire rabies diagnostic force. At present, networking across states or regions among these individuals is rare.

1.3 Reverse-Transcription Polymerase Chain Reaction Assay

Of all molecular methods that seek to detect rabies and rabies-related virus RNAs, tests based on the reverse-transcription polymerase chain reaction (RT-PCR) assay are the most frequently employed (Sacramento, Bourhy, & Tordo, 1991) (see Section 4, following.) Most often, it is applied postmortem to fresh brain material from a suspect animal or human. The technique may also be applied to saliva, skin samples (often collected antemortem from suspect humans), salivary glands, and, indeed, to virtually any other tissue or sample. When this procedure generates a positive amplification result, sequence characterization of the product and comparison with other reference viruses can elucidate the variant type, in what species it is most commonly transmitted, and from which geographic area closely related viruses circulate. The limitations of this method are the lengthy process it requires (up to two days) and the expense, as well as the exquisite sensitivity of the technique, which escalates the risk of cross-contamination and hence erroneous findings.

1.4 Immunohistochemistry on Formalin-Fixed Paraffin Embedded Tissues

If brain material from a potentially rabid animal is fixed in formalin, significant diagnostic delay will occur. The tissue will need time to be completely chemically fixed. Samples of the tissue will need to be trimmed and embedded in paraffin blocks so that sections may be cut with a microtome and placed on microscope slides. These slides may then be subjected to routine histopathology staining, but a specific immunohistochemical procedure will be needed to search for rabies virus antigen. A research technique based on this approach requires significant expertise (Fekadu, Greer, Chandler, & Sanderlin, 1988; Hamir, Moser, Fu, Dietzschold, & Rupprecht, 1995) and can be undertaken in only a few capable laboratories. The process often takes up to a week from the time that tissue in formalin is submitted. Typically, the situations in which this method for diagnosis becomes necessary are complex and often a week or more after the death of an animal or human upon which the diagnosis is sought. In rare cases, RNA extraction can be attempted on chemically fixed material. If the sample has been in formalin for a prolonged period of time, the degradation of RNA and hence sensitivity of the method may be low, although the expected kinetics of this decay are undefined at present. Extraction techniques on paraffin embedded material may be somewhat more rewarding. However, RNA extraction, reverse transcription, and amplification is still challenging and often primers targeting short genetic fragments or approaches using nested or hemi-nested primers may be the most fruitful.

1.5 Other Research Methods

Diagnostic research techniques may include electron microscopy by which virus particles with the characteristic bullet-shape of Rhabdoviruses may be directly observed. The main limitations are the need for highly sophisticated, expensive equipment and expert skill, as well as a potential for low sensitivity because, to confirm negative samples with some confidence, many fields of many samples of a particular specimen may need to be examined, somewhat like searching for a needle in a haystack. Moreover, the characteristic shape does not provide specific information about the genus, genotype, or variant and the species to which it is most likely adapted or the current geographic location of its most closely related viruses.

As described previously, direct virus isolation is possible through intracerebral inoculation of a brain homogenate into young mice or into cell culture. However, if the homogenate contains a pathogen that causes mice to succumb or infects cell cultures, additional methods to confirm the nature of the pathogen are required. The dFA or RT-PCR methods to detect rabies virus or rabies-related viruses are most commonly used.

For research purposes, methods for visualizing various proteins of the virus are often developed. This may include use of specific antibodies that bind to various epitopes on the targeted protein, an association that is then revealed through labeling the primary antibody or using a secondary antibody and a development technique for visualization. Other approaches include use of labeled probes to detect specific rabies virus sequences by *in situ* hybridization. Also, diagnostic or research approaches through real-time PCR are feasible (Hayman et al., 2011a; Nadin-Davis, Sheen, & Wandeler, 2009; Szanto, Nadin-Davis, Rosatte, & White, 2011) (see Section 4 of this chapter). Depending on the precise chemistry employed for real-time PCR, these methods have limitations that could, under certain circumstances, yield false negative or false positive results both of which have significant public health implications.

Another approach to enhanced surveillance is a direct immunohistochemical approach on touch impressions of brain material using a combination of biotinylated anti-nucleocapsid monoclonal antibodies (Lembo et al., 2006). The material is then incubated with streptavidin-peroxidase complex, followed by the AEC peroxidase substrate, and then counterstained with hematoxylin. Examination is by light microscopy. Although in use for research purposes on a limited basis, the primary reagent is not reliably available to qualified personnel and laboratories.

There is a strong desire for immediately available diagnostic (disease detection) and serologic (vaccine response verification) testing, typically at a veterinary practice, humane association, animal control, or wildlife agency, but also for use in humans (Jayakumar & Padmanaban, 1994; Madhusudana, Paul, Abhilash, & Suja, 2004; Perrin, Gontier, Lecocq, & Bourhy, 1992; Vasanth, Madhusudana, Abhilash, Suja, & Muhamuda, 2004; Zanluca et al., 2011), and there are prototype tests being developed towards this end (see Kasempimolporn et al., 2012; Servat et al., 2012). The ultimate problem is that rabies demands exquisite sensitivity so as not to miss a public health threat to humans or their domestic animals. At a minimum, these types of tests would be useful as an initial screening tool. Several strategies could be considered. One approach would be to develop the test method so that it is as specific as possible, meaning that the target yields strongly positive results that would need no further confirmation. All other results would need definitive testing by traditional diagnostic methods. The other approach would be to screen with the highest possible sensitivity, knowing that some true negatives would be identified as potentially positive. Samples with positive results by screening could be accepted as a presumptive positive or be sent for confirmatory or definitive testing. The technology exists for these tests, but the process of refinement, regulatory approval, and then acceptance by the end users will be more complex than with most animal diseases, due to public health implications, as well as the emotion and concern that rabies elicits from many individuals.

2 POSTMORTEM DIAGNOSIS OF RABIES IN ANIMALS

2.1 Indications for Testing

This section describes the policies and test protocols that apply to rabies diagnosis specifically within the USA; while rather similar practices are applied in many different countries, details may vary according to jurisdiction.

The single most important role of diagnostic testing is the protection of human and domestic animal health. In this regard, the indication for rabies testing is directly related to an exposed human or domestic animal. This is passive surveillance in that the submitted animals are selected on the basis of their encounters with humans and animals. Across North America, terrestrial rabies virus vector species such as the raccoon, skunk, and fox species from areas infected with variants adapted to these species will, in most cases, comprise the species with the greatest proportion of positive animals. However, test results on the submitted animals are not representative of population-based testing in the various species. Although active surveillance is necessary to gain good epizootiologic knowledge on rabies, limited information can be enhanced through carefully crafted specimen acceptance policies. An optimal understanding of local and regional transmission patterns will help guide risk assessment for potentially exposed humans and animals, especially in situations where the exposing animal is not available for diagnostic testing or, if it was a dog, cat, or ferret, for observation (Anonymous, 2011). In addition to the testing of animals that may have exposed a human or domestic animal, disease surveillance can be enhanced by the acceptance of uncommonly tested animals, which, although only occasionally available for testing, may be observed with neurologic signs compatible with rabies, such as deer, bear, beaver, or other species. For heightened surveillance, it is also wise to consider testing individuals of a potential rabies virus vector species that succumbed to an illness with signs compatible with rabies but that are from an area not known to be infected with a variant adapted to their species. It is also critically important to continue to test rabies virus vector species from an area infected with a species-adapted variant, so that the intensity of the local epizootic is well understood. This is particularly important in areas where potential rabies control efforts are ongoing through experimental methods, such as oral vaccination or trap-vaccinate-and-release.

Bats comprise a small but significant proportion of the samples upon which diagnostic testing is conducted. The vast majority of these animals are negative for rabies. The speciation of these individuals is extremely useful because the risk of rabies varies according to species, and the variants that are found in these animals are diverse. When rabies is confirmed, the diagnosis guides human post-exposure prophylaxis for potentially exposed persons and the management of potentially exposed domestic animals. The difficulty with human and domestic animal encounters with these animals is that bats are comparatively small-bodied. A potential exposure, such as a bite, from these animals may be ignored by humans because the actual trauma from a bite is so tiny. Indeed, some humans may not fully awaken from slumber when they are bitten by a bat, according to the case histories of some rabid humans.

It is important to emphasize that healthy non-stray dogs, cats, and ferrets that bite a person may be observed for 10 days—*regardless of vaccination status*—in place of euthanasia and testing (Anonymous, 2011). It appears in numerous cases that a biting dog or cat is submitted for rabies testing simply because someone has been bitten and the specter of rabies is raised, when the actual issue may be that an animal is simply no longer wanted by its owner; rabies is not suspected clinically. If a concern about a potential exposure is voiced to health authorities and the animal is euthanized immediately after the biting incident, it must be tested for rabies. Thus diagnostic laboratories often end up testing numerous healthy but unwanted domestic dogs and cats.

2.2 Biosafety

All activities related to the handling of animals and samples for rabies diagnosis should be performed using appropriate biosafety practices to avoid bites, as well as potential exposures to infected tissues or fluids. During sample preparation and the laboratory diagnostic process, the most likely risk of exposure is through accidental penetrating injuries with contaminated laboratory equipment or exposure of mucous membranes or broken skin to infectious tissue or fluids. The highest viral concentrations are found in central nervous system, salivary glands, and saliva, but any innervated tissue may be a source of exposure. People preparing specimens for submission and laboratory personnel involved in diagnostic activities can protect themselves against exposure to rabies through appropriate Personal Protection Equipment (PPE), as detailed in the cornerstone of biosafety practice and policy in the USA, which is the Biosafety in Microbiological and Biomedical Laboratories publication of the Centers for Disease Control and Prevention and the National Institutes of Health (CDC & NIH, 2012). The documents "Laboratory Safety Guidelines" issued by Health Canada (2004) and "Containment Standards for Veterinary Facilities" issued by CFIA (1996) also cover biosafety issues. During necropsy, this often consists of heavy rubber gloves, laboratory gown and waterproof apron, boots, surgical masks, protective sleeves, and a face shield. Fume hoods or biosafety cabinets are useful in providing protection from odor, ectoparasites, and bone fragments, but they are not mandatory. During laboratory activities, PPE often consists of disposable examination gloves, laboratory coats, sharps precautions, eye protection, and so on. During the diagnostic process, all sample materials and slides should be manipulated with attention to preclude creation of an aerosol or airborne droplet spray. Skin, respiratory, and eye protection is necessary for safe removal of brain tissue from animals submitted for testing. The manipulation and staining of slides and cleanup of the microscope and associated areas should be done with care to preclude potential exposures to rabies through glass chips and shards from slides. In addition, pre-exposure immunization against rabies for personnel routinely involved in these activities is recommended, along with regular serological monitoring of rabies antibody titers (Manning et al., 2008).

Pre-exposure vaccination provides priming immunization against rabies so that if an exposure occurs, post-exposure management is relatively simple, just day 0 and day 3 vaccine administrations, without the need for rabies immunoglobulin. Moreover, pre-exposure vaccination may also provide protection against an unrecognized exposure to rabies. Access to rabies diagnostic areas should be restricted to immunized personnel with the area secured and posted accordingly. All sample materials processed in these areas are medical waste and must be disposed of in accordance with guidelines in the U.S. Biosafety in Microbiological and Biomedical Laboratories publication, or similar guidelines for other jurisdictions.

2.3 Animal Confinement or Capture, Preliminary Preparation, Preservation, and Submission of Specimens

Initial capture or confinement of a rabies-suspect animal can be quite problematic. Rabid animals can be unpredictable, sometimes alternating from an apparently weakened neurologic state to one of dangerous aggression. Almost all jurisdiction of animal control is local; this can present substantial difficulties for the potentially exposed person or owner of an exposed domestic animal, particularly if the rabies-suspect animal is a wildlife species. Some local animal control officials may assist with situations involving only stray or rabies-suspect dogs, while some may assist where the problematic animal is any domestic species. It is rare to find a locality with personnel who would respond to any animal control situation whether it involves a domestic species, or bat, skunk, raccoon, fox, or other wildlife species. Individual members of the public often resort to independently confining or somehow capturing a potentially rabid animal and then making arrangements for its decapitation and submission to a laboratory.

For definitive diagnosis, a full cross-section of the brainstem and cerebellum (or alternatively, samples from the hippocampus) must be examined (Anonymous, 2003) (Figure 11.1). Specimens without these tissues are reported as "Unsuitable." Thus, preparation of samples often begins with decapitation of medium-size mammals. The entire carcass of bats and other small mammals are accepted for submission to maximize the amount of tissue available for examination and to facilitate identification of the species. In the case of large animals, the brain is often removed from the calvarium. In cattle, sample collection for rabies and bovine spongiform encephalopathy (BSE) are compatible, as the obex in the caudal medulla, which is required for BSE diagnosis, is directly caudal to the optimal area for rabies testing (Figure 11.1).

The diagnostic specimen must be packaged appropriately to preclude leakage of infectious material or deterioration of the sample due to extreme heat or freeze-thaw cycles. The specimen should be placed within an inner leak-proof container and then a secondary leak-proof

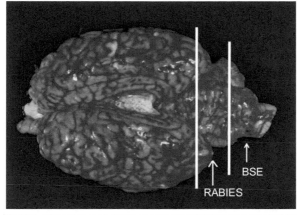

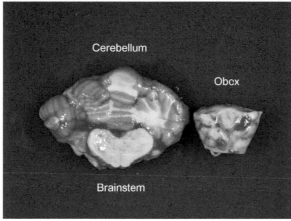

FIGURE 11.1   **Dorsal surface of an animal brain showing the coronal planes of section used for diagnosis of both rabies and bovine spongiform encephalopathy (BSE) (top). Cross section (left) and dorsal view (right) of an animal brain demonstrating the area sampled for rabies diagnosis in relation to the obex, which is appropriate for BSE diagnostic testing (bottom).** *(Photographs contributed by Dr. Jerome Nietfield, Kansas State University College of Veterinary Medicine, Department of Medicine and Pathobiology, Manhattan, KS 66502.)*

container. The double-enclosed specimen is then placed in a leak-proof outer container, which is often a Styrofoam cooler with an outer cardboard pack. Adequate cooling material, such as frozen cold packs, should be placed around the double-enclosed specimen. Absorbent material is mandatory to absorb condensation from cold packs and to contain any possible leakage from the sample. The outside of the container is often labeled with an International Biohazard sticker and "Exempt Animal Specimen" designation or with a UN3373 Biological Substance, Category B sticker. Each laboratory that performs rabies diagnosis typically offers a rabies examination submission form either in hard copy or online. If multiple samples are being submitted at the same time, it is necessary to use a separate form for each specimen. Submission forms should be packed separately from the samples and are typically placed in a sealed plastic bag to ensure that they will not become contaminated during transit or opening of the specimen. Express shipping is often the recommended method for delivery of a sample to testing laboratories. In most cases, specimens must be received by the testing laboratory in the morning on weekdays to be tested on the same day. Results are often available by close of business on the day of arrival and testing at laboratories. However, some specimens may require additional testing by a repeat of the dFA method, by evaluation with other methods, and occasionally perhaps even by sharing with a reference laboratory. This delay may occur as a result of the detection of apparently nonspecific staining, sparse antigen with lack of ability to repeat the finding, and other related observations. Emergency rabies diagnostic evaluation may often be arranged, but the necessity of it is at the discretion of public health professionals in consultation with laboratory personnel. Determinations are often made on a case-by-case basis. Emergency samples are often driven to the laboratory as an alternative to shipping. These guidelines have been established based on the Compendium of Animal Rabies Prevention and Control (2008) and the Advisory Committee on Immunization Practices (2008) which recognizes that "administration of rabies post-exposure prophylaxis is a medical urgency, not a medical emergency."

2.4 Immunofluorescence on Brain Tissue

2.4.1 Necropsy and Dissection

In accordance with the minimum standard guidelines for rabies testing, instruments used for dissection of specimens should be clean and sterile prior to use on each sample. Thus it is most convenient and time-efficient to have enough sets available for expected daily numbers of specimens. Although seemingly obvious, periodic review of practices and emphasis during training should be placed on evaluation of each step in the process of preparing samples for examination in order to minimize opportunities for cross-contamination. Each sample and the material derived from this sample should be prepared individually at each step of the procedure. Initial visual examination of specimens requires the identification of brainstem and cerebellum (and/or hippocampus, if available). Liquefaction of the specimen through decomposition or thermal stress (i.e., excessive heat or one or more freeze-thaw cycles) may preclude gross anatomic identification of these areas. Desiccation may also preclude preparation of the sample for staining with reagents. In these cases, sample results should be reported as untestable, unsuitable, or unsatisfactory for testing. On degraded samples, it may be reasonable and is recommended to continue to attempt testing of the material. Antigen can be quite resilient, and a positive result would still guide human treatment decisions and animal management. However, a negative finding would not be reportable, as there would not be confidence that appropriate areas of sufficient fresh brain material were examined.

Although viral antigen is widespread throughout the brain in most rabid animals, in some specimens from large animals, such as cows and horses, virus may be found only on one side of the brainstem. These observations have led to the recommendation that a complete cross section of the brainstem at the level of the medulla, pons, or midbrain must be examined for a definitive diagnosis. An ideal diagnostic procedure includes the preparation of two sets of a smear or impression of the brainstem and two from cerebellum. Each brain area should be sampled through an impression made in a 15 mm diameter well or a smear of 10 mm^2, as this then allows for examination of 40 separate microscopic fields at 200× magnification. High quality microscope slides with wells of 14 mm to 16 mm diameter are commonly used for rabies tests. In some cases, routine practice used to include marking the areas on a slide for staining after pressing tissue to be examined onto the slide or making a slip smear of the sample. This practice should be avoided because it is possible to cross-contaminate samples through the wax pencil or marking pen. The two sets of slides from each specimen are prepared for staining with two different sources of FITC-conjugated anti-rabies antibodies, as further described in the next section.

An explosion-proof −20 °C freezer is required during the acetone fixation step of the prepared sample slides. Specimens and other reagents may also be stored at this temperature, although −70 °C is necessary for specimens and some reagents for longer term storage. Freezers that automatically defrost cannot be used because the heating cycles that occur in these types of equipment is damaging to reagents and specimens.

Prepared samples (i.e., slide impressions or smears) must be allowed to completely dry at room temperature prior to fixation. This may take up to half an hour. The application of heat to expedite drying of slides is never recommended, as this may degrade antigens in the samples. Slides from an individual test animal may be placed in an individual container for fixation. Slides from multiple animals should never be placed in a container together or with control slides. Slides are fixed in acetone for a minimum of 1 hr to overnight at −20 °C.

2.4.2 *Fluorescent Conjugate Selection, Preparation, and Evaluation*

In the USA, primary rabies diagnosis is typically conducted with reagents that for regulatory purposes are considered ASRs, or "for research purposes only," or "for laboratory use only." The regulation 21 CFR 864.4020(a) defines ASRs as "antibodies, both polyclonal and monoclonal, specific receptor proteins, ligands, nucleic acid sequences, and similar reagents which, through specific binding or chemical reaction with substances in a specimen, are intended for use in a diagnostic application for identification and quantification of an individual chemical substance or ligand in biological specimens." The reagents for rabies diagnosis are classified in Class II of ASRs and thus are subject to a higher level of regulation. To confound matters, the testing is largely performed on animal samples, rather than human, and as such, the activity is considered "public health testing performed on environmental (nonhuman) samples." The Interpretive Guidelines §493.1253(b)(2) for this activity include that, prior to reporting patient test results, the laboratory is responsible for establishing the performance specifications for each test system not subject to FDA clearance or approval, as well as for each test system for which the manufacturer does not provide performance specifications. However, it is unclear whether rabies testing is actually subject to these regulations because the "patient test result" is on an animal, and yet the result will dictate appropriate human medical intervention. Nonetheless, the establishment of method performance specifications should provide evidence that the accuracy, precision, analytical sensitivity, and analytical specificity of the procedure are adequate to meet the clients' needs as determined by the laboratory director and clinical consultant. The terminology of laboratory director and clinical consultant are CLIA roles, but the vast majority of rabies diagnostic laboratories are not under CLIA jurisdiction. While important, these guidelines may not be in practice every day in rabies laboratories throughout the USA.

In Title 21 of the U.S. Code of Federal Regulations, which can be found on the FDA website in Sec. 866.3460, there is specific language about rabies virus immune-fluorescent reagents. They are listed as "devices that consist of rabies virus anti-sera conjugated with a fluorescent dye used to identify rabies virus in specimens taken from suspected rabid animals" and are classified as Class II, and thus it is up to the laboratory to establish performance characteristics. As such, one manufacturer notifies the purchaser that rabies diagnostic conjugates are ASRs and that it is up to the laboratory to establish performance characteristics. In an online catalog description of another source of rabies diagnostic products, the complicated effect of the current regulatory language is in evidence. In the descriptions of products, the intended use is classified as In Vitro Diagnostic (IVD) use (qualified for sale in various parts of the world), ASR, Research Use Only (RUO) or Laboratory Use Only. The list includes: Rabies DFA I Reagent which is available for IVD, CE (CE—can be sold to some countries in Europe); Rabies DFA II Reagent—ASR or IVD, CE, Export (outside of USA) Only; Rabies DFA

III Reagent—ASR; Rabies Mab Typing Set—ASR or IVD, Export Only; Rabies Polyclonal DFA Reagent—ASR or IVD, Export Only; Rabies Diluent—For Laboratory Use; Rabies Diluent with Evan's Blue—For Laboratory Use; Rabies Negative Control (for use with Rabies monoclonal reagents)—RUO; Rabies Negative Control (for use with Rabies polyclonal reagents)—RUO or IVD, CE, Export Only; and Mounting Fluid, Low Glycerol—For Laboratory Use. The ASR regulations under which primary rabies diagnostic reagents fall are confusing and difficult to apply. The imperfections of these rules are indicative of the compromise that was necessary to allow their implementation. There was a clear need to impose some level of regulation on assays using these reagents, but this rule settled for only regulating some ingredients. Nonetheless, the ASR rules may be the first step toward significant regulation of these assays, of which primary rabies diagnosis is only a small fraction.

Each batch of rabies reagents must be titrated to determine a working dilution (Anonymous, 2003). Although the use of a counterstain is optional, the use of Evans Blue, specifically formulated for immunofluorescent assays, provides the advantage of contrast coloring (opposite of the positive immunofluorescent result) of tissue on a slide, quenching of background fluorescence, and also visual confirmation that the test reagent was actually added to a slide. The amount of counterstain added to a test reagent is determined by titration when the working dilution of the conjugate is determined.

In general, the working dilution for each conjugate is determined through initial evaluation of serial twofold dilutions, for example, at 1:10, 1:20, and 1:40 or more. These should be prepared exactly as the reagent will be used for diagnostic samples, for example, with the same diluent, tubes, syringe filters, and counterstain concentration. Two or more diagnosticians should read and record results independently. The results should then be compared to arrive at a consensus for the optimal dilution providing crisp +4 staining, meaning a glaring, apple-green brilliance, with minimal background fluorescence. The working dilution can be more precisely determined by additional examination of the performance of limited dilutions around the end point. Because antigen presentation and antibody avidity and affinity vary with different rabies virus samples, and viral inclusions appear quite differently with different reagents, accurate reagent evaluation requires multiple observations on a diverse collection of positive control material. Control material for conjugate titrations conducted within each laboratory should include animals naturally infected with the most common variant of rabies virus occurring in the region from which samples are submitted and from animals infected with a variant that is more diverse, such as one or more from insectivorous bats. Control material should be stored in aliquots to avoid repeated freeze-thaw cycles and maintained at −40 °C or below.

Stock solutions of testing reagents may be stored as frozen aliquots at a minimum of −20 °C or preferably lower. Reagents at the working dilution must be filtered prior to use and may be stored at +4 °C but discarded if not used within 7 days. With disposable filter units that attach directly to a syringe, the working dilution of the conjugate can be filtered as it is added to test slides. The diluted test reagents ready for dispensing onto test slides can be stored at +4 °C in the syringe with the attached filter unit, as long as reagent is prevented from drying on the filter or tip of the dispensing syringe through sealing with a syringe tip or plastic wrap.

The two sets of slides prepared for staining from each specimen should be evaluated with FITC-conjugated anti-rabies virus antibodies, consisting of two different monoclonal antibody pools or one monoclonal antibody conjugate and one hyperimmune serum conjugate. While hyperimmune serum conjugates consist of the greatest diversity of anti-rabies virus antibodies and hence the broadest potential for reaction with diverse rabies virus and related viruses, these preparations have a higher innate risk of nonspecific reactivity. This risk can be managed through the use of a specificity control reagent. The specificity control for a hyperimmune rabies reagent is an FITC-labeled serum reagent produced in the same animal host as the rabies reagent but directed to an agent other than rabies virus. In contrast, monoclonal antibody reagents rarely manifest nonspecific reactivity. However, they present the risk of not reacting with variants that lack the specific epitope(s) to which the monoclonal antibodies are directed. Thus, it is recommended to use reagents prepared from two different pools of monoclonal antibodies to minimize the risk of non-recognition of any one variant. Specificity controls for monoclonal reagents consist of FITC-labeled mouse monoclonal antibodies that are of the same isotype and protein concentration as the rabies reagents but directed to an agent other than rabies virus.

2.4.3 *Immunofluorescence Test Protocol*

After acetone-fixation, the test slides and control slides are air dried at room temperature. As recommended, two different anti-rabies virus conjugates may be added, one to each set of slides, by dispensing through a syringe fitted with a 0.45 μm low protein binding filter. The slides are incubated for 30 minutes at 37 °C in a high humidity chamber. After staining, the slides are briefly rinsed under a stream of PBS, and then soaked in PBS for 3 to 5 minutes with attention that control slides and slides from each test animal are in separate rinse containers to avoid cross-contamination. The slides are then soaked again in new PBS for a second 3 to 5 minute interval. Slides are carefully blotted to remove excess liquid (no rinsing is necessary), then briefly air dried. A small amount of 20% glycerol-Tris buffered saline pH 9.0 is placed onto coverslips arranged on absorbent paper. The air-dried slides are inverted and placed on the coverslips with mounting media, thus allowing excess mounting media to be wicked into the absorbent paper when light pressure is applied to the back of the slides. Slides should be read by fluorescence microscopy within 2 hours of cover-slipping.

The minimum rabies diagnostic procedure recommends the examination of 40 fields at a magnification of 200× or more for each conjugate and brain area tested by two laboratory diagnosticians. Thus, for basic diagnosis, 40 fields × 2 conjugates × 2 brain areas sampled × 2 observers equals 320 observations for definitive diagnosis. If fluorescence is observed, fluorescing inclusions should be examined at 400× magnification for resolution of very fine dust-like inclusions and recognition of some types of nonspecific staining.

An annual rabies diagnostic proficiency program has been available for voluntary participation in the USA since 1994. Among the diagnostic laboratories enrolled in the recent programs, performance has been acceptable on strongly positive and negative slides; discrepancies occur with very weakly positive specimens (Powell, 1997). There is no mandatory remedial training or evaluation of laboratories that may perform poorly on these challenging samples. As described in previous sections, if the procedure is "modified by the laboratory," meaning any change to the assay that could affect its performance specifications for sensitivity, specificity, accuracy, or precision, and so on, these may be done but at present are not regulated nor subject to oversight. Laboratory modification that could affect performance specifications include but are not limited to: 1) changes in specimen handling instructions; 2) incubation times or temperatures; 3) changes in specimen or reagent dilution (each laboratory determines this internally); 4) using different positive and negative control material (each laboratory determines this internally); 5) using a different antibody reagent (i.e. monoclonal versus polyclonal) (each laboratory determines this internally); 6) changes in or elimination of a procedural step (each laboratory determines this internally).

2.4.4 Laboratory Reporting Practices and Emergency Examinations

At present, the majority of positive rabies results are reported to a state or local health department and the submitter (which is often a veterinarian, humane association, animal control, or related entity) by fax or telephone. State health departments often provide access to monthly summaries of positive cases of rabies, typically to the county level. Although the denominator number of each species tested is critically important (there cannot be any cases if no animals have been tested), these numbers are often more difficult to find and may not be readily available to the public. Rabies prevention efforts would be much enhanced by the compilation of testing data, both negative and positive cases, preferably by a case occurrence location as precise as longitude and latitude. With this information, one would be able to assess the intensity and appropriateness of surveillance among the local rabies reservoir species, and, in comparison, the spillover into non-reservoir species. At the 2012 annual meeting of the Association of Public Health Laboratories, there were a number of commercial entities offering Laboratory Information Management Systems that included rabies data transmission. Electronic automated data management and reporting of results would expedite access to case information and reduce human error.

3 ANTEMORTEM DIAGNOSIS OF RABIES

Antemortem diagnosis of rabies in humans and animals is possible during the clinical illness. The major constraint is the brevity of the clinical period. Typically, each day results in a significant decline in clinical stability, ending invariably with death. In the USA, humans with clinical signs and symptoms suggestive of rabies may be evaluated through a battery of tests, in most cases, conducted at the Centers for Disease Control and Prevention, Atlanta, Georgia. In Canada, such testing is performed by the CFIA Centre of Expertise for Rabies in Ottawa, Ontario, with the exception of the serological analysis, which is done by the Central Laboratory of Public Health Ontario, located in Toronto, Ontario, or the Public Health Agency of Canada National Microbiology Laboratory, located in Winnipeg, Manitoba.

For human antemortem diagnosis of rabies, the best sample appropriate for dFA testing is skin biopsy material from the nape of the neck that includes several innervated hair follicles (Crepin et al., 1998). Initial evaluation consists of cryostat sectioning of the frozen skin biopsy, fixation in acetone, staining with rabies virus specific fluorescent conjugate, and examination under fluorescence microscopy. Positive observations typically occur where the skin follicles are innervated. This evaluation is complete within several hours of receipt of the sample. It is a highly specific test, but a negative finding on this sample with this method cannot definitively rule out rabies; subsequent samples from the patient may ultimately be positive.

Other useful samples are fresh saliva, serum, and cerebrospinal fluid (CSF). Serum and CSF are evaluated for antibodies by two methods. One is the indirect fluorescent antibody test, which can be completed quickly (i.e., within a few hours), but it is not a functional test for antibody; it is a binding test. False positive binding may occur, especially in patients with encephalitis due to unrelated etiologies. Secondly, rabies virus neutralizing antibody reactivity can be assessed in the rapid fluorescent focus inhibition test (RFFIT). This test is a functional assay in that the result reflects activity in the samples, which prevents rabies virus infection of cells (see Chapter 12). The RFFIT takes one full day (20+ hours) to complete. In most patients, antibodies develop rather late in the clinical course, so the patient may remain seronegative for the majority of the investigation. As these initial evaluations are underway, the subcutaneous tissues from the skin biopsy and the saliva are subjected to RT-PCR assays using a number of different primers to maximize sensitivity (see next section). Serial sampling of suspect patients is optimal for definitive diagnosis. If the patient remains alive but ill for more than 2–3 weeks or recovers quickly, the differential diagnosis is less likely to include rabies.

Antemortem diagnosis of rabies in animals is technically feasible but is not practical under the majority of circumstances at this time. Without advanced supportive medical care, a rabid animal will succumb to the disease quickly. By the time diagnostic results would be available on a battery of samples similar to those described for humans, the animal would most likely be moribund or dead. Moreover, negative findings early in the clinical course of a rabid animal may mislead public health management of potentially exposed persons.

4 USE OF MOLECULAR METHODS TO DETECT VIRAL RNA

4.1 Advantages and Disadvantages of Molecular Methods

Molecular methods detect viral RNA using either nucleic acid probes, an amplification strategy, or a combination of both approaches in contrast to the traditional approach of detecting viral protein in the dFA. Due to the difference in the nature of the target molecule in the two types of assay, differences in their performance need to be considered.

In the case of lyssaviruses, including rabies virus, variation at the level of the nucleic acid genome is significantly greater than at the protein level due to genetic code redundancy (Bourhy, Kissi & Tordo, 1993; Kuzmin et al., 2005). Consequently, failure to detect a virus present in a sample (false negative result) is potentially a larger problem using molecular methods, which usually rely on the hybridization of relatively short segments of nucleic acid (oligonucleotide) to the RNA target, compared to antibody-antigen binding strategies that form the basis of the dFA detection method. Since the epitope(s) detected by an antibody are more likely to be conserved than a particular nucleotide sequence, serological methods have an advantage when one requires a broadly reactive test capable of detecting a wide range of lyssaviruses. Moreover, molecular based methods are generally more time consuming and costly than the dFA. In their favor, molecular-based methods are reported to be highly sensitive, especially when an amplification process is included. However, this can exacerbate the potential for false positive results, due to sample contamination that can easily occur when amplification methods such as PCR are applied without rigorous attention to the operational requirements for such assays (Kwok & Higuchi, 1989). Consequently, routine rabies diagnosis on fresh brain tissue, usually applied to animals collected in the terminal stages of disease when levels of viral antigen in the brain are high, still continues to be performed by the dFA, which remains the recommended gold standard postmortem diagnostic test for rabies (OIE, 2011; WHO, 2005). However, there are situations where dFA performance is less than optimal and where molecular methods can, if applied carefully and correctly, provide either a confirmatory or alternate diagnostic capability. The utility and application of various nucleic acid amplification tests were recently reviewed (Wacharapluesadee & Hemachudha, 2010).

The dFA procedure rapidly loses sensitivity when applied to brain tissue that is substantially decomposed. Controlled observations have shown that in such a situation a molecular method of detection can be greatly more sensitive (David et al., 2002; Heaton et al., 1997). Genetic methods have also been found to be superior to the mouse inoculation test for detection of rabies virus in samples maintained under various conditions of storage for long periods (Lopes, Venditti & Queiroz, 2010). Moreover, in some jurisdictions, when an animal has had human contact but is scored as rabies dFA negative, this result must be confirmed with an alternate test; since molecular methods can be completed more quickly than either virus culture or the MIT, and with superior sensitivity (Picard-Meyer et al., 2004) they are potentially useful for the routine confirmation of dFA results. Genetic tests can also be used as an adjunct to the dFA when unexpected or unusual fluorescent staining patterns are observed and confirmation of virus presence is required; sometimes a combination of tests is required to reach a consensus on the disposition of a particular case (McColl et al., 1993). The issue of discordant results between many of these tests can be problematic, however, when RT-PCR is the only method that suggests presence of rabies virus. Distinguishing between a false positive RT-PCR result and a false negative result for the other assays in such a situation may be a highly complex process with no clear-cut resolution. The best recourse is to avoid such situations wherever possible by adherence to the guidelines recommended for the performance of PCRs (Cooper & Poinar, 2000). These include careful processing of tissues using clean, sterile instruments and supplies in each case; use of physically separate areas for performing tissue extraction, PCR and post-PCR analysis; and use of dedicated pipettes in each of these areas to avoid sample cross-contamination.

4.2 Detection of Lyssavirus RNA by Reverse Transcription–Polymerase Chain Reaction

Early molecular methods detected rabies virus RNA present in brain extracts that had been applied to a membrane by hybridization to a probe (see Ermine, Tordo & Tsiang, 1988); however, such methods lacked the required sensitivity for use as diagnostic tools. The emergence of PCR technology (Saiki, 1989) enabled development of highly sensitive diagnostic methods for rabies virus and other lyssaviruses. The principle of the PCR depends on the use of two synthetic oligonucleotides that can hybridize to opposite strands of a dsDNA target and are oriented in such a way that when they prime new DNA synthesis, the newly created DNA strands overlap in sequence. The reaction, catalyzed by a thermostable DNA polymerase, requires repeated thermocycling thus: first, high heat (95 °C) to denature the DNA template into single strands, then a lower temperature (usually 45–60 °C) to allow annealing of the oligonucleotides to their target sequences, and finally an incubation (usually 72 °C) to allow the annealed oligonucleotides to prime DNA synthesis using dNTP substrates. This cycling is repeated for 25–40 cycles and a successful PCR produces a double-stranded DNA product (amplicon) of specific length, defined at its two ends by the primers used in the reaction. The amplification of this product, usually by more than 100,000 fold, is the basis for the assay's exquisite sensitivity. Since its original description, refinement of the technique now allows PCR products of up to several kb to be generated from good quality DNA template. However, to apply this technique to lyssaviruses, the viral RNA must first be converted to a complementary DNA (cDNA) strand using the enzyme reverse transcriptase. This reverse transcription (RT) is often the limiting step when amplifying RNA virus sequences, and it usually precludes amplifying lengths greater than 5 kb in a single reaction.

Once nucleotide sequence information is available for a particular nucleic acid, primers of appropriate sequence can be designed so as to amplify virtually any segment of that nucleic acid. Despite the tremendous versatility inherent to PCR assays, designing broadly reactive PCRs for detection of all members of the *Lyssavirus* genus becomes complicated due to genetic variation between and within species. To develop a robust, broadly cross-reactive assay, very careful primer design is needed, so as to ensure that the primers bind to sequences well conserved across the viral group targeted.

RT-PCR methods for lyssavirus amplification, including detailed protocols, have been described previously (see Nadin-Davis, 1998; Sacramento, Bourhy & Tordo, 1991; Tordo, Bourhy & Sacramento, 1992, 1995). These use a combination of commercially available reagents suitable for general RT-PCR applications and custom-synthesized oligonucleotides of defined sequence. The following provides some general guidelines to consider when developing or applying RT-PCR methods to lyssavirus detection.

4.2.1 Method for RNA Extraction

Before applying RT-PCR to a sample, total RNA or total nucleic acids must first be recovered from the tissue to be tested. One commonly used method utilizes a commercial reagent known as TRIzol® or the equivalent, TriPure®, a phenol/guanidine isothiocyanate solution based on earlier acidic phenol methods of RNA extraction. This reagent rapidly inactivates any nuclease present and quickly dissolves soft tissues such as brain, making it especially suitable for application in rabies diagnosis. After addition of chloroform to the mixture to facilitate a liquid phase separation, RNA recovered in the aqueous phase is readily precipitated by addition of isopropanol. This method is reasonably simple and amenable to moderate throughput in terms of sample numbers.

Other methods that are sometimes used rely on commercially available kits that avoid the use of noxious chemicals and the requirement for RNA precipitation; for example, kits that provide a silica-membrane spin column approach for recovery of total RNA from a wide variety of tissue types.

More recently, the application of magnetic bead technology to nucleic acid purification has resulted in the development of platforms using a 96 well plate format to enable high throughput sample processing. Several companies market instruments employing this technology. It relies on the binding of nucleic acids to magnetic bead particles followed by multiple washing steps to remove other contaminants before elution of highly purified nucleic acid from the particles.

4.2.2 Selection of Target Strand

Since the rabies virus life cycle includes production of full-length negative and positive sense copies of its genome, as well as significant amounts of mRNA, either positive (messenger) or negative (genomic) sense sequences can be targeted for cDNA generation by use of either negative or positive sense primers respectively. This step can be initiated using one or both of the PCR primers. When using a single sequence-specific primer, the sensitivity of the assay may be affected by the sense of the sequence initially targeted in the RT step. Many protocols target the negative sense genomic sequence in consideration of the often less-than-ideal state of diagnostic samples submitted to the laboratory; it is presumed that the encapsulated genomic RNA may be better protected than its corresponding mRNA from degradation occurring due to tissue autolysis, although this issue has never been carefully evaluated. In some situations, where the amount of sample is limited or it is of poor quality, priming of cDNA synthesis may be undertaken using random hexamer primers in place of the sequence specific PCR primer; this may initiate cDNA synthesis at several positions within the RNA target and from both RNA strands. This strategy also facilitates subsequent performance of multiple PCRs targeting distinct viral sequences.

4.2.3 *Primer Design*

A critical factor in determining the specificity of a PCR is the nucleotide sequence of the primer pair used to drive the reaction. For diagnostic purposes, the PCR should ideally be sufficiently broadly reactive to successfully amplify all members of the *Lyssavirus* species likely to be encountered within the region of study. Highly conserved sequences are normally targeted. Since the N gene is one of the more conserved regions of the lyssavirus genome (Le Mercier, Jacob & Tordo,1997), this gene has been the target of virtually all efforts to develop a broadly cross-reactive diagnostic test using PCR methodology. Moreover, the N gene was historically favored for genetic characterization so as to allow direct comparison between the genetic characteristics of these viruses and their antigenic properties, as studied using panels of monoclonal antibodies directed primarily towards the nucleoprotein product of the N gene. Fooks et al. (2009) provide a comprehensive listing of a wide variety of primers used for RT-PCR detection of lyssaviruses and several of these are described below.

Table 11.1 describes some primers that are commonly used for rabies virus detection. In the supplementary material, Tables 11.S1–11.S7 illustrate alignments of these primers with sequences of several lyssaviruses representative of the known diversity of the genus (12 recognized species—see Chapter 4), information that helps to explain why certain primers are more useful than others for amplification of particular species and rabies virus strains. It should be borne in mind that a small number of base mismatches between the primer and its target sequence will not necessarily prevent their annealing. The stringency of primer annealing is dependent on the annealing temperature; the lower the annealing temperature, the less stringent is the annealing process, and the more mismatches can be accommodated. The caveat to this, however, is that as the annealing temperature is lowered, the opportunity for annealing to poorly related sequences rises and substantial nonspecific primer binding occurs. Ultimately, this leads to a highly nonspecific reaction of little value as a diagnostic test. The position of mismatches also impacts on the extent to which they may hinder proper primer annealing. In particular, mismatches at the 3' terminus of the primer or within the three most 3' bases of the primer are often highly detrimental to the PCR since the 3' end of the primer must anneal to its template faithfully in order to prime new DNA synthesis. One strategy often employed to

TABLE 11.1   Listing of Selected Primers used for RT-PCR of Lyssaviruses

Primer Name	Sequence 5′–3′	Location in PV Genome[1]/ Orientation	Reference
N7	ATGTAACACCTCTACAATG	55–73/+	Bourhy, Kissi & Tordo, 1993
JW12	ATGTAACACCYCTACAATG	55–73/+	Heaton, McElhinney & Lowings, 1999
RabN1	AACACCTCTACAATGGATGCCGACAA	59–84/ +	Nadin-Davis, 1998
Nseq0	AACACCTCTACAATGGATGCCGAC	59–82/ +	Nadin-Davis, Casey & Wandeler, 1993
10g	CTACAATGGATGCCGAC	66–82/ +	Smith et al., 1992
RabNfor	TTGTRGAYCAATATGAGTACAA	135–156/+	Nadin-Davis, 1998
BB6	GATCARTATGAGTAYAAATATCC	140–162/+	Black et al., 2002
N165-146	GCAGGGTAYTTRTACTCATA	146–165/−	Wakeley et al., 2005
D017	AGATCAATATGAGTAYAARTAYCC	139–162/+	Foord et al., 2006
JW10 (P)	GTCATTAGAGTATGGTGTTC	617–636/−	Heaton, McElhinney & Lowings, 1999
JW10 (ME1)	GTCATCAATGTGTGRTGTTC	617–636/−	Heaton, McElhinney & Lowings, 1999
JW10 (DLE2)	GTCATCAAAGTGTGRTGCTC	617–636/−	Heaton, McElhinney & Lowings, 1999
JW6 (DPL)	CAATTCGCACACATTTTGTG	641–660/−	Heaton, McElhinney & Lowings, 1999
JW6 (E)	CAGTTGGCACACATCTTGTG	641–660/−	Heaton, McElhinney & Lowings,1999
JW6 (M)	CAGTTAGCGCACATCTTATG	641–660/−	Heaton, McElhinney & Lowings,1999
RabNrev	CCGGCTCAAACATTCTTCTTA	876–896/−	Nadin-Davis, 1998
1087NFdeg	GAGAARGAACTTCARGAATA	1157–1176/+	McQuiston, Yager, Smith & Rupprecht, 2001
1312NBdeg	TTGTTTARAAAYTCAGCRAA	1382–1401/−	McQuiston, Yager, Smith & Rupprecht, 2001
RabN2	ggggagcTCTTATGAGTCACTCGAATATG TCTTG	1399–1425/−	Nadin-Davis, Casey & Wandeler, 1993
304	TTGACAAAGATCTTGCTCAT	1514–1533/−	Smith, 1995
RabN5	GGATTGACRAAGATCTTGCTCAT	1514–1536/−	Nadin-Davis, 1998
N8	AGTCTCTTCAGCCATCTC	1568–1585/−	Bourhy, Kissi & Tordo, 1993

[1]*Bases are numbered according to the PV reference strain, GenBank Accession number NC_001542. Lower case bases (RabN2) do not correspond to viral sequence and were added to increase the oligonucleotide's melting temperature.*

overcome variability within the targeted sequence is to use a combination of primers of different sequences that can anneal to the same position within different target sequences (see the JW6/JW10 primer set in Tables 11.1 and 11.S3). In some cases, a degenerate primer in which two or more bases are inserted into a certain position during synthesis may be employed; alternatively, the base inosine, which has increased flexibility to anneal to all bases, can be inserted into a position in place of any of the four usual bases. The primer 1312NBdeg (Tables 11.1 and 11.S6) is an example of a degenerate primer having three degenerate positions.

Initially, PCRs that amplified the complete coding region of the lyssavirus N gene were developed based on the belief that the 5′ and 3′ coding termini were well conserved and to allow determination of the complete coding sequence of the gene. The N7 primer, which has a 3′ terminus corresponding to the N gene start codon, was successfully used by Bourhy, Kissi, and Tordo (1993) and subsequently by Kissi, Tordo, and Bourhy (1995) to amplify a wide range of lyssaviruses. The only known mismatch within this primer is at base 65 (PV reference strain), at which some sequences contain a C in place of the T. The JW12 primer (Heaton et al., 1997) is identical to N7 except that it contains a C/T degenerate base at this position in order to address this difference. Other primers, RabN1 and Nseq0 (Nadin-Davis, 1998; Nadin-Davis, Casey & Wandeler, 1993) and 10 g (Smith et al., 1992), which have been used extensively for priming at this target sequence, straddle the initiating codon. Although all of these primers successfully amplify most rabies viruses, only the JW12 primer performs well with other lyssaviruses (Foord et al., 2006). In particular, the mismatch of the 3′ A base of primer RabN1 with certain specimens that have a G at this position is problematic, and this primer has been replaced by Nseq0 for routine application (Nadin-Davis, unpublished data). The downstream primers that are often paired with these primers to amplify the complete N gene target either the sequence around the N gene stop codon (RabN2, 1312NBdeg) or the start codon of the P gene (RabN5, 304). The degenerate 1312NBdeg primer is more useful than RabN2, which is no longer in routine use (Nadin-Davis, unpublished data), since it can better accommodate the genetic variation seen at this location. Both 304 (Smith, 1995) and its closely related variant RabN5 (Nadin-Davis, 1998) are widely used due to excellent concordance with their target sequences for almost all lyssaviruses except for the most divergent (Lagos bat virus, Mokola virus and West Caucasian bat virus). Primer N8, which targets downstream sequence at bases 1568–1585, would appear to be better matched to these rabies-related viruses (Bourhy, Kissi & Tordo, 1993).

Several primers that target internal sequence have also been used extensively. The primer pair RabNfor and RabNrev, that amplifies a fragment of 762 bp, was developed to amplify all known rabies virus strains circulating in Canada, but it has since proven useful for amplification of a wide range of rabies viruses (Nadin-Davis, 1998). Tables 11.S2 and 11.S4 illustrate the relatively good match of these primers with most rabies viruses but increased numbers of mismatches are evident when these primers are compared with sequences from other lyssaviruses. Slightly modified versions of these two primers (SuEli+/−) successfully amplified a range of Mexican rabies virus variants (Loza-Rubio et al., 2005). The overlapping primer BB6, described by Black et al. (2002), extends into a less conserved sequence and has not found extensive application but a modified version of this primer, D017, was used to detect a wide range of lyssaviruses (Foord et al., 2006). The genomic sense primer N165–146 targeting this region also appears suitable for a wide variety of lyssaviruses (Wakeley et al., 2005). Two sets of primers (JW6 and JW10) used for hemi-nested PCR (hn PCR) in combination with JW12 are illustrated (Tables 11.1 and 11.S3). In each case, the viral sequence is targeted by three primers, each of slightly different sequence, which when combined attempt to cover the sequence variation found between all lyssavirus species (Heaton et al., 1997). However, the data in Table 11.S3 suggest that mismatches with some rabies virus strains (e.g., Africa 2 dog strain, represented by V461NIG, with the 3′-end of primer JW6 and certain American insectivorous bat strains, represented by V588IBMX with the 3′ proximal base of primer JW10) may result in poor amplification by these primers. While the panel used to evaluate this primer set included 23 rabies viruses (Heaton et al., 1997), more extensive testing would be needed to establish the universal suitability of these primers for all lyssaviruses. Another internal primer, 1087NFdeg (Tables 11.1 and 11.S5), targets bases 1157–1176 near the 3′ terminus of the N gene coding region; use of degenerate positions allows this primer to anneal broadly to rabies viruses. But again, mismatches at the 3′-end may preclude its usefulness for members of the other species.

A few other useful primer sets are illustrated in Table 11.2. The N1/N2 primer pair, originally described by Tordo, Bourhy, and Sacramento (1992), has been used on several occasions to amplify Mokola viruses recovered in South Africa (Nel et al., 2000; Sabeta et al., 2010). The 5′ and 3′ terminal sequences of the lyssavirus genome are highly conserved across the genus and have been used as targets, together with internal primers, for RT-PCR (Table 11.2; Szanto, Nadin-Davis & White, 2008). Other N gene primers that have been used to detect Lagos bat viruses have been described elsewhere (see Kuzmin et al., 2008; Markotter et al., 2006).

It is evident from the preceding that the development of truly universal primers capable of pan-lyssavirus amplification is challenging and that, in practice, combinations of primer pairs are usually used to detect all members of the genus. Consideration must always be given to the nature of the viruses that are likely to be encountered in a certain geographical area. For instance, in the Americas, primers that success-fully amplify all known rabies viruses may be sufficiently broad in scope for most situations because any indigenously acquired lyssavirus infection will be due to classical rabies virus. In other areas, particularly Africa and Europe, the presence of rabies-related viruses dictates the need for assays that will detect all appropriate lyssaviruses. Assays that detect specific sub-sets of the lyssavirus genus have exhibited significant utility. For example, in the UK an assay that employed the JW12 primer for cDNA synthesis followed by PCR with the more selective BB2 primer (genome position 154–174) and primer JW6(E) specifically detected all rabies viruses and both European bat lyssavirus types, EBLV-1 and EBLV-2 (Black et al., 2000). Indeed, the use of primers that match the sequence of only one lyssavirus species or particular rabies virus vari-ants can enable both virus detection and its identification to type in a single assay without the need for downstream amplicon characterization. Using such strategies, multiplex PCRs have been developed to identify multiple rabies virus variants based upon the size of the amplicons pro-duced (Nadin-Davis, Huang & Wandeler, 1996; Nel, Bingham, Jacobs & Jaftha, 1998; Rohde, Neill, Clark & Smith, 1997).

Most laboratories performing PCR on a routine basis will maintain several primer combinations and may apply two or more primer pairs to evaluate a particular specimen, giving due consideration to the source of the sample and the viruses to which the specimen could have been exposed. As additional nucleotide sequence information becomes avail-able for other portions of the lyssavirus genome, other targets for a broadly cross-reactive PCR assay may be identified. Bourhy et al. (2005) designed primers that successfully amplified a section of the L poly-merase gene for a number of Rhabdoviruses representing several genera. These primers targeted sequences encoding highly conserved amino acid motifs required for polymerase function; a PCR targeting this same con-served L gene region has been used for human rabies diagnosis in Africa and Asia (Dacheux et al., 2008). Further studies may yield a pan-lyssavi-rus RT-PCR targeting L gene sequences.

4.2.4 Assay Sensitivity

Success of an RT-PCR is dependent on the integrity of the RNA tem-plate, and while PCR is less sensitive than the dFA to the effects of tissue autolysis (David et al., 2002; Heaton et al., 1997; Rojas Anaya, Loza-Rubio, Banda Ruiz & Hernandez Baumgarten, 2006; Whitby, Johnstone, & Sillero-Zubiri, 1997b), extensive degradation of the sample will ulti-mately lead to false negative results. To maximize the sensitivity of RT-PCR and hence the chance of detecting any lyssavirus sequence pres-ent in a sample, the following strategies will be helpful:

TABLE 11.2   Additional Primers used for Lyssavirus Amplification

Primer Name	Sequence 5'–3'	Position in Genome	Lyssavirus Species Target	References
N1	TTTGAGACTGCTCCTTTTG (+)	587–605	Mokola virus	Tordo, Bourhy & Sacramento, 1992; Nel et al., 2000; Sabeta et al., 2010
N2	GGGTATATCGTAGGATG (−)	1029–1013		
LYS001F[1]	ACGCTTAACGAMAAA (+)	1–15	all	Kuzmin et al., 2008
LYSEND[1]	ACGCTTAACAAAWAAA (−)	5' TERMINAL COMPLEMENT		
RVFor[1,2]	AACTGCAGACTAGTACGCT TAACAACCAGATCAAAG	1–22	Rabies virus	Szanto, Nadin-Davis & White, 2008
RVRev[1,2]	GGATGCATGCGGCCGCACGCT TAACAAATAAAC	5' TERMINAL COMPLEMENT		

[1]*Used in combination with internal primers to amplify the termini of the genome.*
[2]*Underlined bases represent restriction sites added to amplicon termini for cloning and to increase annealing temperature of primers.*

4.2.4.1 TARGETING OF A RELATIVELY SMALL SEQUENCE WINDOW

Amplification of short DNA strands is more efficient than amplification of complete genes, and so the smaller the fragment amplified by PCR, the more sensitive the assay. Indeed, RNA that is significantly degraded may be far more effectively amplified if only short stretches of intact sequence need be present in the sample, and assays that produce products of 200–300 bp (see McQuiston, Yager, Smith & Rupprecht, 2001) may be ideal for this purpose.

4.2.4.2 INCREASING THE NUMBER OF AMPLIFICATION CYCLES

In an ideal reaction, the amount of PCR product doubles after each cycle, so increasing the number of cycles would be expected to significantly increase the product yield. Generally, 25–35 thermocycles of PCR are employed, although this can be increased to 40 or 45. Use of cycle numbers higher than this generally becomes unproductive due to the gradual loss of enzymatic activity of the DNA polymerase employed in the reaction, despite use of thermostable enzymes relatively resistant to high temperature.

4.2.4.3 USE OF NESTED PCR

A nested PCR is one in which the product of a PCR is subjected to a second round of amplification using primers internal to those employed for the first round (Kamolvarin et al., 1993). A hemi-nested PCR (hn PCR) (Heaton et al., 1997; Picard-Meyer et al., 2004) employs one of the first round primers in combination with an internal primer in the second PCR. Nested strategies increase the sensitivity of the assay enormously but at the cost of greatly increasing the chance of a false positive result, unless stringent precautions are taken to prevent carryover contamination of the sample. It has been proposed that the main reason why nested PCRs are sometimes necessary is to compensate for inefficient first-round PCR due to primer mismatches and that the use of well-matched primers for first round PCR should preclude the need for a nested approach in most circumstances (Trimarchi & Smith, 2002).

To demonstrate the utility of nested PCR, the results of an evaluation of several samples from a human case of rabies (Elmgren, Nadin-Davis, Muldoon, & Wandeler, 2002) by first and second round PCR are shown in Figure 11.2. It is apparent that for the first round of PCR (Figure 11.2A), apart from the positive control, the only sample to generate a specific band of the correct size is the saliva sample. However, after the second round (nested) PCR (Figure 11.2B) the eye secretion, saliva, and skin biopsy samples all generated a specific product of size identical to that of the positive control, while all blank samples, the negative control, and the CSF remained negative.

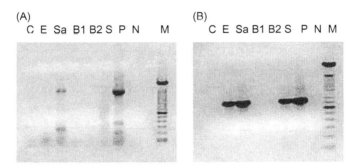

(A)
C E Sa B1 B2 S P N   M

(B)
C E Sa B1 B2 S P N M

FIGURE 11.2 **Diagnosis of human samples for rabies by RT-PCR.** Analysis by gel electrophoresis of first (panel A) and second (panel B) round PCRs of several samples from a human rabies case. First round RT-PCR was performed using primer Nseq0 for RT and primers Nseq0/RabN5 for PCR; the expected product has a size of 1478 bp. Re-amplification of an aliquot of each first round PCR was performed using primers RabNfor/RabNrev that produce an amplicon of 762 bp. The samples tested are as follows: C, CSF; E, eye secretion; Sa, saliva; B1 and B2, water samples extracted and processed in parallel with the tissues; S, skin biopsy. RT-PCR controls included a positive control (P), from a rabies positive skunk, and a water blank as a negative control (N). The marker (M), electrophoresed in parallel with the samples, was a 100 bp DNA ladder (Invitrogen). DNA was detected under UV light after ethidium bromide staining of the agarose gel; an inverted image is presented.

A hn PCR that used JW12 in combination with JW6 (1st round) and JW10 (2nd round) primer cocktails was reported as useful for detection of all major lyssavirus species (Heaton et al., 1997). However, since this study was undertaken, our knowledge of the diversity of the *Lyssavirus* genus has expanded dramatically. Using a panel of viruses representing the current known genetic diversity of the African lyssaviruses, these hnRT-PCR assays were re-evaluated and failed to detect some LBV and MOKV isolates; accordingly, an alternative assay that employed the positive sense primer LYS001F (Table 11.2) in combination with two other novel primers was developed and shown to be more broadly cross-reactive (Coertse, Weyer, Nel & Markotter, 2010). Another set of nested degenerate primers targeting the central region of the N gene sequence have been reported to be suitable for amplification of all lyssaviruses (Vázquez-Morón, Avellón & Echevarría, 2006) but further evaluation of these primers is warranted. Nested PCR can also be employed for selective detection of certain lyssavirus species. For example, an assay that specifically detected EBLV-1 in European bats used a hemi-nested approach in which the first round PCR was performed using primer JW12 and a degenerate version of JW6, whereas the second round of PCR used the EBLV-1 specific reverse primer Jebl1 in combination with JW12 (Picard-Meyer et al., 2004). Distinct primer sets targeting the central region of the N gene were developed for the experimental detection of the Eurasian bat lyssaviruses Aravan, Khujand and Irkut viruses by standard and nested PCRs (Hughes et al., 2006) but use of these tests for routine detection of these viruses remains to be established with further isolates of these species.

4.2.4.4 METHODS OF PCR PRODUCT DETECTION

PCR products are most commonly detected after electrophoresis through agarose by staining with ethidium bromide, a dye that intercalates between DNA bases and that is readily visualized under UV light. This allows the approximate size of the amplicon to be determined with reference to standard DNA markers. Alternative, less toxic dyes are now commercially available but may lack the sensitivity of ethidium bromide. More sensitive methods of detection employ hybridization of the amplicon to rabies-specific oligonucleotide/DNA probes either after Southern transfer from the gel to a membrane (Heaton et al., 1997; Heaton, McElhinney & Lowings, 1999) or by an ELISA-based method (Black et al., 2000; Whitby et al., 1997a). Traditionally, such probes were labeled with radioisotopes, but alternative ligands such as digoxigenin (DIG) or biotin are now more commonly used. Probes increase PCR sensitivity by 10–100 fold over direct ethidium bromide staining. Moreover, they confirm the specific nature of the amplicon (Heaton, McElhinney & Lowings, 1999). In rare instances, false positives have resulted due to the production of nonspecific bands of a size similar to that of the expected product (Trimarchi & Smith, 2002); in such cases only subsequent characterization of the amplicon, by hybridization to a rabies-specific probe or by nucleotide sequencing, can discriminate between specific and nonspecific product. Real-time PCR methods, such as the 5' nuclease assay that incorporates dye-labeled probes as an integral part of the test (see below), provide exquisite sensitivity and specificity.

4.2.5 *Use of Controls and Proficiency Testing*

As in any diagnostic assay use of appropriate controls is essential to proper interpretation of the results. The following controls are strongly recommended:

4.2.5.1 MOCK EXTRACTION CONTROL

A known rabies-negative sample should be processed for RNA extraction in parallel with the sample under investigation so as to control for any inadvertent contamination of the specimens through aerosol generation or reagent contamination. Ideally the negative sample should comprise rabies-negative brain tissue although a water sample can also be used if such tissue is not readily available.

4.2.5.2 POSITIVE AND NEGATIVE SAMPLES AS PCR CONTROLS

Upon PCR set-up, each assay run should include at least one water sample as a negative control and RNA from one or more rabies-positive samples for use as positive controls. Any reaction run in which either the negative or positive controls fail must be considered invalid and the PCR repeated. When a large number of samples are being examined use of a number of negative controls interspersed with the samples is recommended.

4.2.5.3 CONTROL FOR TEMPLATE INTEGRITY

Evaluation of a sample for its suitability for PCR can be a useful control if sample integrity is in question. Smith et al. (2000) described a ribosomal RNA (rRNA) internal control that is suitable for 14 different mammalian species and which can be incorporated into a lyssavirus RT-PCR to assess template quality. A slightly modified version of this method, performed separately from the lyssavirus RT-PCR, has proven useful for evaluation of samples processed at the Rabies Centre of Expertise in Ottawa, Canada (Nadin-Davis, Sheen & Wandeler, 2009, 2012). Use of such controls reduces the chance of obtaining false negative RT-PCR results from lyssavirus infected tissues due to poor sample integrity since any sample which fails to amplify such internal controls should be declared unfit for testing. Such controls could also be invaluable to applied research studies; for example, to investigate conflicting results of studies that explored the utility of RT-PCR for the detection of rabies RNA in brain material stored in 50% glycerol saline (Biswal, Ratho, & Mishra, 2007; Muleya et al., 2012). Indeed, using a rRNA RT-PCR control, poor sample integrity was demonstrated to be a factor in the evaluation of a collection of Indian samples (Nadin-Davis, Sheen & Wandeler, 2012).

Since 2009 the European Union Reference Laboratory has organized an international inter-laboratory trial program for laboratories wishing to evaluate their proficiency in rabies testing (Robardet et al., 2011). These trials occur on an annual basis and consist of panels that evaluate a laboratory's proficiency in dFA, RTCIT, and RT-PCR. It should be noted however that in addition to classical rabies virus the panels include specimens of rabies-related viruses such as the Australian bat lyssavirus (ABLV) and EBLV types not normally encountered by North American laboratories and accordingly these panels are not optimized for a wide clientele.

4.3 Real-Time RT-PCR for Lyssavirus Detection

The development of real-time PCR platforms (see Logan & Edwards, 2004) has provided the technology to develop highly sensitive, rapid and quantitative assays for lyssaviruses which overcome some of the limitations in getting molecular testing regimens accepted for diagnostic application by the international community. While real-time PCR employs the same principles as standard PCR, in which two PCR primers generate an amplicon, the detection of this product is facilitated during the course of the reaction (i.e., in real-time) by linkage to increasing levels of fluorescence that are converted into an amplification plot. Any sample in which the fluorescent signal rises significantly above the applied threshold prior to the end of the run is considered positive. The cycle at which the fluorescence statistically rises significantly above background levels, designated as the critical threshold cycle number (C_T) can be used to estimate the amount of target sequence present in the reaction in comparison to values obtained for a series of standards of known amount.

The high sensitivity of the assay is assured both by the inclusion of fluorescent dyes for product detection and the targeting of relatively short amplicons, between 80–200 bp, that are generated efficiently. In the case of RNA viruses, which require reverse transcription prior to the PCR, both one and two step assays have been described. The two step assay requires a separate RT prior to set-up of the PCR while one step assays employ enzyme mixtures that permit both the RT and PCR to be performed sequentially within the thermocycler. The latter is preferable since it minimizes handling of the sample and thereby maximizes efficiency while also minimizing the chances of sample cross contamination.

The two most popular chemistries that have evolved to support this technology have both been applied to the development of lyssavirus assays.

4.3.1 5′ Nuclease Assay

The chemistry of this assay, originally referred to as TaqMan™, employs a dual-labeled probe (DLP) as an integral part of the reaction. This probe comprises a synthetic oligonucleotide, which binds to internal sequence on one strand of the amplicon, and is labeled at both its 5′-and 3′-ends. The 5′ terminal of the DLP is covalently attached to a reporter dye that fluoresces when irradiated at a certain wavelength. A quencher moiety is covalently attached to the DLP's 3′-end such that emissions by the reporter dye are effectively quenched. When target sequence is present in the PCR, specific product is generated and the DLP binds to its cognate sequence during the annealing step just prior to strand extension from one of the PCR primers. During DNA synthesis, the 5′ nuclease activity of the DNA polymerase degrades the probe, reporter and quencher become dissociated from each other and the presence of PCR product is detected as fluorescence emitted by the reporter dye at a defined wavelength. Levels of this signal rise during the course of the reaction until a plateau is reached once the reagents in the reaction are exhausted. If no amplicon is produced during the PCR, the DLP cannot bind to amplicon and it remains intact, no increase in fluorescent signal is recorded, and the sample is scored as negative.

Whereas this approach provides exquisite specificity to the assay, the design of primers and probes that are broadly cross reactive can be challenging due to nucleotide sequence variation observed between members of the genus. Depending on their location and frequency base mismatches between the target and primer/probe sequences can substantially reduce assay sensitivity or even preclude detection completely.

In a relatively early study, Black et al. (2002) reported an ambitious effort to both detect and discriminate between lyssavirus genotypes 1–6 using TaqMan technology. The assay relied on the previously described JW6 primer set and a newly developed primer BB6, that overlaps in sequence with primer RabNfor, and which generates a product that, at 502 bp, is considered long for a TaqMan assay. In addition 8 different probes, comprising 3 rabies virus specific probes and one probe each for genotypes 2–6, were developed; these targeted different portions of the amplicon over a 261 bp region. Separate assays were undertaken to evaluate samples for each of the 6 genotypes. This complex assay design underscored the difficulty in developing a pan-*lyssavirus* TaqMan method. The authors did report the successful detection and assignment of 106 lyssaviruses using these reagents, with no cross-reactivity with 18 non-lyssavirus isolates detected. Another assay based on primers used in a hnRT-PCR method to detect all African lyssaviruses was converted into a real-time format with addition of a probe and shown to detect all members of a limited cohort representing the diversity of this group (Coertse, Weyer, Nel & Markotter, 2010).

A simplified protocol derived from the study by Black et al. (2002) that detects and discriminates lyssaviruses of genotypes 1, 5 and 6 employs primers JW12 and N165-146 to produce a 111 bp amplicon that is typed using three distinct genotyping probes, each labeled with a different reporter dye (Wakeley et al., 2005). Of 62 lyssaviruses evaluated by this method all were readily detected and typed except for one American bat rabies virus which was only weakly detected, possibly due to three mismatches between the rabies virus probe and the target sequence for this isolate.

In a rather more cautious study on the application of TaqMan technology to rabies virus RNA detection, Hughes, Smith, Hanlon, and Rupprecht (2004) developed a series of assays for several rabies virus strains circulating in North American reservoir hosts. Their studies supported the utility of the quantitative aspects of TaqMan technology but exposed difficulties that might be faced in the design of TaqMan primers and probes to detect a wide range of rabies virus strains, given the genetic diversity that is found within each specific strain. They reported that three historical samples (5% of those tested) were not amplified by any of their primer/probe sets. This was due to several mismatches between the nucleotide sequences of these reagents and the viral target sequences; in particular, mismatches in the center of the probe target sequence were reported to be especially detrimental to the assay. These findings suggest that a simple TaqMan assay could yield significant numbers of false negative results and that large numbers of primer/probe combinations would be required to adequately cover the genetic diversity existing within rabies virus populations.

In an attempt to overcome these expressed shortcomings, Nadin-Davis, Sheen, and Wandeler (2009) explored the utility of three different assays, targeting separate regions of the N gene, to detect all rabies viruses. All of these assays were designed after careful review of the sequence variation observed between members of all known rabies viral lineages. One of these assays (RABVD1) employed a modified version of the assay described by Wakeley et al. (2005); the forward primer was identical to primer JW12, the reverse primer was identical to primer N165-146 except that it included three additional degenerate positions at bases 151, 160 and 163 and the probe corresponded to bases 78 to 111 (positive sense) with some degenerate positions. The RABVD1 reagent set was found to be suitable for detection of all 203 rabies viruses in the study panel; high C_T values yielded by a few samples were due to poor RNA sample integrity as measured by a β-actin internal control assay. The other two reagent sets explored in this study (RABVD2 and RABVD3) that targeted N gene coding region sequences further downstream were also found to be broadly cross-reactive but did fail to detect a few samples in each case. It was concluded that if two separate assays were performed on a single sample, together with an internal control to check for sample integrity, these real-time PCRs were 100% successful in detecting rabies virus but they were not suitable for detection of other lyssaviruses.

In yet another study to develop a TaqMan assay that detects all rabies viruses, Hoffmann et al. (2010) compared the method described by Wakeley et al. (2005) with a second assay that targets sequence slightly downstream within the N gene ORF. While neither assay individually detected all 93 viruses in the study panel, together the two assays were 100% successful.

Other TaqMan assays that target specific groups of lyssaviruses have been described. One such assay uses two biotype specific forward primers, a common reverse primer and two distinct DLPs to detect and discriminate the flying fox and insectivorous biotypes of ABLV (Smith, Northill, Harrower & Smith, 2002). A later optimized version of this method was specific to ABLVs with no cross-reactivity with other lyssaviruses (Foord et al., 2006). Assays to detect and discriminate rabies virus and EBLV-1 in samples from across Poland were reported (Orlowska, Smreczak, Trębas & Zmudziński, 2008) while assays that detect and discriminate Eurasian bat lyssaviruses were applied on an experimental basis only (Hughes et al., 2006).

A TaqMan assay, targeting sequences close to the 3'-end of the N gene ORF, was developed based on rabies virus N gene sequence data acquired over several years during a nationwide survey in Thailand. While this method detected a wide range of field isolates from Thailand and several other Asian countries (Wacharapluesadee et al., 2008) it failed to detect the Challenge Virus Standard (CVS) strain due to significant sequence mismatch of the probe with the target. This study explored the effects of sequence mismatches between the reagents and target and showed that the TaqMan assay was permissive for multiple mismatches, though with some impact on reaction efficiency and sensitivity.

The use of controls is an important aspect of any real-time RT-PCR assay since these assays tend to be more susceptible to inhibitory factors than standard RT-PCRs. Assays that detect 18S rRNA (Coertse, Weyer, Nel & Markotter, 2010) or β-actin mRNA are most often incorporated as internal controls for lyssavirus RNA detection by 5' nuclease assays (Hughes, Smith, Hanlon & Rupprecht, 2004; Nadin-Davis, Sheen & Wandeler, 2009; Wakeley et al., 2005). Hoffmann and colleagues have explored the use of several controls to evaluate RNA extraction and PCR inhibitory effects all of which provided satisfactory results (Hoffmann, Depner, Schirrmeier & Beer, 2006; Hoffmann et al., 2010).

Several studies have reported that TaqMan assays are more sensitive, by up to 100 fold, than even the hnRT-PCR methods previously employed (Foord et al., 2006; Nadin-Davis, Sheen & Wandeler, 2009; Orlowska, Smreczak, Trębas & Zmudziński, 2008; Smith, Northill, Harrower & Smith, 2002). The exception were the studies by Hughes, Smith, Hanlon, and Rupprecht (2004) and Hughes et al. (2006) in which it was reported that their TaqMan assays were equivalent to or higher in sensitivity compared to regular PCRs but lower than that achieved by hnRT-PCR, perhaps indicating less than optimal TaqMan assay design. Few studies have attempted to compare the sensitivity of real-time RT-PCR methods and the dFA. A study by Szanto, Nadin-Davis, Rosatte, and White (2011) suggested that a real-time RT-PCR method that specifically detected the raccoon variant of rabies may be more sensitive than dFA but confirmation of this claim will require more stringently controlled experiments.

4.3.2 Use of Intercalating Dyes

This chemistry employs a DNA-intercalating dye such as SYBR Green to detect the production of all dsDNA within the reaction. As amplicon is produced by the action of the two PCR primers, increasing amounts of the dye intercalate with the product resulting in increasing levels of fluorescence. Whereas this is a relatively inexpensive option for method development the caveat is that unless the reaction is carefully designed and performed the production of nonspecific dsDNA products can potentially yield false positive results. One approach used to try to confirm the identity of products detected in this manner is to determine the melting temperature (T_M) of the amplicon by performance of a melting curve analysis, a process that can readily be automated in most commercial instruments employed for real-time PCR assays.

One report in which this approach was explored employed a two-step protocol in which primer RabN1 (Table 11.1) was used to prime cDNA synthesis followed by real-time PCR using primer O1, corresponding to 10g (Table 11.1), paired with a novel reverse primer, R6, corresponding to bases 201-183 of the rabies virus genome, to generate a 135bp product (Nagaraj et al., 2006). Melting curve analysis discriminated between specific products having a T_M of 78°C and nonspecific primer-dimer products with a T_M of 70–73°C. A comparison of this method with nested PCR on a small number of samples indicated the higher sensitivity of the real-time PCR protocol.

Another protocol employed two primers (JW12 and N165-146) used previously for TaqMan assays with the SYBR Green chemistry in an effort to develop a pan-lyssavirus assay (Hayman et al., 2011a). Both melting curve analysis and gel analysis of the amplicons were performed to guard against false positive results due to generation of nonspecific products. The assay was tested using viral sequences representative of all lyssavirus species and may be of value for the detection of a wide range of lyssaviruses in support of epidemiological studies though it has not yet been validated for routine diagnostic use. This real-time PCR method was also found to be more sensitive than the hnRT-PCR previously described for detection of a broad range of lyssaviruses; indeed, the wide difference in sensitivity between these methods for certain isolates was suggested to be the consequence of poor binding of the hemi-nested primers to target due to sequence mismatches (Coertse, Weyer, Nel & Markotter, 2010; Hayman et al., 2011a).

4.3.3 *Prospects for Future Application of Real-Time PCR*

Despite the many advantages of real-time PCR as a diagnostic tool, some technical aspects of the assay are important to consider when applying this technology. It can be difficult to define the appropriate end point of the assay (i.e., the cycle number beyond which samples cannot be considered positive) given the potential for nonspecific degradation of TaqMan probes at late cycles and the difficulty in correlating results with those of other gold standard tests which may not be as sensitive as real-time PCR methods. To minimize problems related to these and other issues, strict adherence to the "minimal information for publication of quantitative real-time PCR experiments" (MIQE) guidelines for real-time PCR assay development and reporting (Bustin et al., 2009) are encouraged.

Molecular methods do, of course, have enormous value in studies on rabies pathogenesis, and this application is described in more detail elsewhere (see Chapter 8). Standard RT-PCR methods have been used to examine organ distribution of virus in experimental (Charlton, Nadin-Davis, Casey & Wandeler, 1997; Vos et al., 2004) or naturally infected animals (Serra-Cobo, Amengual, Abellán & Bourhy, 2002), but real-time RT-PCR is particularly advantageous in such studies because, by its quantitative nature, it can provide measurements of actual viral load in such samples. Accordingly, real-time RT-PCR has been employed to determine post-mortem levels of virus present in various tissues of infected humans (Panning et al., 2010) or animals (Brookes et al., 2007; Freuling et al., 2009; Johnson et al., 2008), and to determine titer of virus present in saliva samples from patients undergoing experimental treatment (see Nadin-Davis, Sheen & Wandeler, 2009). Panning et al. (2010) also showed that the real-time RT-PCR was a more sensitive method than virus isolation.

4.4 Application of Molecular Methods to Fixed Tissues

Fresh brain material is the ideal tissue for rabies diagnosis, but in some situations brain tissue is fixed prior to suspicion of rabies in the differential diagnosis, and such specimens are occasionally submitted for testing. Molecular techniques have used labeled DNA or RNA probes to detect virus in tissue sections *in situ* using approaches that draw extensively from standard immunohistochemical methods. An *in situ* hybridization method that uses a probe to detect the rabies virus N gene transcript has been developed for application to formalin-fixed tissues received by the Rabies Centre of Expertise, CFIA, Canada (Nadin-Davis, Sheen & Wandeler, 2003); this is occasionally performed to confirm results of immunohistochemical investigations on such samples where direct comparison of the distribution of viral RNAs with viral antigens may facilitate correct interpretation of the observed staining patterns. Refinement of this technique also allows for typing of any virus present using strain-specific probes targeting divergent P gene sequences. Warner, Whitfield, Fekadu, and Ho (1997) also reported on the utility of *in situ* hybridization as a confirmatory test for dFA analysis of fixed tissues. Although commercial kits for labeling of probes are available, there is no commercial source for rabies-specific probes or sequences. So, probes must be developed in-house, with appropriate attention given to the impact that lyssavirus diversity will have on the ability of a probe to anneal to its target sequence. Moreover, *in situ* hybridization methods are quite labor intensive due to the many incubation periods and blocking steps needed, as well as the initial time required for preparation of tissue sections, and so these procedures are performed in exceptional circumstances only.

Other approaches for such sample types have applied PCR protocols to virus RNA detection. In a comparison of the use of dFA and RT-PCR, Kulonen, Fekadu, Whitfield, and Warner (1999) reported that both could accurately detect rabies virus in 12 of 24 paraffin-embedded Finnish samples that had been Carnoy-fixed for between 1–24 hours. The molecular method required RNA extraction from paraffin-embedded tissue sections prior to RT-PCR. Of the two RT-PCRs tested in this study, only the one that generated the shorter (139 bp) product gave results concordant with those of the dFA and the known rabies status of the samples. A PCR that generated a 304 bp amplicon detected only 67% of the positive samples.

Wacharapluesadee, Ruangvejvorachai, and Hemachudha (2006) described a rather different procedure that successfully detected rabies virus sequences in brain tissues fixed in formalin for between one day to one week and then stored as paraffin-embedded tissue for between one month to 16 years at 30 °C. RNA extraction from formalin-fixed paraffin-embedded (FFPE) tissue required tissue deparaffinization with xylene, drying and grinding the tissue followed by treatment with proteinase K before the RNA was recovered using the NucliSense Isolation Reagent (Biomerieux), which employs silica particles in a guanidine thiocyanate solution to bind the nucleic acid during purification. Using a variety of primers for the RT-PCR, they demonstrated that rabies virus N gene sequence could be detected in tissue from different regions of the brain in seven rabies-positive patients, although not all samples could be confirmed by this approach; a higher positivity rate was obtained by an immunohistochemical assay performed in parallel. The most efficient RT-PCR assay was one that detected the smallest product (150 bp), and products larger than 400 bp could not be generated in any sample. The authors did demonstrate, however, that through the production of up to six overlapping PCR amplicons, it was possible to generate nucleotide sequence for the complete N gene of the virus responsible for the infection, thereby indicating the potential value of this sample type for molecular epidemiological studies.

These studies showed that as the length of the formalin fixation increased, generation of RT-PCR products became increasingly difficult, and only short products could be made with any frequency. The fixation and paraffin embedding processes cause degradation and modification of the RNA in the tissue and result in RNA cross-linking to other cellular components, thereby rendering RNA extraction inefficient and adversely affecting reverse transcription; this results in an inability to produce amplicons longer than a few hundred base pairs.

Studies on tissues infected with other RNA viruses suggest possible approaches to improve upon these methods. McKinney et al. (2009) explored the use of commercial products, including the paraffin block RNA isolation kit (Ambion) and the Optimum FFPE RNA isolation kit, for improved RNA extraction from FFPE tissues that had undergone prolonged fixation (21 to 30 days). Use of these kits with a modified protocol employing an extended (24 hr) proteinase K digestion was found to facilitate recovery of RNA from tissues infected with West Nile virus, such that viral sequences could be detected by a real-time PCR protocol (McKinney et al., 2009). To date, such protocols have not been evaluated on rabies-infected FFPE samples, but this approach, in which a real-time PCR that amplifies short target sequences is applied to RNA from FFPE samples using an optimized RNA extraction protocol, is likely to be the best for efficient detection of rabies virus and other RNA viruses in fixed tissues.

4.5 Other Molecular Methods for Rabies Virus Detection

Whereas a number of sensitive amplification methods distinct from PCR have been developed for detection of specific nucleic acid sequences, just a few studies have reported the application of such methods to rabies diagnosis (see Fooks et al., 2009). Wacharapluesadee and Hemachudha (2001) explored the application of nucleic-acid sequence based amplification (NASBA) technology to detect rabies virus RNA.

NASBA employs three enzymes, avian myeloblastosis virus reverse transcriptase, *E. coli* RNase H and T7 RNA polymerase, which, together with a set of primers, support isothermal amplification of an RNA template. In this case, the primers targeted a 180 base viral sequence within the central region of the rabies virus N gene, which was subsequently recognized by a reporter probe binding to internal sequence. With respect to sensitivity, this assay appeared to compare favorably in comparisons with standard RT-PCR. Its application, especially in countries where the costs of acquiring the thermocylers required to perform RT-PCR might be prohibitive, could be further investigated.

Another novel amplification strategy with potential application to tropical countries is the reverse-transcription loop-mediated isothermal amplification (RT-LAMP) method. The LAMP method, first described by Notomi et al. (2000), employs a DNA polymerase with high strand displacement activity (e.g. *Bst* DNA polymerase large fragment), and four primers, two inner and two outer. Each of the two inner primers binds to two distinct sequences, of opposite sense, on the target molecule. During the 1 hr incubation at 65 °C, a complex series of reactions involving all primers generate several stem-loop structures so that the final reaction yields a number of products of various sizes when examined by gel electrophoresis. The specificity of these products can be confirmed by restriction endonuclease analysis, which should generate a small number of digestion products of predicted sizes. Optimal efficient amplification is achieved by targeting DNA sequences <300 bp in length and including additional loop primers in the reaction (Nagamine, Hase & Notomi, 2002). The use of reverse transcriptase together with the DNA polymerase allows the assay to be applied to RNA targets.

This strategy for rabies detection by RT-LAMP has been explored in the Philippines (Boldbataar et al., 2009) and in Brazil where it was shown to discriminate between vampire bat- and dog-associated rabies viruses although other field isolates, including those of insectivorous bats, were not consistently detected (Saitou et al., 2010). Successful application of this technology to the detection of several African rabies viruses was also reported (Hayman et al., 2011b; Muleya et al., 2012). The former study employed two sets of primers that between them detected a small sampling of viruses of both Africa 1 and Africa 2 lineages; they also incorporated the assay into a lateral flow device in which biotinylated primers were employed to facilitate product detection (Hayman et al., 2011b). The sensitivity of RT-LAMP assays appears to compare very favorably with that of a RT-PCR. These isothermal RT-LAMP methods are being promoted as inexpensive assays for rabies virus diagnosis because the technology is not technically demanding, the reagents cost just a few dollars per test, and they do not require expensive thermal cycling equipment. Such assays may be of huge benefit to developing countries, where much laboratory infrastructure is currently lacking and the funding needed to support some of the other diagnostic testing methods, such as dFA or RT-PCR, is unavailable. At a local level such a strategy may be realistic, but some of the challenges inherent to these assays must be borne in mind. The design of broadly reactive RT-LAMP assays that can detect a wide range of lyssaviruses will be very challenging, given that multiple primers within a short stretch of target sequence are needed and especially given the observation that a single mismatch at the end of a primer can critically interfere with the assay (Boldbataar et al., 2009). In Africa in particular, where many of these assays are being evaluated, there is a need to detect and discriminate between rabies virus and rabies-related viruses. None of these assays described to date have been validated against significant numbers of virus samples and much work and refinement is yet needed before they can be considered an appropriate alternative to the dFA or RT-PCR methods.

Another experimental approach to lyssavirus detection involves the use of microarray technology, which, when linked to sequence-independent PCR amplification of template, shows promise for both detection and genotyping applications, and which could be a most useful tool for the identification of novel lyssaviruses (Gurrala et al., 2009). However, extensive refinement to such a method, both to improve sensitivity for testing of clinical samples and to achieve a high level of reproducibility in the microarray printing, would be required before it could be considered as a viable diagnostic alternative.

4.6 Sampling and Sample Storage Issues

Ideally, rabies diagnosis is performed on fresh brain tissue collected within 24 hours and kept refrigerated or alternatively stored frozen during transportation. However, in some jurisdictions adherence to these practices is difficult. Other means of preserving samples so that they remain suitable for diagnosis once they reach the testing laboratory have been explored. Preservation of brain tissue on filter paper has proven to have some value; Wacharapluesadee, Phumesin, Lumlertdacha, and Hemachudha (2003) reported that brain tissue dried onto filter paper and stored at room temperature for up to 222 days could be used for RNA extraction and rabies virus detection, by either RT-PCR or NASBA. Other reports have indicated the value of Whatman FTA cards for the preservation and storage of brain samples until they can be examined by molecular methods (Nadin-Davis, Sheen & Wandeler, 2012; Picard-Meyer, Barrat & Cliquet, 2007). Although such a method may have limited utility for samples where a rapid diagnosis is required, it provides a valuable means of transporting samples destined for epidemiological studies.

A recent study explored the application of a TaqMan RT-PCR assay for detection of rabies virus in non-neural specimens from dogs; these included oral swabs and whisker and hair follicles that are much more easily obtained than brain tissue (Wacharapluesadee et al., 2012). However, the use of such sample types reduced the test sensitivity compared to that obtained with brain tissue, and thus it appears that substitution of brain with these other sample types is not currently a viable alternative for provision of accurate diagnosis.

4.7 Use of Molecular Methods for Antemortem Diagnosis

The versatility of molecular tests for analysis of a wide selection of tissue types make them a most useful tool for antemortem diagnosis, particularly as real-time PCR methods reduce turnaround times. Rapid and accurate diagnosis of human rabies is important with respect to both patient management and identification of the need for postexposure prophylaxis of patient contacts. Moreover, given the extent of global travel these days, the patient's travel history must be considered when determining the nature of lyssaviruses to which the patient could have been exposed. This information will dictate the type of test, including the nature of the reagents, most appropriate to the situation and reinforces the need for pan-lyssavirus diagnostic tests. Most studies on antemortem diagnosis by molecular methods have employed either standard or real-time RT-PCR. Prior to the advent of real-time RT-PCR, Heaton, McElhinney, and Lowings (1999) reported on a rapid PCR thermal cycling technique for human diagnosis and NASBA has also been evaluated for this purpose. As described in Section 3, the sample types that have to date proven most useful for antemortem testing are skin biopsy and saliva samples. Several studies have reported that a testing combination involving dFA applied to skin biopsy tissue and standard or hn RT-PCR applied to saliva and skin biopsy could accurately identify most human cases (Crepin et al., 1998; Dacheux et al., 2008; De Brito et al., 2011; Elmgren, Nadin-Davis, Muldoon & Wandeler, 2002; Macedo et al., 2006; Smith et al., 2003). Some studies that explored the use of saliva only for antemortem rabies testing by molecular methods have shown that, whereas the detection rate can be quite high, it does not reach 100% (Hemachudha & Wacharapluesadee 2004; Nagaraj et al., 2006; Wacharapluesadee & Hemachudha, 2001), either due to sporadic secretion of virus into this fluid and/or presence of very low levels of virus. Dacheux et al. (2008) suggested that a 100% sensitivity rate can be achieved if samples collected from a patient on at least three different days were tested. Testing of serially collected samples from individual patients must thus be considered before rendering a negative diagnosis using this sample type alone. The rate of detection of rabies virus in either urine or CSF samples appears to be consistently lower (Dacheux et al., 2008), although by testing a combination of such fluid samples accurate diagnosis is frequently achieved; interestingly, however, it has been noted that patients with paralytic rabies can give false negative results by such tests, perhaps due to the timing of sample collection and/or the very limited titers of virus present in these sample types in patients exhibiting this form of the disease (Hemachudha & Wacharapluesadee 2004; Wacharapluesadee & Hemachudha, 2002). Moreover, at the Rabies Centre of Expertise in Ottawa, Canada, CSF has consistently been found to be a poor sample type for human rabies diagnosis using molecular techniques (Nadin-Davis, unpublished data). Real-time RT-PCR was able to detect rabies virus intravitam in saliva, corneal swab, and sputum samples taken from a patient infected after organ transplantation (Panning et al., 2010).

Although performance of antemortem diagnosis in domestic animals has not often been considered, Saengseesom, Mitmoonpitak, Kasempimolporn and Sitprija (2007) did report a small study to explore the utility of saliva and CSF samples for antemortem diagnosis in dogs. Again, saliva was the better sample type for virus detection but did not yield a 100% detection level. Accordingly, such tests cannot be considered appropriate for decisions regarding human post-exposure prophylaxis at this time. However, as the animal rights movement gains momentum, pressure to refine antemortem animal testing scenarios may increase in the future. For epidemiological studies of rabies in wildlife, especially in bat populations of Europe where many species are protected, RT-PCR testing for the presence of lyssaviruses in RNA recovered from oral swabs has been useful (Echevarría, Avellón, Juste, Vera, & Ibáñez, 2001); Picard-Meyer et al., 2011).

5 CONCLUSIONS

In summary, rabies testing in North America is performed on more than 100,000 animal specimens by more than 100 laboratories that offer rabies diagnosis. Each positive case has ramifications for a minimum of at least one human or one domestic animal and typically involves three to ten, or even hundreds to tens of thousands of persons at public venues in need of triage for potential exposure to rabies. For example, there were approximately 450 children presented for post-exposure prophylaxis after attending a petting zoo where a sheep became ill and was reported positive for rabies (CDC, unpublished data). Ultimately, the diagnosis appeared to be due to cross-contamination during specimen preparation. In a case with a confirmed rabid goat in an animal petting area at a county fair, more than 25,000 people were in attendance and in need of triage for potential exposure to this animal (CDC, unpublished data). The telephones at the local health department were overwhelmed for more than three days. Similarly, a confirmed-positive, rabid kitten at a popular pet shop with many visitors resulted in over 650 people receiving post-exposure prophylaxis (Noah et al., 1996). Moreover, the opportunity for dog-to-dog transmitted rabies to re-establish itself in the USA through translocation of infected members of its primary reservoir is well illustrated by the following: the recent translocation of dogs (young puppies) incubating canine rabies from Thailand, India, Puerto Rico, and Iraq, into California, Washington state and then ultimately Arkansas, Massachusetts and New Jersey respectively; an estimated vaccinated rate of below 70% of the approximately 80,000 pet dogs in the USA; and the often significant time-delay to identify that these rabies-positive samples are a variant not known to occur in the USA. Rabies is a clear public health threat, but fiscal constraints often jeopardize the infrastructure that is critical to its prevention, of which reliable, accurate, precise, and timely primary and advanced diagnoses are fundamental. Local, state, and federal government entities must defend the population against this rather ancient, easily preventable public health threat. If we are not prepared to shore up our defenses against the potential re-emergence of a known public health threat, how can we be prepared for those infectious diseases that are novel and emergent?

References

Anonymous, (2003). Protocol for postmortem diagnosis of rabies in animals by direct fluorecent antibody testing. *A Minimum Standard for Rabies Diagnosis in the United States*, 1–20.

Anonymous, (2011). Compendium of animal rabies prevention and control, 2011. *Morbidity and Mortality Weekly Reports – Recommendations and Reports*, 60, 1–17.

Anonymous. (6-1-2012). Search of US-based Health Department and rabies laboratory websites.

Biswal, M., Ratho, R., & Mishra, B. (2007). Usefulness of reverse transcriptase-polymerase chain reaction for detection of rabies RNA in archival samples. *Japanese Journal of Infectious Diseases*, 60, 298–299.

Black, E. M., Lowings, J. P., Smith, J., Heaton, P. R., & McElhinney, L. M. (2002). A rapid RT-PCR method to differentiate six established genotypes of rabies and rabies related viruses using TaqMan™ technology. *Journal of Virolological Methods*, 105, 25–35.

Black, E. M., McElhinney, L. M., Lowings, J. P., Smith, J., Johnstone, P., & Heaton, P. R. (2000). Molecular methods to distinguish between classical rabies and the rabies-related European bat lyssaviruses. *Journal of Virolological Methods*, 87, 123–131.

Blanton, J. D., Hanlon, C. A., & Rupprecht, C. E. (2007). Rabies surveillance in the United States during 2006. *Journal of the American Veterinary Medical Association*, 231, 540–556.

Blanton, J. D., Krebs, J. W., Hanlon, C. A., & Rupprecht, C. E. (2006). Rabies surveillance in the United States during 2005. *Journal of the American Veterinary Medical Association*, 229, 1897–1911.

Blanton, J. D., Palmer, D., Christian, K. A., & Rupprecht, C. E. (2008). Rabies surveillance in the United States during 2007. *Journal of the American Veterinary Medical Association*, 233, 884–897.

Blanton, J. D., Palmer, D., Dyer, J., & Rupprecht, C. E. (2011). Rabies surveillance in the United States during 2010. *Journal of the American Veterinary Medical Association*, 239, 773–783.

Blanton, J. D., Palmer, D., & Rupprecht, C. E. (2010). Rabies surveillance in the United States during 2009. *Journal of the American Veterinary Medical Association, 237*, 646–657.

Blanton, J. D., Robertson, K., Palmer, D., & Rupprecht, C. E. (2009). Rabies surveillance in the United States during 2008. *Journal of the American Veterinary Medical Association, 235*, 676–689.

Boldbataar, B., Inoue, S., Suigiura, N., Noguchi, A., Orbina, J. R. C., Demetria, C., et al. (2009). *Japanese Journal of Infectious Diseases, 62*, 187–191.

Bourhy, H., Cowley, J. A., Larrous, F., Holmes, E. C., & Walker, P. J. (2005). Phylogenetic relationshsips among rhabdoviruses inferred using the L polymerase gene. *Journal of General Virology, 86*, 2849–2858.

Bourhy, H., Kissi, B., & Tordo, N. (1993). Molecular diversity of the Lyssavirus genus. *Virology, 194*, 70–81.

Brookes, S. M., Klopfleisch, R., Müller, T., Healy, D. M., Teifke, J. P., Lange, E., et al. (2007). Susceptibility of sheep to European bat lyssavirus type-1 and -2 infection: A clinical pathogenesis study. *Veterinary Microbiology, 125*, 210–223.

Bustin, S. A., Benes, V., Garson, J. A., Hellemans, J., Huggett, J., Kubista, M., et al. (2009). The MIQE guidelines: Minimum information for publication of quantitative real-time PCR experiments. *Clinical Chemistry, 55*, 611–622.

CDC & NIH. (2012). *Biosafety in Microbiological and Biomedical Laboratories 5th ed.*

CFIA. (1996). *Containment Standards for Veterinary Facilities.* Available from <http://www.inspection.gc.ca/english/sci/bio/anima/convet/convete.pdf/> Accessed 20.07.12.

Charlton, K. M., Nadin-Davis, S., Casey, G. A., & Wandeler, A. I. (1997). The long incubation period in rabies: Delayed progression of infection in muscle at the site of exposure. *Acta Neuropathologica, 94*, 73–77.

Coertse, J., Weyer, J., Nel, L. H., & Markotter, W. (2010). Improved PCR methods for detection of African rabies and rabies-related lyssaviruses. *Journal of Clinical Microbiology, 48*, 3949–3955.

Cooper, A., & Poinar, H. N. (2000). Ancient DNA: Do it right or not at all. *Science, 289*, 1139.

Crepin, P., Audry, L., Rotivel, Y., Gacoin, A., Caroff, C., & Bourhy, H. (1998). Intravitam diagnosis of human rabies by PCR using saliva and cerebrospinal fluid. *Journal of Clinical Microbiology, 36*, 1117–1121.

Dacheux, L., Reynes, J. M., Buchy, P., Sivuth, O., Diop, B. M., Rousset, D., et al. (2008). A reliable diagnosis of human rabies based on analysis of skin biopsy specimens. *Clinical and Infectious Diseases, 47*, 1410–1417.

David, D., Yakobson, B., Rotenberg, D., Dveres, N., Davidson, I., & Stram, Y. (2002). Rabies virus detection by RT-PCR in decomposed naturally infected brains. *Veterinary Microbiology, 87*, 111–118.

De Brito, M. G., Chamone, T. L., da Silva, F. J., Wada, M. Y., de Miranda, A. B., Castilho, J. G., et al. (2011). Antemortem diagnosis of human rabies in a veterinarian infected when handling a herbivore in Minas Gerais, Brazil. *Revista do Instituto de Medicina Tropical de São Paulo, 53*, 39–44.

Dimaano, E. M., Scholand, S. J., Alera, M. T., & Belandres, D. B. (2011). Clinical and epidemiological features of human rabies cases in the Philippines: A review from 1987 to 2006. *International Journal of Infectious Diseases, 15*, e495–e499.

Echevarría, J. E., Avellón, A., Juste, J., Vera, M., & Ibáñez, C. (2001). Screening of active lyssavirus infection in wild bat populations by viral RNA detection on oropharyngeal swabs. *Journal of Clinical Microbiology, 39*, 3678–3683.

Elmgren, L. D., Nadin-Davis, S. A., Muldoon, F. T., & Wandeler, A. I. (2002). Diagnosis and analysis of a recent case of human rabies in Canada. *Canadian Journal of Infectious Diseases, 13*, 129–133.

Ermine, A., Tordo, N., & Tsiang, H. (1988). Rapid diagnosis of rabies infection by means of a dot hybridization assay. *Molecular & Cellular Probes, 2*, 75–82.

Expert committee on rabies; second report. (1954). *World Health Organization Technical Report Series. 82*, 1–27.

Fehlner-Gardiner, C., Muldoon, F., Nadin-Davis, S., Wandeler, A., Kush, J., & Jordan, L. T. (2008). Cross-Canada disease report: Laboratory diagnosis of rabies in Canada for calendar year 2006. *Canadian Veterinary Journal, 49*, 359–361.

Fekadu, M., Greer, P. W., Chandler, F. W., & Sanderlin, D. W. (1988). Use of the avidin-biotin peroxidase system to detect rabies antigen in formalin-fixed paraffin-embedded tissues. *Journal of Virological Methods, 19*, 91–96.

Fooks, A. R., Johnson, N., Freuling, C. M., Wakeley, P., Banyard, A., McElhinney, L. M., et al. (2009). Emerging technologies for the detection of rabies virus: Challenges and hopes in the 21st century. *PLoS Neglected Tropical Diseases, 3*, e530.

Foord, A. J., Heine, H. G., Pritchard, L. I., Lunt, R. A., Newberry, K. M., Rootes, C. L., et al. (2006). Molecular diagnosis of lyssaviruses and sequence comparison of Australian bat lyssavirus samples. *Australian Veterinary Journal, 84*, 225–230.

Freuling, C., Vos, A., Johnson, N., Kaipf, I., Denzinger, A., Neubert, L., et al. (2009). Experimental infection of serotine bats (Eptesicus serotinus) with European bat lyssavirus type 1a. *Journal of General Virology, 90*, 2493–2502.

Goldwasser, R. A., & Kissling, R. E. (1958). Fluorescent antibody staining of street and fixed rabies virus antigens. *Proceedings of the Society for Experimental Biology and Medicine, 98*, 219–223.

Gurrala, R., Dastjerdi, A., Johnson, N., Nunez-Garcia, J., Grierson, S., Steinbach, F., et al. (2009). Development of a DNA microarray for simultaneous detection and genotyping of lyssaviruses. *Virus Research, 144*, 202–208.

Hamir, A. N., Moser, G., Fu, Z. F., Dietzschold, B., & Rupprecht, C. E. (1995). Immunohistochemical test for rabies: Identification of a diagnostically superior monoclonal antibody. *Veterinary Record, 136*, 295–296.

Hanlon, C. A., Smith, J. S., & Anderson, G. R. (1999). Recommendations of a national working group on prevention and control of rabies in the United States. Article II: Laboratory diagnosis of rabies. The National working group on rabies prevention and control. *Journal of the American Veterinary Medical Association, 215*, 1444–1446.

Hayman, D. T. S., Banyard, A. C., Wakeley, P. R., Harkess, G., Marston, D., Wood, J. L. N., et al. (2011a). A universal real-time assay for the detection of lyssaviruses. *Journal of Virological Methods, 177*, 87–93.

Hayman, D. T. S., Johnson, N., Horton, D. L., Hedge, J., Wakeley, P. R., Banyard, A. C., et al. (2011b). Evolutionary history of rabies in Ghana. *PLoS Neglected Tropical Diseases, 5*(4), e1001.

Health Canada. (2004). *Laboratory Safety Guidelines* Available from <http://www.phac-aspc.gc.ca/publicat/lbg-ldmbl-04/pdf/lbg_2004_e.pdf/> Accessed 27.06.12.

Heaton, P. R., Johnstone, P., McElhinney, L. M., Cowley, R., O'Sullivan, E., & Whitby, J. E. (1997). Heminested PCR assay for detection of six genotypes of rabies and rabies-related viruses. *Journal of Clinical Microbiology, 35*, 2762–2766.

Heaton, P. R., McElhinney, L. M., & Lowings, J. P. (1999). Detection and identification of rabies and rabies-related viruses using rapid-cycle PCR. *Journal of Virological Methods, 81*, 63–69.

Hemachudha, T., & Wacharapluesadee, S. (2004). Antemortem diagnosis of human rabies. *Clinical Infectious Diseases, 39*, 1085–1086.

Hoffmann, B., Depner, K., Schirrmeier, H., & Beer, M. (2006). A universal heterologous internal control system for duplex real-time RT-PCR assays used in a detection system for pestiviruses. *Journal of Virological Methods, 136*, 200–209.

Hoffmann, B., Freuling, C. M., Wakeley, P. R., Rasmussen, T. B., Leech, S., Fooks, A. R., et al. (2010). Improved safety for molecular diagnosis of classical rabies viruses by use of a TaqMan real-time reverse transcription-PCR "double check" strategy. *Journal of Clinical Microbiology, 48*, 3970–3978.

Hughes, G. J., Kuzmin, I. V., Schmitz, A., Blanton, J., Manangan, J., Murphy, S., et al. (2006). Experimental infection of big brown bats (Eptesicus fuscus) with Eurasian bat lyssaviruses Aravan, Khujand and Irkut virus. *Archives of Virology, 151,* 2021–2035.

Hughes, G. J., Smith, J. S., Hanlon, C. A., & Rupprecht, C. E. (2004). Evaluation of a TaqMan PCR assay to detect rabies virus RNA: Influence of sequence variation and application to quantification of viral loads. *Journal of Clinical Microbiology, 42,* 299–306.

Interpretive Guidelines §493.1253(b)(2). (12-6-1996). <http://www.cms.gov/Regulations-and-Guidance/Legislation/CLIA/downloads/apcsubk1.pdf/>[Interpretive Guidelines §493.1253(b)(2)], 27–54.

Jayakumar, R., & Padmanaban, V. D. (1994). A dipstick dot enzyme immunoassay for detection of rabies antigen. *Zentralblatt fur Bakteriologie, Parasitenkunde, Infektionskrankheiten und Hygiene, 280,* 382–385.

Johnson, N., Vos, A., Neubert, L., Freuling, C., Mansfield, K. L., Kaipf, I., et al. (2008). Experimental study of European bat lyssavirus type-2 infection in Daubenton's bats (Myotis daubentonii). *Journal of General Virology, 89,* 2662–2672.

Kamolvarin, N., Tirawatnpong, T., Rattanasiwamoke, R., Tirawatnpong, S., Panpanich, T., & Hemachudha, T. (1993). Diagnosis of rabies by polymerase chain reaction with nested primers. *Journal of Infectious Diseases, 167,* 207–210.

Kasempimolporn, S., Saengseesom, W., Huadsakul, S., Boonchang, S., & Sitprija, V. (2012). Evaluation of a rapid immunochromatographic test strip for detection of rabies virus in dog saliva samples. *Journal of Veterinary Diagnostic Investigation, 23,* 1197–1201.

Kissi, B., Tordo, N., & Bourhy, H. (1995). Genetic Polymorphism in the Rabies virus nucleoprotein gene. *Virology, 209,* 526–537.

Koprowski, H. (1966). Laboratory techniques in rabies (1st ed.). Mouse inoculation test. *Monograph Series World Health Organization, 23,* 56–68.

Koprowski, H. (1973). Laboratory techniques in rabies: The mouse inoculation test (3rd ed.). *Monograph Series World Health Organization, 23,* 85–93.

Kulonen, K., Fekadu, M., Whitfield, S., & Warner, C. K. (1999). An evaluation of immunofluorescence and PCR methods for detection of rabies in archival carnoy-fixed, paraffin-embedded brain tissue. *Journal of Veterinary Medicine Series B, 46,* 151–155.

Kuzmin, I. V., Hughes, G. J., Botvinkin, A. D., Orciari, L. A., & Rupprecht, C. E. (2005). Phylogenetic relationships of Irkut and West Caucasian bat viruses within the Lyssavirus genus and suggested quantitative criteria based on the N gene sequence for lyssavirus genotype definition. *Virus Research, 111,* 28–43.

Kuzmin, I. V., Niezgoda, M., Franka, R., Agwanda, B., Markotter, W., Deagley, J. C., et al. (2008). Lagos bat virus in Kenya. *Journal of Clinical Microbiology, 46,* 1451–1461.

Kwok, S., & Higuchi, R. (1989). Avoiding false positives with PCR. *Nature, 339,* 237–238.

Le Mercier, P., Jacob, Y., & Tordo, N. (1997). The complete Mokola virus genome sequence: Structure of the RNA-dependent RNA polymerase. *Journal of General Virology, 78,* 1571–1576.

Lembo, T., Niezgoda, M., Velasco-Villa, A., Cleaveland, S., Ernest, E., & Rupprecht, C. E. (2006). Evaluation of a direct, rapid immunohistochemical test for rabies diagnosis. *Emerging Infectious Diseases, 12,* 310–313.

Logan, J. M. J., & Edwards, K. J. (2004). An overview of real-time PCR platforms. In K. Edwards, J. Logan, & N. Saunders (Eds.), *Real-time PCR: An essential guide (pp. 13–29).* Norfolk, UK: Horizon Bioscience.

Lopes, M. C., Venditti, L. L. R., & Queiroz, L. H. (2010). Comparison between RT-PCR and the mouse inoculation test for detection of rabies virus in samples kept for long periods under different conditions. *Journal of Virological Methods, 164,* 19–23.

Loza-Rubio, E., Rojas-Anaya, E., Banda-Ruiz, V. M., Nadin-Davis, S. A., & Cortez-Garcia, B. (2005). Detection of multiple strains of rabies virus RNA using primers designed to target Mexican vampire bat variants. *Epidemiology and Infection, 133,* 927–934.

Macedo, C. I., Carnieli, P., Brandão, P. E., Rosa, E. S., de Oliveira, R. N., Castilho, J. G., et al. (2006). Diagnosis of human rabies cases by polymerase chain reaction of neck-skin samples. *The Brazilian Journal of Infectious Diseases, 10*, 341–345.

Madhusudana, S. N., Paul, J. P., Abhilash, V. K., & Suja, M. S. (2004). Rapid diagnosis of rabies in humans and animals by a dot blot enzyme immunoassay. *International Journal of Infectious Diseases, 8*, 339–345.

Manning, S. E., Rupprecht, C. E., Fishbein, D., Hanlon, C. A., Lumlertdacha, B., Guerra, M., et al. (2008). Human rabies prevention--United States, 2008: Recommendations of the advisory committee on immunization practices. *Morbidity and Mortality Weekly Reports - Recommendations and Reports, 57*, 1–28.

Markotter, W., Randles, J., Rupprecht, C. E., Sabeta, C. T., Taylor, P. J., Wandeler, A. I., et al. (2006). Lagos bat virus, South Africa. *Emerging Infectious Diseases, 12*, 504–506.

McColl, K. A., Gould, A. R., Selleck, P. W., Hooper, P. T., Westbury, H. A., & Smith, J. S. (1993). Polymerase chain reaction and other laboratory techniques in the diagnosis of long incubation rabies in Australia. *Australian Veterinary Journal, 70*, 84–89.

McKinney, M. D., Moon, S. J., Kulesh, D. A., Larsen, T., & Schoepp, R. J. (2009). Detection of viral RNA from paraffin-embedded tissues after prolonged formalin fixation. *Journal of Clinical Virology, 44*, 39–42.

McQuiston, J. H., Yager, P. A., Smith, J. S., & Rupprecht, C. E. (2001). Epidemiologic characteristics of rabies virus variants in dogs and cats in the United States, 1999. *Journal of the American Veterinary Medical Association, 218*, 1939–1942.

Muleya, W., Namangala, B., Mweene, A., Zulu, L., Fandamu, P., Banda, D., et al. (2012). Molecular epidemiology and a loop-mediated isothermal amplification method for diagnosis of infection with a rabies virus in Zambia. *Virus Research, 163*, 160–168.

Nadin-Davis, S. A. (1998). Polymerase chain reaction protocols for rabies virus discrimination. *Journal of Virological Methods, 75*, 1–8.

Nadin-Davis, S. A., Casey, G. A., & Wandeler, A. (1993). Identification of regional variants of the rabies virus within the Canadian province of Ontario. *Journal of General Virology, 74*, 829–837.

Nadin-Davis, S. A., Huang, W., & Wandeler, A. I. (1996). The design of strain-specific polymerase chain reactions for discrimination of the raccoon rabies virus strain from indigenous rabies viruses of Ontario. *Journal of Virological Methods, 57*, 141–156.

Nadin-Davis, S. A., Sheen, M., & Wandeler, A. I. (2003). Use of discriminatory probes for strain typing of formalin-fixed rabies virus-infected tissues by in situ hybridization. *Journal of Clinical Microbiology, 41*, 4343–4352.

Nadin-Davis, S. A., Sheen, M., & Wandeler, A. I. (2009). Development of real-time reverse transcriptase polymerase chain reaction methods for human rabies diagnosis. *Journal of Medical Virology, 81*, 1484–1497.

Nadin-Davis, S. A., Sheen, M., & Wandeler, A. I. (2012). Recent emergence of the Arctic rabies virus lineage. *Virus Research, 163*, 352–362.

Nagamine, K., Hase, T., & Notomi, T. (2002). Accelerated reaction by loop-mediated isothermal amplification using loop primers. *Molecular and Cellular Probes, 16*, 223–239.

Nagaraj, T., Vasanth, J. P., Desai, A., Kamat, A., Madhusudana, S. N., & Ravi, V. (2006). Ante mortem diagnosis of human rabies using saliva samples: Comparison of real time and conventional RT-PCR techniques. *Journal of Clinical Virology, 36*, 17–23.

Negri, A. (1903). Beitrag zum Studium der Aetiologie der Tollwuth. *Zeitschrift fur Hygiene und Infectionskrankheiten, 43*, 507–527.

Nel, L. H., Bingham, J., Jacobs, J. A., & Jaftha, J. B. (1998). A nucleotide-specific polymerase chain reaction assay to differentiate rabies virus biotypes in South Africa. *Onderstepoort Journal of Veterinary Research, 65*, 297–303.

Nel, L., Jacobs, J., Jaftha, J., von Teichman, B., & Bingham, J. (2000). New cases of Mokola virus infection in South Africa: A genotypic comparison of southern African virus isolates. *Virus Genes, 20*, 103–106.

Noah, D. L., Smith, M. G., Gotthardt, J. C., Krebs, J. W., Green, D., & Childs, J. E. (1996). Mass human exposure to rabies in New Hampshire: Exposures, treatment, and cost. *American Journal of Public Health, 86,* 1149–1151.

Notomi, T., Okayama, H., Masubuchi, H., Yonekawa, T., Watanabe, K., Amino, N., et al. (2000). Loop-mediated isothermal amplification of DNA. *Nucleic Acids Research, 28,* e63.

OIE, (2011). *Manual of diagnostic tests and vaccines for terrestrial animals 2011.* Paris, France: World Organisation for Animal Health. (<http://www.oie.int/fileadmin/Home/eng/Health_standards/tahm/2.01.13_RABIES.pdf>).

Orlowska, A., Smreczak, N., Trębas, P., & Zmudziński, J. F. (2008). Comparison of real-time PCR and heminested RT-PCR methods in the detection of rabies virus infection in bats and terrestrial animals. *Bulletin of the Veterinary Institute in Pulawy, 52,* 313–318.

Panning, M., Baumgarte, S., Pfefferle, S., Maier, T., Martens, A., & Drosten, C. (2010). Comparative analysis of rabies virus reverse transcription-PCR and virus isolation using samples from a patient infected with rabies virus. *Journal of Clinical Microbiology, 48,* 2960–2962.

Perrin, P., Gontier, C., Lecocq, E., & Bourhy, H. (1992). A modified rapid enzyme immunoassay for the detection of rabies and rabies-related viruses: RREID-lyssa. *Biologicals, 20,* 51–58.

Picard-Meyer, E., Barrat, J., & Cliquet, F. (2007). Use of filter paper (FTA®) technology for sampling, recovery and molecular characterisation of rabies viruses. *Journal of Virological Methods, 140,* 174–182.

Picard-Meyer, E., Bruyere, V., Barrat, J., Tissot, E., Barrat, M. J., & Cliquet, F. (2004). Development of a hemi-nested RT-PCR method for the specific determination of European bat lyssavirus 1: Comparison with other rabies diagnostic methods. *Vaccine, 22,* 1921–1929.

Picard-Meyer, E., Dubourg-Savage, M. -J., Arthur, L., Barataud, M., Bécu, D., Bracco, S., et al. (2011). Active surveillance of bat rabies in France: A 5-year study (2004–2009). *Veterinary Microbiology, 151,* 390–395.

Powell, J. (1997). Proficiency testing in the rabies diagnostic laboratory. *Abstracts of the eighth annual rabies in the americas conference,* November 2–6, 1997, Kingston, Ontario.

Robardet, E., Picard-Meyer, E., Andrieu, S., Servat, A., & Cliquet, F. (2011). International interlaboratory trials on rabies diagnosis: An overview of results and variation in reference diagnosis techniques (fluorescent antibody test, rabies tissue culture infection test, mouse inoculation test) and molecular biology techniques. *Journal of Virological Methods, 177,* 15–25.

Rohde, R. E., Neill, S. U., Clark, K. A., & Smith, J. S. (1997). Molecular epidemiology of rabies epizootics in Texas. *Clinical and Diagnostic Virology, 8,* 209–217.

Rojas Anaya, E., Loza-Rubio, E., Banda Ruiz, V. M., & Hernandez Baumgarten, E. (2006). Use of reverse transcription-polymerase chain reaction to determine the stability of rabies virus genome in brains kept at room temperature. *Journal of Veterinary and Diagnostic Investigation, 18,* 98–101.

Rudd, R. J., Smith, J. S., Yager, P. A., Orciari, L. A., & Trimarchi, C. V. (2005). A need for standardized rabies-virus diagnostic procedures: Effect of cover-glass mountant on the reliability of antigen detection by the fluorescent antibody test. *Virus Research, 111,* 83–88.

Sabeta, C., Blumberg, L., Miyen, J., Mohale, D., Shumba, W., & Wandeler, A. (2010). Mokola virus involved in a human contact (South Africa). *FEMS Immunology and Medical Microbiology, 58,* 85–90.

Sacramento, D., Bourhy, H., & Tordo, N. (1991). PCR technique as an alternative method for diagnosis and molecular epidemiology of rabies virus. *Molecular and Cellular Probes, 5,* 229–240.

Saengseesom, W., Mitmoonpitak, C., Kasempimolporn, S., & Sitprija, V. (2007). Real-time PCR analysis of dog cerebrospinal fluid and saliva samples for ante-mortem diagnosis of rabies. *Southeast Asian Journal of Tropical Medicine and Public Health, 38,* 53–57.

Saiki, R. K. (1989). The design and optimization of the PCR. In H. A. Erlich (Ed.), *PCR technology: Principles and applications for DNA amplification (pp. 7–16).* New York: Stockton Press.

Saitou, Y., Kobayashi, Y., Hiarano, S., Mochizuki, N., Itou, T., Ito, F. H., et al. (2010). A method for simultaneous detection and identification of Brazilian dog- and vampire bat-related rabies virus by reverse transcription loop-mediated isothermal amplification assay. *Journal of Virological Methods, 168,* 13–17.

Sellers, T. F. (1923). Status of Rabies in the United States in 1921. *American Journal of Public Health (N.Y.), 13,* 742–747.

Serra-Cobo, J., Amengual, B., Abellán, C., & Bourhy, H. (2002). European bat lyssavirus infection in Spanish bat populations. *Emerging Infectious Diseases, 8,* 413–420.

Servat, A., Picard-Meyer, E., Robardet, E., Muzniece, Z., Must, K., & Cliquet, F. (2012). Evaluation of a rapid immunochromatographic diagnostic test for the detection of rabies from brain material of European mammals. *Biologicals, 40,* 61–66.

Smith, I. L., Northill, J. A., Harrower, B. J., & Smith, G. A. (2002). Detection of Australian bat lyssavirus using a fluorogenic probe. *Journal of Clinical Virology, 25,* 285–291.

Smith, J., McElhinney, L., Parsons, G., Brink, N., Doherty, T., Agranoff, D., et al. (2003). Case Report: Rapid Ante-Mortem diagnosis of a human case of rabies imported into the UK from the Philippines. *Journal of Medical Virology, 69,* 150–155.

Smith, J., McElhinney, L. M., Heaton, P. R., Black, E. M., & Lowings, J. P. (2000). Assessment of template quality by the incorporation of an internal control into a RT-PCR for the detection of rabies and rabies-related viruses. *Journal of Virological Methods, 84,* 107–115.

Smith, J. S. (1995). Rabies virus. In P. R. Murray, E. J. Baron, M. A. Pfaller, F. C. Tenover, & P. H. Yolken (Eds.), *Manual of clinical microbiology (pp. 997–1003).* Washington: ASM Press.

Smith, J. S., Orciari, L. A., Yager, P. A., Seidel, H. D., & Warner, C. K. (1992). Epidemiologic and historical relationships among 87 rabies virus isolates as determined by limited sequence analysis. *Journal of Infectious Diseases, 166,* 296–307.

Szanto, A. G., Nadin-Davis, S. A., Rosatte, R. C., & White, B. N. (2011). Re-assessment of direct fluorescent antibody negative brain tissues with a real-time PCR assay to detect the presence of raccoon rabies virus RNA. *Journal of Virological Methods, 174,* 110–116.

Szanto, A. G., Nadin-Davis, S. A., & White, B. N. (2008). Complete genome sequence of a raccoon rabies virus isolate. *Virus Research, 136,* 130–139.

Tordo, N., Bourhy, H., & Sacramento, D. (1992). Polymerase chain reaction technology for rabies virus. In Y. Becker & G. Darai (Eds.), *Frontiers of virology: Diagnosis of human viruses by polymerase chain reaction technology (pp. 389–405).* Berlin: Springer-Verlag.

Tordo, N., Bourhy, H., & Sacramento, D. (1995). Polymerase chain reaction technology for rabies virus. In J. P. Clewley (Ed.), *The Polymerase Chain Reaction (PCR) for human viral diagnosis* (pp. 125–145). Boca Raton, FL: CRC Press.

Trimarchi, C. V., & Smith, J. S. (2002). Diagnostic evaluation. In A. C. Jackson & W. H. Wunner (Eds.), *Rabies (pp. 307–349).* New York: Academic Press.

Vasanth, J. P., Madhusudana, S. N., Abhilash, K. V., Suja, M. S., & Muhamuda, K. (2004). Development and evaluation of an enzyme immunoassay for rapid diagnosis of rabies in humans and animals. *Indian Journal of Pathology and Microbiology, 47,* 574–578.

Vázquez-Morón, S., Avellón, A., & Echevarría, J. E. (2006). RT-PCR for detection of all seven genotypes of Lyssavirus genus. *Journal of Virological Methods, 135,* 281–287.

Vos, A., Müller, T., Neubert, L., Zurbriggen, A., Botteron, C., Pöhle, D., et al. (2004). Rabies in red foxes (Vulpes vulpes) experimentally infected with European bat lyssavirus type 1. *Journal of Veterinary Medicine, series B, Infectious Diseases and Veterinary Public Health, 51,* 327–332.

Wacharapluesadee, S., & Hemachudha, T. (2001). Nucleic-acid sequence based amplification in the rapid diagnosis of rabies. *The Lancet, 358,* 892–893.

Wacharapluesadee, S., & Hemachudha, T. (2002). Urine samples for rabies RNA detection in the diagnosis of rabies in humans. *Clinical and Infectious Diseases, 34,* 874–875.

Wacharapluesadee, S., & Hemachudha, T. (2010). Ante- and post-mortem diagnosis of rabies using nucleic acid-amplification tests. *Expert Review of Molecular Diagnostics, 10,* 207–218.

Wacharapluesadee, S., Phumesin, P., Lumlertdacha, B., & Hemachudha, T. (2003). Diagnosis of rabies by use of brain tissue dried on filter paper. *Clinical Infectious Diseases, 36,* 674–675.

Wacharapluesadee, S., Ruangvejvorachai, P., & Hemachudha, T. (2006). A simple method for detection of rabies viral sequences in 16-year old archival brain specimens with one-week fixation in formalin. *Journal of Virological Methods, 134,* 267–271.

Wacharapluesadee, S., Sutipanya, J., Damrongwatanapokin, S., Phumesin, P., Chamnanpood, P., Leowijuk, C., et al. (2008). Development of a TaqMan real-time RT-PCR assay for the detection of rabies virus. *Journal of Virological Methods, 151,* 317–320.

Wacharapluesadee, S., Tepsumethanon, V., Supavonwong, P., Kaewpom, T., Intarut, N., & Hemachudha, T. (2012). Detection of rabies viral RNA by TaqMan real-time RT-PCR using non-neural specimens from dogs infected with rabies virus. *Journal of Virological Methods, 184,* 109–112.

Wakeley, P. R., Johnson, N., McElhinney, L. M., Marston, D., Sawyer, J., & Fooks, A. R. (2005). Development of a real-time, TaqMan reverse transcription-PCR assay for detection and differentiation of lyssavirus genotypes 1, 5 and 6. *Journal of Clinical Microbiology, 43,* 2786–2792.

Warner, C. K., Whitfield, S. G., Fekadu, M., & Ho, H. (1997). Procedures for reproducible detection of rabies virus antigen, mRNA and genome in situ in formalin fixed tissues. *Journal of Virological Methods, 67,* 5–12.

Whitby, J. E., Heaton, P. R., Whitby, H. E., O'Sullivan, E., & Johnstone, P. (1997a). Rapid detection of rabies and rabies-related viruses by RT-PCR and enzyme-linked immuno-sorbent assay. *Journal of Virological Methods, 69,* 63–72.

Whitby, J. E., Johnstone, P., & Sillero-Zubiri, C. (1997b). Rabies virus in the decomposed brain of an Ethiopian wolf detected by nested reverse transcription-polymerase chain reaction. *Journal of Wildlife Diseases, 33,* 912–915.

World Health Organization (WHO), (2005). *WHO expert consultation on Rabies, 2004.* Geneva: WHO. (First Report: WHO technical report series no. 931).

Young, C. C., & Sellers, T. F. (1927). Laboratory: A new method for staining negri bodies of rabies. *American Journal of Public Health (N.Y.), 17,* 1080–1081.

Zanluca, C., Aires, L. R., Mueller, P. P., Santos, V. V., Carrieri, M. L., Pinto, A. R., et al. (2011). Novel monoclonal antibodies that bind to wild and fixed rabies virus strains. *Journal of Virological Methods, 175,* 66–73.

Measures of Rabies Immunity

Susan M. Moore, Chandra R. Gordon, and Cathleen A. Hanlon

Kansas State University Rabies Laboratory, 2005 Research Park Circle, Manhattan, KS 66502, USA

1 INTRODUCTION

"Measurements are the basis of science.

Therefore the methods used to assess immunological parameters and immunity...... need to be critically reviewed.

Is the chosen parameter accurately measured, is it robust, is it a good correlate of protective immunity.....?" *(Zinkernagel, 2002)*

There are a number of variables to be considered in the assessment of immune status of an individual host or among a population. For a pathogen like rabies virus, important variables include the source and route of potential natural exposure. For vaccination, important variables consist of vaccine type, potency, and virus strain; vaccination route and schedule; and individual host factors. Although perhaps often overlooked, it is essential to have a basic understanding of the laboratory methods used to measure and assess the host's immune status. The precision, accuracy, sensitivity, and specificity of a method must be well defined. Moreover, an "adequate," acceptable, or diagnostic value for each method must be clearly defined so that a particular test result for a patient can be meaningfully interpreted in relation to the patient's history and clinical management. If these parameters are not clearly and objectively defined, conclusions based on test results from various methods may be inherently misleading. If a laboratory method such as the rapid fluorescent focus inhibition test (RFFIT) (Smith, Yager, & Baer, 1973) developed for measuring vaccine response in serum samples is applied for the analysis of biologic products such as human or equine rabies immune globulin (RIG) or rabies virus neutralizing monoclonal antibodies, the method will most likely need modifications and thus also subsequent method validation.

Serology is the study of the immunological properties of blood serum or other bodily fluids. For the most part, serology is the investigation of antibodies in serum, although assessment of immunity may be conducted on cerebrospinal fluid and other sources of fluid. Antibodies are produced by plasma cells which may be specifically activated in response to antigens, such as those from viruses and bacteria, to protect the host. The primary action of an antibody is to bind to antigen. The secondary or effector actions of antibodies include neutralization and opsonization of infectious agents, and activation of other immune mediators (see Figure 12.1). Complement activation and antibody dependent cellular cytotoxity (ADCC) are other effector functions that rely on the binding action of antibodies. Not all antibodies have effector actions. Some antibodies that bind to an antigen may not result in a biological effect because they are not effective in eliciting a secondary effect. Effector actions occur in accordance with the individual characteristics of a specific antibody structure and depend upon the class, subclass, or variable region of an antibody. In a competent host, exposure to an antigen will activate multiple immune cell clones and result in the production of a polyclonal antibody response.

Rabies virus specific antibodies are produced by the immune system in response to infection or vaccination *in vivo*, or by immune cells or molecular methods *in vitro*. The reasons for performing rabies serology can range from diagnosis of infection to investigation of epitope specificity of an anti-rabies virus glycoprotein monoclonal antibody. Characterization of an antibody's affinity, specificity, quantity, and neutralizing function, complement binding function, and class/subclass are achieved by various methods. Many serological techniques developed over the past five decades differ not only in their ability to detect the function, affinity, and specificity of rabies virus antibodies but also in the ease and practicality with which they are performed. To select an appropriate method and appropriately interpret test results, it is essential to understand the specific strengths, weaknesses, and limitations of available methods. Numerous reports indicate that protection against rabies is largely dependent upon the presence of rabies virus neutralizing antibodies (RVNA) (Dietzschold, 1993; Hooper et al., 1998). Thus, assays to detect and quantify RVNA, such as the rapid fluorescent focus inhibition test (RFFIT) (Smith et al., 1973) and the fluorescent antibody virus neutralization test (FAVN) (Cliquet, Aubert, & Sagne, 1998) are the methods recommended for quantitation purposes in rabies serology. Antigen binding assays have proven to be useful for the detection of specific

(A)

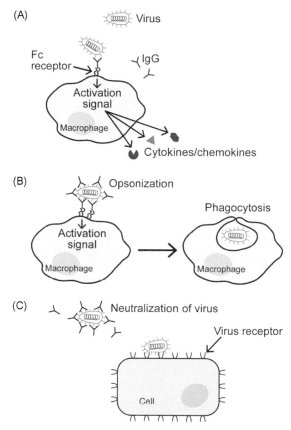

FIGURE 12.1 The effector functions of antibodies include: (A) activation of immune cells such as macrophages to produce cytokines and chemokines through Fc receptor binding; (B) Opsonization of infectious organisms induces phagocytosis of the organisms through Fc receptor binding; and (C) Neutralization of virus though binding of proteins used for attachment and entry of the virus, thereby blocking infection of the cell.

isotypes of rabies virus antibodies, either using whole virions or specific viral proteins as antigen(s). The decision to use a specific assay should start with the purpose of testing and the intended application of results. Other factors to consider are the assay complexity, degree of precision and/or accuracy, specificity and range of detection. In addition, the availability of laboratory materials, instruments, and safety equipment also must be considered. It is critical to understand exactly what aspect of rabies virus specific antibodies are measured, as well as the limitation of the assay, in order to select the best test and also interpret and use the test results in an appropriate manner.

Investigative serology focuses on the detection and measurement of immune components in blood (usually serum), including immunoglobulins of several subclasses directed against specific epitopes. Detection of IgM and IgG classes is dependent upon the time point in the course of the humoral immune response after exposure to an antigen. In the initial or primary antibody response, IgM is produced first in relatively low levels followed by higher levels of IgG after the occurrence of class switching. If the purpose of the assay is to detect the initial response, it should be designed to detect both IgM and IgG. The specificity of the immunoglobulin produced is driven by distinct epitopes present on the rabies viral proteins used to generate the antibodies. Exposure to rabies virus, whether through vaccination with inactivated virus or through exposure, induces the formation of antibodies potentially against all viral proteins, but predominantly against the rabies virus glycoprotein (G) and nucleoprotein (N). Studies of monoclonal antibodies (MAbs) capable of neutralizing rabies virus indicate that these MAbs are directed against a number of epitopes on the G of rabies virus (Tordo, 1996).

Rabies virus neutralization requires a minimum number of antibody molecules per G spike to induce steric hindrance of the virus-receptor-binding activity (Flamand, Raux, Gaudin, & Ruigrok, 1993). Another mechanism may involve conformational changes in the G protein, ultimately resulting in the loss of virion receptor-binding ability (Irie & Kawai, 2002). The humoral immune response elicited by rabies virus vaccination consists of a mixture of polyclonal antibodies that influences a variety of complex neutralization mechanisms.

Specific methods to detect antibodies specific for rabies viral antigens include precipitation, agglutination, immunoelectrophoresis, radioimmunoassay, enzyme-linked immunosorbent assays (ELISA), Western blots, indirect immunofluorescence, immunoelectron microscopy, and serum neutralization assays. All of these assays depend on an antibody–antigen interaction to detect the presence of an antibody. Two basic types of assays are used: 1) assays involving primary binding activity between antibodies and antigens, and 2) functional assays to measure neutralization actions of antibodies. Although other components and products of the immune system are involved, protection from clinical rabies after infection relies heavily on the presence of RVNA. Therefore, methods to detect and quantify antibodies that can functionally neutralize rabies virus are recommended to quantify the level of immunity after rabies vaccination.

2 HISTORY OF REGULATORY STANDARDS

Throughout history, various governmental entities have regulated drug products and related methods for human health assessment, including measures of rabies immunity in relation to diagnosis, qualification of biologic products such as vaccines and rabies immune globulins, and response to vaccination, particularly in at-risk groups. The primary function of regulation has been to ensure the quality and safety of drug products available for human use. Federal regulations began on a large scale in the early twentieth century when the U.S. Congress enacted the Pure Food and Drugs Act of 1906. The Food and Drug Administration (FDA) was born from this law and has oversight over products, marketing of food and drugs to consumers, and the manufacturing practices of food and drug industrial companies.

Most major regulatory standards throughout history arose from disasters. In 1976, Good Laboratory Practices (GLP) were created by FDA in response to a high percentage of studies with flawed data, falsified data, insufficient documentation, inappropriate testing facilities, and instances where experimental animals were subjected to inhumane conditions. In the 1970s, the Organization for Economic Co-operation and Development was created to ensure quality, integrity, and reproducibility of data, a global definition of GLP. Good Manufacturing Practices (GMP) was created in 1972 in response to the Davenport Disaster, in which six people died due to contaminated intravenous fluids. Good Clinical Practices (GCP) began at the end of World War II following inhumane experiments performed on humans, prompting the Nuremberg Code, which outlines the proper means to conduct research. These instructions were the foundation of GCP (Food and Drug Administration, 2009). Each standard has been updated continually to meet the growing demands of ensuring safety and the quality of drug research and clinical trials. Throughout the '70s and '80s, Japan, the UK, and other European countries had developed their own set of GCP guidelines. Different guidelines per country brought into question the validity of clinical trials performed in different countries. In 1996, conferences were held to unify each country's GCP codes of practice, resulting in the "International Conference on Harmonization of Good Clinical Practice" ICH-GCP (Baynes, 2005). The International Conference on Harmonisation of Technical Requirements for Registration of Pharmaceuticals for Human Use (ICH) (1996) *Guideline for Good Clinical Practice E6(R1).)* [Online]. Available from: http://www.ich.org/LOB/media/MEDIA482.pdf (Accessed: 08 October 2010.)

3 REGULATORY REQUIREMENTS

Research is the cornerstone of all scientific endeavors. It is a harnessing of curiosity to solve new or existing problems, prove new ideas, or develop new theories. Research is not defined or controlled. No formal regulatory requirements exist for scientific research. Primary regulations dealing with clinical laboratories like the Clinical Laboratory Improvement Amendments (CLIA) or the College of American Pathologists (CAP) specifically state that they do not have jurisdiction over research. The FDA does not provide regulations for oversight of research testing. Even without regulatory guidance, researchers can and do produce work of the highest quality. However, any technology such as: new test methods, new vaccines, or prophylactic drugs, that may be developed in an academic setting, and eventually transferred to the industrial world must meet criteria of quality in order to have any commercial potential. So can the "assurance of quality" needed for regulatory compliance be found within basic research?

The answer to this question is based upon an understanding of what "assurance of quality" is needed for a researcher to test at a level compliant with "good practices" as described for laboratory, clinical or manufacturing environments, and designated as GLP, GCP or GMP or implying any or all of the three by GxP, in which GxP represents the abbreviations of these titles, where x (a common symbol for a variable) represents the specific regulation. First, we must realize what is meant by "assurance of quality" as defined within a GxP environment. Second, we must understand when to apply the different GxP regulations within the research and development applied to the prevention or treatment of a disease.

The fundamental of any requirement or standard is a complete and operational quality system that will demonstrate that there is overall control of all aspects of testing and quality. Components of a quality system include but are not limited to: the study regulations or quality standards of the regulatory body the testing and standard operational procedures (SOPs), the organization infrastructure with roles and responsibilities (i.e., study director, QA unit, testing personnel, management), trained and competent testing personnel, personnel safety, physical facilities, validated/calibrated equipment, validated testing procedures, certified reference materials, quality control programs (proficiency testing, confirmatory testing, quality control samples), an internal audit program, laboratory information systems, and procedures for recording and archiving of data (Food and Drug Administration, 1996). These standards, if implemented within the laboratory, allow optimal laboratory operations to ensure consistent, reproducible, auditable, and reliable laboratory results of a sufficient quality for regulatory testing. Complete documentation is required to allow reconstruction of the events not only during and right after completion of the study but also 5–10 or more years later. A quality system must be intentionally included in the study or testing process before it starts, and monitored and documented during its performance. Quality cannot be created after the work has been completed.

Drug Development Stages

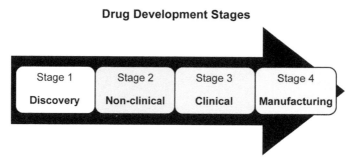

Stage 1	Stage 2	Stage 3	Stage 4
Discovery	Non-clinical	Clinical	Manufacturing

FIGURE 12.2   Drug development is delineated into specific stages, each with its own requirement of regulatory oversight.

With the importance of producing quality products and processes, and the increase in regulatory scrutiny of all aspects of product development and the manufacturing cycle, researchers have to understand what regulatory environment is applicable for each phase (see Figure 12.2) of the development of a product. Researchers need to have an organized approach to applying and/or combining the GXP functions.

In a rabies research environment, when would you need to be concerned with running experiments and testing in accordance with a GxP regulation? All information and data gathered during the basic research stage may be inadmissible in the regulatory findings if not obtained through a quality system in compliance with GxP regulations. The quality system in place while the research is being performed will dictate whether the information and data will be suitable for regulatory evaluation. As detailed in Table 12.1, the applicable regulatory requirements depend upon the type of testing that is conducted.

The intended use of the data to be generated will define the recommended regulatory level of testing for a product or drug. During the first few stages of drug or treatment studies, including basic research to drug discovery, oversight is not provided by any regulatory agency as shown in Figure 12.3. During the preclinical development stage, compliance with GLP is required (Good Laboratory Practise for Nonclinical Laboratory Studies, 2011). GLP is only relevant to nonclinical (human) testing and deals with the organization, processes and conditions under which these studies are planned, performed, monitored, recorded, and reported. The primary purpose of these (nonhuman evaluations) studies is generally safety testing for the drug and/or product. Clinical evaluation in humans would follow successful safety studies. During the human-testing phase, GCP is the basis for quality standards and regulatory compliance. GCP addresses source data for clinical trials and applies to human research studies where the rights, integrity and confidentiality

TABLE 12.1   Regulatory Requirements – Laboratory Testing

Type of Testing	Applicable Regulatory Standard
Drug Product	GMP
Drug Product in Animals	GLP
Human Specimens/Trials	GCP/CLIA
Nonclinical safety studies Development of drugs	GLP
Basic research	Nonregulated Testing
Studies to develop new analytical methods	Nonregulated Testing
Discovery of drug product	Nonregulated Testing
Discovery of disease	Nonregulated Testing

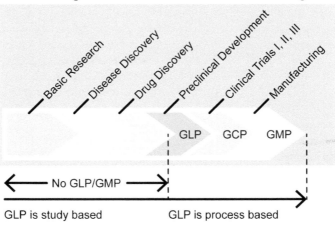

FIGURE 12.3   The intended use of the data to be generated will define the recommended regulatory level of testing for a product or drug.

of trial subjects are protected. Within the United States, GCP used to support human diagnostics and health care are established by Centers for Medicare & Medicaid Services (CMS) regulations and accrediting organizations, such as Clinical laboratory Improvement Amendments (CLIA). GCP must be instituted in all laboratory testing during the clinical trials from Phase I (to demonstrate tolerance of the test drug and to define human pharmacokinetics), through Phase II (where the dose–effect relationship is confirmed), to Phase III (full-scale, often multicenter, clinical efficacy trials in hundreds and thousands of patients). The final stage is manufacturing. Any laboratory testing preformed on the manufactured product, as well as the actual manufacturing of the product, will be completed in compliance with GMP standards. These regulations and instructions are covered in FDA, Code of Regulations Title 21.

Test method development is essential to ensure reproducible and defensible data, and thus, product quality. A test method must be developed and validated for use to analyze samples during the early development of a drug produced. As the drug development progresses from nonclinical study through clinical trials to commercialization, the test method will need to follow a similar progression. Throughout the process of drug research and test development, method validation and revalidation must be considered. Method validation is defined by all regulatory standards as to which components must be tested, the acceptance criteria for each component and the test method. GXP requirements do not specifically state what components of or to what degree method validation must be completed during the phases of drug production. The reasoning behind this is due to the understanding that drug products and test methods will evolve through the course of development. The test method's purpose should be linked to the type of trial and the drug's intended purpose. As an example (GLP and GMP), the FDA regulation 21 CFR 211.194(a)(2) specifically states that users of analytical methods in the U.S. Pharmacopeia/National Formulary (USP/NF) are not required to validate the accuracy and reliability of these methods, but merely verify their suitability under actual conditions of use. However, if testing within a clinical trial setting, CLIA or ICH regulations would be utilized depending upon the requests or needs of the company sponsoring the testing. CLIA specifically states how a test is to be validated for use. FDA regulation CFR Section 493.1253 Standard: Establishment and Verification of Performance Specifications states the performance characteristics: accuracy, precision, analytical specificity, analytical sensitivity to include interfering substances, reportable range of test results for the test system, verify the reference intervals (normal values), determine calibration, and control procedures and documentation of all activities specified above (Anonymous, 1992). In ICH Q7A: "Changes are expected during development, and every change in product, specifications, or test procedures should be recorded adequately" (ICH, 2001). Above all else, method validation must prove the test method is fit for the purpose of the testing to which it is applied. The performance characteristics such as sensitivity, specificity, accuracy, precision, lower limit of detection, linearity, and reproducibility for each testing method must be analyzed. It is the laboratory's responsibility to prove suitability or competency of the test method 'in house' before and during testing. It is possible that a method that functions satisfactorily in one laboratory fails to operate in the same manner in another. It is considered unacceptable for the researcher/laboratorian to use a published 'validated method' without demonstrating their capability in the use of the method (AOAC International, 2002).

The extent of and expectations from early phase method validation are lower than the requirements in the later stages of development. The validation exercise becomes larger and more detailed and collects a larger body of data to ensure that the method is robust and appropriate for use at the commercial site. The final method must be validated for the intended use whether it is to test the actual product or test the product in use. During the research and testing process, when is a partial or full method validation needed? Logically, any assay—whether it is fully or partially validated—must be scientifically sound, suited for its intended purpose and stage of product or drug development, and capable of generating reliable results. Any method that is newly developed must always be fully validated first prior to being used within a laboratory.

To adapt a quality system within the research stage, the technician performing the testing must have the scientific and technical understanding, knowledge of the product (drug, vaccine, etc.), and the ability to execute the quality functions of analytical method validation. All technicians performing the testing must have appropriate training to promote an understanding of the testing principles involved with the method validation, proper documentation of the data as well as understanding of how to interpret the data, and an understanding of the cross-functional relationships of the testing, product, companies, and patients. This means that fundamental quality system components must be applied at the bench level.

4 ASSURING QUALITY RESULTS

The only way to know if a method has the performance characteristics that "fit the purpose" for which it will be used is to define the test method through validation. Just as there are consequences for selecting an unsuitable assay for a given purpose, erroneous conclusions can be made if the capabilities and limitations of the assay are not considered when interpreting results. A method with acceptable accuracy and precision levels for measurement of antibodies in a potency range of 0.1 to 10.0 IU/mL in a serum matrix cannot claim the same accuracy and precision levels for higher potency samples or samples in a different matrix or body fluid without validation experiments to evaluate these adaptions of the method. The method parameters important for a qualitative assay are sensitivity, specificity, and predictive value. In addition to sensitivity and specificity, a quantitative assay requires definition of accuracy (closeness to the true value), precision (repeatability of the measure), linearity, and reportable range. Recent publications are examples of RFFIT or modified RFFIT validations performed for specific purposes—evaluation of clinical trial samples for a human monoclonal antibody combination for the post-exposure treatment of rabies and vaccine potency evaluation (Kostense et al., 2012; Kramer, Bruckner, Daas, & Milne, 2010). Robustness evaluation describes the ability of the method to perform to set criteria during normal laboratory conditions, including normal variations of equipment performance, reagent lots, or between different personnel. Biologic variation must be considered separately from analytical variation. For example, two test results from the same sample may vary solely on the basis of the receptivity of cells to virus infection; cells used last week may have different virus infectivity characteristics than the cells used in subsequent testing. The variation from these types of factors is separate from other sources of variation. For repeat measures of the same sample, there are statistical tools to set the expected variance and for determining what variance is evidence of a significant difference, such as minimum significance ratio (MSR) (Khan & Findlay, 2009). Measurement or detection of rabies virus antibodies can be influenced by interference. Interference can be caused by cross reacting antibodies, nonspecific binding, and matrix effect (hemolysis, lipemia, or "dirty" samples, etc.). Interference can occur not only with the antibody of interest in the sample but also in the interaction of the detected or competing antibodies in the assay. Naturally occurring proteins in samples, such as albumins, fibrinogen, and complement factors, can result in assay interference (Selby, 1999). Results from samples with interfering factors can be misleading if the effects of these interfering factors are not considered. In most cases, interference will occur at low levels and will not cause measurement problems at higher dilutions or in samples with high potency, since specific binding is stronger than the weaker interference reaction. When interference is suspected or needs to be ruled out, samples may be evaluated by an alternative method in which the effect of interfering factors is minimized so that specific activity may be detected and measured.

The lower limit of detection (LOD) is affected by interference and the assay parameters. If the purpose of testing is determination of the presence or absence of rabies virus antibodies, it is critical to define the lowest level of antibodies that an assay can reliably detect. But if the ability to accurately and precisely measure low levels of rabies virus antibodies is important, as in evaluation of passive rabies virus immunoglobulin levels in post-exposure treatment, then defining an assay's lower limit of quantitation (LLOQ) is required. Cut-off values assigned to an assay depend both on the LOD or LLOQ and the purpose of testing. If the application of the rabies serology testing is to identify low levels of

TABLE 12.2   "Fit for Purpose" Method Variations that can be Applied to
Neutralization or Antigen Binding Assays

Neutralization Assay Variables	Antigen Binding Assay Variables

• Strain of challenge virus
• Dose of challenge virus
• Cell type
• Serial dilution scheme
• Detection system Variables
   ◦ Fluorescent-labeled antibody
   ◦ Enzyme-labeled antibody
   ◦ Modified challenge virus
     (e.g., Green Fluorescent Protein
     insert)
	• Source of Antigen–Selection of
	  virus strain
	• Type of Antigen–virus protein(s)
	   ◦ Whole virus
	   ◦ Purified protein
	• Platform – slides, plates, or beads
	• Detection system variables
	   ◦ Species specific or non–species specific
	   ◦ Immunoglobulin specific for class or
	     subclass

rabies virus antibodies and exclude false negative test results, the cut-off level should be low, but this may yield some false positive test results. Conversely, a higher cut-off value (i.e., above a level which might allow some false positive test results) would identify only true positive test results which would be acceptable if the purpose of the method is to reliably identify only those individuals. The trade-off is that a high cut-off level would increase the number of false negatives (i.e., exclude some true positives that are low). The probability of false positive and false negatives is related to the precision of the assay. Assays with a high variability particularly at the cut-off level would exclude some true positive samples with potency values close to the cut-off level and conversely identify some true negatives as positive. Upon repeat testing, these samples could measure either positive or negative.

The matrix of the sample can affect the LOD and LLOQ for a specific method, therefore whenever the sample matrix is altered; re-evaluation of this parameter is required. Any change that impacts the sensitivity of an assay will also change the LOD. Indeed, any change in the procedure or sample may require re-validation to determine the effect on the established performance characteristics. Method variations listed in Table 12.2 (for either binding assays or neutralization assays) can be used to customize a method for certain purposes, such as measurement of antibodies from a particular species, or within a range of potency values, but the changes implemented to customize an assay may also result in changes in the performance characteristics of a method.

Immunity can be measured by different methods. It is natural to compare the results from different methods, and it is important to consider how the comparison is made. Although it is very common to evaluate agreement between methods by a correlation coefficient, conclusions based on this value are improper. According to at least one well-known medical statistician, the best way to conduct a method comparison is to calculate the "mean and standard deviation of the between-method-differences" (Altman, 1991). It is not enough to just generate and examine the data, it is essential to apply appropriate statistical tools. A functional understanding of statistics or collaboration with a statistician is often essential for these exercises. The application of statistics to evaluations of immunoassay performance is a specialized area of competence, and it is of particular importance when the assay will be used to determine acceptance of biologics (Findlay et al., 2000).

As previously mentioned, there are some critical components that are essential to consider, identify, and control to ensure precise and accurate measurements for serum neutralization assays, such as strain and dose of challenge virus, cell type, and reference standards. For results to be comparable over time from the same laboratory and possibly between laboratories, these components must be standardized. Whenever any critical steps or components are changed as may be necessary for a specific purpose, the modified method will require re-validation. A standard reference rabies immune globulin serum (SRIG) provides a defined potency standard in international units per mL (IU/mL). By comparison of the SRIG result to the test sample result, the value of the test sample is standardized and comparable. But if the SRIG used is not the same or not calibrated against a known standard, discrepant results can occur (Yu et al., 2012). The value of the test sample is standardized through comparison with the SRIG result in that assay at that time. It is essential for the SRIG to be precisely described for each batch of test results. If the standards are not identical or not calibrated against a known standard, results cannot be directly compared between assay runs and between laboratories. Standard reference serum of equine source may perform differently than human SRIG such that batches of test results will yield different values depending upon the control serum (Haase, Seinsche, & Schneider, 1985). The potency assigned to a SRIG by one method may not be the same in a different method and cannot be automatically assumed. For example, a control serum at 0.5 IU/mL in the RFFIT method may perform at 0.7 Equivalent units per mL in an ELISA-based assay. If this standard with a known performance by RFFIT was directly applied as the standard for an ELISA and assumed to perform at 0.5 EU/mL, the ELISA results would bias toward the exclusion of some samples which might meet a RFFIT 0.5 IU/mL value. Control serum, such as an SRIG, needs to be fully characterized by a new method and its potency needs to be assigned in units applicable to that particular method (Moore & Hanlon, 2010). A comparison of two international SRIG products in current use, WHO 1st international rabies immune globulin and WHO 2nd international rabies immune globulin, over several years show that reference serum can lose potency over time, with the 1st RIG lower in potency

TABLE 12.3   Comparison of the WHO[a] International Standard Anti-Rabies Immunoglobulin, Human – 1st and 2nd

Laboratory[b]/Year	Difference in Potency (WHO 1st/WHO 2nd)	Method of Testing:
FDA/CDC/KSU		
1997	2.5% lower	RFFIT
KSU		
2006	12% lower	RFFIT
2006	36% lower	FAVN
2006	22% higher	Direct ELISA
2012	19% lower	RFFIT
2012	34% lower	FAVN
2012	15% higher	Direct ELISA

[a]WHO – World Health Organization.
[b]FDA – USA Food and Drug Administration; CDC – US Centers for Disease Control and Prevention; KSU – Kansas State University Rabies Laboratory.

by 2.5% in 1997 to 19% lower in 2012 by RFFIT, yet higher in potency by ELISA (see Table 12.3). This illustrates the importance of calibration and monitoring of the RIG in use in a particular laboratory and for a particular assay. If the challenge virus of an assay is substantially different than the virus source for a vaccine, the serologic results from clinical trials may underestimate responses to the vaccine (Brookes & Fooks, 2006; Moore, Ricke, Davis, & Briggs, 2005). The same is true for antigen binding assays where the virus strain and type (whole or protein) used in the detection system should ideally be the same in order to obtain the most informative results.

Despite the potential negative effect of a change in how a method is performed, there are good reasons to introduce variations to a procedure. These may include the need to measure rabies virus antibodies from a specific species which may require a change in the detection system or the need to measure potency of samples which are beyond the normal linear range of testing and hence may require a pre-dilution to achieve a different range of sample dilutions to be tested. Method validation reveals the robustness and limitations of assay and its performance characteristics. In addition to method validation, conducting continual monitoring of method performance increases the chances that potential problems will be quickly identified. Regular participation in proficiency programs is one way to monitor performance of the method and also assists in the identification of drifts and trends. If good quality control practices are in place, results may be comparable between laboratories even when there are differences in procedure. For example, nine laboratories performing the RFFIT for different purposes (from +/− screening to regulated quantitative measurements) and executing different RFFIT procedures (including SRIG source, virus strain, cell type used) recently participated in a voluntary exchange of an informal proficiency panel of samples. All nine laboratories identified the RVNA negative sample as below their assigned cut-off and had reported values within twofold of the average of all measurements for the remaining five RVNA positive samples. Even if laboratories are following the same protocol and using the same components, the agreement in results for the same sample can vary based on method variables related to environment, personnel training, and equipment performance. In other words, methods can be standardized, but unless the laboratories are adhering to the same quality assurance standards, the results may still demonstrate greater variability than is ideal. Acceptance criteria for precision and accuracy are different depending upon on the type of assay. Cell-based assays, such as serum neutralization, are inherently more variable and thus are allowed greater variability than binding assays. The precision of binding assays is generally expected to be in the range of 5–20%, while cell-based assays may be allowed a precision variability of 30%, and up to 50% (Center for Drug Evaluation and Research, 2001; Chaloner-Larsson, Anderson, & Egan, 1997). In general, for serological titration assays a two-fold difference in replicate measurements is commonly recognized as the upper level of reproducibility (Wood & Durham, 1980). The precision of an assay should be taken into account when reviewing rabies serology results in relation to survival of experimental challenge, inter-laboratory comparisons, and proficiency testing, as well as when establishing acceptable levels for proof of sero-conversion or an adequate response to rabies vaccination.

5 ASSAY SELECTION

One must ask the right question of a method in order to obtain a valid result. Similarly, it is essential to understand the performance characteristics of a particular method and its limitations, in order to determine whether the method can generate appropriate results that can answer a particular question. Rabies virus antibodies are measured for several different reasons. The reasons for these measurements will influence the requirements for method sensitivity, specificity, precision, accuracy, linear range, limit of detection, and robustness of the method. The requirements for specific reagents, instrumentation, or facilities may vary according to the particular methods. The consequence of selecting an improper method can be as simple as getting a result that does not answer an academic question, thus leading the research down a wrong path, or as complex as providing incomplete or misleading information that will be used to make essential health care decisions, whether veterinary or medical, for the prevention of clinical rabies. For example, if there is an encephalitis suggestive of clinical rabies, evaluating a sample from a human or animal with an assay that can only detect IgG antibodies could be insensitive or misleading because IgM antibodies, which are produced before IgG, may remain undetected. Thus, a negative test result would be misleading. Besides the consequence of using an unsuitable method for individual diagnosis, ambiguous results add potentially incorrect information to the body of data compiled for typical antibody responses in rabies patients.

Laboratory tests for rabies virus antibodies are used for research, human vaccination decisions, pet travel permits, wildlife vaccination program evaluations, and pharmaceutical product licensure. No one method will be the ideal fit for all purposes. The method that will "fit" must be defined by the characteristics of rabies virus antibodies that are most important or by the parameter of interest. For example, to research the difference between monoclonal antibodies produced against the glycoprotein of the ERA rabies virus strain, a serum neutralization assay is essential if the ultimate purpose of the monoclonal is therapeutic. The challenge virus used in the serum neutralization assay should be considered. If the purpose of the monoclonal antibody is use as a therapeutic agent, then the challenge virus used should be one that is most closely related to the rabies virus variants that are enzootic in the regions where the biologic is intended for use. Moreover, if the monoclonal antibody is intended for eventual licensure, the laboratory method selected must be an approved, validated method that is recognized by the licensing authority. Conversely, if the purpose of the monoclonal antibody is used in diagnostic testing to differentiate ERA infected brains from brains infected with other strains, then the method best used to illustrate the difference in monoclonal antibodies would be an IFA using ERA infected cells. Below are specific assay requirements that apply to some common reasons for measuring rabies virus antibodies:

- Standardized for comparable results between laboratories and over time (clinical trial testing, human testing for vaccine response either for post-exposure or pre-exposure, oral-bait program evaluation, pet travel)
- Detection of low levels in an initial response to infection or vaccination/ability to measure low levels of IgM and IgG (clinical diagnosis, evaluation of post-exposure treatment, some research purposes)
- Cost effective (to obtain screening results from large numbers of samples)

- Adaptable for detecting difference immunoglobulin subclasses (research and clinical)
- Adaptable for detecting specificities or antibodies from different species (research and surveillance)
- Approved by regulatory authorities (biologic product testing, pet travel)
- Low technology or low level bio-containment facilities (field research or in developing countries)

Consideration of the sample is also a factor in choosing the proper assay. Attempting to measure a F(ab')$_2$ product with an ELISA whose secondary detection system relies on binding to the Fc portion of the immunoglobulin would be futile. A blocking or competitive ELISA or a serum neutralization assay would be a better 'fit' for this purpose because these assays do not rely on the complete structure of the immunoglobulin, only the antigen binding portion, F(ab')$_2$ for detection. In human laboratory medicine, it is not uncommon to screen samples using a sensitive assay to identify positive samples from negative samples and then follow the sensitive screening tool with confirmatory testing with a more specific assay to identify the true positive samples and exclude the false positives. Several methods can be used effectively for screening purposes. Depending on the screening goal, assays such as ELISA using whole virus antigen, lateral flow with a positive or negative readout, and IFA can identify samples that potentially contain rabies virus specific antibodies. Testing with a Western blot technique can confirm the specificity of the antibodies detected in the screening assay or testing with a serum neutralization (SN) method can confirm the neutralizing function of the antibodies. A screening method with lower accuracy (result may not be the true value), but higher precision (repeat measurements are clustered closely although they may not be near the true value) may be more useful for oral baiting surveillance—if it is quick, standardized and simple—than a more accurate method that is cumbersome, time consuming, and more variable. For the purpose of evaluation of oral baiting campaigns, determination of individual "protection" is less important than herd immunity levels and the ability to confidently compare results between laboratories and over time.

6 SERUM NEUTRALIZATION ASSAYS

Rabies serum neutralization assays are distinguished by the ability to detect the neutralization activity of specific antibodies *in vitro* and therefore attempts to measure the potential protective action of these antibodies *in vivo*. Technical performance of rabies virus neutralization assays requires the use of infectious virus and can be labor intensive and time consuming. There are two rabies serum neutralizing assays recognized by the World Health Organization (WHO) and the World Organisation for Animals Health (OIE) to measure RVNA: the rapid fluorescent focus inhibition test (RFFIT), described in 1973 by Smith et al. (1973); and the fluorescent antibody virus neutralization test (FAVN), developed in 1997 by Cliquet et al. (1998). Measurement of RVNA by serum neutralization assays is based on the same principle as the mouse neutralization test (MNT), extensively employed in early rabies serology work. The MNT involves the injection of test serum dilutions in mice followed by a challenge with a standard dose of rabies virus, with the read-out being mortality among the mice (Atanasiu, 1973). Although this is truly a "real" measurement of the protective function RVNA in the serum, the biological variation of individual mouse immunity, as well as possible interference of other immune effectors, made it difficult to standardize the test. With the development of *in vitro* methods such as the RFFIT, improvements in sensitivity and standardization were achieved.

Both the RFFIT and the FAVN tests consist of incubation of dilutions of heat-inactivated serum with a fixed amount of live rabies virus for 60–90 minutes at 37°C. Measurement of residual virus infectivity is accomplished by detection of virus in cell culture via a labeled antirabies virus antibody and subsequent calculation of the quantitative titer by the number microscopic fields containing virus infected cells. The RFFIT method is conducted in multi-chamber slides (see Figure 12.4A). Serum is serially diluted fivefold and tested in each well. Variations of the RFFIT include the use of microtiter plates in place of the slides and the use of twofold or threefold dilutions. The rabies challenge virus should contain 30–100 50% tissue culture infective dose ($TCID_{50}$). After the virus is added to the diluted serum, the slides are incubated at 37°C for 90 minutes, after which baby hamster kidney (BHK) or mouse neuroblastoma (MNA) cells are added to each of the wells. Diethylaminoethyl-Dextran (DEAE-Dextran) has been used, typically at a 0.01 μg/mL concentration, in some variations of the RFFIT to enhance susceptibility of the cells to rabies virus infection (Kaplan, Wiktor, Maes, Campbell, & Koprowski, 1967). The slides are generally incubated at 37°C in a 2–5% CO_2 incubator for 20–24 hours, although the incubation period is extended to 48 hours in some variations of the method conducted in microtiter plates. The wells containing an adherent monolayer of cells are washed, and the cells are fixed with 80% cold acetone. FITC-conjugated anti-rabies virus antibody directed against the rabies virus nucleoprotein (N) is added in order to detect virus-infected cells. In 8-well chamber slides, 20 fields of each well are examined using a fluorescent microscope for the presence of fluorescence in the cells, an absence of which indicates antibodies in the sample neutralized virus and the presence of which indicates a lack of antibodies. The titer of RVNA in the serum sample being analyzed is defined as the dilution at which 50% of the observed microscopic fields contain one or more infected cells. Mathematical calculation using the Reed and Muench formula, Spearman-Karber formula or Probit method will determine the exact quantitative titer of RVNA in the serum sample. Alternatively, the quantitative titer of RVNA can be more simply defined, but with less precision, as the highest serum dilution where 100% viral inhibition occurred, thus indicating that there were no infected cells at that dilution and all subsequent higher dilutions exhibit infected cells (Aubert, 1996; Habel, 1996). Transcribing a serum dilution value into a standardized and more globally recognized measure of IU/mL is achieved by a simple calculation wherein the value from a serum sample being tested is compared to the serum dilution value of a reference serum standard containing a specific amount of RVNA previously tested and verified to be accurate (Velleca & Forrester, 1981). The quality of the test components as well as the skill and expertise of the technician conducting the test, including the analysis of the microscopic readout, can substantially affect the precision of RFFIT test results.

To simplify and reduce the subjectivity of the microscopic counting step, the FAVN method uses four replicates of serum using threefold dilutions in microtiter wells and scores each well as either positive or negative for the presence of rabies virus infected cells after a 48-hour incubation. A direct comparison of the two methods demonstrated no statistically significant differences in results when conducted in a laboratory adhering to good quality assurance standards (Briggs et al., 1998). Precision and repeatability of virus neutralization test results can be controlled by strict adherence to the dose and strain of the challenge virus used and the source of the standard reference serum. Early published reports that compared different laboratory RFFIT results reported that the use of a high infective dose of challenge virus resulted in reduced sensitivity for testing low-titered sera, whereas a low viral dose of challenge virus could result in lower precision when testing high titered sera such as rabies virus immunoglobulin (RIG) preparations (Fitzgerald, Baer, Cabasso, & Vallancourt, 1975). In addition, the use of an equine RIG as the reference standard to determine IU/mL values resulted in significantly different titer results than when a human RIG reference standard was used (Lyng, Bentzon, & Fitzgerald, 1989). Measuring RVNA from people vaccinated with a vaccine prepared with a parent virus strain heterologous to the challenge virus strain in the RFFIT (usually CVS-11) can result in lower titers than if a homologous challenge strain is used (Moore et al., 2005). Rabies virus neutralization tests identify the presence of all classes of immunoglobulin in a sample (both IgM and IgG) and therefore will be able to detect the early production of rabies virus antibody after exposure or vaccination, but they may not be as sensitive as the IFA (Smith, 1991).

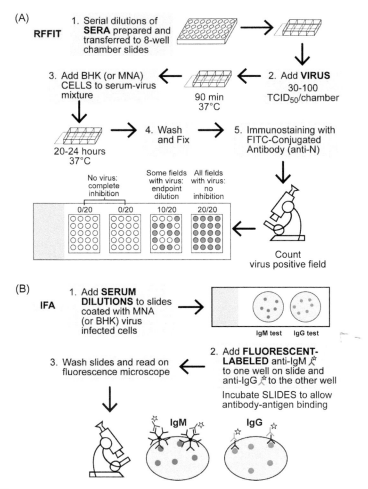

FIGURE 12.4 **(A) RFFIT procedure.** Serum is serially diluted in a 96 well plate and transferred into 8-well chamber slides. The rabies challenge virus is added to the diluted serum, the slides are incubated at 37°C for 90 minutes, after which baby hamster kidney (BHK) or mouse neuroblastoma (MNA) cells are added to each of the wells. The slides are incubated at 37°C in a 2-5% CO_2 incubator for 20–24 hours. The wells containing an adherent monolayer of cells are washed and the cells are fixed with 80% cold acetone. FITC-conjugated anti-rabies virus antibody directed against the rabies virus N is added in order to detect virus-infected cells. In 8-well chamber slides, 20 fields of each well are examined, using a fluorescent microscope, for the presence of fluorescence in the cells indicating the presence of non-neutralized rabies virus. The titer of RVNA in the serum sample being analyzed is defined as the dilution at which 50% of the observed microscopic fields contain one or more infected cells. **(B) IFA technique.** Test serum in added to slides fixed with rabies virus-infected cells. Rabies virus antibodies in the serum bind to antigens on rabies virus proteins present in the infected cells and are subsequently detected by FITC-labeled anti-IgG or anti-IgM. Slides are read on a fluorescence microscope to evaluate the slides for the presence of labeled antibodies. **(C) Competitive ELISA.** A labeled rabies virus antibody competes with the rabies virus antibodies in the test sample. The test serum sample and

(*Continued*)

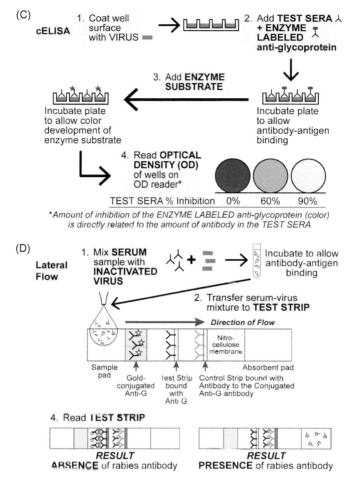

TEST SERA % Inhibition   0%   60%   90%

*Amount of inhibition of the ENZYME LABELED anti-glycoprotein (color)
is directly related to the amount of antibody in the TEST SERA*

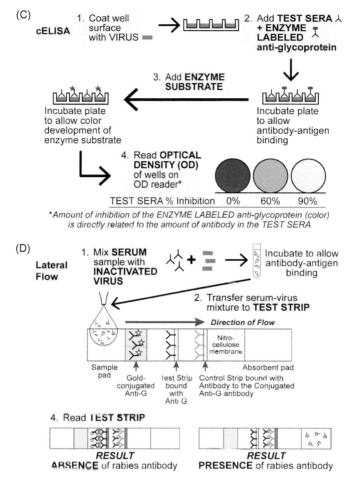

> an enzyme-labeled anti-rabies virus antibody in incubated with the inactivated rabies virus antigen on the well surface. The amount of enzyme-labeled antibody is detected by adding a conjugate for color development; amount of rabies virus antibody in the serum sample is inversely related to intensity of color development. The level of antibody in the test serum can be quantitated by use of a standard curve and an OD reader. (**D**) **Lateral Flow.** The test serum sample is mixed with inactivated virus before adding the mixture to the absorbent material of the test strip. The mixture then flows across to encounter a labeled anti-rabies virus (anti-G) antibody which will bind to any unbound inactivated rabies virus. The mixture continues to flow toward two areas (lines) of the strip bound with detection antibodies; the first detection antibody is specific for the rabies virus (anti-G), and the second detection antibody is specific for the labeled rabies virus (anti-G) antibody. If the mixture contains unbound inactivated rabies virus there will a color development at the first strip, indicating the sample did not contain enough anti-rabies virus antibody to bind the inactivated rabies virus in the mixture. If the test serum sample does contain anti-rabies virus antibody, then the labeled antibody will pass the first strip and be bound by the antibody in the second strip causing a color development at the second strip. The results will be either "presence of rabies virus antibody" (only second strip visible) or 'absence of rabies virus antibody" (first and second strip visible).

Because the virus neutralization testing method depends on the measurement of residual, or 'non-neutralized' rabies virus infecting the cells, the presence of interference factors in the sera or culture media that adversely affects cell (and ultimately viral) growth will mimic virus neutralization by nonspecifically inhibiting viral growth. Any inhibition of viral growth not directly due to neutralizing antibody will give a false positive result. As with ELISA techniques, some steps of the virus neutralization test can be automated, especially when performed using micro-titer plates, for example, the addition of media to the plates, serial dilution of the serum samples, addition of virus and anti-rabies virus conjugate, as well as some of the more tedious steps such as plate washing, allowing the application of these methods to high-throughput testing. Automated reading of FAVN and RFFIT reduces the work time required for the microscopic analysis readout and aids in the minimization of errors (Peharpre et al., 1999). No significant difference was noted when comparison was made between the rabies virus antibody levels reported by automated and non-automated reading, but the automated reading resulted in lower sensitivity. The expense of the equipment required to conduct automated reading of the RFFIT or FAVN, the requirement for a consistent cell monolayer and need for a good quality FITC conjugate limits the practicality of this enhancement, especially for laboratories that do not conduct large numbers of tests. As an alternative to microscopic fluorescence measurement, a microneutralization test (RAMIN), and the indirect immunoperoxidase virus neutralization (IPVN) technique and the modified FAVN, employ a mouse anti-rabies virus antibody and a peroxidase anti-mouse conjugate enabling automated reading by a spectrophotometer (Cardoso, Silva, Albas, Ferreira, & Perri, 2004; Hostnik, 2000; Mannen et al., 1987). In each of these studies, a good correlation was confirmed between traditional rabies virus neutralization methods and the modifications that were made to each test. Other modifications take advantage of molecular techniques to prepare recombinant viruses to use in place of the standard challenge virus, CVS-11, for standardization of fluorescence or for adaptability to detect different specificities of antibodies. Modified CVS-11 expressing green fluorescent protein (GFP) eliminates the need for FITC-conjugated anti-rabies virus antibody (Khawplod et al., 2005). Using this modified CVS-11 in combination with flow cytometry to detect residual virus present after incubation with serum, reportedly increases the sensitivity because each cell is individually assessed for viral infectivity, creating a more precise percentage of viral inhibition (Bordignon et al., 2002). A new method developed by Wright et al. (2008) utilizes pseudotype viruses—lentivirus vectors expressing rabies virus glycoprotein and a reporter (e.g., *lacZ*, GFP) (Wright et al., 2008). By expressing the glycoprotein from different rabies virus strains, a panel of pseudotypes can be used in cross-species comparison studies. Because the pseudoviruses are replication incompetent particles, this method is applicable in areas where high level biocontainment facilities are not available.

7 BINDING ASSAYS

Binding assays are methods that detect or measure immunoglobulin molecules by their ability to bind specifically to their target antigen. This binding can be detected by use of secondary detection systems usually bound to a color development system for visualization or quantitation by optical density (OD) or fluorescent measurement. ELISA assays are the most commonly used binding assay. ELISA assays may be based on indirect, competitive, and blocking approaches. Other binding assays are lateral flow and indirect fluorescent antibody (IFA). Western blots are used to identify the fine specificities of antibodies. Antigen binding assays such as ELISAs and IFAs are rapid, simple, and often do not require manipulation of infectious rabies virus during the assay, although preparation of antigen may involve live virus. These assays rely on the interaction of the antibody and antigen, regardless of the ability of the antibody to neutralize rabies virus, and they are useful for the detection of rabies virus binding antibodies. An assay with whole virus as the target antigen may be useful to identify the presence of rabies virus antibodies specific for the different antigens on the rabies virus. Conversely, purified viral proteins can be used to distinguish the specific composition of antibodies that may be present. Binding assays are able to identify the subclass of rabies virus antibodies, for example, IgM and IgG, by using a conjugated anti-subclass Ig antibody as the secondary antibody.

The IFA technique involves adding test serum to slides fixed with rabies virus-infected cells (see Figure 12.4B). Rabies virus antibodies in the serum bind to antigens on rabies virus proteins present in the infected cells and are subsequently detected by FITC-labeled anti-IgG or anti-IgM. A fluorescence microscope is required to evaluate the slides for the presence of labeled antibodies. Quantification of the antibodies can be accomplished by serial dilution of the serum to determine the antibody titer. Because infected cells are used as the source of rabies virus antigens, the potential exists for both antibodies with specificities to rabies virus antigens and to cellular antigens to be detected. The possibility of detecting antibody binding that is not specific to rabies virus antigen (i.e., autoantibodies, antibodies to cellular antigens, etc.) is important to consider when evaluating the results.

Early ELISA methods, such as the one described by Nicholson and Prestage in 1982 used inactivated whole virus as the antigen and anti-IgG as the secondary antibody (Nicholson & Prestage, 1982). This technique offered greater specificity over the IFA because the only source for binding is whole virus coated to the surface of the well, not in a cell where there are other antigens present as possible targets for interfering antibodies. Modification of the ELISA by Grassi in 1989 improved the assay specificity further by using purified rabies virus glycoprotein (G) antigen allowing detection of anti-rabies G antibodies (Grassi, Wandeler, & Peterhans, 1989). In contrast to early ELISA assays, there is a higher degree of correlation between RFFIT and G-protein ELISA because most neutralizing antibodies are directed against the G protein. The secondary antibody employed by ELISA methods may be species specific, but if staphylococcus Protein A is employed, the method can be applied to samples from a number of species because Protein A binds to the Fc portion of the IgG of many species.

Other types of ELISAs include competitive (cELISA) and blocking ELISA. Both methods involve the use of a labeled rabies virus antibody that is used to either compete with (cELISA) or to detect antigen not blocked by (blocking ELISA) the rabies virus antibodies in the test sample. In the blocking ELISA, incubation of the serum sample with the inactivated rabies virus antigen on the well surface is followed by addition of an enzyme-labeled anti-rabies virus antibody. Any unbound (unblocked) inactivated rabies virus is bound by the labeled antibody. The amount of enzyme-labeled antibody is detected by adding a conjugate for color development; the amount of rabies virus antibody in the serum sample is inversely related to intensity of color development. The level of antibody in the test serum can be quantitated by use of a standard curve and an optical density (OD) reader. Competition for rabies virus binding between the anti-rabies virus antibodies in a serum sample with a labeled anti-rabies virus monoclonal antibody is the basis of the cELISA (see Figure 12.4C). Similar to the blocking ELISA, the labeled antibody is measured and used to determine the level of anti-rabies virus antibody in the sample. Both of these methods can reduce the effect of nonspecific binding because the antibody that is measured is a purified reagent antibody. Use of different reagent monoclonal antibodies in these assays allows detection of various specificities of rabies virus antibodies that may be present in the sample (i.e., use of a labeled anti-rabies virus N will detect anti-N in the sample and use of a labeled anti-rabies virus G will detect anti-G in the sample).

An electrochemiluminescent (ECL) adaption of the blocking ELISA method employs microtiter plates fitted with a series of electrodes at the bottom of the wells. Applying an electrical current across the electrodes causes the generation of a luminescent signal by the chemical energy ligand-binding reactions. Quantitation of the signal converts the measurement to antibody concentration. This method has been applied to measurement of proteins and has the potential for greater sensitivity and faster results compared to the traditional ELISA method (Guglielmo-Viret, Attree, Blanco-Gros, & Thullier, 2005; Ma, Niezgoda, Blanton, Recuenco, & Rupprecht, 2012).

Lateral flow immunoassays to detect and measure antibodies are useful for fieldwork and in areas where a low tech screening method is required. By adapting the concepts of ELISA to an absorbent test strip, the testing process is simplified to progress across a straight line by having the sample interacting with the assay reagents sequentially (see Figure 12.4D). A version of this method for detecting rabies-virus antibody requires an initial step of mixing the serum sample with inactivated virus before adding the mixture to the absorbent material of the test strip. The mixture then flows across to encounter a labeled anti-rabies virus antibody that will bind to any unbound inactivated rabies virus. The mixture continues to flow toward two areas (lines) of the strip bound with detection antibodies; the first detection antibody is specific for the rabies virus, and the second detection antibody is specific for the labeled anti-rabies virus antibody. By using this design, if the mixture contains inactivated rabies virus bound with labeled anti-rabies virus antibody, there will be a color development at the first strip, indicating the sample did not contain enough anti-rabies virus antibody to bind the inactivated rabies virus in the mixture. If the sample contains anti-rabies virus antibody, then the labeled antibody will pass the first strip and be bound by the labeled anti-rabies virus antibody, causing a color development at the second strip. The results will be either positive (only second strip visible) or negative (first and second strip visible) for the presence of anti-rabies virus antibody in the test serum sample. The level of antibody to define positive or negative is set and defined by the design of the assay and can only be altered by concentration or dilution of the serum sample. Its simplicity and portability allows use by operators with a minimal amount of education and training. The lateral flow assay is useful for point-of-care situations where an initial rapid screening result would determine whether or not further action, such as additional testing or vaccination, was necessary.

ELISA methods have several advantages including the fact that they are rapid, require little expertise, do not need high-level biohazard facilities to be performed and several steps of the procedure can be automated (i.e., serial dilution of the sera, addition of reagents, and optical density reading). Additionally, software packages are available to calculate end-point titers or antibody concentration, thus allowing for objective reading and interpretation. The disadvantages of ELISA methods include the restrictive nature of the conjugated antibody or protein A that limits the isotype of immunoglobulin detected. The use of species-specific anti-IgG confines the utility of the assay to a certain species. Additionally, although the use of protein A increases test application to several species, it does not react with all forms of IgG3 and therefore will lead to an underestimation of the level of rabies virus-specific antibodies in serum containing higher proportions of IgG3 antibodies (Carpenter, 1997). The degree of nonspecific binding detected by an ELISA will depend on the purity of the antigen preparation and the efficiency of the coating step because immunoglobulins will nonspecifically adhere to glass, plastic, and also to contaminating material (i.e., mycoplasma) or cell culture components. Quantification of IgG antibodies that bind to rabies virus will *not* precisely demonstrate the level of *protective* virus neutralizing antibodies present in the sera. Therefore, the ELISA method is not appropriate for attempting to measure the amount of RVNA. Reporting of ELISA results in IU/ml is not reflective of this unit of measurement as defined by the WHO, where 1 International Unit of neutralizing activity is present per mg of protein. Therefore, the use of IU/ml to describe an antigen-binding assay for a rabies virus titer result is misleading. Since not all binding antibodies neutralize virus, whether for rabies virus or other pathogens, titers obtained from antigen-binding assays are not biologically identical to RVNA titers. Therefore, ELISA-based test results should only be ordered, used, and interpreted by informed health care providers, whether veterinary or human, or researchers, for the optimal prevention of rabies.

8 DEFINING "ADEQUATE" OR "MINIMUM" RESPONSE TO RABIES VACCINATION

It is well known that vaccination resulting in production of rabies virus neutralizing antibodies (RVNA) serves to prevent rabies in persons who have been in contact with a rabid animal. People who have an increased risk of rabies exposure are vaccinated pre-exposure to provide protection for unnoticed exposures and to reduce the vaccination schedule upon known exposure. This population should have periodic RVNA titer checks to evaluate the need for booster vaccinations. There are two major sources of guidelines in regard to an adequate response to rabies vaccination: the World Health Organization (WHO) Expert Committee on Rabies and the Advisory Committee on Immunization Practices (ACIP). Because the acceptable level given by these two guidelines are different and there is lack of understanding of how these levels were obtained and what they mean, there is confusion in the medical and veterinary fields about how to interpret rabies serology results in regard to booster vaccination decisions. Guidelines for the prevention of rabies, a fatal disease, should be clear and unambiguous. The optimal level for protection in someone who recently completed a post-exposure series may be different than the level that shows continuing immunity in someone who has been pre-exposure vaccinated over two years ago. Health professionals need a clear guideline to reference in making life-saving decisions about rabies treatment, for both pre- and post-exposure situations. The guideline instructions should clearly define the acceptable RVNA level in terms applicable to the recommended laboratory methods and clarify if there are situations where a different level may apply.

Because rabies is preventable by vaccination that produces rabies virus neutralizing antibodies (other immune mediators may be at play but are not readily measured), use of methods to quantitate RVNA are preferred when testing for vaccine response in humans. Currently, the most utilized method for this purpose is the RFFIT (and modified RFFIT methods). Not only should the method for this purpose be confirmed to measure neutralizing antibodies but it should also be standardized to allow comparison to other laboratories and to established guidelines for human vaccination and vaccine manufacturer's instructions. The method needs to provide results that can be related to the guidelines in regards to units of measure and ability to detect the level stated. While no specific RVNA level has been identified as representing absolute protection under all circumstances and in all hosts against all rabies virus variant infections, RVNA levels attained by the majority of subjects in vaccine clinical trials formed the basis for the levels currently recognized as the minimal adequate response in vaccinated humans. The levels recommended by the WHO and the ACIP are different (see Table 12.4). This difference is not new, but a misunderstanding of the different levels is common and indeed present, even in published literature. The ACIP recommends the use of the rapid fluorescent focus inhibition test (RFFIT) because other methods (i.e., ELISA, Lateral Flow) do not measure RVNA specifically and therefore cannot be correlated to the RFFIT. Because applying the WHO level will result in more vaccinated individuals falling within the "need for booster" group and ELISA methods may produce results not correlated with the guideline levels, these important components relied on for vaccination decisions require clarification.

Immunization against rabies with the vaccine produced by Pasteur in 1885, and which remained largely unchanged until the advent of tissue culture vaccines, involved multiple vaccinations and was not without serious consequences (Steele, 1975). In contrast, modern tissue culture vaccines are safe and effective (Wunner & Briggs, 2010). The safety of the vaccine combined with the knowledge of the protective effects of RVNA upon rabies exposure led to the practice of pre-exposure vaccination for people at frequent or continuous risk of exposure (Centers for Disease Control and Prevention, 2008). Initially, it was suggested that this population receive a booster vaccination every two years to ensure ongoing "protection" by RVNA (Centers for Disease Control, 1976). RVNA was

TABLE 12.4   Guidelines for Persons Pre-Exposure Vaccinated and at Risk of Rabies Exposure

Agency/Year	Booster Vaccination Recommended if Level is Below:	Method of Testing:
WHO		
1992	0.5IU/mL	MNT or RFFIT; ELISA only with caution
2005	0.5IU/mL	RFFIT or FAVN; ELISA if RFFIT not available
ACIP		
1976	None, boosters recommended every 2 years	None stated
1980	1:16 titer or booster every 2 years	RFFIT
1984	1:5 titer per CDC; 0.5IU/mL per WHO	RFFIT
1991	1:5 titer[a]	RFFIT
1999	Complete neutralization at a 1:5 serum dilution[b]	RFFIT
2008	Complete neutralization at a 1:5 serum dilution[c]	RFFIT

[a]Recommended response 2–4 weeks after either pre- or post-exposure vaccination is complete neutralization at a 1:25 serum dilution, which is equivalent to the WHO level of 0.5IU/mL.
[b]Recommended response 1–2 weeks after post-exposure vaccination is complete neutralization at a 1:5 serum dilution.
[c]RVNA titer most properly reported according to a standard as IU/mL.

measured in the early days of rabies vaccine development by the mouse neutralization test (MNT), a cumbersome *in vivo* test. Concerns about the adverse effects of too-frequent vaccination and the implementation of a rapid test (RFFIT) for RVNA influenced a change in the recommendation from periodic booster vaccination to boostering only when the RVNA level falls below a level representing an adequate response (Centers for Disease Control, 1980).

The level of 0.5IU/mL was recommended at the Joint WHO/IABS Symposium on the Standardization of Rabies Vaccines for Human Use Produced in Tissue Culture held in Marburg, West Germany, in November 1977 (Bogel, 1978). After the results of several international human rabies vaccine trials were presented, recommendations were given by specific Working Groups. The Working Group for vaccine potency requirements of reduced immunization schedules and pre-exposure vaccination stated: "The group suggests that the serum be tested four weeks after the last inoculation and at that time a minimum value of 0.5IU per ml be attained to demonstrate seroconversion" (Bogel, 1978).

Based on the subsequent reports of rabies virus antibody levels attained after pre- or post-exposure vaccination series, the level of $0.5\,IU/mL$ was accepted as proof of seroconversion [12]. The designated level in the ACIP was also based on RFFIT results from human vaccine trials and the observation that no nonspecific inhibition (false positive) reactions were found in serum dilutions above 1:5 (personal communication, Jean Smith). This led to the conclusion that if a specific RVNA titer result was detected at this level, then seroconversion had been achieved. The level described in the ACIP is approximately $0.1\,IU/mL$ in the RFFIT as originally described (Moore & Hanlon, 2010).

Therefore, both levels, $0.5\,IU/mL$ and $0.1\,IU/mL$, although different by five fold, are based on the same rationale which is the specific detection of RVNA. The difference is the degree of confidence that the designated level can be assured to be a true measurement and not a "false positive." The WHO does not define the assay used and recognizes the differences in testing methods and laboratory capabilities. The variance in rabies serology results was described in early publications on methods; wide differences in RVNA levels were obtained by MNT and RFFIT and by different laboratories (Bogel, 1978; Fitzgerald, Gallagher, Hunter, Spivey, & Seligmann, 1978). The ACIP states that rabies serology should be performed by the RFFIT, thereby designating a single approved method.

Although the ACIP defines a method, it does not define the adequate level of RVNA in standardized terms. "Complete neutralization at a serum dilution of 1:5 in the RFFIT" can represent different titers and IU/mL values in laboratories that perform "modified" RFFIT assays. Besides standardizing the value reported for rabies serology, defining the specific parameters and standard reagents that comprise acceptable methods for RVNA measurement would aid in interpreting results for the determination of the need for booster vaccination.

The ACIP states rabies serology should be performed by the RFFIT, but other rabies serology methods are available, particularly ELISA methods, and may be inadvertently ordered. This may be especially true if samples from humans are collected and sent for "rabies" titer through a commercial human medical laboratory. Understanding the method and then interpretation of the result is essential for the optimal management of humans and animals. Indirect ELISA methods detect and measure the presence of rabies virus specific antibodies based on their binding ability; they do not measure the neutralizing ability of the antibodies (Irie & Kawai, 2002). The immune response to rabies vaccination involves antibody production that is polyclonal. In each individual, a polyclonal response may include clonal antibodies that vary in affinity, avidity, and ability to neutralize the rabies virus. Therefore, the relationship between the level of binding antibodies and neutralizing antibodies cannot be predicted and is not linear. In one individual, an ELISA result may be higher than the RFFIT result and in another individual, the opposite can be true. Applying an "adequate" level of virus neutralizing antibodies, as described by WHO and ACIP, to ELISA results will not be accurate in every individual. The validity of a method is unique to each laboratory and the parameters of a validation should be carefully considered. Method validation documents performance standards and includes identification and verification of the lower limit of quantitation (LOQ). This is the lowest level that will produce accurate and precise results. The LOQ is the level that, by implication based on the history stated above, both WHO and ACIP recognize as an adequate response to rabies vaccination. Performance of a method within each laboratory that generates a rabies serology result and consideration of the rationale behind the two different definitions of adequate rabies vaccine response will need to be considered toward the development of clear language in regard to clinical management of persons or animals at-risk for exposure or following an exposure.

The proportion of individuals in various populations who are below either the WHO or ACIP guidelines of a minimum response to vaccination has not changed from initially published studies. This is illustrated by a review of rabies serology results from sequential student classes at three veterinary colleges during 2005 to 2011, which were tested at the Kansas State University Rabies Laboratory (see Figure 12.5). The percentage of the populations failing to meet the ACIP level ranged from 0–8%, and the percentage of the populations failing to meet the WHO level ranged from 15–25%. These proportions are similar to what is reported in the peer-reviewed literature from clinical vaccine studies of pre-exposure vaccinated individuals where 2–7% of vaccinated people fail to show complete neutralization at a 1:5 serum dilution after 2 years [1], and approximately 75% of persons remain seroconverted ($>0.5\,IU/mL$) for 5–10 years after pre-exposure vaccination (Rodrigues et al., 1987; Strady et al., 1998).

9 CONCLUSIONS

There have been no changes in the past decade in either WHO or ACIP documents of what is considered an adequate level of rabies virus neutralizing antibodies. However, there continues to be a fundamental and relatively widespread misunderstanding of the difference between the two guidelines and between different rabies serology methods that may be available, especially when tests are ordered through large commercial laboratories.

Given the importance of RVNA levels in the prevention of human and animal rabies, guidelines for adequate vaccination should be stated in terms that are readily understood by individuals-at-risk (whether human

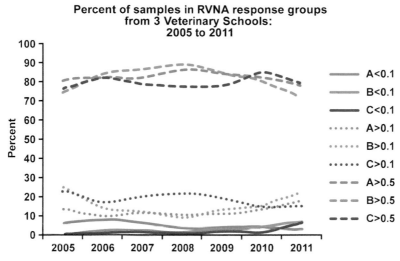

FIGURE 12.5   The RVNA response groups: >0.5 – above the WHO level of 0.5IU/mL (dashed lines), >0.1 above the ACIP level of 0.1IU/mL / below the WIIO level of 0.5IU/mL (dotted lines), and <0.1 below the ACIP level of 0.1IU/mL (solid lines) are represented for samples submitted to the KSU Rabies Laboratory for RFFIT testing from three veterinary schools (designated A, B, and C).

or animal) and health care providers, both veterinary and medical, who will use the recommendations for clinical management of individuals. Besides defining rabies exposure, risk groups, and vaccination schedules and regimens, the guidelines need to clearly state RVNA levels that are considered to be adequate responses to vaccination, both for pre- and post-exposure situations. The levels should be defined in a standard format (IU/mL) and in consideration of methods that are validated for this purpose, giving due consideration to the method's performance characteristic especially in regard to variability at the lower limit of detection. It would be advisable to set a level with lower confidence limits that do not reach below the lower limit of detection.

Across the globe, there are different official regulatory standards and guidelines, both national and international for drugs and related product approvals. Each has a common goal to ensure the integrity of the laboratory data; protect human welfare; and provide safe and effective products. These goals can only be met if all aspects of the testing and phase developments are held to the appropriate regulatory standard and monitored throughout the development process. The implementation of these standards must be from the beginning of the work at bench level to the end of the process, which often continues during clinical use of products intended for the improvement or protection of human and animal health, including direct assessment of, and assessment of host responses to, rabies vaccines and sources of polyclonal and monoclonal products for the prevention of rabies.

References

Altman, D. G. (1991). Statistical analysis of comparison between laboratory methods. *Journal of Clinical Pathology, 44,* 700–701.

Anonymous (1992). *Regulations implementing the Clinical Laboratory Improvement Amendments of 1988 (CLIA). Final Rule.* U.S. Department of Health and Human Services.

AOAC International (2002). AOAC Guidelines for Single Laboratory Validation of Chemical Methods for Dietary Supplements and Botanicals. <http://www.aoac.org/dietsipp6/Dietary-Supplement-web-site/slv_guidelines.pdf/>

Atanasiu, P. (1973). Quantitative assay and potency test of antirabies serum and immuno-globulin. *Monograph Series World Health Organization, 314*–318.

Aubert, M. F. (1996). Methods for the calculation of titres. In F. X Meslin, M. M. Kaplan, & H. Koprowski (Eds.), *Laboratory techniques in rabies (pp. 445–459)* (4th ed.). Geneva: World Health Organization.

Baynes, M. (2005). Introduction to Regulatory World In.

Bogel, K. (1978). Developments in Biological Standardization. *Joint WHO/IABS symposium on the standardization of rabies vaccines for human use produced in tissue culture [rabies III], 40,* 1–288.

Bordignon, J., Comin, F., Ferreira, S. C., Caporale, G. M., Lima Filho, J. H., & Zanetti, C. R. (2002). Calculating rabies virus neutralizing antibodies titres by flow cytometry. *Journal of the Institute of Tropical Medicine of São Paulo, 44,* 151–154.

Briggs, D. J., Smith, J. S., Mueller, F. L., Schwenke, J., Davis, R. D., Gordon, C. R., et al. (1998). A comparison of two serological methods for detecting the immune response after rabies vaccination in dogs and cats being exported to rabies-free areas. *Biologicals, 26,* 347–355.

Brookes, S. M., & Fooks, A. R. (2006). Occupational lyssavirus risks and post-vaccination monitoring. *Developments in Biologicals, 125,* 165–173.

Cardoso, T. C., Silva, L. H., Albas, A., Ferreira, H. L., & Perri, S. H. (2004). Rabies neutralizing antibody detection by indirect immunperoxidase serum neutralization assay performed on chicken embryo related cell line. *Memórias do Instituto Oswaldo Cruz, 99,* 531–534.

Carpenter, A. B. (1997). Enzyme linked immuno-sorbent assay. In N. R Rose, E. C. De Macario, J. Fahey, H. Fiedman, & G. M. Pen (Eds.), *Manual of clinical laboratory immunology (pp. 2–9)* (5th ed.). Boston: Little Brown.

Center for Drug Evaluation and Research (2001). *Guidance for industry bioanalytical method validation.* Rockville, MD.

Centers for Disease Control. (1976). *Rabies recommendations of the public health service advisory committee on immunization practices* (Rep. No. Vol. 25/No. 51). Atlanta, GA.

Centers for Disease Control. (1980). *Rabies prevention recommendation of the immunization practices advisory committee (ACIP)* (Rep. No. Vol. 29/No. 23). Atlanta, GA.

Centers for Disease Control and Prevention. (2008). *Human rabies prevention – united states, 2008 recommendatoins of the advisory committee on immunization practices* (Rep. No. Vol. 57/RR-3). Atlanta, GA.

Chaloner-Larsson, G., Anderson, R., & Egan, A. (1997). *A WHO guide to good manufacturing practice (GMP) requirements. Part: Validation.* Geneva: World Health Organization.

Cliquet, F., Aubert, M., & Sagne, L. (1998). Development of a fluorescent antibody virus neutralisation test (FAVN test) for the quantitation of rabies-neutralising antibody. *Journal of Immunological Methods, 212,* 79–87.

Dietzschold, B. (1993). Antibody-mediated clearance of viruses from the mammalian central nervous system. *Trends in Microbiology*, *1*, 63–66.

Findlay, J. W., Smith, W. C., Lee, J. W., Nordblom, G. D., Das, I., DeSilva, B. S., et al. (2000). Validation of immunoassays for bioanalysis: a pharmaceutical industry perspective. *Journal of Pharmaceutical and Biomedical Analysis*, *21*, 1249–1273.

Fitzgerald, E. A., Baer, G. M., Cabasso, V. F., & Vallancourt, R. F. (1975). A collaborative study on the potency testing of antirabies globulin. *Journal of Biological Standardization*, *3*, 273–278.

Fitzgerald, E. A., Gallagher, M., Hunter, W. S., Spivey, R. F., & Seligmann, E. B., Jr. (1978). Laboratory evaluation of the immune response to rabies vaccine. *Journal of Biological Standardization*, *6*, 101–109.

Flamand, A., Raux, H., Gaudin, Y., & Ruigrok, R. W. (1993). Mechanisms of rabies virus neutralization. *Virology*, *194*, 302–313.

Food and Drug Administration. (1996). Quality System Regulations. 812. Federal Regulations.

Food and Drug Administration. (2009). Milestones in U.S. Food and drug law history. Available at <http://www.fda.gov/AboutFDA/WhatWeDo/History/Milestones/default.htm> Accessed 25.02.13.

Good Laboratory Practice for Nonclinical Laboratory Studies. (2011). 21 CFR part 58. Federal Regulations.

Grassi, M., Wandeler, A. I., & Peterhans, E. (1989). Enzyme-linked immunosorbent assay for determination of antibodies to the envelope glycoprotein of rabies virus. *Jouranl of Clinical Microbiology*, *27*, 899–902.

Guglielmo-Viret, V., Attree, O., Blanco-Gros, V., & Thullier, P. (2005). Comparison of electrochemiluminescence assay and ELISA for the detection of Clostridium botulinum type B neurotoxin. *Journal of Immunological Methods*, *301*, 164–172.

Haase, M., Seinsche, D., & Schneider, W. (1985). The mouse neutralization test in comparison with the rapid fluorescent focus inhibition test: differences in the results in rabies antibody determinations. *Journal of Biological Standardization*, *13*, 123–128.

Habel, K. (1996). Habel test for potency. In F. X Meslin, M. M. Kaplan, & H. Koprowski (Eds.), *Laboratory techniques in rabies* (pp. 369–373) (4th ed.). Geneva: World Health Organization.

Hooper, D. C., Morimoto, K., Bette, M., Weihe, E., Koprowski, H., & Dietzschold, B. (1998). Collaboration of antibody and inflammation in clearance of rabies virus from the central nervous system. *Journal of Virology*, *72*, 3711–3719.

Hostnik, P. (2000). The modification of fluorescent antibody virus neutralization (FAVN) test for the detection of antibodies to rabies virus. *Journal of Veterinary Medicine. B, Infectious Diseases and Veterinary Public Health*, *47*, 423–427.

ICH. (2001). ICH Guidance for Industry Q7A Good Manufacturing Practice for Active Pharmaceutical Ingredients.

Irie, T., & Kawai, A. (2002). Studies on the different conditions for rabies virus neutralization by monoclonal antibodies #1-46-12 and #7-1-9. *The Journal of General Virology*, *83*, 3045–3053.

Kaplan, M. M., Wiktor, T. J., Maes, R. F., Campbell, J. B., & Koprowski, H. (1967). Effect of polyions on the infectivity of rabies virus in tissue culture: construction of a single-cycle growth curve. *Journal of Virology*, *1*, 145–151.

Khan, N. K., & Findlay, J. W. (2009). Assay design. In N. K Khan & J. W. Findlay (Eds.), *Ligand-binding assays: Development, validation, and implementation in the drug development arena (pp. 196–219)*. Hoboken: John Wiley & Sons.

Khawplod, P., Inoue, K., Shoji, Y., Wilde, H., Ubol, S., Nishizono, A., et al. (2005). A novel rapid fluorescent focus inhibition test for rabies virus using a recombinant rabies virus visualizing a green fluorescent protein. *Journal of Virological Methods*, *125*, 35–40.

Kostense, S., Moore, S., Companjen, A., Bakker, A. B., Marissen, W. E., von, E. R., et al. (2012). Validation of the rapid fluorescent focus inhibition test (RFFIT) for rabies virus neutralizing antibodies in clinical samples. *Antimicrobial Agents and Chemotherapy.*

Kramer, B., Bruckner, L., Daas, A., & Milne, C. (2010). Collaborative study for validation of a serological potency assay for rabies vaccine (inactivated) for veterinary use. *Pharmeuropa Bio & Scientific Notes, 2010,* 37–55.

Lyng, J., Bentzon, M. W., & Fitzgerald, E. A. (1989). Potency assay of antibodies against rabies. A report on a collaborative study. *Journal of Biological Standardization, 17,* 267–280.

Ma, X., Niezgoda, M., Blanton, J. D., Recuenco, S., & Rupprecht, C. E. (2012). Evaluation of a new serological technique for detecting rabies virus antibodies following vaccination. *Vaccine.*

Mannen, K., Mifune, K., Reid-Sanden, F. L., Smith, J. S., Yager, P. A., Sumner, J. W., et al. (1987). Microneutralization test for rabies virus based on an enzyme immunoassay. *Journal of Clinical Microbiology, 25,* 2440–2442.

Moore, S. M., & Hanlon, C. A. (2010). Rabies-specific antibodies: measuring surrogates of protection against a fatal disease. *PLoS Neglected Tropical Diseases, 4,* e595.

Moore, S. M., Ricke, T. A., Davis, R. D., & Briggs, D. J. (2005). The influence of homologous vs. heterologous challenge virus strains on the serological test results of rabies virus neutralizing assays. *Biologicals, 33,* 269–276.

Nicholson, K. G., & Prestage, H. (1982). Enzyme-linked immunosorbent assay: a rapid reproducible test for the measurement of rabies antibody. *Journal of Medical Virology, 9,* 43–49.

Peharpre, D., Cliquet, F., Sagne, E., Renders, C., Costy, F., & Aubert, M. (1999). Comparison of visual microscopic and computer-automated fluorescence detection of rabies virus neutralizing antibodies. *Journal of Veterinary Diagnostic Investigation: Official Publicaton of the American Association of Veterinary Laboratory Diagnostians, Inc., 11,* 330–333.

Rodrigues, F. M., Mandke, V. B., Roumiantzeff, M., Rao, C. V., Mehta, J. M., Pavri, K. M., et al. (1987). Persistence of rabies antibody 5 years after pre-exposure prophylaxis with human diploid cell antirabies vaccine and antibody response to a single booster dose. *Epidemiology and Infection, 99,* 91–95.

Selby, C. (1999). Interference in immunoassay. *Annals of Clinical Biochemistry, 36(Pt 6),* 704–721.

Smith, J. (1991). Rabies serology. In G. M Baer (Ed.), *The natural history of rabies (pp. 235–252)* (2nd ed.). Boca Raton: CRC Press.

Smith, J. S., Yager, P. A., & Baer, G. M. (1973). A rapid reproducible test for determining rabies neutralizing antibody. *Bulletin of the World Health Organization, 48,* 535–541.

Steele, J. (1975). History of rabies (1st ed.). In G. M (1975). Baer (Ed.), *The natural history of rabies* (Vol. 1, pp. 1–28). New York: Academic Press.

Strady, A., Lang, J., Lienard, M., Blondeau, C., Jaussaud, R., & Plotkin, S. A. (1998). Antibody persistence following preexposure regimens of cell-culture rabies vaccines: 10-year follow-up and proposal for a new booster policy. *The Journal of Infectious Diseases, 177,* 1290–1295.

Tordo, N. (1996). Characteristics and molecular biology of the rabies virus. In F. X Meslin, M. M. Kaplan, & H. Koprowski (Eds.), *Laboratory techniques in rabies (pp. 28–51)* (4th ed.). Geneva: World Health Organization.

Velleca, W. M., & Forrester, F. T. (1981). *Laboratory methods for detecting rabies.* Atlanta: U.S. Department of Health and Human Services Public Health Service Centers for Disease Control.

Wood, R. J., & Durham, T. M. (1980). Reproducibility of serological titers. *Journal of Clinical Microbiology, 11,* 541–545.

Wright, E., Temperton, N. J., Marston, D. A., McElhinney, L. M., Fooks, A. R., & Weiss, R. A. (2008). Investigating antibody neutralization of lyssaviruses using lentiviral pseudotypes: a cross-species comparison. *The Journal of General Virology, 89,* 2204–2213.

Wunner, W. H., & Briggs, D. J. (2010). Rabies in the 21 century. *PLoS Neglected Tropical Diseases*, 4, e591.

Yu, P. C., Noguchi, A., Inoue, S., Tang, Q., Rayner, S., & Liang, G. D. (2012). Comparison of RFFIT tests with different standard sera and testing procedures. *Virologica Sinica*, 27, 187–193.

Zinkernagel, R. M. (2002). Uncertainties – discrepancies in immunology. *Immunological Reviews*, 185, 103–125.

13

Rabies Vaccines

Deborah J. Briggs[1], Thirumeni Nagarajan[2], and Charles E. Rupprecht[3]

[1]Global Alliance for Rabies Control, 529 Humboldt St., Suite One, College of Veterinary Medicine, Kansas State University, Manhattan, KS 66502, USA, [2]Research and Development Department, Biological E. Limited, Plot No. 1, Phase II, SP Biotech Park, Genome Valley, Kolthur Village, Shameerpet, Hyderabad, India, [3]Global Alliance for Rabies Control, 501 Humboldt St., Suite One, Manhattan, KS 66502, USA

1 INTRODUCTION

Rabies has one of the highest case-fatality rates among infectious diseases, with over 98% of global human deaths attributed to exposure to infected dogs (Hampson et al., 2009). Human and animal rabies vaccines play a critical role in protection against rabies, and together with laboratory-based surveillance, are one quintessential part of a functional feedback system of disease prevention and control (Figure 13.1). Rabies vaccines provide the best service to public health when used in concert with effective risk-management assessment systems that include enhanced surveillance, reliable decentralized diagnostic laboratories, humane animal management based upon the prevalence of disease in a locality, continuing education of public health professionals trained on the use of appropriate pre-exposure prophylaxis (PreP) and post-exposure prophylaxis (PEP) and animal vaccination, enhanced inter-sectorial communication between animal and human health professionals, with appropriate data reporting, sharing, and action. Without an adequate

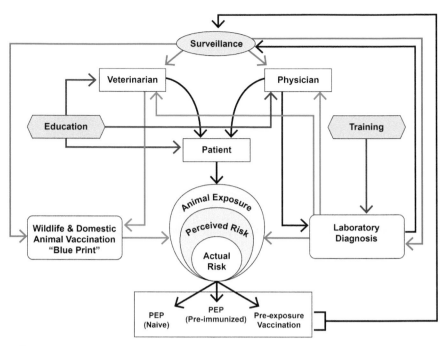

FIGURE 13.1   Components of an integrated risk management assessment system for rabies with the key integration of surveillance, diagnosis, and vaccination.

risk management system in place, PEP is used inappropriately. History has demonstrated that for a singular strategy of governments purchasing an ever needed supply of human rabies vaccine is akin to an 'incurable wound' and increases dramatically the financial burden of disease prevention in regions that can least afford to pay (Zinsstag et al., 2009). Clearly, establishing and enforcing national risk assessment systems to evaluate the use of PEP is cost-beneficial and provides a more effective strategy for the overall prevention and control of rabies.

2   HUMAN AND ANIMAL RABIES VACCINES

2.1   Vaccine Accessibility

Cell culture rabies vaccines (CCVs) are available in all industrialized countries, although supply limitations have been reported occasionally (CDC, 2012). In developing countries, CCV availability may be limited to large metropolitan areas, and these vaccines are often not stocked in rural clinics, nor are they readily affordable to those living in these areas (Hampson et al., 2011). The inability to purchase rabies vaccines in poor populations continues to make rabies largely a disease of poverty. In regions where canine rabies is enzootic and there are no control programs in place, human rabies vaccines are the only defense against contracting rabies. In regions where vaccines are not supplied free by the government, families with limited resources often must sell valuable possessions, including livestock that they depend on for food, to pay for vaccines. Additionally, parents who have several exposed children may need to decide who will receive a vaccine as they cannot afford to pay for PEP for all.

In several countries, particularly in Asia, the administration of PEP is associated with Animal Bite Centers (ABCs) specifically treating patients exposed to suspect or proven rabid animals. The ABC concept was initiated when the original global network of Pasteur Institutes were established in the late 1800s to produce and administer nerve-tissue vaccines (NTVs) produced *in vivo* in the brain tissue of infected animals to exposed patients. Over the decades, the Pasteur Institutes expanded their public health services to local populations to include maladies besides rabies. Production of NTVs became increasingly the responsibility of governments. In most countries, NTVs are being phased out in place of CCVs. Unfortunately, in countries where they are still produced, their use is generally relegated to those that cannot afford to pay for safer and more efficacious CCVs. Presently, three types of NTVs are being produced: Fermi vaccine, produced and administered in Ethiopia; Semple vaccine, produced and administered in Mongolia, Myanmar, and Pakistan; and SMBV, produced in Argentina, Bolivia, Ecuador, El Salvador, Honduras, Peru, Uruguay, and Algeria. For more than a decade, the WHO has recommended that the production of NTVs be stopped and be replaced with more efficacious and safer CCVs. This created a conundrum of how to continue to serve the population most at risk of exposure when, at minimum, start-up funding by governments or co-payment by patients themselves was required. This has served to slow the replacement of NTVs. A few Asian countries have begun to produce CCVs. Establishment of local production facilities has aided accessibility of affordable vaccines, and the replacement of NTVs with CCVs (Supplementary Tables S13.1 and S13.2). Specific recommendations for manufacturing of CCV and a pre-qualification process have been developed (WHO, 2011).

2.2 Vaccine Production

The manufacturing processes for all vaccines are necessarily similar, including production of both human and animal rabies vaccines. The WHO has published standard recommendations for rabies vaccine production (WHO, 2005). Ultimately, it is the responsibility of national governments to approve, license, and monitor human and animal rabies vaccines and production facilities in their own countries.

Whole inactivated rabies virus particles are highly immunogenic and form the basis for rabies vaccines currently used for PreP and PEP of humans and animals. The quality and magnitude of the immune response after vaccination depend largely on the integrity of viral antigens, especially the glycoprotein (G), which must be presented as a membrane-anchored protein in a repetitive rigid form. This can be achieved using whole virus particles for immunization (Dietzschold, Faber, & Schnell, 2003). Only vaccines that are pure, potent, safe, affordable, and capable of providing stable and long-lasting immunity are recommended for mass vaccination of animals because this constitutes the most effective method of controlling and eliminating rabies in animals. The manufacture of affordable inactivated animal rabies vaccine necessitates the choice of inexpensive upstream and downstream processes and improved formulation strategies involving adjuvants that do not significantly add to the cost of production. With the multitude of public health challenges facing resource-poor countries, only affordable rabies vaccines can be purchased and made available to improve accessibility in geographical regions where they are needed most. Wider availability could encourage broader vaccination coverage, which is central for inducing desired levels of herd immunity. Animal rabies vaccines can be broadly classified into inactivated NTVs, inactivated CCVs, and live attenuated or recombinant rabies vaccines. The various aspects of the manufacturing process, such as the choice of cell line, vaccine virus strain, culture system, downstream processes, formulation strategies, and potency testing vary based on the product.

2.3 Production of Modern Rabies Vaccines

Modern rabies vaccines for human and animal use are produced *in vitro*, bypassing the need to infect live animals (Figure 13.2). In 1958, rabies virus was first adapted to grow in hamster kidney cells, which led to considerable progress in cell culture techniques (Sureau, 1987). The growth of virus in cell culture simplified the large-scale production of antigen for both attenuated and inactivated vaccines for human and animal use (Barth et al., 1984; Chapman, Ramshaw, & Crick, 1973). Cell culture technology allows for standardized, safe, effective, economical, and large-scale vaccine production (Regan & Pettriciani, 1987). Improvements in vaccine production techniques during the last few decades led to an increased use of inactivated adjuvanted vaccines for animal immunizations. All modern human rabies vaccines are inactivated, but both inactivated and modified-live or recombinant rabies vaccines are produced for use in animals. Additionally, rabies

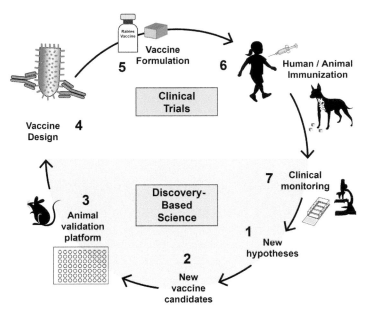

FIGURE 13.2   A modern vaccine development cycle displaying the discovery and clinical development phases in animals and humans.

vaccines differ according to the strain of virus used and the characteristics of the cell substrate chosen for viral replication (Reculard, 1996). Modern CCVs are highly defined, safer, and more potent than NTVs because they use well-defined and characterized cell lines, highly defined media, controlled production processes, and monitoring of the process and product, and have to a large extent replaced NTVs (Tao et al., 2011). Acceptable vaccination coverage can better be achieved if the vaccine cost is reduced (Zinsstag et al., 2009). The trend towards transfer or acquisition of modern cell culture for inactivated animal rabies vaccine production is increasing in developing countries. The process of manufacturing inactivated cell culture rabies vaccine involves a number of complex steps and unit operations, ranging from the selection of cell substrate and virus strain for cultivation in bioreactors to the downstream processing and formulation of final product (Figure 13.3). The overall increase in productivity, without adding significantly to the cost of production, can be achieved by having advanced upstream processes for improved yields and harvest volumes, as well as downstream processes for improved recovery and purification of the products from process and product related impurities in place (Wolff & Reichl, 2011). The seed virus, cell substrate, and viral inactivant vary considerably between different manufacturers. The details of various human and animal rabies vaccines such as name, seed virus, cell substrate, viral

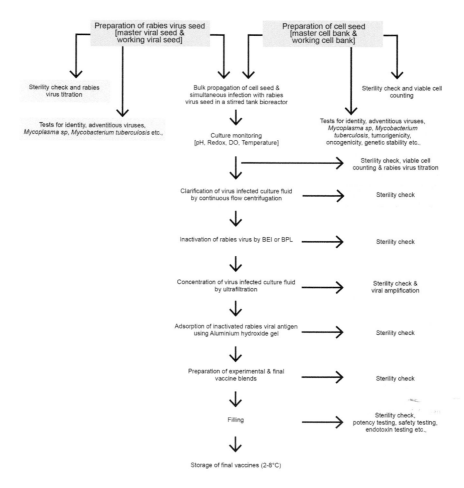

FIGURE 13.3   Generic flowchart of upstream and downstream processes in production of commercial cell culture rabies vaccine for animals.

inactivant, adjuvant, preservative, physical form, and manufacturers available worldwide are listed in Supplementary Tables S13.1 and S13.2.

2.3.1 Vaccine Virus Strains

Many rabies virus strains, such as Challenge Virus Standard (CVS), Flury LEP, Flury HEP, Kelev, Evelyn Rokitniki Abelseth (ERA), Vnukovo-32, Street Alabama Dufferin (SAD), Pasteur Virus (PV), and Pitmann Moore (PM) have all been used for production of inactivated animal rabies vaccines (Reculard, 1996). These strains resemble each other in their ability to confer uniform protection against members of the lyssavirus genotype 1, rabies virus. However, they may differ from each other in their capacity to proliferate to high titers in cell culture. For instance, PV gave the highest titers when BHK-21 cells grown on microcarriers were infected with a multiplicity of infection (MOI) of 0.3 in bioreactors (Kallel et al., 2003). The highest titers obtained for Flury LEP occurred when BHK-21 cells were infected with a MOI of 0.01 in static culture system (Tao et al., 2011). The Flury LEP strain is widely used for making rabies vaccines for humans and animals because it can achieve high titers when grown in cell culture and because of its ability to elicit high immunogenicity (Koprowski & Cox, 1948). The rabies G is a type I membrane protein that exists as a trimer (3×65KDa) and forms a spike extending 8.3nm from the viral membrane, with three domains: ectodomain, transmembrane domain, and cytoplasmic domain (Gaudin, Ruigrok, Tuffereau, Knossow, & Flamand, 1992). The ectodomain of G is known to harbor sites that are responsible for infectivity, neuroinvasiveness, and, when used as a vaccine immunogen, for conferring protection against lethal rabies virus challenge (Langevin & Tuffereau, 2002). The G protein of CVS and ERA rabies virus strains appears as a doublet when resolved by SDS-PAGE under reducing conditions (Langevin & Tuffereau, 2002; Prehaud, Coulon, LaFay, Thiers & Flamand, 2002). The G is a singlet for all other rabies virus strains. Increasing G expression in the viral seed strain may improve inactivated rabies vaccines, not only in terms of immunogenicity but also improved manufacturing capacity. For example, the Flury LEP strain with an additional G gene (rLEP-G) produced strikingly higher levels of G protein in cell culture and showed similar *in vitro* growth properties and bio-safety characteristics. An inactivated vaccine prepared using rLEP-G induced significantly higher antibody titers in vaccinated subjects than were induced by Flury LEP, suggesting that rLEP-G is an improved seed virus candidate for manufacture of inactivated rabies vaccine (Tao et al., 2011). Similarly, an inactivated rabies vaccine prepared using a highly attenu ated triple RV G protein variant of SAD (SPBAANGAS-GAS-GAS) might permit an antigen sparing strategy with reduced production costs and without reducing safety (Faber, Dietzschold, & Li, 2009a).

2.3.2 Cell Substrate

The development of cell culture technology introduced a new era in production of rabies vaccines. Historically, cells derived from animal tissues acted as primary substrates for production of viral vaccines. Cell properties and events linked to their growth have a bearing on the quality of biologics produced in them (WHO, 1998). Although cell substrates, such as primary cells (derived from hamster kidney, fetal bovine kidney, dog kidney, piglet kidney, chick embryo, etc.) and diploid cells can be used for production of rabies vaccines, continuous cell lines possess distinct advantages over primary and diploid cell substrates, such as bulk propagation of virus with ease and lower production costs (Grachev, 1990). Continuous cell lines have the potential of an infinite life span and permit preparation of a well-defined seed lot system consisting of master, working, and extended working cell banks (Barone, Fracchia, Pascuali, & Proglio, 1997). Continuous cell lines are also amenable to cultivation as attached cells in suitable culture systems or suspension in bioreactors. Most animal rabies vaccines produced from continuous cell lines are prepared in BHK-21 (baby hamster kidney origin), some of which can grow in suspension culture, and Vero, and Nil-2 (baby hamster embryo origin) cells that can grow on a solid phase as a monolayer culture (Reculard, 1996). The BHK-21 cell line is widely used for production of viral vaccines because of the significant advantages including a well-documented origin and history (Hay, Caputo, & Chen, 1994). As a non-neuronal cell line, BHK-21 cells are permissive for viral infection and hence are used for preparation of seed for some vaccines (Rudd & Trimarchi, 1987). Animal rabies vaccines produced in BHK-21 cell line have been used for decades and have been proven to be safe and efficacious.

2.3.3 *Production System*

The cell culture vaccine production process is suitable for large-scale manufacture and the process parameters can be ramped up to run routinely and cost effectively. Various culture systems, such as static (e.g., flasks and cell factories™), roller (e.g., smooth walled and ribbed) and bioreactors (e.g., hollow fiber and stirred tank), are available for cultivation of adherent and suspension cells. Static and roller culture systems are used widely for preparation of cell and vaccine virus seeds. However, they are less preferred for animal rabies vaccine manufacture because of high cost, labor intensiveness, difficulty in real-time culture monitoring and control, and multiple handling that would make the culture prone to contamination. Considerable progress has been made in production of inactivated rabies vaccine for animal use during the past three decades, including adoption of bioreactor technology to bulk propagate continuous cell lines for antigen production. The BHK-21 is an adherent cell line, with few exceptions, and requires static or roller culture systems for preparation of cell and virus seeds. A modified solid phase, in the form of microcarriers (e.g., Cytodex 3™) and macrocarriers (e.g., Fibra-Cel® disks), has been developed to overcome limitations associated with static and roller culture systems. Adherent cells have been grown on microcarriers as pseudosuspension culture in a stirred tank bioreactor and used for bulk propagation of rabies vaccine antigens employing a perfusion culture mode (Kallel, et al., 2006). Microcarrier technology offers numerous advantages over traditional techniques, such as high surface area to volume ratio, an efficient monitoring and control of key process parameters, the achievement of high cell densities, and ease of scale-up (Birch, 1999). Similarly, rabies vaccine antigens have been bulk propagated using adherent cells grown on macrocarriers in a packed-bed bioreactor (Hassanzadeh, Zavareh, Shokrgozar, Ramezani, & Fayaz, 2011). Producing animal rabies vaccines using adherent cells grown on microcarriers and macrocarriers might be very expensive. Therefore, it would be ideal to secure adherent cells adapted for growth in suspension culture so that they can be bulk propagated with ease and at a reduced cost using a stirred tank bioreactor (Atanasiu, Ribeiro, & Tsiang, 1972; Chapman et al., 1973). A BHK CZ cell line, a derivative of the BHK-21 cell line, has desirable features for industrial applications, including the ability to grow in suspension and sediment completely in approximately 14 hours, a prerequisite for infecting cells grown in suspension, and the ability to maintain them in a low serum containing medium. Inactivated rabies vaccines for domestic animals that are produced in BHK cells grown in a stirred tank bioreactor are currently marketed worldwide.

2.3.4 Downstream Processing

Regulatory requirements to ensure vaccine purity and safety depend on the particular product application (Wolff & Reichl, 2011). The development process for animal vaccines generally has less stringent regulatory and preclinical trial requirements. Adhering to these requirements constitutes the largest expense in human vaccine development. Development of animal vaccines usually has a shorter time from initial phase to market launch, thus increasing the return on investment required for research and development. The manufacture of inactivated animal rabies vaccines does not involve an elaborate downstream processing, as is the case for human vaccine production, again reducing the time and cost involved. However, all animal rabies vaccines should meet the minimal requirements as required by national regulatory authorities in terms of purity, safety, potency, duration of immunity, efficacy, stability, antigen content, identity, effective inactivation, and sterility (OIE, 2011). Due to less elaborate downstream processing, animal rabies vaccines may contain several undesirable impurities, such as BSA, host cell DNA and viral aggregates. For example, BSA has the risk of increasing immediate type allergic reactions, and sometimes death (Ohmori et al., 2005). Host cell DNA has the potential for transfer of activated cellular and/or viral oncogenes (particularly if the cell substrate is tumorigenic) and production of infectious viruses from the viral nucleic acid and aberrant gene expression by insertion of sequences into sensitive gene control regions (Vitrology, 2008). Nevertheless, inactivated animal rabies vaccines produced in cell culture have been used in the field for more than 2 decades, clearly demonstrating that they are well tolerated by domestic animals when administered parenterally.

2.3.5 Viral Inactivation

Rabies viral particles can be inactivated by employing physical (gamma radiation and UV radiation) and chemical agents [acetyle-thyleneimine (AEI), BEI and β-propiolactone (BPL)]. Chemical agents that act on the viral nucleic acid with little or no effect on the viral protein coat are preferred to physical agents by vaccine manufacturers. However, classical chemical inactivants, such as phenol and formaldehyde, are not recommended because they may reduce the immunogenicity of the vaccine and release toxic or irritant residues (Reculard, 1996). Currently, binary ethylenimine (BEI) and BPL are the inactivants of choice for producing animal rabies vaccines. Preference for BEI as a viral inactivant relates to the fact that it is highly stable, easy to prepare, inexpensive, very effective, less hazardous to handle than many other inactivants, and results in a potent and stable vaccine (Bahnemann, 1990; Mondal, Neelima, Seetha Rama Reddy, Ananda Rao, & Srinivasan, 2005). By comparison, BPL as a viral inactivant is more expensive and less stable than BEI at 37°C. However, it is still widely used for viral inactivation because it is highly effective. In principle, rabies virus can be inactivated either before or after concentration, but care must be exercised to avoid formation of aggregates if virus is inactivated before concentration by ultrafiltration. Viral inactivation is a critical step in the manufacture of rabies vaccine because the presence of virus in the final vaccine can cause infection resulting in death. There have been incidents when rabies vaccines have been recalled from the field because they were suspected to contain infectious virus (CDC, 2004). This emphasizes the significance of optimizing inactivation conditions and monitoring inactivation kinetics. The inactivation process for veterinary rabies vaccine has been validated by studying the inactivation kinetics for BEI (Mondal et al., 2005).

2.3.6 Formulation

The goal of vaccination is to generate a strong immune response to provide long-term protection against infection. In general, inactivated vaccines are formulated with relatively higher dosage of antigens to ensure a high potency in the final product (Habel, 1973). However, inactivated vaccines usually require the addition of a strong adjuvant to be effective because they are unable to infect cells and cause antigen amplification (Montomoli, Piccirella, Khadang, Mennitto, Camerini & De Rosa, 2011). Adjuvanted animal rabies vaccines are usually presented in liquid form, whereas the vaccines meant for human use are presented both in freeze-dried and liquid forms. Ideally, a rabies vaccine should be able to prime a Th1/Th2 balanced and efficient immune response to achieve viral clearance (Ren et al., 2010). Aluminum salt is an adjuvant commonly used in the manufacture of inactivated animal rabies vaccine to improve its efficacy. Although safe and well tolerated, aluminum salt is strongly Th2 biased and relatively weak as an adjuvant. In addition, aluminum salt-based adjuvants are limited in their use, in that they preclude lyophilization or freezing. Obviously, it becomes important to develop a novel adjuvant combined with or as a substitute for aluminum to enhance the efficacy and reduce cost of the existing inactivated animal rabies vaccines. A cytosine phosphate-linked guanine (CpG) oligonucleotide (ODN), which can preferentially activate canine lymphocytes when used in combination with aluminum adjuvant, has been shown to be a potent adjuvant that can induce enhanced antibody titres and Th1 cellular immune response in canine rabies vaccines (Ren et al., 2010). A similar combination of CpG ODN and aluminum adjuvant has been shown experimentally in mice to enable a reduction from five-dose PEP to three-dose PEP (Wang et al., 2008). Recently, rabies vaccines (Rabix and Rabifel) for dogs and cats formulated with Matrix-M™ Vet adjuvant (Isconova, Sweden) have been launched in Russia (VetBioChem, Russia). The Matrix-M™ Vet adjuvant has been shown to elicit both humoral and cellular immune responses with a significantly reduced amount of antigen in comparison with classical adjuvanted inactivated animal rabies vaccines.

2.3.7 Combination Vaccines

Combination vaccines are immunological products intended for (i) immunization against different infectious diseases or (ii) immunization against multifactorial infectious diseases caused by different species, types, or variants of pathogens or (iii) combinations of (i), (ii). Thus, they reduce the number of injections required to prevent several diseases and overcome the constraints of multiple injections. Other potential advantages of combination vaccines include: cost-effective delivery of vaccines; increased safety of immunization by reducing the number of needles and syringes needed; reduction of the number of needle sticks and thus discomfort to the vaccinees; improvement of timely vaccination coverage; reduction of the cost of stocking and administering separate vaccines; reduction of the cost of extra clinic visits; and facilitation of the addition of new vaccines into immunization programs (Di Fabio & de Quadros, 2001). However, combination vaccines might have the drawback of having an unwanted interaction between the vaccine components that may influence reactogenicity, immunogenicity, and stability, thereby complicating laboratory control tests. Presence of more than one component often causes either a diminished or increased response to individual components, compared to the administration of a single specific component. The chemical incompatibility or immunologic interference when different antigens are combined into one vaccine can be difficult to overcome. Therefore, development of combined vaccines is not straightforward, and each antigen included in the vaccine combination must be developed and studied individually in terms of quality, safety, and efficacy. Requirements for new combinations of antigens need to be considered carefully and should not be made solely on assumptions based on the properties of individual components (Corbel, 1994). Several combination vaccines include rabies virus antigens as components. Examples of antigens included in combination rabies vaccines include: foot and mouth virus (FMDV) for cattle, sheep, and goats; leptospira, canine distemper virus, canine hepatitis virus, canine parvovirus, canine parainfluenza, infectious laryngotracheitis virus, and canine adenovirus for dogs; and feline parvovirus, feline panleucopenia virus, and feline calicivirus for cats. Development of combination vaccines is gaining importance and is increasingly recommended because of the concern of repeated clinic visits by reducing the total number of injections and reducing costs compared with administration schedules for monovalent vaccines (Morefield et al., 2008).

2.3.8 Potency Testing

Inactivated rabies vaccine is a product produced by complex batch-wise manufacturing procedures. This implies that vaccine characteristics can vary from batch to batch. Obviously, many variables might influence quality of the product, which makes extensive quality control (QC) of each lot produced essential and mandatory (Hendriksen, 2009). Current QC of inactivated animal vaccines focuses mainly on the potency of the final products in a batch-wise manner. Potency can be defined as the measure of one or several parameters that have been shown to be related directly or indirectly to product efficacy, such as the ability to produce an effective level of protection in the target species (Hendricksen et al., 1998). The potency of inactivated rabies vaccines for animal and human use is conventionally tested by mouse protection test, which was originally developed at the United States National Institutes of Health (NIH). The NIH potency test was adopted by the WHO Expert Committee on Rabies in the 1950s and is now part of many national and international requirements for the evaluation of inactivated rabies vaccines.

Despite the multiplicity of seed viruses used in vaccine production, the strains used for development of most vaccine reference preparations (PM strain) and the virus strain used for challenge (CVS strain) in vaccine potency testing are derivatives of the original PV isolate (WHO, 1992). Many factors inherent in rabies vaccine strain differences and the characteristics of the NIH test influence protection in mice, including: administration route and number of doses; duration between vaccination and challenge; and the age of mice at the time of initial vaccination (Wunderli, Dreesen, Miller, & Baer, 2003). The principle of the NIH test is the immunization of mice with different vaccine concentrations, followed by intracerebral challenge with 5 to 50 LD_{50}/dose of the CVS strain. The protection rates (ED_{50}) of a standard and the test vaccines are calculated, and the potency of the test vaccine is given in International Units (IU). Because different seed virus strains are used for the production of rabies vaccines, difficulties arise from the different strain relationship between the test vaccine virus, as well as the challenge virus strain and the reference vaccine virus strain (Blancou, Aubert, Cain, Selve, Thraenhart, & Bruckner, 1989).

The classical NIH method for potency testing of inactivated rabies vaccines is unacceptably slow and expensive, and it requires large numbers of animals and involves significant pain, distress, inhumane end points and safety issues for laboratory workers. Animal welfare concerns, as well as scientific considerations have led to a "3Rs concept" that comprises the refinement of animal procedures, the reduction of animal numbers, and the replacement of animal models (Romberg et al., 2012). Several methods for rabies vaccine potency testing have been reported as possible replacements for the NIH test including: serologic assays; *in vitro* antigen quantification methods such as ELISA, immunocapture ELISA (IC-ELISA), single radial immunodiffusion (SRID), luciferase immunoprecipitation system (LIPS), and antigen binding tests (ABT); an *in vivo* potency challenge test with refinement (use of analgesics, anesthetics, and humane end points to minimize pain and distress); and reducing number of mice per dilution, and the number of dilutions tested, and elimination of duplicate testing and testing of multiple batches at one time (de Moura, de Araujo, Cabello, Romijn, & Leite, 2009; Kramer, Schildger, Behrensdorf-Nicol, Hanschmann, & Duchow, 2009). New methods and approaches are currently sought that are more humane; use fewer or no animals; are more rapid, less expensive and more precise; and are safer for laboratory workers.

2.4 New Generation Rabies Vaccines

2.4.1 Oral Vaccination

Although administration of conventional rabies vaccines by the parenteral route is highly effective in subjects that can be restrained easily, such as humans or domestic species, logistical problems can limit their obvious extension in harder to capture free-ranging animals. The concept of oral rabies vaccine began in the 1960s. It was conceived at CDC in a prescient, multi-disciplinary inference, recognizing the role of herd immunity in canine rabies prevention; the emerging problem of wildlife rabies; the ethical, ecological, and economic constraints of population reduction as a primary means of disease control; the need for remote delivery of vaccines that could be applied to the buccal mucosa via ingestion of baits; and improvement in production methods and the availability of suitable viral seed candidates, such as the ERA strain (Baer, Abelseth, & Debbie, 1971). The first field trials with ERA rabies virus vaccine (referred to as SAD Bern because of the proprietary use of the actual ERA parent virus already as a commercial vaccine), contained in packets within chicken-head baits, took place in Switzerland, focusing initially upon environmental safety, especially in non-target species such as rodents, and later efficacy for red foxes (Steck, Wandeler, Bichsel, Capt, & Schneider, 1982). Thereafter, Germany used SAD Bern to develop their own seed virus (called SAD B19) and replaced the use of hand-delivery of chicken-head baits with bait components and attractants that could be easily automated, with a greater focus on aerial distribution (Muller, Stohr, Teuffert, & Stohr, 1993). Soon, many European countries were using ORV for prevention and control of wildlife (Aubert, Masson, Artois, & Barrat, 1994). A new vaccine produced from SAD Bern by selection of escape mutants from neutralizing monoclonal antibody pressure resulted in a highly attenuated rabies virus vaccine, SAG2, for improved safety in target species, including dogs (Cliquet & Aubert, 2004). By the end of the 20th century, large portions of Western Europe and southern Ontario, Canada, were becoming free of wildlife rabies virus transmission by the strategic use of ORV (MacInnes et al., 2001; Stohr & Meslin, 1996). However, limitations in effectiveness to certain species, and concerns of residual pathogenicity from first generation rabies viruses remained as hurdles in the field and required fresh biotechnological approaches for additional resolution (Fehlner-Gardiner et al., 2008; Muller et al., 2009).

2.4.2 Recombinant Rabies Vaccines

The production of rabies sub-unit biologics has been a goal in vaccine development for decades, due in large part to biosafety concerns from the high case fatality from lyssaviruses, as well as the search for products that are inexpensive, and more pure, potent, and efficacious than via conventional approaches (Wunner, Dietzschold, Curtis, & Wiktor, 1983). Heterologous expression of lyssavirus genes has been attempted in a wide variety of microorganisms, including bacteria, yeast, baculoviruses, herpesviruses, paramyxoviruses, and plant viruses (Benmaamar, Astray, Wagner, & Pereira, 2009; Ge et al., 2011; Wang et al., 2011). In many of these examples, viral G protein has been expressed, reported to be immunogenic, and may have protected animals against lethal rabies virus challenge under experimental conditions. However, while many of these attempts have generated promising supportive preliminary data, to date, only those approaches using recombinant poxviruses or adenoviruses have passed muster from initial conceptual development towards further Good Manufacturing Practices (GMP) production, actual clinical trials, and commercialization.

2.4.3 Poxviruses

During the 1970s, researchers were exploring the potential utility of recombinant DNA methodology as a fundamental technique to understand basic biochemical processes in prokaryotic and eukaryotic cells, as well as the potential biohazards of combining functional genetic material from different organisms (Berg, Baltimore, Brenner, Roblin, & Singer, 1975). In particular, DNA viruses were examined for their ability to serve as cloning and expression vectors, in part as a way to examine the *in situ* properties of different heterologous viral genes. Poxviruses were one of the first agents to be examined because of their capacity to accommodate foreign genes, focusing upon vaccinia virus as a vector, due in part to the basic knowledge gained from its use in the eradication of smallpox (Panicali & Paoletti, 1982). In 1983, a recombinant vaccinia-rabies glycoprotein (V-RG) virus was developed, with the incorporation of the ERA rabies virus G gene into an attenuated isolate of the vaccinia Copenhagen strain (Wiktor et al., 1984). Shortly thereafter, due in no small part because of an unprecedented outbreak of raccoon rabies in the mid-Atlantic states, preliminary experiments began to test the suitability of V-RG orally in animals (Wiktor, Macfarlan, Dietzschold, Rupprecht, & Wunner, 1985). Considering their importance as rabies reservoirs, both raccoons and red foxes were selected initially for captive studies, in which the safety and efficacy of V-RG was demonstrated (Blancou et al., 1986; Rupprecht et al., 1986). While additional safety studies of V-RG were performed in a wide variety of birds and mammals under laboratory conditions, potential field sites were examined for release of this genetically altered virus (Hable et al., 1992). The first European field trial of V-RG occurred in Belgium (Pastoret et al., 1988). The initial site in North America was an island off the coast of Virginia (Hanlon, Niezgoda, Hamir, Schumacher, Koprowski, & Rupprecht, 1998). Since its development, V-RG has been used successfully in campaigns targeting the red fox rabies in Europe, primarily in Belgium and France (Brochier et al., 1991). In North America, since 1990, more than 100 million doses of V-RG have been distributed in the field, leading to the elimination of coyote rabies, the near elimination of gray fox rabies in Texas, and significant suppression of raccoon rabies in Canada and eastern North America (Slate et al., 2009). Based upon national surveillance in the United States, at least two human infections have been described, both cases attributed to direct contact of V-RG virus and damaged skin (Roess et al., 2012).

Over the past several decades, several other poxviruses have been constructed for the expression of rabies virus genes, but none to date, except V-RG, have been licensed for the prevention and control of rabies in free-ranging wildlife (Weyer, Rupprecht, & Nel, 2009). Recombinant avipoxviruses with an inserted rabies virus G gene have been considered for use in both human and veterinary medicine, resulting in a licensed product for domestic cats (CDC, 2011; Fries et al., 1996).

2.4.4 Adenoviruses

Besides poxviruses, adenoviruses have also been a primary focus for expression of foreign genes (Graham & Prevec, 1991). Adenoviruses in the family *Adenoviridae*, subfamily *Mastadenoviridae*, are non-enveloped, DNA viruses commonly found in mammals (Woods, 2001). These viruses have a predilection for the alimentary and respiratory tracts. Typically, adenoviruses result in mild, sub-clinical infections. Candidate adenoviruses for selection as recombinant vaccines for rabies prevention and control have concentrated upon strains isolated from domestic dogs and primates, both human and nonhuman (Chen et al., 2010; Ertl, 2005; Lees et al., 2002; Tims et al., 2000; Tordo, Foumier, Jallet, Szelechowski, Klonjkowski, & Eloit, 2008; Vos et al., 2001; Xiang et al., 2002; Zhou, Cun, Li, Xiang, & Ertl, 2006).

Canine adenoviruses (CAdVs) are highly contagious viruses that can cause severe and sometimes fatal disease in canids and are endemic among domestic dogs (Decaro, Martella, & Buonavoglia, 2008). The CAdVs are comprised of two major types, 1 and 2. Type 1 virus causes canine hepatitis, affecting the liver of dogs and related species, where cell damage and hemorrhage can occur, leading to death due to shock. An infected dog sheds adenovirus in the feces and urine for prolonged periods. Type 2 CAdV is one of the causes of infectious tracheobronchitis in dogs, known as "kennel cough." Because of the less severe nature of adverse events, CAdV2 has been a primary focus for rabies vaccine development (Hu et al., 2006; Li et al., 2006). Captive studies with dogs, raccoons, and skunks have demonstrated the safety and efficacy of CAdV2 recombinant rabies vaccines by the oral route (Henderson et al., 2009; Zhang, Liu, Fooks, Zhang, & Hu, 2008). Although CAdV2 may appear to be a more ideal candidate than other vectors for oral immunization of free-ranging dogs and wildlife for safety reasons, interference from prior adenovirus exposure may prevent adequate immunization and expression of foreign genes, such as rabies virus. If widespread natural immunity to CAdV among wild carnivores and dogs impedes its development as an oral vaccine in those species, there may be utility for parenteral application in other mammals (Bouet-Cararo et al., 2011; Hu, Liu, Zhang, Zhang & Fooks, 2007; Liu, Zhang, Ma, Zhang, & Hu, 2008).

In a reverse One Health context, it is a bit counterintuitive that human adenoviruses would serve as potential immunogens for carnivores in a capacity better than those isolated from other animals. In humans, adenovirus infections are fairly ubiquitous, and they are not responsible for significant illness in healthy immune-competent people (Echavarria, 2008). Currently, six human adenovirus species are recognized (A–F), with 51 serotypes. Isolates of the C adenovirus species are associated with milder respiratory symptoms, and a majority of people are exposed to adenoviruses during childhood (Sun et al., 2011). In addition, from the standpoint of recombinant vaccines, viral vectors based on human adenovirus type 5 (HAdV-5), a type C adenovirus, have been used in human gene therapy, with few serious events (Douglas, 2007). Despite their apparent safety record, there is an opportunity for human or non-human adenoviruses to be considered as recombinant rabies virus vectors. It remains to be seen if they will be licensed for use in humans, in contrast to veterinary applications (Ertl, 2005).

A HAdV-5 vectored-rabies virus glycoprotein recombinant vaccine (AdRG1.3, known as ONRAB®) has been developed in Canada, produced by Artemis (Shen et al., 2012). The ONRAB® vaccine consists of a replication-competent HAdV-5 vector with an insertion of the ERA rabies virus G gene into the E3 region (Knowles et al., 2009). Insertion of the rabies virus G gene into the HAdV-5 E3 region deletes most activity of the adenovirus E3 gene, involved in evasion of host immune responses (Wold, Tollefson, & Hermiston, 1995). There were no documented environmental or public health impacts associated with the distribution of ONRAB® in Ontario (Rosatte et al., 2009).

Most of the theoretical concerns on the use of adenoviruses as vaccines stem from the perception regarding potential tumorigenicity, or effects after recombination in nature. Adenoviruses have the ability to induce tumors in certain experimental animal models, and transform cells in culture, although no naturally occurring tumors have been documented in animals. The species C (types 1, 2, 5, and 6) viruses are considered non-oncogenic, with little potential for cellular transformation (Graham, Rowe, McKinnon, Bacchetti,, Ruben, & Branton, 1984). The concern about mixing of genetic elements under field conditions stems from the concept that much of adenovirus evolution is driven in part by homologous recombination. Hence, one question is whether genetically modified adenoviruses might encounter others in nature, exchange genetic material, and produce new viruses that could cause disease. As experimental observations in captivity have shown, multiple mammalian species can be infected with HAdV-5, and serologic evidence suggests that humans are occasionally naturally infected with CAdV (Kremer, Boutin, Chillon, & Danos, 2000). However, naturally occurring recombinant adenovirus strains have demonstrated that homologous recombination is restricted to closely related strains or serotypes within the same species and to regions of greatest genetic homology (Munz & Young, 1983). Minimum sequence homology requirements for efficient recombination have been estimated to be more than 200 bp in mammalian cells (Liskay, Letsou, & Stachelek, 1987). Thus, the risk of recombination between ONRAB® and naturally occurring adenoviruses, resulting in more virulent progeny, is felt to be extremely remote.

In addition to important reservoirs such as raccoons and foxes, the extension of ONRAB® in other species may also include skunks, for which the V-RG vaccine may have more limited efficacy. Captive skunks administered ONRAB® orally developed rabies virus neutralizing antibodies (Knowles et al., 2009). Moreover, in Ontario field trials, high-density baiting (300 baits/km^2) appears to have achieved limited rabies virus transmission among skunks. However, other field work has shown that a lower proportion of skunks may develop rabies virus antibody in comparison to raccoons. Additional applied research on both vaccine and bait development will be useful towards resolution of this question.

Recently, a U.S. field trial of the ONRAB® vaccine occurred in West Virginia during 2011. No adverse events have been reported to date. Considering preliminary Canadian laboratory and field research published thus far, no overt adverse human, veterinary, or environmental incidents are anticipated (Knowles et al., 2009; Rosatte et al., 2011). Nevertheless, as with any genetically modified agent, enhanced surveillance, ecological monitoring, and ongoing agricultural and public outreach are necessary to gauge properly the relative risks from environmental release of such viruses, given the opportunity for recombination. Theoretically, recombination between ONRAB® and CAdV or other adenoviruses could result in replication-competent viruses with novel host ranges, tissue tropisms, and disease potential. However, in the case of CAdV types 1 and 2, genomic sequence identity with HAdV-5 is believed to be very low, and for intergenomic recombination to occur, these viruses would have to simultaneously infect the same cell, which should be a very rare event, and in practice, opportunities for co-infection between different adenovirus species are already operative in the environment. Very limited evidence suggests that HAdV-5 can infect canine cells productively (Ternovoi, Le, Belousova, Smith, Siegal & Curiel, 2005). Thus, while the risk of recombination between ONRAB® and other wild-type adenoviruses is not zero, if recombination does occur, it is not likely that such an event would yield a viable, transmissible virus, with enhanced pathogenicity for wildlife, domestic animals, or humans.

Besides direct scrutiny upon wildlife during such limited field trials, plans are under development in the United States to adapt an existing system for detection of suspected adverse events associated with recombinant vaccine. For example, current oversight for human and pet contact with the V-RG virus in the United States consists of a passive surveillance system, in which the public can call a number printed on the vaccine container if they discover a bait. Calls are routed to appropriate designated local personnel to provide basic and biomedical information. Information is collected on the nature of the exposure, including: circumstances of the bait discovery; where the bait was found; whether there was contact with the bait, and how it was retrieved; and whether the vaccine sachet was compromised, resulting in exposure to vaccine virus. These data are collected by federal, state, and local collaborators in participating states using a standardized form at the end of the baiting season. One limitation of this particular system is the delay in aggregation of data until the end of the season, reducing the ability to monitor data in real time. Another is the difference in expected adverse events, such as rash-like illness or vesicular lesions from V-RG, as a poxvirus, in contrast mild respiratory or enteric signs from the adenovirus ONRAB®. Enhanced syndromic surveillance for unusual increases in compatible symptoms on a local level may be warranted for the early detection of any potential shifts in existing rates of illness in localities where ONRAB® would be deployed (Bourgeois, Olson, Brownstein, McAdam, & Mandl, 2006). Given favorable safety, immunogenicity, and efficacy data, this human adenovirus vaccine should be instrumental as another tool in the continued efforts to prevent wildlife rabies in North America and elsewhere, as it is prepared for regulatory review and eventual licensure (Fehlner-Gardiner et al., 2012).

2.4.5 Rhabdoviruses

Historically, unlike the progress in DNA virus recombinant techniques, similar application to the single-stranded negative-sense RNA viruses was quite limited, and rhabdoviruses represented one of the few important families of viral pathogens which could not be manipulated easily from cDNA throughout most of the 20th century. This inability changed with the development of reverse genetics methodology (Schnell, Mebatsion, & Conzelmann, 1994). Such technology permitted a powerful means for studying basic viral pathogenicity, improving the construction of existing vaccines, and creating an opportunity for a new generation of biologics by using these agents as vaccine vectors (Gomme, Wanjalla, Wirblich, & Schnell, 2011; Morimoto et al., 2001; Prehaud et al., 2010; Pulmanausahakul, Li, Schnell, & Dietzschold, 2008; Schnell, McGettigan, Wirblich, & Papaneri, 2010; Wu & Rupprecht, 2008). Compared to the first generation of modified-live rabies viruses based upon the ERA/SAD, LEP/HEP, and other seed strains, most biosafety concerns about the theoretical use of rabies viruses might be alleviated by the precise molecular attenuation of key genes and minimization of replication competence (Dietzschold, Li, Faber, & Schnell, 2008; Du, Tang, Huang, Rodney, Wang, & Liang, 2011; Gomme, Faul, Flomenberg, McGettigan, & Schnell, 2010; Wen et al., 2011). Resulting progeny viruses could be considered for either conventional PreP or PEP scenarios for humans or domestic animals (Cenna et al., 2009; Faber, Li, Kean, Hooper, Alugupalli, & Dietzschold,

2009b; Li et al., 2012; Wu, Franka, Henderson, & Rupprecht, 2011). In addition, because this new generation of rabies viruses is efficacious not only parenterally but also by the oral route, novel candidates may be considered for the immunization of dogs and important wild carnivore reservoirs (Blanton, Self, Niezgoda, Faber, Dietzschold, & Rupprecht, 2007; Dietzschold, Faber, Mattis, Pak, Schnell, & Dietzschold, 2004; Faber et al., 2009b; Faber et al., 2005; Rupprecht et al., 2005; Vos et al., 2011). Moreover, the use of modern reverse genetics not only provides a novel manner for rhabdoviruses to serve as antigenic targets against rabies but also a means for incorporation of foreign genes for toxins, additional zoonoses, and other infectious diseases (Blaney et al., 2011; Faber et al., 2005; Mustafa et al., 2011; Smith et al., 2006; Wanjalla, Faul, Gomme, & Schnell, 2010). Use of the same approach with reproductive antigens suggests that a dual immuno-contraceptive biologic may be designed for a major paradigm shift by simultaneous rabies immunization and population management of dogs and other species (Fayrer-Hosken, 2008; Wu, Franka, Svoboda, Pohl, & Rupprecht, 2009). Despite significant progress in methodology over the past decade, unlike the case for the recombinant poxvirus rabies vaccine, V-RG Raboral, or the recombinant adenovirus rabies vaccine, ONRAB, as of yet, none of these new recombinant rabies virus vaccines have been licensed or tested extensively in the field for either public health, agricultural, or wildlife medicine applications.

2.5 Future Directions

Rabies can be prevented even after exposure has occurred, and yet tens of thousands of human deaths occur annually, amounting to one unnecessary fatality every 10 minutes. The burden of rabies among livestock, wildlife, and domestic animals, especially dogs in canine rabies enzootic countries, is unknown, but must range into the millions of annual cases by even a conservative estimate. Access to efficacious rabies biologics is limited in impoverished regions that incur the greatest rabies burden. Victims in industrialized nations that could afford PEP indicate a lack of awareness as part of the problem. Therefore we conclude this chapter by providing a list of needs that would alleviate the current global burden of rabies in both humans and animals:

- *Simpler vaccine delivery systems.* These would include both improved ID devices and ORVs for humans. Simpler ID devices for CCVs would encourage developing nations to replace NTVs with CCVs, broaden the appeal to expand the use of ID PEP, reduce vaccine costs for resource poor nations, and increase accessibility to CCVs. In many animal species, ORV has already proven to be effective and new ORVs for humans would eliminate needle injury and transmission of other etiological agents in the clinical setting. Alternate safe routes for vaccine delivery outside of IM, ID, and subcutaneous injections require investigation and lessons learned from delivery of other biologics, including pandemic influenza, as a starting point.

- *Paradigm shift in PreP to protect humans and PEP to protect domestic animals*. To date the use of PreP has concentrated upon domestic animals and the use of PEP has only been applicable to humans. However, broader application of PreP for humans is also needed as a paradigm shift to reduce health disparities in particular communities, especially among pediatric populations under frequent risk of viral infection, in which present logistics and economics are not effective based on conventional PEP. Studies into the application of PEP in naïve animal species are also warranted.

- *Reduced vaccination regimens for PreP and PEP*. Current PreP and PEP regimens for humans are long and require several clinic visits. There is a need to investigate the possibility of reducing the time between injections and the total number of doses. Evidence obtained cumulatively from basic clinical responses could be employed to further explore prime-boost strategies and minimum dose-sparing schedules in human PEP.

- *Use of biotechnology to improve rabies biologics*. Considering the revolution in reverse genetics, conducting field trials for recombinant rabies virus vaccines should be considered. Additionally, the use of homologous or heterologous MAbs for PEP should be imminent as an adjunct or total replacement of RIG, especially considering that essential hybridoma technology is more than three decades old.

- *Improved animal welfare applications in production of rabies biologics*. There is an urgent need to improve methods to manage dog populations in regions where canine rabies continues to be endemic and dog fecundity remains unchecked. The use of immuno-contraception for dogs offers the best hope for a humane approach to the problem. Given the complexity of dual measurement of rabies immunogenicity and efficacy together with contraception, limited clinical trials might be preferable to traditional captive studies, once the primary safety and proof of concept of canine rabies immuno-contraception have been demonstrated. Additionally, re-evaluation of the NIH test that requires inoculating hundreds of mice to evaluate the potency of human and animal rabies vaccines is overdue. Eventual universal refinement, reduction and replacement of the use of animals in the NIH test will require careful regulatory assessment on a global basis.

- *A One Health focus on canine rabies elimination*. Considering the eradication of smallpox and rinderpest, and considerations for similar strategies targeting polio, measles, and so on, a World Health Assembly resolution would enable the technical ability and economic resources necessary to achieve the above proposals, with enhanced collaboration among FAO, OIE, WHO, and other principal stakeholders, to remove the single largest factor in the global burden of rabies by creation of herd immunity via proven modern vaccination strategies.

ACKNOWLEDGEMENTS

The authors thank many staff in their home institutions for their contributions to this chapter, as well as multiple colleagues around the world, including: Michael Attlan, Alexander Botvinrin, Sheila de Matos Xavier, Wlamir de Moura, Jacques Delbecque, Anthony Fooks, Sergio Gribencha, Mal Hoover, Rongliang Hu, Ivan Kuzmin, G Kamaraj, Dieter Kneil, R. Kumaran, Fernando Leanes, Francois Meslin, Louis Nel, Gerd Rundstrom, Joanne Maki, Claudius Malerczyk, B. Mohanasubramanian, Anvar Rassouli, Sergio Recuenco, V. A. Srinivasan, Changchun Tu, Marco Vigilato, Xianfu Wu, Naoki Yanagisawa, Andres Velasco-Villa, and Ad Vos.

The views represented in this communication represent that of the authors and not necessarily their institutions.

Use of commercial names is for comparison purposes only and does not represent endorsement.

References

Atanasiu, P., Ribeiro, M., & Tsiang, H. (1972). [Antirabies vaccines from tissue culture obtained with the Pasteur strain. Results of vaccination]. *Annales de l'Institut Pasteur (Paris), 123*(3), 427–441.

Aubert, M. F., Masson, E., Artois, M., & Barrat, J. (1994). Oral wildlife rabies vaccination field trials in Europe, with recent emphasis on France. *Current Topics in Microbiology and Immunology, 187*, 219–243.

Baer, G. M., Abelseth, M. K., & Debbie, J. G. (1971). Oral vaccination of foxes against rabies. *American Journal of Epidemiology, 93*(6), 487–490.

Bahnemann, H. G. (1990). Inactivation of viral antigens for vaccine preparation with particular reference to the application of binary ethylenimine. *Vaccine, 8*(4), 299–303.

Barone, D., Fracchia, S., Pascuali, E., & Proglio, F. (1997, October 5-6). *ICH-4 guidelines for the quality and safety of cell substrate used in the production of pharmaceuticals.* Paper presented at the Italian-German Biotech Forum: Pharmaceutical Biotechnology.

Barth, R., Gruschkau, H., Bijok, U., Hilfenhaus, J., Hinz, J., Milcke, L., et al. (1984). A new inactivated tissue culture rabies vaccine for use in man. Evaluation of PCEC-vaccine by laboratory tests. *Journal of Biological Standardization, 12*(1), 29–46.

Benmaamar, R., Astray, R. M., Wagner, R., & Pereira, C. A. (2009). High-level expression of rabies virus glycoprotein with the RNA-based Semliki Forest Virus expression vector. *Journal of Biotechnology, 139*(4), 283–290. doi:10.1016/j.jbiotec.2008.12.009.

Berg, P., Baltimore, D., Brenner, S., Roblin, R. O., & Singer, M. F. (1975). Summary statement of the Asilomar conference on recombinant DNA molecules. *Proceedings of the National Academy of Sciences of the United States of America, 72*(6), 1981–1984.

Birch, J. R. (1999). Suspension culture, animal cells. In M. C. Flickinger & S. W. Drew (Eds.), *Bioprocess technology: Fermentation, biocatalysis and bioseparation (pp. 2509–2516).* Wiley.

Blancou, J., Aubert, M. F., Cain, E., Selve, M., Thraenhart, O., & Bruckner, L. (1989). Effect of strain differences on the potency testing of rabies vaccines in mice. *Journal of Biological Standardization*, 17(3), 259–266.

Blancou, J., Kieny, M. P., Lathe, R., Lecocq, J. P., Pastoret, P. P., Soulebot, J. P., et al. (1986). Oral vaccination of the fox against rabies using a live recombinant vaccinia virus. *Nature*, 322(6077), 373–375. doi:10.1038/322373a0.

Blaney, J. E., Wirblich, C., Papaneri, A. B., Johnson, R. F., Myers, C. J., Juelich, T. L., et al. (2011). Inactivated or live-attenuated bivalent vaccines that confer protection against rabies and Ebola viruses. *Journal of Virological*, 85(20), 10605–10616. doi:10.1128/JVI.00558-11.

Blanton, J. D., Self, J., Niezgoda, M., Faber, M. L., Dietzschold, B., & Rupprecht, C. (2007). Oral vaccination of raccoons (*Procyon lotor*) with genetically modified rabies virus vaccines. *Vaccine*, 25(42), 7296–7300. doi:10.1016/j.vaccine.2007.08.004.

Bouet-Cararo, C., Contreras, V., Fournier, A., Jallet, C., Guibert, J. M., Dubois, E., et al. (2011). Canine adenoviruses elicit both humoral and cell-mediated immune responses against rabies following immunisation of sheep. *Vaccine*, 29(6), 1304–1310. doi:10.1016/j.vaccine.2010.11.068.

Bourgeois, F. T., Olson, K. L., Brownstein, J. S., McAdam, A. J., & Mandl, K. D. (2006). Validation of syndromic surveillance for respiratory infections. *Annals of Emergency Medicine*, 47(3), 265–271. doi:10.1016/j.annemergmed.2005.11.022.

Brochier, B., Kieny, M. P., Costy, F., Coppens, P., Bauduin, B., Lecocq, J. P., et al. (1991). Large-scale eradication of rabies using recombinant vaccinia-rabies vaccine. *Nature*, 354(6354), 520–522. doi:10.1038/354520a0.

CDC, (2004). Manufacturer's recall of human rabies vaccine. *MMWR Dispatch*, 53, 287–289.

CDC, (2011). Compendium of animal rabies prevention and control, 2011. *MMWR Recommendations and Reports*, 60(RR-6), 1–17.

CDC. (2012). Rabies vaccine supply update: IMOVAX rabies availability Retrieved February 29, 2012, from <http://www.cdc.gov/rabies/resources/news/vaccine_supply/index.html/>.

Cenna, J., Hunter, M., Tan, G. S., Papaneri, A. B., Ribka, E. P., Schnell, M. J., et al. (2009). Replication-deficient rabies virus-based vaccines are safe and immunogenic in mice and nonhuman primates. *Journal of Infectious Diseases*, 200(8), 1251–1260. doi:10.1086/605949.

Chapman, W. G., Ramshaw, I. A., & Crick, J. (1973). Inactivated rabies vaccine produced from the Flury LEP strain of virus grown in BHK-21 suspension cells. *Appl Microbiol*, 26(6), 858–862.

Chen, H., Xiang, Z. Q., Li, Y., Kurupati, R. K., Jia, B., Bian, A., et al. (2010). Adenovirus-based vaccines: Comparison of vectors from three species of adenoviridae. *Journal of Virological*, 84(20), 10522–10532. doi:10.1128/JVI.00450-10.

Cliquet, F., & Aubert, M. (2004). Elimination of terrestrial rabies in Western European countries. *Developmental Biology (Basel)*, 119, 185–204.

Corbel, M. J. (1994). Control testing of combined vaccines: A consideration of potential problems and approaches. *Biologicals*, 22(4), 353–360. doi:10.1006/biol.1994.1054.

de Moura, W. C., de Araujo, H. P., Cabello, P. H., Romijn, P. C., & Leite, J. P. (2009). Potency evaluation of rabies vaccine for human use: The impact of the reduction in the number of animals per dilution. *Journal of Virological Methods*, 158(1-2), 84–92. doi:10.1016/j.jviromet.2009.01.017.

Decaro, N., Martella, V., & Buonavoglia, C. (2008). Canine adenoviruses and herpesvirus. *The Veterinary Clinics of North America. Small Animal Practice*, 38(4), 799–814. doi:10.1016/j.cvsm.2008.02.006. viii.

Di Fabio, J. L., & de Quadros, C. (2001). Considerations for combination vaccine development and use in the developing world. *Clin Infect Dis*, 33(Suppl. 4), S340–345. doi:10.1086/322571.

Dietzschold, B., Faber, M., & Schnell, M. J. (2003). New approaches to the prevention and eradication of rabies. *Expert Review of Vaccines*, 2(3), 399–406. doi:10.1586/14760584.2.3.399.

Dietzschold, B., Li, J., Faber, M., & Schnell, M. (2008). Concepts in the pathogenesis of rabies. *Future Virology*, 3(5), 481–490. doi:10.2217/17460794.3.5.481.

Dietzschold, M. L., Faber, M., Mattis, J. A., Pak, K. Y., Schnell, M. J., & Dietzschold, B. (2004). In vitro growth and stability of recombinant rabies viruses designed for vaccination of wildlife. *Vaccine*, 23(4), 518–524. doi:10.1016/j.vaccine.2004.06.031.

Douglas, J. T. (2007). Adenoviral vectors for gene therapy. *Molecular Biotechnology*, 36(1), 71–80.

Du, J., Tang, Q., Huang, Y., Rodney, W. E., Wang, L., & Liang, G. (2011). Development of recombinant rabies viruses vectors with Gaussia luciferase reporter based on Chinese vaccine strain CTN181. *Virus Research*, 160(1–2), 82–88. doi:10.1016/j.virusres.2011.05.018.

Echavarria, M. (2008). Adenoviruses in immunocompromised hosts. *Clinical Microbiology Reviews*, 21(4), 704–715. doi:10.1128/CMR.00052-07.

Ertl, H. C. (2005). Immunological insights from genetic vaccines. *Virus Research*, 111(1), 89–92. doi:10.1016/j.viruses.2005.03.015.

Faber, M., Dietzschold, B., & Li, J. (2009a). Immunogenicity and safety of recombinant rabies viruses used for oral vaccination of stray dogs and wildlife. *Zoonoses Public Health*, 56(6-7), 262–269. doi:10.1111/j.1863-2378.2008.01215.x.

Faber, M., Lamirande, E. W., Roberts, A., Rice, A. B., Koprowski, H., Dietzschold, B., et al. (2005). A single immunization with a rhabdovirus-based vector expressing severe acute respiratory syndrome coronavirus (SARS-CoV) S protein results in the production of high levels of SARS-CoV-neutralizing antibodies. *The Journal of General Virology*, 86(Pt 5), 1435–1440. doi:10.1099/vir.0.80844-0.

Faber, M., Li, J., Kean, R. B., Hooper, D. C., Alugupalli, K. R., & Dietzschold, B. (2009b). Effective preexposure and postexposure prophylaxis of rabies with a highly attenuated recombinant rabies virus. *Proceedings of the National Academy of Sciences of the United States of America*, 106(27), 11300–11305. doi:10.1073/pnas.0905640106.

Fayrer-Hosken, R. (2008). Controlling animal populations using anti-fertility vaccines. *Reproduction in Domestic Animals*, 2, 179–185.

Fehlner-Gardiner, C., Nadin-Davis, S., Armstrong, J., Muldoon, F., Bachmann, P., & Wandeler, A. (2008). Era vaccine-derived cases of rabies in wildlife and domestic animals in Ontario, Canada, 1989–2004. *Journal of Wildlife Diseases*, 44(1), 71–85.

Fehlner-Gardiner, C., Rudd, R., Donovan, D., Slate, D., Kempf, L., & Badcock, J. (2012). Comparing ONRAB(R) AND RABORAL V-RG(R) oral rabies vaccine field performance in raccoons and striped skunks, New Brunswick, Canada, and Maine, USA. *Journal of Wildlife Diseases*, 48(1), 157–167.

Fries, L. F., Tartaglia, J., Taylor, J., Kauffman, E. K., Meignier, B., Paoletti, E., et al. (1996). Human safety and immunogenicity of a canarypox-rabies glycoprotein recombinant vaccine: An alternative poxvirus vector system. *Vaccine*, 14(5), 428–434.

Gaudin, Y., Ruigrok, R. W., Tuffereau, C., Knossow, M., & Flamand, A. (1992). Rabies virus glycoprotein is a trimer. *Virology*, 187(2), 627–632.

Ge, J., Wang, X., Tao, L., Wen, Z., Feng, N., Yang, S., et al. (2011). Newcastle disease virus-vectored rabies vaccine is safe, highly immunogenic, and provides long-lasting protection in dogs and cats. *Journal of Virological*, 85(16), 8241–8252. doi:10.1128/JVI.00519-11.

Gomme, E. A., Faul, E. J., Flomenberg, P., McGettigan, J. P., & Schnell, M. J. (2010). Characterization of a single-cycle rabies virus-based vaccine vector. *Journal of Virological*, 84(6), 2820–2831. doi:10.1128/JVI.01870-09.

Gomme, E. A., Wanjalla, C. N., Wirblich, C., & Schnell, M. J. (2011). Rabies virus as a research tool and viral vaccine vector. *Advances in Virus Research*, 79, 139–164. doi:10.1016/B978-0-12-387040-7.00009-3.

Grachev, V. P. (1990). World Health Organization attitude concerning the use of continuous cell lines as substrates for production of human virus vaccines. In A. (1990). Mizrahi (Ed.), *Advances in biotechnological processes. Viral vaccines* (Vol. 14, pp. 37–67). Wiley-Liss.

Graham, F. L., & Prevec, L. (1991). Manipulation of adenovirus vectors. *Methods in Molecular Biology, 7,* 109–128.

Graham, F. L., Rowe, D. T., McKinnon, R., Bacchetti, S., Ruben, M., & Branton, P. E. (1984). Transformation by human adenoviruses. *Journal of Cellular Physiology. Supplement, 3,* 151–163.

Habel, K. (1973). *Laboratory techniques in rabies: General considerations in vaccine production* (Vol. 23). World Health Organization. (pp. 189–191).

Hable, C. P., Hamir, A. N., Snyder, D. E., Joyner, R., French, J., Nettles, V., et al. (1992). Prerequisites for oral immunization of free-ranging raccoons (Procyon lotor) with a recombinant rabies virus vaccine: Study site ecology and bait system development. *Journal of Wildlife Diseases, 28*(1), 64–79.

Hampson, K., Cleaveland, S., & Briggs, D. (2011). Evaluation of cost-effective strategies for rabies post-exposure vaccination in low-income countries. *PLoS Neglected Tropical Diseases, 5*(3), e982. doi:10.1371/journal.pntd.0000982.

Hampson, K., Dushoff, J., Cleaveland, S., Haydon, D. T., Kaare, M., Packer, C., et al. (2009). Transmission dynamics and prospects for the elimination of canine rabies. *PLoS Biology, 7*(3), e53. doi:10.1371/journal.pbio.1000053.

Hanlon, C. A., Niezgoda, M., Hamir, A. N., Schumacher, C., Koprowski, H., & Rupprecht, C. E. (1998). First North American field release of a vaccinia-rabies glycoprotein recombinant virus. *Journal of Wildlife Diseases, 34*(2), 228–239.

Hassanzadeh, S. M., Zavareh, A., Shokrgozar, M. A., Ramezani, A., & Fayaz, A. (2011). High vero cell density and rabies virus proliferation on fibracel disks versus cytodex-1 in spinner flask. *Pakistan Journal of Biological Sciences, 14*(7), 441–448.

Hay, R. J., Caputo, J., & Chen, T. R. (1994). *ATCC cell lines and hybridomas* (8th ed., pp. 640).

Henderson, H., Jackson, F., Bean, K., Panasuk, B., Niezgoda, M., Slate, D., et al. (2009). Oral immunization of raccoons and skunks with a canine adenovirus recombinant rabies vaccine. *Vaccine, 27*(51), 7194–7197. doi:10.1016/j.vaccine.2009.09.030.

Hendricksen, C. F., Spieser, J. M., Akkermans, A., Balls, M., Bruckner, L., Cussler, K., et al. (1998). Validation of alternative methods for the potency testing of vaccines. *The report and recommendations of ECVAM workshop 31. Alternatives to laboratory animals,* Vol. 26. (pp. 747–761).

Hendriksen, C. F. (2009). Replacement, reduction and refinement alternatives to animal use in vaccine potency measurement. *Expert Review of Vaccines, 8*(3), 313–322. doi:10.1586/14760584.0.0.313.

Hu, R., Zhang, S., Fooks, A. R., Yuan, H., Liu, Y., Li, H., et al. (2006). Prevention of rabies virus infection in dogs by a recombinant canine adenovirus type-2 encoding the rabies virus glycoprotein. *Microbes and Infection, 8*(4), 1090–1097. doi:10.1016/j.micinf.2005.11.007.

Hu, R. L., Liu, Y., Zhang, S. F., Zhang, F., & Fooks, A. R. (2007). Experimental immunization of cats with a recombinant rabies-canine adenovirus vaccine elicits a long-lasting neutralizing antibody response against rabies. *Vaccine, 25*(29), 5301–5307. doi:10.1016/j.vaccine.2007.05.024.

Kallel, H., Diouani, M. F., Loukil, H., Trabelsi, K., Snoussi, M. A., Majoul, S., et al. (2006). Immunogenicity and efficacy of an in-house developed cell-culture derived veterinarian rabies vaccine. *Vaccine, 24*(22), 4856–4862. doi:10.1016/j.vaccine.2006.03.012.

Kallel, H., Rourou, S., Majoul, S., & Loukil, H. (2003). A novel process for the production of a veterinary rabies vaccine in BHK-21 cells grown on microcarriers in a 20-l bioreactor. *Applied Microbiology and Biotechnology, 61*(5-6), 441–446. doi:10.1007/s00253-003-1245-3.

Knowles, M. K., Nadin-Davis, S. A., Sheen, M., Rosatte, R., Mueller, R., & Beresford, A. (2009). Safety studies on an adenovirus recombinant vaccine for rabies

(AdRG1.3-ONRAB) in target and non-target species. *Vaccine*, 27(47), 6619–6626. doi:10.1016/j.vaccine.2009.08.005.

Koprowski, H., & Cox, H. R. (1948). Studies on chick embryo adapted rabies virus; culture characteristics and pathogenicity. *Journal of Immunology*, 60(4), 533–554.

Kramer, B., Schildger, H., Behrensdorf-Nicol, H. A., Hanschmann, K. M., & Duchow, K. (2009). The rapid fluorescent focus inhibition test is a suitable method for batch potency testing of inactivated rabies vaccines. *Biologicals*, 37(2), 119–126.

Kremer, E. J., Boutin, S., Chillon, M., & Danos, O. (2000). Canine adenovirus vectors: An alternative for adenovirus-mediated gene transfer. *Journal of Virological*, 74(1), 505–512.

Langevin, C., Jaaro, H., Bressanelli, S., Fainzilber, M., & Tuffereau, C. (2002). Rabies virus glycoprotein (RVG) is a trimeric ligand for the N-terminal cysteine-rich domain of the mammalian p75 neurotrophin receptor. *The Journal of Biological Chemistry*, 277(40), 37655–37662. doi:10.1074/jbc.M201374200.

Langevin, C., & Tuffereau, C. (2002). Mutations conferring resistance to neutralization by a soluble form of the neurotrophin receptor (p75NTR) map outside of the known antigenic sites of the rabies virus glycoprotein. *Journal of Virological*, 76(21), 10756–10765.

Lees, C. Y., Briggs, D. J., Wu, X., Davis, R. D., Moore, S. M., Gordon, C., et al. (2002). Induction of protective immunity by topic application of a recombinant adenovirus expressing rabies virus glycoprotein. *Veterinary Microbiology*, 85(4), 295–303.

Li, J., Ertl, A., Portocarrero, C., Barkhouse, D. A., Dietzschold, B., Hooper, D. C., et al. (2012). Postexposure treatment with the live-attenuated Rabies Virus (RV) vaccine TriGAS triggers the clearance of Wild-Type RV from the Central Nervous System (CNS) through the rapid induction of genes relevant to adaptive immunity in CNS tissues. *Journal of Virological*, 86(6), 3200–3210. doi:10.1128/JVI.06699-11.

Li, J., Faber, M., Papaneri, A., Faber, M. L., McGettigan, J. P., Schnell, M. J., et al. (2006). A single immunization with a recombinant canine adenovirus expressing the rabies virus G protein confers protective immunity against rabies in mice. *Virology*, 356(1-2), 147–154. doi:10.1016/j.virol.2006.07.037.

Liskay, R. M., Letsou, A., & Stachelek, J. L. (1987). Homology requirement for efficient gene conversion between duplicated chromosomal sequences in mammalian cells. *Genetics*, 115(1), 161–167.

Liu, Y., Zhang, S., Ma, G., Zhang, F., & Hu, R. (2008). Efficacy and safety of a live canine adenovirus-vectored rabies virus vaccine in swine. *Vaccine*, 26(42), 5368–5372. doi:10.1016/j.vaccine.2008.08.001.

MacInnes, C. D., Smith, S. M., Tinline, R. R., Ayers, N. R., Bachmann, P., Ball, D. G., et al. (2001). Elimination of rabies from red foxes in eastern Ontario. *Journal of Wildlife Diseases*, 37(1), 119–132.

Mondal, S. K., Neelima, M., Seetha Rama Reddy, K., Ananda Rao, K., & Srinivasan, V. A. (2005). Validation of the inactivant binary ethylenimine for inactivating rabies virus for veterinary rabies vaccine production. *Biologicals*, 33(3), 185–189. doi:10.1016/j.biologicals.2005.05.003.

Montomoli, E., Piccirella, S., Khadang, B., Mennitto, E., Camerini, R., & De Rosa, A. (2011). Current adjuvants and new perspectives in vaccine formulation. *Expert Review of Vaccines*, 10(7), 1053–1061. doi:10.1586/erv.11.48.

Morefield, G. L., Tammariello, R. F., Purcell, B. K., Worksham, P. L., Chapman, J., Smith, L. A., et al. (2008). An alternative approach to combination vaccines: Intradermal administration of isolated ocmponents for control of anthrax, botulism, plague and staphylococcal toxic syndrome. *Journal of Immune Based Therapies and Vaccines*, 6(5), 1–11.

Morimoto, K., McGettigan, J. P., Foley, H. D., Hooper, D. C., Dietzschold, B., & Schnell, M. J. (2001). Genetic engineering of live rabies vaccines. *Vaccine*, 19(25-26), 3543–3551.

Muller, T., Batza, H. J., Beckert, A., Bunzenthal, C., Cox, J. H., Freuling, C. M., et al. (2009). Analysis of vaccine-virus-associated rabies cases in red foxes (Vulpes vulpes) after oral rabies vaccination campaigns in Germany and Austria. *Archives of Virology, 154*(7), 1081–1091. doi:10.1007/s00705-009-0408-7.

Muller, T., Stohr, K., Teuffert, J., & Stohr, P. (1993). Experiences with the aerial distribution of baits for the oral immunization of foxes against rabies in eastern Germany. *Deutsche Tierarztliche Wochenschrift, 100*(5), 203–207.

Munz, P. L., Young, C., & Young, C. S. (1983). The genetic analysis of adenovirus recombination in triparental and superinfection crosses. *Virology, 126*(2), 576–586.

Munz, P. L., & Young, C. S. (1984). Polarity in adenovirus recombination. *Virology, 135*(2), 503–514.

Mustafa, W., Al-Saleem, F. H., Nasser, Z., Olson, R. M., Mattis, J. A., Simpson, L. L., et al. (2011). Immunization of mice with the non-toxic HC50 domain of botulinum neurotoxin presented by rabies virus particles induces a strong immune response affording protection against high-dose botulinum neurotoxin challenge. *Vaccine, 29*(28), 4638–4645. doi:10.1016/j.vaccine.2011.04.045.

Ohmori, K., Masuda, K., Maeda, S., Kaburagi, Y., Kurata, K., Ohno, K., et al. (2005). IgE reactivity to vaccine components in dogs that developed immediate-type allergic reactions after vaccination. *Veterinary Immunology and Immunopathology, 104*(3–4), 249–256. doi:10.1016/j.vetimm.2004.12.003.

OIE, (2011). Rabies: *OIE terrestrial manual*. Paris: OIE. (pp. 1–20).

Panicali, D., & Paoletti, E. (1982). Construction of poxviruses as cloning vectors: Insertion of the thymidine kinase gene from herpes simplex virus into the DNA of infectious vaccinia virus. *Proceedings of the National Academy of Sciences of the United States of America, 79*(16), 4927–4931.

Pastoret, P. P., Brochier, B., Languet, B., Thomas, I., Paquot, A., Bauduin, B., et al. (1988). First field trial of fox vaccination against rabies using a vaccinia-rabies recombinant virus. *The Veterinary Record, 123*(19), 481–483.

Prehaud, C., Coulon, P., LaFay, F., Thiers, C., & Flamand, A. (1988). Antigenic site II of the rabies virus glycoprotein: Structure and role in viral virulence. *Journal of Virological, 62*(1), 1–7.

Prehaud, C., Wolff, N., Terrien, E., Lafage, M., Megret, F., Babault, N., et al. (2010). Attenuation of rabies virulence: Takeover by the cytoplasmic domain of its envelope protein. *Sci Signal, 3*(105), ra5. doi:10.1126/scisignal.2000510.

Pulmanausahakul, R., Li, J., Schnell, M. J., & Dietzschold, B. (2008). The glycoprotein and the matrix protein of rabies virus affect pathogenicity by regulating viral replication and facilitating cell-to-cell spread.]. *Journal of Virological, 82*(5), 2330–2338. doi:10.1128/JVI.02327-07.

Reculard, P. (1996). Cell-culture vaccines for veterinary use. In F. M. Meslin, M. M. Kaplan, & H. Koprowski (Eds.), *Laboratory Techniques in Rabies (pp. 314–323)*. Geneva: WHO.

Regan, P. J., & Pettriciani, J. C. (1987). The approach used to establish the safety of veterinary vaccines produced in the BHK-21 cell line. *Developments in Biological Standardization, 68*, 19–25.

Ren, J., Sun, L., Yang, L., Wang, H., Wan, M., Zhang, P., et al. (2010). A novel canine favored CpG oligodeoxynucleotide capable of enhancing the efficacy of an inactivated aluminum-adjuvanted rabies vaccine of dog use. *Vaccine, 28*(12), 2458–2464. doi:10.1016/j.vaccine.2009.12.077.

Roess, A. A., Rea, N., Lederman, E., Dato, V., Chipman, R., Slate, D., et al. (2012). National surveillance for human and pet contact with oral rabies vaccine baits, 2001–2009. *Journal of the American Veterinary Medical Association, 240*(2), 163–168. doi:10.2460/javma.240.2.163.

Romberg, J., Lang, S., Balks, E., Kamphuis, E., Duchow, K., Loos, D., et al. (2012). Potency testing of veterinary vaccines: The way from in vivo to in vitro. *Biologicals, 40*(1), 100–106. doi:10.1016/j.biologicals.2011.10.004.

Rosatte, R. C., Donovan, D., Davies, J. C., Allan, M., Bachmann, P., Stevenson, B., et al. (2009). Aerial distribution of ONRAB baits as a tactic to control rabies in raccoons and striped skunks in Ontario, Canada. *Journal of Wildlife Diseases, 45*(2), 363–374.

Rosatte, R. C., Donovan, D., Davies, J. C., Brown, L., Allan, M., von Zuben, V., & Bachmann, P. (2011). High-density baiting with ONRAB(R) rabies vaccine baits to control Arctic-variant rabies in striped skunks in Ontario, Canada. *J Wildl Dis*, 47(2), 459–465.

Rudd, R. J., & Trimarchi, C. V. (1987). Comparison of sensitivity of BHK-21 and murine neuroblastoma cells in the isolation of a street strain rabies virus. *Journal of Clinical Microbiology & Infectious Diseases*, 25(8), 1456–1458.

Rupprecht, C. E., Hanlon, C. A., Blanton, J., Manangan, J., Morrill, P., Murphy, S., et al. (2005). Oral vaccination of dogs with recombinant rabies virus vaccines. *Virus Research*, 111(1), 101–105. doi:10.1016/j.viruses.2005.03.017.

Rupprecht, C. E., Wiktor, T. J., Johnston, D. H., Hamir, A. N., Dietzschold, B., Wunner, W. H., et al. (1986). Oral immunization and protection of raccoons (Procyon lotor) with a vaccinia-rabies glycoprotein recombinant virus vaccine. *Proceedings of the National Academy of Sciences of the United States of America*, 83(20), 7947–7950.

Schnell, M. J., McGettigan, J. P., Wirblich, C., & Papaneri, A. (2010). The cell biology of rabies virus: Using stealth to reach the brain. *Nature Reviews Microbiology*, 8(1), 51–61. doi:10.1038/nrmicro2260.

Schnell, M. J., Mebatsion, T., & Conzelmann, K. K. (1994). Infectious rabies viruses from cloned cDNA. *EMBO Journal*, 13(18), 4195–4203.

Shen, C. F., Lanthier, S., Jacob, D., Montes, J., Beath, A., Beresford, A., et al. (2012). Process optimization and scale-up for production of rabies vaccine live adenovirus vector (AdRG1.3). *Vaccine*, 30(2), 300–306. doi:10.1016/j.vaccine.2011.10.095.

Slate, D., Algeo, T. P., Nelson, K. M., Chipman, R. B., Donovan, D., Blanton, J. D., et al. (2009). Oral rabies vaccination in north america: Opportunities, complexities, and challenges. [Review]. *PLoS Neglected Tropical Diseases*, 3(12), e549. doi:10.1371/journal.pntd.0000549.

Smith, M. E., Koser, M., Xiao, S., Siler, C., McGettigan, J. P., Calkins, C., et al. (2006). Rabies virus glycoprotein as a carrier for anthrax protective antigen. *Virology*, 353(2), 344–356. doi:10.1016/j.virol.2006.05.010.

Steck, F., Wandeler, A., Bichsel, P., Capt, S., & Schneider, L. (1982). Oral immunisation of foxes against rabies. A field study. *Zentralblatt fur Veterinarmedizin B*, 29(5), 372–396.

Stohr, K., & Meslin, F. M. (1996). Progress and setbacks in the oral immunisation of foxes against rabies in Europe. *The Veterinary Record*, 139(2), 32–35.

Sun, C., Zhang, Y., Feng, L., Pan, W., Zhang, M., Hong, Z., et al. (2011). Epidemiology of adenovirus type 5 neutralizing antibodies in healthy people and AIDS patients in Guangzhou, southern China. [Research Support, Non-U.S. Gov't]. *Vaccine*, 29(22), 3837–3841. doi:10.1016/j.vaccine.2011.03.042.

Sureau, P. (1987). Rabies vaccine production in animal cell cultures. *Advances in Biochemical Engineering/Biotechnology*, 34, 11–128.

Tao, L., Ge, J., Wang, X., Wen, Z., Zhai, H., Hua, T., et al. (2011). Generation of a recombinant rabies Flury LEP virus carrying an additional G gene creates an improved seed virus for inactivated vaccine production. *Virology Journal*, 8, 454. doi:10.1186/1743-422X-8-454.

Ternovoi, V. V., Le, L. P., Belousova, N., Smith, B. F., Siegal, G. P., & Curiel, D. T. (2005). Productive replication of human adenovirus type 5 in canine cells. *Journal of Virological*, 79(2), 1308–1311. doi:10.1128/JVI.79.2.1308-1311.2005.

Tims, T., Briggs, D. J., Davis, R. D., Moore, S. M., Xiang, Z., Ertl, H. C., et al. (2000). Adult dogs receiving a rabies booster dose with a recombinant adenovirus expressing rabies virus glycoprotein develop high titers of neutralizing antibodies. *Vaccine*, 18(25), 2804–2807.

Tordo, N., Foumier, A., Jallet, C., Szelechowski, M., Klonjkowski, B., & Eloit, M. (2008). Canine adenovirus based rabies vaccines. *Developmental Biology (Basel)*, 131, 467–476.

Vitrology. (2008). Vitrology Biotech Retrieved February 15 2012. from

Vos, A., Conzelmann, K. K., Finke, S., Muller, T., Teifke, J., Fooks, A. R., et al. (2011). Immunogenicity studies in carnivores using a rabies virus construct with a site-directed deletion in the phosphoprotein. *Advance in Prevention Medicine, 2011,* 898171. doi:10.4061/2011/898171.

Vos, A., Neubert, A., Pommerening, E., Muller, T., Dohner, L., Neubert, L., et al. (2001). Immunogenicity of an E1-deleted recombinant human adenovirus against rabies by different routes of administration. *The Journal of General Virology, 82*(Pt 9), 2191–2197.

Wang, X., Bao, M., Wan, M., Wei, H., Wang, L., Yu, H., et al. (2008). A CpG oligodeoxynucleotide acts as a potent adjuvant for inactivated rabies virus vaccine. *Vaccine, 26*(15), 1893–1901. doi:10.1016/j.vaccine.2008.01.043.

Wang, X., Liu, J., Wu, X., Yu, L., Chen, H., Guo, H., et al. (2011). Oral immunisation of mice with a recombinant rabies virus vaccine incorporating the heat-labile enterotoxin B subunit of Escherichia coli in an attenuated Salmonella strain. *Research in Veterinary Science* doi:10.1016/j.rvsc.2011.09.015.

Wanjalla, C. N., Faul, E. J., Gomme, E. A., & Schnell, M. J. (2010). Dendritic cells infected by recombinant rabies virus vaccine vector expressing HIV-1 Gag are immunogenic even in the presence of vector-specific immunity. *Vaccine, 29*(1), 130–140. doi:10.1016/j.vaccine.2010.08.042.

Wen, Y., Wang, H., Wu, H., Yang, F., Tripp, R. A., Hogan, R. J., et al. (2011). Rabies virus expressing dendritic cell-activating molecules enhances the innate and adaptive immune response to vaccination. *Journal of Virological, 85*(4), 1634–1644. doi:10.1128/JVI.01552-10.

Weyer, J., Rupprecht, C. E., & Nel, L. H. (2009). Poxvirus-vectored vaccines for rabies–a review. *Vaccine, 27*(51), 7198–7201. doi:10.1016/j.vaccine.2009.09.033.

WHO, (1992). WHO expert committee on rabies. *World Health Organization Technical Report Series, 824,* 1–84.

WHO, (1998). *Requirements for the use of animal cells as in vitro substrates for the production of biologicals.* Geneva: WHO. WHO technical report series 878 (pp. 19–52).

WHO, (2005). *WHO expert committee on biological standardization.* Geneva: WHO. (pp. 154).

WHO. (2011). *WHO prequalified vaccines,* from <http://www.who.int/immunization_standards/vaccine_quality/rabies/en/index.html/>

Wiktor, T., Macfarlan, R. I., Dietzschold, B., Rupprecht, C., & Wunner, W. H. (1985). Immunogenic properties of vaccinia recombinant virus expressing the rabies glycoprotein. *Annales de l'Institut Pasteur. Virology, 136,* 6.

Wiktor, T. J., Macfarlan, R. I., Reagan, K. J., Dietzschold, B., Curtis, P. J., Wunner, W. H., et al. (1984). Protection from rabies by a vaccinia virus recombinant containing the rabies virus glycoprotein gene. *Proceedings of the National Academy of Sciences of the United States of America, 81*(22), 7194–7198.

Wold, W. S., Tollefson, A. E., & Hermiston, T. W. (1995). E3 transcription unit of adenovirus. *Current topics in Microbiology and Immunology, 199*(Pt 1), 237–274.

Wolff, M., & Reichl, U. (2011). Downstream processing of cell culture-derived virus particles. *Review of Vaccines, 10*(10), 1451–1475.

Woods, L. W. (2001). Adenovirus diseases. In E. S. Williams & I. K. Barker (Eds.), *Infectious diseases of wild animals (pp. 558).* Ames, Iowa: Iowa State University Press.

Wu, X., Franka, R., Henderson, H., & Rupprecht, C. E. (2011). Live attenuated rabies virus co-infected with street rabies virus protects animals against rabies. *Vaccine, 29*(25), 4195–4201. doi:10.1016/j.vaccine.2011.03.104.

Wu, X., Franka, R., Svoboda, P., Pohl, J., & Rupprecht, C. E. (2009). Development of combined vaccines for rabies and immunocontraception. *Vaccine, 27*(51), 7202–7209. doi:10.1016/j.vaccine.2009.09.025.

Wu, X., & Rupprecht, C. E. (2008). Glycoprotein gene relocation in rabies virus. *Virus Research, 131*(1), 95–99. doi:10.1016/j.virusres.2007.07.018.

Wunderli, P. S., Dreesen, D. W., Miller, T. J., & Baer, G. M. (2003). Effect of heterogeneity of rabies virus strain and challenge route on efficacy of inactivated rabies vaccines in mice. *American Journal of Veterinary Research, 64*(4), 499–505.

Wunner, W. H., Dietzschold, B., Curtis, P. J., & Wiktor, T. J. (1983). Rabies subunit vaccines. *The Journal of General Virology, 64*(Pt 8), 1649–1656.

Xiang, Z., Gao, G., Reyes-Sandoval, A., Cohen, C. J., Li, Y., Bergelson, J. M., et al. (2002). Novel, chimpanzee serotype 68-based adenoviral vaccine carrier for induction of antibodies to a transgene product. *Journal of Virology, 76*(6), 2667–2675.

Zhang, S., Liu, Y., Fooks, A. R., Zhang, F., & Hu, R. (2008). Oral vaccination of dogs (Canis familiaris) with baits containing the recombinant rabies-canine adenovirus type-2 vaccine confers long-lasting immunity against rabies. *Vaccine, 26*(3), 345–350. doi:10.1016/j.vaccine.2007.11.029.

Zhou, D., Cun, A., Li, Y., Xiang, Z., & Ertl, H. C. (2006). A chimpanzee-origin adenovirus vector expressing the rabies virus glycoprotein as an oral vaccine against inhalation infection with rabies virus. *Molecular Therapy, 14*(5), 662–672. doi:10.1016/j.ymthe.2006.03.027.

Zinsstag, J., Durr, S., Penny, M. A., Mindekem, R., Roth, F., Menendez Gonzalez, S., et al. (2009). Transmission dynamics and economics of rabies control in dogs and humans in an African city. *Proceedings of the National Academy of Sciences of the United States of America, 106*(35), 14996–15001. doi:10.1073/pnas.0904740106.

Next Generation of Rabies Vaccines

Zhi Quan Xiang and Hildegund C.J. Ertl

The Wistar Institute, 3601 Spruce St, Philadelphia, PA 19104, USA

1 INTRODUCTION

Rabies is a simple, negative-stranded RNA virus that encodes five structural proteins, that is, the nucleoprotein (NP), the glycoprotein (G), the phosphoprotein (P), the matrix protein (M), and the polymerase (L). Correlates of protection are well defined, and virus-neutralizing antibodies (VNAs) present in serum at titers of or above 0.5 international units (IU)/mL provide protection (Wunderli, Shaddock, Schmid, Miller, & Baer, 1991). VNAs are solely directed against the G protein. For post-exposure prophylaxis (PEP), active immunization has to be combined with passive immunization with a rabies virus specific immunoglobulin (RIG). Rabies has the highest fatality rate of all known pathogens, and, with a handful of exceptions, humans that develop symptomatic rabies inevitably die. Although efficacious vaccines are available, mortality due to rabies remains high, and the disease claims the lives of an estimated 55,000 humans each year. Worldwide rabies virus is mainly transmitted by dogs, and fatal infections are disproportionately high in developing countries and in children. Although ignorance of appropriate treatment, such as wound cleaning and disinfection, or use of alternative ineffective treatments such as traditional herbal medicines may contribute to the high death rate, economic factors play a major role. PEP is costly, as it requires four injections given sequentially, and is also time consuming. In addition, RIG, which is required for protection upon severe exposure, is expensive and in very short supply and thus habitually underutilized. The development of alternative, less expensive vaccines, which would ideally reduce the need for RIG, is thus warranted. Such vaccines could replace the vaccines used for PEP. Alternatively, provided they are inexpensive and induce sustained immunity after a single dose, they could be used for childhood vaccination in highly endemic areas.

A number of vaccine platforms that have been evaluated for prevention or PEP of rabies pre-clinically are discussed in this chapter. The most promising are vaccines based on live attenuated rabies virus for PEP and recombinant vaccines based on replication-defective adenoviruses for preventive vaccinations. Other innovative approaches, such as edible vaccines that grow in plants, continue to face technical challenges.

2 HISTORY OF RABIES VACCINES

Rabies, a disease whose name originates from the Sanskrit word of *rabhas*, which translates into "to do violence," has been known as a fatal disease since at least 2300 BC. There was and still is no cure for rabies. Avoidance of rabid dogs was the only means of prevention until a vaccine was developed and tested in 1885 by Louis Pasteur in a young boy named Joseph Meisner who had been bitten by a rabid dog. The vaccine consisted of a crude solution of dried spinal cord tissue from rabbits infected with a rabies virus stock, which originated from a rabid cow and then was serially passaged 80 times in rabbits. The boy, who received 13 injections of this vaccine, survived without reported sequelae, and the vaccine, with some modifications, continued to be used until vaccines, in which rabies virus was inactivated by chemical methods, replaced it. The most commonly used inactivated rabies vaccine was developed in 1911 by Lieutenant-Colonel Sir David Semple in India. He used virus from the brains of rabies virus-infected sheep that were inactivated with phenol. Semple-type vaccines, based on rabies virus derived from nerve tissue of infected animals, were used for decades (Sureau, 1988). Although in general the treatment prevented rabies, it did so at a cost. Immune responses to the nerve tissue present in the vaccine preparations caused debilitating and sometimes fatal neurological disease in approximately one out of 80–200 vaccine recipients. To reduce this side effect, rabies vaccines were subsequently generated from brains of suckling mice or newborn rats with not yet fully matured nervous systems.

Vaccines derived from 7-day-old embryonated duck eggs were developed in the 1960s (Powell & Culbertson, 1959). Advances in mammalian tissue culture allowed Hilary Koprowski, Stanley Plotkin, and Tadeusz Wiktor to develop a human diploid cell line WI-38-derived, inactivated rabies vaccine (Wiktor, Plotkin, & Koprowski, 1978), called human diploid cell vaccine (HDCV), which was licensed in 1976 in Europe and in 1980 in the United States. This and similar vaccines, as well as vaccines derived from duck or chick eggs, are currently used worldwide for post-exposure or preventive vaccination against rabies.

Passive immunization with RIG in combination with active immunization was first tested in 1953 in a village in Iran that was attacked by a rabid wolf (Habel & Koprowski, 1955). The clinical trial confirmed that treatment with RIG is essential to prevent vaccine failures in cases of severe exposures to rabies virus.

The initial rabies vaccines were developed without means to assess the likely efficacy of new vaccines. In the 1960s, potency tests were firmly established for pre-clinical testing of rabies vaccines (Habel, 1966).

3 CURRENT RABIES VACCINE REGIMENS

Rabies vaccinations follow guidelines formulated by expert panels of the World Health Organization (WHO) and fine-tuned by local regulatory agencies such as the Centers for Disease Control and Prevention in the United States. WHO strongly discouraged the use of Semple-type vaccines, and by 2004 their use was discontinued in most countries.

Pre-exposure immunization is only given to individuals at very high risk for exposure, such as veterinarians, animal handlers, rabies laboratory workers, and international travelers who are likely to come in contact with animals in countries where rabies is prevalent. The initial immunization consists of three doses given on days 0, 7, and 21 or 28. For humans at continued risk, periodic testing for antibody titers is recommended. Booster immunizations with a single dose of the vaccine are indicated once titers fall below 0.5 international units (IU)/mL, which is felt to be the minimal titer of rabies virus-specific antibodies that reliably provides protection to humans and animals.

In most cases, humans are vaccinated after exposure. Initial cleaning and disinfecting of the bite wound are essential to remove remaining virus-containing saliva. Immunization should be initiated as soon as possible in previously unvaccinated individuals. In the United States, 4 doses of vaccine are given on days 0, 3, 7, and 14. The first dose is combined with RIG that should be given into and around the bite site. Residual RIG should be given intramuscularly away from the site of active immunization to reduce interference with the vaccine that upon contact with specific antibodies may be neutralized and thus unable to elicit a B cell response. The recommended dose of human RIG is 20 mg/kg; equine RIG should be used at 40 mg/kg. Higher doses should not be used as they may interfere with active immunization. Previously vaccinated individuals require 2 doses of vaccine on days 0 and 3 without RIG. Active immunization can be given intramuscularly into the deltoid muscle or at lower doses intradermally (but not recommended for PEP). Other countries may use modified regimens or may omit the use of RIG for minor scratches, abrasions, or bites.

4 INCIDENCE AND RISK FOR RABIES AND VACCINE FAILURES

Due to mandatory dog rabies vaccination combined, in part, with wildlife immunization programs, rabies has become rare in the Americas and Europe but remains common in Asia and Africa with an estimated annual mortality rate of 55,000 humans per year. Most cases (99%) are caused by rabid dogs, and approximately 40% of humans receiving post-exposure prophylaxis (PEP) are children below the age of 15. A total of 3.3 billion humans are at risk to contract rabies and each year more than 15 million humans receive rabies PEP. It has been estimated that in some countries up to 40% of humans require PEP during childhood, which suggests that in such areas preventive childhood vaccination should be considered to reduce the incidence of rabies and the cost of PEP.

Appropriate wound cleaning and the full course of PEP in most cases prevent the development of symptomatic rabies. Bites by bats, which often cause minor lesions, can be overlooked and thus left untreated; they are the major cause of rabies in the United States (Blanton, Palmer, Dyer, & Rupprecht, 2011). In developing countries, the cost of PEP, which in Africa and Asia exceeds the monthly income of an average family, discourages its use. Vaccine failures after full PEP are rare in healthy individuals but have been reported more frequently in humans with immunodeficiencies (Tantawichien, Jaijaroensup, Khawplod, & Sitprija, 2001).

5 CORRELATES OF PROTECTION

Most of the rabies virus proteins, such as the G, N, and NP proteins, contain epitopes for recognition by $CD4^+$ or $CD8^+$ T cells (Desmezieres et al., 1999; Ertl et al., 1989; Larson, Wunner, Otvos, & Ertl, 1991). VNAs are exclusively directed to the G protein. The 65–67 kd G protein, which contains ~500 amino acids, has 2–6 potential sites for N-glycosylation, 12–16 conserved cysteine residues, 2–3 hydrophobic heptad repeats, a transmembrane domain, and a short cytoplasmic domain (Schneider & Diringer, 1976). Its structure has not yet been resolved by X-ray crystal-lography, although it is known that the protein is displayed in form of non-covalently linked trimers on the virion's surface. The G protein is a major determinant for pathogenicity and single amino acid substitution can strongly attenuate the virulence of rabies virus. The G protein also attaches to cellular receptors and as such determines the virus's tropism for neurons. Most importantly, the G protein expresses, depending on the strain, 3–5 conformation-dependent binding sites each with multiple partially overlapping epitopes for VNAs (Luo et al., 1998). Additional linear epitopes have been defined (Mansfield, Johnson, & Fooks, 2004;

Zhao et al., 2008). Protection against rabies virus strongly correlates with titers of circulating VNAs. Titers of 0.5 IU/mL determined by a validated and standardized assay such as a rapid immunofluorescent focus inhibition test (RFFIT) (Gelosa & Borroni, 1990) suffice to prevent an infection. CD8$^+$ T cells do not contribute to protection, whereas CD4$^+$ T cells are indirectly involved by their essential role in providing help for induction of long-lived plasma cells producing affinity-matured antibodies (Xiang, Knowles, McCarrick, & Ertl, 1995).

Molecular analyses suggest that currently circulating rabies viruses have evolved within the last 1,500 years (Nadin-Davis & Real, 2011). Rabies virus, like all RNA viruses, show high mutation rates and genetically distinct isolates characterize outbreaks in distinct regions or species. Accordingly, most monoclonal neutralizing antibodies fail to inhibit a wide range of rabies virus isolates, and cross-reactive monoclonal antibodies appear to be rare (Muller et al., 2009). This may provide a challenge in switching from RIG to a monoclonal antibody cocktail for global use in PEP. Notwithstanding, rabies vaccines based on a limited number of strains, such as the challenge virus standard (CVS) or Pasteur virus (PV) strain, are efficacious worldwide, as is RIG, which is induced by the same vaccine strains, demonstrating that vaccines to rabies virus do not have to be updated periodically or do not require adjustments to regionally prevalent strains.

6 EXPERIMENTAL VACCINES FOR RABIES

The continued high incidence of rabies in developing countries, the economic burden of PEP, and the globally limited availability of RIG warrant development of novel cost-effective rabies vaccines for preventive vaccination or PEP. Just about every vaccine prototype, ranging from peptides to plant-derived vaccines, has been evaluated and in many cases has shown efficacy in experimental animals. In this section, we will briefly discuss some of the more promising prototypes and their potential for use in humans. Monoclonal antibodies either derived from human B cells or from mouse hybridomas, the latter genetically modified to humanize the antibodies constant region, are being developed to eventually replace RIG (Smith, Wu, Franka, & Rupprecht, 2011) and are not discussed. Vaccines are typically divided into live-attenuated and inactivated vaccines. This division does not suffice for most of the modern experimental vaccines, which are either based on genetic attenuation or modifications that enhance immunogenicity or on individual viral proteins, so-called subunit vaccines, which in case of rabies vaccines are typically based on the viral G protein or parts thereof. Subunit vaccines can consist of synthetic peptides or proteins carrying crucial B cell epitopes. Proteins can be isolated from a variety of expression systems, including mammalian cells, yeast or plants. Subunit vaccines can also reflect genetic vaccines, which encode the rabies virus G protein. In addition, with advances in our knowledge of the interplay between innate and adaptive immune responses and the identification of pathogen recognition receptors as crucial initiators of inflammatory reactions (Coban, Ishii, & Akira, 2009) that are a prerequisite for activation of T and B cells, novel adjuvants are being developed that may enhance the immunogenicity of vaccines. Such adjuvants can be added to a vaccine, or their sequences can be directly incorporated into a vaccine. Traditional rabies vaccines are delivered systemically to humans. Alternative routes of immunization, such as intranasal or oral routes, are being explored.

6.1 Inactivated "Enhanced" Traditional Rabies Vaccines

Inactivated rabies vaccines are not potently immunogenic, and, therefore, require multiple doses until protective VNA titers are achieved. Rabies vaccines used in the United States, for example, Imovax Rabies, a HDCV produced by Sanofi Pasteur and RabAvert, a purified chick embryo cell vaccine from Novartis, do not contain adjuvant such as alum. In fact, studies in experimental animals indicate that addition of alum does not improve the vaccine's immunogenicity (Lin & Perrin, 1999). Formulations containing rabies vaccine mixed with CpG-oligodeoxynucleotides, which trigger innate immune responses through Toll-like receptor 9, showed enhanced and accelerated VNA responses in animals (Wang et al., 2008).

6.2 Protein and Peptide Vaccines

Linear B cell epitopes have been identified in the rabies virus G proteins that can be expressed by synthetic peptide vaccines. Peptide vaccines, although exceptionally safe, are commonly poorly immunogenic and induce very narrow B cell responses. Indeed, one study with a peptide expressing a linear epitope of the rabies virus glycoprotein only induced low titers of VNAs that failed to neutralize an escape mutant (Niederhauser et al., 2008). Mimotopes of the rabies virus G protein isolated from a random constrained hexapeptide phage display library and corresponding to antigenic site III induced a VNA response in mice. This study did not address the potential problem of viral escape or the overall breadth of the antibody response (Houimel & Dellagi, 2009). G and N protein peptides expressed by the coat protein of alfalfa mosaic virus grown in tobacco plants or spinach induced an immune response in mice (Yusibov et al., 2002). Spinach-derived virus, when tested in rabies vaccine-immune humans, elicited a weak recall response. Immunization of mice with spinach expressing peptides of the G protein was also poorly immunogenic and provided limited protection against subsequent challenge (Yusibov et al., 2002). Overall, the high variability of the viral G protein contraindicates vaccines that induce antibodies to single epitopes.

The rabies virus G protein has been purified from a number of expression systems and used for immunization of animals. Yeast-derived protein failed to induce protective immune responses in mice (Klepfer et al., 1993), most likely reflecting poor folding of the final product. Baculovirus-derived G protein expressed in insect cells was found to be immunogenic (Fu et al., 1993), but the extensive purification that would be required for its use in humans would likely render this approach cost-ineffective. G protein has been expressed in plants such as maize, which upon ingestion induced a detectable VNA response and protection against challenge in mice (Loza-Rubio, Rojas, Gomez, Olivera, & Gomez-Lim, 2008). Again, the structural complexity of the rabies virus G protein, the requirement for its correct folding to elicit broad VNA responses, and the need for extensive purifications prior to injection into humans make it unlikely that protein-based rabies vaccines will replace current vaccines. The use of edible vaccines such as spinach or maize genetically engineered to express G protein remains an attractive option that continues to face experimental challenges.

6.3 Genetic Vaccines

Genetic vaccines to rabies virus carry the G protein gene that is expressed once the vaccine carrier has transduced or infected a cell. The advantage of genetic vaccines is that they are highly versatile. Furthermore, the mammalian expression system allows for faithful expression of the protein and its correct folding, and, depending on the system, genetic vaccines can be highly immunogenic and cost-effective. The disadvantage of genetic vaccines is that production of immunogenic levels of the transgene product takes time. The incubation period of rabies varies tremendously, ranging from a few days to several years. Upon severe exposures, which typically result in shorter incubation times, rapid induction of protective VNA titers is of the essence. This may not be achieved by genetic vaccines, which nevertheless provide attractive alternatives for preventive vaccination.

6.4 DNA Vaccines

DNA vaccines are very easy to construct, and they are well-tolerated in humans. Numerous studies with DNA vaccine to rabies virus reported induction of protective VNA titers using various routes of immunization in experimental animals, ranging from mice, cats and dogs, to monkeys (Bahloul et al., 2003; Bahloul et al., 2006; Lodmell, Ewalt, Parnell, Rupprecht, & Hanlon, 2006; Lodmell, Parnell, Bailey, Ewalt, & Hanlon, 2002; Lodmell, Parnell, Weyhrich, & Ewalt, 2003; Tesoro Cruz et al., 2008; Tesoro Cruz, Hernandez Gonzalez, Alonso Morales, & Aguilar-Setien, 2006; Xiang, Spitalnik et al., 1995). Some studies reported protection with DNA vaccines given to already infected animals (Bahloul et al., 2003; Lodmell et al., 2002; Tesoro Cruz et al., 2008). Unfortunately, in humans, DNA vaccines were found to be poorly immunogenic. Addition of genetic adjuvants (Xiang & Ertl, 1995), prime boost regimens in which DNA vaccines are typically used for priming followed by a booster immunization with a recombinant viral vector (Lodmell & Ewalt, 2000) or novel delivery methods such a injection of DNA followed by electroporation (Sardesai & Weiner, 2011), increase the immunogenicity of DNA vaccine in animals and for some antigens in humans as well. To what degree such modifications affect the vaccine's safety remains to be investigated in more depth. Prime-boost regimens, although highly effective, may not be suited for PEP, and their cost-effectiveness for preventive vaccination is unlikely. The suitability of electroporation, which increases immune responses by enhancing transduction rates, for immunization in developing countries remains to be explored.

Efficacy was also reported for Sindbis virus-based DNA vaccines (Saxena et al., 2008), which, unlike conventional DNA vaccine, generate self-replicate RNA transcripts and thus achieve superior protein-expression levels.

6.5 Viral Vector Vaccines

Viral vector vaccines carry an expression cassette encoding the vaccine antigen within their genome. Viral vector vaccines are by definition infectious vaccines, as production of the vaccine antigen is achieved *in situ* upon infection of cells. Some viral vectors are based on attenuated viruses, such as vaccinia virus or Modified Vaccinia Anchora (MVA), while others, such as adenoviral vectors, are genetically altered to render them replication-defective. The most commonly explored viral vectors for rabies G protein are based on attenuated poxviruses or different strains of E1-deleted and, hence, replication-defective adenoviruses.

6.6 Recombinant Poxviruses

A vaccinia virus expressing the full-length G protein of the Evelyn Rokitnicki Abelseth (ERA) strain was one of the first viral vectors that was produced and tested. The vaccine termed VR-G showed adequate efficacy in animals after oral administration and is now being used for vaccination of wildlife animals (Brochier et al., 1996). Due to its high reactogenicity, it is not suited for use in humans. Vectors based on the more attenuated MVA are less immunogenic and fail to elicit an immune response after oral application (Weyer, Rupprecht, Mans, Viljoen, & Nel, 2007).

6.7 Recombinant Adenoviruses

In most adenoviral vaccine vectors derived from human or simian serotypes, the E1 domain, which encodes polypeptides essential for viral replication, is deleted, thus rendering the virus replication-defective. An expression cassette is then cloned into the deleted E1 domain. Alternatively, a foreign gene can be cloned into the E3 domain, which encodes polypeptides that are not essential for viral replication. E3-deleted vectors remain replication competent. Adenoviruses are ubiquitous DNA viruses, which can be found in many species, although individual serotypes are species-specific. Fifty-seven different serotypes can infect humans; 25 serotypes have been isolated from simians, 9 from bovines, 6 from sheep and 2 from dogs. In theory all of these adenoviruses can be vectored, although thus far efforts have focused on human serotypes HAdV-5, -26, and -35 (commonly referred to as AdHu5, AdHu26, etc.); simian serotypes SadV-23, -24, and -25 (also called chimpanzee adenovirus (AdC) 6, 7, or 68; canine adenovirus serotype 2 (CAV-2); and bovine adenovirus serotype 3 (BAdV-3). Human serotype vectors, such as AdHu5, are unsuited for clinical use, as most humans become naturally infected with multiple serotypes of AdHu viruses and, consequently, develop VNAs, which interfere with active immunization (Chen et al., 2010). VNAs to adenoviruses derived from other species are rare in humans, thus favoring their development as vaccine vectors (Chen et al., 2010; Xiang et al., 2006). Adenovirus vectors, similar to replicating adenoviruses, persist at very low levels in a transcriptionally active form and thereby maintain sustained levels of transgene product-specific immune responses (Tatsis et al., 2007). A number of different E1-deleted adenovirus vectors have had preclinical testing, including those based on HAdV-5 (Xiang, Yang, Wilson, & Ertl, 1996), -26 (Chen et al., 2010), SAdV-23, 24, and 25 (Lasaro & Ertl, 2009). Vectors are highly immunogenic and provide protection to subsequent challenges. E3-deleted replication-competent CAV-2 vectors were tested in dogs, cats, sheep, swine, raccoons, and skunks, in which they induced protective immunity upon various routes of immunization (Bouet-Cararo et al., 2011; Henderson et al., 2009; Hu, Liu, Zhang, Zhang, & Fooks, 2007; Liu, Zhang, Ma, Zhang, & Hu, 2008; Zhang, Liu, Fooks, Zhang, & Hu, 2008).

Although live adenovirus vectors may raise safety concerns for use in humans, dogs are routinely vaccinated with an attenuated CAV-2 vaccine, suggesting that a combination CAV-2 rabies vaccine for canines is feasible. As mentioned above, adenoviral vectors are highly immunogenic, but immune responses develop with a delay, thus precluding their use for PEP. As fairly low doses can induce sustained protective titers of VNAs, their use for large-scale preventive vaccination should be explored.

6.8 Newcastle Disease Virus (NDV) Vectors

NDV is an avian paramyxovirus that can be vectored by reverse genetics. Insertion of full-length rabies virus G protein gene in between the M and P protein genes results in a vector that expresses the rabies virus G protein on its surface, and, in addition, encodes the protein upon infection of cells (Ge et al., 2011). Rabies G protein expressing NDV vectors are immunogenic and induce protective immunity in dogs and cats. Clinical development of such vectors may be problematic, especially in the United States, where NDV is viewed as a select agent due to its threat to the poultry industry.

6.9 Live Attenuated Rabies Virus Vaccines

In general, live vaccines are more immunogenic than killed vaccines, which relates to superior induction of innate immunity and higher antigenic loads for stimulation of adaptive immune responses by the former. Reverse genetics allows for attenuation of rabies virus through manipulation of its genome and thus for the development of potentially safe, attenuated live rabies vaccines. Rabies virus can be attenuated by several means. Deletion of the P protein, which serves as a cofactor for the viral polymerase, prevents viral replication (Cenna et al., 2009). Furthermore, the P protein subverts immune responses by inhibiting IFN type I signaling thus attenuating rabies virus-specific immune responses. P protein-deleted rabies viruses, although they express markedly less G protein on the surface of infected cells compared to wild-type virus induce more potent immune responses compared to inactivated virus (Cenna et al., 2009). Responses can be further improved by incorporating a second G protein gene in between the genes encoding the M protein and L protein (Cenna et al., 2008). The P protein-deleted virus lacks pathogenicity even after direct intracerebral injection into immunodeficient mice. Growth kinetics of recombinant virus show a strong reduction in titers, which may provide problems for cost-effective manufacturing.

M protein-deleted rabies virus, which shows growth kinetics similar to those of P protein-deleted rabies virus, express higher levels of

G protein on *in vitro* infected cells and accordingly induces more potent VNA responses in mice (Cenna et al., 2009). In nonhuman primates, the M protein-deleted rabies virus given at two doses induces better antibody responses, including VNA responses, compared to two doses of HDCV (Cenna et al., 2009). M protein-deleted virus fails to cause disease in immunodeficient mice.

Deletion of the G protein encoding gene and its transcomplementation by packaging cell lines results in a single cycle rabies virus, which induces a VNA response, albeit at markedly lower titer compared to a replication-competent control rabies virus (Gomme, Faul, Flomenberg, McGettigan, & Schnell, 2010).

Although vaccines based on attenuated rabies virus may be suitable for prevention or PEP, several problems remain to be addressed. Foremost, reduced growth may cause manufacturing problems and thus render the vaccine too costly. Secondly and equally important, deletion mutants that are grown in transcomplementing cell lines or injected into infected humans may undergo recombination, which may reverse their attenuation. Such events would most likely be very rare and thus hard to detect in experimental animals but nevertheless pose potential safety issues that need to be addressed.

Incorporation of genes that encode granulocyte-macrophage colony-stimulating factor, macrophage-derived chemokines, or macrophage inflammatory protein 1a into attenuated rabies virus was shown to increase the vaccines immunogenicity (Wen et al., 2011).

7 SUMMARY

Rabies, which is well controlled in developed countries, has become a neglected disease. Dog rabies remains enzoonotic through most of the developing world, where it causes an estimated 55,000 human deaths each year, a number that is likely an underestimate due to misdiagnosis and underreporting. Efficacious vaccines that could markedly reduce human fatalities are available, but as they are costly and of poor immunogenicity, requiring multiple doses, they will continue to be underutilized. Alternative vaccines are required. A number of vaccines have undergone pre-clinical testing, but, of those, only live rabies vaccines attenuated through gene deletion are likely to be suitable for PEP where rapid induction of protective immune responses are essential. A number of other vaccine platforms, especially genetic vaccines based on highly immunogenic carriers such as E1-deleted adenoviruses, may be suitable for preventive childhood vaccination and should be explored further.

References

Bahloul, C., Ahmed, S. B., B'Chir, B. I., Kharmachi, H., Hayouni el, A., & Dellagi, K. (2003). Post-exposure therapy in mice against experimental rabies: A single injection of DNA vaccine is as effective as five injections of cell culture-derived vaccine. *Vaccine, 22*(2), 177–184.

Bahloul, C., Taieb, D., Diouani, M. F., Ahmed, S. B., Chtourou, Y., B'Chir, B. I., et al. (2006). Field trials of a very potent rabies DNA vaccine which induced long lasting virus neutralizing antibodies and protection in dogs in experimental conditions. *Vaccine, 24*(8), 1063–1072.

Blanton, J. D., Palmer, D., Dyer, J., & Rupprecht, C. E. (2011). Rabies surveillance in the United States during 2010. *Journal of the American Veterinary Medical Association, 239*(6), 773–783.

Bouet-Cararo, C., Contreras, V., Fournier, A., Jallet, C., Guibert, J. M., Dubois, E., et al. (2011). Canine adenoviruses elicit both humoral and cell-mediated immune responses against rabies following immunisation of sheep. *Vaccine, 29*(6), 1304–1310.

Brochier, B., Aubert, M. F., Pastoret, P. P., Masson, E., Schon, J., Lombard, M., et al. (1996). Field use of a vaccinia-rabies recombinant vaccine for the control of sylvatic rabies in Europe and North America. *Revue Scientifique et Technique, 15*(3), 947–970.

Cenna, J., Hunter, M., Tan, G. S., Papaneri, A. B., Ribka, E. P., Schnell, M. J., et al. (2009). Replication-deficient rabies virus-based vaccines are safe and immunogenic in mice and nonhuman primates. *Journal of Infectious Diseases, 200*(8), 1251–1260.

Cenna, J., Tan, G. S., Papaneri, A. B., Dietzschold, B., Schnell, M. J., & McGettigan, J. P. (2008). Immune modulating effect by a phosphoprotein-deleted rabies virus vaccine vector expressing two copies of the rabies virus glycoprotein gene. *Vaccine, 26*(50), 6405–6414.

Chen, H., Xiang, Z. Q., Li, Y., Kurupati, R. K., Jia, B., Bian, A., et al. (2010). Adenovirus-based vaccines: Comparison of vectors from three species of adenoviridae. *Journal of Virology, 84*(20), 10522–10532.

Coban, C., Ishii, K. J., & Akira, S. (2009). Immune interventions of human diseases through toll-like receptors. *Advances in Experimental Medicine and Biology, 655*, 63–80.

Desmezieres, E., Jacob, Y., Saron, M. F., Delpeyroux, F., Tordo, N., & Perrin, P. (1999). Lyssavirus glycoproteins expressing immunologically potent foreign B cell and cytotoxic T lymphocyte epitopes as prototypes for multivalent vaccines. *Journal of General Virology, 80*, 2343–2351. Pt 9.

Ertl, H. C., Dietzschold, B., Gore, M., Otvos, L., Jr., Larson, J. K., Wunner, W. H., et al. (1989). Induction of rabies virus-specific T-helper cells by synthetic peptides that carry dominant T-helper cell epitopes of the viral ribonucleoprotein. *Journal of Virology, 63*(7), 2885–2892.

Fu, Z. F., Rupprecht, C. E., Dietzschold, B., Saikumar, P., Niu, H. S., Babka, I., et al. (1993). Oral vaccination of racoons (Procyon lotor) with baculovirus-expressed rabies virus glycoprotein. *Vaccine, 11*(9), 925–928.

Ge, J., Wang, X., Tao, L., Wen, Z., Feng, N., Yang, S., et al. (2011). Newcastle disease virus-vectored rabies vaccine is safe, highly immunogenic, and provides long-lasting protection in dogs and cats. *Journal of Virology, 85*(16), 8241–8252.

Gelosa, L., & Borroni, G. (1990). Serological determination of rabies antibodies in vaccinated subjects. *Microbiologica, 13*(3), 257–262.

Gomme, E. A., Faul, E. J., Flomenberg, P., McGettigan, J. P., & Schnell, M. J. (2010). Characterization of a single-cycle rabies virus-based vaccine vector. *Journal of Virology, 84*(6), 2820–2831.

Habel, K. (1966). Laboratory techniques in rabies. Habel test for potency. *Monograph Series of the World Health Organ, 23*, 140–143.

Habel, K., & Koprowski, H. (1955). Laboratory data supporting the clinical trial of anti-rabies serum in persons bitten by a rabid wolf. *Bulletin of the World Health Organization, 13*(5), 773–779.

Henderson, H., Jackson, F., Bean, K., Panasuk, B., Niezgoda, M., Slate, D., et al. (2009). Oral immunization of raccoons and skunks with a canine adenovirus recombinant rabies vaccine. *Vaccine, 27*(51), 7194–7197.

Houimel, M., & Dellagi, K. (2009). Peptide mimotopes of rabies virus glycoprotein with immunogenic activity. *Vaccine, 27*(34), 4648–4655.

Hu, R. L., Liu, Y., Zhang, S. F., Zhang, F., & Fooks, A. R. (2007). Experimental immunization of cats with a recombinant rabies-canine adenovirus vaccine elicits a long-lasting neutralizing antibody response against rabies. *Vaccine, 25*(29), 5301–5307.

Klepfer, S. R., Debouck, C., Uffelman, J., Jacobs, P., Bollen, A., & Jones, E. V. (1993). Characterization of rabies glycoprotein expressed in yeast. *Archives of Virology, 128*(3-4), 269–286.

Larson, J. K., Wunner, W. H., Otvos, L., Jr., & Ertl, H. C. (1991). Identification of an immuno-dominant epitope within the phosphoprotein of rabies virus that is recognized by both class I- and class II-restricted T cells. *Journal of Virology, 65*(11), 5673–5679.

Lasaro, M. O., & Ertl, H. C. (2009). New insights on adenovirus as vaccine vectors. *Molecular Therapy, 17*(8), 1333–1339.

Lin, H., & Perrin, P. (1999). [Influence of aluminum adjuvant to experimental rabies vaccine]. *Chinese Journal of Experimental and Clinical Virology, 13*(2), 133–135.

Liu, Y., Zhang, S., Ma, G., Zhang, F., & Hu, R. (2008). Efficacy and safety of a live canine adenovirus-vectored rabies virus vaccine in swine. *Vaccine, 26*(42), 5368–5372.

Lodmell, D. L., & Ewalt, L. C. (2000). Rabies vaccination: Comparison of neutralizing antibody responses after priming and boosting with different combinations of DNA, inactivated virus, or recombinant vaccinia virus vaccines. *Vaccine, 18*(22), 2394–2398.

Lodmell, D. L., Ewalt, L. C., Parnell, M. J., Rupprecht, C. E., & Hanlon, C. A. (2006). One-time intradermal DNA vaccination in ear pinnae one year prior to infection protects dogs against rabies virus. *Vaccine, 24*(4), 412–416.

Lodmell, D. L., Parnell, M. J., Bailey, J. R., Ewalt, L. C., & Hanlon, C. A. (2002). Rabies DNA vaccination of non-human primates: Post-exposure studies using gene gun methodology that accelerates induction of neutralizing antibody and enhances neutralizing antibody titers. *Vaccine, 20*(17–18), 2221–2228.

Lodmell, D. L., Parnell, M. J., Weyhrich, J. T., & Ewalt, L. C. (2003). Canine rabies DNA vaccination: A single-dose intradermal injection into ear pinnae elicits elevated and persistent levels of neutralizing antibody. *Vaccine, 21*(25-26), 3998–4002.

Loza-Rubio, E., Rojas, E., Gomez, L., Olivera, M. T., & Gomez-Lim, M. A. (2008). Development of an edible rabies vaccine in maize using the Vnukovo strain. *Developments in Biologicals, 131*, 477–482.

Luo, T. R., Minamoto, N., Hishida, M., Yamamoto, K., Fujise, T., Hiraga, S., et al. (1998). Antigenic and functional analyses of glycoprotein of rabies virus using monoclonal antibodies. *Microbiology and Immunology, 42*(3), 187–193.

Mansfield, K. L., Johnson, N., & Fooks, A. R. (2004). Identification of a conserved linear epitope at the N terminus of the rabies virus glycoprotein. *Journal of General Virology, 85*(Pt 11), 3279–3283.

Muller, T., Dietzschold, B., Ertl, H., Fooks, A. R., Freuling, C., Fehlner-Gardiner, C., et al. (2009). Development of a mouse monoclonal antibody cocktail for post-exposure rabies prophylaxis in humans. *PLoS Neglected Tropical Diseases, 3*(11), e542.

Nadin-Davis, S. A., & Real, L. A. (2011). Molecular phylogenetics of the lyssaviruses–insights from a coalescent approach. *Advances in Virus Research, 79*, 203–238.

Niederhauser, S., Bruegger, D., Zahno, M. L., Vogt, H. R., Peterhans, E., Zanoni, R., et al. (2008). A synthetic peptide encompassing the G5 antigenic region of the rabies virus induces high avidity but poorly neutralizing antibody in immunized animals. *Vaccine*, 26(52), 6749–6753.

Powell, H. M., & Culbertson, C. G. (1959). Action of rabies vaccine derived from embryonated duck eggs against street virus. *Proceedings of the Society for Experimental Biology and Medicine*, 101, 801–803.

Sardesai, N. Y., & Weiner, D. B. (2011). Electroporation delivery of DNA vaccines: Prospects for success. *Current Opinion in Immunology*, 23(3), 421–429.

Saxena, S., Dahiya, S. S., Sonwane, A. A., Patel, C. L., Saini, M., Rai, A., et al. (2008). A Sindbis virus replicon-based DNA vaccine encoding the rabies virus glycoprotein elicits immune responses and complete protection in mice from lethal challenge. *Vaccine*, 26(51), 6592–6601.

Schneider, L. G., & Diringer, H. (1976). Structure and molecular biology of rabies virus. *Current Topics in Microbiology and Immunology*, 75, 153–180.

Smith, T. G., Wu, X., Franka, R., & Rupprecht, C. E. (2011). Design of future rabies biologics and antiviral drugs. *Advances in Virus Research*, 79, 345–363.

Sureau, P. (1988). History of rabies: Advances in research towards rabies prevention during the last 30 years. *Reviews of Infectious Diseases*, 10(Suppl. 4), S581–S584.

Tantawichien, T., Jaijaroensup, W., Khawplod, P., & Sitprija, V. (2001). Failure of multiple-site intradermal postexposure rabies vaccination in patients with human immunodeficiency virus with low CD4+ T lymphocyte counts. *Clinical Infectious Diseases*, 33(10), E122–124.

Tatsis, N., Fitzgerald, J. C., Reyes-Sandoval, A., Harris-McCoy, K. C., Hensley, S. E., Zhou, D., et al. (2007). Adenoviral vectors persist in vivo and maintain activated CD8+ T cells: Implications for their use as vaccines. *Blood*, 110(6), 1916–1923.

Tesoro Cruz, E., Feria Romero, I. A., Lopez Mendoza, J. G., Orozco Suarez, S., Hernandez Gonzalez, R., Favela, F. B., et al. (2008). Efficient post-exposure prophylaxis against rabies by applying a four-dose DNA vaccine intranasally. *Vaccine*, 26(52), 6936–6944.

Tesoro Cruz, E., Hernandez Gonzalez, R., Alonso Morales, R., & Aguilar-Setien, A. (2006). Rabies DNA vaccination by the intranasal route in dogs. *Developments in Biologicals*, 125, 221–231.

Wang, X., Bao, M., Wan, M., Wei, H., Wang, L., Yu, H., et al. (2008). A CpG oligodeoxynucleotide acts as a potent adjuvant for inactivated rabies virus vaccine. *Vaccine*, 26(15), 1893–1901.

Wen, Y., Wang, H., Wu, H., Yang, F., Tripp, R. A., Hogan, R. J., et al. (2011). Rabies virus expressing dendritic cell-activating molecules enhances the innate and adaptive immune response to vaccination. *Journal of Virology*, 85(4), 1634–1644.

Weyer, J., Rupprecht, C. E., Mans, J., Viljoen, G. J., & Nel, L. H. (2007). Generation and evaluation of a recombinant modified vaccinia virus Ankara vaccine for rabies. *Vaccine*, 25(21), 4213–4222.

Wiktor, T. J., Plotkin, S. A., & Koprowski, H. (1978). Development and clinical trials of the new human rabies vaccine of tissue culture (human diploid cell) origin. *Developments in Biological Standardization*, 40, 3–9.

Wunderli, P. S., Shaddock, J. H., Schmid, D. S., Miller, T. J., & Baer, G. M. (1991). The protective role of humoral neutralizing antibody in the NIH potency test for rabies vaccines. *Vaccine*, 9(9), 638–642.

Xiang, Z. Q., & Ertl, H. C. (1995). Manipulation of the immune response to a plasmid-encoded viral antigen by coinoculation with plasmids expressing cytokines. *Immunity*, 2(2), 129–135.

Xiang, Z. Q., Knowles, B. B., McCarrick, J. W., & Ertl, H. C. (1995). Immune effector mechanisms required for protection to rabies virus. *Virology*, 214(2), 398–404.

Xiang, Z. Q., Li, Y., Cun, A., Yang, W., Ellenberg, S., Switzer, W. M., et al. (2006). Chimpanzee adenovirus antibodies in humans, sub-Saharan Africa. *Emerging Infectious Diseases*, 12(10), 1596–1599.

Xiang, Z. Q., Spitalnik, S. L., Cheng, J., Erikson, J., Wojczyk, B., & Ertl, H. C. (1995). Immune responses to nucleic acid vaccines to rabies virus. *Virology, 209*(2), 569–579.

Xiang, Z. Q., Yang, Y., Wilson, J. M., & Ertl, H. C. (1996). A replication-defective human adenovirus recombinant serves as a highly efficacious vaccine carrier. *Virology, 219*(1), 220–227.

Yusibov, V., Hooper, D. C., Spitsin, S. V., Fleysh, N., Kean, R. B., Mikheeva, T., et al. (2002). Expression in plants and immunogenicity of plant virus-based experimental rabies vaccine. *Vaccine, 20*(25-26), 3155–3164.

Zhang, S., Liu, Y., Fooks, A. R., Zhang, F., & Hu, R. (2008). Oral vaccination of dogs (Canis familiaris) with baits containing the recombinant rabies-canine adenovirus type-2 vaccine confers long-lasting immunity against rabies. *Vaccine, 26*(3), 345–350.

Zhao, X. L., Yin, J., Chen, W. Q., Jiang, M., Yang, G., & Yang, Z. H. (2008). Generation and characterization of human monoclonal antibodies to G5, a linear neutralization epitope on glycoprotein of rabies virus, by phage display technology. *Microbiology and Immunology, 52*(2), 89–93.

Public Health Management of Humans at Risk

Louise H. Taylor[1], Peter Costa[1], and Deborah J. Briggs[1,2]

[1]Global Alliance for Rabies Control, 529 Humboldt St. Suite One, Manhattan, Kansas, 66502, USA, [2]College of Veterinary Medicine, Kansas State University, Manhattan, Kansas, 66502, USA

1 INTRODUCTION

Human to human transmission of rabies is extremely rare, with almost all human infections occurring as a result of "spillover" events caused by the bite of an infected animal. Each rabies virus variant tends to circulate within a specific species of animal, and the close relationship between dogs and humans explains why over 99% of all global human rabies deaths occur after an exposure to an infected dog. More than 3.3 billion people live at daily risk of exposure to canine rabies, and more than 15 million people receive post-exposure prophylaxis (PEP) after having been exposed to a suspected rabid animal every year (WHO, 2011c). Most human rabies deaths occur in Africa and Asia, and over 50% of these deaths take the lives of children under 15 years of age in the poorest segments of the population (WHO, 2010). Currently, Asia is reported to be the continent most affected by rabies, carrying the highest public health burden, with most cases occurring in India, Bangladesh, China, Indonesia, and Pakistan (Clifton, 2011; Knobel et al., 2005), but effective rabies surveillance in both Asia and Africa is lacking.

In industrialized nations where canine rabies has been largely eliminated, wildlife, particularly bats, serve as a primary reservoir species for rabies viruses and are responsible for the majority of human rabies deaths. The number of reported rabies cases in wildlife is increasing;

linked in part to urban expansion that has brought wildlife into closer proximity with humans (Haider, 2008). In several countries of Latin America, rabies transmitted by vampire bats is a serious threat to indigenous people living in traditional housing in remote regions (Carvalho-Costa, Tedesqui, de Jesus Nascimento Monteiro, & Boia, 2012).

2 HUMAN DEATHS ARE PREVENTABLE

Rabies is a deadly disease with a mortality rate approaching 100%, giving it the dubious distinction of the highest case fatality rate of any infectious disease of humans. To date, few rabies victims have survived after the onset of clinical symptoms (Franka & Rupprecht, 2011; Willoughby et al., 2005). However, PEP is highly effective at preventing rabies when it is administered promptly after an exposure has occurred as recommended by WHO (WHO, 2005). Virtually every human rabies death could have been prevented if the bite victims had known about the risks of disease and if they had sought and received prompt PEP. This highlights the need for increased educational awareness about the importance of wound care and appropriate medical attention as soon as possible after an exposure occurs (Rupprecht, 2009).

In parts of Asia, particularly in Thailand, increased access and utilization of human rabies vaccine as part of the treatment protocol after nearly every animal bite has significantly reduced human mortality from rabies, crediting Thailand as the Asian country with the steadiest decline in human rabies cases in recent history (Robertson et al., 2011), decreasing over the past decade from around 200 to about 20 cases annually (Fu, 2008). Where access to timely PEP is not an option, such as in remote regions of the Amazon, pre-exposure prophylaxis (PrEP) should be considered to protect communities at high risk of exposure and infection. (Gomez-Benavides, Laguna-Torres, & Recuenco, 2010). However, the most effective and sustainable method to reduce the public health risk of transmitting rabies to humans is to eliminate the disease at the source of infection, and there are many examples of locations where rabies in dogs or wildlife has been eliminated and others where significant progress has been made (Cleaveland, Kaare, Knobel, & Laurenson, 2006; Rupprecht et al., 2008; Slate et al., 2009). Canine rabies has been eliminated in many countries, including Japan (1957), Malaysia (1954), Taiwan (1961), and several European countries (1920s–1960s). The United States was declared canine rabies free in 2007 (CDC, 2007). Considerable progress has been made in Latin America, where the number of rabies cases in dogs was reduced from 15,686 in 1982 to 1,131 in 2003. As the incidence of canine rabies was reduced, the number of human cases decreased from 355 to 27 during the same period. In 2008, only 6 out of 48 countries and territories in Latin America reported human cases of canine rabies, with 11 in total (Ruiz & Chavez, 2010). These successes were achieved through mass canine vaccination programs focused on eliminating the circulation of rabies virus in dogs. The policy of mass culling of stray dogs or wildlife hosts without vaccination has been shown to be ineffective in numerous situations, is not recommended by WHO, and should not be promoted (Clifton, 2011; Rupprecht et al., 2008; WHO, 2005).

In some areas, notably islands, elimination of canine rabies can be achieved within a short time period through implementation of well-organized programs, including government and community support. In October 2010, the Province of Bohol, Philippines, celebrated two years without a human or animal case of rabies after an intensive program beginning in 2007 (Lapiz et al., 2012). This program involved highly effective multi-sectorial collaboration, integrated education of schoolchildren, and a strong emphasis on training community volunteers to prepare for and implement rabies prevention and control activities. The Bohol rabies prevention program, recently recognized by the president of the Philippines as a model for other public health initiatives, achieved the first elimination of canine rabies in the Asian region in 30 years (Galing Pook Foundation, 2012).

3 FACTORS RESPONSIBLE FOR THE CONTINUATION OF HUMAN RABIES DEATHS

3.1 Awareness

A lack of awareness about how to prevent rabies represents one of the most important contributing factors adding to the burden of human rabies. Recent estimates suggest that more than 55,000 people succumb to rabies annually and 28,000 of these deaths are children under the age of 15. The Disability Adjusted Life Years (DALYs) lost due to rabies is estimated to be 1.74 million each year (Knobel et al., 2005). Almost every one of these deaths could have been prevented if the exposed patient would have received appropriate PEP in a timely manner. Improving rabies educational awareness among health care providers as well as the general public could help to reduce the number of human deaths. For example, the importance of wound cleansing as the first step in PEP is often overlooked, and in resource-poor countries, rabies immune globulin (RIG) is rarely administered to patients with Category III wounds (transdermal bites, scratches and licks on broken skin, or contamination of mucous membranes with saliva) where it is indicated. A recent survey reviewing the anti-rabies prevention activities in medical personnel from six bite treatment centers in India reported that personnel in only one out of the centers washed the wounds of patients and personnel in only two centers administered RIG (Ichhpujani et al., 2008).

Traditional healers are still consulted by dog bite victims in resource-poor countries and often provide explanations of rabies as being caused by evil spirits and witchcraft. They may offer ineffective remedies such as rubbing chili or ash into the wound and providing false assurance to exposed patients, who then do not seek effective treatment (Satapathy, Sahu, Behera, Patnaik, & Malini, 2005).

Children are at increased risk of contracting rabies because they are more likely to be bitten by a dog due to their curiosity, lack of inhibition, limited knowledge about dog behavior, inability to protect themselves from an attack, and, not realizing the danger, they may not tell their parents of a minor exposure (Tenzin et al., 2011). For these reasons, increasing educational awareness in children about the risk of rabies exposure resulting from dog bites is critical, and they should be a primary target of rabies prevention efforts. Without a basic understanding of how to recognize an exposure, potential rabies victims cannot be expected to seek appropriate care. That rabies virus continues to circulate in bats presents a challenge to the public health management of humans at risk because many people do not relate exposure to bats as a potential danger, leading to under-reporting of the problem.

3.2 Free Roaming Unvaccinated Dogs

Most human rabies cases occur after a bite from a rabid dog. Although the exact number is unknown, the World Society for the Protection of Animals indicates that there are 375 million free-roaming dogs globally, almost all of which remain unvaccinated (WSPA, 2012). It has become increasingly clear that canine rabies endemic countries that have high numbers of dog bites also report high numbers of human rabies cases. In India, for example, it is estimated that more than 17 million people are bitten annually, and stray dogs far outnumber pet dogs (Sudarshan, Mahendra, & Narayan, 2001). The relationship between humans and dogs varies throughout the world, and free roaming dogs may have owners, or be fed and cared for by the community. Whatever their ownership status, free-roaming unvaccinated dogs in rabies endemic regions pose a public health threat to communities and environments, especially in areas where poor sanitation provides sufficient food for dog populations to thrive and even increase. Where food sources continue to be available, Animal Birth Control (ABC) programs often cannot address the scale of the problem (Sudarshan et al., 2001).

3.3 Accessibility to Vaccine

Effective rabies vaccines were developed well over a century ago, and RIG was recommended for use as part of PEP in the 1950s (WHO, 1950).

However, these life-saving biologicals, especially RIG, still remain inaccessible in many places where they are needed most (Hampson, Cleaveland, & Briggs, 2011; Hampson et al., 2008). Bite victims may need to travel several miles over a number of days in order to reach one or more health clinics where rabies vaccine is available. Once they find the vaccine, it is often very expensive compared to their household income. Families living in resource-poor countries have had to sell valuable livestock that they depend upon for food in order to pay for the rabies vaccine and the cost of travel to reach the clinic. This situation often has a long-term impact on household finances.

3.4 Political Support

After more than 5,000 years, rabies remains a serious public health threat, marginalized by competing national health priorities, inadequate resources and sheer lack of political will (Bourhy, Dautry-Varsat, Hotez, & Salomon, 2010). Rabies is a neglected disease substantiated by the persistent threat of exposure yet lack of sustained funding for prevention and control efforts. Nonexistent surveillance systems, resulting in an absence of reported human deaths, contribute to the reluctance of public officials, governments, and other key stakeholders to financially support rabies prevention and control efforts. That most human rabies deaths occur in impoverished populations in canine rabies endemic countries where rabies may not be a notifiable disease makes the true burden of rabies difficult to quantify. Due to the near 100% fatality rate of rabies and the high cost of health care, rabies patients are usually sent home to die. This method of care is unfortunately the most cost-effective way for impoverished families to care for their loved ones. The lack of reliable data on disease burden, limited awareness amongst clinicians and policy makers, diagnostic deficiencies, and fragmented disease surveillance systems continues to ensure that rabies remains low or nonexistent on the health agenda of most developing countries (Molyneux et al., 2011). Additionally, the lack of trained clinicians and absence of diagnostic laboratories leads to missed diagnoses or misdiagnoses. For example, a study in Malawi reported that 10.5% of the deaths in children initially attributed to cerebral malaria were in fact caused by rabies (Mallewa et al., 2007). This is an indication of the number of human rabies deaths that remain unrecorded or are misdiagnosed, adding to the inaccuracies in data and ultimately rendering rabies as a low public health priority (Hampson et al., 2008).

4 TOOLS TO PREVENT HUMAN RABIES DEATHS

Worldwide elimination of canine rabies is possible (Rupprecht et al., 2008). All of the tools to prevent human rabies are available, including robust educational materials; safe and effective biologicals; and various comprehensive guidelines, protocols, and standard operating procedure documents. For decades, rabies elimination was perceived to be an impossible feat for many countries. However, given renewed prominence under the banner of One Health, policy makers are beginning to recognize its importance to both human and animal health. Recently, various countries and multinational associations have set goals to improve rabies prevention and control, including reducing the number of human deaths to zero, including the Association of Southeast Asian Nations, the Philippines, and Canada, the United States, and Mexico (Gongal & Wright, 2011; PCHRD, 2012; Slate et al., 2009).

Increased awareness about rabies and grassroots efforts and advocacy through the initiatives like the World Rabies Day (WRD) campaign are leading to a change in the way rabies is perceived by policy makers. In the 21st century, rabies is beginning to be viewed as a winnable battle as more and more countries understand what is required to prevent human rabies and are outfitting themselves with the knowledge and tools necessary to begin planning their own national rabies control programs.

4.1 Legislative framework

National governments are responsible for the public health legislation in a country, including strategic plans and laws to prevent rabies (Partners for Rabies Prevention, 2010c). These include mandatory animal vaccination laws, collaboration between human and animal public health departments, improved educational awareness, improving sanitary conditions, humane animal welfare, as well as community involvement and responsibility (Lapiz et al., 2012). Making rabies a notifiable disease at a national level is the first step in building risk assessment systems that can help improve diagnosis and surveillance. National and provincial legislation is critical for local leaders to have support to enforce stray dog control and vaccination of pets. Without a legal framework to build a rabies prevention program, community members may be reluctant to take responsibility for vaccinating their pets and keeping them under their control.

4.2 Risk Assessment Systems

Although the first line of defense to prevent rabies in an exposed human patient is to implement PEP in a timely manner, human antirabies biologicals are only part of the solution to reduce the risk of rabies. Establishing risk assessment systems, including reliable diagnostic laboratories, improved surveillance systems, and continued training of public health experts about when to appropriately administer PEP is essential. By understanding the rabies epidemiology in the area, the vaccination history of the patient, and the potential risk of contracting rabies, medical personnel can help save lives as well as determine when not to administer PEP. Without risk assessment systems, exposed patients may not receive PEP when it is required to save their lives, or PEP may be given unnecessarily to unexposed patients, resulting in an escalation of the financial burden of rabies. It is only through building this type of infrastructure that a clear understanding of the epidemiology of rabies in a region can be understood. In regions where rabies is endemic, all animal bites should be evaluated by a medical professional, and bites incurred from an animal species belonging to the reservoir host species for circulating rabies viruses should always be considered a potential exposure to rabies. Determining the origin of rabies viruses through the use of genetic analyses has been useful in understanding the epidemiology of rabies as it can provide an insight into how to structure control programs, including where independent cycles of virus transmission are occurring and where elimination programs need to be targeted (Hayman et al., 2011; Nanayakkara, Smith, & Rupprecht, 2003).

Rabies endemic countries can monitor the epidemiology of rabies by establishing a system of satellite diagnostic laboratories under the direction of a national diagnostic reference facility. The national reference laboratory should be responsible for training personnel, maintaining quality assurance, and compiling data. Collaboration with international WHO Collaborative Centers can assist in the transfer of technology, including state-of-the-art diagnostic methods. Rabies surveillance can then be built on the diagnostic framework of a country. Laboratory results from the satellite diagnostic facilities should be sent to the national laboratory, where they can be compiled and distributed to all stakeholders involved in rabies prevention and control efforts.

4.3 Human Rabies Prevention

Preventing rabies in humans involves three strategies, including: eliminating the possibility of exposure; administering PrEP to those at increased risk of exposure; and administering PEP to patients that have been exposed. Rabies is present on every continent except Antarctica, and therefore poses a risk to most of the human population, especially those living in regions where the virus continues to circulate in the terrestrial animal population. This does not mean that every person living in a region where terrestrial rabies is circulating should receive PrEP no matter what their level of risk. Rather it means that rabies education awareness should remain a high priority for the general public living in rabies endemic regions and those living at high risk with limited or no access to biologicals should be considered for PrEP.

4.3.1 Preventing Exposures Through Improved Education

Rabies prevention begins with education. Without a clear understanding of how to prevent rabies by all stakeholders, the burden of rabies is unlikely to be reduced. Although there are numerous rabies virus variants circulating in a number of different animal reservoirs, the educational messages to prevent, recognize, and treat potential exposures are similar throughout the world regardless of individual, organizational, societal or environmental differences. First and foremost, education must focus on the fact that rabies viruses are maintained in animals but can be transmitted to humans, illustrating that elimination is therefore only possible through interventions targeting animal reservoirs and that routine vaccination of animals in close contact with humans, particularly dogs, is a critical component to prevent human rabies (Zinsstag et al., 2007). Secondly, emphasis should be placed on bite prevention. A potential exposure to rabies should be a consideration for every patient who has experienced an animal bite in rabies endemic regions. In regions with high endemicity, increased scrutiny of the details surrounding the biting incident is vital. Education about avoiding unknown animals that are likely to be unvaccinated and how to treat animals in a safe and responsible manner is essential in reducing the number of biting incidents, and hence, the number of potential rabies exposures. Finally, the lack of awareness about attending to bite wounds and needing to seek appropriate medical care after an exposure merits greater prominence through education. This knowledge gap is a major contributing factor to the ongoing burden of human rabies, particularly as immediate and thorough wound washing is largely regarded as an effective method of directly inactivating rabies virus that may have been infiltrated at wound sites that occurred during an exposure (Tenzin et al., 2011). Numerous infections can result from animal bites, and therefore the prompt cleansing of all wounds is warranted in every case. Educational efforts focusing on the need to thoroughly wash all wounds could be included with other public health hygiene initiatives already ongoing in many nations.

4.3.2 Pre-Exposure Vaccination

PrEP consists of a series of three doses of vaccine, given either intramuscularly (IM) or intradermally (ID) (Table 15.1). One dose of vaccine is administered on each of days 0, 3, and either day 21 or day 28 (WHO, 2011b). In the event of a subsequent exposure, PrEP reduces the doses of vaccine needed for PEP and eliminates the need for RIG, significantly reducing the costs.

TABLE 15.1   Vaccination Regimes Against Rabies. Pre-Exposure and Post-Exposure Vaccination Regimens Currently Recommended by the World Health Organization[a] and the Advisory Committee on Immunization Practices.[b] Regimens that are Recommended for Intradermal (ID) and/or Intramuscular (IM) are Listed

Vaccination	Number of Doses of Vaccine	Number of Clinical Visits Required	Route of Vaccine Administration	Schedule of Injections (days)
Pre-exposure				
Routine	3	3	ID[a]	0, 7, 21 or 28
			IM[a,b]	
Post-exposure				
Essen	5	5	IM[a,b]	0, 3, 7, 14, 28
Zagreb	4	3	IM[a,b]	0 (2 doses in each deltoid), 7, 21
Reduced 4-dose	4	4	IM[b]	0, 3, 7, 14
Modified Thai Red Cross	8	4	ID[a]	0, 3, 7, 28 (2 doses on each day)
Post-exposure for previously vaccinated persons				
Two-dose	2	2	IM[a,b]	0, 3
Four-dose	4	1	ID[a]	0 (2 doses above each deltoid)

PrEP should be administered to all persons at increased risk of rabies. This includes personnel working in rabies research laboratories and diagnostic laboratories, veterinarians, animal control workers, and individuals whose vocation or hobbies (e.g., caving) put them at increased risk of exposure. Additionally, in high risk regions where access to antirabies biologicals is difficult, PrEP should be considered as a valuable tool to save lives. For example, in the Amazon region where there are no medical clinics and vampire bat rabies continues to take the lives of unprotected indigenous populations, PrEP would be a valuable asset to improve rabies prevention activities (Carvalho-Costa et al., 2012). There are also remote populations living in high risk regions in Asia and Africa where dog rabies is uncontrolled and access to rabies biologicals is difficult to impossible where PrEP would help to save lives, especially those of children. Currently, the Philippine government has endorsed the administration of PrEP to children living in remote regions of the country where canine rabies poses a high risk and access to biologicals is limited.

4.3.3 Post-Exposure Prophylaxis

PEP must be administered to every patient that has been exposed to rabies because when administered in a timely manner according to WHO recommendations, PEP can save almost every exposed patient. In fact, there have only been a handful of human rabies deaths reported in patients that received PEP in a timely manner as per WHO protocol (Deshmukh & Yemul, 1999; Hemachudha et al., 1999; Tantawichien, Jaijaroensup, Khawplod, & Sitprija, 2001; Wilde, 2007). An exposure to rabies occurs when rabies virus enters an open wound or mucous membrane. This usually occurs through the trauma inflicted during the bite of a rabid animal. PEP consists of three components: appropriate wound care; administration of rabies immune globulin (RIG); and administration of a cell culture-based rabies vaccine (CCV). Wound care is often overlooked as the primary treatment in patients exposed to rabies, especially in resource poor countries. All wounds should be washed with soap and water for at least 15 minutes followed by administration of a virucidal antiseptic (WHO, 1950).

RIG is administered into and around the wound sites of patients who have not previously received PrEP or PEP with a CCV. RIG is administered as "passive immunity" and provides exposed patients with an immediate supply of rabies neutralizing antibodies prior to their ability to produce their own neutralizing antibodies in response to rabies vaccine. RIG can be administered up to seven days after a patient has received rabies vaccine. Currently, both heterologous and homologous RIGs are being produced globally. Heterologous equine rabies immune globulin (ERIG) is administered at a volume of 40 IU/Kg body weight and homologous human rabies immune globulin (HRIG) is administered at 20 IU/Kg body weight. Both products should be administered by injecting as much of the volume as possible into and around the wound sites to inactivate any virus that may have been injected at the time of the exposure. Any remaining RIG is administered by deep intramuscular injection at a site away from the vaccination site (WHO, 1950). In the case of multiple wounds, RIG should be diluted to a sufficient volume to be able to infiltrate all wounds (WHO, 2005). There is currently a global shortage of RIG, and even where it is available the cost is often prohibitively high. Ongoing clinical trials investigating the use of MAbs to replace the current RIGs will hopefully provide a solution to the lack of availability (Bakker et al., 2008).

For IM administration, WHO recommends that vaccine be administered intramuscularly (IM) into the upper arm (deltoid muscle) in adults and into the anterolateral thigh in small children. There are currently three IM vaccine regimens for PEP. Two are recommended by WHO, and a third reduced four dose regimen has recently been recommended by the Advisory Committee for Immunization Practices (ACIP) in the United States (Rupprecht et al., 2010; WHO, 1950, 2010) (Table 15.1). The Essen five-dose IM regimen is administered as one dose on each of days 0, 3, 7, 14, and 28. The ACIP recommendation has dropped the fifth dose on day 28 and one dose of vaccine is given on each of days 0, 3, 7, and 14 (Rupprecht et al., 2010). WHO also recommends a four dose IM regimen, known as the Zagreb or "2-1-1" PEP regimen. It is administered as two IM doses on day 0, one dose given into each deltoid, and one dose given on each of days 7 and 21. The WHO also recommends one intradermal (ID) PEP regimen known as the modified Thai red cross regimen. The vaccine is administered ID as two doses, one above each deltoid, on days 0, 3, 7, and 28. The ID method of delivery requires a lower dose of vaccine and can provide cost savings in high throughput clinics (Hampson et al., 2011).

There are three recommended PEP regimens for exposed patients that have previously received a complete PEP or PrEP series with a CCV. Two of the regimens are administered as a two-dose series on day 0 and 3 and can be administered either IM or ID. A third PEP regimen is administered as a four-dose ID regimen on day 0 (WHO, 2010).

4.4 Controlling Rabies in the Animal Reservoir

As mentioned previously, the best solution to reduce the risk of rabies exposure to humans is to eliminate the disease in the animal reservoir. Eliminating or greatly reducing the incidence of canine rabies in Latin America has had a direct impact on the number of human deaths (Belotto, 2004; Belotto, Leanes, Schneider, Tamayo, & Correa, 2005). Ensuring adequate funding for vaccination of dogs to save human lives is often challenging, as sometimes the Ministry of Health and the Ministry of Agriculture may be unwilling to recognize dogs as their responsibility.

Reducing rabies in specific species of wildlife is also possible, as has been demonstrated through the oral rabies vaccination program aimed at vaccinating the red fox population in Europe, the coyote population in the Southern United States, and the raccoon population in the eastern United States (Baer, Abelseth, & Debbie, 1971; Blancou, 2008; Blancou et al., 1986; Rupprecht, Dietzschold, Cox, & Schneider, 1989) (see Chapter 18). Controlling rabies in bat populations continues to be difficult due to lack of appropriate biologicals. In the Americas, bat rabies is a major threat to humans, and there are more human rabies cases attributed to infected bats than to other reservoirs. In Latin America, vampire bats pose a serious threat to humans as well as to livestock.

Investment in animal rabies control programs will save human lives and should be designed to reduce the cost of PEP in a sustainable manner. This means that to be effective, rabies prevention and control programs require an integrated approach including: local community involvement; communication management; dog population control; mass dog vaccination; dog bite management; veterinary quarantine; enforcement of rabies prevention legislation; and improved diagnostic capability, surveillance and monitoring.

5 COMMUNICATION, AWARENESS, AND ADVOCACY

Health communication is defined as the "study and use of communication strategies to inform and influence individual and community decisions that enhance health" (U.S. Department of Health and Human Services, 2000). Health communications are becoming increasingly important to public health practice and are a key component of effective and sustainable rabies prevention and control programs. People must be informed about how to protect themselves against rabies, and messages to that end must be clear, concise, timely, and targeted. The recent increased incorporation of health communicators into public health activities focused on rabies prevention has brought new evidence-based approaches to improve awareness. These elements of communication, rooted mainly—although not entirely—in health education, have resulted in renewed enthusiasm among anti-rabies advocates at all levels of society.

Rabies prevention from the perspective of a health communicator concerns getting the right messages to the right people at the right time. The World Rabies Day (WRD) campaign was created in part to fill the long-standing gap between scientific discovery involving rabies and improving public health through increased awareness; essentially giving a voice to data. In order to practice healthy behaviors, the general population should understand the significance of the public health messages they are receiving. It is therefore information, not data, that must be communicated for target audience knowledge acquisition. Communication is relevant to everyone, and much can be learned about rabies prevention through effective communication. For example, simply observing and recording how children in a particular community interact with dogs can help significantly in the development of localized bite prevention messages. Similarly, understanding the role livestock plays in the provision of food, fur, finance, and transport can help generate messages on the long-term cost benefit of animal vaccination. Relevant information exchange is the most important and cost-effective way for lay people to understand how to prevent rabies.

5.1 Communicating Effectively

It is important to realize that public health information is received and processed through individual and social filters. These pre-existing beliefs and norms can have a dramatic effect on how messages are perceived and interpreted. Since a single message will not be related to or be understood by all people, targeted communications must flow through appropriate channels to intended audiences from credible sources (Table 15.2). Rabies communications therefore need to take into account who needs to

TABLE 15.2   Benefits and Limitations of Communication Channels

Type of Outreach	Example Activities	Benefits	Limitations
Interpersonal communications	School presentations Conference or Symposium	Viewed as credible Allows for 2-way discussion Good for educational initiatives	Can be expensive Requires time Audience limited
Community outreach events	Parades, runs, walks Town-hall meetings Workplace campaigns	May be familiar, trusted, influential, motivational Can reach a large audience in one place	Requires fair amount of time and coordination from several partners
Newspaper	Feature articles Advertisements Letters to the editor	Can convey health information more thoroughly than radio/TV	Needs to be newsworthy Exposure limited to one day Article placement varies
Radio	News Advertisements (PSAs) Talk shows	May be main form of media in some locales Can direct messages toward target audiences	Topic must fit station's format Difficult for audience to retain or pass on information
Television	News Advertisements (PSAs) Talk shows	Visual and audio format is good for demonstrating prevention behaviors	May be expensive May not be available Messages can get "lost"
Internet	Web sites E-mail listservs Social networking Webinars	Large number of people Can be updated quickly Can be tailored/ customized	May not be available Requires that target audience is connected and looking for health information

receive information (audience); what information needs to be communicated (message); how the information will be transmitted (channel); and from whom or where the message will originate (source). These considerations are essential components of communications (Rimal & Lapinski, 2009). For example, to illustrate the importance of dog vaccination and address any local misconceptions about the safety of vaccinating dogs, a well-known public official might be filmed taking his or her dog to be vaccinated on WRD. The video recording could then accompany a news broadcast on national television with the public official discussing the importance of vaccination, dog bite prevention, and PEP.

In addition to these fundamental considerations, specific characteristics of effective health communications include accuracy, clarity, consistency, credibility, relevance, and correct tone and appeal (Parvanta, Nelson, Parvanta, & Harner, 2011). Messages must be current, clear, and simple, using the best available and most up-to-date technical data in balance with understandable information. Credibility is vital and audiences must be able to relate to or identify in some way with the spokesperson delivering the information. In many cases, the most effective public health spokespersons are those who themselves have suffered or are presently suffering from a particular disease or condition. Rabies, however, is an exception to this rule, given the disease's terminal nature. In the case of rabies, credibility can be built by partnering with established and trusted organizations, authorities, and individuals to achieve a common mission while operating in a neutral, genuine and transparent manner. Another facet of effective health communications is relevance. Combined with tone and appeal, materials and messages are relevant if they are meaningful to your audience. Relevance can be assessed during pretesting with target intended audiences. For example, when developing a poster for the prevention of rabies spread by the Arabian red fox in Saudi Arabia, special care must be taken to use appropriate and correct imagery, language, color, sizing and format. An image of a red fox that is not species correct may immediately result in disregard for the material, despite the value of any linked educational messages. Such nuances can significantly help or hinder the acceptance of health messages and are often discovered and remedied during pretesting. Appreciation and utilization of these core elements and characteristics of effective health communications will greatly assist in the development and delivery of rabies prevention and control efforts and are focal points of consideration when developing communications plans.

5.2 Increasing Awareness About Rabies

There are numerous audiences who need to know about how to prevent rabies and each will require specific and applicable messages.

In order to help determine what messages need to reach which audiences, it is most helpful to develop a communications plan. A communications plan for rabies can be developed for any societal level. A micro plan might target an individual locality (e.g., Bali, Indonesia) and focus on specific objectives and agents of change, whereas a macro plan might focus on aggregate health behavior change across multinational audiences. There are eight interrelated steps in developing a rabies communications plan, starting with a scientific assessment to identify the important points, potential issues, challenges, and barriers to change that may affect rabies outreach, as well as what role communication can play in addressing each problem. Step two defines the purpose of the outreach by identifying why communication is necessary and what recipients are supposed to do once they receive the information. The third step is to identify and segment target audiences to ensure that individual messages best resonate with specific recipients since not everyone will understand or relate to the same message. The fourth step entails developing and testing messages with subject-matter experts and intended audience(s) to improve messaging and better understand format, context, and delivery preferences. In step five, message format and dissemination channels are evaluated to understand which communication media (newspaper, radio, etc.) are most utilized and most trusted by target audience members. Step six focuses on determining when the message(s) should be transmitted to have the greatest impact, and step seven is the implementation of the communications activities. Lastly, step eight is the evaluation of the effort and its impact.

5.3 The World Rabies Day Campaign

World Rabies Day (WRD) is a health communications initiative that was launched by the Global Alliance for Rabies Control in 2007 to specifically address the widespread lack of awareness, misconceptions, and ignorance about rabies on all societal levels (WRD, 2012). Observed annually on September 28, WRD offers a unique opportunity to increase awareness for rabies prevention and control (Goswami, 2010). Since its inception, 150 countries have participated in WRD, resulting in the mass delivery of rabies prevention information to nearly 200 million people worldwide (Figure 15.1). Observance of WRD has been adopted by numerous multinational public health organizations and networks, including the Association of Southeast Asian Nations (ASEAN), the Asian, African, Middle East and Eastern European Rabies Expert Bureaus (AREB, AfroREB, MEEREB, respectively), South and East African Rabies Group (SEARG), Rabies in Asia Foundation (RIA), PAHO, WHO, OIE, CDC and numerous others. WRD plays a key role in the targeted

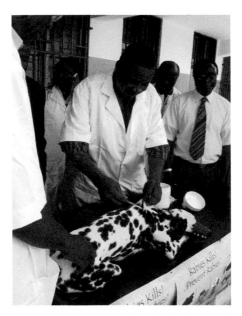

FIGURE 15.1   **World Rabies Day activities have helped to send educational messages to approximately 190 million people living at risk of rabies.** In Ghana, a government official is vaccinating a dog to kick off the canine rabies vaccination program. *Photo credit: Dr Richard Suu-ire.*

distribution of correct and actionable information regarding rabies. The WRD website (GARC, 2012) serves as a central point source for rabies education material and has catalyzed revitalization of national rabies control programs, facilitated development of a global rabies network, created new organizational partnerships, and led to new donor programs for rabies control (Cleaveland, Costa, Lembo, & Briggs, 2010). WRD is an established observance listed on the United Nations calendar of International Days (United Nations, 2012) and is a functioning "One Health" effort, with numerous international partners from the human and animal health fields successfully working together towards the common goal of preventing human and animal rabies.

6   INVOLVING COMMUNITIES IN RABIES CONTROL EFFORTS

6.1   The Roles of Different Organizations and Individuals

International health organizations such as the WHO, OIE, and FAO have the ability to assess current knowledge and feasibility of disease control methods, produce guidelines, and define research priorities. Their reports and guidelines are invaluable in setting international standards and establishing best practices to achieve disease control (WHO, 2005, 2010). They make recommendations to governments regarding public health issues and work with donors to fund large-scale interventions for improving public health. There are increasing numbers of successful large-scale disease control programs, such as the eradication of smallpox and rinderpest and the massive reduction of polio and measles (Hinman, 1999; Moss & Strebel, 2011; Roeder, 2011; Smits, 2009). These programs have generally been directed through the efforts of international health agencies, with significant external funding, driven by the belief that freedom from disease is a public good and even a human right (Hinman, 1999). However, there is a need to involve the citizens at risk, as well as international agencies and national governments. Published examples indicate that disease control programs that are conducted in a heavy-handed manner with excessive emphasis on a top down, "one-strategy-fits-all" approach often result in failures or delays (Atkinson, Vallely, Fitzgerald, Whittaker, & Tanner, 2011; Easterly, 2006).

There are many examples of community based non-governmental organizations (NGOs) funded by private donations that are working on preventing rabies. These are often not documented in the scientific literature and their cost-effectiveness is seldom analyzed, but these types of programs can often be the only rabies control efforts being conducted in an area. In Sierra Leone, for example, animal health clubs have increased responsible pet ownership practices in the community and have improved rabies vaccination dramatically in an environment having little infrastructure (FAO, 2010). The Blue Cross of India's ABC program in Chennai, India, has been involved in vaccinating and sterilizing the local stray dog population for several years. As a result, they proved to the local government that through their activity the number of animal bites and human rabies cases decreased (Blue Cross of India, 2009). In the 2008 Bali rabies outbreak, the Bali Animal Welfare Association campaigned for and helped to implement a culturally sensitive vaccination campaign to replace the culling program that had been ongoing. The demonstration of their success eventually led to the scaling up of dog vaccination programs to cover the entire island (WSPA, 2011). NGOs, particularly local ones, are often praised for their ability to innovate solutions based on the needs, desires, and constraints of their own communities (Easterly, 2006; Mukhopadhyay, 2007). In Sri Lanka, the Blue Paw Trust has pioneered the concept of managed dog zones, where workers in enclosed compounds, such as hospitals and schools, adopt the stray dogs, making sure that they are fed, healthy, and vaccinated against rabies (Blue Paw Trust, 2009).

6.2 Integrating the Support of all Partners

Coordinated global diseases control efforts must include the political commitment, infrastructure, and planning that can only be achieved by a "top down" approach, as exemplified by the eradication of polio initiative (Aylward, Acharya, England, Agocs, & Linkins, 2003). However, the full participation of the communities is also critical in order to meet targets, control costs, and eventually realize the success of the entire program. The community should be considered as a critical partner in the design as well as the implementation of control programs (Atkinson et al., 2011; Kilpatrick, 2009). However, successful locally designed and implemented control efforts may be difficult to scale up, especially considering their limited funding sources. Clearly, in order to be successful, it is necessary to incorporate both top-down and bottom-up approaches to insure that the program will have a maximum chance of success.

The Bohol Rabies Prevention and Elimination Program in the Philippines is a recent example of a successful and sustainable rabies prevention and control program that utilized an integrated approach to achieve rabies free status for the island within two years (Lapiz et al., 2012). The program was initiated by highly integrated health, veterinary, and legal departments at the provincial level. In addition, the program was fully aligned with Philippine national policy, followed international guidelines, and involved a close-working partnership with the community, beginning with the first planning stages. The Bohol project developed targeted information for education outreach, enhanced surveillance, and sustained animal vaccination campaigns, including training thousands of teachers to help with education campaigns and community volunteers to increase rabies surveillance in all villages. The incorporation of rabies education into the elementary class curriculum provided opportunities for routine and sustained rabies prevention information to be disseminated to all students living on the island. During the first 18 months, community groups were also trained and prepared for mass dog vaccination campaigns and dog registration schemes that served as the cornerstones of the program (Lapiz et al., 2012).The fostering and investing of competent communities (a concept derived from the need for community readiness for effective HIV/AIDS control) is often overlooked in the race to scale up interventions and achieve short-term successes. However, building effective community support is critical to insure long-term success and sustainability (Atkinson et al., 2011). The three-year effort required for the Bohol project included participation from the local communities, policy makers, a dedicated group of intersectoral partners, and key opinion leaders and international rabies experts.

6.3 Developing Programs with Communities

In order to gain the support of communities, it is critical to understand the geography and culture of the region involved in the rabies control and prevention program. Local consultations are invaluable during the strategic design process. In developing programs aimed at eliminating canine rabies, the relationship between humans and dogs must be understood. Building on the local indigenous knowledge by asking community members to design their own methods of participation can be a very helpful tool, even if it is slower than may be ideal (Atkinson et al., 2011; Kilpatrick, 2009). A detailed review of published examples of community participation in disease control efforts has identified key factors that may influence the participation of individuals, households, communities, and governments in control efforts (Atkinson et al., 2011). One important observation is that the most vulnerable members of society are usually least capable of participating in prevention efforts, and empowerment of these people can have a substantial impact on improving the chances of success.

Village leaders, community group heads, teachers, and local health professionals are in the best position to influence the community and to leverage community support for programs. It is helpful to include an education component specifically aimed at improving awareness and support of the local leaders as a component of the program. In addition, assessing and communicating the value of intervention for the leader's community will help align the expectations of the health service and the community and help convince village leaders to be strong advocates for change (Attree et al., 2011; Kilpatrick, 2009). In Bohol and Kenya, village-based volunteers and community elders continue to play critical roles in rabies control efforts and have received paralegal training to enable them to fairly enforce disease control legislation in their communities (FAO, 2011; Lapiz et al., 2012). There are increasing examples of rabies prevention education being incorporated into various subjects in the school curriculum to reach the highest-risk group, children (Lapiz et al., 2012).

Rabies control programs should be integrated into the existing public health infrastructure. If these programs are strengthened, including staff training that provides advantages beyond the disease effort currently considered, investments can help to benefit the management of several diseases, thus reducing overall costs (Aylward et al., 2003; Smits, 2009). Adequate training at a local level has been highlighted as a necessity to insure correct implementation of rabies control efforts in Latin America and Asia (Mukhopadhyay, 2007; Ruiz & Chavez, 2010). In Colombia, public health experts trained to facilitate communication between the public and hospitals have played a vital role in urban rabies prevention, promoting participation in vaccination campaigns and supporting community surveillance for rabies exposures (Partners for Rabies Prevention, 2010a). Additionally, community-based animal health workers have played an important cost-saving role in dog vaccination campaigns in Tanzania (Kaare et al., 2009). Health care and veterinary professionals are often involved in efforts to educate communities, and they also serve as advocates to their governments for increased rabies prevention and control measures. One innovative scheme in India has involved rabies experts at Kempegowda Institute of Medical Sciences, animal health experts from the College of Veterinary Medicine in Bangalore, and animal welfare experts at Karuna Animal Welfare Association in Karnataka India. This 'Adopt-a-village' program brought the experts to rural villages to train teachers and women involved in mother–child health care programs and empowered local village members to participate in rabies control efforts that helped to prevent human rabies deaths in their own community (GARC, 2010).

Recent years have seen a strengthening of rabies expert groups, both regionally based rabies expert groups (Aylan et al., 2011; Dodet et al., 2008; Dodet & Asian Rabies Expert, 2006) and international collaborations such as the Global Alliance for Rabies Control and the Partners for Rabies Prevention (Lembo et al., 2011). These experts, due to their training and expertise, play an important and powerful role in helping to design control strategies and advocating for improved resources for rabies control at national and international levels.

7 SUSTAINABILITY

7.1 The Goal is Elimination of Risk

The timely administration of PEP and effective anti-rabies biologicals has helped save millions of lives. However, a long-term sustainable rabies prevention and control program must include an integrated approach aimed at preventing humans being exposed to rabid animals, with the goal of eventually eliminating rabies at the source of infection. Thailand, Vietnam, and other countries have concentrated very heavily on expanding the provision of PEP, which has reduced human cases of rabies dramatically (Hemachudha, 2005). Without investing in increased vaccination coverage for dogs, the prime source of infection, this approach has resulted in a decrease in human cases, yet a dramatic and continually rising expenditure for human biologicals. The sustainability of this type of approach has been rightly questioned (Clifton, 2011; Hemachudha, 2005; Kasempimolporn, Jitapunkul, & Sitprija, 2008). Without elimination of the reservoir of infection in animals, humans will continue to be exposed and PEP costs will remain high in perpetuity.

The ultimate goal of disease control is eradication, the reduction of worldwide incidence of infection to zero, such that intervention measures are no longer needed. Eradication thus represents the ultimate in sustainability and social justice (Hinman, 1999). To date this has been achieved for only one human disease, smallpox, and one animal disease, rinderpest. Global eradication of polio is within reach, and measles could soon follow (Moss & Strebel, 2011). Due to the circulation of rabies viruses in various species of wildlife, particularly in bats where effective control methods do not yet exist, complete eradication of rabies is not feasible in the near future (Rupprecht et al., 2008). However, as the vast majority of human cases result from the bites of rabid domestic dogs and occur in regions where domestic dogs are the only maintenance host, mostly in Africa and Asia (Knobel et al., 2005), canine rabies elimination is a realistic and valuable goal (Rupprecht et al., 2008).

Effective elimination of canine rabies virus in dogs has the advantage of moving past large-scale canine vaccination programs to control strategies based on the establishment of effective surveillance and emergency response systems. This transition is critical for realizing the greatest public health cost benefit and can only be achieved after concerted surveillance efforts suggest that no residual reservoir exists and that an established infrastructure is in place to respond to any potential reintroduction of disease. For small areas with relatively easily protected borders, this transition can occur within a short time span (2 to 4 years for Bohol) (Lapiz et al., 2012). In KwaZulu, Natal, a 5-year plan is underway (WHO, 2011a). Longer periods of time are required for large geographic regions, such as Latin America, where a concomitant wildlife reservoir of rabies poses a threat of re-introduction into the domestic dog population (Ruiz & Chavez, 2010).

The benefits of a regional approach are clear. In 1983, the governments of the Pan American Health Organization (PAHO) region endorsed a political decision to eliminate dog-transmitted rabies in humans (Ruiz & Chavez, 2010). Nearly US$ 40 million annually was allocated by governments for this purpose, and more than 100 national and regional laboratories were established to help improve surveillance throughout the continent (Rupprecht et al., 2008). Recognizing that rabies viruses do not respect borders, instituting international agreements and identifying potential weaknesses in such agreements is important. For example, in South America, Brazil has donated millions of doses of canine vaccine to several other countries with limited resources, such as Bolivia and Haiti, where a lack of financing and infrastructure have slowed progress (Rupprecht et al., 2008).

7.2 Integrating Sustainability from the Start

Even if elimination is not achieved (and may never have been intended), large-scale integrated canine rabies prevention and control programs that are properly financed and implemented will save human lives and can reduce the demand for—and therefore the cost of—PEP dramatically (Cleaveland, Kaare, Tiringa, Mlengeya, & Barrat, 2003; FAO, 2011).

With all program approaches, maintaining the effort is vital if advances in the fight against rabies are not to be lost. Infected dogs can travel long distances with their owners (Johnson, Freuling, Horton, Muller, & Fooks, 2011) and considerable distances without their help, making outbreaks a constant threat, especially when rabies vaccination levels are low.

In reviewing successful elimination programs, Rupprecht et al. (2008) identified several factors that contributed to their success: (i) an initial assessment of the potential costs and benefits of widespread rabies prevention programs; (ii) long-term, national, rather than temporary, local plans; (iii) dog population control programs integrated with mass vaccination, but not used alone; (iv) a detailed understanding of the local dog–human relationship and how that affects participation; (v) culturally appropriate education; (vi) where wildlife rabies persists, dog vaccination is always recommended, and is often tied to compulsory dog registration; (vii) monitoring and surveillance to assess progress is vital.

Recent data published by the scientific community has shown that elimination is feasible, even in places where it was perceived to be impossible (Lembo et al., 2010). The effective reproduction rate (Ro) of rabies virus is not very high, meaning that transmission can be managed through effective vaccination programs (Hampson et al., 2009). In a detailed analysis of perceived barriers to elimination in Africa, Lembo and colleagues found that a lack of awareness of the burden of rabies, uncertainties about required levels of vaccination, accessibility of dogs for vaccination, and limited implementation and surveillance resources were not insurmountable hurdles to rabies prevention and control (Lembo et al., 2010). Tools to prevent rabies are available, and guidance is now freely accessible, through the Canine Rabies Blueprint, published online in 2010 (Partners for Rabies Prevention, 2012). This resource has compiled practical advice from experienced field projects and relevant guidelines from international health agencies to ensure that best practices in rabies control and case studies are available globally to anyone who needs them (Lembo, 2012).

There is an increasing commitment of international health organizations to control rabies through their individual organizations and together through collaboration of efforts such as the Partners for Rabies Prevention (PRP) (Lembo et al., 2011). One of the early goals of the PRP was to develop a detailed roadmap to identify the gaps that needed to be filled. Progress is being made. Over the last 5 years, achievements of this group include the initiation of the WRD campaign, the development of the Canine Rabies Blueprint, and securing funding for model rabies control program initiatives. Currently, studies are underway to reassess the global burden of rabies and develop models to demonstrated the economic impact of rabies and the cost-effectiveness of control programs.

Against a background of many health and other development needs in developing countries, strong, locally relevant economic arguments for rabies control are needed so that governments can make informed budget choices. Recent reassessments of the true burden of rabies in developing countries are aimed at persuading governments that prevention of canine rabies has benefits in terms of saving lives (Aylan et al., 2011; Knobel et al., 2005; Mahendra et al., 2010). This process is made easier where rabies is a legally notifiable (reportable) disease and effective reporting systems are in place.

There is a paucity of actual data on the cost-effectiveness of rabies control strategies, even in developed countries. Recent work has quantified the cost of different vaccination campaign designs and their ability to achieve desired levels of coverage in Tanzania and Chad (Kayali, Mindekem, Hutton, Ndoutamia, & Zinsstag, 2006). In the absence of real data, modeling approaches can predict the long-term cost savings of investment in canine vaccination. For the Philippines, an assessment in 1990 suggested that rabies elimination costs would be recovered 4–11 years after the initiation of a one-year elimination campaign (Fishbein et al., 1991). In Chad, investments in a canine vaccination program in addition to the standard use of PEP for human cases was predicted to be cost saving after 5–7 years (Zinsstag et al., 2009). Modeling approaches have also be used to assess the most cost effective ways to reduce PEP costs, strongly supporting the use of the ID vaccine route (Gongal & Wright, 2011; Hampson et al., 2011).

7.3 Sources of Funding and Cost Recovery

Initial investment in dog vaccination programs may seem to be overwhelming. Therefore, helping governments to find methods to reduce costs is important. In some cases, external funding has helped to support the strengthening of the infrastructure and to conduct mass vaccination with the understanding that maintenance of the rabies-free state will continue to be supported by national government (Lapiz et al., 2012; WSPA, 2011). The strength of government commitment to fund the maintenance phase will influence potential external funders, and the ongoing costs of a maintenance phase need to be assessed and presented so that governments can make informed decisions. Often, the financial responsibility for rabies prevention and control is unclear. Veterinary services may be expected to implement the operational part of canine rabies control with little or no economic benefit to the veterinary sector, yet the health ministry may benefit through cost savings on PEP and lives saved. Thus innovative strategies should be established where costs are shared between the Ministry of Health and the Ministry of Agriculture so that fair and adequate long-term funding for rabies prevention and control programs is ensured (Lembo et al., 2010).

7.3.1 Volunteers

Involving locally based community volunteers can be a strong asset in rabies prevention and control programs (Mukhopadhyay, 2007). In the Bohol Rabies Project, a volunteer program was established that trained 'rabies watchers' in every village, responsible for providing advice to bite victims and monitoring the movement of stray dogs. Across the island, thousands of teachers volunteered their time and talent to develop a curriculum that integrated rabies education into their subjects to ensure that children knew how to avoid dog bites and what to do if bitten (Lapiz et al., 2012). In the Adopt-a-Village project in India, women were trained to help their own communities with correct advice for dog bite treatment (GARC, 2010). In Kenya, village elders have been integrated into the rabies control activities and fulfill a vital role in monitoring stray dog populations, unvaccinated dogs, and bite incidents (FAO, 2011). In certain districts of Thailand, volunteers have been trained to conduct animal vaccinations and utilize chemical contraception to better serve their own community. They work in cooperation with municipal authorities during regional mass campaigns, and to capture, vaccinate, and chemically castrate stray dogs (Rupprecht et al., 2008). In Turkey, schoolchildren have been essential and enthusiastic participants in vaccination campaigns, leading vaccinators around their communities to locate both owned and stray dogs (Partners for Rabies Prevention, 2010b). In many parts of the world, NGOs—particularly animal welfare groups—are already involved in rabies control efforts and could be approached to help reduce governmental costs for labor during mass vaccination campaigns or elsewhere. This is being exemplified in Bali (WSPA, 2011).

Expectations of volunteers need to be well-managed in order to empower volunteers and to keep them involved but not overburdened by their responsibilities. Well-defined duties and time frames for engagement can help set realistic expectations (Aylward et al., 2003). When a good balance between engagement and responsibility is achieved, volunteer participation can lead to benefits to society beyond the immediate control project. One excellent example includes the schistosomiasis elimination campaign in China that invested in mass literacy classes with the aim of providing broader benefits for the development of communities (Atkinson et al., 2011). Maintaining the commitment to controlling rabies and periodic retraining and education of the community, including a provision for rewarding accomplishments, is critical for long-term success. The use of financial incentives to motivate volunteers can be useful but can easily be abused and can set unhelpful precedents. However, in certain circumstances, for example where vigilance after elimination is critical to maintaining success, financial rewards for volunteers reporting disease occurrence could be considered as a way to supplement governmental surveillance and reporting efforts (Atkinson et al., 2011). Integration of other related disease control programs can also help to decrease costs. In Kenya, for example, rabies vaccination rates were increased when farmers were asked to bring their dogs for vaccination when they brought their cattle for livestock health programs (FAO, 2011).

7.3.2 *Charging for Vaccination*

Ideally, dog vaccinations and human PEP are provided free of charge, to ensure maximum participation and life-saving PEP to those most at risk of rabies. In some countries, such as Sri Lanka, PEP is provided free, but this is not the norm. As rabies vaccination campaigns increase awareness, numbers of patients seeking PEP should be expected to rise initially and unnecessary administration of PEP can dramatically reduce the cost effectiveness of programs. In most resource-poor countries, the cost of PEP is the responsibility of the bite victim, a huge financial burden on families that often decreases the number of people seeking PEP. A more equal approach, where governments subsidize PEP costs, may be a way to achieve the balance, and models have been developed that investigate cost sharing between the government and the public (Hampson et al., 2011).

In most industrialized countries, pet owners assume the cost and are legally bound to vaccinate their pets against rabies. For developing countries where rabies is endemic, disease control strategies differ In some areas, charging for vaccination clearly results in lower vaccination rates than when vaccination campaigns are conducted free of charge (Durr et al., 2009). However, in Bohol, an enforced dog registration and vaccination campaign at a small cost to owners has been very successful and will fund help to fund the project in perpetuity. (Lapiz et al., 2012).

Including financial benefits of vaccination can also be helpful to increase uptake, even if the costs are incurred by the owners. In the Bohol project, owners of registered and vaccinated dogs are entitled to free PEP (Lapiz et al., 2012). In Kenya, legislation to enforce dogs owners to pay for PEP for bite victims if they were bitten by their unvaccinated dog has led to an increase in dog vaccination, now seen as an insurance policy (FAO, 2011).

8 CONCLUSIONS

We know that rabies is a totally preventable disease and that building sustainable rabies prevention and control programs is possible. There are excellent examples of regions that have managed to eliminate canine rabies, and others that have significantly reduced the disease in wildlife. Consequently, the burden of human and animal rabies has been dramatically decreased. Human rabies prevention starts with education of all sectors. The public must know how to recognize the threat, health professionals need to know how to respond appropriately to exposure events, and governments need to know the real impact of the disease in their countries and the potential benefits of increased control efforts. The recent application of health communications approaches to rabies has already reaped benefits. The WRD campaign and the Blueprint for canine rabies elimination and human rabies prevention has increased the access to educational materials and information about how to develop effective communication plans for those who need them.

Encouraged by the recent success stories, rabies experts in affected regions are calling for coordinated approaches to rabies control (Dodet, 2007; Dodet & Asian Rabies Expert, 2006), with some countries and regions forming plans for elimination (Gongal & Wright, 2011).

The ultimate goal in rabies control is to eliminate the virus in the animal reservoirs. This has been demonstrated and is an achievable goal for canine rabies in most, if not all, countries where it is endemic. Elimination of canine rabies would reduce the burden of human deaths globally by an estimated 99%. Even if complete elimination of canine rabies is not possible at the current time, long-term sustainable control efforts through dog vaccination would increase the number of human lives saved from rabies as well as reduce the fear associated with rabies in communities living with dog-mediated rabies. However, without building in a plan for sustainability, the benefits of mass vaccination are short lived and benefits will not outweigh costs. National—ideally, supra-national—policies and commitments are called for, but persuading governments to commit to long-term financing to ensure sustainability is a challenge. Strategies to share costs between external donors, local NGOs, the public, and the government should be investigated on a case-by-case basis to find the most practical solution for ensuring success.

Rabies control requires a multi-sectorial approach, combining the strengths of all stakeholders, from international organizations to individual community members, and successful programs are excellent examples of a 'One Health' approach to public health. Support from governments is critical, but without participation from the community, efforts will not succeed. Additionally, the input and support of local communities should not be overlooked when it comes to designing and implementing programs. Numerous publications outline specific methods to prevent rabies, but exactly how these methods are implemented in each region will vary according to the local circumstances. The most effective recent examples of control have used innovative models to increase intersectoral collaborations beyond the basic veterinary and medical field into education and legislation spheres to foster community ownership and participation that achieve maximum efficiency and sustainability.

References

Atkinson, J. A., Vallely, A., Fitzgerald, L., Whittaker, M., & Tanner, M. (2011). The architecture and effect of participation: A systematic review of community participation for communicable disease control and elimination. Implications for malaria elimination. *Malaria Journal, 10*, 225. doi:10.1186/1475-2875-10-225.

Attree, P., French, B., Milton, B., Povall, S., Whitehead, M., & Popay, J. (2011). The experience of community engagement for individuals: A rapid review of evidence. *Health & Social Care Community, 19*(3), 250–260. doi:10.1111/j.1365-2524.2010.00976.x.

Aylan, O., El-Sayed, A. F., Farahtaj, F., Janani, A. R., Lugach, O., Tarkhan-Mouravi, O., et al. (2011). Report of the first meeting of the middle east and eastern europe rabies expert bureau, istanbul, turkey (june 8–9, 2010). *Advance Preventive Medicine, 2011*, 812515. doi:10.4061/2011/812515.

Aylward, R. B., Acharya, A., England, S., Agocs, M., & Linkins, J. (2003). Global health goals: Lessons from the worldwide effort to eradicate poliomyelitis. *Lancet, 362*(9387), 909–914. doi:10.1016/S0140-6736(03)14337-1.

Baer, G. M., Abelseth, M. K., & Debbie, J. G. (1971). Oral vaccination of foxes against rabies. *American Journal of Epidemiology, 93*(6), 487–490.

Bakker, A. B., Python, C., Kissling, C. J., Pandya, P., Marissen, W. E., Brink, M. F., et al. (2008). First administration to humans of a monoclonal antibody cocktail against rabies virus: Safety, tolerability, and neutralizing activity. *Vaccine, 26*(47), 5922 5927. doi:10.1016/j.vaccine.2008.08.050.

Belotto, A., Leanes, L. F., Schneider, M. C., Tamayo, H., & Correa, E. (2005). Overview of rabies in the americas. *Virus Research, 111*(1), 5–12. doi:10.1016/j.virusres.2005.03.006.

Belotto, A. J. (2004). The pan american health organization (paho) role in the control of rabies in latin america. *Developmental Biology (Basel), 119*, 213–216.

Blancou, J. (2008). The control of rabies in eurasia: Overview, history and background. *Developmental Biology (Basel), 131*, 3–15.

Blancou, J., Kieny, M. P., Lathe, R., Lecocq, J. P., Pastoret, P. P., Soulebot, J. P., et al. (1986). Oral vaccination of the fox against rabies using a live recombinant vaccinia virus. *Nature, 322*(6077), 373–375. doi:10.1038/322373a0.

Blue Cross of India. (2009). The success of the abc-ar programme in india. Retrieved March 7th, 2012, 2012, from <http://bluecrossofindia.org/abc.html/>.

Blue Paw Trust. (2009). Dog managed zones. Retrieved March 7th, 2012, from <http://www.bluepawtrust.org/?page_id = 183/>.

Bourhy, H., Dautry-Varsat, A., Hotez, P. J., & Salomon, J. (2010). Rabies, still neglected after 125 years of vaccination. *PLoS Neglected Tropical Diseases, 4*(11), e839. doi:10.1371/journal.pntd.0000839.

Carvalho-Costa, F. A., Tedesqui, V. L., de Jesus Nascimento Monteiro, M., & Boia, M. N. (2012). Outbreaks of attacks by hematophagous bats in isolated riverine communities in the brazilian amazon: A challenge to rabies control. *Zoonoses and Public Health, 59*(4), 272–277. doi:10.1111/j.1863-2378.2011.01444.x.

CDC. (2007). Us declared canine-rabies free. Retrieved March 26th, 2012, from <http://www.cdc.gov/media/pressrel/2007/r070907.htm/>.

Cleaveland, S., Costa, P., Lembo, T., & Briggs, D. (2010). Catalysing action against rabies. *The Veterinary Record, 167*(11), 422–423. doi:10.1136/vr.c4775.

Cleaveland, S., Kaare, M., Knobel, D., & Laurenson, M. K. (2006). Canine vaccination–providing broader benefits for disease control. *Veterinary Microbiology, 117*(1), 43–50. doi:10.1016/j.vetmic.2006.04.009.

Cleaveland, S., Kaare, M., Tiringa, P., Mlengeya, T., & Barrat, J. (2003). A dog rabies vaccination campaign in rural africa: Impact on the incidence of dog rabies and human dog-bite injuries. *Vaccine, 21*(17–18), 1965–1973.

Clifton, M. (2011). How to eradicate canine rabies: A perspective of historical efforts. *Asian Biomedicine, 5*(4), 559–568.

Deshmukh, R. A., & Yemul, V. L. (1999). Fatal rabies encephalitis despite post-exposure vaccination in a diabetic patient: A need for use of rabies immune globulin in all post-exposure cases. *The Journal of the Association of Physicians of India, 47*(5), 546–547.

Dodet, B. (2007). Advocating rabies control in asia. *Vaccine, 25*(21), 4123–4124. doi:10.1016/j.vaccine.2007.03.002.

Dodet, B., Africa Rabies Expert, Bureau, Adjogoua, E. V., Aguemon, A. R., Amadou, O. H., Atipo, A. L., et al. (2008). Fighting rabies in africa: The africa rabies expert bureau (afroreb). *Vaccine, 26*(50), 6295–6298. doi:10.1016/j.vaccine.2008.04.087.

Dodet, B., & Asian Rabies Expert, Bureau, (2006). Preventing the incurable: Asian rabies experts advocate rabies control. *Vaccine, 24*(16), 3045–3049.

Durr, S., Mindekem, R., Kaninga, Y., Doumagoum Moto, D., Meltzer, M. I., Vounatsou, P., et al. (2009). Effectiveness of dog rabies vaccination programmes: Comparison of owner-charged and free vaccination campaigns. *Epidemiology and Infection, 137*(11), 1558–1567. doi:10.1017/S0950268809002386.

Easterly, W. (2006). *The white man's burden: Why the west's efforts to aid the rest have done so much ill and so little good.* US: Penguin Press.

FAO. (2010). "Dogs of war": Animal health clubs champion rabies prevention to protect livelihoods and lives in sierra leone. from <http://www.fao.org/ag/againfo/home/en/news_archive/AGA_in_action/2010_Animal_Health_Clubs_3.html/>.

FAO. (2011). Rabies control in kisumu, kenya. Retrieved March 7th 2012, 2012, from <http://www.fao.org/ag/againfo/home/en/news_archive/2011_Rabies_Control_in_Kisumu.html/>.

Fishbein, D. B., Miranda, N. J., Merrill, P., Camba, R. A., Meltzer, M., Carlos, E. T., et al. (1991). Rabies control in the republic of the philippines: Benefits and costs of elimination. *Vaccine, 9*(8), 581–587.

Franka, R., & Rupprecht, C. E. (2011). Treatment of rabies in the 21st century: Curing the incurable? *Future Microbiology, 6*(10), 1135–1140. doi:10.2217/fmb.11.92.

Fu, Z. F. (2008). The rabies situation in Far East Asia. *Developmental Biology (Basel), 131*, 55–61.

Galing Pook Foundation. (2012). Galing pook outstanding local governance programmes 2011 (pp. 24–25). Manila Philippines.

GARC. (2010). Project reports. Retrieved March 28th, 2012, from <http://rabiescontrol.net/what-we-do/project-reports.html/>.

GARC. (2012). World rabies day website. from <http://www.worldrabiesday.org/>.

Gomez-Benavides, J., Laguna-Torres, V. A., & Recuenco, S. (2010). [the real significance of being bitten by a hematophagous bat in indigenous communities in the remote peruvian amazon]. *Revista Peruana de Medicina Experimental y Salud Pública, 27*(4), 657–658.

Gongal, G., & Wright, A. E. (2011). Human rabies in the who southeast asia region: Forward steps for elimination. *Advances in Preventive Medicine, 2011*, 383870. doi:10.4061/2011/383870.

Goswami, A. (2010). Report of the sixth areb meeting, november 2009, at manila, philippines. Issue i, july 2010. *APCRI Journal, 12*(1), 16–20.

Haider, S. (2008). Rabies: Old disease, new challenges. *Canadian Medical Association Journal*, *178*(5), 562–563. doi:10.1503/cmaj.071709.

Hampson, K., Cleaveland, S., & Briggs, D. (2011). Evaluation of cost-effective strategies for rabies post-exposure vaccination in low-income countries. *PLoS Neglected Tropical Diseases*, *5*(3), e982. doi:10.1371/journal.pntd.0000982.

Hampson, K., Dobson, A., Kaare, M., Dushoff, J., Magoto, M., Sindoya, E., et al. (2008). Rabies exposures, post-exposure prophylaxis and deaths in a region of endemic canine rabies. *PLoS Neglected Tropical Diseases*, *2*(11), e339. doi:10.1371/journal.pntd.0000339.

Hampson, K., Dushoff, J., Cleaveland, S., Haydon, D. T., Kaare, M., Packer, C., et al. (2009). Transmission dynamics and prospects for the elimination of canine rabies. *PLoS Biology*, *7*(3), e53. doi:10.1371/journal.pbio.1000053.

Hayman, D. T., Johnson, N., Horton, D. L., Hedge, J., Wakeley, P. R., Banyard, A. C., et al. (2011). Evolutionary history of rabies in ghana. *PLoS Neglected Tropical Diseases*, *5*(4), e1001. doi:10.1371/journal.pntd.0001001.

Hemachudha, T. (2005). Rabies and dog population control in thailand: Success or failure? *Journal of the Medical Association of Thailand*, *88*(1), 120–123.

Hemachudha, T., Mitrabhakdi, E., Wilde, H., Vejabhuti, A., Siripataravanit, S., & Kingnate, D. (1999). Additional reports of failure to respond to treatment after rabies exposure in thailand. *Clinical Infectious Diseases*, *28*(1), 143–144. doi:10.1086/517179.

Hinman, A. (1999). Eradication of vaccine-preventable diseases. *Annual Review of Public Health*, *20*, 211–229. doi:10.1146/annurev.publhealth.20.1.211.

Ichhpujani, R. L., Mala, C., Veena, M., Singh, J., Bhardwaj, M., Bhattacharya, D., et al. (2008). Epidemiology of animal bites and rabies cases in india. A multicentric study. *Journal of Communication Disorders*, *40*(1), 27–36.

Johnson, N., Freuling, C., Horton, D., Muller, T., & Fooks, A. R. (2011). Imported rabies, european union and switzerland, 2001–2010. *Emerging Infectious Diseases*, *17*(4), 753–754. doi:10.3201/eid1706.101154.

Kaare, M., Lembo, T., Hampson, K., Ernest, E., Estes, A., Mentzel, C., et al. (2009). Rabies control in rural africa: Evaluating strategies for effective domestic dog vaccination. *Vaccine*, *27*(1), 152–160. doi:10.1016/j.vaccine.2008.09.054.

Kasempimolporn, S., Jitapunkul, S., & Sitprija, V. (2008). Moving towards the elimination of rabies in thailand. *Journal of the Medical Association of Thailand*, *91*(3), 433–437.

Kayali, U., Mindekem, R., Hutton, G., Ndoutamia, A. G., & Zinsstag, J. (2006). Cost-description of a pilot parenteral vaccination campaign against rabies in dogs in n'djamena, chad. *Tropical Medicine & International Health*, *11*(7), 1058–1065. doi:10.1111/j.1365-3156.2006.01663.x.

Kilpatrick, S. (2009). Multi-level rural community engagement in health. *The Australian Journal of Rural Health*, *17*(1), 39–44. doi:10.1111/j.1440-1584.2008.01035.x.

Knobel, D. L., Cleaveland, S., Coleman, P. G., Fevre, E. M., Meltzer, M. I., Miranda, M. E., et al. (2005). Re-evaluating the burden of rabies in africa and asia. *Bulletin of the World Health Organization*, *83*(5), 360–368. doi: /S0042-96862005000500012.

Lapiz, M.D., Miranda, M.E., Romulo, G.G., Daguro, L.I., Paman, M.D., Madrinan, F.P., et al. (2012). Implementation of an intersectoral program to eliminate human and canine rabies: The Bohol Rabies Prevention and Elimination Project. *PLoS NTD*. doi:10.1371/journal.pntd.0001891.

Lembo, T. (2012). The blueprint for rabies prevention and control: A novel operational toolkit for rabies elimination. *PLoS Neglected Tropical Diseases*, *6*(2), e1388. doi:10.1371/journal.pntd.0001388.

Lembo, T., Attlan, M., Bourhy, H., Cleaveland, S., Costa, P., de Balogh, K., et al. (2011). Renewed global partnerships and redesigned roadmaps for rabies prevention and control. *Veterinary Medicine International*, *2011*, 923149. doi:10.4061/2011/923149.

Lembo, T., Hampson, K., Kaare, M. T., Ernest, E., Knobel, D., Kazwala, R. R., et al. (2010). The feasibility of canine rabies elimination in africa: Dispelling doubts with data. *PLoS Neglected Tropical Diseases*, *4*(2), e626. doi:10.1371/journal.pntd.0000626.

Mahendra, B. J., Madhusudana, S. N., Sampath, G., Datta, S. S., Ashwathnarayana, D. H., Venkatesh, G. M., et al. (2010). Immunogenicity, safety and tolerance of a purified duck embryo vaccine (pdev, vaxirab) for rabies post-exposure prophylaxis: Results of a multicentric study in india. *Human Vaccines*, 6(9).

Mallewa, M., Fooks, A. R., Banda, D., Chikungwa, P., Mankhambo, L., Molyneux, E., et al. (2007). Rabies encephalitis in malaria-endemic area, malawi, africa. *Emerging Infectious Diseases*, 13(1), 136–139.

Molyneux, D., Hallaj, Z., Keusch, G. T., McManus, D. P., Ngowi, H., Cleaveland, S., et al. (2011). Zoonoses and marginalised infectious diseases of poverty: Where do we stand? *Parasites & Vectors*, 4, 106. doi:10.1186/1756-3305-4-106.

Moss, W. J., & Strebel, P. (2011). Biological feasibility of measles eradication. *Journal of Infectious Diseases*, 204(Suppl. 1), S47–53. doi:10.1093/infdis/jir065.

Mukhopadhyay, A. (2007). South asia's health promotion kaleidoscope. *Promotion & Education*, 14(4), 238–243.

Nanayakkara, S., Smith, J. S., & Rupprecht, C. E. (2003). Rabies in sri lanka: Splendid isolation. *Emerging Infectious Diseases*, 9(3), 368–371.

Partners for Rabies Prevention. (2010a). An example of the use of health promoters in dog vaccination campaigns. Retrieved March 7th 2012, 2012, from <http://www.rabiesblueprint.com/An-example-of-the-use-of-health/>.

Partners for Rabies Prevention. (2010b). House-to-house rabies vaccination campaigns using schoolchildren in istanbul, turkey. Retrieved March 7th, 2012, 2012, from <http://www.rabiesblueprint.com/House-to-house-rabies-vaccination/>.

Partners for Rabies Prevention. (2010c). Legislation section of blueprint for rabies prevention and control. Retrieved March 28th, 2012, from <http://www.rabiesblueprint.com/3-2-Legislation/>.

Partners for Rabies Prevention. (2012). Blueprint for rabies prevention and control. from <http://www.rabiesblueprint.com/>.

Parvanta, C. F., Nelson, D. E., Parvanta, S. A., & Harner, R. N. . (2011). *Essentials of public health communication*. Sudbury, MA, USA: Jones & Bartlett.

PCHRD. (2012). Rabies-free philippines, still possible. Philippine council for health research and development. Retrieved March 30th, 2012, from <http://www.pchrd.dost.gov.ph/index.php/news/576-rabies-free-philippines-still-possible/>.

Rimal, R. N., & Lapinski, M. K. (2009). Why health communication is important in public health. *Bulletin of the World Health Organization*, 87(4) 247–247a.

Robertson, K., Lumlertdacha, B., Franka, R., Petersen, B., Bhengsri, S., Henchaichon, S., et al. (2011). Rabies-related knowledge and practices among persons at risk of bat exposures in thailand. *PLoS Neglected Tropical Diseases*, 5(6), e1054. doi:10.1371/journal.pntd.0001054.

Roeder, P. L. (2011). Rinderpest: The end of cattle plague. *Preventive Veterinary Medicine*, 102(2), 98–106. doi:10.1016/j.prevetmed.2011.04.004.

Ruiz, M., & Chavez, C. B. (2010). Rabies in latin america. *Neurological Research*, 32(3), 272–277. doi:10.1179/016164110X12645013284257.

Rupprecht, C. E. (2009). Bats, emerging diseases, and the human interface. *PLoS Neglected Tropical Diseases*, 3(7), e451. doi:10.1371/journal.pntd.0000451.

Rupprecht, C. E., Barrett, J., Briggs, D., Cliquet, F., Fooks, A. R., Lumlertdacha, B., et al. (2008). Can rabies be eradicated? *Developments in Biologicals (Basel)*, 131, 95–121.

Rupprecht, C. E., Briggs, D., Brown, C. M., Franka, R., Katz, S. L., Kerr, H. D., et al. (2010). Use of a reduced (4-dose) vaccine schedule for postexposure prophylaxis to prevent human rabies: Recommendations of the advisory committee on immunization practices. *MMWR Recommendations and Report*, 59(RR-2), 1–9.

Rupprecht, C. E., Dietzschold, B., Cox, J. H., & Schneider, L. G. (1989). Oral vaccination of raccoons (procyon lotor) with an attenuated (sad-b19) rabies virus vaccine. *Journal of Wildlife Diseases*, 25(4), 548–554.

Satapathy, D. M., Sahu, T., Behera, T. R., Patnaik, J. K., & Malini, D. S. (2005). Socio-clinical profile of rabis cases in anti-rabies clinic, m. K. C. G. Medical college, orissa. *Indian Journal of Public Health*, 49(4), 241–242.

Slate, D., Algeo, T. P., Nelson, K. M., Chipman, R. B., Donovan, D., Blanton, J. D., et al. (2009). Oral rabies vaccination in north america: Opportunities, complexities, and challenges. *PLoS Neglected Tropical Diseases*, 3(12), e549. doi:10.1371/journal.pntd.0000549.

Smits, H. L. (2009). Prospects for the control of neglected tropical diseases by mass drug administration. *Expert review of Anti-Infective Therapy*, 7(1), 37–56. doi:10.1586/14787210.7.1.37.

Sudarshan, M. K., Mahendra, B. J., & Narayan, D. H. (2001). A community survey of dog bites, anti-rabies treatment, rabies and dog population management in bangalore city. *Journal of Communication Disorders*, 33(4), 245–251.

Tantawichien, T., Jaijaroensup, W., Khawplod, P., & Sitprija, V. (2001). Failure of multiple-site intradermal postexposure rabies vaccination in patients with human immunodeficiency virus with low cd4+ t lymphocyte counts. *Clinical Infectious Diseases*, 33(10), E122–124. doi:10.1086/324087.

Tenzin, (2011)., Dhand, N. K., Gyeltshen, T., Firestone, S., Zangmo, C., Dema, C., et al. (2011). Dog bites in humans and estimating human rabies mortality in rabies endemic areas of bhutan. *PLoS Neglected Tropical Diseases*, 5(11), e1391. doi:10.1371/journal.pntd.0001391.

U.S. Department of Health and Human Services, (2000). (2nd ed.) *Healthy people 2010. With understanding and improving health and objectives for improving health* (Vol. 1–2). Washington, DC, USA: U.S. Government Printing Office.

United Nations. (2012). United nations observances. Retrieved March 26th, 2012, from <http://www.un.org/en/events/observances/days.shtml/>.

WHO. (1950). Who expert consultation on rabies: First report *Technical report series* 28. Geneva.

WHO. (2005). Who expert consultation on rabies (2004: Geneva, switzerland) *WHO Technical Report Series; 931*. Geneva: World Health Organization.

WHO, (2010). Rabies vaccines: Who position paper–recommendations. *Vaccine*, 28(44), 7140–7142. doi:10.1016/j.vaccine.2010.08.082.

WHO. (2011a). Celebrating one year without a reported human case of rabies in kwazulunatal, south africa. Retrieved March 7th, 2012, 2012, from <http://www.who.int/rabies/Celebrating_one_year_rabies_free_KwaZuluNatal/en/>.

WHO. (2011b) *The immunological basis for immunization series: Rabies* (pp. 1–23). Geneva.

WHO. (2011c). Rabies fact sheet, no 99. from <http://www.who.int/mediacentre/factsheets/fs099/en/>.

Wilde, H. (2007). Failures of post-exposure rabies prophylaxis. *Vaccine*, 25(44), 7605–7609. doi:10.1016/j.vaccine.2007.08.054.

Willoughby, R. E., Jr., et al., Tieves, K. S., Hoffman, G. M., Ghanayem, N. S., Amlie-Lefond, C. M., Schwabe, M. J., et al. (2005). Survival after treatment of rabies with induction of coma. *The New England Journal of Medicine*, 352(24), 2508–2514. doi:10.1056/NEJMoa050382.

WSPA. (2011). Rabies cases on decline in bali as first round of vaccination completed. Retrieved March 7th, 2012, 2012, from <http://www.wspa-international.org/latest-news/2011/rabies-decline-bali-vaccination-completed.aspx/>.

WSPA. (2012). Stray animals. Retrieved May 2nd, 2102, from <http://www.wspa-international.org/wspaswork/dogs/strayanimals/Default.aspx/>.

Zinsstag, J., Durr, S., Penny, M. A., Mindekem, R., Roth, F., Menendez Gonzalez, S., et al. (2009). Transmission dynamics and economics of rabies control in dogs and humans in an african city. *Proceedings of the National Academy of Sciences of the United States of America*, 106(35), 14996–15001. doi:10.1073/pnas.0904740106.

Zinsstag, J., Schelling, E., Roth, F., Bonfoh, B., de Savigny, D., & Tanner, M. (2007). Human benefits of animal interventions for zoonosis control. *Emerging Infectious Diseases*, 13(4), 527–531.

Therapy of Human Rabies

Alan C. Jackson

Departments of Internal Medicine (Neurology) and Medical
Microbiology University of Manitoba, Winnipeg,
Manitoba R3A 1R9, Canada

Preventative therapy for rabies after exposures is highly effective if current recommendations are followed (Manning et al., 2008). Unfortunately, the therapy of rabies has proved to be disappointing. A number of approaches have been unsuccessful. Therapy with human leukocyte interferon in three patients with high-dose intraventricular and systemic (intramuscular) administration was not associated with a beneficial clinical effect, but this therapy was not initiated until between 8 and 14 days after the onset of symptoms (Merigan et al., 1984). Similarly, antiviral therapy with intravenous ribavirin (16 patients given doses of 16–400 mg) was unsuccessful in China (Kureishi, Xu, Wu, & Stiver, 1992). An open trial of therapy with combined intravenous and intrathecal administration of either ribavirin (one patient) or interferon-alfa (three patients) (Warrell et al., 1989) was also unsuccessful. Anti-rabies virus hyperimmune serum of either human or equine origin has been administered intravenously and by the intrathecal route (Basgoz & Frosch, 1998; Emmons et al., 1973; Hattwick, Corey, & Creech, 1976; Hemachudha et al., 2003), but there was no clear beneficial effect. In some cases, survival has been prolonged for a few weeks with critical care measures.

In 2001, a conference was held that included physicians who have experience in the management of human rabies and researchers with expertise in rabies pathogenesis. The opinions of the participants were published in a viewpoint article in *Clinical Infectious Diseases*, including therapeutic options when aggressive therapy is considered desirable (Jackson et al., 2003). Patients in good health with relatively early disease and access to adequate resources and facilities were felt to be potential candidates for an aggressive approach. It was felt that a

Rabies: Scientific Basis of the Disease and its Management
DOI: http://dx.doi.org/10.1016/B978-0-12-396547-9.00016-X **575**

combination of specific therapies should be considered, including rabies vaccine, rabies immune globulin, monoclonal antibodies (in the future), ribavirin, interferon-α and ketamine, which is a dissociative anesthetic agent that is a noncompetitive antagonist of the N-methyl-D-aspartate (NMDA) receptor. Previous studies performed *in vitro* and in an experimental animal model suggested ketamine may be a useful therapeutic agent (Lockhart, Tordo, & Tsiang, 1992; Lockhart, Tsiang, Ceccaldi, & Guillemer, 1991). Recovery has occurred in one patient who received therapy with ketamine and other agents (Hu, Willoughby, Jr., Dhonau, & Mack, 2007; Willoughby, Jr. et al., 2005), but, disappointingly, this therapeutic approach was subsequently unsuccessful in at least 26 cases (Table 16.1) and perhaps many other cases in which there is no available documentation.

TABLE 16.1   Cases of Human Rabies with Treatment Failures that Used the Main Components of the "Milwaukee Protocol"

Case No.	Year of Death	Age and Sex of Patient	Virus Source	Country	Reference
1	2005	47 male	kidney and pancreas transplant (dog)	Germany	(Maier et al., 2010)
2	2005	46 female	lung transplant (dog)	Germany	(Maier et al., 2010)
3	2005	72 male	kidney transplant (dog)	Germany	(Maier et al., 2010)
4	2005	unknown	dog	India	(Bagchi, 2005)
5	2005	7 male	vampire bat	Brazil	_[a]
6	2005	20–30 female	vampire bat	Brazil	_[a]
7	2006	33 male	dog	Thailand	(Hemachudha et al., 2006)
8	2006	16 male	bat	USA (Texas)	(Houston Chronicle, 2006)
9	2006	10 female	bat	USA (Indiana)	(Christenson et al., 2007)
10	2006	11 male	dog (Philippines)	USA (California)	(Aramburo et al., 2011; Christenson et al., 2007)
11	2007	73 male	bat	Canada (Alberta)	(McDermid et al., 2008)

(Continued)

TABLE 16.1   (Continued)

Case No.	Year of Death	Age and Sex of Patient	Virus Source	Country	Reference
12	2007	55 male	dog (Morocco)	Germany	(Drosten, 2007)
13	2007	34 female	bat (Kenya)	The Netherlands	(van Thiel et al., 2009)
14	2008	5 male	dog	Equatorial Guinea	(Rubin et al., 2009)
15	2008	55 male	bat	USA (Missouri)	(Pue et al., 2009; Turabelidze et al., 2009)
16	2008	8 female	cat	Colombia	(Juncosa, 2008)
17	2008	15 male	vampire bat	Colombia	(Badillo et al., 2009)
18	2009	37 female	dog (South Africa)	Northern Ireland	(Hunter et al., 2010)
19	2009	42 male	dog (India)	USA (Virginia)	(Troell et al., 2010)
20	2010	11 female	cat	Romania	(Luminos et al., 2011)
21	2011	41 female	dog (Guinea-Bissau)	Portugal	(Santos et al., 2012)
22	2011	25 male	Dog (Afghanistan)	USA (Massachusetts)	(Javaid et al., 2012)
23	2012	63 male	brown bat	USA (Massachusetts)	(Greer et al., 2013)
24	2012	9 male	marmoset	Brazil	(NE 10, 2012)
25	2012	41 male	dog (Dominican Republic)	Canada (Ontario)	(Branswell, 2012)
26	2012	29 male	dog (Mozambique)	South Africa	(IAfrica.com, 2012; Times Live, 2012)

[a]Personal communication from Dr. Rita Medeiros, University of Para, Belem, Brazil.
(Updated from Jackson, A. C.: Therapy in human rabies, in Research Advances in Rabies, Alan C. Jackson (ed.), Advances in Virus Research 79:365–375, 2011; Copyright Elsevier.)

1 HUMAN CASES WITH RECOVERY FROM RABIES

Survival from rabies has been well documented in only seven patients (Table 16.2), and all but one of these patients received rabies immunization prior to the onset of clinical disease. The first recovery from rabies, which has been the only case without significant neurological sequelae,

TABLE 16.2 Cases of Human Rabies with Recovery

Location	Year	Age of Patient	Transmission	Immunization	Outcome	Reference
USA	1970	6	Bat bite	Duck embryo vaccine	Complete recovery	(Hattwick et al., 1972)
Argentina	1972	45	Dog bites	Suckling mouse brain vaccine	Mild sequelae	(Porras et al., 1976)
USA	1977	32	Laboratory (vaccine strain)	Pre-exposure vaccination	Severe sequelae	(Tillotson et al., 1977a, 1977b)
Mexico	1992	9	Dog bites	Post-exposure vaccination (combination)	Severe sequelae[a]	(Alvarez et al., 1994)
India	2000	6	Dog bites	Post-exposure vaccination (combination)	Severe sequelae[b]	(Madhusudana et al., 2002)
United States	2004	15	Bat bite	None	Mild to moderate sequelae	(Willoughby et al., 2005; Hu et al., 2007)
Brazil	2008	15	Vampire bat bite	Post-exposure vaccination	Severe sequelae	(Ministerio da Saude in Brazil, 2008)

[a]Patient died less than 4 years after developing rabies with marked neurological sequelae (L. Alvarez, personal communication).
[b]Patient died about 2 years after developing rabies with marked neurological sequelae (S. Mahusudana, personal communication).

occurred in 1970 (Hattwick, Weis, Stechschulte, Baer, & Gregg, 1972). Matthew Winkler, a 6-year-old boy from Ohio, was bitten on his left thumb by a big brown bat (*Eptesicus fuscus*), which was later shown to be rabid. Vaccination was initiated with duck embryo rabies vaccine beginning 4 days after the bite. Shortly after completing the multidose therapy (20 days after the bite), he became ill with fever and meningeal signs. His CSF showed 125 white cells/μL (75% mononuclear cells and 25% polymorphonuclear leukocytes), and the CSF protein was elevated. He developed abnormal behavior and later lapsed into a coma. He had focal neurological signs and seizures and developed cardiac and respiratory complications. He subsequently showed progressive improvement and apparently had a good neurologic recovery. A brain biopsy was consistent with encephalitis. His serum neutralization titer against rabies virus peaked at 1:63,000 at 3 months. This titer was much higher than has been observed secondary to vaccination. He also had very high titers of neutralizing antibodies in the

CSF, which have not been observed with vaccination. Rabies virus was not isolated from brain tissue, CSF, or saliva, probably as a result of viral neutralization related to the high antibody levels.

The second case with recovery was a 45-year-old woman who sustained multiple deep bites to her left arm from a dog in Argentina in 1972 (Porras et al., 1976). The dog developed neurologic signs and died 4 days later. The patient received 14 daily doses of suckling mouse brain rabies vaccine beginning 10 days after the bites, which were followed by two booster doses. Twenty-one days after the bites (at the time of her twelfth vaccine dose), she developed left-arm paresthesias, which subsequently spread and became accompanied by pain; vaccination was continued. She was admitted to hospital with quadriparesis and hyperreflexia 31 days after the bites. She had limb weakness, tremor in her upper extremities (greater on the left), cerebellar signs (asynergia, ataxia, dysmetria and dysdiadochokinesia), generalized myoclonus, and hyperreflexia in her lower extremities. Prominent cerebellar signs are unusual in rabies, despite the characteristic infection of neurons in the cerebellum, including Purkinje cells and deep cerebellar nuclei. Her CSF showed 5 cells/μL, and CSF protein was mildly elevated at 0.65 g/L. Her serum neutralization titer against rabies virus peaked at 1:640,000 at about 3 months, and she also had very high titers of neutralizing antibodies in the CSF. Rabies virus was not isolated from her saliva or CSF and corneal impression smears were negative for rabies virus antigen. Neurologic deterioration occurred shortly after she received each of the two booster doses of rabies vaccine and included altered mental status, generalized seizures, dysphagia, and quadriparesis. She showed neurological improvement over the next few months. Thirteen months after the onset of her symptoms, her recovery was reported as "nearly complete." However, there was no description of her residual neurological deficits (Porras et al., 1976). The unusual neurologic features of this patient, as well as the clinical worsening after booster doses of the suckling mouse brain rabies vaccine were administered, raise the question of whether encephalomyelitis due to the rabies vaccine played a significant role in this patient's clinical picture.

The third case occurred in a 32-year-old laboratory technician in New York in 1977 who was pre-immunized with duck embryo rabies vaccine (Tillotson, Axelrod, & Lyman, 1977a; Tillotson, Axelrod, & Lyman, 1977b). About 5 months prior to his illness, he had a rabies virus neutralizing antibody titer of 1:32. He worked with live rabies virus vaccine strains, and he was likely exposed to an aerosol of rabies virus about two weeks prior to the onset of his illness. He experienced initial malaise, headache, fever, chills, and nausea and then lethargy with intermittent delirium. He was admitted to hospital in Albany, New York, 6 days after the onset of his symptoms with expressive aphasia, hyperreflexia, and primitive reflexes. CSF showed 230 white cells/μL (95% mononuclear cells), and

CSF protein was elevated at $1.17\,g/L$. The day after admission to the hospital, he deteriorated and went into a deep coma. His serum neutralizing antibody titer increased from 1:32 to 1:64,000 and subsequently increased to 1:175,000 over a 10-day period during his illness (Tillotson et al., 1977a). He also developed a high titer of CSF antibodies. Rabies virus antigen was not detectable in a skin biopsy or in corneal impression smears. Four months after the onset of his illness, he was ambulatory, but he had residual aphasia and spasticity (Tillotson et al., 1977a). This was the first report of a case of rabies in a pre-immunized individual and only the fourth well-documented case with transmission due to airborne exposure to the virus.

The fourth case, from 1992, occurred in a 9-year-old boy in Mexico (Alvarez et al., 1994). This boy sustained severe facial bites from a dog and received local wound treatment. On the day after the bites, vaccination was initiated with VERO rabies vaccine, but passive immunization with rabies immune globulin was not given. Nineteen days after the bites, he developed fever and dysphagia. He subsequently had a variety of abnormal neurological signs and convulsions. He never developed hydrophobia or inspiratory spasms. He was admitted to hospital and subsequently became comatose. CSF showed 184 cells/μL (65% mononuclear cells). He required mechanical ventilation for several days. Rabies virus was not isolated from saliva, and rabies virus antigen was not found in either a skin biopsy or corneal impression smears. His peak serum neutralizing antibody titer was 1:34,800 (39 days after the bite), and he had a very high CSF antibody titer. He had severe neurologic sequelae, including quadriparesis and visual impairment. Although he recovered for a period, he died almost 4 years later (L. Alvarez, personal communication).

The fifth case was a 6-year-old girl who was bitten on the face and hands by a dog in India, and the dog died 4 days later (Madhusudana, Nagaraj, Uday, Ratnavalli, & Kumar, 2002). She received three doses of rabies purified chick embryo cell vaccine (PCECV) on days 0, 3, and 7, but no local wound treatment was given and rabies immune globulin was not administered. She developed clinical features of rabies 14 days after the bites, which included fever, dysphagia to liquids, and visual hallucinations. A rare neurologic complication to the PCECV was considered, and she was given methylprednisolone and one dose of rabies human diploid cell vaccine. She subsequently developed hypersalivation and focal motor seizures, and she became comatose. A MR scan showed T_2-weighted hyperintense signals in the cerebral cortex, basal ganglia, and brainstem. She had a CSF pleocytosis. Her peak serum neutralizing antibody titer was 1:312,000 (7800 IU/mL) after 110 days of illness, and she had a CSF antibody titer of 1:182,000 (4550 IU/mL) at this time. Rabies virus was not isolated, and both skin biopsies and corneal tests were negative for rabies virus antigen. She had severe neurologic sequelae, including rigidity and involuntary movements of her limbs, and she had frequent opisthotonic postures. She died about 2 years later (S. Madhusudana, personal communication).

The sixth case occurred in Wisconsin in 2004 (Willoughby, Jr. et al., 2005). A previously healthy 15-year-old female was bitten by a bat on her left index finger while attending a church service, and she subsequently released the bat. The wound was washed with peroxide, but she did not seek medical attention at that time. About one month after the bite, she developed numbness and tingling of her left hand, and over the next 3 days, she developed diplopia related to bilateral partial sixth-nerve palsies, unsteadiness, and nausea and vomiting. MRI brain was normal. On her fourth day of illness, CSF showed 23 white cells/μL (93% lymphocytes), and CSF protein was mildly elevated at 50 mg/dL. She subsequently developed fever (38.8°C), nystagmus, left arm tremor, and hypersalivation, and at about that time the history of the bat bite was obtained. The patient was transferred to a tertiary care hospital in Milwaukee 5 days after the onset of neurologic symptoms. A repeat MRI scan was normal. Neutralizing anti-rabies virus antibodies were detected in serum and CSF on the first hospital day (initially 1:102 and 1:47, respectively) and subsequently increased (to 1:1183 and 1:1300, respectively). Nuchal skin biopsies were negative for rabies virus antigen, rabies virus RNA was not detected in the skin biopsies or in saliva by RT-PCR, and viral isolation on saliva was negative. The patient was intubated and put into a drug-induced coma, which included the noncompetitive NMDA antagonist ketamine at 48 mg/kg/day as a continuous infusion and intravenous midazolam for 7 days. There was a deliberate attempt to maintain a burst-suppression pattern on her electroencephalogram, and supplemental phenobarbital was given. She also received intravenous ribavirin and amantadine 200 mg per day administered enterally. She improved and was discharged from the hospital with neurologic deficits, and she has subsequently shown further progressive neurologic improvement (Hu et al., 2007).

The seventh and most recent case was a 15-year-old boy from Brazil (Ministerio da Saude in Brazil, 2008). He was aggressively attacked by a hematophagous bat and developed symptoms of rabies 29 days later on October 6, 2012. Prior to the onset of symptoms, he received four doses of rabies vaccine. A skin biopsy was positive for rabies virus RNA, and a vampire bat variant was identified. On October 11, 2012, he was intubated and treated with therapeutic (induced) coma and other therapies ("Milwaukee protocol") were initiated. The boy survived rabies, apparently with fairly severe neurological sequelae.

There are also two reported cases with rabies virus antibodies, but without neutralizing anti-rabies virus antibodies in serum and cerebrospinal fluid. The first was a female who lacked the typical clinical features of rabies and did not require intensive care (Holzmann-Pazgal et al., 2010). She had fever, headache, nuchal rigidity, disorientation, and limb weakness, and had a cerebrospinal fluid pleocytosis and enlarged lateral ventricles on MR imaging. She developed only a low titer of rabies virus neutralizing antibodies in sera (up to 1:14) after receiving human rabies immune globulin and one dose of rabies vaccine and no detectable neutralizing antibodies in CSF (Holzmann-Pazgal et al., 2010). Other diagnostic tests for rabies (detection of rabies virus antigen and RNA) were negative. The second case was an 8-year-old female from California (Wiedeman et al., 2012). She experienced sore throat and vomiting. Later, over a few days, she developed swallowing difficulties. A few days later, she developed abdominal pain and neck and back pain, and then on the next day she had sore throat and abdominal pain and was noted to be confused. She deteriorated rapidly and required endotracheal intubation. CSF showed 6 leukocytes/μL with a protein of 62 mg/dL. Over the next few days, she developed ascending flaccid paralysis, decreased level of consciousness, and fever. MR imaging of brain showed multiple T2 and FLAIR signal abnormalities in cortical and subcortical regions and in the periventricular white matter. Electrophysiological studies were consistent with a demyelinating and predominantly motor polyneuropathy. She had rabies virus-specific IgG and IgM in her serum and CSF, but she did not develop rabies virus neutralizing antibodies. All other diagnostic tests for rabies were also negative. She showed progressive improvement after just over two weeks in hospital and she was discharged home after another 5 weeks in rehabilitation. Neither of these cases had typical clinical features of rabies. The atypical clinical features, plus the lack or minimal development of rabies virus neutralizing antibodies, suggest that it is unlikely that these two patients recovered from rabies. The exact etiology and pathogenetic mechanisms involved in the illnesses of these two patients remains elusive, and they should not be considered rabies survivors.

2 FUTURE PROSPECTS FOR THE AGGRESSIVE MANAGEMENT OF RABIES IN HUMANS

Recovery of the preceding seven patients with rabies has inspired physicians to aggressively manage patients with rabies in critical care units. In part, because of the mild neuropathological changes in rabies, the hope was that patients, even when previously unimmunized, could be maintained through the acute phase of their illness, and, if they could avoid medical complications, then perhaps they could clear the viral infection and recover. Overall, this approach has proved to be disappointing (Bhatt, Hattwick, Gerdsen, Emmons, & Johnson, 1974; Emmons et al., 1973; Gode, Raju, Jayalakshmi, Kaul, & Bhide, 1976; Lopez et al., 1975; Rubin, Sullivan, Summers, Gregg, & Sikes, 1970; Udwadia et al., 1989). However, the sixth case from Milwaukee has provided some optimism that aggressive therapy may become much more effective in the future. This case was the first documented survivor who did not receive rabies vaccine prior to onset of clinical rabies. As discussed in the accompanying editorial (Jackson, 2005), it is unknown if therapy with one or more specific agents given played an important role in the outcome of this case. The induction of coma per se has not been shown to be useful in the management of infectious diseases of the nervous system, and to date there is no evidence supporting this approach in rabies or other forms of viral encephalitis. Hence, this approach should not become routine for the management of rabies at this time. There is now increased doubt about the efficacy of ketamine therapy in rabies virus infection. More recent studies on ketamine in both rabies virus-infected primary neuron cultures and in experimental rabies in mice have shown a lack of efficacy of ketamine (Weli, Scott, Ward, & Jackson, 2006). Unlike other viral infections of the nervous system, including Sindbis virus encephalomyelitis (Darman et al., 2004; Nargi-Aizenman et al., 2004; Nargi-Aizenman & Griffin, 2001) and human immunodeficiency virus infection (Kaul & Lipton, 2007), there is not yet any established experimental evidence supporting excitotoxicity in rabies. Even where there is strong experimental evidence of excitotoxicity in animal models, multiple clinical trials in humans have shown a lack of efficacy of neuroprotective agents in stroke (Ginsberg, 2009). Hence, a strong neuroprotective effect of a therapy given to a single patient without a clear scientific rationale is highly unlikely to be responsible for a favorable outcome.

The presence of neutralizing anti-rabies virus antibodies early in a patient's clinical course was probably an important factor contributing to the favorable outcome. This occurs in less than 20% of all patients with rabies. The presence of neutralizing anti-rabies virus antibodies is a marker of an active adaptive immune response that is essential for viral clearance (Chapter 10). Because six survivors of rabies had received rabies vaccine prior to the onset of their disease, this suggests that an early immune response is associated with a positive outcome. Although the seventh case above is an exception (rabies virus RNA was detected in the skin biopsy), usually all of the other diagnostic laboratory tests in rabies survivors are negative for rabies virus antigen and RNA in fluids and tissues (without testing of brain tissues). This may be because viral clearance was so effective that centrifugal spread of the infection to peripheral organ sites was reduced or rapid clearance occurred through immune-mediated mechanisms.

Bat rabies viruses are likely less neurovirulent than canine and perhaps other variants that are responsible for human cases of rabies (Lafon, 2005), and human rabies cases due to canine rabies virus variants may have a less favorable outcome. A previous survivor of rabies, who received rabies vaccine prior to the onset of disease, had an excellent neurological recovery and was also infected with a bat rabies virus (Hattwick et al., 1972). It will never be known whether the causative bat rabies virus variant in the Milwaukee case was, in some way, attenuated and different from previously isolated bat rabies virus variants because viral isolation was not successful.

An aggressive approach to therapy of rabies will require the full resources of a critical care unit and have a high risk of failure. The following should all be considered a "favorable" factors for initiating an aggressive therapeutic approach: 1) therapy with dose(s) of rabies vaccine prior to the onset of illness, 2) young age, 3) healthy and immunocompetent individual, 4) rabies due to a bat rabies variant (e.g., in contrast to a canine variant), 5) early presence of neutralizing anti-rabies virus antibodies in serum and CSF, and 6) mild neurological disease at the time of initiation of therapy. New approaches to treating human rabies need to be developed rather than repeating ineffective therapies. It remains highly doubtful that the Milwaukee Protocol will prove to be useful in the management of human rabies in light of the fact that at least 26 patients received similar therapeutic approaches with fatal outcomes (Table 16.1). Repetition of this therapy will likely impede progress in moving the development of new effective therapies for rabies forward. Finding an effective neuroprotective drug is highly unlikely with a "trial and error" approach, in light of the fact that so many clinical trials failed to show efficacy for a neuroprotective drug for acute stroke (Ginsberg, 2009). The most effective "neuroprotective" therapy to date for an acute brain insult is therapeutic hypothermia, in which the body temperature is reduced by a variety of cooling methods in order to reduce neuronal injury and improve clinical outcomes. Efficacy has been established in Australian (Bernard et al., 2002) and European (The Hypothermia After Cardiac Arrest Study Group, 2002) studies for patients who remain unconscious after witnessed cardiac arrest due to ventricular fibrillation. Efficacy for hypothermia for traumatic brain injury has not yet been established (Christian, Zada, Sung, & Giannotta, 2008), but it continues to be an area of active investigation. Hypothermia decreases cerebral metabolism, production of reactive oxygen species, lipid peroxidation, and inflammatory response activity, which, at least in part, may explain its beneficial effects. There are generalized methods of inducing hypothermia and also regional methods that can be applied to the head and neck, which include use of a cooling helmet (Wang et al., 2004) and intranasal cooling (Busch et al., 2010; Castren et al., 2010). Intranasal cooling involves spraying an inert evaporative coolant via nasal prongs that rapidly evaporates after contact with the nasopharynx, and it has the advantage of reducing the temperature more rapidly. The regional methods are associated with less systemic adverse effects and would also be expected to have a reduced effect on a natural or rabies vaccine-induced systemic immune response, which is important for viral clearance in rabies virus infection. Rabies virus replication is generally fairly efficient at lower-than-normal body temperatures (e.g., 33°C), particularly with infection of an epithelial cell line with a bat rabies virus variant (Morimoto et al., 1996). However, there may be reduced viral spread due to inhibitory effects of hypothermia on fast axonal transport (Bisby & Jones, 1978) and trans-synaptic spread, as well as other beneficial and neuroprotective effects. Under natural conditions, hibernation of rabies vectors likely results in "suspension" of viral replication (Sulkin, Allen, Sims, Krutzsch, & Kim, 1960) and inhibition of viral spread by marked inhibition of axonal transport (Bisby & Jones, 1978) due to very low body temperatures (e.g., below 5 to 10°C). In contrast, therapeutic mild (34°C) or moderate (30°C) hypothermia maintained for periods of 24 to 72 hours would be expected to be associated with much more modest, but potentially beneficial effects. Entirely new approaches need to be taken for the aggressive management of human rabies, which may combine a variety of different therapeutic approaches, including, for example, hypothermia and antiviral and other therapeutic agents. More work is needed to identify new efficacious therapeutic agents and more basic research is needed to improve our understanding of basic mechanisms underlying rabies pathogenesis in humans and animals. Hopefully, in the future, with early clinical diagnosis of rabies, good clinical outcomes can be achieved with the development of effective therapy.

References

Alvarez, L., Fajardo, R., Lopez, E., Pedroza, R., Hemachudha, T., Kamolvarin, N., et al. (1994). Partial recovery from rabies in a nine-year-old boy. *The Pediatric Infectious Disease Journal, 13,* 1154–1155.

Aramburo, A., Willoughby, R. E., Bollen, A. W., Glaser, C. A., Hsieh, C. J., Davis, S. L., et al. (2011). Failure of the Milwaukee Protocol in a child with rabies. *Clinical Infectious Diseases, 53*(6), 572–574.

Badillo, R., Mantilla, J. C., & Pradilla, G. (2009). Human rabies encephalitis by a vampire bat bite in an urban area of Colombia (Spanish). *Biomédica, 29*(2), 191–203.

Bagchi, S. (2005, July 4). Coma therapy. *The Telegraph,* Calcutta.

Basgoz, N., & Frosch, M. P. (1998). Case records of the Massachusetts General Hospital: A 32-year-old woman with pharyngeal spasms and paresthesias after a dog bite. *New England Journal of Medicine, 339*(2), 105–112.

Bernard, S. A., Gray, T. W., Buist, M. D., Jones, B. M., Silvester, W., Gutteridge, G., et al. (2002). Treatment of comatose survivors of out-of-hospital cardiac arrest with induced hypothermia. *New England Journal of Medicine, 346*(8), 557–563.

Bhatt, D. R., Hattwick, M. A. W., Gerdsen, R., Emmons, R. W., & Johnson, H. N. (1974). Human rabies: Diagnosis, complications, and management. *American Journal of Diseases of Children, 127*, 862–869.

Bisby, M. A., & Jones, D. L. (1978). Temperature sensitivity of axonal transport in hibernating and nonhibernating rodents. *Experimental Neurology, 61*(1), 74–83.

Branswell, H. (2012, April 19). Testing suggests Toronto rabies case infected in Dominican Republic. *The Globe and Mail*, Toronto.

Busch, H. J., Eichwede, F., Fodisch, M., Taccone, F. S., Wobker, G., Schwab, T., et al. (2010). Safety and feasibility of nasopharyngeal evaporative cooling in the emergency department setting in survivors of cardiac arrest. *Resuscitation, 81*(8), 943–949.

Castren, M., Nordberg, P., Svensson, L., Taccone, F., Vincent, J. L., Desruelles, D., et al. (2010). Intra-arrest transnasal evaporative cooling: A randomized, prehospital, multicenter study (PRINCE: Pre-ROSC IntraNasal Cooling Effectiveness). *Circulation, 122*(7), 729–736.

Christenson, J. C., Holm, B. M., Lechlitner, S., Howell, J. F., Wenger, M., Roy-Burman, A., et al. (2007). Human rabies–Indiana and California, 2006. *Morbidity and Mortality Weekly Report, 56*(15), 361–365.

Christian, E., Zada, G., Sung, G., & Giannotta, S. L. (2008). A review of selective hypothermia in the management of traumatic brain injury. *Neurosurgical Focus, 25*(4), E9.

Darman, J., Backovic, S., Dike, S., Maragakis, N. J., Krishnan, C., Rothstein, J. D., et al. (2004). Viral-induced spinal motor neuron death is non-cell-autonomous and involves glutamate excitotoxicity. *Journal of Neuroscience, 24*(34), 7566–7575.

Drosten, C. (2007). Rabies – Germany (Hamburg) ex Morocco. *ProMED-mail, 20070419.1287.* Available at <http://www.promedmail.org/> Accessed 07.08.12.

Emmons, R. W., Leonard, L. L., DeGenaro, F., Jr., et al., Protas, E. S., Bazeley, P. L., Giammona, S. T., et al. (1973). A case of human rabies with prolonged survival. *Intervirology, 1*(1), 60–72.

Ginsberg, M. D. (2009). Current status of neuroprotection for cerebral ischemia: Synoptic overview. *Stroke, 40*(*Suppl. 3*), S111–S114.

Gode, G. R., Raju, A. V., Jayalakshmi, T. S., Kaul, H. L., & Bhide, N. K. (1976). Intensive care in rabies therapy. Clinical observations. *Lancet, 2*, 6–8.

Greer, D. M., Robbins, G. K., Lijewski, V., Gonzales, R. G., & McGuone, D. (2013). Case records of the Massachusetts General Hospital: Case 1-2013: a 63-year-old man with paresthesias and difficulty swallowing. *New England Journal of Medicine, 368*(2), 172–180.

Hattwick, M. A., Corey, L., & Creech, W. B. (1976). Clinical use of human globulin immune to rabies virus. *Journal of Infectious Diseases, 133*(*Suppl*), A266–A272.

Hattwick, M. A. W., Weis, T. T., Stechschulte, C. J., Baer, G. M., & Gregg, M. B. (1972). Recovery from rabies: A case report. *Annals of Internal Medicine, 76*, 931–942.

Hemachudha, T., Sunsaneewitayakul, B., Desudchit, T., Suankratay, C., Sittipunt, C., Wacharapluesadee, S., et al. (2006). Failure of therapeutic coma and ketamine for therapy of human rabies. *Journal of Neurovirology, 12*, 407–409.

Hemachudha, T., Sunsaneewitayakul, B., Mitrabhakdi, E., Suankratay, C., Laothamathas, J., Wacharapluesadee, S., et al. (2003). Paralytic complications following intravenous rabies immune globulin treatment in a patient with furious rabies (Letter). *International Journal of Infectious Diseases, 7*(1), 76–77.

Holzmann-Pazgal, G., Wanger, A., Degaffe, G., Rose, C., Heresi, G., Amaya, R., et al. (2010). Presumptive abortive human rabies – Texas, 2009. *Morbidity and Mortality Weekly Report, 59*(7), 185–190.

Houston Chronicle. (2006). Rabies, human – USA (Texas). *ProMED-mail, 20060513.1360.* Available at <http://www.promedmail.org/> Accessed 07.08.12.

Hu, W. T., Willoughby, R. E., Jr., Dhonau, H., & Mack, K. J. (2007). Long-term follow-up after treatment of rabies by induction of coma (Letter). *New England Journal of Medicine, 357*(9), 945–946.

Hunter, M., Johnson, N., Hedderwick, S., McCaughey, C., Lowry, K., McConville, J., et al. (2010). Immunovirological correlates in human rabies treated with therapeutic coma. *Journal of Medical Virology, 82*(7), 1255–1265.

IAfrica.com. (2012). Rabies – South Africa (03): (Kwazulu-Natal), human ex Mozambique. *ProMED-mail, 20120608.1160328.* Available at <http://www.promedmail.org/> Accessed 07.08.12.

Jackson, A. C. (2005). Recovery from rabies (Editorial). *New England Journal of Medicine, 352*(24), 2549–2550.

Jackson, A. C., Warrell, M. J., Rupprecht, C. E., Ertl, H. C. J., Dietzschold, B., O'Reilly, M, et al. (2003). Management of rabies in humans. *Clinical Infectious Diseases, 36*(1), 60–63.

Javaid, W., Amzuta, I. G., Nat, A., Johnson, T., Grant, D., Rudd, R. J., et al. (2012). Imported human rabies in a U.S. Army soldier – New York, 2011. *Morbidity and Mortality Weekly Report, 61*(17), 302–305.

Juncosa, B. (2008). Hope for rabies victims: Unorthodox coma therapy shows promise. First a U.S. girl – and now two South American kids survive onset of the deadly virus. Available at <http://www.scientificamerican.com/> Accessed 07.08.12.

Kaul, M., & Lipton, S. A. (2007). Neuroinflammation and excitotoxicity in neurobiology of HIV-1 infection and AIDS: Targets for neuroprotection. In J. O. Malva, A. C. Rego, R. A. Cunha, & C. R. Oliveira (Eds.), *Interaction between neurons and glia in aging and disease (pp. 281–308).* New York: Springer Science.

Kureishi, A., Xu, L. Z., Wu, H., & Stiver, H. G. (1992). Rabies in China: Recommendations for control. *Bulletin of the World Health Organization, 70,* 443–450.

Lafon, M. (2005). Bat rabies–the Achilles heel of a viral killer? *Lancet, 366*(9489), 876–877.

Lockhart, B. P., Tordo, N., & Tsiang, H. (1992). Inhibition of rabies virus transcription in rat cortical neurons with the dissociative anesthetic ketamine. *Antimicrobial Agents and Chemotherapy, 36,* 1750–1755.

Lockhart, B. P., Tsiang, H., Ceccaldi, P. E., & Guillemer, S. (1991). Ketamine-mediated inhibition of rabies virus infection *in vitro* and in rat brain. *Antiviral Chemistry and Chemotherapy, 2,* 9–15.

Lopez, M., Neves, J., Moreira, E. C., Reis, R., Tafuri, W. L., Pittella, J. E., et al. (1975). Human rabies. 1. intensive treatment. *Revista Do Instituto de Medicina Tropical de Sao Paulo, 17*(2), 103–110.

Luminos, M., Barboi, G., Draganescu, A., Streinu Cercel, A., Staniceanu, F., Jugulete, G., et al. (2011). Human rabies in a Romanian boy – an ante mortem case study. *Rabies Bulletin Europe, 35*(2), 5–10.

Madhusudana, S. N., Nagaraj, D., Uday, M., Ratnavalli, E., & Kumar, M. V. (2002). Partial recovery from rabies in a six-year-old girl (Letter). *International Journal of Infectious Diseases, 6*(1), 85–86.

Maier, T., Schwarting, A., Mauer, D., Ross, R. S., Martens, A., Kliem, V., et al. (2010). Management and outcomes after multiple corneal and solid organ transplantations from a donor infected with rabies virus. *Clinical Infectious Diseases, 50*(8), 1112–1119.

Manning, S. E., Rupprecht, C. E., Fishbein, D., Hanlon, C. A., Lumlertdacha, B., Guerra, M., et al. (2008). Human rabies prevention–United States, 2008: Recommendations of the Advisory Committee on Immunization Practices. *Morbidity and Mortality Weekly Report, 57*(RR-3), 1–28.

McDermid, R. C., Saxinger, L., Lee, B., Johnstone, J., Noel Gibney, R. T., Johnson, M., et al. (2008). Human rabies encephalitis following bat exposure: Failure of therapeutic coma. *Canadian Medical Association Journal, 178*(5), 557–561.

Merigan, T. C., Baer, G. M., Winkler, W. G., Bernard, K. W., Gibert, C. G., Chany, C., Collaborative Group, (1984). Human leukocyte interferon administration to patients with symptomatic and suspected rabies. *Annals of Neurology, 16,* 82–87.

Ministerio da Saude in Brazil. (2008). Rabies, human survival, bat – Brazil: (Pernambuco). *ProMED-mail, 20081114.3599.* Available at <http://www.promedmail.org/> Accessed 07.08.12.

Morimoto, K., Patel, M., Corisdeo, S., Hooper, D. C., Fu, Z. F., Rupprecht, C. E., et al. (1996). Characterization of a unique variant of bat rabies virus responsible for newly emerging human cases in North America. *Proceedings of the National Academy of Sciences of the United States of America, 93*(11), 5653–5658.

Nargi-Aizenman, J. L., & Griffin, D. E. (2001). Sindbis virus-induced neuronal death is both necrotic and apoptotic and is ameliorated by N-methyl-D-aspartate receptor antagonists. *Journal of Virology, 75*(15), 7114–7121.

Nargi-Aizenman, J. L., Havert, M. B., Zhang, M., Irani, D. N., Rothstein, J. D., & Griffin, D. E. (2004). Glutamate receptor antagonists protect from virus-induced neural degeneration. *Annals of Neurology, 55*(4), 541–549.

NE 10. (2012). Rabies, human – Brazil (03): (Ceara). *ProMED-mail, 20120314.1070531.* Available at <http://www.promedmail.org/> Accessed 07.08.12.

Porras, C., Barboza, J. J., Fuenzalida, E., Adaros, H. L., Oviedo, A. M., & Furst, J. (1976). Recovery from rabies in man. *Annals of Internal Medicine, 85,* 44–48.

Pue, H. L., Turabelidze, G., Patrick, S., Grim, A., Bell, C., Reese, V., et al. (2009). Human rabies – Missouri, 2008. *Morbidity and Mortality Weekly Report, 58*(43), 1207–1209.

Rubin, J., David, D., Willoughby, R. E., Jr., et al., Rupprecht, C. E., Garcia, C., Guarda, D. C., et al. (2009). Applying the Milwaukee Protocol to treat canine rabies in Equatorial Guinea. *Scandinavian Journal of Infectious Diseases, 41*(5), 372–375.

Rubin, R. H., Sullivan, L., Summers, R., Gregg, M. B., & Sikes, R. K. (1970). A case of human rabies in Kansas: Epidemiologic, clinical, and laboratory considerations. *Journal of Infectious Diseases, 122,* 318–322.

Santos, A., Cale, E., Dacheux, L., Bourhy, H., Gouveia, J., & Vasconcelos, P. (2012). Fatal case of imported human rabies in Amadora, Portugal, August 2011. *Eurosurveillance, 17*(12), 20130.

Sulkin, S. E., Allen, R., Sims, R., Krutzsch, P. H., & Kim, C. (1960). Studies on the pathogenesis of rabies in bats. II. Influence of environmental temperature. *Journal of Experimental Medicine, 112,* 595–617.

The Hypothermia After Cardiac Arrest Study Group, (2002). Mild therapeutic hypothermia to improve the neurologic outcome after cardiac arrest. *New England Journal of Medicine, 346*(8), 549–556.

Tillotson, J. R., Axelrod, D., & Lyman, D. O. (1977a). Follow-up on rabies – New York. *Morbidity and Mortality Weekly Report, 26,* 249–250.

Tillotson, J. R., Axelrod, D., & Lyman, D. O. (1977b). Rabies in a laboratory worker – New York. *Morbidity and Mortality Weekly Report, 26,* 183–184.

Times Live. (2012). Rabies – South Africa (02): (Kwazulu-Natal), human ex Mozambique. *ProMED-mail, 20120528.1147931.* Available at <http://www.promedmail.org/> Accessed 07.08.12.

Troell, P., Miller-Zuber, B., Ondrush, J., Murphy, J., Fatteh, N., Feldman, K., et al. (2010). Human rabies—Virginia, 2009. *Morbidity and Mortality Weekly Report, 59*(38), 1236–1238.

Turabelidze, G., Pue, H., Grim, A., & Patrick, S. (2009). First human rabies case in Missouri in 50 years causes death in outdoorsman. *Missouri Medicine, 106*(6), 417–419.

Udwadia, Z. F., Udwadia, F. E., Katrak, S. M., Dastur, D. K., Sekhar, M., Lall, A., et al. (1989). Human rabies: Clinical features, diagnosis, complications, and management. *Critical Care Medicine, 17,* 834–836.

van Thiel, P. P., de Bie, R. M., Eftimov, F., Tepaske, R., Zaaijer, H. L., van Doornum, G. J., et al. (2009). Fatal human rabies due to Duvenhage virus from a bat in Kenya: Failure of treatment with coma-induction, ketamine, and antiviral drugs. *PLoS Neglected Tropical Diseases, 3*(7), e428.

Wang, H., Olivero, W., Lanzino, G., Elkins, W., Rose, J., Honings, D., et al. (2004). Rapid and selective cerebral hypothermia achieved using a cooling helmet. *Journal of Neurosurgery, 100*(2), 272–277.

Warrell, M. J., White, N. J., Looareesuwan, S., Phillips, R. E., Suntharasamai, P., Chanthavanich, P., et al. (1989). Failure of interferon alfa and tribavirin in rabies encephalitis. *British Medical Journal, 299,* 830–833.

Weli, S. C., Scott, C. A., Ward, C. A., & Jackson, A. C. (2006). Rabies virus infection of primary neuronal cultures and adult mice: Failure to demonstrate evidence of excitotoxicity. *Journal of Virology, 80*(20), 10270–10273.

Wiedeman, J., Plant, J., Glaser, C., Messenger, S., Wadford, D., Sheriff, H., et al. (2012). Recovery of a patient from clinical rabies – California, 2011. *Morbidity and Mortality Weekly Report, 61*(4), 61–65.

Willoughby, R. E., Jr., et al., Tieves, K. S., Hoffman, G. M., Ghanayem, N. S., Amlie-Lefond, C. M., Schwabe, M. J., et al. (2005). Survival after treatment of rabies with induction of coma. *New England Journal of Medicine, 352*(24), 2508–2514.

Dog Rabies and Its Control

Darryn L. Knobel[1], Tiziana Lembo[2], Michelle Morters[3], Sunny E. Townsend[2], Sarah Cleaveland[2], and Katie Hampson[2]

[1]Department of Veterinary Tropical Diseases, Faculty of Veterinary Science, University of Pretoria, Onderstepoort, 0110 South Africa, [2]Boyd Orr Centre for Population and Ecosystem Health, Institute of Biodiversity, Animal Health and Comparative Medicine, College of Medical, Veterinary and Life Sciences, University of Glasgow, Glasgow G12 8QQ UK, [3]Cambridge Infectious Diseases Consortium, Department of Veterinary Medicine, Cambridge University, Madingley Road, Cambridge CB3 0ES UK

1 INTRODUCTION

Domestic dogs are the major reservoir of rabies virus (RABV) throughout Africa and Asia (Chapter 3, this volume). The association between the bite of a "mad" dog and human rabies has been recognized since antiquity (reviewed by Neville, 2004), and rabid dogs are still responsible for the vast majority (>90%) of human deaths from rabies worldwide (WHO, 1999). The control of the disease in domestic dogs thus has important implications for public health, particularly in Africa and Asia where canine rabies is endemic.

1.1 Historical Perspectives and the Current Situation

In the 19th century, muzzling and dog movement restrictions were used to successfully eliminate dog rabies in parts of Europe. By the early 1900s, animal rabies vaccines were being developed in Japan, with the first mass dog vaccinations in 1921 (Umeno & Doi, 1921). Over the second half of the 20th century, widespread successes in eliminating dog

Rabies: Scientific Basis of the Disease and its Management
DOI: http://dx.doi.org/10.1016/B978-0-12-396547-9.00017-1

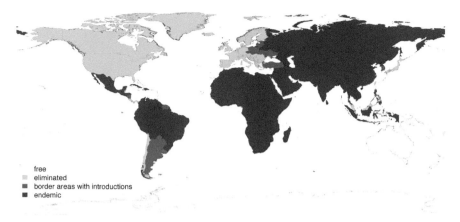

free
eliminated
border areas with introductions
endemic

FIGURE 17.1   Countries with rabies circulating endemically in domestic dog populations (dark red), countries with sporadic canine rabies cases (pale red) and countries that have been freed from canine rabies through concerted control efforts (grey) or have historically been rabies-free (white).

rabies through concerted mass vaccination efforts were seen: dog rabies was eliminated from Western Europe (King, Fooks, Aubert, & Wandeler, 2004), Japan (Shimada, 1971), and several countries in the Americas, with targets currently being set for elimination of dog rabies from this region as a whole (Schneider et al., 2007). During periods of colonial occupation in Africa, rabies was successfully controlled in some countries through mass vaccinations, supported by strict implementation of tie-up orders (Shone, 1962), but rabies has since re-emerged as these control measures lapsed. Vaccination campaigns and strictly enforced culling eliminated rabies from Malaysia in the 1960s (Tan, Abdul, Mohd, & Beran, 1972; WHO, 1988), and recent decades have seen improved control measures result in continual reductions in rabies incidence in other parts of Asia (Hahn, 2009). However, recent resurgences have also occurred: for example, in China re-emergence has been attributed to a relaxation of control measures, use of ineffective local animal vaccines Hu, Fooks, Zhang, Liu, & Zhang, 2008), and poor awareness (Wu, Franka, Svoboda, Pohl, & Rupprecht, 2009). Neighboring Nepal and Bhutan have also reported increases in incidence (Tenzin, Sharma, Dhand, Timsina, & Ward, 2010), and since the late 1990s rabies has spread to many previously rabies-free islands in Indonesia, including Flores in 1997 (Bingham, 2001; Windiyaningsih, Wilde, Meslin, Suroso, & Widarso, 2004), Maluku and Ambon in 2003, Bali in 2008 and Nias islands in 2010 (Lembo, Craig, Miles, Hampson, & Meslin, 2013). Figure 17.1 shows the current and historical distribution of canine rabies, illustrating successful elimination programs in more developed and relatively isolated countries.

1.2 The Burden of Human Disease

More than 99% of all human deaths from rabies occur in Africa and Asia (WHO, 1999), where the disease is responsible for some 55,000 human deaths and 1.74 million disability-adjusted life years (DALYs) annually (Knobel et al., 2005). These figures are derived from a probability model using estimates of the incidence of bite injuries from suspected rabid dogs, the site of bite injuries, and probability of receiving post-exposure prophylaxis (PEP). Empirical data on the incidence of human deaths derived from detailed contract-tracing studies in Tanzania (Hampson et al., 2008) and a large multi-centric study in India (Sudarshan et al., 2007) generated consistent estimates of human rabies incidence, equivalent to 1.5–5 deaths/100,000/year in endemic areas. In Africa, this is approximately 100 times higher than indicated by official reports.

Each year, an estimated 7 million people are bitten worldwide by suspect rabid dogs (Coleman, Fèvre, & Cleaveland, 2004; Knobel et al., 2005), leading to a high demand for expensive PEP. This places a substantial economic burden on the public health sector in Africa and Asia (with direct medical costs estimated at $196 million/year), but also results in high costs to bite victims and their care givers, who need to travel to health facilities on multiple occasions to obtain the complete course of post-exposure vaccines required. Indirect (patient-borne) costs are estimated to be $289 million/year, contributing to more than 50% of the total medical costs associated with PEP administration (Knobel et al., 2005). These high costs have disproportionate impacts on the poor, who struggle to raise the cash needed in a timely fashion, and hence are more likely to incur delays in obtaining prompt PEP, sometimes with fatal consequences (Hampson et al., 2008).

Although the costs of dog rabies relate mainly to countries in Africa and Asia, other countries also incur costs as a result of dog rabies, even where the disease has been eliminated. For example, in the UK, where terrestrial rabies has been eliminated for many decades, PEP use is now increasing (Health Protection Agency, 2011). For rabies-free countries, vaccine costs arise as a result of (a) pre-exposure vaccination of travelers, (b) the need for PEP in travelers returning with a recent history of exposure, and (c) inhabitants exposed to rabid dogs following new incursions or illegal importation. A study of PEP use in travelers returning home to Australia, New Zealand, and France indicated that only 11% of bitten travelers received both vaccine and rabies immunoglobulin at the time of exposure, and many only received appropriate PEP on return to clinics in their home country, which invariably resulted in delays and high risks (Gautret et al., 2008). In France, substantial costs are still incurred for PEP use, with peaks of demand linked with increased public awareness following cases of imported human rabies, as well as the illegal importation of rabid dogs from North Africa (Lardon et al., 2010).

Although limited data are available for other components of the economic burden, costs associated with livestock rabies are also unlikely to be trivial (Knobel et al., 2005; Lembo et al., 2010). Furthermore, a major component of economic loss that has not yet been quantified relates to productivity losses due to premature human rabies deaths, which is likely to be substantial given the high incidence of disease in young people.

Domestic dogs are responsible for the maintenance of the virus and transmission to the vast majority of human cases in endemic areas in Africa and Asia. Control of the disease in humans should include a strong focus of control in this reservoir population. Dog rabies control measures have the objective of decreasing the burden of human rabies, and of eventual elimination in endemic areas.

2 EPIDEMIOLOGICAL THEORY OF DOG RABIES CONTROL

An understanding of some of the theoretical concepts of the epidemiology of infectious diseases can help in the effective design and implementation of dog rabies control programs. A number of these key concepts are reviewed in this section.

2.1 Key Epidemiological Parameters

The basic reproductive number (R_0) of an infectious agent such as rabies virus is defined as the average number of secondary infections produced by an infected individual in an otherwise susceptible host population (Anderson & May, 1991). R_0 determines whether a pathogen can persist in such a population, and it is valuable for assessing control options. When R_0 is less than 1, on average each infectious individual infects less than one other individual, and the pathogen will die out in the population. In contrast, when R_0 exceeds 1 there is an exponential rise in the number of cases over time, and an epidemic results. R_0 is consistently estimated to be between 1 and 2 from rabies outbreaks in dog populations around the world (Hampson et al., 2009), which is relatively close to the extinction threshold of 1.

A closely related concept is that of the effective reproductive number (R_e), when transmission occurs in a population that is not entirely susceptible due to implemented control efforts. For dog rabies, R_e is determined by the number of susceptible dogs bitten by each infected dog during its infectious period, and the probability that those bitten dogs go on to develop rabies (i.e. become infectious themselves). The number of dogs bitten may depend upon genetic, behavioral, environmental, and anthropogenic factors, whereas the probability of developing rabies once bitten depends upon vaccination status (since there is no natural immunity to rabies), as well as other intrinsic factors including viral dose, location of bite(s), and degree of tissue injury. The aim of control measures is to reduce transmission so that R_e is reduced below the threshold of 1. For rabies, these control measures can therefore operate by reducing either the number of dogs bitten or the probability that bitten dogs develop rabies.

Historically, prior to the advent of effective vaccines, domestic dog rabies control measures focused on reducing the number of susceptible dogs bitten, through movement restrictions and culling rabid, bitten and "stray" dogs (Bögel, 2002; Fooks, Roberts, Lynch, Hersteinsson, & Runolfsson, 2004; Meldrum, 1988; Muir & Roome, 2005; Tierkel, 1959; WHO, 1987). The low value of R_0 and cultural context of strict confinement and muzzling of dogs probably contributed to the success of these measures in isolated locations such as the UK, but few other successes were reported using only these measures. However, with the advent of effective animal vaccines, mass vaccination has become the mainstay of successful dog rabies control and a more ethically and culturally acceptable measure. The control of infectious diseases through mass vaccination is based on the concept of herd immunity, when the vaccination of a proportion of the population (or "herd") provides protection for individuals who are not vaccinated. That proportion of the population that needs to be vaccinated to achieve herd immunity and thus control disease depends on R_0. For rabies, low values of R_0 suggest that the critical vaccination coverage, P_{crit}, required to control disease should be roughly between 20 and 40% (i.e., 20–40% of the dog population should be immune at any time in order to prevent sustained outbreaks of rabies, although short chains of transmission can still occur in such partially vaccinated populations; Hampson et al., 2009). By comparison, some other infectious diseases that have been successfully controlled by mass vaccination (e.g., measles or rinderpest) have considerably higher values of R_0, and require coverages well above 90%. These figures thus suggest that dog rabies is amenable to control by mass vaccination.

2.2 Factors Influencing the Effectiveness of Control Efforts

Although only low levels of vaccination coverage are theoretically required to control dog rabies, in practice the achievement of this goal is hampered by factors responsible for declines in the proportion of the vaccinated population over time. Host demography plays a major role in influencing the long-term effectiveness of vaccination efforts, because coverage levels decline as vaccinated dogs die and new susceptible dogs

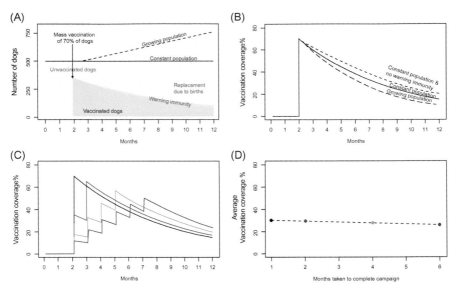

FIGURE 17.2   **Factors affecting vaccination coverage.** A) A population model showing
the change in vaccinated individuals in a stable population (solid line) following a campaign
implemented in month 2, whereby 70% of the population is vaccinated, and the corresponding change in a growing population (dashed line). Vaccinated dog numbers (green region)
decline due to deaths (cross-hatched region) and waning immunity (solid blue region)
whilst B) coverage declines due to the replacement of the deaths of vaccinated dogs by susceptible pups as well as the waning immunity. Deaths occur at the same rate in both vaccinated and unvaccinated animals, therefore increases in dog lifespan only affect coverage
by enabling dogs to reproduce for longer. However, if increased survival results in lower
rates of reproduction then coverage may be stabilized, suggesting that research into the factors regulating dog populations is merited. Parameters used in the model are: birth rate =
1.5 year^{-1} in the constant population and 2 year^{-1} in the growing population, death rate =
1.5 year^{-1}, immunity wanes at a rate of 0.333 year^{-1}. C) The effect of the speed of delivery
of campaigns on vaccination coverage: faster but otherwise equivalent campaigns achieve
a higher coverage than slower campaigns (see the campaigns implemented in 1, 2, 3 or
4 months – black, red, green, and blue respectively) but coverage in the population is lower
at the end of the year due to births and loss of immunity, whilst D) average coverage over
the entire year is only marginally affected by conducting a slower campaign.

are born (Figure 17.2). The movement of dogs by people similarly affects
coverage, particularly if large numbers of unvaccinated dogs are brought
into an area. In most dog rabies-endemic areas, dog birth rates are high,
and therefore coverage levels decline rapidly. If coverage falls below
the target threshold of 20–40%, rabies transmission can be sustained
and outbreaks can persist. This phenomenon is particularly acute when
mass dog vaccination is conducted as "pulsed" campaigns (rather than
through the ongoing vaccination of new dogs born into the population),
and it is influenced by the interval between campaigns (Figure 17.2).

   The vaccination coverage that should be reached during any one campaign will therefore depend not only on the critical threshold coverage

to be maintained (P_{crit}), but also on the dog population dynamics, the duration of vaccine-induced immunity, and the inter-campaign intervals because of the declines in coverage that occur over time. WHO-approved dog vaccines provide immunity lasting for at least 2 years and usually longer. Empirically, consecutive annual vaccinations that achieve 70% coverage in dog populations have repeatedly proven effective in eliminating endemic canine rabies (Coleman & Dye, 1996). On the contrary, campaigns that do not reach 60–70% of the dog population can be effective, but are often not successful in controlling rabies in the long term, because of these coverage declines in between campaigns (Hampson et al., 2009). By incorporating demographic parameters from dog populations in northern Tanzania, Hampson et al. (2009) estimated that annual campaigns should aim to vaccinate 60% of the dog population to avoid coverage falling below P_{crit}. Indeed, a target of 70% coverage would appear to be sufficient to maintain herd immunity, even in populations with the highest demographic rates and should remain a universal target for programs that aim to eliminate rabies.

Other factors besides dog demography may impact the sustainability of control measures and prospects of dog rabies elimination. Declines in rabies incidence can occur relatively quickly given sufficient effort: two consecutive years of high-coverage annual campaigns led to progressive declines in rabies incidence in the Americas (Schneider et al., 2007). However, the period over which sustained coverage needs to be maintained to achieve elimination is generally unknown, although it is likely to be influenced by factors related to host population size and isolation, the speed of recruitment of susceptible hosts, spatial (Selhorst et al., 2005) and human-related factors, heterogeneities in coverage, as well as epidemic stochasticity (Townsend et al., 2013). In particular, the spatial distribution of host populations and interventions play a key role in disease dynamics and the efficacy of control measures. In contiguous landscapes, local transmission frequently introduces infections from dog populations with endemic rabies into neighboring areas, seriously impeding any localized vaccination efforts (Beyer et al., 2011). In more isolated or island populations, imported infections are much rarer and mainly occur as a consequence of human-mediated transport of incubating or infectious dogs. Furthermore, the probability that imported infections spark new epidemics depends upon corresponding vaccination efforts (campaign frequency and spatial distribution, and coverage achieved) and demographically-driven declines in coverage.

2.3 Dog Population Management

Dog population management is a broad term encompassing interventions that usually target free-roaming domestic dogs and aim to stabilize or reduce population size or density, adjust population structure, improve the overall health, or alter the behavior of the dog population. Interventions include culling or sterilization programs, sometimes combined with rabies vaccinations (Reece & Chawla, 2006; Totton, 2009; Totton et al., 2010). There can be many positive aspects to sterilization programs, but we now restrict our discussion to issues related to rabies control. It is important to note that canine rabies can very effectively be controlled by vaccination programs alone.

2.3.1 Culling

Culling, or the widespread killing of hosts regardless of infection status, is conducted on the assumption that, if host densities are sufficiently reduced, infectious diseases will be unable to invade, or persist (Morters et al., 2012). Although the assumption that rabies transmission rates increase with increasing host densities is intuitively appealing, there is no evidence that this is the case in domestic dog populations (Hampson et al., 2009; Morters et al., 2012). The assumption that contact rate scales with host density ignores the complexity and heterogeneity of animal and human behavior. This is reflected in the evidence of the ineffectiveness of culling to control rabies. As an example, nearly 300,000 dogs were culled in Flores, Indonesia, over a four-year period in response to a rabies outbreak on that island in 1997. Rabies was still endemic in 2004, even though the total dog population had been reduced by around half to approximately 400,000 dogs (Windiyaningsih et al., 2004). Similarly, culling failed to control canine rabies in Korea (Lee et al., 2001) and Israel (Kaplan, Goor, & Tierkel, 1954), whereas subsequent vaccination programs in both countries controlled the disease. Historically, some culling efforts may have succeeded through strict authoritarian policies, by removing almost 100% of the dog population (Shone, 1962). However, there is no evidence that culling operations today have any significant effects on controlling rabies, and enforcing such extreme levels of population reduction would now be logistically impractical and culturally unacceptable.

Two further points should be considered with regards to culling dogs to control rabies. Firstly, although culling programs may lead to the removal of infected animals, the low prevalence of canine rabies means that untargeted culling will usually remove far more healthy than sick dogs. Selective removal of diseased and hence potentially infectious dogs will usually still be carried out by communities, even in the absence of culling programs. Indeed, euthanasia of animals suspected to be infected with rabies should be an essential part of rabies control programs. Secondly, a variable degree of culling of free-roaming dogs, historically regarded as "strays," has often been undertaken alongside mass vaccination programs (Cheuk, 1969; Ernst & Fabrega-G, 1989; Larghi, Arrosi, Nakajata-A, & Villa-Nova, 1988; Wells, 1954). However, although culling can reduce the number (but not the proportion) of dogs to be vaccinated to achieve herd immunity, the vast majority of free-roaming dogs in most societies globally are owned (Butler & Bingham, 2000; Cleaveland & Dye, 1995; Kaare et al., 2009; Windiyaningsih et al., 2004; WHO/WSPA, 1990) and in reasonable health, and so culling these animals is ethically questionable. Culling can thus result in unintended negative consequences for the community often provoking considerable upset, whereas such dogs can usually be vaccinated without any of these negative repercussions.

2.3.2 Sterilization Programs

Sterilization programs aim to reduce the birth rate in a population by permanently or temporarily removing the ability of individuals to reproduce. Currently, surgical sterilization through the removal of the ovaries and uterus in females (ovariohysterectomy) or testes in males (castration or orchidectomy) is the primary method to sterilize dogs, although there is ongoing research into safe, effective, and affordable chemical and immunological means (Alliance for Contraception in Dogs and Cats, 2010).

Limited data suggest that sterilizations at the population level may stabilize or gradually reduce population size or density over time scales of several years (Reece & Chawla, 2006; Totton, 2009; Totton et al., 2010). However, as discussed in Section 2.3.1, there is no evidence that rabies transmission increases with domestic dog population density and that reductions in host density by sterilization (or culling) reduce the incidence of rabies. Although reductions in population density may plausibly reduce the number of dogs that require vaccination, timely reductions in density may be constrained by resources and population dynamics.

Theory suggests that birth control could help to maintain levels of vaccination coverage by reducing birth rates and increasing the longevity of sterilized dogs. Yet, empirical evidence for this is equivocal. Based on the limited data currently available (Totton, 2009; Totton et al., 2010) the effects of sterilizations on population structure and the maintenance of vaccination coverage are unclear. Where these programs are undertaken, there is no evidence that sterilization increases the life expectancy of dogs. Critically, there are no definitive studies providing evidence of how dog populations are regulated, nor what drives the demand for new animals. Given that most dogs, including roaming animals, are owned, the effect of sterilizations on herd immunity may be hampered by two possible responses to reductions in the local availability of pups, coupled with an ongoing demand for dogs. Firstly, concomitant reductions in pup mortality might result from reduced competition for food (although there is no empirical evidence that free-roaming dogs compete at the population level for food to survive), reductions in surplus pups being dumped or improved care of pups produced from any remaining unsterilized bitches. Secondly, pups or dogs may be sought from outside the population. Migration of people with their dogs may also affect population structure and vaccination coverage. It is unclear what effect these compensatory mechanisms could have on vaccination coverage. If sterilizations precipitate an influx of dogs from external sources, it might result in an increase in the spread of rabies between populations or within very large populations. Arguably, a local supply of healthy pups might actually help to limit the spread of rabies. What is clear is that the influence of human factors on population structure, vaccination coverage and the spread of rabies are currently poorly understood and warrant further research.

Sterilizations to reduce the number of pups and sub-adults in the population has been advocated on the basis that a proportionally higher incidence of rabies occurs in dogs under 12 months of age (Belcher, Wurapa, & Atuora, 1976; Beran 1991; Bhatia, Bhardwaj, & Sehgal, 1988; Kayali et al., 2003; Malaga, Lopez Nieto, & Gambirazio, 1979; Mitmoonpitak, Tepsumethanon, & Wilde, 1998) and an increased public health risk is often reported because of close contact with puppies (Widdowson, Morales, Chaves, & McGrane, 2002; WHO 1998). This may be the effect of comparatively lower vaccination coverage in this sub-group (Belcher et al., 1976; Malaga et al., 1979; Widdowson et al., 2002) rather than a higher rate of contact (Mitmoonpitak, Wilde, & Tepsumethanon, 1997; Mitmoonpitak et al., 1998). Dogs younger than 12 weeks of age are not often vaccinated (Chomel et al., 1988; Mitmoonpitak et al., 1998) on the assumption that maternal antibody may prevent adequate immune responses (this issue is discussed in more detail in Section 3.2). Rather than try to reduce the proportion of these age groups in the population through sterilizations, it may be more efficient to simply vaccinate them as a means to maintain herd immunity.

Sterilization programs have only been shown to reduce the incidence of rabies when combined with rabies vaccinations (Reece & Chawla 2006). However, as all sterilized dogs are also vaccinated, it is unclear to what extent sterilization itself impacts on rabies control. There are no data on the effect of sterilizations as a sole modality on disease incidence and currently no scientific evidence to support sterilizations as an essential component of rabies control programs. Rather, for the purposes of rabies control, available resources should be directed towards repeat vaccination campaigns. For sterilization programs with the primary aim of addressing animal welfare concerns, animals should still be vaccinated at the time of sterilization to increase vaccination coverage. Similarly, legislation regulating the number of owned dogs as a means to control rabies (either to reduce recruitment or population densities) are also sometimes implemented (Lapiz et al., 2012), with differing degrees of enforcement. All these measures may have positive benefits if they increase awareness of rabies and responsible dog ownership, but direct impacts on reducing transmission are likely to be negligible if not carried out as part of a sustained, high-coverage vaccination program.

3 PRACTICAL ASPECTS OF DOG RABIES CONTROL

For infection to be eliminated from a given area, rabies control strategies must target the animal population mostly responsible for viral maintenance and transmission to humans (the reservoir host). Once rabies has been eliminated in the reservoir species ("attack/elimination" component), strategies need to be implemented to ensure that freedom from rabies is maintained ("maintenance" component).

3.1 Planning Phase

The successful accomplishment of the attack and maintenance components of a canine rabies elimination program requires careful planning ("preparatory" component). Preparatory steps should include (a) those aimed at increasing awareness at the national and community level to ensure commitment of politicians and policymakers (hence legislative and financial support) and inter-sectoral dialogue, and community participation in dog rabies control operations and responsible dog ownership; (b) those directed at establishing local capacity for rabies surveillance, prevention and control, including training of relevant professionals; and (c) surveys focusing on estimating dog population sizes to inform vaccination programs.

Emphasis has been placed on the need to conduct detailed surveys for planning purposes prior to the start of canine rabies control programs. While these surveys offer the opportunity to collect useful data on differences in dog demographics, relationships between dogs and people, levels of supervision and accessibility, and local community attitudes (Ratsitorahina et al., 2009; Robinson, Miranda, Miranda, & Childs, 1996; Suzuki et al., 2008), they require specialized personnel, are costly to implement, and may considerably delay the implementation of control programs, as well as divert resources from dog vaccination. The issue of inaccessibility of dogs to vaccination is often perceived as a critical impediment to successful dog rabies control, and it is therefore viewed as an essential aspect to determine through preliminary surveys. Ecological studies from a range of rabies endemic settings have however demonstrated that, although most domestic dogs are allowed to roam freely, at least one household claims some degree of responsibility for them, and that most owners are willing to restrain their dogs for vaccination (Kaare et al., 2009; Kayali et al., 2003; Kongkaew, Coleman, Pfeiffer, Antarasena, & Thiptara, 2004; Perry, 1993; Ratsitorahina et al., 2009; Robinson et al., 1996; Suzuki et al., 2008; Wandeler, Matter, Kappeler, & Budde, 1993). Thus, in most circumstances, detailed surveys to determine these factors may not be required.

Initial determination of the dog population size in a given community on the other hand is important in the preparatory and evaluation stages in order to determine campaign logistics (e.g., number of dogs to vaccinate to reach the recommended coverage, estimating supplies needed, etc.), and evaluate the intervention in terms of vaccination coverage. Complete dog counts are rarely available, and, although techniques to estimate dog population sizes exist, they are not without problems. Rapid estimates can be made prior to program implementation (e.g., based on expert opinion or through simple household surveys to establish the mean number of owned dogs per household and dog-to-human ratios) and refined subsequently in combination with surveys for estimation of vaccination coverage. Household surveys potentially capture much of the owned population but methods are also available to assess the number of free-roaming dogs through counts along selected representative routes or capture-mark-recapture methods consisting of temporarily marking dogs (e.g., during vaccination campaigns) and estimating the total number of free-roaming dogs from the observed ratio of marked to unmarked dogs determined during "visual recapture" efforts (WSPA, 2007). There are, however, several issues with all these methods. Bias may be introduced by multiple extrapolations that may produce unrealistic estimates of dog population sizes. Counts obtained through these techniques are based on random selection of sub-regions, households, or routes, and general dog population estimates therefore result from extrapolating counts from a sample to whole regions or countries, despite large variation in dog ownership patterns. Similarly, estimates of the owned dog populations are generally extrapolated from the total human population or number of households known through national population censuses. Human population censuses are generally carried out at wide time intervals with associated potential errors in dog size estimates resulting from population growth. In addition, the design, implementation and analysis of data generated from these surveys are not trivial.

Despite these caveats, and given the importance of reliable denominator data to evaluate the effectiveness of any given intervention efforts, relatively accurate determination of dog population sizes should be viewed as an essential part of planning and implementing dog rabies elimination programs. An option would be to focus on complete dog census efforts that could be integrated within nationwide censuses of the human population, which are routinely carried out by census bureau offices.

3.2 Implementation of Mass Dog Vaccination Campaigns

The "attack/elimination" component of canine rabies control programs aims at preventing the occurrence of human rabies by reducing and eventually eliminating virus transmission in the dog population. The feasibility of rabies elimination through mass dog vaccination is now widely advocated as a realistic prospect also for rabies endemic areas of the less developed world (Dodet, Meslin, & Heseltine, 2001; Lembo et al., 2010; Rupprecht et al., 2008; WHO 1992 & 2004). There are several approaches for mass dog vaccination, but given the evidence of general accessibility of dogs for parenteral vaccination (Kayali et al., 2003; Kongkaew et al., 2004; Lembo et al., 2010; Perry, 1993; Ratsitorahina et al., 2009; Robinson et al., 1996; Suzuki et al., 2008; Wandeler et al., 1993), here we will focus on approaches that apply to free-roaming dogs that vary in accessibility and handleability. For highly accessible rural communities and dogs that are easy to handle, central-point vaccinations, consisting of mobile teams that set up temporary vaccination posts in central village locations, are generally the most cost-effective strategy (Kaare et al., 2009). This, along with continual vaccination at fixed vaccination posts in well-recognized sites within the community (e.g., private or government veterinary clinics), is also a suitable approach in urban settings. In dispersed communities, or where dogs are difficult to handle, alternative delivery strategies may be required to achieve high coverage. In very remote communities combined approaches using central point and house-to-house vaccination conducted by either mobile teams of permanent staff or trained local community-based animal health workers have proven very effective in achieving high coverage and eliminating rabies from given areas (Kaare et al., 2009, Lembo et al., 2010), although they require considerable investment in labor and capital and are operationally difficult (Kaare et al., 2009). The use of trained dog-catchers with nets has been used to attain high coverage in Bali, Indonesia, where dog densities are very high, and most dogs are not used to being restrained (Putra et al., in press).

Oral delivery strategies alone or combined with parenteral vaccination can in various settings improve coverage, especially where dogs are not readily handleable (Guzel, Leloglu, & Vos, 1998; Matter et al., 1995; Matter et al., 1998; WHO, 2007). Yet, oral vaccination has not been operationalized in any dog rabies elimination program and currently available commercial vaccine baits prequalified by WHO are limited and much more expensive than injectables. Locally produced baits could make oral vaccination more affordable (Estrada, Vos, De Leon, & Muller, 2001;

WHO, 2007), but they would need to overcome high regulatory scrutiny both in terms of quality of biologicals and human safety, the latter being of particular concern in areas where dogs and humans live in close proximity and the human population comprises high numbers of immune suppressed individuals, as is the case in many rabies-endemic areas due to the HIV pandemic.

An essential issue to be addressed in the design and implementation of dog rabies control programs is that of vaccination of young pups. A lack of inclusion of pups in vaccination efforts can result in insufficient coverage being attained (Kaare et al., 2009). Despite the widespread perception that vaccine-induced active immunity may be affected by the presence of maternally derived antibodies in young pups, evidence from a range of settings demonstrates that vaccination of dogs younger than 3 months is safe and confers protection independently of the health condition of dogs (Barrat et al., 2001, Chappuis, 1998). For example, in village settings in Tanzania, young pups responded well to a single vaccine dose, with no indication of interference (McNabb 2008).

Aspects such as improving community participation in dog rabies control programs and more specifically the inclusion of pups in vaccination campaigns can be addressed through education and awareness. Ensuring adequate engagement of local communities is undoubtedly a key step for successful dog rabies control. The input of other disciplines, dedicated to developing methods most compatible with local circumstances, to inform and influence individual and community decisions that enhance health (i.e., global health communications) is therefore increasingly recognized as essential in global rabies elimination strategies. To this effect, initiatives have been developed that have had considerable impacts worldwide (Costa, Briggs, & Dedmon, 2010; Costa, Briggs, Tumpey, Dedmon, & Coutts, 2009). A major factor that is likely to compromise community participation and that can be a difficult challenge to overcome is that of costs. Charging for vaccination is usually counterproductive in poorer communities without an effective regulatory culture as turnout may be too low to achieve adequate vaccination coverage (Durr et al., 2009; see also Section 3.5).

Given that dog rabies control programs often operate under logistical and financial constraints, an important question relates to the design of vaccination campaigns that maximize the effectiveness of control efforts in terms of both reducing the occurrence of rabies, and the time and intensity of effort required to achieve this goal. The design of vaccine delivery in pulses is an important determinant of the effectiveness of vaccination campaigns (Nokes & Swinton, 1997), with factors such as pulse frequency and pulse size (number of vaccine doses) having a considerable influence on longer-term impacts of interventions (Beyer et al., 2012). When only limited vaccination resources are available, achieving sparse coverage throughout a large population will have only negligible impacts, whereas concentrating vaccination effort in a particular community, preferably one that is isolated, will generate greater gains (Beyer et al., 2012).

Although the rabies elimination phase should focus on eliminating infection at its source, the integration of dog rabies control (through mass dog vaccination) and human rabies prevention (through human prophylaxis) is required to avoid unnecessary human deaths when dog rabies is still prevalent. Efforts should be made towards correct and cost-effective utilization of costly human biologicals by public health authorities and enhanced awareness about prevention behaviors amongst communities. Progressive declines of canine rabies should ultimately result in reduced demand for human biologicals (see Section 3.5).

3.3 Measuring the Impact of Mass Dog Vaccination Campaigns

3.3.1 Vaccination Coverage

Achieving high coverage is the most important aim of any vaccination program, since demographic processes cause coverage levels to rapidly decline and the logistic challenge of carrying out campaigns means that it takes time to achieve effective coverage (Figure 17.2). It is therefore vital to reach as large a proportion of the population as possible: that is, dogs of all ages, dogs that are free-roaming, and dogs in all communities.

To determine whether sufficient dogs have been vaccinated, coverage needs to be measured. The ratio of vaccinated dogs to the total dog population is the most direct estimate of coverage, but requires reliable denominator data on the dog population, which are often problematic to obtain (Section 3.1). Coverage can also be estimated from post-vaccination questionnaire surveys as the proportion of vaccinated to unvaccinated dogs in households, or from direct observation of marked (vaccinated) and unmarked (unvaccinated) dogs as discussed above. If the latter method is used, identification of vaccinated dogs is required. In situations where permanent identification of dogs is constrained by a lack of resources, temporary forms of identification (e.g., colored tags or plastic collars) can be applied and are often very popular in local communities, providing motivation for owners to participate in vaccination campaigns (Kaare et al., 2009). Both methods rely on a sample of the population being surveyed, with potential problems arising from extrapolations, as mentioned above.

A common misperception is that serological surveillance is an alternative method for measuring coverage. The safety and efficacy of WHO pre-qualified cell-culture vaccines currently used for parenteral immunization of dogs are widely recognized, but serological studies can provide information about immunogenicity of alternative vaccines (e.g., locally produced vaccines/baits) and/or effectiveness of vaccine delivery (e.g., correct vaccine administration and maintenance of cold chain). However, there are several reasons why serology is unsuitable for estimating coverage. Firstly, gold standard serology assays are of extremely limited availability in rabies-endemic areas because they require BSL3/SAPO4 high containment laboratories (as live or recombinant virus is used), are difficult to perform and are prohibitively expensive (Meslin & Kaplan, 1996). Alternative neutralization assays to assess seroconversion using lentiviral pseudotypes have been developed that remove the need for high containment laboratories and may be more suitable for a wider range of settings (Wright et al., 2008 & 2009), but high-standard infrastructure for cell culture is still needed. Secondly, quantification of serological responses to vaccination requires determining baseline and 4-week post-vaccination antibody titres, making dog sampling strategies operationally cumbersome and very expensive. Thirdly, the minimum rabies virus neutralizing antibody titre of 0.5 IU per mL of serum considered satisfactory for the international transfer of vaccinated dogs should not be used to assess vaccine efficacy in dogs immunized in the context of mass campaigns in which lower values should be expected.

3.3.2 Surveillance

To evaluate the effectiveness of intervention efforts, adequate surveillance measures need to be established so that the rabies situation can be determined at the start of the control program and impacts of intervention can be monitored through time. Surveillance is also essential to detect new incursions that could compromise maintenance of freedom in areas where rabies has been eliminated.

International standards for rabies surveillance and recognition of freedom from rabies require laboratory confirmation of cases postmortem, hence effective systems for sample collection and submission, and prompt diagnosis (Meslin & Kaplan, 1996). However, limited capacity for sample collection and diagnosis is an enduring problem in many dog rabies-endemic countries. Therefore, laboratory-confirmed cases may not necessarily provide a good measure of incidence. In many countries, national statistics are still based on clinical signs and are largely incomplete (Dodet et al., 2008).

Over the past 40 years, the direct fluorescent antibody technique (DFA) has been the global standard for rapid rabies diagnosis (Meslin & Kaplan, 1996), but in many parts of the world infrastructure and capacity (i.e., expertise for fluorescence microscopy) for performing this test are still inadequate. Simplified techniques for sample collection, preservation, and diagnosis are now available, which have shown potential to increase in-country capabilities for rabies surveillance. A direct rapid immunohistochemical test (dRIT) based on light microscopy (Niezgoda & Rupprecht,

2006) shows complete concordance with the DFA and performs well on samples preserved under field conditions (Durr et al., 2008; Lembo et al., 2006; Tao et al., 2008). Other simple techniques have been described to successfully allow rapid rabies screening, including enzyme immunoassays (Vasanth, Madhusudana, Abhilash, Suja, & Muhamuda, 2004), dot blot enzyme immunoassays (Madhusudana, Paul, Abhilash, & Suja, 2004) and lateral-flow immunodiagnostic test kits (Nishizono et al., 2008), but these remain research tools. Serological techniques based on lentiviral pseudotypes have potential for more widespread use in rabies-endemic areas (Wright et al., 2008 & 2009), but, while they are likely to help generate improved rabies epidemiological data, they are not diagnostic of rabies infection.

Medical records of animal-bite injuries from suspect rabid animals can provide useful epidemiological information both in terms of rabies incidence and human exposures (Hampson et al., 2009) and for making inferences about spatial transmission dynamics and the efficacy of control measures (Beyer et al., 2011). Mobile phone network access offers opportunities for real-time reporting/detection of cases and animal bite injuries, and studies are in place to evaluate this as a tool to enhance rabies surveillance in remote communities. In general, channels of reporting and communication need to be improved, and increased deployment of field diagnostic tools and cheap and user-friendly means for reporting cases should be prioritized.

3.4 Maintaining Rabies-Free Status

Once achieved, the economic and ethical incentives for maintaining freedom from rabies through sustained (potentially targeted) control (e.g., use of a cordon sanitaire) and effective surveillance are extremely strong, as demonstrated by the devastating effects of emerging epidemics in previously rabies-free areas (Putra et al., in press; Windiyaningsih et al., 2004). There is currently very little research available to advocate what kind of strategic approaches, such as targeted ring vaccination, would be most effective to eliminate residual foci in the final stages of an elimination program or for responding to new incursions. However, much can be learned from countries that have achieved and maintained freedom from rabies.

After achieving rabies elimination, the island nations of Britain and Japan both suffered, but were able to control incursions during periods of wartime. Swift dog muzzling and dog confinement contained an imported case to Britain in 1921 (Fooks et al., 2004) and mass vaccination freed Japan from rabies following WWII (Shimada, 1971). These nations have since maintained freedom from rabies through stringent quarantine systems, with Japan enforcing mandatory dog registration and vaccination (Takahashi-Omoe, Omoe, & Okabe, 2008). Occasional introductions of dog rabies from North Africa to Europe have had significant economic ramifications (Lardon et al., 2010), but all have so far been contained without significant further spread. Similarly, despite periodic reintroductions, a vaccination "cordon-sanitaire" maintained at the Malaysia-Thailand border (Tan et al., 1972; WHO, 1988), has also kept Malaysia effectively free from rabies, with only one isolated and probably imported dog rabies case reported outside the buffer zone in 1996 (Hussin, 1997). Across continental geopolitical boundaries, maintaining freedom from rabies has proven more difficult. Although rabies in North America has been largely controlled since the 1960s through mass vaccination campaigns followed by a long period of compulsory pet vaccination (Blanton, Hanlon, & Rupprecht, 2007; Held, Tierkel, & Steele, 1967; Korns & Zeissig, 1948), the USA was only declared free of dog rabies in 2007 after concerted transboundary collaborations to prevent importations (Blanton et al., 2007).

Geographic isolation, high levels of surveillance, local capacity/infrastructure for rapid mobilization, continued political commitment and inter-sectoral cooperation, and enforced legislation have been important factors in keeping countries rabies-free (Takahashi-Omoe et al., 2008). In contiguous landscapes, an assessment of the rabies situation in neighboring areas is also of great importance, ideally followed by the establishment of rabies control and prevention efforts in these jurisdictions through liaison and collaborations between key stakeholders. Although contiguous international boundaries undoubtedly pose considerable challenges to maintaining freedom from infection, phylodynamic studies from north Africa indicate little mixing of viral sequences from Algeria, Tunisia, and Morocco, and evidence for only a few long-distance introductions, which are most likely to be the result of human-mediated spread (Talbi et al., 2010). The importance of human-mediated dispersal in north Africa gives grounds for cautious optimism that geopolitical boundaries may represent more of a barrier to canine rabies dispersal than may be expected from geographic features alone, and that zoosanitary border controls are likely to be an effective means of preventing re-introduction.

3.5 Economics of Dog Vaccination for Rabies Control

To ensure that an area is continuously maintained free from rabies, careful consideration also needs to be given to building sustainability in established programs. Financial sustainability needs to address factors affecting the cost-effectiveness of different rabies control strategies, as well as operational issues associated with the design and implementation of dog vaccination campaigns. For prevention of human rabies deaths, rabies control strategies that incorporate mass vaccination of domestic dogs have the potential to be more cost-effective than strategies relying on administration of human PEP alone, with dog vaccination strategies typically becoming more cost-effective within 5–6 years of the onset of mass dog vaccination campaigns (Bögel & Meslin, 1990; Zinsstag et al., 2009). These economic benefits can be explained by the high costs of PEP in comparison with delivery of dog vaccination (typically US$1.5-US$4/ dog – Bögel & Meslin, 1990; Kaare et al., 2009; Kayali, Mindekem, Hutton, Ndoutamia, & Zinsstag, 2006), as well as the escalating costs of PEP over time as dog populations and rabies incidence continue to rise. However, in cost-effectiveness models, there is often an assumption of a linear relationship between dog rabies incidence, human exposure, and demand for human PEP, and it is clear that this relationship may vary in different settings. For example, the incidence of suspected rabid dog bites reported at local health facilities declined rapidly in rural Tanzania following the implementation of annual dog vaccination campaigns and decline in dog rabies incidence (Cleaveland, Kaare, Tiringa, Mlengeya, & Barrat, 2003; Lembo et al., 2010). In contrast, dog vaccination has been associated with increased demand for PEP in some parts of Asia (Hahn, 2009). Critical factors are likely to include education and awareness (of both bite victims and clinicians), availability and cost of human PEP, the size of national health budgets, and a society's tolerance of risk in clinical decision-making regarding administration of PEP. A further important factor relates to costs of different regimens of PEP, with most cost-effectiveness analyses including only intramuscular regimens. However, efforts are being made to introduce intradermal regimens across Asia and Africa, which are highly efficacious (Chapter 13) and have the potential for major cost savings (Hampson, Cleaveland, & Briggs, 2011).

The recent implementation of several large-scale rabies elimination demonstration projects in Asia and Africa provides an opportunity to explore cost-effectiveness and sustainability strategies in more detail. Ongoing projects in the Philippines, South Africa, and Tanzania are likely to generate valuable comparative information on cost-effectiveness and sustainability of dog rabies control from settings that differ in terms of the level of established infrastructure, the degree of government support and community engagement, and strategies for vaccine delivery (including centralized campaigns, house-to-house delivery, and community-led initiatives).

References

Alliance for Contraception in Dogs and Cats. (2010). *Proceedings of the fourth international symposium on non-surgical methods of pet population control, April 8–10, 2010, Dallas, TX.* Retrieved from <http://www.acc-d.org/4thSymposiumProceedings/>.

Anderson, R. M., & May, R. M. (1991). *Infectious diseases of humans: Dynamics and control.* Oxford: Oxford University Press.

Barrat, J., Blasco, E., Lambot, M., Cliquet, F., Brochier, B., Renders, C., et al. (2001). Is it possible to vaccinate young canids against rabies and to protect them?: In *Proceedings of the sixth Southern and Eastern African rabies group/world health organization meeting, Lilongwe, Malawi*. Lyon: Éditions Fondation Marcel Mérieux. pp. 151–163.

Belcher, D. W., Wurapa, F. K., & Atuora, D. O. C. (1976). Endemic rabies in Ghana. *American Journal of Tropical Medicine and Hygiene, 25*, 724–729.

Beran, G. W. (1991). Urban rabies. In G. M. Baer (Ed.), *The natural history of rabies (pp. 427–443)* (2nd ed.). Boca Raton, FL: CRC Press.

Beyer, H., Hampson, K., Cleaveland, S., Kaare, M., Lembo, T., & Haydon, D. T. (2011). Metapopulation dynamics of rabies and the efficacy of vaccination. *Proceedings of the Royal Society B-Biological Sciences, 278*, 2182–2190.

Beyer, H. L., Hampson, K., Lembo, T., Cleaveland, S., Kaare, M., & Haydon, D. T. (2012). The implications of metapopulation dynamics on the design of vaccination campaigns. *Vaccine, 30*, 1014–1022.

Bhatia, R., Bhardwaj, M., & Sehgal, S. (1988). Canine rabies in and around Delhi–a 16 year study. *Journal of Communicable Diseases, 20*, 104–110.

Bingham, J. (2001). Rabies on Flores Island, Indonesia: Is eradication possible in the near future?. In B. Dodet, F. -X. Meslin, & E. Heseltine (Eds.), *Proceedings of the fourth international symposium on rabies control in Asia, Hanoi, 5–9 March 2001 (pp. 148–155)*. Montrouge: John Libbey Eurotext.

Blanton, J. D., Hanlon, C. A., & Rupprecht, C. E. (2007). Rabies surveillance in the United States during 2006. *Journal of the American Veterinary Medical Association, 231*, 540–556.

Bögel, K. (2002). Control of dog rabies. In A. C. Jackson & W. H. Wunner (Eds.), *Rabies (pp. 428–443)*. San Diego, CA: Elsevier Science.

Bögel, K., & Meslin, F. -X. (1990). Economics of human and canine rabies elimination–guidelines for programme orientation. *Bulletin of the World Health Organization, 68*, 281–291.

Butler, J. R. A., & Bingham, J. (2000). Demography and dog-human relationships of the dog population in Zimbabwean communal lands. *The Veterinary Record, 147*, 442–446.

Chappuis, G. (1998). Neonatal immunity and immunisation in early age: Lessons from veterinary medicine. *Vaccine, 16*, 1468–1472.

Cheuk, H. (1969). A review of the history and control of rabies in Hong Kong. *Agricultural Sciences in Hong Kong, 1*, 141–174.

Chomel, B., Chappuis, G., Bullon, F., Cardenas, E., de Beublain, T. D., Lombard, M., et al. (1988). Mass vaccination campaign against rabies: Are dogs correctly protected? The Peruvian experience. *Reviews of Infectious Diseases, 10*, S697–702.

Cleaveland, S., & Dye, C. (1995). Maintenance of a microparasite infecting several host species: Rabies in the Serengeti. *Parasitology, 111*, S33–S47.

Cleaveland, S., Kaare, M., Tiringa, P., Mlengeya, T., & Barrat, J. (2003). A dog rabies vaccination campaign in rural Africa: Impact on the incidence of dog rabies and human dog-bite injuries. *Vaccine, 21*, 1965–1973.

Coleman, P. G., & Dye, C. (1996). Immunization coverage required to prevent outbreaks of dog rabies. *Vaccine, 14*, 185–186.

Coleman, P. G., Fèvre, E. M., & Cleaveland, S. (2004). Estimating the public health impact of rabies. *Emerging Infectious Diseases, 10*, 140–142.

Costa, P., Briggs, D. J., Tumpey, A., Dedmon, R., & Coutts, J. (2009). World rabies day outreach to Asia: Empowering people through education. *Asian Biomedicine, 3*, 451–457.

Costa, P., Briggs, D., & Dedmon, R. (2010). World rabies day (September 28, 2010): The continuing effort to "make rabies history". *Asian Biomedicine, 4*, 671.

Dodet, B., Meslin, F. -X., & Heseltine, E. (Eds.), (2001). *Proceedings of the fourth international symposium on rabies control in Asia, Hanoi, 5–9 March 2001*. Montrouge: John Libbey Eurotext.

Dodet, B., Adjogoua, E. V., Aguemon, A. R., Amadou, O. H., Atipo, A. L., Baba, B. A., et al. (2008). Fighting rabies in Africa: The Africa rabies expert bureau (AfroREB). *Vaccine, 26*, 6295–6298.

Durr, S., Naissengar, S., Mindekem, R., Diguimbye, C., Niezgoda, M., Kuzmin, I., et al. (2008). Rabies diagnosis for developing countries. *PLoS Neglected Tropical Diseases, 2*, e206.

Durr, S., Mindekem, R., Kaninga, Y., Moto, D. D., Meltzer, M. I., Vounatsou, P., et al. (2009). Effectiveness of dog rabies vaccination programmes: Comparison of owner-charged and free vaccination campaigns. *Epidemiology and Infection, 137*, 1558–1567.

Ernst, S. N., & Fabrega-G, F. (1989). Epidemiology of rabies of Chile 1950–1986: A descriptive study of laboratory-confirmed cases. *Revista de Microbiologia, 20*, 121–127.

Estrada, R., Vos, A., De Leon, R., & Muller, T. (2001). Field trial with oral vaccination of dogs against rabies in the Philippines. *BMC Infectious Diseases, 1*, 23.

Fooks, A. R., Roberts, D. H., Lynch, M., Hersteinsson, P., & Runolfsson, H. (2004). Rabies in the united Kingdom, Ireland and Iceland. In A. A. King, A. R. Fooks, M. Aubert, & A. I. Wandeler (Eds.), *Historical perspective of rabies in europe and the Mediterranean Basin (pp. 25–32)*. Paris: OIE (World Organisation for Animal Health).

Gautret, P., Shaw, M., Gazin, P., Soula, G., Delmont, J., Parola, P., et al. (2008). Rabies postexposure in returned injured travelers from France, Australia, and New Zealand: A retrospective study. *Journal of Travel Medicine, 15*, 25–30.

Guzel, N., Leloglu, N., & Vos, A. (1998). Evaluation of a vaccination campaign of dogs against rabies, including oral vaccination, in Kusadasi, Turkey. *Etlik Veteriner Mikrobiyoloji Dergisi, 9*, 121–134.

Hahn, N. T. H. (Ed.). (2009). *Proceedings of the second international conference of rabies in Asia (RIA) foundation, 9–11 September, Hanoi, Viet Nam, 2009*. Retrieved from <http://www.rabiesinasia.org/vietnam/riacon2009report.pdf/>.

Hampson, K., Dobson, A., Kaare, M., Dushoff, J., Magoto, M., Sindoya, E., et al. (2008). Rabies exposures, post-exposure prophylaxis and deaths in a region of endemic canine rabies. *Plos Neglected Tropical Diseases, 2*, e339.

Hampson, K., Dushoff, J., Cleaveland, S., Haydon, D. T., Kaare, M., Packer, C., et al. (2009). Transmission dynamics and prospects for the elimination of canine rabies. *Plos Biology, 7*, e1000053.

Hampson, K., Cleaveland, S., & Briggs, D. (2011). Evaluation of cost-effective strategies for rabies post-exposure vaccination in low-income countries. *PLoS Neglected Tropical Diseases, 5*, e982.

Health Protection Agency. (2011). Demand for post-exposure rabies vaccine trebles. Retrieved from <http://www.hpa.org.uk/NewsCentre/NationalPressReleases/2011PressReleases/110913Demandforpostexprabiesvaccinetrebles/>.

Held, J. R., Tierkel, E. S., & Steele, J. H. (1967). Rabies in man and animals in the United States, 1946–1965. *Public Health Reports, 82*, 1009–1018.

Hu, R. L., Fooks, A. R., Zhang, S. F., Liu, Y., & Zhang, F. (2008). Inferior rabies vaccine quality and low immunization coverage in dogs (*Canis familiaris*) in China. *Epidemiology and Infection, 136*, 1556–1563.

Hussin, A. (1997). Malaysia: Veterinary aspects of rabies control and prevention. In B. Dodet & F. -X. Meslin (Eds.), *Rabies control in Asia (pp. 167–170)*. Paris: Elsevier.

Kaare, M., Lembo, T., Hampson, K., Ernest, E., Estes, A., Mentzel, C., et al. (2009). Rabies control in rural Africa: Evaluating strategies for effective domestic dog vaccination. *Vaccine, 27*, 152–160.

Kaplan, M. M., Goor, Y., & Tierkel, E. S. (1954). A field demostration of rabies control using chicken-embryo vaccine in dogs. *Bulletin of the World Health Organization, 10*, 743–752.

Kayali, U., Mindekem, R., Yemadji, N., Vounatsou, P., Kaninga, Y., Ndoutamia, A. G., et al. (2003). Coverage of pilot parenteral vaccination campaign against canine rabies in N'Djamena, Chad. *Bulletin of the World Health Organization, 81*, 739–744.

Kayali, U., Mindekem, R., Hutton, G., Ndoutamia, A. G., & Zinsstag, J. (2006). Cost-description of a pilot parenteral vaccination campaign against rabies in dogs in N'Djaména, Chad. *Tropical Medicine and International Health, 11*, 1058–1065.

King, A. A., Fooks, A. R., Aubert, M., & Wandeler, A. I. (Eds.), (2004). *Historical perspective of rabies in europe and the Mediterranean Basin*. Paris: OIE (World Organisation for Animal Health).

Knobel, D. L., Cleaveland, S., Coleman, P. G., Fèvre, E. M., Meltzer, M. I., Miranda, M. E. G., et al. (2005). Re-evaluating the burden of rabies in Africa and Asia. *Bulletin of the World Health Organization, 83*, 360–368.

Kongkaew, W., Coleman, P., Pfeiffer, D. U., Antarasena, C., & Thiptara, A. (2004). Vaccination coverage and epidemiological parameters of the owned-dog population in Thungsong district, Thailand. *Preventive Veterinary Medicine, 65*, 105–115.

Korns, R. F., & Zeissig, A. (1948). Dog, fox, and cattle rabies in New-York State–evaluation of vaccination in dogs. *American Journal of Public Health, 38*, 50–65.

Lapiz, S. M. D., Miranda, M. E. G., Garcia, R. G., Daguro, L. I., Paman, M. D., Madrinan, F. P., et al. (2012). Implementation of an intersectoral program to eliminate human and canine rabies: the Bohol Rabies Prevention and Elimination Project. *PLoS Neglected Tropical Diseases, 6*, e1891.

Lardon, Z., Watier, L., Brunet, A., Bernède, C., Goudal, M., Dacheux, L., et al. (2010). Imported episodic rabies increases patient demand for and physician delivery of antirabies prophylaxis. *PLoS Neglected Tropical Diseases, 4*, e723.

Larghi, O. P., Arrosi, J. C., Nakajata-A, J., & Villa-Nova, A. (1988). Control Of urban rabies. In J. B. Campbell & K. M. Charlton (Eds.), *Developments in veterinary virology: Rabies (pp. 407–422)*. Dordrecht, The Netherlands: Kluwer Academic.

Lee, J. H., Lee, M. J., Lee, J. B., Kim, J. S., Bae, C. S., & Lee, W. C. (2001). Review of canine rabies prevalence under two different vaccination programmes in Korea. *The Veterinary Record, 148*, 511–512.

Lembo, T., Niezgoda, M., Velasco-Villa, A., Cleaveland, S., Ernest, E., & Rupprecht, C. E. (2006). Evaluation of a direct, rapid immunohistochemical test for rabies diagnosis. *Emerging Infectious Diseases, 12*, 310–313.

Lembo, T., Hampson, K., Kaare, M., Ernest, E., Knobel, D., Kazwala, R., et al. (2010). The feasibility of canine rabies elimination in africa: Dispelling doubts with data. *Plos Neglected Tropical Diseases, 4*, e626.

Lembo, T., Craig, P., Miles, M. A., Hampson, K., & Meslin, F. X. (2013). Zoonoses prevention, control and elimination in dogs. In C. N. L. Macpherson, F. X. Meslin & A. I. Wandeler (Eds.), *Dogs, Zoonoses and public health* (2nd ed.) (pp. 205–258). Wallingford, Oxon, UK: CAB International.

Madhusudana, S. N., Paul, J. P., Abhilash, V. K., & Suja, M. S. (2004). Rapid diagnosis of rabies in humans and animals by a dot blot enzyme immunoassay. *International Journal of Infectious Diseases, 8*, 339–345.

Malaga, H., Lopez Nieto, E., & Gambirazio, C. (1979). Canine rabies seasonality. *International Journal of Epidemiology, 8*, 243–245.

Matter, H. C., Kharmachi, H., Haddad, N., Benyoussef, S., Sghaier, C., Benkhelifa, R., et al. (1995). Test of three bait types for oral immunization of dogs against rabies in Tunisia. *American Journal of Tropical Medicine and Hygiene, 52*, 489–495.

Matter, H. C., Schumacher, C. L., Kharmachi, H., Hammami, S., Tlatli, A., Jemli, J., et al. (1998). Field evaluation of two bait delivery systems for the oral immunization of dogs against rabies in Tunisia. *Vaccine, 16*, 657–665.

McNabb, S. (2008). *Welfare and health assessment of Tanzanian dogs and its relationship to immunological response to rabies vaccination*. (Unpublished MSc thesis). University of Edinburgh, Edinburgh.

Meldrum, K. C. (1988). Rabies contingency plans in the United Kingdom. *Parassitologia, 30*, 97–103.

Meslin, F. X., & Kaplan, M. M. (1996). An overview of laboratory techniques in the diagnosis and prevention of rabies and in rabies research. *Laboratory techniques in rabies* (4th ed.). (Eds.), pp. 9–27.

Mitmoonpitak, C., Wilde, H., & Tepsumethanon, V. (1997). Current status of animal rabies in Thailand. *Journal of Veterinary Medical Science, 59*, 457–460.

Mitmoonpitak, C., Tepsumethanon, V., & Wilde, H. (1998). Rabies in Thailand. *Epidemiology and Infection, 120*, 165–169.

Morters, M. K., Restif, O., Hampson, K., Cleaveland, S., Wood, J. L. N., & Conlan, A. J. K. (2012). Evidence-based control of canine rabies: A critical review of population density reduction. *Journal of Animal Ecology* doi:10.1111/j.1365-2656.2012.02033.x.

Muir, P., & Roome, A. (2005). Indigenous rabies in the UK. *Lancet, 365*, 2175.

Neville, J. (2004). Rabies in the ancient world. In A. A. King, A. R. Fooks, M. Aubert, & A. I. Wandeler (Eds.), *Historical perspective of rabies in europe and the Mediterranean Basin (pp. 1–14)*. Paris: OIE (World Organisation for Animal Health).

Niezgoda, M., & Rupprecht, C. E. (2006). Standard operating procedure for the direct rapid immunohistochemistry test for the detection of rabies virus antigen: *National laboratory training network course*. Atlanta, GA: US Department of Health and Human Services & Centers for Disease Control and Prevention. pp. 1–16.

Nishizono, A., Khawplod, P., Ahmed, K., Goto, K., Shiota, S., Mifune, K., et al. (2008). A simple and rapid immunochromatographic test kit for rabies diagnosis. *Microbiology and Immunology, 52*, 243–249.

Nokes, D. J., & Swinton, J. (1997). Vaccination in pulses: A strategy for global eradication of measles and polio? *Trends In Microbiology, 5*, 14–19.

Perry, B. D. (1993). Dog ecology in Eastern and Southern Africa–implications for Rabies control. *Onderstepoort Journal of Veterinary Research, 60*, 429–436.

Putra, A. A. G., Hampson, K., Girardi, J., Hiby, E., Knobel, D. L., Mardiana, I. W., et al. (in press). Response to a rabies epidemic in Bali, Indonesia, 2008–2009. *Emerging Infectious Diseases.*

Ratsitorahina, M., Rasambainarivo, J. H., Raharimanana, S., Rakotonandrasana, H., Andriamiarisoa, M. P., Rakalomanana, F. A., et al. (2009). Dog ecology and demography in Antananarivo, 2007. *BMC Veterinary Research, 5*, 21.

Reece, J. F., & Chawla, S. K. (2006). Control of rabies in Jaipur, India, by the sterilisation and vaccination of neighbourhood dogs. *The Veterinary Record, 159*, 379–383.

Robinson, L. E., Miranda, M. E., Miranda, N. L., & Childs, J. E. (1996). Evaluation of a canine rabies vaccination campaign and characterization of owned-dog populations in the Philippines. *Southeast Asian Journal of Tropical Medicine and Public Health, 27*, 250–256.

Rupprecht, C. E., Barrett, J., Briggs, D., Cliquet, F., Fooks, A. R., Lumlertdacha, B., et al. (2008). Can rabies be eradicated? *Developments in Biologicals (Basel), 131*, 95–121.

Schneider, M. C., Belotto, A., Adé, M. P., Hendrickx, S., Leanes, L. F., Rodrigues, M. J., et al. (2007). Current status of human rabies transmitted by dogs in Latin America. *Cadernos de Saúde Pública, 23*, 2049–2063.

Selhorst, T., Muller, T., Schwermer, H., Ziller, M., Schluter, H., Breitenmoser, U., et al. (2005). Use of an area index to retrospectively analyze the elimination of fox rabies in European countries. *Environmental Management, 35*, 292–302.

Shimada, K. (1971). The last rabies outbreak in Japan. In Y. Nagano & F. M. Davenport (Eds.), *Rabies (pp. 11–28)*. Baltimore, MA: University Park.

Shone, D. K. (1962). Rabies in southern Rhodesia: 1900 to 1961. *Journal of the South African Veterinary Medical Association, 33*, 567–580.

Sudarshan, M. K., Madhusudana, S. N., Mahendra, B. J., Rao, N. S. N., Ashwath Narayana, D. H., Abdul Rahman, S., et al. (2007). Assessing the burden of human rabies in India: Results of a national multi-center epidemiological survey. *International Journal of Infectious Diseases, 11*, 29–35.

Suzuki, K., Pereira, L. A., Frias, R., Lopez, R., Mutinelli, L. E., & Pons, E. R. (2008). Rabies-vaccination coverage and profiles of the owned-dog population in Santa Cruz de la Sierra, Bolivia. *Zoonoses and Public Health, 55*, 177–183.

Takahashi-Omoe, H., Omoe, K., & Okabe, N. (2008). Regulatory systems for prevention and control of rabies, Japan. *Emerging Infectious Diseases, 14*, 1368–1374.

Talbi, C., Lemey, P., Suchard, M. A., Abdelatif, E., Elharrak, M., Nourlil, J., et al. (2010). Phylodynamics and human-mediated dispersal of a zoonotic virus. *PLoS Pathogens, 6*, e1001166.

Tan, D. S. K., Abdul, W. M. A., Mohd, N. K., & Beran, G. (1972). An outbreak of rabies in West Malaysia in 1970 with unusual laboratory observations. *Medical Journal of Malaysia, 27*, 107–112.

Tao, X. Y., Niezgoda, M., Du, J. L., Li, H., Wang, X. G., & Liang, G. D. (2008). The primary application of direct rapid immunohistochemical test to rabies diagnosis in China. *Zhonghua Shi Yan He Lin Chuang Bing Du Xue Za Zhi, 22*, 168–170.

Tenzin, (2010)., Sharma, B., Dhand, N. K., Timsina, N., & Ward, M. P. (2010). Reemergence of rabies in Chhukha district, Bhutan, 2008. *Emerging Infectious Diseases, 16*, 1925–1930.

Tierkel, E. S. (1959). Rabies. In C. A. Brandly & E. L. Jungherr (Eds.), *Advances in veterinary science (pp. 183–226)*. New York, NY: Academic Press.

Totton, S. C. (2009). *Stray dog population health and demographics in Jodhpur, India, following a spay/neuter/rabies vaccination program*. (Unpublished doctoral dissertation), University of Guelph, Canada.

Totton, S. C., Wandeler, A. I., Zinsstag, J., Bauch, C. T., Ribble, C. S., Rosatte, R. C., et al. (2010). Stray dog population demographics in Jodhpur, India following a population control/ rabies vaccination program. *Preventive Veterinary Medicine, 97*, 51–57.

Townsend, S. E., Lembo, T., Cleaveland, S., Meslin, F-X., Miranda, M. E., Putra, A. A. G., et al. (2013). Surveillance guidelines for disease elimination: a case study of canine rabies. *Comparative Immunology, Microbiology and Infectious Diseases* doi:10.1016/j. cimid.2012.10.008.

Umeno, S., & Doi, Y. (1921). A study on the anti-rabic inoculation of dogs and the results of its practical application. *The Kitasato Archives of Experimental Medicine, 4*, 89–108.

Vasanth, J. P., Madhusudana, S. N., Abhilash, K. V., Suja, M. S., & Muhamuda, K. (2004). Development and evaluation of an enzyme immunoassay for rapid diagnosis of rabies in humans and animals. *Indian Journal of Pathology and Microbiology, 47*, 574–578.

Wandeler, A. I., Matter, H. C., Kappeler, A., & Budde, A. (1993). The ecology of dogs and canine rabies: A selective review. *Revue Scientifique et Technique de l' Office International des Epizooties, 12*, 51–71.

Wells, C. W. (1954). The control of rabies in Malaya through compulsory mass vaccination of dogs. *Bulletin of the World Health Organization, 10*, 731–742.

WHO, (1987). *Guidelines for dog rabies control*. Geneva: WHO.

WHO, (1988). *Report of a WHO consultation on dog ecology studies related to rabies control (WHO/Rabies Research/88.25)*. Geneva: WHO.

WHO, (1990)., & WSPA, *Guidelines for dog population management*. Geneva: WHO. (WHO/ ZOON/90.166).

WHO, (1992). WHO expert committee on rabies, 8th report. *WHO Technical Report Series, 824*, 1–84.

WHO, (1998). *Field application of oral rabies vaccines for dogs*. Geneva: WHO. (WHO/EMC/ ZDI/98.15).

WHO, (1999). *World survey of rabies No. 34 for the year 1998*. Geneva: WHO. (WHO/CDS/ CSR/APH/99.6).

WHO, (2004). WHO expert consultation on rabies, 1st report. *WHO Technical Report Series, 931*, 1–88.

WHO, (2007). *Guidance for research on oral rabies vaccines and field application of oral vaccination of dogs against rabies*. Geneva: WHO.

Widdowson, M. -A., Morales, G. J., Chaves, S., & McGrane, J. (2002). Epidemiology of urban canine rabies, Santa Cruz, Bolivia, 1972–1997. *Emerging Infectious Diseases, 8*, 458–461.

Windiyaningsih, C., Wilde, H., Meslin, F. -X., Suroso, T., & Widarso, H. S. (2004). The rabies epidemic on Flores Island, Indonesia (1998–2003). *Journal of the Medical Association of Thailand, 87*, 1389–1393.

Wright, E., Temperton, N. J., Marston, D. A., McElhinney, L. M., Fooks, A. R., & Weiss, R. A. (2008). Investigating antibody neutralization of lyssaviruses using lentiviral pseudotypes: A cross-species comparison. *Journal of General Virology, 89,* 2204–2213.

Wright, E., McNabb, S., Goddard, T., Horton, D. L., Lembo, T., Nel, L. H., et al. (2009). A robust lentiviral pseudotype neutralisation assay for in-field serosurveillance of rabies and lyssaviruses in Africa. *Vaccine, 27,* 7178–7186.

WSPA. (2007). Surveying roaming dog populations: Guidelines on methodology. Retrieved from <http://www.icam-coalition.org/downloads/Surveying%20roaming%20dog%20 populations%20-%20guidelines%20on%20methodology.pdf/>.

Wu, X. F., Franka, R., Svoboda, P., Pohl, J., & Rupprecht, C. E. (2009). Development of combined vaccines for rabies and immunocontraception. *Vaccine, 27,* 7202–7209.

Zinsstag, J., Dürr, S., Penny, M. A., Mindekem, R., Roth, F., Gonzalez, S., et al. (2009). Transmission dynamics and economics of rabies control in dogs and humans in an African city. *Proceedings of the National Academy of Sciences, USA, 106,* 14996–15001.

Rabies Control in Wild Carnivores

Richard C. Rosatte

Ontario Ministry of Natural Resources, Wildlife Research and
Development Section, Trent University, DNA Building,
Peterborough, Ontario K9J 7B8, Canada

1 INTRODUCTION

The control of rabies in wildlife is impacted by many variables, including the variant of rabies virus, the vector species that is being targeted, vector ecology, as well as the landscape features of the rabies infected area. In addition, public attitudes toward control tactics, the effectiveness of the wildlife rabies control strategy, and human practices such as wildlife relocation need to be considered when planning for rabies control (Rosatte, 2011; Rosatte & MacInnes, 1989). Historically, wildlife rabies control has evolved from strictly a culling operation to a more comprehensive approach involving the use of several tactics such as point infection control, vaccination by parenteral injection, and oral immunization with vaccine baits (Johnston & Tinline, 2002; Rosatte, 2011; Rosatte et al., 2001). By far, the most successful rabies control programs globally are those that have utilized a multi-agency approach (Brochier et al., 1996, 2001; Rosatte, 2011). In this chapter, the traditional and contemporary approaches to wildlife rabies control will be examined.

2 HISTORICAL AND CONTEMPORARY ASPECTS OF RABIES CONTROL IN WILDLIFE

2.1 North America

Trapping of foxes, coyotes (*Canis latrans*), and/or poisoning of skunks were methods used while attempting to control rabies in those species in

Rabies: Scientific Basis of the Disease and its Management
DOI: http://dx.doi.org/10.1016/B978-0-12-396547-9.00018-3

some areas of North America during the 1940s to the 1960s (Ballantyne & O'Donoghue, 1954; Rosatte, Pybus, & Gunson, 1986). Many of those operations were unsuccessful. Research was initiated in the late 1960s in North America to develop an oral rabies vaccine for red foxes (*Vulpes vulpes*) (Baer, Abelseth, & Debbie, 1971; Black & Lawson, 1970; Winkler, 1992; Winkler, McLean, & Cowart, 1975). In Ontario, Canada, oral rabies vaccination (ORV) using baits containing the attenuated Evelyn-Rokitnicki-Abelseth (ERA®) strain of rabies virus was later proven to be effective for the control of rabies in red foxes in rural and urban habitats (Lawson et al., 1997; MacInnes et al., 2001; Rosatte et al., 1993, 2007d; Rosatte, 2011; Rosatte, Power, & MacInnes, 1992). Following this, the vaccinia-rabies glycoprotein (V-RG®) recombinant vaccine in baits was shown to be successful in Texas for controlling rabies in coyotes and gray foxes (*Urocyon cinereoargenteus*) (Farry, Henke, Beasom, & Fearneyhough, 1998; Fearneyhough et al., 1998; Sidwa et al., 2005). In fact, the elimination of rabies in coyotes using vaccine baits resulted in the United States being declared free of canine rabies in 2007 (Slate et al., 2009). To prevent re-emergence of the disease, each year a 30 km to 65 km wide buffer zone of V-RG® vaccine baits is placed along the Mexico/Texas border. However, gray fox rabies re-emerged in Texas during 2008 and 2009. Deyoung et al. (2009) using genetics concluded that gray foxes in Texas have a potential for long distance dispersals >30 km. In view of this, they suggested that a vaccination zone 16 to 24 km wide would be too narrow to control rabies in gray foxes. Control is currently on-going using 32 km wide vaccine-baiting strips (Slate et al., 2009).

Although ERA® was effective for the control of rabies in red foxes in Ontario, it was not effective in striped skunks (*Mephitis mephitis*) and raccoons (*Procyon lotor*) (MacInnes et al., 2001; Nadin-Davis, Muldoon, & Wandeler, 2006). In addition, ERA® caused vaccine-induced rabies in red foxes, raccoons, striped skunks, rodents, and one bovine calf (Black & Lawson, 1980; Fehlner-Gardiner et al., 2008; Rosatte, 2011). During 2010, Ontario replaced ERA® baits with a newly developed recombinant vaccine called ONRAB® rabies vaccine for the control of rabies in red foxes in that province (Knowles et al., 2009a,b; Lutze-Wallace, Sapp, Sidhu, & Wandeler, 1995; Rosatte et al., 2009a, 2011a; Yarosh, Wandeler, Graham, Campbell, & Prevec, 1996).

Raccoons and skunks are currently the primary terrestrial reservoirs of rabies (raccoon variant) in eastern North America. The raccoon variant of rabies is currently enzootic over an approximate 1 million km^2 area in the eastern United States. Genetic tracking revealed that the raccoon variant of rabies virus spread from the United States into Canada and was reported in Ontario during 1999, in New Brunswick during 2000, and in Quebec in 2006 (Szanto, Nadin-Davis, Rosatte, & White, 2011a; Wandeler & Salsberg, 1999). Ontario used a combination of population reduction,

Trap-Vaccinate-Release (TVR) and V-RG® baits to control and eliminate raccoon rabies with the last case being reported in 2005 (Rosatte et al., 2001, 2008, 2009b). In fact, during 1994 to 2007, a total of 96,621 raccoons and 7,967 skunks were captured and vaccinated against rabies using TVR in Ontario (Rosatte et al., 2009b; Sobey et al., 2010). Population reduction and TVR were used in New Brunswick and Quebec during the 2000s to control raccoon rabies (Rosatte, 2011). TVR was also employed to control raccoon rabies in Ohio and Massachusetts, as well as an outbreak of bat-variant rabies in skunks in Flagstaff, Arizona, during the 2000s (Algeo et al., 2008; Engeman, Christensen, Pipas, & Bergman, 2003; Leslie et al., 2006; Rosatte, 2011; Slate et al., 2009). In Ontario, proactive TVR in the Niagara region was replaced by the hand and aerial distribution of ONRAB® vaccine baits beginning in 2009 (Rosatte, 2011).

Baits containing V-RG® vaccine have been used in several eastern states from Maine to Florida and Canada in an attempt to control raccoon rabies (Boulanger et al., 1996; Fehlner-Gardiner et al., 2012; Olson, Mitchell, & Werner, 2000; Rosatte et al., 2008; Roscoe et al., 1998; Rupprecht, Hamir, Johnston, & Koprowski, 1988; Russell, Smith, Childs, & Real, 2005; Weyer, Rupprecht, & Nel, 2009). An oral rabies vaccination program was established in Ohio to prevent the westward spread of raccoon rabies (Ramey, Blackwell, Gates, & Siemons, 2008). Unfortunately, the performance of V-RG® in the field was below expectations, with vaccine efficacy in raccoons in Ontario being a disappointing 7 to 28% (Rosatte et al., 2008). Similarly, efficacy was 30% in Maine (Fehlner-Gardiner et al., 2012). Only 16% of raccoons were seropositive following V-RG® baiting in Erie County, New York, during 2002–2005 (Boulanger et al., 2008) and in Ohio only 12% of raccoons sero-converted following distribution of V-RG® (Ramey et al., 2008). In addition, raccoons fed V-RG® baits in captivity in Ontario did not respond well to challenge with rabies virus (Brown et al., 2011). Although raccoon rabies was controlled and eliminated in Ontario, ONRAB® vaccine baits are currently being used proactively to prevent raccoon rabies from again entering the province (Rosatte, 2011; Rosatte et al., 2009a, 2011a).

ONRAB® is an oral vaccine containing human adenovirus type 5 recombinant virus expressing the rabies virus glycoprotein that was developed in Ontario, Canada. Vaccine efficacy of ONRAB® during field trials with raccoons in Ontario ranged from 79 to 90% at bait densities of 75–400/km² (Rosatte et al., 2009a). A comparative trial using ONRAB® and V-RG® in Maine and New Brunswick yielded a seropositive level of 74% in raccoons in ONRAB® baited areas (Fehlner-Gardiner et al., 2012). ONRAB® was also used in Quebec to control raccoon rabies during the late 2000s (Boyer, Canac-Marquis, Guerin, Mainguy, & Pelletier, 2011; Rosatte, 2011). The last case of raccoon rabies in Quebec (as of October 2011) was in 2009. In view of this, efforts are being focussed on preventing a re-invasion of raccoon rabies from Vermont into Quebec (Rees, Belanger, Lelievre, Cote, & Lambert, 2011a). An ONRAB® field trial occurred in south-eastern West Virginia during September 2011 as the disease still persists in raccoons and skunks in the eastern United States (Rosatte, 2011; Sattler et al., 2009; Slate et al., 2009).

Rabies in skunks is currently enzootic in the prairie provinces of Canada, the north-central United States, and California. It is thought that rabies in skunks in those areas may have originated historically from spill-over from long-term epizootics in dogs (Velasco-villa et al., 2008). Skunks have also played a major role in the maintenance of Arctic variant rabies virus in southwestern Ontario during the 2000s, especially since rabies in red foxes was controlled (MacInnes et al., 2001; Nadin-Davis et al., 2006; Rosatte, 2011). Skunks have also been reported with the raccoon variant of rabies in eastern North America (Rosatte et al., 2006). Unfortunately, V-RG® has not been effective in skunks using similar doses as used for raccoons (Briggs, Briggs, & Howard, 1988; Hanlon, Niezgoda, Morrill, &, Rupprecht, 2002; Rosatte et al., 2008). However, population reduction in combination with TVR or vaccination by injection with conventional animal vaccine(s) has been used successfully to control rabies in skunks (and raccoons) in urban and rural areas of Ontario, Canada (Rosatte, Howard, Campbell, & MacInnes, 1990; Rosatte, Power, & MacInnes, 1991; Rosatte, Power, & MacInnes, 1992; Rosatte et al., 1993, 2001, 2007b, 2009b).

Street Alabama Dufferin (SAD) B19, a modified live rabies virus vaccine has been tested orally in skunks (Vos, Pommerening, Neubert, Kachel, & Neubert, 2002). In addition, the attenuated SAG-2 rabies virus vaccine (a variant of the SAD vaccine strain) also proved to be effective when given orally to skunks as well as raccoons and arctic foxes (*Alopex lagopus*). That vaccine shows promise as a candidate vaccine for wildlife rabies control in the United States (Hanlon et al., 2002; Follmann, Ritter, & Hartbauer, 2004). A canine adenovirus recombinant rabies vaccine was effective in laboratory experiments and produced protective immunity in skunks as well as raccoons (Henderson et al., 2009). Currently (2011), ONRAB® is being used to control rabies in striped skunks in Ontario (Rosatte, 2011; Rosatte et al., 2009a, 2011a). ONRAB® has been shown to be stable and safe in a number of target and non-target species with low recovery rates of the virus from tissues, feces, and oral cavities (Knowles et al., 2009a,b).

Currently, rabies in Canada exists at low levels primarily due to the success of wildlife rabies control programs in Ontario, Quebec, New Brunswick, and Newfoundland. In fact, during 2010 there were only 123 cases reported in Canada with 49% of those in skunks (primarily in Manitoba and Saskatchewan), 39% in bats, and 5% in foxes. There were no rabid raccoons reported in Canada during 2010 (http://www.inspection.

gc.ca/). However, rabies is still rampant in the United States, with 6,154 rabid animals being reported during 2010 (Blanton, Palmer, Dyer, & Rupprecht, 2011). Thirty-seven percent (2,246) of the U.S. cases were in raccoons, 24% (1,448) in skunks, 23% (1,430) in bats, and 7% (429) in foxes. Currently, the raccoon variant of rabies is responsible for the majority of rabies cases in the United States (Blanton et al., 2011).

2.2 Europe

In the past, thousands of foxes were culled in an effort to control rabies throughout much of Europe (Matouch & Polak, 1982). Depopulation of foxes generally only resulted in a transient lull in the prevalence of rabies (Aubert, 1999). Due to this, parenteral vaccination programs targeting foxes were initiated in Switzerland and Germany (Aubert, Masson, Artois, & Barral, 1994). However, this method was viewed as impractical because too few foxes were captured in a limited area (Aubert et al., 1994). Significant advances were made in 1977, when the first field trial using baits containing oral rabies vaccine was initiated in Switzerland. In 1978, Steck, Wandeler, Bichsel, Capt, & Schneider (1982a) attempted to halt an advancing fox rabies epizootic in the Rhone Valley area of Switzerland using chicken-head baits containing SAD-Berne rabies vaccine. By 1985, the disease was under control in Switzerland (Aubert et al., 1994; Steck et al., 1982a,b).

Field trials with baits containing oral rabies vaccine for the control of rabies in foxes were initiated in West Germany in 1983 (Schneider & Cox, 1988). By 1987, complete elimination of fox rabies was achieved over large areas of West Germany (Schneider, Cox, Muller, & Hohnsbeen, 1988). In the mid to late 1980s and early 1990s, fox rabies control programs were initiated using oral baits containing SAD-B19 vaccine in Austria, Belgium, Czechoslovakia, East Germany, France, Hungary, Italy, Luxembourg, The Netherlands, and Slovenia (Brochier et al., 2001; Curk & Carpenter, 1994; Gerletti, Guidali, Scherini, & Tosi, 1991; Irsara, Bressan, & Mutinelli, 1990; Pastoret, Kappeler, & Aubert, 2004). Baits containing SAD-B19 vaccine were also utilized beginning in 1988 in Finland to control rabies in raccoon dogs (*Nyctereutes procyonoides*) (Nyberg et al., 1992; Pastoret et al., 2004; Westerling, 1991).

The SAD-B19 and SAD-Berne vaccines were found to be pathogenic for a variety of rodents and other species. In addition, 6 vaccine induced rabies cases in foxes were documented in Germany during 2001 to 2006 (Muller et al., 2009). Therefore, research was initiated to develop more attenuated and safer vaccines. The V-RG® recombinant vaccine was shown to be effective in foxes, raccoon dogs, and dogs (Cliquet et al., 2008; Kieny et al., 1984). That vaccine was successfully used to control rabies in foxes in Belgium and France beginning in 1989 (Brochier et al., 2001;

Pastoret & Brochier, 1999). In 1991, the SAD-Berne vaccine was replaced by a SAG-1 vaccine for use in controlling rabies in foxes in Switzerland (Aubert et al., 1994). SAG-1 was also used in France beginning in 1990 (Artois, Cliquet, Barrat, & Schumacher, 1997; Aubert et al., 1994; Coulon et al., 1992). A SAD mutant (SAG$_2$) has also been developed for rabies control in Europe (Lambot et al., 2001). During the 2000s, Rabidog® baits containing SAG$_2$ vaccine were used to control rabies in raccoon dogs and red foxes in Estonia (Niin, Laine, Guiot, Demerson, & Cliquet, 2008).

Oral rabies vaccination of foxes and raccoon dogs has reduced rabies cases throughout western Europe (Artois, 2003; Brochier et al., 2001; Matouch & Vitasek, 2005; Pastoret & Brochier, 1999; Thulke, Eisinger, Selhorst, & Muller, 2008; Zanoni, Kappeler, Muller, Wandeler, & Breitenmoser, 2000). In Germany and Austria alone, more than 97 million vaccine baits (SAD B19 and SAD P5/88) have been distributed since the 1980s (Muller et al., 2009). However, re-infections of fox rabies have occurred from time to time in France, Belgium, Italy, and Germany due to a lack of geographic isolation, cross-country contamination, and budgetary restrictions (Aubert et al., 1994; Brochier et al., 2001; Chautan, Pontier, & Artois, 2000; Hostnik, Rihtaric, Grom, Malovrh, & Toplak, 2011). In addition, since rabies programs were initiated there have been increases in fox population density (which is conducive to rabies infection and spread) in all European countries. In some areas, such as Poland, fox numbers have nearly doubled with densities increasing from 1.3 to 2 foxes/km^2 (Goszczynski, Misiorowska, & Juszko, 2008). Oral rabies vaccination campaigns were reinitiated in many western European countries, including Poland during 2005, to bring the disease under control (Muller et al., 2009; Smreczak & Zmudzinski, 2009). In response to a re-emergence of rabies in northeastern Italy, SAD B19 vaccine baits were distributed by helicopter at a density of 25–30 baits/km^2 over a 9000 km^2 area during the winters of 2008–2010. About 77% of the foxes sampled were immunized, and rabies is declining (Capello et al., 2010). Currently, much of Western Europe is rabies free (including Finland, The Netherlands, Switzerland, France, Belgium, Luxembourg, Czech Republic), but the ultimate goal is to eliminate rabies from the entire European Union (Smith et al., 2008; Hostnik et al., 2011). However, it is critical that strategies such as ring vaccination (baits are distributed around a rabies case location) are ready to be re-implemented should an outbreak occur. In view of this, a model is being used to assess the feasibility of different emergency tactics based on risk and economics (Thulke et al., 2008).

Rabies is currently on the increase (>6500 cases in 2005) in Central and Eastern European countries such as Lithuania, Latvia, Estonia, Poland, Slovenia, Romania, Bulgaria, Turkey, and Belarus, primarily due to an increase in fox populations (Johnson et al., 2008; Matouch, 2008;

Zienius, Bagdonas, & Dranseika, 2003; Hostnik et al., 2011). About 69% of the rabies cases in eastern and central Europe are in wildlife, primarily in foxes (Matouch, 2008). However, an invasive species from East Asia, the raccoon dog, is becoming increasingly important as a rabies vector (Johnson et al., 2008; Singer et al., 2009). Raccoon dogs now occupy an approximate 1.4 million km^2 area of Europe (Sutor, 2008). Fortunately, the Balkan Mountains play a role as a partial barrier in limiting the southern spread of rabies in south-eastern Europe (Johnson et al., 2007). However, there has been an increase in rabies in raccoon dogs in the Baltic States, especially Lithuania (Holmala & Kauhala, 2009). As a result, rabies control strategies have had to be re-designed to achieve more effective results (Artois, 2003; Aubert, 1999). This included the use of modeling to determine if rabies could persist in raccoon dogs with a resultant recommendation that rabies vaccination programs begin immediately after raccoon dogs emerge from hibernation (Singer et al., 2009).

The first oral rabies vaccination program in Estonia targeting raccoon dogs and foxes occurred in 2005. As a result, rabies cases decreased dramatically in Estonia by 2006 (Laine, Niin, & Partel, 2008). During 2002 to 2008, a 30 to 50 km wide vaccination belt in Slovenia was implemented along the southern border with Croatia. In 2008, most of Slovenia received vaccine baits due to an outbreak of rabies in Italy (Hostnik et al., 2011). Unfortunately, elimination of rabies from the poorer counties of Eastern Europe will be a challenge, primarily due to economics. Therefore, the current recommendation is to maintain a buffer zone of vaccinated animals to protect the rabies free area of Western Europe (Smith et al., 2008).

2.3 Asia

Rabies in canine species is endemic in most of south and south-east Asia, and it is speculated that wildlife rabies only plays a very minor role in the maintenance of the disease (Wilde et al., 2005). As a result, recent research has focussed on methods to control the disease in dogs in areas such as India (Totton, Wandeler, Ribble, Rosatte, & McEwen, 2011). The primary vectors of rabies in areas of northern Asia such as the former Soviet Union are arctic foxes, red foxes, and raccoon dogs (Kuzmin et al., 2004). Candidate vaccines under consideration for arctic regions include a SAG$_2$ lyophilized product (Follmann et al., 2004; Kovalev, Sedov, Shashenko, Osidse, & Ivanovsky, 1992). Metlin, Rybakov, & Mikhalishin (2009) recommend that federal and regional oral rabies vaccination programs be implemented in the Russian Federation countries. However, intensive control programs are generally lacking in northern areas of Asia.

Although dogs account for the majority of rabies cases in China, the ferret badger (*Melogale moschata*) is becoming increasingly important as a wildlife rabies vector, especially since the mid 1990s (Zhang et al., 2009).

In fact, there was an epizootic of rabies in ferret badgers in southeast China during 2007/08. Ferret badgers are also becoming a significant public health threat and have been implicated in human rabies cases (Zhang et al., 2009). Unfortunately, to date, no practical rabies vaccine has been developed for use in wildlife in China.

Since the 1970's, sylvatic rabies, primarily in red foxes and jackals (*Canis sp*), accounted for nearly 50% of the reported rabies cases in Israel with foxes being the primary reservoir (Yakobson, King, Sheichat, Eventov, & David, 2008). Linhart et al. (1997) evaluated the use of vaccine-laden baits for controlling rabies in red foxes and golden jackals (*Canis aureus*) in Israel. The results indicated that ORV with baits for the control of rabies in the Middle East was a feasible option. Yakobson, Manalo, Bader, Perl, & Haber (1998) recommended the implementation of an oral rabies vaccination program for wildlife, which included extension of the program to include Israel, Egypt, Jordan, and Palestine (Yakobson et al., 1998). ORV with vaccine baits was initiated in Israel in 1998 and the disease was nearly eliminated by 2005 (Jacobson et al., 2008). However, an outbreak of rabies occurred in stray dogs during 2005/06. Attempts were made to control the disease in dogs using wildlife ORV techniques, but the program was not effective (Jacobson et al., 2008).

Turkey, which contains territory in Europe (12%) as well as in Asia (88%), experienced an increase in fox rabies cases during the early to mid 2000s, and the disease is currently endemic (Johnson et al., 2003, 2010). In view of this, an oral immunization proposal for the control of fox rabies in Turkey was designed and submitted for approval. ORV was approved and initiated in Turkey in an attempt to control the disease. It is believed that foxes were contracting rabies from rabid dogs (Johnson et al., 2010). Rabies is also endemic in Iran, with the primary reservoir being wolves (Janani et al., 2008). In many cases, wolves have transmitted rabies to humans. An oral rabies vaccination program has been proposed for Iran (Janani et al., 2008).

2.4 Africa

Rabies still kills thousands of people annually in Africa. Dogs are the primary hosts that maintain the disease throughout much of that continent, however, wildlife rabies is becoming increasingly important due to the threat of endangered species extinction and the emergence of new wildlife hosts (Cleaveland, 1998; Lembo et al., 2010). Sylvatic rabies has been reported in a number of African countries in African wild cats (*Felis libyca*) in Botswana, in jackals (*Canis adustus*) and hyena (*Hyaena brunnea*) in Zimbabwe, in yellow mongoose (*Cynictus penicillata*) and black-backed jackals (*Canis mesomelas*) in South Africa, to name a few (Bingham, Foggin, Wandeler, & Hill, 1999a; Bingham, Schumacher, Aubert, Hill, & Aubert, 1997; Bingham, Schumacher, Hill, & Aubert, 1999b; Brown, 2011; Keightley, Struthers, Johnson, & Barnard, 1987; Zulu, Sabeta, & Nel, 2009). In fact, although a primary reservoir of rabies in South Africa is domestic dogs, black-backed jackals and mongooses are also becoming emerging host species for rabies in that country (Zulu et al., 2009; Brown, 2011). It is speculated that jackals could sustain rabies independent of domestic dogs. Therefore, effective rabies control strategies in South Africa will need to include wildlife vaccination tactics targeting jackals and mongoose, as well as domestic dog vaccination campaigns (Brown, 2011; Zulu et al., 2009).

Rabies is a threat to the wild dog population in parts of Namibia, and thousands of kudu have also succumbed to the disease (Laurenson, Van Heerden, Stander, & Van Vuuren, 1997). Unfortunately, wildlife rabies control (as well as dog rabies control) is not a priority in Africa due to a lack of human and monetary resources (Cleaveland, 1998). Some research has been initiated with respect to oral rabies vaccination of species such as jackals in Zimbabwe (Bingham et al., 1997). In addition, endangered African wild dogs (*Lycaon pictus*) were administered inactivated rabies vaccine by dart or via parenteral vaccination during the early 1990s in the Serengeti region of Tanzania following an outbreak of rabies that threatened to eliminate the population (Gascoyne et al., 1993). An oral vaccination system for free-ranging wild dogs was developed during the early 2000s (Knobel, Toit, & Bingham, 2002). In Ethiopia, an outbreak of rabies occurred in the highly endangered Ethiopian wolf (*Canis simensis*) during 2003. Control was attempted by live-capture and parental vaccination and the outbreak was controlled (Knobel et al., 2008). The use of ORV is being considered for use in Ethiopian wolves (Office of International Epizootics, 2003; Knobel et al., 2008).

Despite the diversity of wildlife species in Africa, some of which are capable of being rabies vectors, there is no coordinated effort to organize wildlife rabies control programs, either locally or nationally. The feeling in some areas such as the Serengeti is that the disease in species such as jackals and mongoose can be controlled by interrupting the rabies cycle from domestic animals to wildlife by targeting dogs during rabies control operations (Lembo et al., 2008).

2.5 Central and South America Including the West Indies

The domestic dog accounted for more than 80% of the reported rabies cases in Latin America during the 1990s (De Mattos et al., 1999). Rabies in vampire bats (*Desmodus rotundus*) is also prevalent in Central and South America and has been implicated in human deaths as well as transmission of the disease to wildlife species (Almeida, Martorelli, Aires, Barros, & Massad, 2008; Delpietro, Lord, Russo, & Gury-Dhomen, 2009). Rabies in mongooses (*Herpestes sp*) is a public health concern on some Caribbean islands such as Cuba, Grenada, and Puerto Rico (Diaz, Papo, Rodriguez, & Smith, 1994; Everard & Everard, 2006). A field trial was initiated to evaluate baits for delivery of oral rabies vaccine to the Indian mongoose (*Herpestes auropunctatus*) on the island of Antigua, West Indies (Linhart et al., 1993). In that study, baiting stations showed promise for vaccinating mongooses (Linhart et al., 1993). More recently, a genetically engineered oral rabies vaccine (SPBNGA-S) showed promise for immunizing Asian mongooses (*Herpestes javanicus*) (Blanton et al., 2006).

Sylvatic rabies is maintained in certain geographic areas of Mexico, and wildlife rabies control measures in Latin American countries such as Brazil and Mexico were recommended, including identification of the primary wildlife reservoirs of rabies and the implementation of oral rabies vaccination programs for wildlife (De Mattos et al., 1999; Diaz et al., 1994). However, the extent of rabies infection in wildlife species is not well documented (Almeida et al., 2001; Sato et al., 2004), and wildlife rabies control programs are still nonexistent (other than vampire bat control).

3 THE CONCEPT OF CONTROLLING RABIES IN WILDLIFE

The culling of wildlife was generally not a successful rabies control tactic in many areas of the world. In other situations, it was not an acceptable practice (Debbie, 1991). In view of that, researchers focused on developing alternate methods for rabies control. During the 1980s, Rosatte et al. (1990) demonstrated that it was possible to immunize free-ranging skunks and raccoons against rabies in Ontario by injection of inactivated rabies vaccine in live-trapped animals. Baer (1988), one of the pioneers of modern day rabies research, tried to orally vaccinate foxes with the existing injectable rabies vaccines Flury LEP and HEP. His research proved that the concept of oral rabies vaccination was possible. Many prototype baits (without vaccine) were first field tested by ground and aerial distribution in small-scale field trials at the same time that oral rabies vaccines were being developed (Baer et al., 1971; Black & Lawson, 1970; Linhart et al., 1997). Instead of using an actual vaccine, chemical biomarkers were put into the baits to test the feasibility of reaching the percentage of animals necessary to provide herd immunity for a free-ranging population of rabies vectors (Linhart et al., 1997). This early work showed that the concept of mass baiting of a wild species was feasible, which led to the first release in the wild of an oral vaccine targeting red foxes in Switzerland in 1977 (Steck et al., 1982a,b). It also showed the need for the integration of a wide range of technical resources to implement large-scale ORV programs.

4 INITIATION OF WILDLIFE RABIES CONTROL PROGRAMS

Historically, the initiation of wildlife rabies control programs was stimulated by epizootics, but sometimes they were implemented due to unexpected events. In Ontario, it is interesting that the site of the first well documented human death from wildlife rabies in 1819 (Jackson, 1994) precipitated the Canadian wildlife rabies control program. In 1967, at the same location, a young girl died of rabies contracted from the bite of a stray cat. Her death lead to public demand for the Province of Ontario to initiate a major control program against fox rabies, which had been epizootic in the Province since 1954 (Johnston & Beauregard, 1969). That program continues to have strong provincial support and has been successful in nearly eliminating arctic strain rabies from the province (MacInnes et al., 2001; Rosatte, Tinline, & Johnston, 2007f; Rosatte, 2011; Rosatte et al., 2011a).

The implementation of wildlife rabies control programs requires the careful integration of a complex series of events and disciplines. To be effective, industry, government, and science must function together as a coherent system. Modification or omission of any component of the control system may alter the efficiency and outcome of the control program. Timing is important with regard to both the initiation and continuation of a control program to achieve and maintain a rabies-free status in an area. In the past, rabies epizootics have flourished in the absence of timely control efforts or appropriate technology. One can only surmise that the northward spread of raccoon rabies in the ridge and valley topography of Pennsylvania in the late 1980s and subsequently into New York and New England could have been stopped with vaccine barriers in the valleys had newly developed vaccines and the oral baiting technology, tested and proven in Ontario in the early 1990s, been utilized earlier (Johnston et al., 1988; Rupprecht et al., 1986). Unfortunately, even when appropriate control technologies are at hand, lack of political will or financial resources have often wasted precious time, allowing epizootics to gain hold and advance into naïve animal populations. Indeed, even with support, budget restraints force trade-offs between investment in prevention weighed against the risk of invasion and subsequent higher costs. Programs that are interrupted due to budgetary or other reasons can exacerbate existing ecological and epidemiological conditions and prolong epizootics. Conversely, timely, efficient programs have proved effective in halting advancing epizootic fronts in areas such as the United

States and Canada (Rosatte, 2011; Rosatte et al., 2009a,b; Slate et al., 2009). The key to being prepared for rabies epizootic control is a well designed contingency plan that denotes control tactics for a given epizootic scenario, and having rapid access to the resources to implement the plan quickly (Rosatte et al., 1997; Whitney, Johnston, Wandeler, Nadin-Davis, & Muldoon, 2005).

5 DIAGNOSIS OF SUSPECT WILDLIFE

The diagnosis of rabies is the basis for the control of the disease in wildlife. Paramount to the initiation of any wildlife rabies control program is a delineation of the problem in wildlife. The diagnosis of suspect cases and identification of the strains of rabies virus that are isolated, the species involved, and where in the local ecology and epizootiology these species fit in relation to humans, pets, and domestic animals are important to determine. A comprehensive and continuing diagnostic system that focuses on secondary as well as primary vector species surveillance is imperative if wildlife rabies is to be controlled and eventually eradicated. Furthermore, it is critical that a sufficient sample of the vector population be obtained to portray a realistic picture of the epidemiological situation. New diagnostic tests such as the direct rapid immunohistochemical test (dRIT) make it possible to test surveillance specimens in the field and it is currently being used as a quick, reliable test in the United States and Canada (Lembo et al., 2006; Rosatte, 2011). In addition, sensitive real-time PCR assays have been developed that are capable of detecting even minute amounts of rabies virus in rabies-suspect specimens (Szanto, Nadin-Davis, Rosatte, & White, 2011b).

6 VECTOR SPECIES BIOLOGY IN RELATION TO RABIES EPIDEMIOLOGY

To manage wildlife rabies scientifically, a good understanding of the behavioral ecology of wild mammals that are the vectors in rabies epizootics is critical (Childs et al., 2000; Guerra et al., 2003; Rosatte, Wandeler, Muldoon, & Campbell, 2007g). Ecological studies of the primary and secondary vectors of rabies are a necessity to understand how the vector ecology impacts the epizootiology of the disease (Artois & Aubert, 1991; Broadfoot, Rosatte, & O'Leary, 2001; Rosatte & Allan, 2009; Rosatte et al., 2005; Rosatte, Kelly, & Power, 2011b; Rosatte & Lariviere, 2003; Totton, Rosatte, Tinline, & Bigler, 2004; Totton, Tinline, Rosatte, & Bigler, 2002; Voigt, Tinline, & Broekhoven, 1985). Patterns of rabies vector dispersal are critical to determining patterns of disease spread (Cullingham, Pond, Kyle, Rosatte & White, 2008).

The genetic composition and diversity of the vector species will also dictate the degree of immune response the animal makes when challenged with disease in the wild, thus determining its resistance or susceptibility to rabies virus infection (Srithayakumar, Castillo, Rosatte, & Kyle, 2011). Also critical is an understanding of the behavior of rabies vector species when they become rabid. For example, in Texas, rabid skunks were active during the day time and frequently attacked pets including entering dog pens (Oertti, Wilson, Sidwa, & Rohde, 2009). Rabid skunk prevalence peaked in that state during March/April (Oertti et al., 2009). It was also determined that rabid raccoon movements in Ontario were not different than non-rabid raccoons (Rosatte et al., 2006). To control rabies in wildlife, it is important to understand the relationship among the potential wildlife hosts, including vector metapopulation structure, the genetic composition of the vector species, and human demographic and environmental features, particularly if oral vaccination is proposed as a control method (Jones, Curns, Krebs, & Childs, 2003; Srithayakumar et al., 2011).

7 TRANSPORTATION OF WILDLIFE

The translocation of native and exotic wildlife species at the global level is increasing annually for purposes ranging from restoration of extirpated species to endangered species restoration (Woodford & Rossiter, 1993). The United States is the largest importer of wildlife that represents a source of pathogens such as rabies. During 2000 to 2005, a staggering 246,772 mammals representing 190 genera were imported into the United States (Pavlin, Schluegel, & Daszak, 2009). Rabies was a risk in 78 of those genera. Obviously, changes to importation laws to minimize the risk of disease importation are urgently needed.

Species relocations often occur without proper disease risk assessments being completed, and it is becoming clear that species translocations by humans are the origin of disease-infected wildlife (Rosatte, Donovan, Allan, Bruce, & Davies, 2007a; Rosatte & MacInnes, 1989). For example, raccoons are known to have arrived on the island of Newfoundland, where they are not native, via transport trucks ferried from Nova Scotia (H. Whitney personal communication). In New York, raccoons are a common occurrence at dumpster and trash transfer sites and have often been observed tumbling from newly arriving trash trucks (B. Laniewicz, personal communication). Incredibly, mean raccoon and opossum (*Didelphis virginiana*) movement via hitching rides on transport trucks and trains from the United States and Quebec into Ontario averaged 479 km and 688 km, respectively, during 1998 to 2005 (Rosatte et al., 2007a). In addition, relocation of wildlife can result in extraordinary movements leading to the spread of infectious diseases over large areas (Rosatte & MacInnes, 1989; Roscoe et al., 1998). Therefore, jurisdictions are encouraged to implement legislation regulating the importation, distribution, and relocation of wildlife. It is also imperative that a proper risk assessment be undertaken before wild animal translocations occur.

8 POINT INFECTION CONTROL

The primary objective of population reduction is to reduce the density of the animal vector population below that which is required for successful transmission of rabies to a susceptible individual. The aim is also to remove incubating and clinical animals from the population, as vaccination will not work on those animals (Rosatte et al., 2001, 2007e). Unfortunately, birth rates in primary vectors, such as foxes and raccoons, are high, resulting in rapid population growth and recovery following control (Bogel, Arata, Moegle, & Knorpp, 1974; Rosatte, 2000; Rosatte et al., 2007e). This means that vector removal will have to be applied annually until the disease is eliminated. In Denmark, population reduction (gassing and shooting) was used to control fox rabies during the 1960s, and euthanasia of foxes was used extensively in Czechoslovakia for the control of rabies in the 1970s and 1980s (Matouch & Polak, 1982). An Arctic fox population was also reduced in an area of Alaska during 1994 in response to a rabid arctic fox attacking two humans (Ballard, Follmann, Ritter, Robards, & Cronin, 2001).

Population reduction may have its place as a component of control technology when applied to a newly established point of infection that occurs in a naive population. However, vaccination strategies (both parenteral and oral) in addition to population reduction have proven to be a more effective control strategy and has been termed Point Infection Control (PIC) (Rosatte et al., 2001, 2007e, 2009a) (see Figure 18.2 in Rosatte et al., 2007e). In this situation, incubating and clinically rabid animals are removed from the immediate rabies outbreak area by population reduction. Vaccination of vector species beyond this zone provides a buffer to contain any dispersing infected animals that may have been missed during culling operations.

PIC has been successful in containing and eliminating a new focus of raccoon-strain rabies that appeared in eastern Ontario in 1999 after moving across the St. Lawrence River from adjacent New York State (Rosatte, MacInnes, Williams, & Williams, 1997; Rosatte et al., 2001, 2009b). During 1999 to 2005, 8,311 raccoons and 1,449 skunks using 576,359 trap-nights were captured and humanely euthanized. In addition, 31,844 raccoons and 4,435 skunks were vaccinated using TVR in eastern Ontario (Rosatte et al., 2009b). Raccoon rabies has not been reported in Ontario since September 2005 (to October 2011) (Rosatte et al., 2009b).

A 700 km^2 TVR operation was in place in the Niagara Falls, Ontario, area between 1994 and 2008 to prevent raccoon rabies from spreading to Ontario from New York State (Rosatte et al., 1997, 2009b). A total of 64,778 raccoons and 3,532 skunks were vaccinated using TVR during that 14 year operation (Figure 18.1). That program was so successful that the area has remained free of reported cases of raccoon rabies (to 2011) despite the disease being enzootic in nearby New York State. However, TVR in Niagara was replaced with the aerial and ground distribution of ONRAB$^®$ baits in 2009.

Although PIC is effective in some situations, such as on a new focus of wildlife rabies, generally, the population reduction and TVR portions of the strategy are only feasible for areas smaller than about 2000 km^2 as the logistics of these types of operations becomes difficult for larger areas. In an enzootic situation, where several thousands of square kilometres of area are infected, the most feasible rabies control tactic is the aerial distribution of vaccine baits. However, PIC may still be used to contain "hot spots" within large enzootic areas, especially in situations where

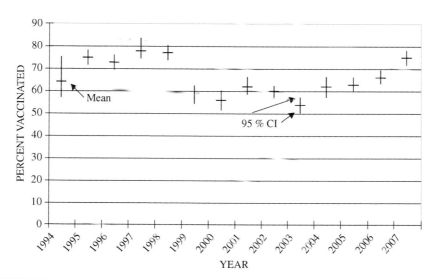

FIGURE 18.1   **Percentage of the raccoon population that was vaccinated against rabies during a Trap-Vaccinate-Release (TVR) program in Niagara Falls, Ontario during 1994–2007.** The TVR area was about 700 km^2. No rabies cases were detected in the area during 1994–2011 despite rabies being present on the New York side of the Niagara River.

secondary vectors do not respond to the vaccine used to immunize the primary vector (Robbins et al., 1998; Rosatte et al., 2001; 2009b).

9 PERCENTAGE OF A VECTOR POPULATION THAT HAS TO BE CULLED OR IMMUNIZED IN ORDER TO CONTROL RABIES

Regardless of whether the tactic employed is culling, vaccination by injection, or oral vaccination with baits, a certain percentage of the vector population must receive treatment so that rabies is either controlled or eliminated. In one Ontario culling program, 8,311 raccoons and 1,449 skunks were euthanized during 1999 to 2005 (Rosatte et al., 2009b). In some of the target areas, about 90% of the target populations were removed and rabies control was successful. During the TVR portion of those programs about 43 to 83% of the raccoon populations were vaccinated against rabies by injection, again with overall successful results and raccoon rabies was eliminated (Rosatte et al., 2009b).

Mathematically derived values for rabies immunity (e.g. 70%) that result in rabies control or eradication are most likely conservative. Thulke & Eisinger (2008) suggest that low immunization levels in foxes (e.g. 50%) provide a reasonable chance of controlling rabies in foxes in Europe. However, the question still remains "at what immunization level is rabies controlled in a wildlife population?" Part of the answer was revealed in an experiment conducted on Wolfe Island, Ontario, an area that was free of raccoon rabies up until 1999. During 1996 to 1998, 52 to 55% of the raccoon population on the island was vaccinated against rabies by injection (TVR). However, during 1999, the percent of the population that was immunized declined to 39%, and an outbreak of 6 rabies cases in raccoons occurred (Rosatte et al., 2007c). The disease disappeared by 2001 when 52% of the raccoon population was immunized. This suggests that the critical level of immunity to control raccoon rabies is above 39%.

In another Ontario study, rabies was controlled in skunks when 36 to 62% of the sera samples were cELISA positive at an inhibition value of at least 16% (Rosatte et al., 2011a). This suggests that in some cases, a population immunity level of lower than 50% may result in the successful control of rabies in skunks. In another study in Texas, V-RG® baits were used to control rabies in gray foxes and coyotes with resultant rabies antibody levels of 61 and 63%, respectively, in those two species (Slate et al., 2009). Based on the above data, if at least 50 to 60% of a wildlife population (i.e., foxes, raccoons, striped skunks) is immunized, there should be a high probability of rabies being controlled and possibly eliminated.

10 BAIT DEVELOPMENT FOR DELIVERY OF ORAL RABIES VACCINE

Most materials thought to be attractive to a target species have been tested as potential baits to deliver oral rabies vaccine to rabies vectors (Linhart et al., 1997). In addition to the safety of the rabies vaccine itself, safety of the entire bait material matrix is critical. The release of a prion-bearing component could have severe repercussions in a large-scale field trial. For this reason, Ontario changed from an animal fat-based bait to a vegetable-based bait during the late 2000s (Rosatte, 2011).

The shape and consistency of the bait can cause deleterious consequences. Heavy, hard baits could be lethal when dropped from the air. Sharp-cornered vaccine packages may become lodged in the throat of ingesting animals if swallowed, and there is the possibility of human contacts with rabies vaccine baits that must be considered (Rupprecht et al., 2001). The mass-production capability and cost of baits are also critical determinants when developing baits for the delivery of rabies vaccine to wildlife (MacInnes, 1988).

Four main types of rabies vaccine baits are currently in use (or were in the past) in large-scale ORV programs in North America. The first type is the Raboral V-RG® bait manufactured by Merial, Inc., Athens, Georgia. It contains the V-RG® vaccine produced by Merial, which is inserted into a fishmeal polymer cube (Rupprecht et al., 1986; Rupprecht, Hanlon, & Slate, 2004; Slate et al., 2005, 2009) manufactured by Bait-Tek, Inc., Orange, Texas. The second bait is the Ontario Bait manufactured by Artemis Technologies, Inc., Guelph, Ontario, Canada, that was used until 2009. It contained either the attenuated ERA® vaccine (Lawson & Bachmann, 2001) or the same vaccine (Merial V-RG®) as used in the Raboral V-RG® bait (Bachmann et al., 1990; MacInnes et al., 2001; Rosatte & Lawson, 2001; Rosatte et al., 2001; 2008). The third bait type is a matrix-coated sachet containing the Merial V-RG® vaccine. These baits (except the Merial sachet) contain tetracycline-HCl as a biomarker. The fourth and newest vaccine-bait is ONRAB® that was developed in Ontario for the control of rabies in raccoons, skunks, and red foxes (Rosatte, 2011; Rosatte et al., 2009a, 2011a). Beginning in 2012, ONRAB® baits for use in Ontario will not contain tetracycline, as vaccine efficacy and bait acceptance studies in that province have been completed and published (Rosatte et al., 2009a; 2011a).

There have been many other bait developments in North America and Europe for the various carnivore species during the last decade (Bruyère, Vuillaume, Cliquet, & Aubert, 2000; Follmann et al., 2004; Masson et al., 1999; Olson et al., 2000; Robbins et al., 1998; Rosatte et al., 2011a; Rosatte, Lawson, & MacInnes, 1998). In Europe, several different types of baits have been used for delivery of ORV to rabies vector species. Those include the chicken head and fishmeal/fish-oil baits which contain the V-RG® vaccine for fox rabies control in Belgium and Tubingen baits containing SAD B19 vaccine for fox rabies control in Germany, as well as for the control of rabies in raccoon dogs and foxes in Finland (Schneider & Cox, 1988; Westerling, 1991). Linhart et al. (1997) found that fishmeal baits were feasible for delivery of ORV to foxes and jackals in Israel.

11 ORAL RABIES VACCINATION INITIATION/CONSIDERATIONS

Oral rabies vaccination is a total system and has been defined as an attempt at wildlife rabies control intended to protect human health and prevent economic losses. Ideally, candidate vaccines will immunize all primary vector species in an area (e.g., raccoons, skunks, foxes, and coyotes in the United States) (Weyer et al., 2009). If the success of the ORV program is to be evaluated after treatment, specimens of the target species should be tested before treatment to establish levels of existing virus neutralizing antibody (VNA) (Hanlon et al., 1989; Rosatte & Gunson, 1984) as well as biomarker levels (Fearneyhough et al., 1998; Rosatte and Lawson, 2001; Sidwa et al., 2005) in the target species population. If a recombinant vaccine is to be used, a survey of related viruses present in the target population should be initiated before and after bait distribution to determine the potential for recombination between the vaccine virus and naturally occurring virus present in the target population (e.g., animal poxviruses present in an area to be baited with V-RG®) (Boulanger et al., 1996). As well, the safety of the vaccine virus should be tested in non-target species that may be present in the target area (Knowles et al., 2009a,b). Legal and liability issues also need to be examined before releasing a biologic agent into the environment.

11.1 Biomarkers in Vaccine-Baits

Ideally, a biomarker should be homogeneous with the vaccine but not impair vaccine efficacy. It should also be detectable for months or years following bait ingestion by an animal. To date, an ideal marker that is compatible with vaccine, long-lasting, inexpensive, and safe has not been developed. Tetracyclines form a long-term mark in bones and teeth, and they have proven useful in verifying not only bait contact but also the actual time-specific bait ingestion regimens using counts of fluorescent lines in tooth dentin and cementum (Figure 18.2). The age of the animal can also be determined from a count of the cementum growth zones in

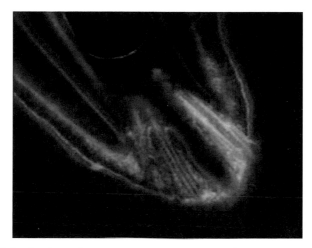

FIGURE 18.2   **Section of a raccoon canine tooth showing yellow fluorescent lines from ingestion of ONRAB® rabies vaccine baits containing 100 mg of tetracycline as a biomarker.** Photo by Andrew Silver, OMNR.

the tooth. When yearly biomarker lines in cementum from multi-year baitings are correlated with VNA levels, they can indicate the duration of detectable levels of VNA and any anamnestic rise in VNA from yearly boosters (Johnston & Tinline, 2002).

When an area is baited, most animals that eat baits will be "tagged" with the biomarker, and this has proven useful for tracking animal dispersal from a baited area (Rosatte et al., 2007f). However, tetracyclines can be found in wildlife at low levels derived from agricultural sources, usually less than 5% (Rosatte unpublished). This has been a concern in some programs (Linhart et al., 1997), but if specimens of the target species are examined for ambient levels prior to baiting, this can be taken into account (Fearneyhough et al., 1998; Johnston et al., 1999; Sidwa et al., 2005: Whitney et al., 2005). There is, however, a need for unique, long-lasting marker materials that can be used in combination with tetracycline to produce unique time-specific life-long "biomarks" in teeth. Such biomarkers will further help to monitor the proportion of a population that has been exposed to vaccine, and in the absence of declining VNA evidence, to determine when to rebait the area (Rosatte et al., 2007f; Sidwa et al., 2005). Specimens submitted for routine rabies diagnosis can also be used to establish the sex-age structure of the infected vector populations (Johnston & Beauregard, 1969), and are of value to establish pre-baiting levels of tetracycline, and for post-ORV monitoring of biomarkers in animals that have dispersed from the baited area. Unfortunately, tetracycline has been found to be an unreliable biomarker in some species such as skunks (even though some skunks ingest baits, for some unknown reason tetracycline is not deposited into their teeth) (Rosatte et al., 2009a) and there is need for development of more reliable markers (Rosatte et al., 2007f).

11.2 Bait Density

The control of rabies ultimately depends on establishing herd immunity through animals contacting vaccine baits and sero-converting. Therefore, bait density usually must correlate with animal density in some positive fashion (Rosatte & Lawson, 2001). This includes the density of all bait-consuming species, not just the target vector. Non-target species such as opossum may consume a considerable proportion of baits intended for raccoons. In urban habitats, raccoon density may be extremely high; for example, in Washington D.C. raccoon density was 67–333/km^2 (Hadidian, Prange, Rosatte, Riley, & Gehrt, 2010). High bait densities will be required to reach a substantial portion of the population. In Scarborough, Ontario, where raccoon density ranged from 37–94/km^2 (Broadfoot et al., 2001), raccoon acceptance of baits was 74% when bait density was 200/km^2 (Rosatte & Lawson, 2001). Roscoe et al. (1998) used fishmeal polymer V-RG® baits at a density 64/km^2 in the Cape May area of New Jersey to control raccoon rabies during 1992–1994. Tetracycline was detected in 73% of the sampled raccoons and 61% of the raccoons tested sero-converted (Roscoe et al., 1998). However, following distribution of V-RG® baits in Ohio at 70 baits/km^2, bait acceptance by raccoons was 57%, but only 12% had rabies antibody. It was concluded that raccoon density in the area was too high for the selected bait density (Ramey et al., 2008).

In some situations, bait density does not correlate with animal density. Such is the case for striped skunks in Ontario. Skunk density in rural habitats of southern Ontario is approximately 1–2/km^2 (Rosatte & Lariviere, 2003; Rosatte, Sobey, Dragoo, & Gehrt, 2010c). Given the low density of skunks in Ontario, one would expect that few vaccine baits would have to be distributed to reach a substantial portion of the skunk population. However, this is not the case, due to the foraging habits and limited home ranges and movements of skunks (Rosatte et al., 2011b; Rosatte & Lariviere, 2003). In fact, 300 ONRAB® baits/km^2 were required to immunize 20–62% of the skunks (variation based on the ELISA inhibition cut-off) during experiments in Ontario in 2008 and 2009 (Rosatte et al., 2011a) (Figure 18.3).

In Newfoundland, under severe winter snow conditions, Whitney et al. (2005) eradicated an invading epizootic of arctic-strain rabies in red foxes using a density of 35 baits/km^2. Metlin et al. (2009) recommend baiting for foxes in Russia at a density of 25–30 baits/km^2 for at least 6 years in duration and for at least 2 years after the last rabies case. Given

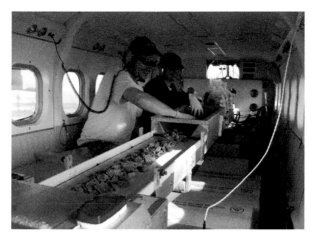

FIGURE 18.3   Ontario Ministry of Natural Resources staff loading ONRAB® rabies vaccine baits onto the baiting machine in a Twin Otter aircraft in Ontario, Canada. Photo by Mark Gibson, OMNR.

the wide variation in results in a small number of trials, it is imperative when planning ORV programs, to consider other factors that will maximize success such as the methods of vaccine placement, timing of vaccination, spatial variations in animal density, and bait design (Bachmann et al., 1990; Olson et al., 2000; Robbins et al., 1998; Rosatte et al., 2007f).

11.3 Method of Vaccine Bait Placement

In urban areas of Ontario, such as Toronto and Niagara Falls, hand placement of vaccine baits was used for the control of fox and raccoon rabies (Rosatte, 2011; Rosatte et al., 2007d, 1992). Roscoe et al. (1998) and Olson et al. (2000) used helicopters and motor vehicles to distribute baits for raccoon rabies control in New Jersey and Florida, respectively. In the New Jersey program, raccoon habitats or ecotones were targeted, rather than random dispersion of baits.

In Ontario, Bachmann et al. (1990) found that aerial distribution of baits was more cost effective than ground distribution. A variety of aircraft including Cessna, Turbo Beaver, Bell helicopter, and Twin Otter have been used to distribute baits for the control of rabies in wildlife in Ontario (Bachmann et al., 1990; MacInnes et al., 2001; Rosatte & Lawson, 2001; Rosatte et al., 1993, 2001, 2011a). Aircraft were also used for a massive ORV program in Texas for the control of the disease in coyotes (Fearneyhough et al., 1998; Sidwa et al., 2005). In addition, hand placement, airplanes and helicopters were used to distribute V-RG® baits for the control of rabies in foxes in Belgium (Brochier et al., 1996; 2001).

Control programs using aircraft typically distribute baits along flight lines with fixed spacing. Optimum spacing must ultimately match the movement behavior/territoriality of the target species and the pattern of distribution on the ground. In Ontario, spacing of flight lines during baiting operations is 2.0 km for foxes, 0.75 km for raccoons, and 0.5 or 0.25 km for skunks (MacInnes et al., 2001; Rosatte et al., 2007f, 2011a; Rosatte, 2011). Rosatte et al. (2011a) found that bait acceptance in skunks was higher (14 to 26%) where flight lines were placed 250 meters apart (compared to 500 meters) during 2009 experiments in Ontario.

11.4 Time of Year for Vaccination Campaigns

Much experimentation has been conducted with respect to the appropriate time of year to distribute rabies vaccine baits in order to maximize the percentage of the target population that is vaccinated against rabies. Fall ORV baiting campaigns resulted in the control of the arctic variant of rabies in Ontario (MacInnes et al., 2001; Rosatte et al., 2007f; Rosatte, 2011). Foxes can be baited during any time of the year, as they are active year-round, even in the northern regions of Canada, which experience harsh winters. In fact, >600,000 ERA® vaccine baits were distributed during severe winter snow conditions in Newfoundland to combat an outbreak of rabies in foxes during 2002 to 2004 (Nadin-Davis et al., 2008; Whitney et al., 2005). However, species such as raccoons and skunks den during the winter in northern climates; therefore, winter baiting is not feasible. For the control of raccoon rabies in Ontario, PIC and TVR programs were successful when deployed during the summer, as capture success for raccoons and skunks was higher during the summer than during the spring or fall (Rosatte et al., 2001). ORV campaigns using V-RG® to control raccoon rabies were also initiated during the summer in Ontario to allow time for assessment of vaccine bait uptake for raccoons prior to winter denning (Rosatte et al., 2008). However, bait acceptance for raccoons was higher in September than in June (Rosatte et al., 2008). During some experiments, the optimal time for baiting raccoons (based on bait acceptance) in southern Ontario was August, while bait uptake was higher for skunks during September (Rosatte et al., 2007a,b,c, 2009a, 2011a; Rosatte & Lawson, 2001; Rosatte, 2011). Both summer and fall baiting campaigns have been used in many U.S. states to combat raccoon rabies. For example, Roscoe et al. (1998) deployed V-RG® baits during the spring and fall in New Jersey to control raccoon rabies.

In most areas of Europe, ORV campaigns for the successful control of fox rabies occurred during the spring (March to May) and autumn (September to October) (Vos, 2003). In France, spring and fall ORV campaigns have been successful for the control of fox rabies, but summer campaigns met with lower success (Masson et al., 1999). In Germany,

Vos et al. (2001) concluded that if the objective is to only vaccinate adult foxes, then baits should be distributed during the first half of March. If juveniles are to be vaccinated, baits should not be deployed before the end of May in previously vaccinated areas, as many young foxes would have maternally transferred immunity, which could interfere with vaccination via baits (Bruyère et al., 2000; Muller et al., 2001a,b; Vos et al., 2001). Selhorst et al. (2001) concluded that June is the recommended time for vaccine-bait distribution for the control of fox rabies in Europe through modeling. However, as foxes are territorial, Vos (2003) concluded that the optimal time for bait distribution in Europe to target territorial foxes should be late autumn (November) or early winter (December).

Regardless of the time of year for baiting, consideration needs to be given to the effect of ambient temperature on the vaccine, as well as the effects of storage on the titer of the vaccine (Lawson & Bachmann, 2001; Selhorst et al., 2001). In Arctic regions, extreme temperatures may require the use of lyophilized oral vaccines for immunizing vector species such as Arctic foxes (Follmann et al., 2004; Kuzmin et al., 2004). However, in Newfoundland, under severe winter snow conditions, Whitney et al. (2005) used frozen or semi-frozen ERA® vaccine in Ontario baits to reach 66% of red foxes and eradicate an invading epizootic of Arctic-variant rabies.

12 IMPORTANCE OF RABIES VECTOR ECOLOGY FOR RABIES CONTROL PLANNING

In order to implement an effective and cost efficient oral rabies vaccination program, it is critical to have basic knowledge of the ecology of the target species. For example, if raccoons are the target species for rabies control, it is advantageous to know the home range and density of that species in specific habitats. In some urban habitats of North America, raccoon densities can be in excess of $100/km^2$ (Hadidian et al., 2010). However, in rural habitats, raccoon densities usually average $<10/km^2$ (Rosatte et al., 2010a,b). Obviously, many more vaccine-baits per unit area will be needed to control rabies where vector densities are high. In addition, sedentary species such as skunks will require high densities of baits to vaccinate a significant portion of the population as their movements and their ability to find baits are restricted (due to being less mobile) compared to species such as foxes and raccoons (Rosatte & Lariviere, 2003; Rosatte et al., 2011a,b).

For rabies control programs to be effective, vaccine-baits should be placed within the home range of the target animal. If each member of the target species only consumed one vaccine-bait and vector population density was constant, there would be no problem. Unfortunately, this is not the case. For example, Blackwell et al. (2004) found that 3.3 baits/raccoon were consumed on average (bait density of 75/km^2) with a 23% non-target uptake where raccoon density was about 25 raccoons/km^2. However, there are many non-target species which consume baits and target vector density is usually variable (Blackwell et al., 2004; Boyer et al., 2011). In addition, the spatial distribution of raccoons changes over time, making it difficult to design baiting strategies with a single animal density in mind.

Rabies vector species occupy specific habitats during certain periods of the year, and targeting those habitats during rabies control operations may increase the chances of success (Boyer et al., 2011). Currently, the only way to determine the parameters of bait density and distribution pattern for a particular area is by empirical experimentation. Fortunately, there are a growing number of successful rabies control programs from which to draw experience, but each geographic area and vector species possess unique factors that can impact results (Boyer et al., 2011; Tinline & MacInnes, 2004).

There should be a relationship between the home range distribution of the target species and spatial areas where rabies vaccine baits are distributed. If the baits are not placed within that bait-available home range, the animal may not encounter a bait and have the opportunity for vaccination. For example, in Ontario, raccoon home-ranges are about 3 to 4 km^2 in size (Rosatte et al., 2010a,b), whereas striped skunk ranges are about 1 to 3 km^2 depending on the habitat (Rosatte & Lariviere, 2003; Rosatte et al., 2010c, 2011b). The key is to ensure that sufficient baits are distributed with narrow enough flight-lines so that each target animal will encounter at least one vaccine-bait. In addition, a sufficient number of vaccine-baits must be distributed to allow for bait-eating competitors. However, if too many baits are distributed, the baiting system is not cost-efficient (Rosatte et al., 2007f).

The distance that vectors move within and beyond their home range during the baiting period is also an important consideration during rabies control planning. Cullingham et al. (2008), using genetic parentage analysis, showed that about 85% of raccoons in a study along the Ontario/New York border moved <3 km. Only about 4% moved >20 km. Cullingham, Kyle, Pond, Rees, & White (2009), using genetic analyses, suggested that differing permeability of rivers in Ontario affected raccoon movement between New York and Ontario and in turn affected rabies incidence in Ontario. Rosatte et al. (2010a,b) using radio-telemetry in Ontario showed that raccoons generally travel <10 km annually. These types of data can be used to design the area of coverage for control programs in order to vaccinate the majority of a rabies vector population while minimizing rabies control costs. These data can also be used to conduct a risk analysis assessment of a rabies control operation. For example, if 95% of raccoons travel less than 10 km annually with only 5% travelling 20 to 50 km, one can implement 10 km radial control programs, thereby playing the odds that rabies infected raccoons will not travel beyond the 10 km control zone. This type of approach could result in substantial savings compared to a 50 km radial program. For example, a 10 km radial rabies control program @ $200/km^2 would cost $62,800.00. By comparison, a 50 km radial program at the same rate would cost $1.6 million. The question would be, is it worth risking the possibility of rabies spreading beyond the 10 km radial control zone to save $1.5 million?

Another related vector ecology factor for consideration is the variability in home-range size between sex and age classes in the vector population. Depending on the time of year, adults usually have larger home ranges than young animals, and yet the young are usually the largest cohort in a carnivore population (Rosatte, 2000; Rosatte et al., 2001, 2010c). Vos, Selhorst, Schroder, & Mulder (2008) found that bait uptake for young foxes in Europe was poor in summer due to their limited ranging behavior and concluded it was not economically feasible to attempt to immunize the juvenile cohort of a fox population during the summer. The bait distribution parameters therefore should be tailored to reach this cohort, provided the bait drop is after the young are immunocompetent and the young are capable of moving considerable distances to find baits (Lawson et al., 1997; Müller et al., 2001a,b; Olson et al., 2000; Vos et al., 2008).

Most bait uptake studies have shown that more than 50% of baits are consumed within 1 week and more than 80% by 1–3 weeks (Bachmann, et al., 1990; Blackwell et al., 2004; Boyer et al., 2011; Linhart et al., 1997). Therefore, the extent of the target animal's movements inside its home range is critical to bait uptake during this short period. If the home range is small and movement is limited (e.g., the movement of female raccoons during the spring perinatal period), distribution patterns with widely spaced lines may miss many individuals. Home-range size and movement information from telemetry studies during the bait-available period can be incorporated into simulation models and geographic information system (GIS) flight-planning programs to further enhance bait distribution success.

13 LARGE-SCALE VACCINE-BAIT DISTRIBUTION TECHNOLOGY

The development of vaccine bait distribution technology has progressed from hand-dropped test baits to systems in many countries capable of air dropping thousands of baits per day (Fearneyhough et al.,

1998; Robbins et al., 1998; Rosatte & Lawson, 2001; Sidwa et al., 2005). In fact, more than 153 million rabies vaccine baits were distributed in the United States and Canada during 1985 to 2010 for the control of rabies in foxes, raccoons, coyotes, and skunks. Currently, one of the larger scale operations is the aerial distribution of V-RG® baits in the United States to control variants of the rabies virus unique to raccoons, gray foxes, and coyotes (Sidwa et al., 2005; Slate et al., 2005; 2009). During 2003, an approximate 180,000 km² area was treated with more than 10 million rabies-vaccine baits (Slate et al., 2005).

The Ontario oral rabies vaccination system has progressed to the point where about 176,000 ONRAB® baits can be aerially distributed per day, with landscape coverage capabilities in excess of 300 baits/km² and flight line spacing of 0.25 km (Rosatte, 2011; Rosatte et al., 2011a). Between 1985 and 2010, 23.5 million rabies vaccine baits were distributed in Ontario. In programs such as this, there are three principal aspects for consideration with respect to delivery of rabies vaccine-baits: (1) ground versus airborne distribution, (2) automation of air-borne bait delivery systems, and (3) organization of the bait delivery team. Ground distribution includes placing baits by hand or throwing baits along roadways from moving vehicles. Airborne distribution ranges from hand baiting by helicopter to automated dropping systems with fixed-wing aircraft. The decision to use a particular method depends largely on the scale of the operation, budget considerations, target species, and the ability to select habitats. In practice, hand baiting has been useful for urban areas and specific habitat types. Airborne delivery by helicopter or fixed-wing aircraft is best to cover large areas uniformly (Rosatte et al., 2007f).

The use of global positioning systems (GPS) in navigation technology in the 1990s has greatly enhanced aircraft baiting operations. GPS allows aircraft to follow pre-programmed flight lines, and, since ground speed can be calculated, the speed of the baiting machine can be continuously adjusted to match the prescribed target drop rate per kilometer (Rosatte et al., 2007f). Furthermore, the X,Y location of baits at the drop point can be recorded and compared with the distribution plan. In practice, Ontario has found that the system permits accurate accounting (+/–1–2%) on a bait drop and prevents costly errors in distribution, especially when millions of baits are being dropped. In addition to using GPS-based flight control, Ontario has developed software (FPLAN) to pre-plan flight routes and upload waypoints to the GPS navigation system (Rosatte et al., 2007f). This software has been used successfully in a number of large-scale ORV programs in North America. Texas, the USDA, and Ontario have also cooperated to add crew-management database software to coordinate a team of 30–60 personnel (Rosatte et al., 2007f).

14 SURVEILLANCE PRIOR TO, DURING, AND AFTER A RABIES CONTROL PROGRAM

Surveillance and direct fluorescent antibody (DFA) diagnosis are the critical tools that can be utilized to determine the extent of a rabies situation. These are key tactics that should be employed as rabies approaches an area, during rabies control operations, as well as following the control or elimination of rabies in an area. The level of surveillance should be determined by the intensity of the epizootic/enzootic situation. For example, in Ontario, during a raccoon rabies outbreak in 1999–2004, less than 1% (21/9,760) of raccoons and skunks (live-trapped) submitted by the Ontario Ministry of Natural Resources (OMNR) for rabies diagnosis were rabid (Rosatte et al., 2001, 2009b). This implied that very intensive rabies surveillance programs had to be implemented in order to maintain an early warning system for the rapid implementation of control programs. However, in other areas, the prevalence of raccoon rabies was quite high, as the epizootic was very intense (Winkler & Jenkins, 1991). In those situations, fewer animals had to be submitted to confirm that rabies was present in an area. Unfortunately, intensive rabies surveillance programs are lacking in most countries due to economic reasons. Some countries only test animals for rabies that have contacted or bitten humans. Furthermore, many animals that are involved in human biting incidents escape and are not tested.

Modeling showed that the worst case scenario of rabies resurgence occurs when rabies has persisted at low levels despite control efforts and has remained undetected by surveillance. In addition, as prolonged vaccination programs eventually make continuance uneconomical, surveillance during the termination stage of the control program is imperative. Unfortunately, many wildlife rabies control programs relax surveillance once the prevalence of the disease decreases. However, surveillance should be pursued aggressively both during field control operations as well as following rabies control to ensure the disease does not become re-established. This will allow for early detection of the disease and a rapid control response (Rosatte et al., 1997). Quebec, which has been free of raccoon rabies for 2 years (2010–2011), is targeting certain habitats that are likely to contain rabid animals during their rabies surveillance programs (Rees, Pond, Tinline, & Belanger, 2011b).

Following an ORV bait drop there are many questions that arise as to the success and efficiency of the project. How many target vectors contacted the baits, and how many were immunized by the vaccine? Under controlled laboratory conditions, efficacy trials can show that a particular vaccine-bait combination will immunize a high percentage of a vector species. However, once released into the environment, there is no control over this vaccine-baiting system (Rosatte et al., 2007f). To evaluate success and to provide information for improving the design of subsequent control efforts, experience has demonstrated that many aspects of surveillance are important, including vaccine-bait distribution parameters. These are the controllable parameters of vaccine-bait distribution, such as the maintenance of the cold chain storage prior to dropping, the values of bait density and dispersal pattern over the target habitat, and the time of year of placement (Rosatte et al., 2007f).

Also of importance are post-drop surveillance sampling techniques. Although not controllable because of the variation in field sampling procedures, these variables can be standardized for all projects and should include (a) a brain tissue sample for DFA diagnosis of a sample of normal animals as well as any suspect rabid animals from the treated area, (b) a serum sample for the detection and titre of VNA, and (c) a maxillary bone sample including the canine tooth to determine the animal's age and the presence or absence of the bait biomarker (Johnston et al., 1999; Johnston & Tinline, 2002; Rosatte et al., 2007f).

Also of importance during post-drop surveillance sampling is the size of sample required for a proper statistical analysis of the data. It is imperative to collect an adequate sample of target and non-target animals from the baited area before as well as after the bait drop. A statistically adequate sample of animals can be determined through statistical modeling, but questions unique to field samples remain (Rosatte et al., 2007f). In laboratory experiments under controlled conditions with uniform test animals, an optimal sample size may be 25 individuals or less. However results from the Texas ORV program have verified that at least 100 specimens are required to give a confident evaluation of coyote and gray fox populations from the wild (Rosatte et al., 2007f; Sidwa et al., 2005).

Another consideration during surveillance operations is the time to commence collection of post-drop surveillance specimens. If feasible, post-drop surveillance specimens should not be collected until 6 weeks following the last day of a bait drop. This allows at least 4 weeks for bait contact and vaccine ingestion and at least 2 weeks for the formation of detectable VNA (Brown et al., 2011). Past experience indicates that 50% of baits are contacted by 7 days, 80% by 2 weeks, and up to 100% by 3–4 weeks (Bachmann et al., 1990). Collections taken too early, that is, less than 6 weeks post drop, will include specimens that have contacted a bait and are tetracycline-positive but VNA negative due to lack of time to develop detectable antibody, which can take 1 to 5 weeks (Brown et al., 2011; Rosatte et al., 2007f) (Figure 18.2).

Also of importance during post-drop surveillance activities is the monitoring of vaccine efficacy. Once out of the production plant, rabies vaccine baits are open to the vagaries of storage temperatures and field distribution. When on the ground, each vaccine-bait is subject to the local microclimatic changes, which may affect the attractiveness of the bait to the target vector species and the ability of the vaccine to successfully immunize (Rosatte et al., 2007f). All of these influences will be different for each individual vaccine and ORV program. In order to verify the efficacy of the vaccine for the duration of the "bait-available period" of approximately 4 weeks, a protocol can be established to harvest vaccine-baits from the field at various intervals to monitor vaccine titer (Rosatte et al., 2007f). Serum VNA is normally used in laboratory experiments to verify vaccine efficacy, but these titers often fail to indicate the true immune status in a wild population. If the serum sample is taken too soon after baiting, sero-conversion may not have yet occurred; if taken months or years later, the antibody titer may have declined below detectable levels (Brown et al., 2011). For these reasons, incorporation of a time-specific biomarker such as tetracycline into the bait along with the vaccine can give additional evidence of vaccine bait contact and the potential duration of immunity, (Fearneyhough et al., 1998; Johnston et al., 1999; Linhart et al., 1997; Olson et al., 2000; Olson & Werner, 1999; Rosatte & Lawson, 2001; Rosatte et al., 2007f; Sidwa et al., 2005).

Tetracycline is a calciphilic marker that persists for the life of the animal in hard tissues such as teeth and bone. With multi-year baitings, as animals grow older, there is a buildup of tetracycline in the population, whereas VNA levels in these older animals decline unless they are rebosted. This can result in the false conclusion that vaccine efficacy is declining in the face of apparent increase in bait uptake as indicated by total tetracycline positivity. To overcome this phenomenon, analysis of yearly tetracycline line profiles in canine teeth in comparison to the annual dentin and cementum growth zones (age lines) can verify the true tetracycline-positive to VNA-positive ratio (see Figures 18.2 and 18.3 in Johnston & Tinline, 2002) (Rosatte et al., 2007f).

The standardization of post-drop surveillance sampling techniques is another critical consideration. The VNA tests for rabies antigen include the rapid fluorescent focus inhibition test (RFFIT), the fluorescent antibody virus neutralization test (FAVN), and the enzyme-linked immunosorbent assay (ELISA). These have undergone various degrees of standardization over time. Concerning VNA titer levels in ORV programs, there is the need to establish a standard level in International Units (IUs) that will be taken as evidence of ORV vaccination (Rosatte et al., 2007f). This is needed particularly for comparative purposes among different ORV systems. Various studies report VNA levels ranging from 0.012 IU or 0.05 IU, up to a presumed human WHO standard protective level of 0.5 IU. In field specimens, correlation between VNA levels and challenge protection are hard to pin down due to variation in the sampling time after vaccine contact (Brown et al., 2011; Rosatte et al., 2007f). Therefore, a VNA minimum titer standard needs to be established. To date, there has been virtually no standardization of the tetracycline biomarker technique from either the tissue sampling or interpretation points of view. The tetracycline line count has only been employed to advantage in a few studies. Here again, there is room for standardization of methods among laboratories so that all field results will be comparable among programs (Rosatte et al., 2007f).

15 THE ONTARIO RABIES MANAGEMENT PROGRAM—A MODEL OF A MULTI-FACETED APPROACH TO WILDLIFE RABIES CONTROL

The Ontario rabies management program is an example where all of the variables discussed in the previous sections have been put together in an efficient, effective operation. The system has evolved from using ERA® modified live-virus vaccine baits for fox rabies control beginning in 1989, to the use of population reduction, TVR, and V-RG® baits for the control of rabies in raccoons and skunks during 1999 to 2007, to the use of a single vaccine-bait, ONRAB® during 2009–2011 for the control of rabies in all three species (MacInnes et al., 2001; Rosatte et al., 2007f, 2008, 2009a, 2011a; Rosatte, 2011) (Figure 18.3). As each of the three species has a different ecology, with differing home ranges, spatial distribution, and habits, a different design is used, depending on whether the control response is for rabies cases in foxes, raccoons, or skunks.

As red foxes in Ontario exist at low densities (about 1–2/km²), have fairly large home ranges, and are fairly mobile, bait density can be fairly low compared to baiting for raccoons and skunks (Voigt, 1987; Rosatte & Allan, 2009; Rosatte, 2001). In Ontario, if the operational response is for a case of rabies in foxes, bait density is 20/km² with flight line spacing of about 2.0 km between baiting lines (MacInnes et al., 2001; Rosatte et al., 2011a) (Figure 18.4). Baiting usually occurs once per year during September. ONRAB® vaccine-baits are distributed within a 50 km radius of the fox case (Figure 18.4). Baiting continues for 2 years after the last reported case. This tactic has been so successful that no cases of rabid foxes were reported in Ontario during 2010 and 2011 (Figure 18.5).

Raccoons exist at fairly high densities in Ontario (about 5 to 10/km²), are fairly mobile, and have home ranges that are smaller than foxes but larger than skunks (Hadidian et al., 2010; Rosatte, 2000; Rosatte et al., 2010a,b). In view of this, bait density has to be higher than that for foxes. After much experimentation, it was decided that the most cost effective and efficient baiting density for raccoons is about 75 ONRAB® baits/km² (Rosatte et al., 2008, 2009a) (Table 18.1) (Figure 18.4). Flight line spacing is about 0.75 km, with baiting being done in either August or September. As with foxes, a 50 km radial area is treated in response to raccoon rabies cases (Rosatte et al., 2008). As raccoon rabies is present in nearby New

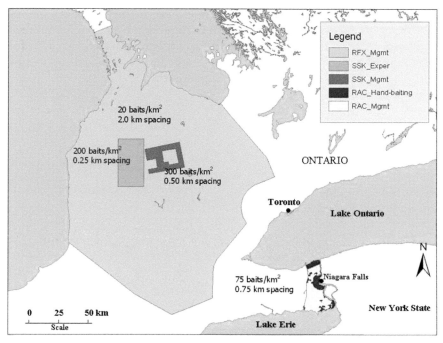

FIGURE 18.4 Map of the Ontario response to Arctic variant rabies in red foxes (20 ONRAB® baits/km²) and striped skunks (300 baits/km²), and proactive baiting (75 baits/km²) to prevent raccoon variant rabies from entering Ontario from New York State. RFX = red fox; SSK = striped skunk; RAC = raccoon. Figure designed by Lucy Brown, OMNR.

York State and is threatening to again enter Ontario, proactive baiting with ONRAB® targeting raccoons is currently still occurring (August 2011) in Ontario, even though the last case was in 2005 (Figure 18.6). Using this tactic, as well as PIC and TVR, raccoon rabies was eliminated from Ontario in 2005 but is still enzootic in neighboring New York State (Figure 18.6).

Striped skunks are fairly sedentary, with small home ranges in Ontario when compared with foxes or raccoons. Densities are fairly low and are about 1–2/km² (Rosatte & Lariviere, 2003; Rosatte et al., 2010c, 2011b). As skunks exist at low densities, one would think that few vaccine-baits would be needed to immunize a substantial portion of the skunk population. This is not the case. Skunks move very slowly, do not travel much, and spend much of their time with their noses to the ground in search of insects and grubs. As a result of their sedentary behavior, flight line spacing has to be very narrow (0.25 km to 0.5 km), and a significant number of baits (about 300/km²) must be distributed in order to reach a significant portion of the population (there is much competition for baits from other more active species, such as foxes and

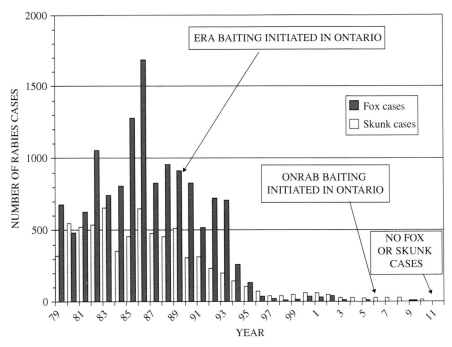

FIGURE 18.5   Fox and skunk rabies cases in Ontario during 1979 to October 2011 with respect to the initiation of ERA® and ONRAB® vaccine baiting campaigns.

raccoons). After much experimentation, it was determined that the most effective bait density for striped skunks in Ontario is 300 ONRAB® baits/km² (Rosatte et al., 2009a, 2011a) (Table 18.1) (Figure 18.4). Flight line spacing at 0.25 km achieves good bait acceptance and immunity in skunks; however, 0.5 km spacing is also acceptable (Rosatte et al., 2011a). Aerial distribution of baits targeting skunks usually occurs in August or September. For a single rabid skunk case, a 64 km² area (8 km by 8 km) is treated with ONRAB® baits. If multiple skunk cases occur in a single focal area, the entire area where cases occur is treated with baits, while maintaining a 4 km wide buffer zone of baits around the outside perimeter of all cases (Figure 18.4). This tactic appears to be working, as there were no rabid skunks reported in Ontario during 2011 (to October, 2011) (Figure 18.5).

Traditionally, Ontario has responded to rabies cases by distributing baits once per year—usually in the early fall. However, beginning in 2010, it was decided to take a more aggressive approach and respond to cases as they occur. The idea was that if cases are treated immediately, the disease would have less time to spread. For example, if a case occurs in March and the area is not baited until September, the disease has

TABLE 18.1   Bait Acceptance and Serological Response in Raccoons and Skunks Following Distribution of ONRAB® Rabies Vaccine Baits in Southwestern Ontario During 2006–2010[a]

Year	Species	Bait Density (/km[b])	Flight Line Spacing (km)	Bait Acceptance (% tetra +)	Serology (% cELISA +)[b]
2007	raccoon	75	0.5	80	76
2007	raccoon	75	0.5	59	na
2007	raccoon	75	0.5	62	na
2006	raccoon	150	0.5	74	66
2006	raccoon	150	0.5	62	na
2006	raccoon	300	0.5	77	81
2006	raccoon	300	0.5	75	na
2007	raccoon	400	0.5	87	84
2007	raccoon	400	0.5	83	na
2007	raccoon	400	0.5	87	na
2007	skunk	75	0.5	32	29
2006	skunk	150	0.5	na	49
2006	skunk	150	0.5	na	17
2010	skunk	200	0.25	na	45
2006	skunk	300	0.5	na	51
2006	skunk	300	0.5	na	37
2008	skunk	300	0.5	na	60
2009	skunk	300	0.5	na	36
2009	skunk	300	0.25	na	62
2007	skunk	400	0.5	45	37

[a]Data from Rosatte et al., 2009a and Rosatte et al., 2011a. Sample size is 1,536 raccoons and 1,132 striped skunks. na = not applicable. Bait acceptance is based on the percentage of sectioned teeth that were positive for tetracycline. Serological response is based on the percentage of sera samples that were positive for rabies antibody using a cELISA.
[b]Serology values are based on samples collected from all plots baited at a particular density e.g. the serology value of 76 is based on raccoons collected from all 3 plots baited at 75 baits/km^2 during 2007.

6 months to spread before treatment occurs. As skunks and raccoons den during periods of inclement weather (Hadidian et al., 2010; Rosatte & Lariviere, 2003: Rosatte et al., 2010a,b,c, 2011b), in Ontario, treatment of cases in those species only occurs during March to late November/early December. However, as red foxes remain active year round in Ontario,

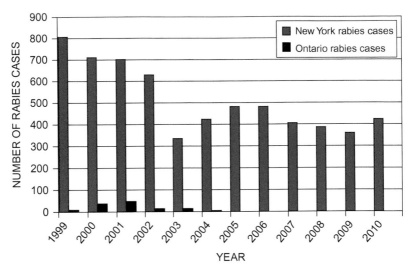

FIGURE 18.6   **Raccoon rabies cases in New York State and Ontario, Canada, during 1999 to 2010.** New York rabies data acquired from http://www.wadsworth.org/rabies.

treatment may occur during any month if required. This approach to rabies control appears to be working, as there were only 10 terrestrial wildlife rabies cases (all skunks) in the province during 2010 (Figure 18.5). There was only one rabid cow and no wildlife cases (except in bats) in Ontario during 2011 (to October, 2011). This is a dramatic decline from the 3,526 cases reported in 1986 when Ontario had the distinction of being labeled the "rabies capital of North America" (Rosatte, 1988).

16 THE COSTS AND BENEFITS OF WILDLIFE RABIES CONTROL

Distributing rabies vaccine-baits over large areas is very expensive (Kreindel et al., 1998; Meltzer, 1996; Selhorst et al., 2001). Costs are variable and dependent on bait density and vaccine cost, but they average about $150.00 to $300.00 Cdn/km^2 (Rosatte et al., 2007f; 2008). Ontario used vaccine baits to control rabies in red foxes during 1989 to 2011 (MacInnes et al., 2001; Rosatte, 2011; Rosatte et al., 2007d,f). Prior to control in 1989, annual human post-exposure treatments (PET) and fox rabies cases averaged 2,248 and 1,861, respectively. In addition, annual indemnity payments for livestock losses due to rabies averaged $247,000.00 annually. During 1990 to 2000, after baiting was initiated, there was a 35, 66, and 41% reduction in PETs, rabies cases, and indemnity payments, respectively (Shwiff, Nunan, Kirkpatrick, & Shwiff, 2011).

Financial benefits of the Ontario program ranged from $35.5 million to $98.4 million (Shwiff et al., 2011).

In North America, when fox rabies was controlled in Ontario, human post-exposure treatments and rabies cases declined, resulting in significant economic benefits (MacInnes et al., 2001). The same scenario can be assumed when raccoon rabies was controlled in Ontario using point infection control strategies (Rosatte et al., 2001). However, vaccinating raccoons with vaccine baits requires 4–8 times as many baits per unit area as does the control of fox rabies (MacInnes et al., 2001; Rosatte & Lawson, 2001; Rosatte et al., 2001, 2009a). Meltzer & Rupprecht (1998) have questioned the benefits of using oral rabies vaccination to eliminate raccoon rabies where the disease is enzootic. However, Kemere, Liddel, Evangelou, Slate, & Osmek (2002) through modeling, have suggested that the net benefits to be gained would be substantial ($48M to $496M US) if raccoon rabies were eliminated using ORV. Shwiff et al. (2009) also predicted that under certain scenarios, the control of skunk rabies in California would be economically feasible using oral vaccination tactics. In Europe, Eisinger & Thulke (2008) using a model, predicted that lower effort (i.e., lower bait densities) while controlling fox rabies via oral vaccination would save one-third of the resources currently being used to combat the disease. The question of the economic feasibility of wildlife rabies control is controversial, to say the least, and is an area that warrants further study.

An example of baiting large geographic areas occurred in 1995–96 when Texas initiated two ORV programs to control epizootics in coyotes and gray foxes (Fearneyhough et al., 1998). Implementation of these two programs averaged $3.8 M per year. This is in contrast to the initial projected cumulative cost of $63 M from 1994 to 2004 for resulting post exposure prophylaxis alone. Other than PET biologicals, this estimate did not include any additional medical care, lost productivity, human suffering, or impact on agriculture and tourism (Rosatte et al., 2007f). By 2000, the number of cases in South Texas had declined to zero. Given this success and the cost-effectiveness of these programs, the South Texas Coyote ORVP has continued in maintenance mode from 2000 to 2005 at a cost of $1.2 M annually, and the Gray Fox ORVP will be continued in order to close in on the focus of that epizootic in west Texas (Sidwa et al., 2005).

The containment and elimination of raccoon-strain rabies in eastern Ontario has been successful using a PIC strategy (Rosatte et al., 2001, 2009b). There have been no cases since September 2005 (to October 2011). Total costs to distribute V-RG baits for raccoon rabies control were about $200Cdn/km^2 and costs for population reduction (PR) and TVR programs were about $500 Cdn/km^2; however, areas treated by PR and TVR were much smaller than those treated by ORV, making overall costs for the former two strategies less costly than ORV (Rosatte et al., 2001).

Rabies only moved 50 km in 6 years (from 1999 to 2005) in Ontario due to control efforts. Without control, modeling suggests the disease would have progressed at least 180–240 km by 2005 and would have cost the government an additional \$8M–\$12M/yr in rabies-associated costs (Rosatte et al., 2001, 2007f).

17 VACCINE BAITING EFFICIENCY AND BAITING SUCCESS GUIDELINES

Oral vaccination is a relatively inefficient method of vaccinating an animal rabies vector, that is, many vaccine doses are required per animal vaccinated versus one dose per animal by parenteral injection (Rosatte et al., 2007f). Thulke et al. (2004) suggested there is no overall optimal strategy for deploying ORV over large areas, and increasing bait density does not necessarily improve acceptance by the target species. Rosatte et al. (2009a) also found that increasing bait density for raccoons in Ontario did not necessarily increase bait acceptance after a certain threshold of bait density. In their study, raccoon bait acceptance was 74% in a plot baited at 150 baits/km^2, yet bait acceptance in a plot baited at 300 baits/km^2 was only 75%. Doubling bait density only resulted in a one percent increase in bait acceptance. However, bait acceptance by raccoons increased dramatically to 87% when bait density was increased to 400 baits/km^2 (Rosatte et al., 2009a).

With limited budgets for rabies control, efficient use of vaccine baits is imperative for the operating agency. Variables affecting baiting success are many and include the geographic area to be baited, target species and target species density, bait-eating competitor density, habitat variability, and year-to-year climate and phenological changes (Rosatte et al., 2007f). Results to date indicate that for a vaccine baiting system to be successful in limiting rabies spread and at the same time to be cost efficient, the following criteria (as noted in Rosatte et al., 2007f) that must be met include (a) to become immune, an individual animal of the target species needs to consume only one bait containing an efficacious vaccine; (b) for an individual animal to eat only one bait, the bait must be dropped within its active home range during the bait-available period. Also, enough baits must be dropped within the home range so that at least one bait is found by the target individual before all the baits are eaten by other competitors; (c) the target vector population is able to resist continuing a rabies epizootic if approximately 50–70% of the population finds a vaccine bait, becomes immune, and maintains that immunity for an extended period (Voigt et al., 1985); and (d) a cost-efficient program will achieve the 50–70% population bait consumption level using only an average of one bait per individual of the target species (Johnston & Tinline, 2002; Rosatte et al., 2007f).

A potential advantage of producing a more effective bait is that lower bait densities can be used in oral baiting, and costs will be reduced. Two other potential methods of reducing oral baiting costs are to target habitat and to partition large areas into bait and no-bait zones (Rosatte et al., 2007f). Targeting habitat is problematic for species like raccoons because they are generalists and can survive in a wide variety of habitats at highly variable densities. An alternative strategy would be to take advantage of natural barriers (mountains, large rivers, and lakes) to delineate areas for intensive baiting and areas for "emergency" baiting depending on the presence or absence of rabies (Rosatte et al., 2007f). In an epizootic situation, natural barriers can be used selectively to complement oral vaccine barriers and reduce or eliminate bait distribution behind or within the barriers. In an enzootic situation, the control of rabies in red foxes in Ontario proved to be successful by distributing vaccine baits in "Rabies Units" in geographical sequence, breaking the cycle in one unit before moving on to another (MacInnes et al., 2001; Rosatte, 2011; Tinline & MacInnes, 2004).

Guidelines for baiting success have been developed and refined using three variables from post-drop surveillance results: (1) the age of the animal, (2) the VNA titer level, and (3) the count of tetracycline biomarker lines in canine teeth (Rosatte et al., 2007f) (Figure 18.2). The presence of tetracycline fluorescence in bones or teeth indicates an animal has contacted a bait, but the proportion of a population that is tetracycline-positive does not by itself indicate baiting efficiency if over baiting is occurring; that is, there has been more than one bait contact per animal. However, a count of tetracycline biomarker lines per animal, determined by ultraviolet–fluorescence microscopy (Johnston et al., 1999), can show if baits are being distributed uniformly, or whether some animals are getting too few or too many baits. Specimens up to 1 year of age give the best count of tetracycline biomarker lines because rapid tooth growth separates daily ingestions of single baits more clearly than in older animals where growth is slow. In older animals, multiple ingestions often appear clumped together into a single line or band of fluorescence (Rosatte et al., 2007f).

18 MODELING

Recently, researchers have employed mathematical models to assist with the design and development of rabies vaccination programs (Johnston & Tinline, 2002; Rosatte et al., 2007f). The proportion of the target population (foxes in Europe) that needed to be vaccinated to achieve herd immunity to halt the spread of rabies was examined using modeling (Anderson, Jackson, May, & Smith, 1981). Modeling was also used to

study the spatial ecology of the vector (red foxes) and the rabies pathogen in Europe with a conclusion that decreasing vaccine-bait densities below that which are currently being used would save a significant
amount of resources (Eisinger & Thulke, 2008). Haydon, Laurenson, &
Sillero-Zubiri (2002) suggested that, in the endangered Ethiopian wolf
population, a population that is dispersed and isolated, herd immunity
could be achieved if only 20–40% of the population were vaccinated.

Recuenco, Eidson, Cherry, Kulldorff, & Johnson (2008) examined the
characteristics of raccoon rabies in New York and concluded that models could be used to select priority areas for rabies control operations
based on risk and costs for control. In addition, Gordon et al. (2004) have
developed an epidemiological model to estimate the risk of infection of
secondary species in areas epizootic/enzootic for raccoon rabies. Models
were also used by researchers investigating the impact of animal density
on the survival of rabies in a contained raccoon population. It was discovered that persistence of rabies occurred at medium vector densities
(5–10 animals per km^2). However, the disease disappeared at lower vector densities (Johnston & Tinline, 2002; Rosatte et al., 2007f).

Coyne et al. (1989) investigated the relative merits of culling and/or
vaccination in the United States, and concluded that vaccination or vaccination plus culling would be required to control rabies if it became
established in the host population. However, deterministic models do
not reveal how spatial behaviors such as territoriality, dispersal, and
daily interaction would affect the patterns of spread and the persistence of rabies (Johnston & Tinline, 2002; Rosatte et al., 2007f). Other
researchers developed spatial and stochastic models of disease spread
(Voigt et al., 1985). For example, field and laboratory experiments in
Ontario using the aerial distribution of ERA® vaccine baits at 20 baits/
km^2 have indicated that only 50–60% of red foxes would be immune to
rabies (Lawson & Bachmann, 2001; MacInnes, 1988). These values are on
the border between eradication and persistence according to Anderson
et al. (1981). However, the Ontario fox rabies model demonstrated that
immunity levels of 50–60% had the highest probability of eradicating fox
rabies in Eastern Ontario if the vaccination program began after a peak
in rabies incidence (Voigt et al., 1985). Following this design, rabies was
controlled in red foxes in eastern Ontario (MacInnes et al., 2001; Rosatte
et al., 2007f).

18.1 Modeling and the use of Habitat and Landscape Barriers for Rabies Control

Many researchers have used simulation models to study the various
aspects of the spread of infectious disease (Johnston & Tinline, 2002;
Rosatte et al., 2007f). Rees et al. (2011a,b) used the Ontario Rabies Model

to examine how the configuration of habitats can impact rabies control operations. They postulated that breaches in vaccine barriers are more likely if the structure of the habitat increases. Smith, Brendan, Waller, Childs, & Real (2002) developed a spatial model of raccoon rabies spread through Connecticut that implied that major rivers acted as semi permeable barriers to rabies spread. Russell et al. (2005) showed the spread of raccoon rabies was impacted by rivers in Ohio following a breach in a vaccination barrier in 2004. Similarly, Rees et al. (2011b) also demonstrated that landscape barriers such as rivers and topography can reduce rabies spread. In support of this is the fact that raccoon rabies was first reported in SE Alabama in 1975 but to date (2011) has not spread west across the Alabama River. The river appears to be serving as a partial barrier to rabies spread. However, poor raccoon habitat west of the river may also be playing some role in limiting the westward spread of raccoon rabies (Arjo, Fisher, Armstrong, Boyd, & Slate, 2008).

Natural barriers should be an integral part of rabies control planning and modelers (Broadfoot et al., 2001) have examined control responses that should be implemented when rabies reappears or breaches a barrier (Rosatte et al., 2007f). In Alabama, the plan is to use the Alabama River as a barrier to raccoon movement and distribute rabies vaccine baits in hardwood forests and around water bodies where raccoon densities are highest east of the river (Arjo et al., 2008).

18.2 Modeling and Evaluation of the Efficacy of Wildlife Rabies Control Tactics

Smith & Wilkinson (2003) employed a model to evaluate culling, oral vaccination, and fertility control as strategies for the control of rabies in foxes. They suggested that the best strategy to control a point source wildlife rabies outbreak would include an area of culling in the center of the disease focus, followed by an outer ring of vaccination or an outer ring of vaccination and fertility control. Modeling was also used to determine the efficacy of controlling raccoon rabies in the United States using oral rabies vaccine baits (Ma, Blanton, Rathban, Recuenco, & Rupprecht, 2010). The model output indicated that ORV should be continued in that country. Neilan & Lenhart (2010) also developed a spatio-temporal rabies epidemic model to determine optimal raccoon rabies control strategies. The model provided insights as to when, where, and to what degree rabies control tactics should be implemented.

Eisinger et al. (2005) evaluated the concept of point infection control by ring vaccination for the control of sylvatic rabies. Rosatte et al. (2001) actually controlled and eliminated an outbreak of raccoon rabies in Ontario, Canada, with point infection control strategies during the early 2000s. That resulted in a $6M to $10M annual savings in rabies associated costs (Rosatte, 2011; Rosatte et al., 2009a). Following that, a landscape genetic model was used by Rees et al. (2009) to study the variables affecting raccoon rabies spread in southern Ontario.

During the last two decades, there has been renewed interest in other disease control methods, such as contraception and comparing combinations of strategies using modeling (Johnston & Tinline, 2002; Rosatte et al., 2007f; Selhorst et al., 2001). Smith & Cheeseman (2002) demonstrated that culling or lethal control can be more effective than vaccination for diseases such as rabies in isolated populations and suggested that permanent contraception would increase the chances of disease eradication. A contraceptive known as GonaCon is being considered to control raccoon populations in the United States. (Bender et al., 2009). Barlow (1996) suggested that culling was likely to be more effective than vaccination but equally as effective as sterilization.

The advantage of simulation models are that they are inexpensive (compared to field experiments), and they allow for controlled experimentation. The primary disadvantage of models is that they are only as good as the data that are acquired from the field and input to the model. However, simulation models force researchers to make explicit assumptions and clearly indicate the unknowns about the variables in rabies vector ecology. This provides direction for additional field research and represents a gain from the modeling exercise (Johnston & Tinline, 2002; Rosatte et al., 2007f).

19 CONTINGENCY PLANNING

Contingency or management planning is critical for the rapid control of disease outbreaks. Ontario had a contingency plan for raccoon rabies 6 years before the disease was reported in the province (Rosatte et al., 1997). This proactive approach allowed staff to implement a control tactic 24 hours after the first case of raccoon rabies was confirmed (Rosatte et al., 2001). Ontario's contingency plan now includes a control response to wildlife rabies cases as soon as they are confirmed. Although Australia is free of reported wildlife rabies, experiments were initiated regarding oral rabies vaccination of foxes. This tactic is being considered should rabies occur in the wildlife population (Marks & Bloomfield, 1999). In Great Britain, contingency plans are in place that involve the distribution of poison baits as well as ORV should an outbreak occur in foxes and/or badgers (*Meles meles*) (Harris, Cheeseman, Smith, & Trewhella, 1990).

In Ohio, the deployment of ORV was successful for controlling raccoon rabies from 1998–2003; however, during 2004, there was a significant outbreak in eastern Ohio and the disease now threatens to expand westward (Russell et al., 2005; Slate et al., 2009). Neighboring states are currently discussing contingency plans (Slate et al., 2005). At a larger scale, the North American Rabies Management Plan was signed in 2008 to build long-term rabies management goals in Canada, the United States, and Mexico (Slate et al., 2009). Similarly, a Canadian Rabies Management Plan was signed in 2009 to provide rabies management goals at the provincial and territorial level (Rosatte, 2011).

20 CONCLUSIONS

The success of wildlife rabies control programs, over the short term, has indirectly assisted managers in securing public funds for additional control efforts. For the medium and long term, especially in areas where rabies control has been successful, the danger is that there is no perceived need for control and therefore no need for funds to perpetuate a state of preparedness (Rosatte et al., 2007f). Furthermore, it is imperative to understand that oral rabies vaccination with baits is not a panacea for wildlife rabies control in all situations worldwide. In some situations, managers have made decisions to continue with tactics that have failed, instead of modifying the tactic or implementing new untried tactics. Managers as well as researchers must take an adaptive management approach to meet the needs of wildlife rabies control in any given area (Rosatte et al., 2007f). For example, the use of ORV is currently the most feasible tactic for wildlife rabies control over large areas in species such as foxes and raccoons. However, ORV is currently only feasible for rabies control in skunks in small areas due to the high densities of vaccine baits required and resultant high costs (Rosatte et al., 2011a). In other jurisdictions, where effective vaccine-baits for species such as raccoons and skunks are not yet available, or have not yet received approval for use, alternate strategies will have to be employed, such as population reduction and TVR, until effective oral rabies vaccines are ready for field application. Additional research is also needed so that vaccine baits that are effective in northern climates can be modified for use in warmer southern climates. However, the most important advancement needed is more affordable baits, as currently vaccine baits account for 85 to 90% of ORV program costs (Rosatte et al., 2007f; Rosatte, 2011).

Diagnostic and assessment techniques should be standardized to allow comparisons between various control programs. Adaptive management also means that rabies control programs must be treated as ongoing experiments to enable a fair assessment of the success/failure of the program. This approach also implies the need for continuing surveillance programs both during and after rabies control program operations, with appropriate sample sizes to allow for a valid assessment of the situation. If rabies is not completely eliminated from an area, the disease may perpetuate at low levels and form a reservoir of rabies to fuel future outbreaks. Adaptive management also requires inter-agency cooperation and a well-designed communication plan to ensure continued cooperation and public understanding and support of a rabies control program (Rosatte et al., 2007f).

ACKNOWLEDGEMENTS

David Johnston and Roland Tinline were co-authors for the Rabies 2nd edition chapter on Rabies Control in Wild Carnivores, and some of their ideas and writings are cited in this chapter as Rosatte et al., 2007f. The chapter was reviewed by Dr. J. C. Davies, manager, OMNR, Wildlife Research and Development Section, Peterborough, Ontario, Canada.

References

Algeo, T., Chipman, R., Bjorldand, B., Chandler, M., Wang, X., Slate, D., et al. (2008). Anatomy of the Cape Cod oral rabies vaccination program. In R. Timm & M. Madon (Eds.), *Twenty third vertebrate pest conference proceedings (pp. 264–269)*. Davis, Ca: University of California Publishers.

Almeida, M., Martorelli, L., Aires, C., Barros, R., & Massad, E. (2008). Vaccinating the vampire bat (*Desmodus rotundus*) against rabies. *Virus Research, 137*, 275–277.

Almeida, M., Massas, E., Aguiar, E., Martorelli, L., & Joppert, A. (2001). Neutralizing anti-rabies antibodies in urban terrestrial wildlife in Brazil. *Journal of Wildlife Diseases, 37*, 394–398.

Anderson, R. M., Jackson, H. C., May, R. M., & Smith, M. (1981). Population dynamics of fox rabies in Europe. *Nature, 289*, 765–771.

Arjo, W., Fisher, C., Armstrong, J., Boyd, F., & Slate, D. (2008). Effects of natural barriers and habitat on the western spread of raccoon rabies in Alabama. *Journal of Wildlife Management, 72*, 1725–1735.

Artois, M. (2003). Wildlife infectious disease control in Europe. *Journal of Mountain Ecology, 7*, 89–97.

Artois, M., & Aubert, M. (1991). Foxes and rabies in Lorraine: A behavioural-ecology approach. *Hystrix, 3*, 149–158.

Artois, M., Cliquet, F., Barrat, J., & Schumacher, C. (1997). Effectiveness of SAG1 Oral vaccine for long-term protection of red foxes (*Vulpes vulpes*) against rabies. *The Veterinary Record, 140*, 57–59.

Aubert, M. (1999). Costs and benefits of rabies control in wildlife in France. *Revue Scientific et Technique de L'Office International des Epizootics, 18*, 533–543.

Aubert, M., Masson, E., Artois, M., & Barrat, J. (1994). Oral wildlife rabies vaccination field trials in Europe with recent emphasis on France. In C. Rupprecht, B. Dietzschold, & H. Koprowski (Eds.), *Lyssaviruses (pp. 219–243)*. Berlin: Springer-Verlag.

Bachmann, P., Bramwell, R. N., Fraser, S. J., Gilmore, D. H., Johnston, D. H., Lawson, ... et al. (1990). Wild carnivore acceptance of baits for delivery of liquid rabies vaccine. *Journal of Wildlife Diseases, 26*, 486–501.

Baer, G. M. (1988). Oral rabies vaccination: An overview. *Reviews of Infectious Diseases, 10(Suppl. 4)*, S644–S648.

Baer, G. M., Abelseth, M. K., & Debbie, J. G. (1971). Oral vaccination of foxes against rabies. *American Journal of Epidemiology, 93*, 487–490.

Ballantyne, E., & O'Donoghue, J. (1954). Rabies control in Alberta. *Journal of the American Veterinary Medical Association, 125*, 316–326.

Ballard, W. B., Follmann, E., Ritter, J., Robards, M., & Cronin, M. (2001). Rabies and canine distemper in an arctic fox population in Alaska. *Journal of Wildlife Diseases, 37*, 133–137.

Barlow, N. (1996). The ecology of wildlife disease control: Simple models revisited. *Journal of Applied Ecology, 33*, 303–314.

Bender, S., Bergman, D., Wenning, K., Miller, L., Slate, D., Jackson, F., et al. (2009). No adverse effects of simultaneous vaccination with the immuno-contraceptive GonaCon and a commercial rabies vaccine on rabies virus neutralizing antibody production in dogs. *Vaccine, 27*, 7210–7213.

Bingham, J., Foggin, C., Wandeler, A., & Hill, F. (1999a). The epidemiology of rabies in Zimbabwe. Rabies in jackals (*Canis adustus* and *Canis mesolelas*). *Onderstepoort Journal of Veterinary Research, 66*, 11–23.

Bingham, J., Schumacher, C., Aubert, M., Hill, F., & Aubert, A. (1997). Innocuity studies of SAG-2 oral rabies vaccine in various Zimbabwean wild non-target species. *Vaccine, 15*, 937–943.

Bingham, J., Schumacher, C. L., Hill, F. W., & Aubert, A. (1999b). Efficacy of SAG-2 oral rabies vaccine in two species of jackal (*Canis adustus* and *Canis mesomelas*). *Vaccine, 17*, 551–558.

Black, J. G., & Lawson, K. F. (1970). Sylvatic rabies studies in the silver fox (*Vulpes vulpes*): Susceptibility and immune response. *Canadian Journal of Comparative Medicine, 34*, 309–311.

Black, J. G., & Lawson, K. F. (1980). The safety and efficacy of immunizing foxes (*Vulpes vulpes*) using baits containing attenuated rabies virus vaccine. *Canadian Journal of Comparative Medicine, 44*, 169–176.

Blackwell, B., Seamans, T., White, R., Patton, Z., Bush, R., & Cepek, J. (2004). Exposure time of oral rabies vaccine baits relative to baiting density and raccoon population density. *Journal of Wildlife Diseases, 40*, 222–229.

Blanton, J., Meadows, A., Murphy, S., Manangan, J., Hanlon, C., Faber, … et al. (2006). Vaccination of small Asian mongoose (*Herpestes javanicus*) against rabies. *Journal of Wildlife Diseases, 42*, 663–666.

Blanton, J., Palmer, D., Dyer, J., & Rupprecht, C. (2011). Rabies surveillance in the United States during 2010. *Journal of the American Veterinary Medical Association, 239*, 773–783.

Bogel, K., Arata, A., Moegle, H., & Knorpp, F. (1974). Recovery of reduced fox populations in rabies control. *Zentralbl Veterinarmed, 21*, 401–412.

Boulanger, D., Crouch, A., Brochier, B., Bennett, M., Clement, J., Gaskell, R., et al. (1996). Serological survey for orthopoxvirus infection of wild mammals in areas where a recombinant rabies virus is used to vaccinate foxes. *The Veterinary Record, 138*, 247–249.

Boulanger, J., Bigler, L., Curtis, P., Lein, D., & Lembo, A., Jr (2008). Evaluation of an oral vaccination program to control raccoon rabies in a suburbanized landscape. *Human-Wildlife Conflicts, 2*, 212–224.

Boyer, J., Canac-Marquis, P., Guerin, D., Mainguy, J., & Pelletier, F. (2011). Oral vaccination against raccoon rabies: Landscape heterogeneity and timing of distribution influence wildlife contact rates with the ONRAB vaccine bait. *Journal of Wildlife Diseases, 47*, 593–602.

Briggs, D., Briggs, J., & Howard, D. (1988). Prevalence of rabies in the striped skunk (*Mephitis mephitis*) in Kansas from 1966 to 1986. *Transactions of the Kansas Academy of Science, 91*, 123–131.

Broadfoot, J. D., Rosatte, R. C., & O'Leary, D. T. (2001). Raccoon and skunk population models for urban disease control planning in Ontario, Canada. *Ecological Applications, 11*, 295–303.

Brochier, B., Aubert, M. F., Pastoret, P. P., Masson, E., Schon, J., Lombard, M., et al. (1996). Field use of a vaccinia–rabies recombinant vaccine for the control of sylvatic rabies in

Europe and North America. *Revue Scientifique et Technique de L'Office International des Epizootics*, 15, 947–970.

Brochier, B., Deschamps, P., Costy, F., Leuris, J., Villers, M., Peharpre, D., et al. (2001). Elimination of sylvatic rabies in Belgium by oral vaccination of the red fox (*Vulpes vulpes*). *Annales de Medecine Veterinaire*, 145, 293–305.

Brown, K. (2011). Rabies epidemiologies: The emergence and resurgence of rabies in twentieth century South Africa. *Journal of the History of Biology*, 44, 81–101.

Brown, L., Rosatte, R., Fehlner-Gardiner, C., Knowles, K., Bachmann, P., Davies, J. C., et al. (2011). Immunogenicity and efficacy of two rabies vaccines in wild-caught, captive raccoons. *Journal of Wildlife Diseases*, 47, 192–194.

Bruyère, V., Vuillaume, P., Cliquet, F., & Aubert, M. (2000). Oral rabies vaccination of foxes with one or two delayed distributions of SAG2 baits during the spring. *Veterinary Research*, 31, 339–345.

Capello, K., Mulatti, P., Gagliazzo, L., Guberti, V., Citterio, C., DeBenedictis, P., et al. (2010). Impact of emergency oral rabies vaccination of foxes in north-eastern Italy, 28 December 2009 – January 2010: preliminary evaluation. *Eurosurveillance*, 15, 1–4.

Chautan, M., Pontier, D., & Artois, M. (2000). Role of rabies in recent demographic changes in red fox (*Vulpes vulpes*) populations in Europe. *Mammalia*, 64, 391–410.

Childs, J. E., Curns, A. T., Dey, M. E., Real, L. A., Feinstein, L., Bjørnstad, O. N., et al. (2000). Predicting the local dynamics of epizootic rabies among raccoons in the United States. *Proceedings of the National Academy of Sciences, USA*, 97, 13666–13671.

Cleaveland, S. (1998). Epidemiology and control of rabies: The growing problem of rabies in Africa. *Transactions of the Royal Society of Tropical Medicine and Hygiene*, 92, 131–134.

Cliquet, F., Barrat, J., Guiot, A., Cael, N., Boutrand, S., Maki, J., et al. (2008). Efficacy and bait acceptance of vaccinia vectored rabies glycoprotein vaccine in captive foxes (*Vulpes vulpes*), raccoon dogs (*Nystereutes procyonoides*) and dogs (*Canis familiaris*). *Vaccine*, 26, 4627–4638.

Coulon, P., Lafay, F., LeBlois, H., Tuffereau, C., Artois, M., Blancou, J., et al. (1992). The SAG1, a new attenuated oral rabies vaccine. In K. Bogel, F. -X. Meslin, & M. Kaplan (Eds.), *Wildlife Rabies Control (pp. 105–111)*. Kent, England: Wells Medical.

Coyne, M. J., Smith, G., & McAllister, F. E. (1989). Mathematical model for the population biology of rabies in raccoons in the mid-Atlantic states. *American Journal of Veterinary Research*, 12, 2148–2154.

Cullingham, C., Kyle, C., Pond, B., Rees, E., & White, B. (2009). Differential permeability of rivers to raccoon gene flow corresponds to rabies incidence in Ontario, Canada. *Molecular Ecology*, 18, 43–53.

Cullingham, C., Pond, B., Kyle, K., Rosatte, R., & White, B. (2008). Combining direct and indirect genetic methods to estimate dispersal for informing wildlife disease management decisions. *Molecular Ecology*, 17, 4874–4886.

Curk, A., & Carpenter, T. (1994). Efficacy of the first oral vaccination against fox rabies in Slovenia. *Revue Scientific et Technique de L'Office International des Epizootics*, 13, 763–775.

De Mattos, C., Mattos, C., Loza-Rubio, E., Aguilar-Setien, A., Orciari, L., & Smith, J. (1999). Molecular characterization of rabies virus isolates from Mexico: Implications for transmission dynamics and human risk. *American Journal of Tropical Medicine and Hygiene*, 61, 587–597.

Debbie, J. (1991). Rabies control in terrestrial wildlife by population reduction. In G. Baer (Ed.), *The Natural History of Rabies (pp. 477–484)* (2nd ed.). Boca Raton, Florida: CRC Press.

Delpietro, H., Lord, R., Russo, R., & Gury-Dhomen, F. (2009). Observation of sylvatic rabies in Northern Argentina during outbreaks of paralytic cattle rabies transmitted by vampire bats *Desmodus rotundus*. *Journal of Wildlife Diseases*, 45, 1169–1173.

Deyoung, R., Zamorano, A., Mesenbrink, B., Campbell, T., Leland, B., Moore, G., et al. (2009). Landscape-genetic analysis of population structure in the Texas gray fox oral rabies vaccination zone. *Journal of Wildlife Management*, 73, 1292–1299.

Diaz, A., Papo, S., Rodriguez, A., & Smith, J. (1994). Antigenic analysis of rabies-virus isolates from Latin America and the Caribbean. *Journal of Veterinary Medicine, 41*, 153–160.

Eisinger, D., & Thulke, H. (2008). Spatial pattern formation facilitates eradication of infectious diseases. *Journal of Applied Ecology, 45*, 415–423.

Eisinger, D., Thulke, H., Selhorst, T., & Muller, T. (2005). Emergency vaccination of rabies under limited resources – combating or containing? *BMC Infectious Diseases, 5*, 10–26.

Engeman, R., Christensen, K., Pipas, M., & Bergman, D. (2003). Population monitoring in support of a rabies vaccination program for skunks in Arizona. *Journal of Wildlife Diseases, 39*, 746–750.

Everard, C., & Everard, J. (2006). Mongoose rabies in the Caribbean. *Annals of the New York Academy of Sciences*, doi:10.1111/j.1749-6632.1992.tb19662.x.

Farry, S. C., Henke, S. E., Beasom, S. L., & Fearneyhough, M. G. (1998). Efficacy of bait distributional strategies to deliver canine rabies vaccines to coyotes in southern Texas. *Journal of Wildlife Diseases, 34*, 23–32.

Fearneyhough, M. G., Wilson, P. J., Clark, K. A., Smith, D. R., Johnston, D. H., Hicks, B. N., et al. (1998). Results of an oral rabies vaccination program for coyotes. *Journal of the American Veterinary Medical Association, 212*, 498–502.

Fehlner-Gardiner, C., Nadin-Davis, S., Armstrong, J., Muldoon, F., Bachmann, P., & Wandeler, A. (2008). ERA vaccine-induced cases of rabies in wildlife and domestic animals in Ontario, Canada, 1989–2004. *Journal of Wildlife Diseases, 44*, 71–85.

Fehlner-Gardiner, C., Rudd, R., Donovan, D., Slate, D., Kempf, L., & Badcock, J. (2012). Comparison of ONRAB® and Raboral V-RG oral rabies vaccine field performance in raccoons and striped skunks in New Brunswick, Canada, and Maine, USA. *Journal of Wildlife Diseases, 48*, 157–167.

Follmann, E., Ritter, D., & Hartbauer, D. (2004). Oral vaccination of captive arctic foxes with lyophilized SAG2 rabies vaccine. *Journal of Wildlife Diseases, 40*, 328–334.

Gascoyne, S., King, A., Laurenson, M., Borner, M., Schildger, B., & Barrat, J. (1993). Aspects of rabies infection and control in the conservation of the African wild dog (*Lycaon pictus*) in the Serengeti region, Tanzania. *Onderstepoort Journal of Veterinary Research, 60*, 415–420.

Gerletti, G., Guidali, F., Scherini, G., & Tosi, G. (1991). Management of the fox (*Vulpes vulpes*) in Lombardy region (Northern Italy) in relation to rabies. *Hystrix, 3*, 191–195.

Gordon, E., Curns, A., Krebs, J., Rupprecht, C., Real, L., & Childs, J. (2004). Temporal dynamics of rabies in a wildlife host and the risk of cross-species transmission. *Epidemiology and Infection, 132*, 515–524.

Goszczynski, J., Misiorowska, M., & Juszko, S. (2008). Changes in density and spatial distribution of red fox dens and cub numbers in central Poland following rabies vaccination. *Acta Theriologica, 53*, 121–127.

Guerra, M., Curns, A., Rupprecht, C., Hanlon, C., Krebs, J., & Childs, J. (2003). Skunk and raccoon rabies in the eastern United States: Temporal and spatial analysis. *Emerging Infectious Diseases, 9*, 1143–1150.

Hadidian, J., Prange, S., Rosatte, R., Riley, S., & Gehrt, S. (2010). Raccoons (*Procyon lotor*). In S. Gehrt, S. Riley, & B. Cypher (Eds.), *Urban Carnivores, Ecology, Conflict and Conservation (pp. 35–48)*. Baltimore, Maryland: The Johns Hopkins University Press.

Hanlon, C., Niezgoda, M., Morrill, P., & Rupprecht, C. (2002). Oral efficacy of an attenuated rabies vaccine in skunks and raccoons. *Journal of Wildlife Diseases, 38*, 420–427.

Hanlon, C. L., Hayes, D. E., Hamir, A. N., Snyder, D. E., Jenkins, S. R., Hable, C. P., et al. (1989). Proposed field evaluation of a rabies recombinant vaccine for raccoons (*Procyon lotor*): Site selection, target species characteristics, and placebo baiting trials. *Journal of Wildlife Diseases, 25*, 555–567.

Harris, S., Cheeseman, C., Smith, G., & Trewhella, W. (1990). Rabies contingency planning in Britain. In P. O'Brien & G. Berry (Eds.), *Wildlife Rabies Contingency Planning in Australia (pp. 63–77)*. Canberra, Australia: Australian Government Publishing Service.

Haydon, D., Laurenson, M., & Sillero-Zubiri, C. (2002). Integrating epidemiology into population viability analysis: Managing the risk posed by rabies and canine distemper to the Ethiopian wolf. *Conservation Biology, 16,* 1372–1385.

Henderson, H., Jackson, F., Bean, K., Panasak, B., Niezgoda, M., Slate, D., et al. (2009). Oral immunization of raccoons and skunks with a canine adenovirus recombinant rabies vaccine. *Vaccine, 27,* 7194–7197.

Holmala, K., & Kauhala, K. (2009). Habitat use of medium sized carnivores in southeast Finland – key habitats for rabies spread. *Annales Zoologici Fennici, 46,* 233–246.

Hostnik, P., Rihtaric, D., Grom, J., Malovrh, T., & Toplak, I. (2011). Maintenance and control of the vaccination belt along neighbouring rabies infected area. *Acta Veterinaria, 61,* 163–174.

Irsara, A., Bressan, G., & Mutinelli, F. (1990). Sylvatic rabies in Italy: Epidemiology. *Journal of Veterinary Medicine, B37,* 53–63.

Jackson, A. C. (1994). The fatal illness of the Fourth Duke of Richmond in Canada: Rabies. *Annals of the Royal College of Physicians and Surgeons Canada, 27,* 40–41.

Janani, A., Fayaz, A., Simani, S., Farahtaj, F., Eslami, N., Howaizi, N., et al. (2008). Epidemiology and control of rabies in Iran. *Developmental Biology, 131,* 207–211.

Johnson, N., Black, C., Smith, J., Un, H., McElhinney, L., Aylan, O., et al. (2003). Rabies emergence among foxes in Turkey. *Journal of Wildlife Diseases, 39,* 262–270.

Johnson, N., Fooks, A., Valtchovski, R., & Muller, T. (2007). Evidence for trans-border movement of rabies by wildlife reservoirs between countries in the Balkan Peninsula. *Veterinary Microbiology, 120,* 71–76.

Johnson, N., Freuling, C., Vos, A., Un, H., Valtchovski, R., Turcitu, M., et al. (2008). Epidemiology of rabies in southeast Europe. *Developmental Biology, 131,* 189–198.

Johnson, N., Un, H., Fooks, A., Freuling, C., Muller, T., Aylan, O., et al. (2010). Rabies epidemiology and control in Turkey. *Epidemiology and Infection, 138,* 305–312.

Johnston, D. H., & Beauregard, M. (1969). Rabies epidemiology in Ontario. *Bulletin of the Wildlife Disease Association, 5,* 357–370.

Johnston, D. H., Joachim, D. G., Bachmann, P., Kardong, K. V., Stewart, R. E. A., Dix, … et al. (1999). Aging furbearers using tooth structure and biomarkers. In M. Novak, J. A. Baker, M. E. Obbard, & B. Malloch (Eds.), *Wild Furbearer Management and Conservation in North America (pp. 228–243)* (CD Edition). Sault Ste Marie, Ontario: Ontario Fur Managers Federation.

Johnston, D. H., & Tinline, R. R. (2002). Rabies control in Wildlife. In A. Jackson & W. Wunner (Eds.), *Rabies (pp. 445–471).* San Diego, California: Academic Press.

Johnston, D. H., Voigt, D. R., MacInnes, C. D., Bachmann, P., Lawson, K. F., & Rupprecht, C. E. (1988). An aerial baiting system for the distribution of attenuated or recombinant rabies vaccines for foxes, raccoons and skunks. *Reviews of Infectious Diseases, 10,* S660–664.

Jones, M., Curns, A., Krebs, J., & Childs, J. (2003). Environmental and human demographic features associated with epizootic raccoon rabies in Maryland, Pennsylvania, and Virginia. *Journal of Wildlife Diseases, 39,* 23–32.

Keightley, A., Struthers, J., Johnson, S., & Barnard, B. (1987). Rabies in South Africa: 1980–1984. *South African Journal of Science, 83,* 466–472.

Kemere, P., Liddel, M., Evangelou, P., Slate, D., & Osmek, S. (2002). Economic analysis of a large scale oral vaccination program to control raccoon rabies. In L. Clark (Ed.), *Human Conflicts with Wildlife Economic Considerations (pp. 109–116).* Fort Collins, Colorado: National Wildlife Research Center.

Kieny, M., Lathe, R., Drillien, R., Spehner, D., Skory, S., Schmitt, D., et al. (1984). Expression of rabies virus glycoprotein from a recombinant vaccinia virus. *Nature, 312,* 163–166.

Knobel, D., Fooks, A., Brooks, S., Randall, D., Williams, S., Argaw, K., et al. (2008). Trapping and vaccination of endangered Ethiopian wolves to control an outbreak of rabies. *Journal of Applied Ecology, 45,* 109–116.

Knobel, D., Toit, J., & Bingham, J. (2002). Development of a bait and baiting system for delivery of oral rabies vaccine to free-ranging African wild dogs (*Lycaon pictus*). *Journal of Wildlife Diseases*, 38, 352–362.

Knowles, M. K., Nadin-Davis, S. A., Sheen, M., Rosatte, R., Mueller, R., & Beresford, A. (2009a). Safety studies on an adenovirus recombinant vaccine for rabies (AdRG1.3-ONRAB®) in target and non-target species. *Vaccine*, 27, 6619–6626.

Knowles, M. K., Roberts, D., Craig, S., Sheen, M., Nadin-Davis, S., & Wandeler, A. (2009b). In vitro and in vivo genetic stability studies of a human adenovirus type 5 recombinant rabies glycoprotein vaccine (ONRAB®). *Vaccine*, 27, 2662–2668.

Kovalev, N., Sedov, V., Shashenko, A., Osidse, D., & Ivanovsky, E. (1992). An attenuated oral vaccine for wild carnivores in the USSR. In K. Bogel, F. -X. Meslin, & M. Kaplan (Eds.), *Wildlife Rabies Control (pp. 112–114)*. Kent, England: Wells Medical.

Kreindel, S. M., McGuill, M., Meltzer, M., Rupprecht, C., & DeMaria, A. (1998). The cost of rabies post-exposure prophylaxis: One state's experience. *Public Health Reports*, 113, 247–251.

Kuzmin, I., Botvinkin, A., McElhinney, M., Smith, J., Orciari, L., Hughes, G., et al. (2004). Molecular epidemiology of terrestrial rabies in the former Soviet Union. *Journal of Wildlife Diseases*, 40, 617–631.

Laine, M., Niin, E., & Partel, A. (2008). The rabies elimination program in Estonia using oral rabies vaccination of wildlife – preliminary results. *Developmental Biology*, 131, 239–247.

Lambot, M., Blasco, E., Barrat, J., Cliquet, F., Brochier, B., Renders, C., et al. (2001). Humoral and cell-mediated immune responses of foxes (*Vulpes vulpes*) after experimental primary and secondary oral vaccination using SAG$_2$ and V-RG vaccines. *Vaccine*, 19, 1827–1835.

Laurenson, K., Van Heerden, J., Stander, P., & Van Vuuren, M. J. (1997). Seroepidemiological survey of sympatric domestic and wild dogs (*Lycaon pictus*) in Tsumkwe District, north-eastern Namibia. *Onderstepoort Journal of Veterinary Research*, 64, 313–316.

Lawson, K. F., & Bachmann, P. (2001). Stability of attenuated live virus rabies vaccine in baits targeted to wild foxes under operational conditions. *Canadian Veterinary Journal*, 42, 368–374.

Lawson, K. F., Chiu, H., Crosgrey, S. J., Matson, M., Casey, G. A., & Campbell, J. B. (1997). Duration of immunity in foxes vaccinated orally with ERA vaccine in a bait. *Canadian Journal of Veterinary Research*, 61, 39–42.

Lembo, T., Hampson, K., Haydon, D., Craft, M., Dobson, A., Dushoff, J., et al. (2008). Exploring reservoir dynamics: A case study of rabies in the Serengeti ecosystem. *Journal of Applied Ecology*, 45, 1246–1257.

Lembo, T., Hampson, K., Kaare, M., Ernest, E., Knobel, D., Kazwala, R., et al. (2010). The feasibility of canine rabies elimination in Africa: Dispelling doubts with data. *Public Library of Science Neglected Tropical Diseases*, 4(2), e626. doi:10.1371/journal.pntd.0000626.

Lembo, T., Niezgoda, M., Velasco-Villa, A., Cleaveland, S., Ernest, E., & Rupprecht, C. (2006). Evaluation of a direct, rapid, immunohistochemical test for rabies diagnosis. *Emerging Infectious Diseases*, 12, 310–313.

Leslie, M., Messenger, S., Rohde, R., Smith, J., Cheshier, R., Hanlon, C., et al. (2006). Bat associated rabies virus in skunks. *Emerging Infectious Diseases*, 12, 1274–1277.

Linhart, S., Creekmore, T., Corn, J., Whitney, M., Snyder, B., & Nettles, V. (1993). Evaluation of baits for oral rabies vaccination of mongooses: Pilot field trials in Antigua, West Indies. *Journal of Wildlife Diseases*, 29, 290–294.

Linhart, S., King, R., Zamir, S., Naveh, U., Davidson, M., & Perl, S. (1997). Oral rabies vaccination of red foxes and golden jackals in Israel; preliminary bait evaluation. *Revue Scientific et Technique de L'Office International des Epizootics*, 16, 874–880.

Lutze-Wallace, C., Sapp, T., Sidhu, M., & Wandeler, A. (1995). In vitro assessments of the genetic stability of a live recombinant human adenovirus vaccine against rabies. *Canadian Journal of Veterinary Research*, 59, 157–160.

Ma, X., Blanton, J., Rathban, S., Recuenco, S., & Rupprecht, C. (2010). Time series analysis of the impact of oral vaccination on raccoon rabies in West Virginia, 1990–2007. *Vector-borne and Zoonotic Diseases, 10*, 801–809.

MacInnes, C. (1988). Control of Wildlife rabies: The Americas. In J. Campbell & K. Charlton (Eds.), *Rabies (pp. 381–405)*. Boston: Kluwer Academic.

MacInnes, C. D., Smith, S. M., Tinline, R. R., Ayers, N. R., Bachmann, P., Ball, D., et al. (2001). Elimination of rabies from red foxes in eastern Ontario. *Journal of Wildlife Diseases, 37*, 119–132.

Marks, C., & Bloomfield, T. (1999). Bait uptake by foxes (*Vulpes vulpes*) in urban Melbourne: The potential of oral vaccination for rabies control. *Wildlife Research, 26*, 777–787.

Masson, E., Bruyère-Masson, V., Vuillaume, P., Lemoyne, S., & Aubert, M. (1999). Rabies oral vaccination of foxes during the summer with the VRG vaccine bait. *Veterinary Research, 30*, 595–605.

Matouch, O. (2008). The rabies situation in Eastern Europe. *Developmental Biology, 131*, 27–35.

Matouch, O., & Polak, L. (1982). Rabies epizootiology and control in Czechoslovakia. *Comparative Immunology Microbiology and Infectious Diseases, 5*, 303–307.

Matouch, O., & Vitasek, J. (2005). Elimination of rabies in the Czech Republic by oral vaccination of foxes. *WHO Rabies Bulletin Europe, 29(1)*, 10–15.

Meltzer, M. I. (1996). Assessing the cost and benefits of an oral vaccine for raccoon rabies: A possible model. *Emerging Infectious Diseases, 2*, 343–349.

Meltzer, M., & Rupprecht, C. (1998). Economics of the prevention and control of rabies. Part 2: Rabies in dogs, livestock and wildlife. *Pharmacoeconomics, 14*, 481–498.

Metlin, A., Rybakov, S., & Mikhalishin, V. (2009). Oral vaccination of wild carnivores against rabies. *Veterinary Medicine Journal, 9*, 18–25.

Muller, T., Batza, H., Becket, A., Banzenthal, C., Cox, J., Freuling, … et al. (2009). Analysis of vaccine-virus associated rabies cases in red foxes (*Vulpes vulpes*) after oral rabies vaccination in Germany and Alaska. *Archives of Virology, 154*, 1081–1091.

Müller, T. F., Schuster, P., Vos, A. C., Selhorst, T., Wenzel, U. D., & Neubert, A. M. (2001a). Effect of maternal immunity on the immune response of young foxes to oral vaccination with SAD B19. *American Journal of Veterinary Research, 62*, 1154–1158.

Muller, T., Vos, A., Selhorst, T., Stiebling, U., Tackmann, K., Schuster, P., et al. (2001b). Is it possible to orally vaccinate juvenile red foxes against rabies in spring campaigns? *Journal of Wildlife Diseases, 37*, 791–797.

Nadin-Davis, S., Muldoon, F., & Wandeler, A. (2006). Persistence of genetic variants of the Arctic fox strain of rabies virus in southern Ontario. *Canadian Journal of Veterinary Research, 70*, 11–19.

Nadin-Davis, S., Muldoon, F., Whitney, H., & Wandeler, A. (2008). Origins of the rabies virus associated with an outbreak in Newfoundland. *Journal of Wildlife Diseases, 44*, 86–98.

Neilan, R., & Lenhart, S. (2010). Optimal vaccine distribution in a spatio temporal epidemic model with an application to rabies and raccoons. *Journal of Mathematical Analysis and Applications, 378*, 603–619.

Niin, E., Laine, M., Guiot, A., Demerson, J., & Cliquet, F. (2008). Rabies in Estonia before and after the first campaigns of oral vaccination of wildlife with SAG2 vaccine bait. *Vaccine, 26*, 3556–3565.

Nyberg, M., Kulonen, K., Neuvonen, E., Ek-Kommonen, C., Nuorgam, M., & Westerling, B. (1992). An epidemic of sylvatic rabies in Finland – Descriptive epidemiology and results of oral vaccination. *Acta Veterinaria Scandinavica, 33*, 43–57.

Oertti, E., Wilson, P., Sidwa, T., & Rohde, R. (2009). Epidemiology of rabies in skunks in Texas. *Journal of the American Veterinary Medical Association, 234*, 616–620.

Olson, C., & Werner, P. (1999). Oral rabies vaccine contact by raccoons and nontarget species in a field trial in Florida. *Journal of Wildlife Diseases, 35*, 687–695.

Olson, C. A., Mitchell, K. D., & Werner, P. A. (2000). Bait ingestion by free-ranging raccoons and nontarget species in an oral rabies vaccine field trial in Florida. *Journal of Wildlife Diseases*, 36, 734–743.

Pastoret, P., & Brochier, B. (1999). Epidemiology and control of fox rabies in Europe. *Vaccine*, 17, 1750–1754.

Pastoret, P., Kappeler, A., & Aubert, M. (2004). European rabies control and its history. In A. King, M. Fooks, A. Aubert, & A. Wandeler (Eds.), *Historical perspective of rabies in Europe and the Mediterranean basin (pp. 337–350)*. Paris, France: OIE. World Organization for Animal Health.

Pavlin, B., Schloegel, L., & Daszak, P. (2009). Risk of importing zoonotic diseases through wildlife trade, United States. *Emerging Infectious Diseases*, 15, 1721–1726.

Ramey, P., Blackwell, B., Gates, R., & Siemons, R. (2008). Oral rabies vaccination of a northern Ohio raccoon population: Relevance of population density and prebait serology. *Journal of Wildlife Diseases*, 44, 553–568.

Recuenco, S., Eidson, M., Cherry, B., Kulldorff, M., & Johnson, G. (2008). Factors associated with endemic raccoon (*Procyon lotor*) rabies in terrestrial mammals in New York State, USA. *Preventative Veterinary Medicine*, 86, 30–42.

Rees, E., Belanger, D., Lelievre, F., Cote, N., & Lambert, L. (2011a). Targeted surveillance of raccoon rabies in Quebec, Canada. *Journal of Wildlife Management*, 75, 1406 1416.

Rees, E., Pond, B., Cullingham, C., Tinline, R., Ball, D., Kyle, C., et al. (2009). Landscape modeling spatial bottlenecks: Implications for raccoon rabies disease spread. *Biology Letters*, 5, 387–390.

Rees, E., Pond, B., Tinline, R., & Belanger, D. (2011b). Understanding effects of barriers on the spread and control of rabies. In A. C. (2011). Jackson (Ed.), *Research Advances in Rabies: Advances in Virus Research (Vol. 79, pp. 421–447)*. London, UK: Academic Press, Elsevier.

Robbins, A. H., Borden, M. D., Windmiller, B. S., Niezgoda, M., Marcus, L. C., O'Brien, S., et al. (1998). Prevention of the spread of rabies to wildlife by oral vaccination of raccoons in Massachusetts. *Journal of the American Veterinary Medical Association*, 213, 1407–1412.

Rosatte, R. (2000). Management of raccoons (Procyon lotor) in Ontario, Canada: Do human intervention and disease have significant impact on raccoon populations? *Mammalia*, 64, 369–390.

Rosatte, R., & Allan, M. (2009). The ecology of red foxes, *Vulpes vulpes*, in metropolitan Toronto, Ontario: Disease management implications. *Canadian Field Naturalist*, 123, 215–220.

Rosatte, R., Allan, M., Warren, R., Neave, P., Babin, T., Buchanan, L., et al. (2005). Movements of two rabid raccoons (*Procyon lotor*) in eastern Ontario, Canada. *Canadian Field Naturalist*, 119, 453–454.

Rosatte, R., Donovan, D., Allan, M., Bruce, L., & Davies, C. (2007a). Human-assisted movements of raccoons, *Procyon lotor*, and opossums, *Didelphis virginiana*, between the United States and Canada. *Canadian Field Naturalist*, 121, 212–213.

Rosatte, R., Howard, D., Campbell, J., & MacInnes, C. (1990). Intramuscular vaccination of skunks and raccoons against rabies. *Journal of Wildlife Diseases*, 26, 225–230.

Rosatte, R., Kelly, P., & Power, M. (2011b). Home range, movements, and habitat utilization of striped skunks (*Mephitis mephitis*) in Scarborough, Ontario, Canada: Disease management implications. *Canadian Field Naturalist*, 125, 27–33.

Rosatte, R., & Lariviere, S. (2003). Skunks. In G. Feldhamer, B. Thompson, & J. Chapman (Eds.), *Wild Mammals of North America: biology, management and conservation (pp. 692–707)* (2nd ed.). Baltimore: Johns Hopkins University Press.

Rosatte, R., & Lawson, K. (2001). Acceptance of baits for delivery of oral rabies vaccine to raccoons. *Journal of Wildlife Diseases*, 37, 730–739.

Rosatte, R., MacDonald, E., Sobey, K., Donovan, D., Bruce, L., Allan, M., et al. (2007c). The elimination of raccoon rabies from Wolfe Island, Ontario: Animal density and movements. *Journal of Wildlife Diseases*, 43, 242–250.

Rosatte, R., MacInnes, C., Power, M., Johnston, D., Bachmann, P., Nunan, C., et al. (1993). Tactics for the control of wildlife rabies in Ontario, Canada. *Revue Scientific et Technique de L'Office International des Epizootics, 12*, 95–98.

Rosatte, R., MacInnes, C., Williams, R., & Williams, O. (1997). A proactive prevention strategy for raccoon rabies in Ontario, Canada. *Wildlife Society Bulletin, 25*, 110–116.

Rosatte, R., Power, M., & MacInnes, C. (1991). Ecology of urban skunks, raccoons, and foxes in Metropolitan Toronto. In L. Adams & D. Leedy (Eds.), *Wildlife conservation in Metropolitan environments (pp. 31–38).* Columbia, Maryland: National Institute for Urban Wildlife.

Rosatte, R., Pybus, M., & Gunson, J. (1986). Population reduction as a factor in the control of skunk rabies in Alberta. *Journal of Wildlife Diseases, 22*, 459–467.

Rosatte, R., Ryckman, M., Ing, K., Proceviat, S., Allan, M., Bruce, L., et al. (2010a). Density, movements, and survival of raccoons in Ontario, Canada: Implications for disease spread and management. *Journal of Mammalogy, 91*, 122–135.

Rosatte, R., Ryckman, M., Meech, S., Proceviat, S., Bruce, L., Donovan, D., et al. (2010b). Home range, movements, and survival of rehabilitated raccoons (*Procyon lotor*) in Ontario, Canada. *Journal of Wildlife Rehabilitation, 30*, 7–12.

Rosatte, R., Sobey, K., Donovan, D., Allan, M., Bruce, L., Buchanan, T., et al. (2007e). Raccoon density and movements after population reduction to control rabies. *Journal of Wildlife Management, 71*, 2373–2378.

Rosatte, R., Sobey, K., Donovan, D., Bruce, L., Allan, M., Silver, A., et al. (2006). Behaviour, movements, and demographics of rabid raccoons in Ontario, Canada: Management implications. *Journal of Wildlife Diseases, 42*, 589–605.

Rosatte, R., Sobey, K., Dragoo, J., & Gehrt, S. (2010c). Striped skunks and allies (*Mephitis* spp). In S. Gehrt, S. Riley, & B. Cypher (Eds.), *Urban Carnivores, Ecology, Conflict and Conservation (pp. 97–108).* Baltimore, Maryland: The Johns Hopkins University Press.

Rosatte, R., Tinline, R., & Johnston, D. (2007f). Rabies control in wild carnivores Jackson & W. Wunner (Eds.), *Rabies (pp. 595–634)* (2nd ed.). San Diego, California: Academic Press.

Rosatte, R., Wandeler, A., Muldoon, F., & Campbell, D. (2007g). Porcupine quills in raccoons as an indicator of rabies, distemper, or both diseases: Disease management implications. *Canadian Veterinary Journal, 48*, 299–300.

Rosatte, R. C. (1988). Rabies in Canada – history, epidemiology and control. *Canadian Veterinary Journal, 29*, 362–365.

Rosatte, R. C. (2001). Long distance movement by a coyote and a red fox in Ontario: Implications for disease-spread. *Canadian Field Naturalist, 116*, 129–131.

Rosatte, R. C. (2011). Evolution of wildlife rabies control tactics. In A. C. (2011). Jackson (Ed.), *Research Advances in Rabies: Advances in Virus Research (Vol. 79*, pp. 397–419). London, UK: Academic Press, Elsevier.

Rosatte, R. C., Allan, M., Bachmann, P., Sobey, K., Donovan, D., Davies, J. C., et al. (2008). Prevalence of tetracycline and rabies virus antibody in raccoons, skunks, and foxes following aerial distribution of V-RG baits to control raccoon rabies in Ontario, Canada. *Journal of Wildlife Diseases, 45*, 772–784.

Rosatte, R. C., Donovan, D., Allan, M., Bruce, L., Buchanan, T., Sobey, K., et al. (2007b). Rabies in vaccinated raccoons from Ontario, Canada. *Journal of Wildlife Diseases, 43*, 300–301.

Rosatte, R. C., Donovan, D., Allan, M., Bruce, L., Buchanan, T., Sobey, K., et al. (2009b). The control of raccoon rabies in Ontario Canada: Proactive and reactive tactics, 1994–2007. *Journal of Wildlife Diseases, 45*, 772–784.

Rosatte, R. C., Donovan, D., Allan, M., Howes, L., Silver, A., Bennett, K., et al. (2001). Emergency response to raccoon rabies introduction in Ontario. *Journal of Wildlife Diseases, 37*, 265–279.

Rosatte, R. C., Donovan, D., Davies, J. C., Allan, M., Bachmann, P., Stevenson, B., et al. (2009a). Aerial distribution of ONRAB® baits as a tactic to control rabies in raccoons and striped skunks in Ontario, Canada. *Journal of Wildlife Diseases, 45*, 363–374.

Rosatte, R. C., Donovan, D., Davies, J. C., Brown, L., Allan, M., von Zuben, V., et al. (2011a). High density baiting with ONRAB® rabies vaccine baits to control Arctic-variant rabies in striped skunks in Ontario, Canada. *Journal of Wildlife Diseases, 47*, 459–465.

Rosatte, R. C., & Gunson, J. R. (1984). Presence of neutralizing antibodies to rabies virus in striped skunks from areas free of skunk rabies in Alberta. *Journal of Wildlife Diseases, 20*, 171–176.

Rosatte, R. C., & MacInnes, C. D. (1989). Relocation of city raccoons. In A. J. Bjugstad, D. W. Uresk, & R. H. Hamre (Eds.), *Ninth great plains wildlife damage control workshop proceedings (pp. 87–92)*. Ft. Collins, Colorado: Great Plains Agricultural Council Publication 127. (USDA Forest Service General Technical Report RM-171).

Rosatte, R. C., Lawson, K., & MacInnes, C. (1998). Development of baits to deliver oral rabies vaccine to raccoons in Ontario. *Journal of Wildlife Diseases, 34*, 647–652.

Rosatte, R. C., Power, M., Donovan, D., Davies, J. C., Allan, M., Bachmann, P., et al. (2007d). Elimination of arctic variant rabies in red foxes, metropolitan Toronto. *Emerging Infectious Diseases, 13*, 25–27.

Rosatte, R. C., Power, M. J., & MacInnes, C. D. (1992). Trap-vaccinate-release and oral vaccination techniques for rabies control in urban skunks, raccoons and foxes. *Journal of Wildlife Diseases, 28*, 562–571.

Roscoe, D. E., Holste, W. C., Sorhage, F. E., Campbell, C., Niezgoda, M., Buchannan, R., et al. (1998). Efficacy of an oral vaccinia–rabies glycoprotein recombinant vaccine in controlling epidemic raccoon rabies in New Jersey. *Journal of Wildlife Diseases, 34*, 752–763.

Rupprecht, C., Blass, L., Smith, K., Orciari, L., Niezgoda, M., Whitfield, S., et al. (2001). Human infection due to recombinant vaccinia-rabies glycoprotein virus. *New England Journal of Medicine, 345*, 582–586.

Rupprecht, C., Hamir, A., Johnston, D., & Koprowski, H. (1988). Efficacy of vaccinia-rabies glycoprotein recombinant virus vaccine in raccoons (Procyon lotor). *Reviews of Infectious Diseases, 10*(Suppl. 4), S803–S809.

Rupprecht, C. E., Wiktor, T. J., Johnston, D. H., Hamir, A. N., Dietzschold, B., Wunner, W., et al. (1986). Oral immunization and protection of raccoons (Procyon lotor) with a vaccinia–rabies glycoprotein recombinant virus vaccine. *Proceedings of the National Academy of Sciences, USA, 83*, 7947–7950.

Russell, C., Smith, D., Childs, J., & Real, L. (2005). Predictive spatial dynamics and strategic planning for raccoon rabies emergence in Ohio. *PLoS Biology, 3*, 382–388.

Sato, G., Itou, T., Shoji, Y., Miura, Y., Mikami, T., Ito, M., et al. (2004). Genetic and phylogenetic analysis of glycoprotein of rabies virus isolated from several species in Brazil. *Journal of Veterinary Medical Science, 66*(7), 747–753.

Sattler, A., Krogwold, R., Wittum, T., Rupprecht, C., Algeo, T., Slate, D., et al. (2009). Influence of oral rabies vaccine bait density on rabies seroprevalence in wild raccoons. *Vaccine, 27*, 7187–7193.

Schneider, L., & Cox, J. (1988). Eradications of rabies through oral vaccination. The German field trial. In P. Pastoret, B. Brochier, I. Thomas, & J. Blancou (Eds.), *Vaccination to control rabies in foxes (pp. 22–38)*. Luxembourg: Commission of the European Communities.

Schneider, L., Cox, J., Muller, W., & Hohnsbeen, K. (1988). Current oral rabies vaccination in Europe, an interim balance. *Reviews of Infectious Diseases, 10*, S654–S659.

Selhorst, T., Thulke, H., & Muller, T. (2001). Cost-efficient vaccination of foxes (*Vulpes vulpes*) against rabies and the need for a new baiting strategy. *Preventive Veterinary Medicine, 51*, 95–109.

Shwiff, S., Sterner, R., Hale, R., Jay, M., Sun, B., & Slate, D. (2009). Benefit cost scenarios of potential oral rabies vaccination for skunks in California. *Journal of Wildlife Diseases, 45*, 227–233.

Shwiff, S. A., Nunan, C., Kirkpatrick, K., & Shwiff, S. S. (2011). A retrospective economic analysis of the Ontario red fox oral rabies vaccination programme. *Zoonoses Public Health, 58*, 169–177.

Sidwa, T., Wilson, P., Moore, G., Oertli, E., Hicks, B., Rohde, R., et al. (2005). Evaluation of oral rabies vaccination programs for control of rabies epizootics in coyotes and gray foxes: 1995–2003. *Journal of the American Veterinary Medical Association, 227*, 785–792.

Singer, A., Kauhala, K., Holmala, K., & Smith, G. (2009). Rabies in north-eastern Europe-the threat from invasive raccoon dogs. *Journal of Wildlife Diseases, 45*, 1121–1137.

Slate, D., Algeo, T., Nelson, K., Chipman, R., Donovan, D., Blanton, J., et al. (2009). Oral rabies vaccination in North America: Opportunities Complexities and Challenges. *PLOS Neglected Tropical Diseases, 3(12)*, e549. doi:10.1371/journalpntd.0000549.

Slate, D., Rupprecht, C., Rooney, J., Donovan, D., Lein, D., & Chipman, R. (2005). Status of oral rabies vaccination in wild carnivores in the United States. *Virus Research, 111*, 68–76.

Smith, D. L., Brendan, L., Waller, L., Childs, J., & Real, L. (2002). Predicting the spatial dynamics of rabies epidemics on heterogeneous landscapes. *PNAS, 99*, 3668–3672.

Smith, G., & Cheeseman, C. (2002). A mathematical model for the control of diseases in wildlife populations: Culling, vaccination and fertility control. *Ecological Modeling, 150*, 45–53.

Smith, G., Thulke, H., Fooks, A., Artois, M., MacDonald, D., Eisinger, D., et al. (2008). What is the future of wildlife rabies control in Europe? *Developments in Biologicals, 131*, 283–289.

Smith, G., & Wilkinson, W. (2003). Modeling control of rabies outbreaks in red fox populations to evaluate culling, vaccination, and vaccination combined with fertility control. *Journal of Wildlife Diseases, 39*, 278–286.

Smreczak, M., & Zmudzinski, J. (2009). Rabies in Poland in 2007. *Medycyna Weterynaryjna, 65*, 617–620.

Sobey, K., Rosatte, R., Bachmann, P., Buchanan, T., Bruce, L., Donovan, D., et al. (2010). Field evaluation of an inactivated vaccine to control raccoon rabies in Ontario, Canada. *Journal of Wildlife Diseases, 46*, 818–831.

Srithayakumar, V., Castillo, S., Rosatte, R., & Kyle, C. (2011). MHC class II DRB diversity in raccoons (*Procyon lotor*) reveals associations with raccoon rabies (Lyssavirus). *Immunogenics, 63*, 103–113.

Steck, F., Wandeler, A., Bichsel, P., Capt, S., Hafliger, U., & Schneider, L. (1982b). Oral immunisation of foxes against rabies. Laboratory and field studies. *Comparative Immunology Microbiology and Infectious Diseases, 5*, 165–171.

Steck, F., Wandeler, A., Bichsel, P., Capt, S., & Schneider, L. (1982a). Oral immunisation of foxes against rabies. A field study. *Zentralbl Veterinärmed, 29*, 372–396.

Sutor, A. (2008). Dispersal of the alien raccoon dog *Nyctereutes procyonoides* in southern Brandenburg, Germany. *European Journal of Wildlife Research, 54*, 321–326.

Szanto, A., Nadin-Davis, S., Rosatte, R., & White, B. (2011a). Genetic tracking of the raccoon variant of rabies virus in eastern North America. *Epidemics, 3*, 76–87.

Szanto, A., Nadin-Davis, S., Rosatte, R., & White, B. (2011b). Re-assessment of direct fluorescent antibody negative brain tissues with a real-time PCR assay to detect the presence of raccoon rabies virus RNA. *Journal of Virological Methods, 174*, 110–116.

Thulke, H., & Eisinger, D. (2008). The strength of 70%: revision of a standard threshold of rabies control. *Developments in Biologicals, 131*, 291–298.

Thulke, H., Eisinger, D., Selhorst, T., & Muller, T. (2008). Scenario- analysis evaluating emergency strategies after rabies re-introduction. *Developmental Biology, 131*, 265–272.

Tinline, R., & MacInnes, C. (2004). Ecogeographic patterns of rabies in southern Ontario based on time series analysis. *Journal of Wildlife Diseases, 40*, 212–221.

Totton, S., Rosatte, R., Tinline, R., & Bigler, L. (2004). Seasonal home ranges of raccoons, Procyon lotor, using a common feeding site in rural eastern Ontario: Rabies management implications. *Canadian Field-Naturalist, 118*, 65–71.

Totton, S., Tinline, R., Rosatte, R., & Bigler, L. (2002). Contact rates of raccoons (*Procyon lotor*) at a communal feeding site in rural eastern Ontario. *Journal of Wildlife Diseases, 38*, 313–319.

Totton, S., Wandeler, A., Ribble, C., Rosatte, R., & McEwen, S. (2011). Stray dog population health in Jodhpur, India in the wake of an animal birth control (ABC) program. *Preventive Veterinary Medicine, 98*, 215–220.

Velasco-villa, A., Reeder, S., Orciari, L., Yager, P., Franka, R., Blanton, J., et al. (2008). Enzootic rabies elimination from dogs and re-emergence in wild terrestrial carnivores, United States. *Emerging Infectious Diseases, 14*, 1849–1854.

Voigt, D. (1987). Red fox. In M. Novak, J. Baker, M. Obbard, & B. Malloch (Eds.), *Wild furbearer management and conservation in North America (pp. 379–393).* North Bay, Ontario: Ontario Trappers Association publishers.

Voigt, D. R., Tinline, R. R., & Broekhoven, L. H. (1985). Spatial simulation model for rabies control. In P. J. Bacon (Ed.), *Population Dynamics of Rabies in Wildlife (pp. 311–349).* London: Academic Press.

Vos, A. (2003). Oral vaccination against rabies and the behavioural ecology of the red fox (Vulpes vulpes). *Journal of Veterinary Medicine, 50*, 477–483.

Vos, A., Muller, T., Selhorst, T., Schuster, P., Neubert, A., & Schluter, H. (2001). Optimizing spring oral vaccination campaigns against rabies. *Deutsche Tierarztliche Wochenschrift, 108*, 55–59.

Vos, A., Pommerening, E., Neubert, L., Kachel, S., & Neubert, A. (2002). Safety studies of the oral rabies vaccine SAD B19 in striped skunk (*Mephitis mephitis*). *Journal of Wildlife Diseases, 38*, 428–431.

Vos, A., Selhorst, T., Schroder, R., & Mulder, J. (2008). Feasibility of oral rabies vaccination campaigns of young foxes *Vulpes vulpes* against rabies in summer. *European Journal of Wildlife Research, 54*, 763–766.

Wandeler, A., & Salsberg, E. (1999). Raccoon rabies in eastern Ontario. *Canadian Veterinary Journal, 40*, 731.

Westerling, B. (1991). Rabies in Finland and its control 1988–90. *Suomen Riista, 37*, 93–100.

Weyer, J., Rupprecht, C., & Nel, L. (2009). Poxvirus vectored vaccines for rabies- a review. *Vaccine, 27*, 7198–7201.

Whitney, H., Johnston, D.H., Wandeler, A., Nadin-Davis, S., & Muldoon, F. (2005). Elimination of rabies from the Island of Newfoundland: 2002–2004. *Rabies in the Americas conference abstract* (p. 80). Ottawa.

Wilde, H., Khawplod, P., Khamoltham, T., Hemachudha, T., Tepsumethanon, V., Lumlerdacha, B., et al. (2005). Rabies control in South and Southeast Asia. *Vaccine, 23*, 2284–2289.

Winkler, W., & Jenkins, S. (1991). Raccoon rabies. In G. Baer (Ed.), *The Natural History of Rabies (pp. 325–340)* (2nd ed.). Boca Raton, Florida: CRC Press.

Winkler, W., McLean, R., & Cowart, J. (1975). Vaccination of foxes against rabies using ingested baits. *Journal of Wildlife Diseases, 11*, 382–388.

Winkler, W. G. (1992). A review of the development of the oral vaccination technique for immunizing wildlife against rabies. In K. Bögel, F. -X. Meslin, & M. Kaplan (Eds.), *Wildlife Rabies Control (pp. 82–96).* Kent, England: Wells Medical.

Woodford, M., & Rossiter, P. (1993). Disease risks associated with wildlife translocation projects. *Revue Scientific et Technique de L'Office International des Epizootics, 12*, 115–135.

Yakobson, B., King, A., Sheichat, N., Eventov, B., & David, D. (2008). Assessment of the efficacy of oral vaccination of livestock guardian dogs in the framework of oral rabies vaccination of wild canids in Israel. *Developments in Biologicals, 131*, 151–156.

Yakobson, B., Manalo, D., Bader, K., Perl, S., & Haber, A. (1998). An epidemiological retrospective study of rabies diagnosis and control in Israel, 1948–1997. *Israel Journal of Veterinary Medicine, 53*, 114–126.

Yarosh, O., Wandeler, A., Graham, F., Campbell, J., & Prevec, L. (1996). Human adenovirus type 5 vectors expressing rabies glycoprotein. *Vaccine, 14*, 1257–1264.

Zanoni, R., Kappeler, A., Muller, U., Wandeler, A., & Breitenmoser, U. (2000). Rabies-free status of Switzerland after 30 years of fox rabies. *Schweizer Archiv Fur Tierheilkunde, 142*, 423–429.

Zhang, S., Tang, Q., Wu, X., Liu, Y., Zhang, F., Rupprecht, C., et al. (2009). Rabies in ferret badgers, southern China. *Emerging Infectious Diseases, 15,* 946–949.

Zienius, D., Bagdonas, J., & Dranseika, A. (2003). Epidemiological situation of rabies in Lithuania from 1990–2000. *Veterinary Microbiology, 93,* 91–100.

Zulu, G., Sabeta, C., & Nel, L. (2009). Molecular epidemiology of rabies: Focus on domestic dogs (*Canis familiaris*) and black-backed jackals (*Canis mesomelas*) from northern South Africa. *Virus research, 140,* 71–78.

19

Blueprint for Rabies Prevention and Control

Tiziana Lembo

Boyd Orr Centre for Population and Ecosystem Health,
Institute of Biodiversity, Animal Health and Comparative Medicine,
College of Medical, Veterinary and Life Sciences, University of Glasgow,
Glasgow G12 8QQ, UK, and Partners for Rabies Prevention, Global
Alliance for Rabies Control, 529 Humboldt St., Suite One, Manhattan,
Kansas 66502, USA

1 INTRODUCTION

Despite many developments in advanced tools and methods to combat rabies reflected in successful efforts in many parts of the world (Fooks et al., 2004; Blanton, Hanlon, & Rupprecht 2007; Schneider et al. 2007), canine rabies remains a major problem in most developing countries (Knobel et al., 2005). In recent years, achieving the ultimate goal of canine rabies elimination has gained increasing international attention (Rupprecht et al., 2008) and has attracted substantial investment from entrepreneurs and governments (World Health Organization 2007). Additionally, studies on the feasibility and potential cost benefits of canine rabies control have helped to make such activities more relevant and attractive for endemic countries (Lembo et al., 2010).

Although existing expertise and tools are adequate to achieve this goal (World Health Organization 2001, 2005), those involved in rabies control efforts on the ground still encounter many challenges. In order to provide governments and practitioners with the necessary support to develop their own national canine rabies elimination programs, a need was identified to gather together and translate practice-based evidence and research findings into user-friendly and widely accessible guidelines that are all

Rabies: Scientific Basis of the Disease and its Management
DOI: http://dx.doi.org/10.1016/B978-0-12-396547-9.00019-5

671

available through one reference point. Global rabies experts from the "Partners for Rabies Prevention" (PRP) have therefore developed an online standard operating procedure, or "blueprint," namely the "Blueprint for Rabies Prevention and Control" (available at http://www.rabiesblueprint .com). One section of this document includes information to help governments plan, implement, evaluate, and sustain canine rabies elimination programs (www.caninerabiesblueprint.org). The methods described in the Canine Rabies Blueprint are also the best way to control canine rabies, and, even if elimination is not anticipated, the benefits in protecting human health are still significant. This chapter describes the Blueprint for Rabies Prevention and Control with a focus on the Canine Rabies Blueprint website, providing some background information regarding this initiative as well as a description of the website. The following phase of the Blueprint development has expanded the information to that necessary to control and eliminate terrestrial wildlife rabies, focusing on fox rabies initially (www .foxrabiesblueprint.org), which will be described elsewhere.

2 PRINCIPLES AND FORMAT

It is widely recognized that a truly multidisciplinary effort is necessary for a one health approach to rabies prevention and control (Lembo et al., 2011). Based on this principle and through effective partnerships within the PRP, the Rabies Blueprint initiative has linked several disciplines required for developing comprehensive guidelines, including medicine, veterinary medicine, virology, epidemiology, immunology, ecology, biology, economics, animal welfare, diagnostics, education outreach, and health communication.

Through the combined expertise of a wide range of contributors, the Canine Rabies Blueprint integrates all the necessary components to achieve elimination of canine rabies, including information gathered from previously published guidelines by international health and animal welfare organizations, scientific findings, and country-specific successful experiences. These guidelines are provided in a question-and-answer format, addressing the most common questions on rabies prevention and control through key take-home messages and referring to case studies in a range of settings. Links to additional information, such as global recommendations, scientific articles, reports, specific standard operating procedures, and other websites are also provided.

Since the Rabies Blueprint is aimed at guiding all personnel concerned with the development and implementation of rabies control programs, from government officials to animal and human health professionals and field personnel, the document presents information using simple language and an intuitive format. In particular, to maximize usage of this tool, the

Rabies Blueprint has an openly accessible Internet-based format developed at low resolution and with contents also available as downloads.

To promote usage worldwide, the Canine Rabies Blueprint, originally developed in English, has been translated into French, Spanish, Portuguese, Russian, and Arabic, with additional translations undertaken (Persian) and some others under consideration (e.g. Chinese).

3 STEPS REQUIRED FOR CANINE RABIES ELIMINATION

The Canine Rabies Blueprint uses a practical approach to guiding readers through a detailed plan to prevent human rabies by means of eliminating canine rabies both in areas where canine rabies is endemic or has been re-introduced. This plan includes a number of key steps that are summarized in this section, while referring the reader to the appropriate sections of the Rabies Blueprint or chapters of this book for more specific information on particular aspects of rabies prevention and control. The steps with links to the relevant sections of the website are provided in Table 19.1. A visual representation (Figure 19.1) also helps guide users through the process of designing, implementing and evaluating an elimination strategy. All information can also be accessed through five main sections listed within the left navigation bar or the Site Map of the Canine Rabies Blueprint website: "Introduction," "Roles and responsibilities," "Infrastructure, legislative framework, costs and funding," "Communications plan," and "Operational activities."

3.1 Epidemiologic Assessment

Identifying key aspects of the epidemiology of rabies in a given area is an essential first step in order to design and implement appropriate control strategies compatible with local circumstances. An epidemiologic assessment, which is entirely reliant on an effective surveillance system, should address issues related to the animal species responsible for rabies maintenance in the area of interest in order to target control measures accordingly.

3.2 Agencies that Need to be Involved

In the "Roles and Responsibilities" section of the Rabies Blueprint, relevant agencies that should be involved in rabies prevention and control activities are identified and their roles and responsibilities are defined. Given that the burden of rabies affects multiple sectors and countries, this section recognizes the importance of engaging a range of country-based governmental and non-governmental organizations, institutions,

TABLE 19.1   Key Steps of a Canine Rabies Control and Elimination Program

Section	Information Provided	Link
Epidemiologic assessment	Information required before designing and implementing canine rabies elimination programs	http://www.rabiesblueprint.com/5-1-What-do-we-need-to-know-before
Agencies to be involved	Roles and responsibilities of agencies that should be involved in rabies prevention and control efforts	http://www.rabiesblueprint.com/Roles-and-Responsibilities
Infrastructure	Minimum infrastructure required for dog vaccination, rabies surveillance, and human rabies prevention	http://www.rabiesblueprint.com/3-1-Infrastructure
Legislation	Legislative measures relevant to rabies prevention and control	http://www.rabiesblueprint.com/3-2-Legislation
Costs and funding	Resource requirements, costs, sources of funding and budget determination	http://www.rabiesblueprint.com/3-3-Costs-and-Funding
Surveillance	Importance of surveillance in intervention programs and commonly used surveillance methods specific to rabies	http://www.rabiesblueprint.com/5-4-2-Why-is-epidemiological
Determining dog population sizes	Commonly used techniques for estimating the number of owned or roaming dogs	http://www.rabiesblueprint.com/5-4-1-What-techniques-are
Training needs	Personnel to be involved in rabies management programs and type of training required	http://www.rabiesblueprint.com/5-3-Who-do-we-need-to-train-and-in
Raising awareness	Guidelines for development of country-specific communications campaigns for rabies prevention and control adaptable to the local situation	http://www.rabiesblueprint.com/Communications-plan
Supplies needed	Supplies required for dog vaccination, sterilization, human prophylaxis, and rabies surveillance	http://www.rabiesblueprint.com/5-2-What-do-we-need-to-buy
Human prophylaxis	Guidelines for human rabies prevention	http://www.rabiesblueprint.com/5-5-What-are-we-going-to-do-human
Sustainability mechanisms	Guidelines for creating long-term sustainability strategies	http://www.rabiesblueprint.com/5-7-1-How-do-we-ensure
Canine rabies control	Guidelines for canine rabies control (attack/elimination phase)	http://www.rabiesblueprint.com/5-4-What-are-we-going-to-do-dog
Evaluation	Indicators of success that need to be determined	http://www.rabiesblueprint.com/5-6-Evaluation
Maintaining freedom	Strategies to ensure maintenance of freedom (maintenance phase)	http://www.rabiesblueprint.com/5-4-17-Our-programme-has-been

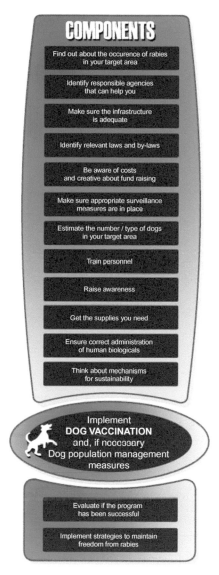

FIGURE 19.1   **Diagram of the components of a canine rabies control and elimination program as provided on the Canine Rabies Blueprint website (www.caninerabiesblueprint.org).** Relevant information can be accessed by clicking on the headings of the diagram.

and agencies, as well as international bodies in rabies elimination efforts, and it emphasizes the importance of building political commitment and ensuring cooperation and dialogue amongst all those involved. The establishment of a formal body (either an interministerial committee or

Task Force) responsible for coordinating activities including all parties is advocated as an ideal strategy.

3.3 Infrastructure

Basic infrastructure necessary to establish comprehensive rabies prevention and control programs is described in a subsequent section, which covers minimum infrastructure required to carry out dog vaccination, rabies surveillance and diagnostics at national and local levels, and human immunization, including requirements for personnel.

3.4 Legislative Frameworks

This section provides details on how a legislative framework relevant to rabies prevention and control may contribute to enhance rabies management efforts. The importance of making rabies a notifiable disease to more accurately evaluate the rabies burden in a given area is highlighted. Legislative measures that have relevance for rabies control are also described, such as responsible dog ownership (including mandatory registration and identification, and vaccination requirements), management of unowned dogs, control of dog movements/relocations, tie-up orders, regulations to control waste disposal, and abandonment legislation.

3.5 Costs and Funding Opportunities

In this section, the Rabies Blueprint provides some indication of resource requirements and costs associated with rabies control and prevention initiatives, based on published studies to which links are provided. In particular, information is available about costs involved in dog vaccination programs, dog sterilization programs, and administration of human post-exposure prophylaxis, as well as sources of funding that might be available for dog rabies control and budget estimations.

3.6 Surveillance

Setting up nationwide surveillance networks is an essential component of rabies elimination programs in order to have access to baseline epidemiologic information prior to the commencement of any intervention, evaluate impacts of intervention, and uncover potential new cases, or rabies affected areas, once the disease has been eliminated. Aspects related to the collection of field- and laboratory-based data through basic gold standard rabies diagnostic techniques are addressed in the Rabies Blueprint, with emphasis on the need for establishing motivation for reporting cases and collection of samples and building capacity for laboratory-based surveillance, including training of field personnel and laboratory employees. Links to standard operating procedures for various diagnostic tests are provided, including techniques developed for use under conditions with little laboratory infrastructure.

3.7 Determination of Dog Population Sizes

In order to define appropriate strategies for canine rabies control in a given area and to evaluate the effectiveness of intervention efforts in terms of vaccination coverage, there are techniques for obtaining information on the size and accessibility of owned and unowned dog populations, which are described here. These include commonly used census techniques such as household surveys, indicator counts, and capture-mark-recapture methods.

3.8 Training Requirements

Training of different categories of personnel, such as rabies surveillance staff and personnel involved in rabies control and prevention efforts, is critical for adequate implementation of elimination strategies. The objectives of the training will vary depending on specific activities (e.g., dog vaccination, epidemiological surveillance systems, emergency responses to outbreaks, etc.) and are covered in specific sections of the Rabies Blueprint.

3.9 Raising Awareness

This section recognizes the importance of raising awareness about rabies prevention and control as part of any comprehensive canine rabies elimination program (Dodet et al. 2008; Matibag et al. 2009) and the role that health communications (the study and use of methods to inform and influence individual and community decisions that enhance health) can play in this respect (Brownson & Nelson 2002).

A template for developing an effective rabies communications strategy that can be adapted to the cultural, political, and behavioral needs of any location is also provided. This communications plan includes eight stages: 1) assessing the science and identifying potential issues, challenges, and barriers to change that may affect the communication outreach; 2) defining the purpose of the communication, which includes identifying targeted goals and measurable objectives that should be adapted towards any national needs; 3) targeting or segmenting the communication outreach to the specific audience at highest risk; 4) developing and pretesting concepts, messages, and materials to ensure that messages resonate with target populations and to improve uptake of prevention/control behaviors; 5) choosing appropriate media channels; 6) determining the best timing for release; 7) implementing the communications effort; and 8) evaluating the communications initiative and its impact, and making refinements.

3.10 Supplies Required

Materials and supplies to be acquired for dog vaccination, sterilization programs, human immunization, and field- and laboratory-based surveillance are listed in detail in this section.

3.11 Guidelines for Human Rabies Prevention

In recognition of the importance of integrating dog rabies control and human rabies prevention as part of the initial stages of an elimination program, approaches to reduce human rabies incidence through adequate and cost-effective utilization of human biologics are described in detail here. The human rabies prevention section provides both bite victims and health care personnel with information on how to recognize and assess exposures, products and regimens available for human prophylaxis, preventive immunization of people deemed at high risk, management of animal bite wounds, and immunization of individuals exposed to suspect rabid animals.

3.12 Mechanisms for Sustainability

Careful attention needs to be given to building sustainability into any rabies control program, so that the benefits accrued and investments made are not lost. Maintenance of a rabies-free status once elimination has been achieved or the planning of emergency responses to potential re-incursions are vital. Aspects to be considered and that are covered in this section include: political stability; maintenance of features established as part of the elimination phase such as surveillance, legislative and financial mechanisms, capacity for dog vaccination and policies for human rabies prevention; ensuring that level of commitment and inter-sectoral and inter-institutional dialogue are high amongst relevant parties; collaborative efforts involving neighboring countries; and continuous community engagement.

3.13 Guidelines for Canine Rabies Control

This section describes the core phase of a rabies elimination program. The main focus is on the practicalities of achieving rabies elimination through mass dog vaccination campaigns, which include strategies available for dog vaccination, planning of campaigns on the ground, community sensitization techniques, frequency of campaigns, and estimation of vaccination coverage. In addition, critical aspects of an effective operational response in case of reintroduction of rabies into an area after a period of absence are described.

Activities required to manage the dog population—which, however, is not always necessary—are also discussed, such as promotion of responsible dog ownership, reproduction control, temporary/permanent removal of dogs (shelters, foster homes, capture and release, euthanasia), and habitat control.

3.14 Mechanisms for Evaluation

Indicators necessary to determine the effectiveness of the canine rabies control program are described, including determining vaccination coverage and assessing impact on dog and human rabies cases. Emphasis is placed on the need for a high level of surveillance and for using appropriate diagnostic tools for confirmation of vaccination success.

3.15 Maintaining Freedom from Rabies

Once strategies required to eliminate canine rabies from an area are discussed, the Rabies Blueprint guides the reader through techniques necessary to ensure that the rabies-free status is maintained (maintenance phase), emphasizing the importance of continuing features established during the elimination phase as well as implementing activities specific to this phase, such as, for example, establishing close liaisons with neighboring jurisdictions and ensuring sustainable financing mechanisms.

4 CONCLUSIONS AND FUTURE DEVELOPMENTS

There are a number of features that render the Rabies Blueprint a powerful toolkit, including the online, easy-to-use format of the document, which compiles all existing material on rabies prevention and control resulting from global expertise into one website developed in simple language and freely accessible to a wide range of users. The economical online design also ensures easy and regular updating, unlike many existing paper-based guidelines, and it allows impact and performance to be monitored as well as feedback to be received from users. Since its inception in 2010, the Rabies Blueprint has met with substantial success, to date exceeding 154,500 visitors from the Americas, Europe, Asia, Africa, and Oceania. The initiative is expected to continue growing in interest and local impacts, especially due to the expansion to a multilingual interface. Future key areas of development include the regular incorporation of revised or new recommendations and rabies resources, as well as upgrading the website to include components focusing on rabies wildlife reservoir species.

ACKNOWLEDGEMENTS

The Partners for Rabies Prevention are an informal public–private partnership comprising a group of stakeholders working in the field of rabies prevention and control (http://www.rabiescontrol.net/about-us/partners/partners-for-rabies-prevention.html). Contributors to the Rabies Blueprint initiative include: Animal Health and Veterinary Laboratories Agency (Anthony Fooks); Canadian Food Inspection Agency (Alexander Wandeler); Centers for Disease Control and Prevention (Abbigail Tumpey, Andrés Velasco-Villa, Sergio Recuenco, Nadia Gallardo-Romero); Dodet Bioscience (Betty Dodet); European Commission (Moritz Klemm); Food and Agriculture Organization (Katinka de Balogh, Olivier Adier, Toni Ettel); French Food Safety Agency (Florence Cliquet); Friedrich-Loeffler-Institute (Thomas Müller, Conrad Freuling); Global Alliance for Rabies Control (Deborah Briggs, Peter Costa, Mary Elizabeth Miranda, Robert Dedmon, Louise Taylor, Kim Doyle, Charles Rupprecht); IDT Biologika GmbH (Adrian Vos); Institut Pasteur (Hervé Bourhy, Noël Tordo); Kedrion Biopharma (Ferdinando Borgese); KwaZulu - Natal Rabies Project (Kevin Le Roux); Merial (Carolin Schumacher, Joanne Maki); Novartis (Dieter Gniel); Pan American Health Organization (Fernando Leanes, Marco Antonio Natal Vigilato); Sanofi Pasteur (Michaël Attlan); Thai Red Cross Society (Boonlert Lumlertdacha); Universitá di Torino (Natalia Cediel); University of Glasgow (Tiziana Lembo, Sarah Cleaveland, Katie Hampson); University of Pretoria (Louis Nel, Melinda Hergert); World Health Organization (François-Xavier Meslin); World Organization for Animal Health (Lea Knopf); World Society for the Protection of Animals (Elly Hiby, Esmee Russell).

References

Blanton, J. D., Hanlon, C. A., & Rupprecht, C. E. (2007). Rabies surveillance in the United States during 2006. *Journal of the American Veterinary Medical Association*, 231, 540–556.

Brownson, R. C., & Nelson, D. E. (2002). *Communicating public health information effectively: A guide for practitioners*. Washington, DC: American Public Health Association.

Dodet, B., Goswami, A., Gunasekera, A., de Guzman, F., Jamali, S., Montalban, C., et al. (2008). Rabies awareness in eight Asian countries. *Vaccine, 26*, 6344–6348.

Fooks, A. R., Roberts, D. H., Lynch, M., Hersteinsson, P., & Runolfsson, H. (2004). Rabies in the United Kingdom, Ireland and Iceland. In A. A. King, A. R. Fooks, M. Aubert, & A. I. Wandeler (Eds.), *Historical perspective of rabies in Europe and the Mediterranean Basin (pp. 25–32)*. Paris: World Organisation for Animal Health.

Knobel, D. L., Cleaveland, S., Coleman, P. G., Fevre, E. M., Meltzer, M. I., Miranda, M. E. G., et al. (2005). Re-evaluating the burden of rabies in Africa and Asia. *Bulletin of the World Health Organization, 83*, 360–368.

Lembo, T., Attlan, M., Bourhy, H., Cleaveland, S., Costa, P., de Balogh, K., et al. (2011). Renewed global partnerships and re-designed roadmaps for rabies prevention and control. *Veterinary Medicine International*, 923149.

Lembo, T., Hampson, K., Kaare, M. T., Ernest, E., Knobel, D., Kazwala, R. R., et al. (2010). The feasibility of canine rabies elimination in Africa: Dispelling doubts with data. *Plos Neglected Tropical Diseases, 4,* e626.

Matibag, G. C., Ohbayashi, Y., Kanda, K., Yamashina, H., Kumara, W. R., Perera, I. N., et al. (2009). A pilot study on the usefulness of information and education campaign materials in enhancing the knowledge, attitude and practice on rabies in rural Sri Lanka. *Journal of Infection in Developing Countries, 3,* 55–64.

Rupprecht, C. E., Barrett, J., Briggs, D., Cliquet, F., Fooks, A. R., Lumlertdacha, B., et al. (2008). Can rabies be eradicated? *Towards the elimination of rabies in Eurasia, 131,* 95–121.

Schneider, M. C., Belotto, A., Adé, M. P., Hendrickx, S., Leanes, L. F., Rodrigues, M. J., et al. (2007). Current status of human rabies transmitted by dogs in Latin America. *Cadernos de Saúde Pública, 23,* 2049–2063.

World Health Organization, (2001). *Strategies for the control and elimination of rabies in Asia. Report of a WHO interregional consultation, 17–21 July 2001.* Geneva: World Health Organization.

World Health Organization, (2005). WHO Expert Consultation on rabies. *WHO Technical Report Series, 931,* 1–88.

World Health Organization. (2007). Bill & Melinda Gates Foundation fund WHO-coordinated project to control and eventually eliminate rabies in low-income countries. Available at <http://www.who.int/rabies/bmgf_who_project/en/index.html> Accessed 01.01.13.

Future Developments and Challenges

Alan C. Jackson

Departments of Internal Medicine (Neurology) and Medical Microbiology
University of Manitoba, Winnipeg, Manitoba R3A 1R9 Canada

1 INTRODUCTION

In recent years, there has been considerable progress in tackling the worldwide rabies problem in certain geographic regions, although, unfortunately, setbacks have occurred in other regions. Many countries in both Asia and Africa still remain a very long way off from making major progress in reducing the number of human deaths due to rabies. Elimination of endemic canine rabies using mass vaccination is the key component in the prevention of rabies in humans. Rabies biologicals remain expensive and continue to have limited availability in resource-poor and resource-limited settings, and this is unlikely to improve in the near future. Neural tissue-derived rabies vaccines are disappearing, and only a few countries still continue to use them. This final chapter will speculate on areas that may become the focus of future research efforts, as well as strategies to reduce the human death toll due to this ancient and dreaded disease.

2 PATHOGENESIS

Major advances in our understanding of the basic mechanisms involved in the pathogenesis of rabies have been slow. Work from my own research laboratory in an experimental mouse model of rabies has focused on the role of neuronal process degeneration, affecting dendrites and axons of infected neurons, which explains why only mild histo-pathological changes are observed in rabies (Scott, Rossiter, Andrew, &

Jackson, 2008). There is accumulating evidence from my laboratory that this neuronal process degeneration is caused by oxidative stress (Jackson, Kammouni, Zherebitskaya, & Fernyhough, 2010), which also has a demonstrated role in a number of neurodegenerative diseases, including diabetic neuropathy. Very recent evidence indicates that the oxidative stress is likely due to rabies virus-induced mitochondrial dysfunction. Confirmation will be important to show that this mechanism is also important in natural rabies and, in particular, in human rabies. Further work is needed to gain an improved understanding of the basic mechanisms involved in how neuronal infection produces clinical disease and death. Understanding the role of up-regulation and down-regulation of host genes, and also host microRNAs in good animal models, will be a difficult task, but this will be useful in order to understand basic mechanisms in rabies pathogenesis.

For logistical reasons, most studies in rabies pathogenesis continue to utilize rodent models, but the use of more natural models such as dog or bat models with the rabies virus variant associated with the species will be important in the future in gaining a better understanding of natural rabies. The barrier is that this research is very expensive, and these models will not be able to quickly address fundamental questions that are hypothesis-based, which makes funding of this research a difficult challenge. Reagents for use in dogs or bats are also more limited than in rodents, and there are greater limitations on the available numbers of animals for studies and in the variety of genetic modifications available for research studies, including transgenic animals. It is clear that more information is needed about the very early events before any clinical features or symptoms develop in rabies and also about the mechanisms involved in producing clinical disease with a fatal outcome.

A limited understanding of the basic mechanisms responsible for clinical disease and death in rabies has made therapeutic efforts much more difficult. Further research is essential in good animal models in order to provide this critical information to assist in the development of therapeutic efforts to successfully move the field forward. A "trial and error" approach on the basis of current available information is very unlikely to prove fruitful and result in the availability of effective therapy for human rabies.

3 EPIDEMIOLOGY

Inadequate surveillance of human and animal rabies is a widespread problem worldwide. The lack of reporting of cases or of reporting requirements plus the lack of available confirmatory laboratory diagnostic evaluations in many regions are contributory. The lack of accurate surveillance data likely contributes to control efforts being under-resourced because the full extent of the rabies problem is undocumented and not realized. In the United States, animal surveillance is passive, which of course naturally greatly underestimates the extent of animal rabies.

Human rabies is not a major public health problem in North America, although a small number of sporadic cases continue to occur, typically due to unrecognized bat exposures and "imported" cases from countries with endemic dog rabies. Endemic bat rabies in insectivorous bats will likely continue for centuries, and there is really no solution in sight even with anticipated scientific advances that may occur. When humans are bitten by bats and the exposure is not recognized, there is no opportunity at all to initiate effective post-exposure rabies prophylaxis measures. Widespread pre-exposure immunization of the entire human population would depend on a major advance in immunizations, in which individuals could be vaccinated against a large variety of pathogens with a multivalent vaccine that has an extremely low incidence of adverse effects. Otherwise, the risk of developing adverse effects from the vaccine would greatly outweigh the very low risk of an unrecognized rabies exposure.

The number of human cases of rabies in Latin America has markedly decreased with effective control of dog rabies that utilized mass vaccination (Schneider et al., 2011). Recent cases of rabies in the indigenous population of the Amazon region of Brazil, Peru, and Ecuador have been recognized to be due to vampire bat rabies variants (da Rosa et al., 2006; Schneider et al., 2009), which are also responsible for economic losses in the cattle industry in Latin America (Lee, Papes, & Van Den Bussche, 2012).

Asia and Africa continue to carry the vast majority of the worldwide burden of rabies. About three billion people live in regions with endemic dog rabies and are at potential risk of a dog exposure. Impoverished people and particularly children carry a very disproportional burden of the disease. The number of global human cases has been estimated at 55,000–75,000 deaths per year. However, these are only very crude estimates. Accurate numbers of cases are not available, for example, in India, which has by far the greatest number of cases of human rabies in any country in the world, and in most countries in Africa. A recent resurgence of rabies has occurred in China, which peaked in 2007 with just over 3,300 deaths per year (Hu, Tang, Tang, & Fooks, 2009). There has been some recent progress (now about 2,000 deaths per year), but major improvements in the death rate will depend on essential governmental commitment and investment. An outbreak of human rabies developed on the island of Bali, Indonesia, in 2008, where there have been more than 100 human deaths. The situation is improving and the yearly human death toll is shrinking in Bali. There have also been outbreaks on other islands in Indonesia, including Nias and Flores.

4 THERAPY OF HUMAN RABIES

There has been little recent progress in developing effective therapy for human rabies. Unfortunately, repetition of the ineffective Milwaukee Protocol (Willoughby, Jr. et al., 2005) has proved to be a major impediment in moving forward (see Chapter 16). Repeated failures and the absence of any well-documented success after the index survivor clearly indicate that new approaches are needed. It has proved very difficult to get this message across to physicians who are faced with real cases and are not familiar with the complex treatment issues in rabies. No real progress will be made in the future until physicians are willing to take new approaches instead of repeating what is easy and has repeatedly failed in the past.

A lack of effective neuroprotective therapies for acute diseases of the nervous system has hampered progress in many neurological diseases, including stroke, hypoxic-ischemic (global) brain insults (e.g., cardiac arrest), and traumatic brain and spinal cord injury. Therapeutic advances have been very slow for these other neurological diseases, despite considerable resources and numerous research investigations, including studies in animal models and in clinical trials. In comparison, research efforts on rabies have been miniscule. Many other viral brain infections also lack effective therapies, although there are notable exceptions, such as the therapy of herpes simplex encephalitis with intravenous acyclovir.

5 RABIES VIRUS AS A BIOTHREAT AGENT

Rabies virus is classified as a Category C agent for bioterrorism, which falls in the third highest priority among pathogens, and it is considered an emerging threat for disease. Rabies virus is subject to rapid inactivation in the environment and is much less hardy than many other pathogens, which reduces its capacity as a weapon. Deliberate release of the agent in a reservoir host in a rabies-free area could produce the threat of substantial economic losses (Fooks, Johnson, & Rupprecht, 2009). Rabies virus can be aerosolized, although it would be difficult to produce exposures of large numbers of humans via an airborne route. Practically everyone is familiar with rabies, and any perceived threat of the disease could easily produce fear in a population, even without a threat of large-scale human fatalities (Fooks et al., 2009). Limitations on the amount of rabies vaccine and human rabies immune globulin available for administration could compound the situation, even if the actual risk of a significant exposure were low. In addition, an incident with mass human exposures could seriously deplete the supply of rabies biologicals (rabies vaccine and human rabies immune globulin). Post-exposure vaccination is also of reduced potential benefit because of the possibility of rapid spread of rabies virus into the brain via an olfactory pathway. Similarly, administration of human rabies immune globulin may be of more limited therapeutic benefit with aerosol exposures.

6 PREVENTION OF HUMAN RABIES

Effective prevention of human rabies will require that the disease receive appropriate recognition by the governments of countries in which it is an important problem. A comprehensive strategy is needed to combat the disease, and rabies must gain the attention from government officials that it deserves. Control of canine rabies is fundamental. Intergovernmental agencies need to cooperate, and the necessary resources need to be allocated. The World Health Organization should play a key role in efforts to combat rabies as a global problem. Without being on the "radar," rabies will not receive the attention or the resources that it deserves. The World Health Assembly is the decision-making body of the World Health Organization (WHO). The assembly meets yearly and passes resolutions that are very influential on the activities of the WHO. There have been no recent resolutions at all concerning rabies; in 2011 and 2012 there were resolutions concerning a variety of infectious diseases, including human immunodeficiency virus infection, dra cunculiasis (guinea-worm disease), cholera, polio, and schistosomiasis. Hopefully, a resolution will be passed in the near future concerning the elimination of canine rabies. A recent report indicates that canine rabies elimination is feasible in Africa, largely through mass vaccination of domestic dogs, which has been proven to be highly effective in preventing human rabies deaths (Lembo et al., 2010).

The high cost of rabies biologicals continues to be a major impediment to the widespread use of both pre-exposure and post-exposure rabies prophylaxis measures. Nerve-tissue rabies vaccines are no longer being used in most countries because of their association with an unacceptable risk of neuroparalytic complications. Human rabies vaccines (cell culture-derived) are expensive, with a North American price of about US $250 per dose with four doses required for post-exposure prophylaxis, plus the cost of human rabies immune globulin for administration to one individual of at least US $1,000 (Dr. Deborah Briggs, personal communication). Hence, the resulting cost for rabies biologicals alone is at least US $2,000 for one exposed individual, excluding associated medical costs.

A study performed over 10 years ago in California showed a cost of about US $3,700 for a suspected rabies exposure (direct costs of $2,600 plus indirect costs of $1,100) (Shwiff et al., 2007). Clearly, costs of this magnitude are completely out of reach for individuals in many countries of the world. New, low-cost rabies vaccines for human use are urgently needed. Because development costs are high, especially with the required associated clinical trials, we cannot expect the availability of new effective rabies vaccines of low cost within the next decade. Intradermal rabies vaccination regimens have lowered the dose of vaccine required for vaccination of each individual, resulting in very significant cost savings. Clinical studies are now evaluating post-exposure vaccination schedules over much shorter periods, including a preliminary study from Thailand with a schedule showing good results in which the vaccine was given over a period of only one week (4-site intradermal vaccination given on days 0, 3, and 7) (Shantavasinkul et al., 2010). This schedule reduces the number of clinic visits and would be much more convenient and less costly in situations where patients need to travel to a distant location to receive the therapy. More research is needed evaluating vaccination schedules that would result in cost savings.

Because children may have unrecognized rabies exposures and are at greater risk than adults and suffer a disproportionate burden of rabies (accounting for about half of all cases), one could argue that all children in countries with endemic canine rabies should receive pre-exposure rabies immunization. This would simplify the post-exposure prophylaxis measures required and would also likely provide some protection against inapparent and unreported exposures. One can safely assume that only a tiny minority of this highly susceptible population currently receives pre-exposure immunization, and economic factors are the main barrier.

Human rabies immune globulin (HRIG) is said to be virtually unavailable in most canine rabies-endemic countries (Wilde, 2012) and is unaffordable for most individuals because of its high cost. Equine rabies immune globulin manufactured in France, India, China, and Thailand, is available in many rabies-endemic countries (Wilde, 2012) and is much cheaper than HRIG. A cocktail of two humanized monoclonal antibodies has been developed for post-exposure rabies prophylaxis, and the phase I studies look very encouraging (Bakker et al., 2008; de Kruif et al., 2007). It can be expected that the necessary clinical trials will be expensive, and the cost of the cocktail will be lower than HRIG. However, it is not clear that the cost will still be within reach of many people in the world.

The Advisory Committee on Immunization Practices in the United States recommends initiation of post-exposure rabies prophylaxis in situations in which a bat exposure cannot be excluded, including, for example, a situation with an adult sleeping in a room in which a bat was present that subsequently escaped (Manning et al., 2008). Canadian researchers have now seriously cast doubt on the wisdom of this recommendation by demonstrating the very low risk of developing rabies and the high cost of rabies biologicals with associated risks of adverse effects. De Serres et al. (2009) have estimated that the cost of preventing a single case of rabies from a bedroom exposure would be nearly 2 billion Canadian dollars for biological products and virologic analysis. This excludes the costs of health care workers. This indicates that the current recommendations should be reconsidered in the context of a complete risk analysis.

7 CONTROL OF ANIMAL RABIES

Outbreaks and resurgences of rabies will continue to occur in unexpected places in the world. Resources will be mobilized when the problems become large, leading to subsequent improvements in local situations. However, in the absence of effective policies backed with appropriate funding, there will be an endless cycle of rabies outbreaks in different locations in the future. An important advance in the control of canine rabies is the development of a blueprint for rabies prevention and control (Lembo, 2012) (see Chapter 19). This greatly facilitates organization of what needs to be done in controlling canine rabies.

There are many barriers to controlling canine rabies, including political, economic, religious, and cultural barriers. Some of the barriers result from lack of awareness of the burden and effects of rabies, a lack of funding, a lack of commitment from public health and veterinary services, and practical difficulties in bringing different authorities together to develop and implement an effective plan of action (Senior, 2012). Thailand is an example of a country that made a major reduction of human deaths from hundreds of deaths to less than 10 deaths per year by developing accessible effective post-exposure rabies prophylaxis rather than by controlling canine rabies. However, about half a million people continue to require post-exposure rabies prophylaxis each year in Thailand (Senior, 2012), which are ongoing costs. Countries in Latin America largely reduced human deaths transmitted by dogs by controlling canine rabies with a regional elimination program, which included mass vaccination of dogs. This resulted in a reduction of human and canine cases by about 90% over a period of 20 years (Belotto, Leanes, Schneider, Tamayo, & Correa, 2005; Schneider et al., 2011). In 2007, there were only 14 human deaths, mostly occurring in Bolivia and Haiti. When will resource-poor or resource-limited countries in Asia and Africa likely be able to control or eliminate canine rabies? Unfortunately, rabies has not yet been deemed a priority in light of the numerous other health problems and issues requiring funding. Is there a research advance that would make it easier to control canine rabies in these countries? A highly effective oral rabies vaccine for dogs would be helpful in accessing stray dogs and would potentially facilitate vaccination of a difficult to reach population. Mass vaccination of dogs is of fundamental importance. Ineffective measures of dog rabies control, such as culling of dogs, continue to be used all too frequently.

Oral vaccination programs and associated rabies control methods have proved to be highly effective in control of wildlife rabies in some countries (see Chapter 18). Canada has been at the forefront of these efforts and has effectively controlled rabies in red foxes in Ontario and very effectively battled incursions of raccoon rabies from the United States into Ontario, New Brunswick, and Quebec and prevented the development of endemic raccoon rabies in Canadian provinces. In contrast, raccoon rabies gradually spread up the entire eastern coast of the United States over a period of decades with no major concerted efforts to prevent this geographical expansion, with resulting ongoing costs for post-exposure prophylaxis measures in humans and animals. Although the costs may be high, prevention of the development of endemic rabies in a region may result in a long-term overall financial benefit. Hence, decisions concerning management of wildlife rabies need to be proactive.

References

Bakker, A. B., Python, C., Kissling, C. J., Pandya, P., Marissen, W. E., Brink, M. F., et al. (2008). First administration to humans of a monoclonal antibody cocktail against rabies virus: Safety, tolerability, and neutralizing activity. *Vaccine, 26*(47), 5922–5927.

Belotto, A., Leanes, L. F., Schneider, M. C., Tamayo, H., & Correa, E. (2005). Overview of rabies in the Americas. *Virus Research, 111*(1), 5–12.

da Rosa, E. S. T., Kotait, I., Barbosa, T. F. S., Carrier, M. L., Brandão, P. E., Pinheiro, A. S., et al. (2006). Bat-transmitted human rabies outbreaks, Brazilian Amazon. *Emerging Infectious Diseases, 12*(8), 1197–1202.

de Kruif, J., Bakker, A. B., Marissen, W. E., Kramer, R. A., Throsby, M., Rupprecht, C. E., et al. (2007). A human monoclonal antibody cocktail as a novel component of rabies postexposure prophylaxis. *Annual Review of Medicine, 58*, 359–368.

De Serres, G., Skowronski, D. M., Mimault, P., Ouakki, M., Maranda-Aubut, R., & Duval, B. (2009). Bats in the bedroom, bats in the belfry: Reanalysis of the rationale for rabies postexposure prophylaxis. *Clinical Infectious Diseases, 48*(11), 1493–1499.

Fooks, A. R., Johnson, N., & Rupprecht, C. E. (2009). Rabies. In A. D. T. Barrett & L. R. Stanberry (Eds.), *Vaccines for biodefense and emerging and neglected diseases* (pp. 609–630). London: Elsevier.

Hu, R., Tang, Q., Tang, J., & Fooks, A. R. (2009). Rabies in China: An update. *Vector-Borne and Zoonotic Diseases, 9*(1), 1–12.

Jackson, A. C., Kammouni, W., Zherebitskaya, E., & Fernyhough, P. (2010). Role of oxidative stress in rabies virus infection of adult mouse dorsal root ganglion neurons. *Journal of Virology, 84*(9), 4697–4705.

Lee, D. N., Papes, M., & Van Den Bussche, R. A. (2012). Present and potential future distribution of common vampire bats in the Americas and the associated risk to cattle. *PLoS One, 7*(8), e42466.

Lembo, T. (2012). The blueprint for rabies prevention and control: A novel operational toolkit for rabies elimination. *PLoS Neglected Tropical Diseases, 6*(2), e1388.

Lembo, T., Hampson, K., Kaare, M. T., Ernest, E., Knobel, D., Kazwala, R. R., et al. (2010). The feasibility of canine rabies elimination in Africa: Dispelling doubts with data. *PLoS Neglected Tropical Diseases, 4*(2), e626.

Manning, S. E., Rupprecht, C. E., Fishbein, D., Hanlon, C. A., Lumlertdacha, B., Guerra, M., et al. (2008). Human rabies prevention--United States, 2008: Recommendations of the advisory committee on immunization practices. *Morbidity and Mortality Weekly Report, 57*(RR-3), 1–28.

Schneider, M. C., Aguilera, X. P., Barbosa da Silva, J. J., Ault, S. K., Najera, P., Martinez, J., et al. (2011). Elimination of neglected diseases in Latin America and the Caribbean: A mapping of selected diseases. *PLoS Neglected Tropical Diseases, 5*(2), e964.

Schneider, M. C., Romijn, P. C., Uieda, W., Tamayo, H., da Silva, D. F., Belotto, A., et al. (2009). Rabies transmitted by vampire bats to humans: An emerging zoonotic disease in Latin America? *Revista Panamericana Salud Publica, 25*(3), 260–269.

Scott, C. A., Rossiter, J. P., Andrew, R. D., & Jackson, A. C. (2008). Structural abnormalities in neurons are sufficient to explain the clinical disease and fatal outcome in experimental rabies in yellow fluorescent protein-expressing transgenic mice. *Journal of Virology, 82*(1), 513–521.

Senior, K. (2012). Global rabies elimination: Are we stepping up to the challenge? *Lancet Infectious Diseases, 12*, 366–367.

Shantavasinkul, P., Tantawichien, T., Wilde, H., Sawangvaree, A., Kumchat, A., Ruksaket, N., et al. (2010). Postexposure rabies prophylaxis completed in 1 week: Preliminary study. *Clinical Infectious Diseases, 50*(1), 56–60.

Shwiff, S. A., Sterner, R. T., Jay, M. T., Parikh, S., Bellomy, A., Meltzer, M. I., et al. (2007). Direct and indirect costs of rabies exposure: A retrospective study in southern California (1998–2002). *Journal of Wildlife Diseases, 43*(2), 251–257.

Wilde, H. (2012). Rabies postexposure vaccination: Are antibody responses adequate? (Editorial). *Clinical Infectious Diseases, 55*(2), 206–208.

Willoughby, R. E., Jr., et al., Tieves, K. S., Hoffman, G. M., Ghanayem, N. S., Amlie-Lefond, C. M., Schwabe, M. J., et al. (2005). Survival after treatment of rabies with induction of coma. *New England Journal of Medicine, 352*(24), 2508–2514.

Index

Note: Page numbers followed by "*f*" and "*t*" refer to figures and tables respectively.